D1803752

HIGH QUALITY BEAMS

Related Titles from the AIP Conference Proceedings
Subseries on Accelerators, Beams, and Instrumentation

588 Physics with an Electron Polarized Light-Ion Collider: Second Workshop, EPIC 2000
Edited by Richard G. Milner, October 2001, 0-7354-0028-8

581 Physics of, and Science with, the X-Ray Free-Electron Laser: 19th Advanced ICFA Beam Dynamics Workshop
Edited by C. Pellegini, S. Chattopadhyay, M. Cornacchia, and I. Lindau, August 2001, 0-7354-0022-9

576 Application of Accelerators in Research and Industry: Sixteenth International Conf.
Edited by J. L. Duggan and I. L. Morgan, July 2001, 0-7354-0015-6

572 Electron Beam Ion Sources and Traps and Their Applications: 8th International Symp.
Edited by Krsto Prelec, June 2001, 0-7354-0011-3

571 PHOTON 2000: International Conf. on the Structure and Interactions of the Photon
Edited by A. J. Finch, July 2001, 0-7354-0010-5

569 Advanced Accelerator Concepts: Ninth Workshop
Edited by Patrick L. Colestock and Sandra Kelley, June 2001, CD-ROM included, 0-7354-0005-9

546 Beam Instrumentation Workshop 2000: Ninth Workshop
Edited by Kenneth D. Jacobs and R. Coles Sibley III, December 2000, 1-56396-975-0

521 Synchrotron Radiation Instrumentation: Eleventh US National Conference
Edited by Piero Pianetta, John Arthur, and Sean Brennan, May 2000, 1-56396-941-6

512 Nuclear Physics at Storage Rings: Fourth International Conference: STORI99
Edited by Hans-Otto Meyer and Peter Schwandt, June 2000, 1-56396-928-9

468 Nonlinear and Collective Phenomena in Beam Physics — 1998 Workshop: International Committee on Future Accelerators
Edited by Swapan Chattopadhyay, Max Cornacchia, and Claudio Pellegrini, April 1999, 1-56396-862-2

To learn more about these titles, or the AIP Conference Proceedings Series, please visit the webpage **http://proceedings.aip.org**

HIGH QUALITY BEAMS

Joint US-CERN-JAPAN-RUSSIA Accelerator School

St. Petersburg and Moscow, Russia   1-14 July 2000

EDITORS
S. I. Kurokawa
KEK, Tsukuba, Japan

S. Y. Lee
Indiana University, USA

J. Miles
CERN, Geneva, Switzerland

E. A. Perevedentsev
Budker Institute of Nuclear Physics, Novosibirsk, Russia

Melville, New York, 2001
AIP CONFERENCE PROCEEDINGS ■ VOLUME 592

Editors:

S. I. Kurokawa
KEK, Accelerator Laboratory
1-1 Oho
Tsukuba, Ibaraki 305-0801
JAPAN

S. Y. Lee
Indiana University
Department of Physics, SW117
Bloomington, IN 47405
USA
E-mail: shylee@indiana.edu

J. Miles
CERN
M07510
1211 Geneva 23
SWITZERLAND

E-mail: John.Miles@cern.ch

E. A. Perevedentsev
The Budker Institute of Nuclear Physics
Laboratory 11
Acad. Lavrentiev prospect 11
630090 Novosibirsk
RUSSIA

E-mail: E.A.Perevedent@inp.nsk.su

The article on pp. 205–230 was authored by a U. S. Government employee and is not covered by the below mentioned copyright.

Authorization to photocopy items for internal or personal use, beyond the free copying permitted under the 1978 U.S. Copyright Law (see statement below), is granted by the American Institute of Physics for users registered with the Copyright Clearance Center (CCC) Transactional Reporting Service, provided that the base fee of $18.00 per copy is paid directly to CCC, 222 Rosewood Drive, Danvers, MA 01923. For those organizations that have been granted a photocopy license by CCC, a separate system of payment has been arranged. The fee code for users of the Transactional Reporting Service is: 0-7354-0034-2/01/$18.00.

© 2001 American Institute of Physics

Individual readers of this volume and nonprofit libraries, acting for them, are permitted to make fair use of the material in it, such as copying an article for use in teaching or research. Permission is granted to quote from this volume in scientific work with the customary acknowledgment of the source. To reprint a figure, table, or other excerpt requires the consent of one of the original authors and notification to AIP. Republication or systematic or multiple reproduction of any material in this volume is permitted only under license from AIP. Address inquiries to Office of Rights and Permissions, Suite 1NO1, 2 Huntington Quadrangle, Melville, N.Y. 11747-4502; phone: 516-576-2268; fax: 516-576-2450; e-mail: rights@aip.org.

L.C. Catalog Card No. 2001095515
ISBN 0-7354-0034-2
ISSN 0094-243X
Printed in the United States of America

Contents

Preface vii

Accelerators: Their Role, History, Status, Prospects, and Practical Applications 1
    A. Skrinsky

Linear Beam Dynamics and Beyond 6
    E. A. Perevedentsev

Quality Limitations of Hadron Beams 24
    J. Gareyte

Quality Limits for Electron Rings 44
    T. Kasuga

Review of Cooling 53
    V. V. Parkhomchuk

Intrabeam Scattering 66
    J. Struckmeier

Emittance Dilution in Proton Rings 85
    F. Willeke

Emittance Preservation in Linear Accelerators 118
    M. Minty

Luminosity and the Beam-Beam Interaction 163
    J. T. Seeman

Beam-beam Interaction in Linear Collider 185
    K. Yokoya

Wake and Impedance 205
    G. V. Stupakov

Longitudinal Single-Bunch Instabilities 231
    M. Migliorati and L. Palumbo

Transverse Mode Coupling Instabilities 260
    J. Gareyte

Polarized Beams in Accelerators and Storage Rings 279
    Y. M. Shatunov

Vlasov Equation and Landau Damping 317
    D. V. Pestrikov

Transverse Instabilities 339
    D. V. Pestrikov

Head-Tail Instabilities 356
    D. V. Pestrikov

Beam Feedback Systems 374
    M. Tobiyama

TUTORIALS

Space Charge 390
    N. A. Vinokurov

Space-Charge Effects in Circular Accelerators 405
   S. Machida
Insertion and Crossing Region Design 435
   U. Wienands and P. Beloshitsky
Tutorial on Linear Colliders 494
   F. Zimmermann
Beam Quality Control for Linear Colliders 552
   P. Logatchov

SEMINARS

Future Colliders 566
   E. Keil
High Power DC Electron Accelerators of the ELV Type 580
   R. A. Salimov
Muon Collider(s): Basics, Status, Problems, Prospects 601
   A. Skrinsky
The Antihydrogen and Positronium Problem in Particle Physics 616
   I. N. Meshkov

Author Index 633

Preface

Since 1985 the US Particle Accelerator School and the CERN Accelerator School have jointly organized a series of special topical courses on the frontiers of Beam Physics and Accelerator Technology. The titles, location, and dates of these schools are:

Nonlinear Dynamics	Santa Margherita di Pula, Sardinia
	January 31 - February 5, 1985
New Acceleration Methods and Techniques	South Padre Island, Texas
	October 23-29, 1986
Observation, Diagnosis and Correction	Anacapri, Italy
	October 20-26, 1988
Beam Intensity Limitations	Hilton Head Island, South Carolina
	November 7-14, 1990
Factories with e^+e^- Rings	Benelmadena, Spain
	October 29 - November 4, 1992

The proceedings of these courses were published in the Lecture Notes in Physics series of Springer-Verlag as volumes 247, 296, 343, 400 and 425.

In 1993, the KEK Particle Accelerator School (KEKPAS) was formed and joined us to form the US-CERN-Japan Joint Accelerator School. This new collaboration resulted in two programs:

Frontiers of Accelerator Technology	Maui, Hawaii
	November 3-9, 1994
Radio Frequency Engineering for Particle Accelerators	Hayama and Tsukuba, Japan
	September 9-18, 1996

The proceedings of these programs were published in 1996 and 1998 respectively by the World Scientific Publishing Company.

In 1996, the Russia Accelerator School was invited to join the collaboration which then became the Joint US-CERN-Japan-Russia Accelerator School, to be referred to as the JAS. The JAS collaboration led to the following two programs:

Beam Measurement	Montreux and CERN, Switzerland
	May 11-20, 1998
High Quality Beams	St. Petersburg and JINR, Dubna
	July 3-14, 2000

The proceedings of the *Beam Measurement* school was published by the World Scientific Publishing Company. The *High Quality Beams* school is the subject of this proceedings.

The guideline in the choice of the topics listed has been to present training courses for scientists and engineers working in the ever-expanding field of accelerators and beams. The course starts with basic theory and continues to the current development of advanced topics presented by established experts. In this way, all students, engineers, and scientists working in the accelerator field have the opportunity to learn and discuss their ideas and problems with experts.

This volume is the proceedings of the latest Joint Accelerator Physics and Technology on *High Quality Beams*. The international school was hosted by the Joint Institute for Nuclear Research (JINR) and the Russia Accelerator School (RAS). The program was held on a river-boat en-route from St. Petersburg to the Joint Institute for Nuclear Research (JINR), Dubna, Moscow Region from July 1-14, 2000. The program committee members were J. Galayda, S. Holmes, S.I. Kurokawa, S.Y. Lee, I. Meshkov, E. Perevedentsev, T. Shintake, G. Shirkov, A. Skrinsky, S. Turner, and E. Wilson. The resulting lectures on topics of high quality beams are listed in the Table of Contents. Seminar topics of general interests are also included.

On behalf of JAS2000, we express our sincere thanks to the local organizing committee members: I. Meshkov, G. Shirkov, N. Tokareva, N. Dokalenko, V. Katrasev, E. Shirkova, L. Soboleva, T. Stepanova, O. Strekalovsky, and V. Zhabitsky at the JINR and RAS, whose efforts were essential to the success of the school. The financial support of our sponsors is gratefully acknowledged. The greatest praise must go to the lecturers, and demonstrators for their teaching and papers for the publication of this proceedings. Finally, we thank all participants of the school for their efforts to make this school a success.

S.I. Kurokawa, KEKPAS, Tsukuba, Japan
S.Y. Lee, USPAS, Fermilab, and Indiana University, USA
J. Miles, CERN, Geneva, Switzerland
E. Perevedentsev, Russia Accelerator School, Novosibirsk, Russia

April 2001

Accelerators: Their Role, History, Status, Prospects, and Practical Applications

A. Skrinsky

Budker Institute of Nuclear Physics, Novosibirsk, 630090 Russia

In this lecture I will present a sketch of the important and interesting filed of charged particle acceleration. This field, speaking globally, is a direct expansion of the long-term drive of mankind towards deeper understanding the microstructure of matter and – now it is possible to add – towards understanding the origin and development of the Universe.

Just a hundred years ago the first elementary particle was discovered directly – the electron. The discovery was made by J.J. Thomson using an accelerator of the very first generation – the cathode-ray tube, which produced (in modern language) electrons of a few tens of keV, and J.J. succeeded in measuring the sign and charge-to-mass ratio of the species produced. Since that time, a high fraction of the discoveries in nuclear and elementary particle physics have been based on successes in the accelerator field. The other, and quite important, sources of experimental information in the field were "natural accelerators" – cosmic rays and the decay products of radioactive elements.

1. The first inventor of accelerators was probably the Norwegian engineer and physicist R. Wideroe: in the early 1920s he proposed and patented the induction accelerator (later called the betatron), and the linear RF accelerator – the first devices aimed at accelerating particles to energies much higher than the high voltage used in them.

At that time, it was already understood that an atom consists of a compact positively charged and massive nucleus plus a cloud of negatively charged and very light electrons (the Rutherford-Bohr model). Moreover, Rutherford succeeded in producing an artificial nuclear transformation, using a radioactive source.

The real accelerators that started to be used in nuclear studies in the early 1930s were the electrostatic accelerator (named Van-de-Graaff after its inventor), the Cockcroft-Walton rectifier, and the Lawrence cyclotron. With these, in many labs a lot of nuclear properties and nuclear reactions were studied and new isotopes discovered, including the discovery of the neutron (the positron, and then the muon, initially considered to be the carrier of nuclear forces, were discovered in cosmic rays).

2. The accelerator field started to grow especially actively in the process of nuclear weapon and nuclear energy development. The scale of all the devices, including accelerators, grew rapidly to its practical limits. In the 1940s and 1950s the important

steps in the accelerator field were the invention of autophasing and of strong focusing. In parallel, the long-awaited synchrotron radiation was observed at electron accelerators. Soon after, it was understood that this spectacular phenomenon brings completely new aspects in dynamics of electrons in cyclic accelerators: damping of all the oscillations and corresponding beam compression and, at the same time, quantum fluctuations of synchrotron radiation which limit electron beam compression.

It is interesting, that pions and even kaons and hyperons were discovered during these years with the use of cosmic rays, but much deeper and more detailed information on their properties and interactions was obtained using accelerators. The same is happening now with the neutrino oscillations.

3. A new stage in high energy physics started with the development of colliding beams. When the energy of an accelerated particle E_{acc} becomes bigger than its rest-frame energy M_{tar} at relativistic energies, the effective reaction energy E_{reac} that is equal to the center of mass energy E_{CoM}, becomes a smaller and smaller fraction of accelerated particle energy:

$$E_{reac} = E_{CoM} = \sqrt{2E_{acc}M_{tar}c^2},$$

whereas if two particles of full energy E_{coll} collide head-on, the reaction energy is

$$E_{reac} = 2E_{coll}.$$

In 1956 a proton-proton collider, based on a special kind of ring that can store same-sign particles moving in opposite directions, and an electron-electron collider, based on storing and compressing beams with the use of synchrotron radiation cooling, were proposed. Many groups in several labs throughout the world started work on colliding beams and two of them – the Princeton-Stanford group and the Novosibirsk group – succeeded in carrying out electron-electron experiments in 1965.

Before the success of these collider experiments, the first storage ring, AdA (Frascati), proved the possibility of maintaining an accelerated beam for hours, and accomplished the observation of bremsstrahlung in electron-positron collisions (1964).

A more difficult problem – a study of electron-positron annihilation at high energy – was solved at Novosibirsk in 1967. Since that time, colliders have become the leading tool in particle physics, bringing out interesting findings and in many cases key results.

The scale of electron-positron colliders grew rapidly, culminating in the LEP collider built and operated at CERN, with a circumference around 30 km, a top energy of 106 GeV per beam, 3 GeV per turn energy loss due to synchrotron radiation compensated by a huge RF system using superconducting cavities, etc. But LEP is probably the last step to higher energy in electron-positron collisions because of the catastrophic rise in synchrotron radiation enargy losses in cyclic machines.

4. It was understood at Novosibirsk already in the 1960s that the way to reach hundreds of GeV for e^+e^- collisions is to use linear colliders; this approach avoids the catastrophic rise in energy losses in cyclic colliders. A conceptually complete physical project for such a collider was presented in Novosibirsk in 1978. Then, a single-pass collider for 50 GeV per beam energy, based on the Stanford linear accelerator, was put into practical and effective operation. Now, several labs (SLAC, KEK, DESY, and

CERN, with the participation of some others, including the BINP) are developing technologies and components for a 500 GeV per beam – and even higher – future linear collider.

5. The race for higher and higher energies is not the only way to raise the potential of electron-positron colliders. The sharp rise in luminosity in the already explored energy domains was – and is – a very important direction. The Novosibirsk pre-Factory VEPP-2M was a successful and productive collider, for 25 years the absolute leader in its energy range of up to 1.4 GeV total. The new B-Factories are now operating efficiently at SLAC and KEK, mostly for CP violation studies in the B-meson sector; the dedicated Phi-Factory at Frascati is progressing towards carving a new way to study CP violation in the familiar K-meson sector. Very slowly, because of a lack of state support, the Charm/Tau Factory project at Novosibirsk (which includes, in addition to super-high luminosity, highly monochromatic and longitudinally polarized options) is progressing also.

6. An important direction in the accelerator field is the arranging of polarized beams in accelerators, storage rings and colliders. It is reasonable to inject polarized proton, deuteron and, in most cases, electron beams into accelerators directly from the polarized beam sources. For positrons, the only way used up to now is their polarization in storage rings due to the magneto-dipole radiation of their magnetic moments (because of the strong rest-frame magnetic field for high energy positrons); this option is useful for electrons also. Some degree of muon beam polarization appears effortless as the result of spin-momentum correlation in the process of the decay of pions – muon producers. To obtain a high degree of muon polarization requires special pion/muon beam gymnastics.

In linear accelerators it is easy to prevent loss of the degree of polarization for any particles and for any direction of polarization (either transverse or longitudinal). For a cyclic accelerator/collider, the most sustainable polarization is the one perpendicular to the median plane (the plane of the orbit). In this case, the depolarization occurs due to the action of spin resonances, and the fight against depolarization is a developed field in beam dynamics. Reaching a good degree of polarization at high energies requires the use of Siberian Snakes – special spin rotators that set the spin precession frequency to one half for any energy.

The important aim of arranging in a collider reactions with the helicities needed in an experiment (when the particles in the interaction region have the proper sign of longitudinal polarization) requires the installation of proper spin rotators, but in principle this is not less stable than transverse polarization.

An interesting and useful application of transversely polarized beams and their resonant depolarization via an external RF depolarizer in a collider, is the ultimately precise calibration of beam particle energy (in principle, up to 10^{-7} accuracy). This high accuracy knowledge allows measurement with ultimate precision of the masses of particles produced in the annihilation of electrons and positrons (and, possibly in the future, of protons and antiprotons). This method (developed and applied for kaons and for φ-, Ψ-, Ψ'-, Υ-, Υ'- and Υ''-resonances at Novosibirsk, and then used at CERN for the Z-boson – to measure their masses with an accuracy up to 2×10^{-5}) allowed the establishment of the high precision mass scale for elementary particles in the range of 1 GeV to 100 GeV.

7. The first proton-proton collider, ISR, successfully operated at CERN in 1972-1983 with energies up to 62 GeV total. But much more exciting results were obtained at the first proton-antiproton collider at CERN (1981-1986; energy range up to 700 GeV total). The success was due to the development of cooling methods effective for heavy particles, antiprotons in this case. The first method, electron cooling, was proposed at Novosibirsk (1965) for storing high intensity proton beams, and developed at Novosibirsk for storing and compressing antiprotons to arrange proton-antiproton collisions. The idea looks simple: in some straight section of a proton (antiproton) storage ring let a parallel low temperature electron beam move with the same mean velocity, and upon some relaxation time the initially high temperature of antiprotons, and hence the beam size, will go down. The idea was published in 1966 and attracted a lot of interest in the world community, but nobody outside Novosibirsk tried to develop this method till 1974, when the first electron cooling was reached at Novosibirsk. The second method for cooling beams of heavy particles, stochastic cooling, was proposed and developed at CERN. This method was used in proton-antiproton colliders at CERN (and then at Fermilab).

The methods are complementary: if you need to cool secondary particles with initially low intensity and big amplitudes (the 3-dimensional phase-space density of the beam is low), use stochastic cooling; but to reach very low temperature and high phase-space density it is much more efficient to use electron cooling. This became especially clear when it was realized from experiment and theory, that the effective transverse electron temperature is very much suppressed by the guiding longitudinal magnetic field, and the electron longitudinal temperature after acceleration also becomes very low, so that the effective temperature becomes thousands of times lower than the cathode temperature, of the order of 1 K°! And such temperatures of high energy ion beams are now more or less routinely achieved in experiments.

These cooling methods have been and are being used in many labs around the world. A new prospect was opened with the developing of high luminosity nuclei-nuclei and electron-nuclei colliders under continuous electron cooling; cooling is used to suppress multiple intra-beam scattering and to raise the beam-beam limits.

8. Ionization cooling presents interesting and important prospects. Here the dissipative force is caused by ionization energy losses of charged particles traveling in some material. This has been considered since the middle 60s, before electron and stochastic cooling. But it immediately became clear that this is not applicable to protons and other hadrons because of the high cross-sections of strong interactions, which in practically interesting cases kill the beams faster than they are cooled down. For electrons and positrons, again, the radiation losses in the matter kill the beam, not cool it. The only possible, and extremely interesting, application of ionization cooling is to cool beams of muons; for muon colliders (muon-muon, muon-proton, etc.) and for neutrino (v_e, v_μ) factories, see my paper for some aspects of this field in these Proceedings.

The "natural" field for the muon collider seems to be 1 TeV per beam and higher – lepton colliders are very important for understanding the physics in this new energy region, but linear colliders become excessively expensive and not as clean as at much lower energies.

9. One of the directions of accelerator development, especially for linear colliders, is the effort to achieve higher and higher accelerating gradients. If we consider the usual accelerator structures, where the electromagnetic field is formed by metallic boundaries, the highest possible field for normal-conducting materials (in pulsed regime) is limited by the action of the electric field normal to the surface, which can produce cold emission and then discharge. This effect limits the accelerating gradient with a value of 100 MeV/m at a frequency of about 10 GHz even for perfectly machined surfaces. The shift to 30 GHz can give 200 MeV/m. These figures are used in the projects of future linear colliders. For superconducting materials, an additional limitation is due to the critical magnetic field, which limits currently achieved values at 30 MeV/m (albeit a superconducting linear collider can provide much cleaner conditions for elementary particle physics experiments).

To achieve much higher accelerating gradients we need to switch to a plasma-formed electromagnetic field (at such gradients, because of field emission and the formation of streamer-like objects, the effective boundary is no longer a solid, but a plasma-type substance!).

The really high gradients were achieved in laser-plasma devices, but for a few laser wavelengths only. A really high energy accelerator should imply not only macroscopic plasma structures, but devices many meters and even kilometers long, excited properly (providing proper phase velocity, proper focusing, etc.) to be competitive with modern and future linacs. It has been shown (for high gradients – theoretically only!) that such beam-driven plasma wakefield structures could be feasible and practically interesting, especially for the acceleration of unstable particles – not only muons, but charged pions and kaons – and for electron-based high luminosity photon-photon colliders.

10. Up to now, our focus has been on colliders, but the growth of scale and energies of accelerators used for fixed-target experiments through all these years has also produced many important results. In the future, the main fraction of neutrino physics and an important fraction of hadron spectroscopy and low energy hadron physics in general will be based on fixed-target experiments, and this will require more and more advanced accelerators. In many cases, the same accelerators are used for fixed-target experiments and as a core part of the injection chain for colliders.

11. Accelerators, first developed as the key tool for basic research in nuclear and elementary particle physics, are now important also for technology, medicine, environment protection, etc. Here we mention only high power electron beam accelerators used for plastic materials processing, fast developing accelerator-based cancer therapy, single-use medical instrument sterilization, grain and food sterilization, and sewage water and flue gas cleaning, and accelerator-type devices for basic and applied research in many non-nuclear sciences (synchrotron radiation and neutron sources, electron beam based lasers). This important aspect of the accelerator field deserves much broader and more detailed consideration.

Being a part of high energy physics, accelerator physics and technology is an important, fast developing, and interesting field, fruitful in applications.

Linear Beam Dynamics and Beyond

E.A. Perevedentsev

Budker Institute of Nuclear Physics,
Novosibirsk 630090 Russia

I INTRODUCTION

The objective of this lecture is to give a survey of the tools needed for linear lattice analysis. Although known for a long time [1] and widely available in textbooks, e.g. [2–6], linear beam dynamics presented in an entire view may be useful in the framework of this School to provide reference material for subsequent lectures and tutorial courses.

Among different formulations of the betatron motion theory, we prefer the complex Floquet function formalism [2], and the examples of linear and nonlinear dynamics problems presented below are intended to illustrate its efficiency.

II EQUATIONS OF MOTION

We consider the design orbit in a circular accelerator as a closed planar curve made up of arcs, each with a constant radius of curvature r_0. Each bend in the orbit is caused by a sector magnet, see Fig. 1a. The cylindrical coordinates r, θ, z are applied in each sector magnet (Fig. 1b), and its magnetic field $\mathbf{B} = (B_r, B_\theta, B_z)$ is two-dimensional:

$$B_\theta = 0, \quad \frac{\partial B_{r,z}}{\partial \theta} = 0.$$

We start with the relativistic equation of motion in the horizontal plane

$$\gamma m \ddot{r} = \gamma m \frac{v_\theta^2}{r} + \frac{e}{c} B_z v_\theta \qquad (1)$$

with the centrifugal and Lorentz forces on the right side. Here e and m are the particle's charge and mass, $\mathbf{v} = (v_r, v_\theta, v_z)$ is its velocity, c is the speed of light, $\beta = v/c$ and $\gamma = (1 - v^2/c^2)^{-1/2}$ are the relativistic factors. On the design orbit the vertical field B_0 is related to the nominal momentum p_0,

$$p_0 c = \gamma_0 \beta_0 m c^2 = -e B_0 r_0, \qquad (2)$$

where the subscript zero indicates nominal particle parameters. Consider a particle's trajectory in a neighborhood close to the design orbit and use the development around the design orbit

$$v_{z,r} \ll v_\theta \approx \beta_0 c, \quad x = r - r_0 \ll r_0.$$

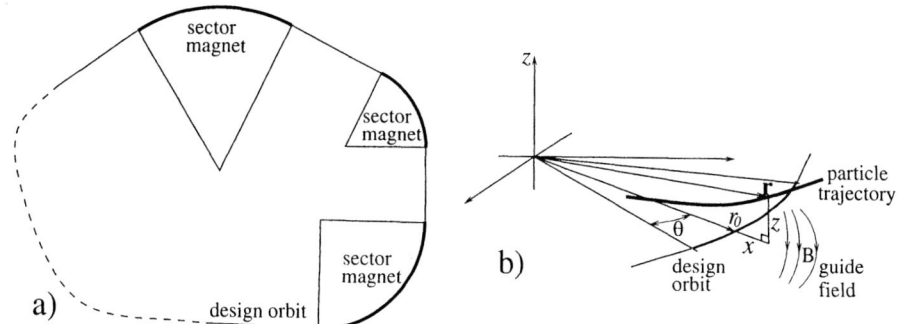

FIGURE 1. a) Design orbit formed by sector magnets; b) accelerator coordinates x, z describe particle trajectories using the design orbit as a reference.

Instead of equations of motion it is convenient to use equations of trajectories, changing the independent variable in Eq. (1) from time t to the path s along the design orbit, $ds = v_\theta dt \approx v_0 dt$,

$$r'' = \frac{1}{r} + \frac{eB_z}{\gamma\beta mc^2}, \qquad (3)$$

where the prime indicates the derivative over s. Development to first order on the right side of Eq. (3), taking account of the particle's momentum offset $p = p_0 + \Delta p$, and relation (2) yields

$$\frac{1}{r_0 + x} + \frac{e(B_0 + \partial B_z/\partial r\, x + \ldots)}{\gamma_0\beta_0 mc^2(1 + \Delta p/p_0)} \approx -\frac{x}{r_0^2} + \frac{e}{p_0 c}\frac{\partial B_z}{\partial r}x + \frac{1}{r_0}\frac{\Delta p}{p}.$$

Thus we have obtained the linearized equation of horizontal motion,

$$x'' + K_x x = \frac{1}{r_0}\frac{\Delta p}{p}, \qquad (4)$$

with the horizontal focusing function K_x in terms of the orbit curvature, and the guide field gradient calculated on the design orbit,

$$K_x = \frac{1}{r_0^2} - \frac{e}{p_0 c}\frac{\partial B_z}{\partial r}. \qquad (5)$$

Similar development for the equation of vertical motion

$$\gamma m\ddot{z} = -\frac{e}{c}B_r v_\theta,$$

taking account of

$$B_r = 0 + \frac{\partial B_r}{\partial z}z + \ldots \approx \frac{\partial B_z}{\partial r}z, \qquad (\mathrm{rot}\mathbf{B})_\theta = \frac{\partial B_r}{\partial z} - \frac{\partial B_z}{\partial r} = 0,$$

yields the linearized equation of vertical motion

$$z'' + K_z z = 0, \tag{6}$$

where K_z is the vertical focusing function,

$$K_z = \frac{e}{p_0 c} \frac{\partial B_z}{\partial r}. \tag{7}$$

A Dispersion

Consider a closed trajectory for an off-momentum particle with $p = p_0 + \Delta p$, and call it the off-momentum orbit.

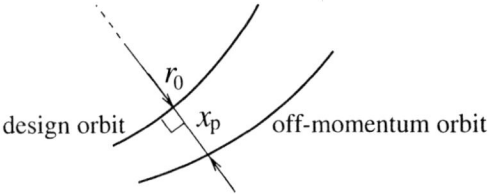

FIGURE 2. Deviation x_p of the closed orbit for off-momentum particles from the design orbit.

This orbit deviates from the design orbit by x_p, as shown in Fig. 2,

$$x_p = D_x \frac{\Delta p}{p} \tag{8}$$

where D_x is called the (horizontal) dispersion. It should to be found, after substituting Eq. (8) into Eq. (4), as a periodic particular solution of the linearized horizontal equation,

$$D_x'' + K_x D_x = \frac{1}{r_0}, \tag{9}$$

where the focusing function $K_x(s)$ and the design orbit curvature radius $r_0(s)$ are periodic functions with the period C_0 of the design orbit circumference.

Dispersion results in first-order path lengthening along the off-momentum orbit,

$$ds\left(1 + \frac{D_x}{r_0}\frac{\Delta p}{p}\right) - ds = \frac{D_x}{r_0}\frac{\Delta p}{p} ds. \tag{10}$$

Integration of Eq. (10) around the machine gives for the off-momentum orbit circumference C,

$$C - C_0 = \Delta C = \frac{\Delta p}{p} \oint \frac{D_x}{r_0} ds \equiv \alpha_p C_0 \frac{\Delta p}{p}, \tag{11}$$

where we introduced the momentum compaction factor

$$\alpha_p = \frac{1}{C_0} \oint \frac{D_x}{r_0} ds.$$

In modern strong-focusing machines $\alpha_p \ll 1$ (a simple estimate is $\alpha_p \approx 1/Q_x^2$, where Q_x is the horizontal betatron oscillation tune).

For the off-momentum particle revolution period T we take into account both the off-momentum orbit lengthening, Eq. (11), and the deviation of the off-momentum velocity $\beta = \beta_0 + \Delta\beta$ from the nominal one,

$$T = \frac{C}{\beta c} = \frac{C_0(1 + \Delta C/C_0)}{\beta_0(1 + \Delta\beta/\beta_0)} \approx T_0 \left(1 + \frac{\Delta C}{C} - \frac{\Delta\beta}{\beta}\right). \tag{12}$$

Using the momentum compaction α_p and relating the particle's velocity to its momentum,

$$\frac{\Delta p}{p} = \frac{\Delta(\gamma m v)}{p} = \frac{1}{\gamma m v} \Delta\left(\frac{mv}{\sqrt{1-v^2/c^2}}\right) = \gamma^2 \frac{\Delta v}{v},$$

we can express the deviation of the revolution frequency ω_0 for the off-momentum particle from Eq. (12),

$$\frac{\Delta\omega_0}{\omega_0} = -\frac{T - T_0}{T_0} = -\left(\alpha_p - \frac{1}{\gamma^2}\right)\frac{\Delta p}{p} \equiv \eta \frac{\Delta p}{p},$$

where we defined the slippage factor η, which plays an essential role in the longitudinal dynamics of particles, considered elsewhere [4].

We deal hereafter only with the transverse motion of on-momentum particles.

III HILL'S EQUATION

Modern focusing systems of circular accelerators are composed of complicated combinations of focusing magnets. To reveal the general properties of the transverse motion, we need to study first the linearized equations with variable focusing functions $K(s)$,

$$x'' + K(s)\, x = 0, \tag{13}$$

with the single restriction that this function is apparently periodic with the machine orbit circumference C, i.e. $K(s + C) = K(s)$. This equation is called Hill's equation; our Eqs. (4) and (6) belong to this type.

A Constant Focusing

Consider first the simple special case of $K = \text{const}$. For $K > 0$, we can take the cosine trajectory,

$$C(s) = \cos\sqrt{K}s\,, \quad \begin{cases} C(0) = 1 \\ C'(0) = 0 \end{cases}$$

and the sine trajectory,

$$S(s) = \sin\sqrt{K}s\,, \quad \begin{cases} S(0) = 0 \\ S'(0) = 1 \end{cases}$$

as a complete set of two linearly independent particular solutions of Hill's equation, Eq. (13), which yields simple harmonic oscillations in this special case.

For $K < 0$, we should replace the above solutions by the respective hyperbolic functions,

$$C(s) = \cosh\sqrt{-K}s\,, \quad \begin{cases} C(0) = 1 \\ C'(0) = 0 \end{cases}, \qquad S(s) = \sinh\sqrt{-K}s\,, \quad \begin{cases} S(0) = 0 \\ S'(0) = 1 \end{cases},$$

and this motion is locally unstable, the deviations from the design orbit grow exponentially.

The chosen set of solutions provides for convenient expression of the general solution to Eq. (13), with given initial displacement x_0 and initial slope x'_0 of a trajectory in the matrix form,

$$\begin{pmatrix} x(s) \\ x'(s) \end{pmatrix} = \begin{pmatrix} C(s) & S(s) \\ C'(s) & S'(s) \end{pmatrix} \begin{pmatrix} x_0 \\ x'_0 \end{pmatrix} \equiv T \begin{pmatrix} x_0 \\ x'_0 \end{pmatrix}. \tag{14}$$

Here matrix T is called the transfer matrix, and we see that transport of a trajectory specified by its initial conditions is just a linear transformation.

B Alternating-Gradient Focusing

Since Hill's equation is a second-order linear equation, its general solution in the form given by Eq. (14) holds for arbitrary $K(s) \neq \text{const}$. In this case the cosine and sine trajectories are to be found first, by solving the differential equation with the appropriate initial conditions. In practice, we can recommend approximation of $K(s)$ by step functions, to whatever detail needed, as shown in Fig. 3, then using solutions with $K(s) = \text{const}$ at each interval.

When these intervals are concatenated, the continuity of the solution $x(s)$ (and of the trajectory slope $x'(s)$) is preserved by multiplication of the respective transfer matrices:

$$\begin{pmatrix} x(s) \\ x'(s) \end{pmatrix} = T_3(s|s_2)\left(T_2(s_2|s_1)\left(T_1(s_1|s_0)\begin{pmatrix} x_0 \\ x'_0 \end{pmatrix}\right)\right) = T_3(s)T_2T_1\begin{pmatrix} x_0 \\ x'_0 \end{pmatrix}.$$

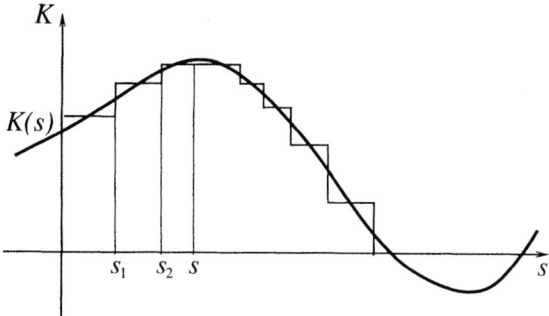

FIGURE 3. Approximation of $K(s)$ by step functions.

The resulting transfer matrix of an arbitrary focusing system is thus available,

$$T(s|s_0) = \ldots T_3 T_2 T_1 = \begin{pmatrix} \mathcal{C}(s) & \mathcal{S}(s) \\ \mathcal{C}'(s) & \mathcal{S}'(s) \end{pmatrix}. \tag{15}$$

For any T, $\det T = $ const, being the Wronskian of Hill's equation. Our specific choice of the initial conditions provides for $\det T = 1$.

C One-Period Matrix $M(s)$ and Stability

Now we introduce the one-period transfer matrix $M(s)$,

$$M(s) = T(s + C|s), \tag{16}$$

which transports the solution forward by one period, see Fig. 4.

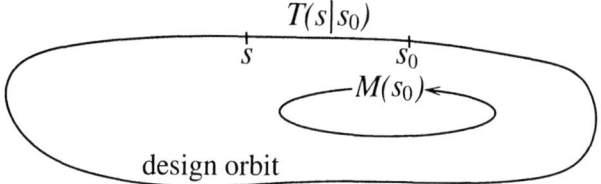

FIGURE 4. Transformation by a one-period matrix $M(s_0)$.

For stable motion, we should have limited values of x and x' when applying M repeatedly to any initial condition. Transport over N periods is given by M^N, therefore stability requires that eigenvalues λ of M must be limited, $|\lambda| \leq 1$. Otherwise, with $|\lambda| > 1$, λ^N means a possibility of unlimited growth of displacements.

Rewriting M via its matrix elements,

$$M = \begin{pmatrix} m_{11} & m_{12} \\ m_{21} & m_{22} \end{pmatrix},$$

we find the eigenvalues of M from the characteristic equation

$$\det(M - \lambda I) = 0,$$

or, explicitly,

$$\begin{vmatrix} m_{11} - \lambda & m_{12} \\ m_{21} & m_{22} - \lambda \end{vmatrix} = \lambda^2 - (m_{11} + m_{22})\lambda + \det M = 0.$$

Using $\det M = 1$ and denoting the trace of M, $m_{11} + m_{22} = \operatorname{tr} M$, we solve this equation for λ,

$$\lambda_{1,2} = \frac{1}{2}\operatorname{tr} M \pm i\sqrt{1 - \left(\frac{1}{2}\operatorname{tr} M\right)^2} \equiv \cos\mu \pm i\sin\mu = e^{\pm i\mu}, \tag{17}$$

where $\cos\mu = \frac{1}{2}\operatorname{tr} M$. From $\det M = 1$ we immediately have $\lambda_1 \lambda_2 = 1$, which means that assuming $|\lambda_1| < 1$ we would have $|\lambda_2| > 1$, i.e. growing displacements become possible with certain initial conditions. Thus, the stability condition $|\lambda| \leq 1$ is reduced to $|\lambda| = 1$ only.

In other words, $\operatorname{Im}\mu = 0$ in Eq. (17), or $|\cos\mu| \leq 1$. Finally, the stability condition can be expressed in terms of matrix M,

$$-2 \leq \operatorname{tr} M \leq 2. \tag{18}$$

Stable solutions of Hill's equation are called betatron oscillations, and the meaning of parameter μ is the phase advance of these oscillations over one period of the focusing structure.

D Twiss Parametrization

The one-period matrix M may be conveniently represented via the identity matrix I and the zero-trace matrix J, $\operatorname{tr} J = 0$, composed of the so-called Twiss parameters,[1]

$$M = I\cos\mu + J\sin\mu = \begin{pmatrix} \cos\mu + \alpha\sin\mu & \beta\sin\mu \\ -\gamma\sin\mu & \cos\mu - \alpha\sin\mu \end{pmatrix}. \tag{19}$$

Among the matrix elements of J,

$$J = \begin{pmatrix} \alpha & \beta \\ -\gamma & -\alpha \end{pmatrix}, \tag{20}$$

there are only two independent parameters, since the relation $\det M = 1$ bounds these matrix elements, $\gamma\beta - \alpha^2 = 1$, or $\det J = 1$. Hence, $J^2 = -I$, and the matrix exponent form of M follows, $M = \exp(\mu J)$.

---

[1] Conventionally denoted as β, γ, the Twiss parameters should not be confused with the relativistic factors.

The matrix elements of $M(s)$ are apparently periodic functions of s, $M(s+C) = M(s)$, and so are the Twiss functions $\beta(s)$, $\alpha(s)$ and $\gamma(s)$.

Provided $M(s_0)$ is known, transformation to another point s is given by the transfer matrix $T(s|s_0)$, see Fig. 4,

$$M(s) = T M(s_0) T^{-1}. \tag{21}$$

Exercise. Proof of transformation (21) is left to the reader as an exercise.
Exercise. Prove that parameter μ does not depend on s.

This transformation of M is in fact a linear transformation of its matrix elements, therefore a linear transformation of the Twiss functions from s_0 to s. Sometimes it is convenient to represent this transformation by a 3×3 matrix, called Steffen's matrix, acting on a 3-vector (β, α, γ),

$$\begin{pmatrix} \beta(s) \\ \alpha(s) \\ \gamma(s) \end{pmatrix} = \begin{pmatrix} t_{11}^2 & -2t_{11}t_{12} & t_{12}^2 \\ -t_{11}t_{21} & t_{11}t_{22} + t_{12}t_{21} & -t_{12}t_{22} \\ t_{21}^2 & -2t_{21}t_{22} & t_{22}^2 \end{pmatrix} \begin{pmatrix} \beta_0 \\ \alpha_0 \\ \gamma_0 \end{pmatrix}, \tag{22}$$

where t_{ik} are the matrix elements of $T(s|s_0)$.

Exercise. Derive the elements of Steffen's matrix from Eq. (21).

Next we need to derive differential equations for the Twiss parameters. First define matrix D,

$$D = \begin{pmatrix} 0 & 1 \\ -K & 0 \end{pmatrix}, \tag{23}$$

which contains the focusing function K and serves for rewriting Hill's equation in matrix form,

$$X' = \frac{d}{ds} \begin{pmatrix} x \\ x' \end{pmatrix} = \begin{pmatrix} 0 & 1 \\ -K & 0 \end{pmatrix} \begin{pmatrix} x \\ x' \end{pmatrix} = DX. \tag{24}$$

The differential equation for the transfer matrix T has the same form. Indeed,

$$\frac{d}{ds} T = \begin{pmatrix} C' & S' \\ C'' & S'' \end{pmatrix} = \begin{pmatrix} C' & S' \\ -KC & -KS \end{pmatrix} = \begin{pmatrix} 0 & 1 \\ -K & 0 \end{pmatrix} \begin{pmatrix} C & S \\ C' & S' \end{pmatrix},$$

or

$$T' = DT. \tag{25}$$

However, the differential equation for a one-period matrix $M(s) = T(s+C|s)$ is quite different. To derive it, we start from Eq. (21),

$$M = M(s) = T M_0 T^{-1},$$

rewrite it as

$$MT = TM_0,$$

and differentiate with respect to s,
$$M'T + MT' = T' M_0 .$$
Hence, using Eq. (25), we get
$$M'T + MDT = DTM_0 .$$
Multiplying on the right by T^{-1}, and using Eq. (21), we obtain
$$M' + MD = DTM_0 T^{-1} = DM ,$$
or, finally,
$$M' = DM - MD . \tag{26}$$
Substituting here the Twiss form of M, Eq. (19), we find a set of differential equations for the Twiss functions,
$$\beta' = -2\alpha ,$$
$$\alpha' = K\beta - \gamma ,$$
$$\gamma' = 2K\alpha . \tag{27}$$
Elimination of α and substitution of $\gamma = (1+\alpha^2)/\beta$ yields a rather cumbersome equation for the β-function alone,
$$\frac{1}{2}\beta\beta'' - \frac{1}{4}\beta'^2 + K\beta^2 = 1 . \tag{28}$$
Note that these equations should be solved with periodic boundary conditions, since the Twiss functions are periodic.

Fortunately, equations for some more convenient functions, $w(s) = \sqrt{\beta(s)}$, look much better. Substituting
$$\beta = w^2, \quad \alpha = -\beta'/2 = -ww', \quad \gamma = w'^2 + \frac{1}{w^2}, \tag{29}$$
into $\alpha' = K\beta - \gamma$ in Eq. (27),
$$-(ww')' = -ww'' - w'^2 = Kw^2 - w'^2 - \frac{1}{w^2},$$
we get a nice equation for w,
$$w'' + Kw = \frac{1}{w^3}, \tag{30}$$
again with periodic boundary conditions.

Exercise. Show that the general solution of Eq. (30) in a focusing-free section ($K = 0$) has the form
$$w(s) = \sqrt{\beta_0 + \frac{(s - s_0)^2}{\beta_0}},$$
s_0 and β_0 being the constants of integration. What is their meaning?

Exercise. Find the general solution of Eq. (30) in a constant-focusing section ($K = $ const). Compare the result with that given by the transformation (22).

E Eigenvectors of $M(s)$

Now we find the eigenvectors $F^T = (f, f')$ of the one-period matrix $M(s)$, using its Twiss form Eq. (19) and knowing its eigenvalues $\lambda_{1,2} = e^{\pm i\mu}$. From $MF = e^{\pm i\mu} F$,

$$\begin{pmatrix} \cos\mu + \alpha\sin\mu & \beta\sin\mu \\ -\gamma\sin\mu & \cos\mu - \alpha\sin\mu \end{pmatrix} \begin{pmatrix} f_\pm \\ f'_\pm \end{pmatrix} = e^{\pm i\mu} \begin{pmatrix} f_\pm \\ f'_\pm \end{pmatrix},$$

we have

$$\frac{f'_\pm}{f_\pm} = \frac{\pm i - \alpha}{\beta}. \tag{31}$$

Note that the eigenvector components are functions of s and obey Hill's equation, Eq. (13). Substituting $\alpha = -\beta'/2$ from Eq. (27) on the right side of Eq. (31), we obtain an expression of the eigenvector components via the Twiss parameters, in the form of a differential equation,

$$\frac{f'_\pm}{f_\pm} = \frac{\beta'}{2\beta} \pm \frac{i}{\beta}.$$

Integration yields a fundamental relation of the eigenvector to the β-function,

$$f_\pm(s) = f_0 \sqrt{\beta(s)} \exp\left[\pm i \int^s \frac{ds'}{\beta(s')}\right], \tag{32}$$

where f_0 is the integration constant. Using freedom of normalization, we choose $f_0 = 1$ and get the complex-conjugate pair of eigenvectors, substituting Eq. (32) into Eq. (31),

$$\begin{pmatrix} \beta \\ \pm i - \alpha \end{pmatrix} \frac{e^{\pm i\psi}}{\sqrt{\beta}}, \quad \psi = \int^s \frac{ds'}{\beta(s')}. \tag{33}$$

Note that the initial phase in Eq. (33) is still left a free parameter. Using Eqs. (29), we can also write this complex-conjugate pair of normalized eigenvectors F, F^* in terms of w and w',

$$F = \begin{pmatrix} f \\ f' \end{pmatrix} = \begin{pmatrix} w \\ w' + i/w \end{pmatrix} e^{i\psi}, \quad \psi' = \frac{1}{w^2}. \tag{34}$$

F The Floquet Theorem

Theorem. For Hill's equation

$$x'' + K(s)\,x = 0$$

where the focusing function is periodic, $K(s+C) = K(s)$, there exist normal solutions $f(s)$,

$$f'' + K(s)\,f = 0,$$

for which advance by one period means multiplication by a phase factor,
$$f(s+C) = e^{i\mu} f(s).$$

Indeed, the above constructed eigenvectors of M, with $f(s) = w(s)e^{i\psi(s)}$, whose absolute value is a periodic function, are transformed by M when advanced by one period, and this transformation is reduced to multiplication by the eigenvalue $e^{i\mu}$ of M,

$$\begin{pmatrix} f \\ f' \end{pmatrix}_{s+C} = M \begin{pmatrix} f \\ f' \end{pmatrix}_s = e^{i\mu} \begin{pmatrix} f \\ f' \end{pmatrix}_s.$$

Moreover, the phase advance is related to the amplitude function w,

$$\psi(s+C) - \psi(s) = \mu = \oint d\psi = \oint \frac{ds}{w^2}. \tag{35}$$

These normal solutions $f(s) = w(s)e^{i\psi(s)}$ are often called Floquet functions. We will call $F^T = (f, f')$, given by Eq. (34), the Floquet vector. Together with its complex conjugate, they form a complete basis. Any solution of Hill's equation can be decomposed in this basis,

$$\begin{pmatrix} x \\ x' \end{pmatrix} = \frac{A}{2} \begin{pmatrix} f \\ f' \end{pmatrix} + \frac{A^*}{2} \begin{pmatrix} f^* \\ f^{*\prime} \end{pmatrix} = \text{Re}[AF]. \tag{36}$$

Using the normalization condition in the Wronskian form,

$$\begin{vmatrix} f & f^* \\ f' & f^{*\prime} \end{vmatrix} = e^{i\psi} e^{-i\psi} \begin{vmatrix} w & w \\ w' + i/w & w' - i/w \end{vmatrix} = -2i, \tag{37}$$

we rewrite the determinant as a skew-scalar product with the help of matrix S,

$$S = \begin{pmatrix} 0 & 1 \\ -1 & 0 \end{pmatrix}. \tag{38}$$

Then

$$\begin{vmatrix} f & f^* \\ f' & f^{*\prime} \end{vmatrix} = (f, f') \begin{pmatrix} 0 & 1 \\ -1 & 0 \end{pmatrix} \begin{pmatrix} f^* \\ f^{*\prime} \end{pmatrix} = F^T S F^* = -2i, \tag{39}$$

while $F^T S F = 0$. These relations help to find the decomposition constant A in Eq. (36) where multiplication on the left by $F^{*T} S$ yields:

$$A = \frac{1}{i} F^{*T} S X = -i e^{-i\psi} \begin{vmatrix} w & x \\ w' - i/w & x' \end{vmatrix}. \tag{40}$$

From the fact that A is a constant determined by the initial conditions of the trajectory, follows the Courant-Snyder invariant,

$$|A|^2 = (wx' - w'x)^2 + \frac{x^2}{w^2} = \gamma x^2 + 2\alpha x x' + \beta x'^2 \equiv \epsilon. \tag{41}$$

When the solution $x(s)$ is propagated in an AG focusing lattice, the quadratic form remains constant because of appropriate variation of the Twiss functions. The physical meaning of this invariant is that it is proportional to the action variable in the particle motion.

G Pseudo-Harmonic Oscillations

From Eq. (36), using Eqs. (33) and (41), we arrive at the pseudo-harmonic form of the solutions to Hill's equation,

$$x(s) = \sqrt{\epsilon \beta(s)} \cos \psi(s), \tag{42}$$

$$x'(s) = -\sqrt{\frac{\epsilon}{\beta(s)}} \left(\sin \psi(s) + \alpha(s) \cos \psi(s) \right). \tag{43}$$

Exercise. Derive Eq. (43) from Eq. (42) by differentiation, using the relation

$$\psi(s) = \int^s \frac{ds'}{\beta(s')}.$$

Exercise. Find the betatron oscillation tune Q,

$$Q = \frac{\mu}{2\pi} = \oint \frac{ds}{\beta(s)}.$$

Exercise. Show by straightforward substitution that the solution given by Eq. (42) satisfies Hill's equation, Eq. (13).

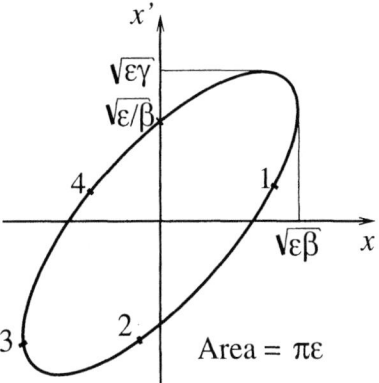

FIGURE 5. Elliptic phase-space trajectory of the betatron oscillation.

Figure 5 shows the phase space of the betatron oscillation, illustrating the meaning of the Twiss parameters. The Courant-Snyder quadratic form, Eq. (41), defines the ellipse with area $\pi \epsilon$. Being a locus of points 1, 2, ... representing one-period mapping, the ellipse is often called a phase-space trajectory of the betatron oscillation. From Fig. 5 and Eq. (42) we conclude that $w(s) = \sqrt{\beta(s)}$ is the envelope function enclosing all betatron trajectories with given $|A|$.

The pseudo-harmonic oscillation is related to the simple harmonic oscillation by a linear transformation, following from Eqs. (42) and (43),

$$\begin{pmatrix} x \\ x' \end{pmatrix} = \begin{pmatrix} \sqrt{\beta} & 0 \\ -\alpha/\sqrt{\beta} & 1/\sqrt{\beta} \end{pmatrix} \begin{pmatrix} \sqrt{\epsilon}\cos\psi \\ -\sqrt{\epsilon}\sin\psi \end{pmatrix}. \tag{44}$$

and by the change of independent variable from s to ψ. The new variables are called the normalized variables.

H Perturbation of Hill's Equation

Let us add to the nominal Hill equation, Eq. (13), an additional term $g(x,s)$ on the right side,

$$x'' + K(s)\,x = g(x,s), \tag{45}$$

or, in matrix form,

$$\frac{d}{ds}\begin{pmatrix} x \\ x' \end{pmatrix} = \begin{pmatrix} 0 & 1 \\ -K & 0 \end{pmatrix}\begin{pmatrix} x \\ x' \end{pmatrix} + \begin{pmatrix} 0 \\ g \end{pmatrix}$$

or,

$$X' = DX + G. \tag{46}$$

The solution may be sought in the same form, Eq. (36), as before, since the nominal Floquet vectors provide for a complete basis. However, the amplitude A is no longer constant. Differentiating Eq. (40) for the amplitude A,

$$A' = \frac{1}{i}(F^{*\prime})^T S\,X + \frac{1}{i}F^{*T} S\,X'$$

using Eq. (46) for X' and the nominal Hill equation, Eq. (13), for the Floquet vectors, $F^{*\prime} = DF^*$, we have

$$A' = \frac{1}{i}F^{*T}\left(\left(D^T S + SD\right)X + SG\right).$$

Since $D^T S + SD = 0$, we finally get the equation for the variation of A caused by the perturbation

$$A' = -i\,F^{*T} S\,G. \tag{47}$$

Putting $A = |A|\,e^{i\phi}$, where both the absolute value and the phase of A are variable because of perturbation on the right side of Hill's equation, Eq. (45), we rewrite Eq. (47) as

$$|A|' + i\phi'\,|A| = -ie^{-i(\psi+\phi)}\sqrt{\beta}\,g(x,s), \tag{48}$$

where on the right side we should put $x = |A|\,\sqrt{\beta}\cos(\psi+\phi)$, in order to complete the change of variables from x, x' to the complex amplitude $A = |A|\,e^{i\phi}$. Note that no assumptions have been made about the smallness of $g(x,s)$, and Eqs. (47) and (48) are exact equations.

IV APPLICATION OF THE FLOQUET FORMALISM

Below we present a number of problems in linear and nonlinear transverse dynamics, which are solved by means of the Floquet formalism, to illustrate its efficiency.

A Transfer Matrix in Terms of Twiss Parameters

Problem. Consider a section of a circular accelerator lattice with specified values of the Twiss parameters on its entrance and exit, i.e. β_i, α_i and β_f, α_f, respectively. Find the transfer matrix T of the section.

To solve this problem, we write the transformation of the Floquet vector performed by this optics,

$$T \begin{pmatrix} f \\ f' \end{pmatrix}_i = e^{i\phi} \begin{pmatrix} f \\ f' \end{pmatrix}_f, \qquad (49)$$

where the arbitrary phase factor follows from freedom in the eigenvector normalization, see Eq. (33). The meaning of ϕ is the betatron oscillation phase advance provided by this optics; it is a free parameter.

Introducing $w_{i,f} = \sqrt{\beta_{i,f}}$ and $w'_{i,f} = -\alpha_{i,f}/w_{i,f}$, we compose matrix W from column Floquet vectors,

$$W = \begin{pmatrix} w & w \\ w' + i/w & w' - i/w \end{pmatrix},$$

then from Eq. (49) we obtain a matrix equation for unknown T,

$$T W_i = W_f \begin{pmatrix} e^{i\phi} & 0 \\ 0 & e^{-i\phi} \end{pmatrix}.$$

The solution is

$$T(f|i) = W_f \begin{pmatrix} e^{i\phi} & 0 \\ 0 & e^{-i\phi} \end{pmatrix} W_i^{-1}.$$

Calculation of the right side yields, for the matrix elements of $T(f|i)$,

$$T(f|i) = \begin{pmatrix} t_{11} & t_{12} \\ t_{21} & t_{22} \end{pmatrix},$$

$$t_{11} = \frac{w_f}{w_i}(\cos\phi - w_i w'_i \sin\phi) = \sqrt{\frac{\beta_f}{\beta_i}}(\cos\phi + \alpha_i \sin\phi),$$

$$t_{12} = w_f w_i \sin\phi = \sqrt{\beta_f \beta_i} \sin\phi,$$

$$t_{21} = \left(\frac{w'_f}{w_i} - \frac{w'_i}{w_f}\right)\cos\phi - \left(w'_f w'_i + \frac{1}{w_f w_i}\right)\sin\phi$$

$$= -\frac{1}{\sqrt{\beta_f \beta_i}}\left((\alpha_f - \alpha_i)\cos\phi + (1 + \alpha_f \alpha_i)\sin\phi\right),$$

$$t_{22} = \frac{w_i}{w_f}(\cos\phi + w_f w'_f \sin\phi) = \sqrt{\frac{\beta_i}{\beta_f}}(\cos\phi - \alpha_f \sin\phi),$$

where i, f indicate the entrance and exit of the optical section.

B Propagation of Mismatched β-Function

Problem. At injection, $s = 0$, initial β_i and β'_i are mismatched with the nominal values β_0 and β'_0 of the lattice. Trace $\beta(s)$ along the lattice.

Using the initial Twiss parameters, we construct the initial Floquet vector,

$$F_i = \begin{pmatrix} w_i \\ w'_i + i/w_i \end{pmatrix},$$

and decompose F_i in the Floquet basis with the nominal w_0 and w'_0,

$$F_i = c_1 F_0 + c_2^* F_0^*, \quad F_0 = \begin{pmatrix} w_0 \\ w'_0 + i/w_0 \end{pmatrix}. \tag{50}$$

To find the constants, we multiply this decomposition on the left by F^{*T}. Then, using the normalization relations, see Eqs. (37) and (39),

$$F^{*T} S F = -(F^T S F^*)^T = (F^T S F^*)^* = 2i,$$

$$F^T S F = F^{*T} S F^* = 0,$$

we obtain for the decomposition constants,

$$c_1 = \frac{1}{2i} F_0^{*T} S F_i, \quad c_2 = \frac{1}{2i} F_0^{*T} S F_i^*.$$

Decomposition (50) holds for any s downstream of the injection point at $s = 0$. Propagation of this initial Floquet vector is then determined by known functions of the nominal lattice, $w_0(s)$ and $\psi_0(s)$:

$$\begin{pmatrix} w(s) \\ w'(s) + \frac{i}{w(s)} \end{pmatrix} e^{i\psi(s)} = c_1 \begin{pmatrix} w_0(s) \\ w'_0(s) + \frac{i}{w_0(s)} \end{pmatrix} e^{i\psi_0(s)} + c_2^* \begin{pmatrix} w_0(s) \\ w'_0(s) - \frac{i}{w_0(s)} \end{pmatrix} e^{-i\psi_0(s)},$$

where $\psi(0) = \psi_0(0) = 0$. The mismatched β-function is

$$\beta(s) = w_0^2(s) \left| c_1 e^{i\psi_0(s)} + c_2^* e^{-i\psi_0(s)} \right|^2 = \beta_0(s) \left(|c_1|^2 + |c_2|^2 + 2\mathrm{Re}\left[c_1 c_2 e^{2i\psi_0(s)} \right] \right).$$

A $\cos(2\psi_0(s) + \varphi)$ term emerges from the right side, indicating the beat of the β-function at twice the betatron tune.

C Amplitude-Dependent Tuneshift

Problem. Consider now a nonlinear perturbation $g(x,s) = q_m(s)x^m$ on the right side of Hill's equation, Eq. (45). Find the resulting correction to the betatron tune.

To solve this problem, we put $g(x,s) = q_m(s)x^m$ in Eq. (48) for the complex amplitude $A = |A|\, e^{i\phi}$,

$$|A|' + i\phi'\,|A| = -ie^{-i(\psi+\phi)}\sqrt{\beta}q_m(s)x^m. \tag{51}$$

The periodic function $q_m(s)$ here can be represented by its Fourier series. From Eq. (51) we see that the phase ϕ obeys the equation

$$\phi' = -\frac{q_m}{|A|}x^m\sqrt{\beta}\cos(\psi+\phi),$$

where we should substitute $x = |A|\sqrt{\beta}\cos(\psi+\phi)$,

$$\phi' = -q_m\,|A|^{m-1}\,\beta^{(m+1)/2}\cos^{m+1}(\psi+\phi). \tag{52}$$

The average of the right side over fast oscillations[2] is non-vanishing for odd m. The phase ϕ on the right side may be kept constant while averaging, if the perturbation is small (which means that ϕ' is also small).

Starting with the case of perturbation of linear focusing, $m=1$, we get from Eq. (52), after averaging,

$$\phi' = -\frac{1}{2}q_1(s)\beta(s).$$

Integration of ϕ' over the orbit circumference yields contribution $\Delta\mu$ to the phase advance μ from the perturbation. Thus we can obtain the betatron tuneshift from additional focusing,

$$\Delta Q = \frac{\Delta\mu}{2\pi} = \frac{1}{2\pi}\oint \phi'ds = -\frac{1}{4\pi}\oint q_1(s)\beta(s)\,ds.$$

For the cubic nonlinearity $m=3$, averaging of the right side of Eq. (52) results in

$$\overline{\cos^4(\psi+\phi)} = \frac{3}{8}. \tag{53}$$

Integration of ϕ' given by Eq. (52) with $m=3$ yields the tuneshift

$$\Delta Q = \frac{\Delta\mu}{2\pi} = \frac{1}{2\pi}\oint \phi'ds = -\frac{3|A|^2}{16\pi}\oint q_3(s)\beta^2(s)ds,$$

where use has been made of Eq. (53). The found amplitude-dependent tuneshift is due to nonlinearity of the betatron motion, and the Twiss β-function squared is the weight function of the cubic perturbation. For mth-order nonlinearity the weight function will be $\beta^{(m+1)/2}$.

---

[2] The averaging is performed with the assumption that the betatron tune stays well apart from (higher-order) resonances.

D Dynamic Aperture Limitation by a Single Nonlinear Kick

Consider a nonlinear one-period map formed by a linear optics section, with a betatron phase advance of $2\pi Q$, followed by a thin nonlinear element which can be treated in the kick approximation. A fixed point of this map (other than $x = 0$, $x' = 0$) is the location in the phase space of the system, where the onset of stochasticity occurs. Thus the position of the fixed point(s) may provide a rough estimate of the available dynamic aperture, i.e. the phase space area around the origin where the regularity of motion is preserved.

Problem. Taking an example with a single sextupole kick $k_2 x^2$, as shown in Fig. 6, find the position of the period-one fixed point. In other words, the fixed point gives initial conditions for a special trajectory, i.e. the periodic one, to be found here.

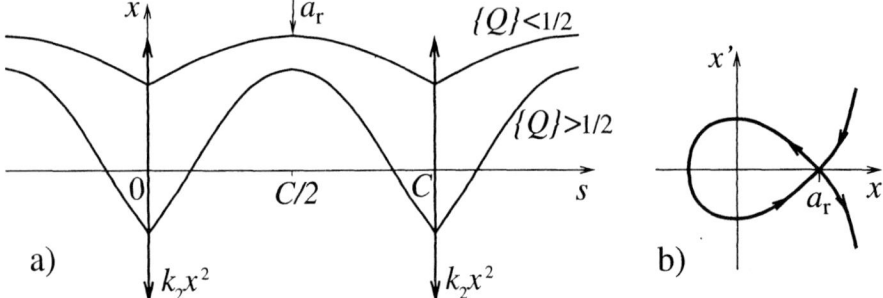

FIGURE 6. a) Periodic trajectories corresponding to the period-one fixed points of the nonlinear transformation; b) phase-space trajectory crossing the period-one fixed point a_r.

It is convenient to work in the normalized betatron variables, Eq. (44). We write the equation of the periodic trajectory on the right side of the kick, for $s > 0$,

$$x = a\cos(\psi - \pi Q),$$
$$x' = -a\sin(\psi - \pi Q),$$

and on the left side, for $s < 0$,

$$x = a\cos(\psi + \pi Q),$$
$$x' = -a\sin(\psi + \pi Q).$$

Continuity of x at $s = 0$ is provided by the form of these expressions. The slope x' is changed by the kick,

$$x'|_{+0} - x'|_{-0} = 2a_r \sin \pi Q = k_2 x_r^2 = k_2 a_r^2 \cos^2 \pi Q,$$

and this equation determines the position a_r of the fixed point,

$$a_r = \frac{2 \sin \pi Q}{k_2 \cos^2 \pi Q}.$$

The solution is markedly tune-dependent, and the neighborhood of the integer resonance should be avoided. Turning from the normalized variables back to normal ones, we see from Eq. (44) that the strength k_2 of the sextupole kick scales as $\beta^{3/2}$, where β is the value of the β-function at the kick location. Therefore, the dynamic aperture scales as $\beta^{-3/2}$.

V CONCLUSION

A versatile formalism is available (in different forms) to fully support linear lattice analysis and to simplify the formulation of nonlinear dynamics problems.

REFERENCES

1. Courant E.D. and Snyder H.S., "Theory of the Alternating-Gradient Synchrotron," *Annals of Physics* **3**, 1-48 (1958).
2. Kolomensky A.A. and Lebedev A.N., *Theory of Cyclic Accelerators*, North-Holland, Amsterdam, 1966.
3. Bruck H., *Accélérateurs Cirulaires des Particules*, Presses Universitaires, Paris, 1966.
4. Sands M., *The Physics of Electron Storage Rings, An Introduction*, SLAC Report 121 (1971).
5. Edwards D.A. and Syphers M.J., *An Introduction to the Physics of High Energy Accelerators*, Wiley, New York, 1993.
6. Lee S.Y., *Accelerator Physics*, World Scientific, 1999.

QUALITY LIMITATIONS OF HADRON BEAMS

Jacques Gareyte

CERN, 1211 Geneva 23, Switzerland

Abstract. The parameters that characterize the quality of hadron beams in the different accelerators or storage rings most commonly used or in project are reviewed. Because of the absence of a natural damping mechanism, hadron beams require careful manipulations. Some of those are described, in particular the processes used to inject from one machine into another. Among the phenomena that have to be carefully controlled in order to reach the desired performance, special attention is given to space-charge and slow instabilities arising from nonlinearities.

INTRODUCTION

The parameters used to judge the quality of a particle beam delivered by an accelerator or circulating in a storage ring are essentially its intensity and its size. Many different phenomena limit our ability to achieve beams of high intensity and small size, particularly if we want high intensity and small size simultaneously. Therefore we usually have to make compromises, sacrificing some parameters to optimize other, more important ones.

This is particularly true of hadron beams. Hadrons are heavy particles which, at the energies attained nowadays, radiate very little and therefore essentially obey Liouville's theorem: their density in phase space is at best constant, and in practice tends to decrease during acceleration or storage processes because of imperfect manipulations or imperfectly controlled interactions. On the one hand, extreme care is required to preserve the quality of hadron beams all along the accelerator's chain; on the other hand, the quasi absence of dissipative processes allows one to perform extremely delicate and sophisticated manipulations on hadron beams in order to adjust in the best way their parameters to the requirements of the users. These two aspects of hadron beam handling are illustrated in this report.

Concerning the different effects that can destroy the quality of hadron beams, we will concentrate on those not treated elsewhere in this course, for instance space charge and nonlinear motion. Other effects, like intra-beam scattering or collective instabilities, which are treated in detail in specialized courses, will at most be mentioned.

QUALITIES REQUIRED IN DIFFERENT MACHINES

The beam parameters most often used, with their units when appropriate, are shown in Table 1.

TABLE 1. Definition of parameters

Parameter	Symbol	Unit
Beam current	I	A
Number of particles per bunch	N	
Total number of particles	N_t	
Transverse R.M.S. size	σ_T	m
Longitudinal R.M.S. size	σ_t	s
R.M.S. energy spread	σ_E	eV
Normalized transverse emittance	$\varepsilon_{T,n} = \beta\gamma(\sigma_T^2/\beta_T)$	m
Longitudinal emittance	$\varepsilon_\ell = 4\pi\sigma_E\sigma_t$	eV.s

Here β_T is the value of the betatron function amplitude at the location where σ_T is measured, and β and γ are the usual relativistic factors. The definitions above are not necessarily logical, but they are widely used in the contemporary accelerator literature.

We will now review different classes of machines, stressing in each case, which are the most important parameters and illustrating the presentation with examples.

Classical Proton Accelerator

This machine usually sends its beams onto material targets, in order to produce secondary particles like electrons, muons, kaons, neutrinos, etc. The main figure of merit is the total number of accelerated particles N_t. Other parameters are much less important, although they have to stay within certain bounds.

Let us take the example of the CERN SPS, which accelerates protons up to 450 GeV.

Its parameters are shown in Table 2.

TABLE 2. SPS parameters

Parameter	
N_t	$4.8\ 10^{13}$
$\varepsilon_{T,n}$	$\sim 10\ 10^{-6}$ m
ε_ℓ	0.2-2 eV.s

Here the aim is to reach the largest possible N_t while keeping beam losses below a predetermined limit in order to control the radioactivity of the machine. At injection energy (14 GeV in the SPS) the transverse emittance must be small enough to fill the aperture of the vacuum chamber without losing more than 1 or 2 % of beam. There is no incentive to reduce it further, since high density beams are more prone to collective instabilities. During acceleration a moderate increase of ε_T can even be allowed, provided the beam dimension, which tends to shrink because of adiabatic damping, remains adequate for high energy beam manipulations (ejection, beam splitting, targeting).

The longitudinal emittance increases from 0.2 to 2 eV.s during acceleration. This is due to numerous longitudinal collective instabilities, and is harmless since it produces no beam losses (the longitudinal acceptance is sufficiently large) and provides a natural way of stabilizing the beam at high energy during ejection.

As we will see later, when the SPS is used as injector for the LHC, such a blow-up is unacceptable and must be suppressed.

Injectors

It is perhaps in the injectors that the demand on beam quality is the most severe. Here all parameters listed in Table 1 may be important, in particular the ratio N/ε, the beam density. In the SPS used as LHC injector, the total intensity is somewhat less than the maximum achieved in fixed-target mode, but the emittances, both transverse and longitudinal, have to be kept below very strict limits. A comprehensive program aimed at upgrading the machine towards this goal is being pursued.

Colliders

The figure of merit of a collider is its luminosity

$$L = \frac{N^2 k f_{rev}}{4\pi \sigma_T^2} \tag{1}$$

where k is the number of bunches and f_{rev} the revolution frequency. In terms of transverse emittance this formula becomes

$$L = \frac{\gamma}{4\pi} \frac{1}{\beta_T} \left[\frac{N}{\varepsilon_{T,n}} \right] \left[\frac{N}{\Delta t} \right] \tag{2}$$

where $\Delta t = 1/kf_{rev}$, the bunch spacing in seconds, is limited by the performance of the physics detectors and the beam crossing arrangements.

The first bracket represents the beam transverse density. It is proportional to the beam-beam parameter $\xi = \dfrac{r_p}{4\pi} \dfrac{N}{\varepsilon_{T,n}}$, which is limited by the physics of the beam-beam interaction (in the LHC $\xi \leq 0.0035$). Here r_p is the classical particle radius $r_p = e^2 / 4\pi \varepsilon_0 mc^2$.

The second bracket is proportional to the beam current I. To obtain a high value of the luminosity one must provide a large number of particles per bunch N, and a transverse emittance $\varepsilon_{T,n}$, that just matches the beam-beam limit. For the LHC, in order to reach $L = 10^{34} \text{cm}^{-2}\text{s}^{-1}$, the design value, with $\Delta t = 25\text{ns}$, one must provide $N = 10^{11}$ with $\varepsilon_{T,n} = 3.75 \, 10^{-6}\text{m}$.

The late SSC as well as the VLHC, the Very Large Hadron Collider now under consideration, have $\varepsilon_{T,n} = 1.10^{-6}\text{m}$. This is because they aim at operating below the beam-beam limit, at relatively modest luminosity. In this case it is interesting to have a very small emittance, which helps reduce the cost of the machine.

The longitudinal emittance ε_ℓ does not appear in the formula for luminosity; it can therefore be chosen to optimize other aspects of the machine. We want the bunch length $\sigma_s = c\,\sigma_t$ to be smaller than the value β_T of the betatron function at the collision point (otherwise luminosity would be affected) and the momentum spread at injection to be small enough to preserve single particle stability. This favors small ε_ℓ. On the other hand a large ε_ℓ reduces transverse emittance growth due to intra-beam scattering and helps fight collective instabilities. The best compromise in the LHC is $\varepsilon_\ell = 1$ eV.s at injection and $\varepsilon_\ell = 2.5$ eV.s in collision. The longitudinal emittance is increased during acceleration in a controlled way.

Drivers

These machines have been the subject of considerable work recently. They are medium energy (typically 1 to 15 GeV) high intensity hadron accelerators (most of them proton accelerators) used to generate short, intense pulses of neutrons (spallation neutron sources) or neutrinos (neutrino factories) or to drive subcritical nuclear fission reactors or nuclear waste burners. A very large number of particles per pulse and a high repetition rate are required, together with the possibility for some of them to compress the bunch down to a length of a nanosecond or so. This last demand can be met only with bunches of reasonably small longitudinal emittance. In these machines the transverse emittance is not critical, in fact it has to be large in order to overcome space-charge problems.

Coolers

These are low energy, proton, antiproton, or ion storage rings, in which cooling techniques are used to increase the density N/ε to extreme values. They constitute exceptions in our list since cooling processes (either stochastic cooling or electron cooling) are globally non-Liouvillian.

SOURCES AND LINACS

In order to be able later-on in the process of acceleration to manipulate beams to meet various user requirements, it is essential to start with beams of the largest possible density. Here we list typical average performances currently attainable, which can then be compared to requirements at succeeding stages.

The Sources

Typically sources reach transverse normalised emittances $\varepsilon_{T,n}$ of the order of 0.2 10^{-6} m. The corresponding intensity depends on the type of particles; it goes up to 200 mA for proton sources, but reaches only 40 mA for H$^-$ sources. We will see later-on how these low intensity H$^-$ beams can be used to give, through charge-exchange injection, better performances than the more intense H$^+$ beams.

Linacs

The source accelerates the beams to about 50 keV and injects them into a linear accelerator, which in turn brings the energy up to a few hundred MeV or even in some cases 1 or 2 GeV. There is an unavoidable increase of transverse emittance in the linac due to space-charge effects. These effects will be described later. The blow-up is more pronounced for high beam intensity. In the CERN 50-MeV proton linac, with I = 170 mA, the emittance at exit is 1.2 10^{-6} m. It has increased sixfold along the linac, for reasons which are not all clear, but probably mainly due to space-charge. In the CERN SPL, the 2.2-GeV superconducting linac now under study, the H$^-$ intensity is 40 mA and the emittance at exit is expected to be 0.6 10^{-6} m, which is only 3 times the entrance value. Comparing these numbers to the stringent collider requirements (LHC: $\varepsilon_{T,n}$ = 3.75 10^{-6} m, VLHC: $\varepsilon_{T,n}$ = 10^{-6} m) we see that they are adequate but there is not much margin to allow for emittance dilution in the successive synchrotrons of the injection chain.

TRANSVERSE SPACE CHARGE

Direct Space Charge

Particles of the same charge are subject to mutual repulsion due to Coulomb interaction. However, since these particles travel side by side at velocity βc, there is an attractive component and the total force is [1]:

$$F = e(E_r - \beta c B_\phi) \quad (3)$$

where E_r is the average radial electric field and B_ϕ the average azimuthal magnetic field induced by all other particles of the beam on the test particle, as shown in Fig 1.

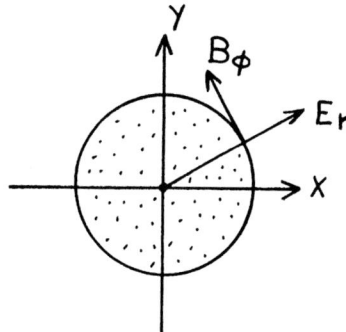

FIGURE 1. Space-charge fields.

Let us envisage first the case of a cylindrical beam with constant particle density $\rho = \dfrac{I}{\pi a^2 \beta c}$, where I is the current and a the beam radius.

Applying Gauss' and Ampere's laws at radius $r = a$ we can write respectively

$$2\pi r\, E_r = \frac{1}{\varepsilon_0} \pi r^2 \rho, \qquad 2\pi r\, B_\phi = \mu_0 \pi r^2 \rho \beta c$$

from which, remembering that $\varepsilon_0 \mu_0 c^2 = 1$, we get,

$$F = \frac{Ie}{2\pi \varepsilon_0 a^2 \beta c}\left[1 - \beta^2\right] r . \qquad (4)$$

Since $1 - \beta^2 = 1/\gamma^2$, we see that because of the partial cancellation of the electrostatic and magnetic forces, the direct space-charge interaction diminishes fast with increasing energy. It mainly affects beams at injection energy.

Beam Transport with Space Charge

At each position s along the accelerator, particles are subjected to a space-charge defocusing force $k_{sc}(s)$, which is superimposed on the externally applied focusing $k(s)$. The transverse equation of motion becomes [2]

$$x'' + \left[k(s) + k_{sc}(s)\right] x = 0 . \qquad (5)$$

Writing $x'' = \frac{1}{\beta^2 c^2}\ddot{x} = \frac{1}{\beta^2 c^2}\frac{F_x}{m\gamma}$, where $\ddot{x}$ is the second-order time derivative of x, we find

$$k_{sc} = -\frac{2r_p I}{ea^2(s)\beta^3\gamma^3 c}. \qquad (6)$$

In a circular accelerator of radius R and betatron function $\beta_T(s)$ the space-charge force induces a tune shift:

$$\Delta Q_{sc} = \frac{1}{4\pi}\int_0^{2\pi R} k_{sc}(s)\beta_T(s)ds = \frac{r_p RI}{e\beta^3\gamma^3 c}\left\langle\frac{\beta_T(s)}{a^2(s)}\right\rangle.$$

The quantity $A_{T,n} = \beta\gamma a^2(s)/\beta_T(s)$ is proportional to the normalized beam emittance $\varepsilon_{T,n}$. For a hadron beam obeying Liouville's theorem this quantity is invariant along the circumference as well as during acceleration.

Therefore, using also $I = \frac{Ne\beta c}{2\pi R}$, where N is the number of particles in the machine, we get

$$\Delta Q_{sc} = -\frac{r_p N}{2\pi A_{T,n}\beta\gamma^2}, \qquad (7)$$

which is the classical formula for space-charge detuning. This formula was derived for a continuous beam with a uniform current along the machine, and a uniform particle distribution in each transverse slice of the beam. It can however be generalized to real bunches of particles.

Space Charge in Real Particle Bunches

Most of the time particles are gathered in bunches by the radio-frequency accelerating system. The particle density is usually maximum at the centre of the bunch and decreases smoothly to zero at the ends. The space-charge detuning is proportional to the local density, and formula (7) applies if we replace N by $2\pi R d N(s)/ds$.

Figure 2 shows the space-charge tune spread in the tune diagram of the CERN PS Booster: central particles of the bunch suffer a negative tune shift of the order of $\Delta Q_{sc} = 0.5$ in the vertical direction (point A of the "necktie") while particles at the ends of the bunch see practically no effect so that their tune is the unperturbed tune of the machine at point C. Experience shows that betatron resonances of order 2 (on integer or half-integer tunes) which are excited by quadrupolar errors in the guide fields, and of order 3 (on 1/3 integer) which are excited by sextupolar errors, produce an increase of particle amplitudes leading to beam losses. Therefore the usual recipe in accelerator design is to restrict the maximum tune shift to about 0.3: this allows us to locate the whole beam in-between resonances of order 2 and 3. In the PS Booster, we could increase the tune shift up to 0.5 after careful compensation of the third order resonances using sextupole correctors.

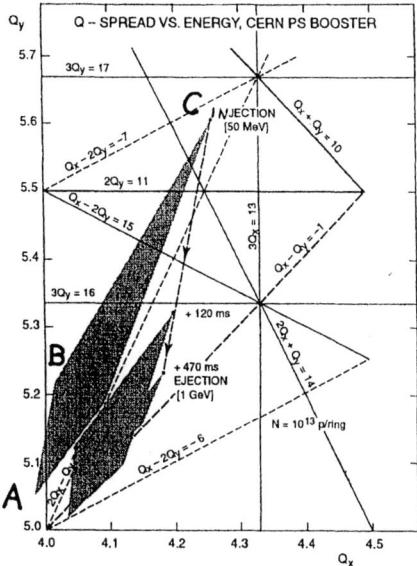

FIGURE 2. Example of space-charge tune spread in a space-charge-limited synchrotron: betatron tune diagram and areas covered by direct space-charge tune spreads at injection, intermediate, and extraction energies for the CERN Proton Synchrotron Booster. During acceleration, space charge gets weaker and the 'necktie' area shrinks, enabling the external machine tunes to move the 'necktie' to an area clear of betatron resonances.

Another complication arises because the transverse distribution in a slice of the beam is usually non-uniform. The uniform distribution used up to now is the projection, in the real transverse plane, of a "shell" distribution in the four-dimensional phase space x, p_x, y, p_y; this is the Kapchinski-Vladimirski (K.V.) distribution [3]. A "real" distribution is usually approximated by a Gaussian distribution. It has the remarkable property that a Gaussian distribution in the four-dimensional phase space generates a Gaussian distribution also in the real transverse plane. Figure 3 summarizes the situation for the K.V. and the Gaussian distributions.

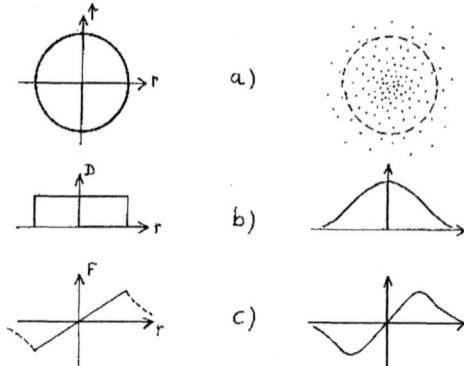

FIGURE 3. Particle distributions. Left: K.V.   Right: Gaussian.
   a)   Distribution in 4-dimensional phase space.
   b)   Density in real space.
   c)   Space charge force.

When the distribution is non-uniform, we can still apply Gauss' and Ampere's laws to evaluate the force at a distance r from the beam centre, but now we have to integrate from zero to r to find the charge. The result for a Gaussian beam is

$$F = \frac{Ie}{2\pi\varepsilon_o \beta c\gamma^2} \frac{1}{r}\left(1 - e^{-\frac{r^2}{2\sigma^2}}\right). \tag{8}$$

For small values of r the force increases linearly with r, but at large amplitude it saturates and decreases to zero. As a consequence particles with large betatron amplitudes suffer less tune shift than small-amplitude ones. This introduces an additional source of tune spread among the particles of a bunch. On Fig. 2, point B represents particles that are longitudinally at the bunch centre but have large betatron amplitudes. This is important for the following discussion.

The tune depression at the centre of a Gaussian beam is given by

$$\Delta Q_{sc} = -\frac{r_p N}{4\pi \varepsilon_{T,n} \beta \gamma^2}. \tag{9}$$

The Envelope Equation

The "naïve" approach outlined above is well justified experimentally in circular machines but is not self-consistent and would lead to inaccurate results in the case of linear accelerators or transport lines, where space-charge forces are often much stronger.

This is because the force depends on the beam size. If the space-charge force displaces the tune towards a half integer resonance, the beam size grows, which in turn decreases the force. A self-consistent treatment can be made in the case of a uniform distribution (K.V. distribution [3]) by considering the envelope equation

$$r_x'' + k(s)r_x - \frac{\varepsilon_x^2}{r_x^3} - \frac{4 r_p I}{ec\beta^3\gamma^3(r_x + r_y)} = 0. \tag{10}$$

Here $r_{x,y}$ is the horizontal, vertical extension of the envelope, $r_x = \sqrt{\beta_x \varepsilon_x}$, and $\varepsilon_x = \varepsilon_{x,n}/\beta\gamma$ is the (unnormalized) emittance. The first three terms of the left side are the usual envelope equation, and the fourth term represents the space-charge effect.

Stationary solutions of this equation give the "matched" solutions. For a given intensity, any distribution in the strength of the focusing elements which satisfy this equation allows transport of the beam without deterioration. The intensity limit above which this is no longer possible is the space-charge limit. In linear accelerators or transfer lines this limit is much higher than in circular machines.

If a matched solution exists but the beam is injected unmatched (this happens for instance if the beam intensity varies from pulse to pulse), the beam envelope oscillates coherently around the matched solution. There are two coherent modes:

- the zero mode in which r_x and r_y oscillate in phase, and
- the π mode in which they oscillate in antiphase.

The π mode is the more interesting. Its frequency is

$$Q_\pi = 2(Q_0 - 0.75\Delta Q_{sc})$$

where ΔQ_{sc} is the incoherent tune shift calculated above. In the absence of space charge the frequency of the π mode of the envelope is twice the unperturbed machine tune Q_0. At high intensity the depression of the envelope frequency is less than twice the incoherent space-charge detuning. Therefore the space-charge limit due to half-integer resonances (they happen when Q_π=integer) is larger than that naively calculated by considering single particles.

The solutions of the envelope equation outlined above can be found analytically for the K.V. distribution. For a real distribution, for instance a Gaussian one, the analysis of the problem is much more complicated. However it has been shown [4] that in this case the envelope equation can still be applied if one considers only the second moment (R.M.S.) of the distribution. There is a π mode oscillation of the R.M.S. size with a frequency depression equal (for round beams) to that of the K.V. distribution.

However the fact that the Gaussian distribution has tails extending well beyond the R.M.S. amplitude has very interesting consequences.

The first consequence is that a mismatched beam "filaments," which means that because of the tune spread among the particles the coherent oscillation energy of the envelope is transferred to single particles, and the beam R.M.S. increases.

The second consequence is more subtle but may be more dangerous, because it is not "self healing" as the R.M.S. blow-up tends to be; this is the formation of a halo. This halo, although it contains a small fraction of the particles, is dangerous in high intensity machines like drivers, because it leads to particle losses in aperture restrictions and consequent activation of the machine. Its formation can be understood qualitatively by looking back at the tune diagram of Fig. 2. We have seen that with a realistic transverse distribution the particles with large transverse amplitudes have a reduced tune depression and are situated approximately at point B. This region corresponds to half the frequency of the R.M.S. oscillation. Therefore if a beam is injected mismatched, it oscillates coherently and part of the corresponding energy can be transferred resonantly to single particles, increasing their amplitude and creating a halo.

Minimizing Space-Charge Effects

As we have seen, the largest tune depression occurs at the dense centre of particle bunches. In order to minimize this effect one should distribute particles inside the bunch so as to approach a uniform distribution in all dimensions. In the transverse dimensions, we will see later how "injection painting" can be used to approach the K.V. distribution. In the longitudinal direction, two techniques are used to create "rectangular "bunches.

The first technique consists in creating a "hollow" bunch in longitudinal phase space (similar to the K.V. distribution). Before RF capture, while the beam is circulating debunched, one introduces adiabatically empty buckets into the beam by sweeping the frequency of a high harmonic RF cavity, as shown in Fig. 4.

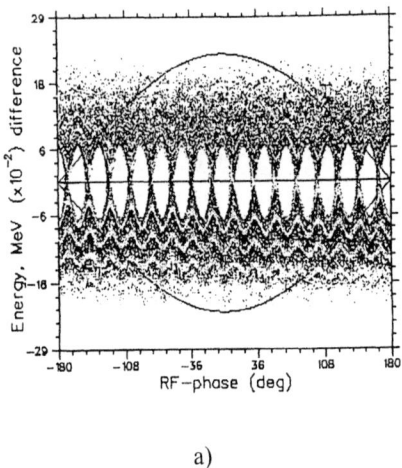

a)

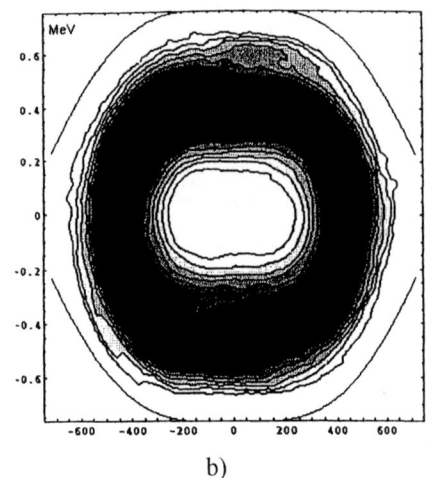

b)

FIGURE 4. Creation of hollow bunch.
a) Deposition of empty buckets (simulation).
b) Tomographic reconstruction of longitudinal phase space after adiabatic capture.

Subsequent adiabatic capture with the accelerating cavity produces the "hollow" bunch on the right [5]. Its projection onto the time axis gives an approximately rectangular distribution.

The second technique uses a harmonic RF cavity excited in antiphase with the main cavity; the combination results in an accelerating wave that has a flat part at the bunch centre. As a consequence particles tend to cluster towards the bunch edges, creating a quasi-rectangular bunch as seen in Fig. 5 [6].

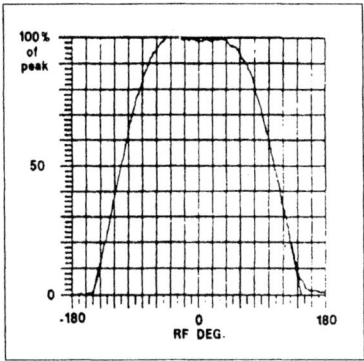

FIGURE 5. Flat-topped bunch obtained with second harmonic RF in CERN PS Booster.

INJECTION FROM LINAC TO SYNCHROTRON

Monoturn Injection

The simplest scheme consists of injecting a linac pulse with a length equal to the ring circumference. This allows preservation of the linac emittance $\varepsilon_{T,n}$ but requires a linac of high intensity. Moreover, often the injected beam would be too dense and would be destroyed by space-charge effects. This scheme is seldom used.

Multiturn Injection

The principle is shown in Fig.6a [7]. The beam is injected continuously over many turns, and during this time the closed orbit at the location of the injection septum is moved inwards (from points 0 to 6 between turn 0 and 6), so that after a complete betatron oscillation the injected beam can clear the septum. Figure 6b shows the resulting phase space after, in this case, 13 turns. The emittance in the plane of injection has been considerably increased. A sizable dilution is created by the "shadows" of the septum magnet, which remain in the phase space.

An improvement, used at the CERN PS Booster, consists of transferring to the vertical plane some of the horizontal oscillation amplitude of the injected beam through linear coupling. The effect is shown in Fig. 6a: one can clear the septum with a reduced change in orbit.

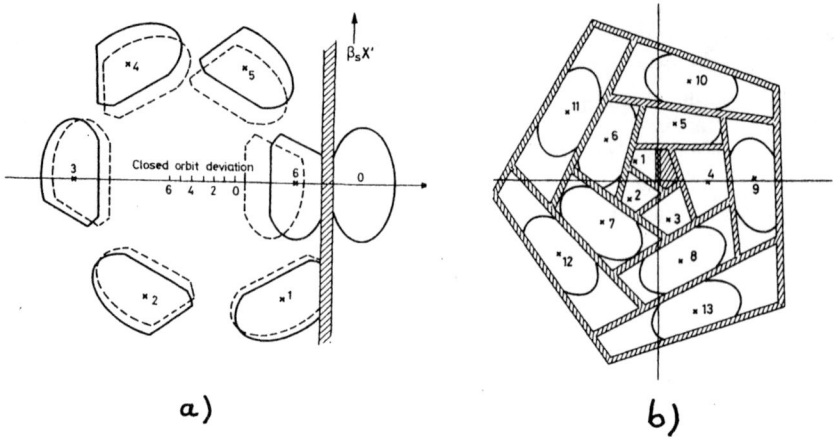

FIGURE 6. Multiturn injection.
    a) Without (continuous line) and with (dashed) horizontal-vertical coupling.
    b) Dilution due to shadow of septum.

Charge-Exchange Injection

The drawbacks of multiturn injection (large dilution) can be overcome by accelerating H⁻ ions in the linac and stripping them of their two electrons to get H⁺ inside the injection magnet, as shown in Fig. 7. With this non-Liouvillian process one can theoretically accumulate many turns in the same phase-space area [8]. Exploiting this property, one can use a linac with moderate intensity, in which it is possible to obtain a small emittance. It is then easy, using varying orbit bumps, to paint the phase space to optimize the injected beam properties. In particular one can obtain an approximation of the K.V. distribution by injecting successively at large amplitude and small angle, then large angle and small amplitude, and large horizontal, small vertical amplitude and vice versa. In this way one can populate (approximately) a "shell" in the four-dimensional transverse phase space.

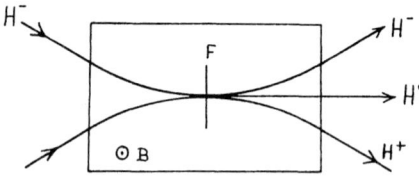

FIGURE 7. Charge-exchange injection. H⁻ ions are stripped in the foil F.

This is a very powerful method which theoretically allows one to reach the best possible performance. As an example, the injection into the CERN neutrino factory driver under design is made over 600 turns. The initial emittance of the H$^-$ linac beam is $0.6 \cdot 10^{-6}$ m, and the final one in the accumulator, which is dictated entirely by space-charge considerations, is $50 \cdot 10^{-6}$ m.

Cooling Injection

Cooling is another non-Liouvillian process that can be used to augment considerably the performance of a linac-to-synchrotron injection process. As an example, the process at the CERN lead ion accumulator [9] is as follows:

- Inject 35 turns while ramping energy by 4‰ and decreasing the horizontal orbit bump from 4 cm to 0 (multiturn injection – the value of the dispersion is large at the septum therefore ramping the energy helps in the process).
- Cool in 0.1 s (electron-cooling.
- Transfer the cooled beam to a "parking orbit."
- Repeat 12 times.

In this way, one can inject a beam 120 times more intense than with monoturn injection.

TRANSFER BETWEEN SYNCHROTRONS

In a long chain of injectors like that of the LHC, it is imperative to minimize the particle losses and emittance dilutions at each transfer. This is obtained by careful matching of the beam properties.

Injection offsets are measured with position monitors (at least two) and corrected by steering devices in the injection line. The remaining effects are reduced further by active feedback systems, which must damp coherent oscillations faster than the filamentation process induced by tune spread.

Mismatch between the focusing functions in the receiving machine and the optical properties of the injected beam can also lead to dilution. Recently it has become possible, for instance using Optical Transition Radiation screens and Charge-Coupled Device cameras, to measure shape oscillations. These can be minimized by adjusting quadrupoles.

The value of the dispersion function D must also be carefully adjusted to avoid longitudinal-transverse emittance transfers.

LONGITUDINAL SPACE CHARGE

Let us consider again a continuous cylindrical beam with uniform density circulating in a concentric vacuum pipe. As we have seen above, for reasons of symmetry, electromagnetic fields generated in the vacuum by this beam are transverse to the beam direction. The addition of a concentric, perfectly conducting wall does not change this property. Therefore in this case there is no longitudinal effect of the space charge. On the contrary, if the line density $\lambda(s)$ varies along the beam, the space charge results in a longitudinal electric field, which can easily be calculated [10] by applying Stokes' law to the small circuit of Fig. 8:

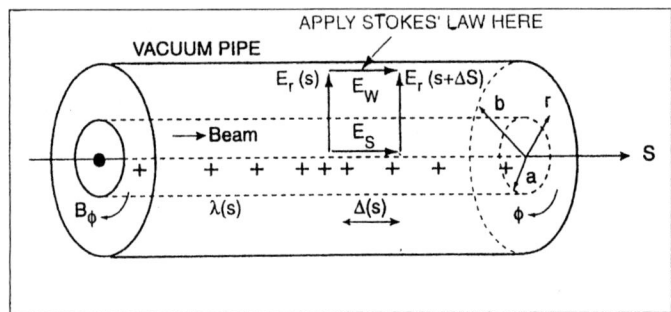

FIGURE 8. Calculation of longitudinal space charge.

$$\oint_{LINE} \vec{E} \cdot d\vec{l} = -\frac{\partial}{\partial t} \int_{SURFACE} \vec{B} \cdot d\vec{\sigma} = -\frac{\partial}{\partial t} \Delta s \int_0^b B_\phi \, dr \qquad (11)$$

with

$$E_r = \frac{e\lambda}{2\pi\varepsilon_0} \frac{1}{r}, \quad B_\phi = \frac{\mu_0 e\lambda \beta c}{2\pi} \frac{1}{r}, \quad r \geq a,$$

$$E_r = \frac{e\lambda}{2\pi\varepsilon_0} \frac{r}{a^2}, \quad B_\phi = \frac{\mu_0 e\lambda \beta c}{2\pi} \frac{r}{a^2}, \quad r \leq a,$$

and using

$$\frac{\partial \lambda}{\partial t} = \frac{\partial \lambda}{\partial s} \frac{ds}{dt} = \beta c \frac{\partial \lambda}{\partial s}$$

and

$$g_0 = 1 + 2\ln\frac{b}{a}$$

one gets

$$E_s = \frac{eg_0}{4\pi\varepsilon_0}(1-\beta^2)\frac{\partial \lambda}{\partial s} + E_w = -\frac{eg_0}{4\pi\varepsilon_0}\frac{1}{\gamma^2}\frac{\partial \lambda}{\partial s} + E_w. \qquad (12)$$

For a perfectly conducting wall, $E_w = 0$ and therefore the longitudinal space charge field is

$$E_s = -\frac{eg_0}{4\pi\varepsilon_0} \frac{1}{\gamma^2} \frac{\partial \lambda}{\partial s}. \tag{13}$$

A parabolic bunch of length $2\beta c\tau$ has

$$\lambda(s) = \frac{3}{4} \frac{N}{\beta c \tau} \left(1 - \frac{s^2}{\beta^2 c^2 \tau^2}\right).$$

In this simple example the field grows linearly from the centre to the edges of the bunch and this reduces (or increases above transition energy) the focusing effect of the RF voltage.

Integrating over the ring we find the equivalent voltage seen by the particle at the bunch extremity:

$$V_{s.c} = \frac{3}{2}\left[\frac{g_0}{2c\varepsilon_0 \beta\gamma^2}\right]\frac{R}{\beta c}\frac{Ne}{\tau^2}.$$

The quantity in brackets is the "space-charge impedance." During accumulation in low energy rings, the space-charge voltage must stay smaller than the RF voltage at the bunch extremity. This puts a lower limit on the necessary RF voltage.

The requirement becomes more stringent during bunch compression. This technique is used to produce the very short bunches needed in drivers. A long bunch with a small momentum spread is injected into a large RF bucket far from matching conditions (Fig. 9). In a quarter of a synchrotron oscillation the bunch rotates to the upright position and is ejected from the machine there. At the end of the process the space-charge-induced voltage is maximum and, if too important, may hamper the compression. As an illustration, the voltage developed at the extremities of a $N = 10^{13}$ p bunch at 2 GeV with $\sigma_t = 1$ ns (close to the CERN neutrino factory driver parameters) is 15 MV.

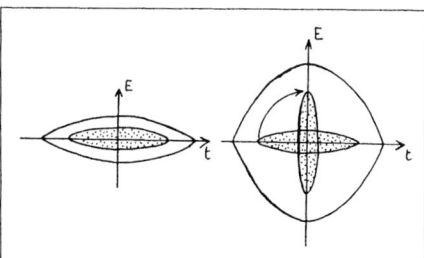

FIGURE 9 Bunch compression.

WEAK CHAOS

Now we turn our attention to a problem of single particle instability induced by unwanted nonlinear components of the guide field. Magnet builders have learned how to minimize these effects. However, in the case of hadron storage rings, the particles have to circulate for enormous distances: in the LHC, storage time is 10 hours, which means 4.10^8 revolutions or 10^{10} km. The trajectories of these particles, which see practically no damping at this time scale, are extremely sensitive to small nonlinear forces. The problem of transition from regular to chaotic motion is nowadays widely studied for the solar system and many other dynamical systems. We try here to give a terse account of studies done in accelerators, and their conclusions.

Nonlinear fields are essentially responsible for two effects that conspire to destabilize in long-term particle motion.

In the first place they induce tune spread in the beam. This is caused mainly by the systematic errors, those that are the same in all magnets around the ring.

In the second place they excite nonlinear, high-order resonances. This is due mainly to the random errors, those that vary from magnet to magnet and thus generate a rich spectrum of azimuthal harmonics.

High-order resonances are very dense in the Q_x, Q_y tune diagram. They occur whenever $\ell Q_x + m Q_y = p$, where ℓ, m, p are integers, $\ell + m = n$ being the order of the resonance and p the azimuthal harmonic of the multipolar error of order n which excites this resonance.

Because of the tune spread one cannot prevent some particles from crossing resonances, especially those that have a large betatron amplitude and therefore a large tune shift. Moreover the basic parameters of the machine cannot be completely fixed, and are subject to some jitter. For instance it is very difficult in practice to control residual tune modulations due to the imperfections of the quadrupole power supplies to better than $\Delta Q = 10^{-4}$. This is sufficient to make particles repeatedly cross a large number of resonances, which may in turn induce weakly chaotic motion leading to slow diffusion of particles towards large amplitude, and eventually to particle losses.

This problem has been investigated both by experiments in existing machines and by computer simulations. We will give an example of each approach [11].

In an experiment on the CERN SPS, the machine, which is otherwise very linear, was perturbed in a controlled way using strong sextupolar lenses. The tune, the diffusion rate, and the onset of particle losses were measured versus the initial particle amplitude. The results were compared to computer simulation in which the tune shift and the onset of chaotic motion were evaluated. Simulation results were in good agreement with experimental observations, as can be seen in Fig. 10.

Figure 11 shows the results of a computer simulation of the above experiment which evaluates, turn after turn, the oscillation amplitude of 32 particles for up to 3 million turns [12]. All particles start with the same initial amplitude, which corresponds to the lower edge of the chaotic zone shown in Fig. 10. In a linear system, the motion would be regular forever and one would observe just one horizontal line in Fig. 11: all amplitudes would stay constant. On the contrary, we observe that the amplitudes of individual particles jitter and slowly move apart. In addition, from time to time a particle escapes towards large amplitudes and is lost out of the machine. This behavior is typical of chaotic motion. The largest particle amplitude below which the motion remains regular is called the dynamic aperture.

Hadron colliders must be carefully designed to provide a sufficiently large dynamic aperture for the beam.

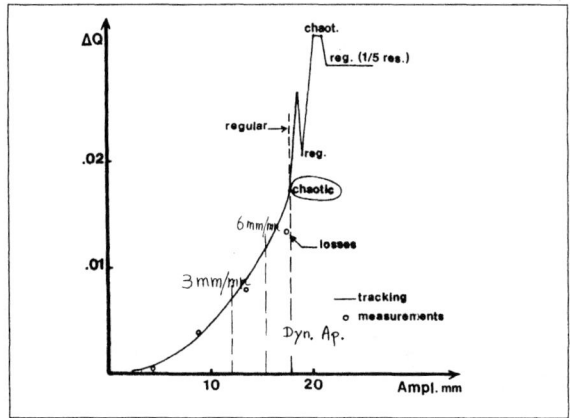

FIGURE 10. SPS experiment on nonlinear motion. Tune shift with amplitude (line is from simulation, dots are measurements). Indications of diffusion rate (in mm/min, measured) onset of losses (measured) and chaotic motion (from simulation).

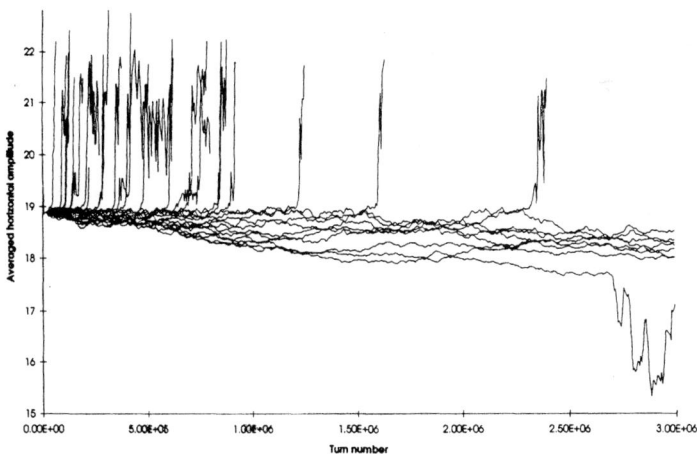

FIGURE 11. SPS experiment. Simulation of 32 particles over 3 million turns, in the chaotic region.

EXAMPLE OF HADRON BEAM MANIPULATION

Hadron beams are delicate objects, to be manipulated with care, because they keep a memory of all treatments inflicted on them. However, and for essentially the same reason, they lend themselves to beautiful experiments, such as the triple bunch splitting which is now used to prepare the beam for the LHC. Figure 12 shows how a long bunch is split into three by adiabatically raising and lowering, as required, the voltage in three RF systems of different frequencies [13].

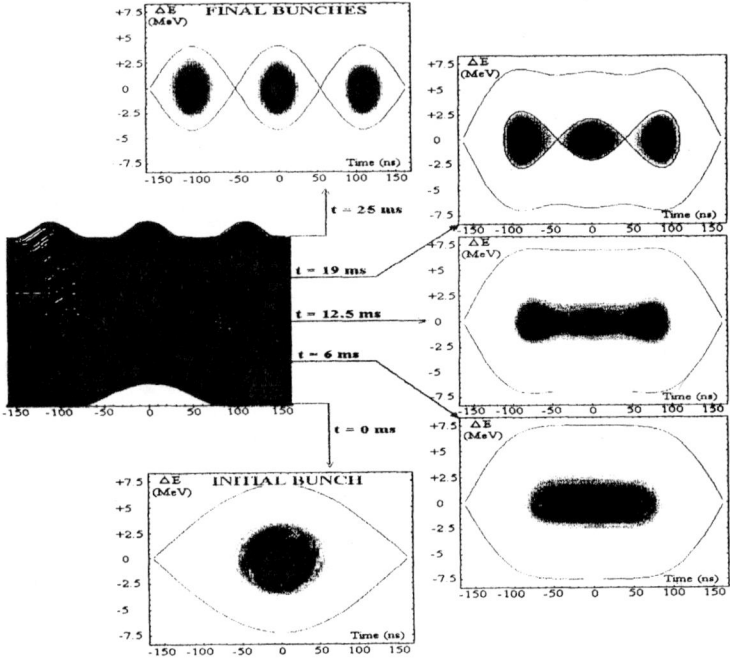

FIGURE 12. Triple bunch splitting in the CERN PS. Evolution of bunch profile (left) and tomographic reconstruction of phase-space density.

CONCLUSION

Hadron beams are used in a large variety of applications, ranging from proton drivers to high energy colliders. Obtaining and preserving the beam quality needed in these machines require enormous amounts of ingenuity. Whereas half a century of studies were necessary to reach today's performance, the field is still blooming and the considerable efforts now engaged hold the promise of continuing progress.

REFERENCES

1. Schindl, K., *"Space Charge"*, The Joint US-CERN-Japan-Russia School on Particle Accelerators, Course on "Beam Measurements", Montreux Switzerland, 1998. Also CERN/PS 99-012 (DI).
2. Courant, E. and Snyder, H., *"Theory of the Alternating-Gradient Synchrotron"*, Ann. Phys. 3,1, 1958.
3. Kapchinsky, I. And Vladimirsky, V.V., *"Limitations of Proton Beam Current in a Strong Focussing Linear Accelerator Associated with the Beam Space Charge"*, Proceedings of the Conference on High Energy Accelerators and Instrumentation, CERN, 1959, p.274.
4. Sacherer, F.J., *"Transverse Space Charge Effects in Circular Accelerators"*, Thesis, University of California, Berkely, 1968.
5. Blas, A. et al., *"New Technique for Bunch Shape Flattening"*, Proceedings, Particle Accelerator Conference, New York, USA, 29 March - 2 April 1999. Also CERN/PS 99-036 (RF).
6. Gelato, G. et al., *"Progress in Space Charge Limited* Machines ", Proceedings, Particle Accelerator Conference, Washington D.C., USA, 16-19 March 1987. Also CERN/PS 87-36 (BR).
7. Van der Stock, P.D.V., "Multiturn Injection into the CERN Proton Synchrotron Booster", Thesis, University of Amsterdam, The Netherlands, 1981. Also CERN PS/BR 81-28.
8. Rees, G.H., *Injection*, CERN Accelerator School, Fifth General Accelerator Physics Course, 7-18 September 1992, University of Jyväskylä, Finland, CERN 94-01.
9. Bosser, J. et al., "Results on Lead Ion Accumulation in LEAR for the LHC", 6th Epac, June 22-26, Stockholm, Sweeden, 1998. Also CERN/PS 98-033 (CA).
10. Chao, A.W., *Physics of Collective Beam Instabilities in High Energy Accelerators"*, Wiley Series in Beam Physics and Accelerator Technology, 20.
11. Gareyte, J., Hilaire, A, Schmidt, F., "Dynamic Aperture and Long Term Particle Stability in the Presence of Strong Sextupoles in the CERN SPS, " PAC, Chicago, USA, March 20-23, 1989. Also CERN SPS/89-2 (AMS).
12. Fisher, W, Giovannozzi, M, Schmidt, F, *"The Dynamic Aperture Measurement at the CERN SPS"*, Physical Review E, 55,. 3307, 1997. Also CERN SL 95-96 (AP).
13. Garoby, R. Hancock, S. Vallet, J.L., "Demonstration of Bunch Triple Splitting in the CERN PS", 7th EPAC, 26-30 June 2000, Vienna, Austria. Also CERN PS 2000-038 (RF).

Quality Limits for Electron Rings

Toshio Kasuga

Institute of Materials Structure Science
High Energy Accelerator Research Organization (KEK)
1-1 Oho, Tsukuba 305-0801, Japan

Abstract. Longitudinal phase-space issues that limit the performance of electron storage rings, especially synchrotron radiation sources, are discussed. Single-bunch operation, a phenomenon that deteriorates single-bunch purity and measures of purity deterioration are given. The importance of the longitudinal emittance and low-alpha operation is also discussed.

1 INTRODUCTION

Users of the first-generation light sources that were originally built for particle physics wanted only high current and stable beams that they could handle with their experimental apparatus. When they started to have their own dedicated light sources, they wanted beams whose six-dimensional phase-space areas were as small as possible. The requirement for transverse emittances became important first. For example, many of the first- and the second-generation light sources have horizontal emittances larger than 100 nm.rad and horizontal-vertical coupling coefficients of several percent. However, machines with emittances smaller than several nm.rad and coupling coefficients less than 1% or 0.1% have been achieved in third-generation light sources. Design technique became sophisticated with the development of computer-aided design, and straight sections with almost ideal machine functions for insertion devices have been realized. Several lecturers at this school have discussed the importance of transverse phase space; therefore we focus mainly on phenomena concerning the longitudinal phase space of electron storage rings for SR sources.

Long beam lifetimes and stable beams are essential for synchrotron radiation users. However, small six-dimensional phase-space areas, high beam currents, long lifetimes, and stability do not go together. Many phenomena that destroy beam stability, such as instabilities due to wakefields, the influence of charged particles trapped around the beam, and beam-beam interactions have also been discussed at this school. Even if we could build an ideal vacuum system, beam lifetimes would be limited by the Touschek effect, especially in low energy and low emittance machines. This effect also limits single-bunch purity, as we will discuss in detail in the next section.

Steadiness around closed orbits is essential in low emittance machines. Low emittance means that the phase-space area of the optical image of the beam is small. Therefore, even small movements of the image due to wavering of orbits can affect experiments that require high brilliance. Stability of beam profiles is also important for the same reason. Stabilization of circulating beams in light source accelerators is a never-ending theme for accelerator physicists. The many practical issues including the vibration of buildings, magnetic fields induced by other accelerators near the light source, etc. For example, at KEK, there are six circular accelerators in a 2x1-km site: PF, PF-AR, LER and HER of the B-Factory, ATF, and PS. When the magnet system of the B-Factory is being started up, deformation of several microns is observed in the close orbit in PF. However I will not discuss the stability of circulating beams further.

Table 1 shows the parameters of several Japanese light sources to be discussed. PF and PF-AR are dedicated light sources at the Institute of Materials Structure Science of KEK. PF-AR was originally constructed as a booster for the TRISTAN collider and was recently converted into a dedicated single-bunch synchrotron radiation source. UVSOR is a 0.75-GeV dedicated light source mainly for molecular science and its related fields at the Institute for Molecular Science. A low-alpha and short-bunch lattice has been tested recently in this machine. SPring-8 is one of the largest third-generation light sources in the world. New Subaru of Himeji Institute of Technology was recently commissioned at the site of SPring-8. The lattice of New Subaru was designed to enable to operation in low-alpha or negative-alpha modes. A racetrack type compact electron storage ring was commissioned at Hiroshima University recently.

2 SINGLE-BUNCH OPERATION

Recently, demand for single-bunch operation, in which only one of many buckets is filled with an intense bunch, has gradually increased. Single-bunch operation is contradictory to many experiments that require high current beams because there are some difficulties in storing an intense beam in one bucket. However, in KEK PF, several days a year are devoted to single-bunch operation, and PF-AR is essentially a single-bunch machine, as mentioned before.

Satellite bunches, namely undesirable bunches around the main bunch, affect time-resolved experiments. Therefore, experimenters require excellent single-bunch purity. Since an impurity level of 10^{-8} is easily detectable in photon counting experiments, we must keep the impurity level as low as possible. Table 2 shows the present required impurity levels, namely better than 10^{-4} to 10^{-9}. Even though an electron gun can produce a beam whose length is less than a bucket width, electrons due to field emission from an electron gun or a linac leak into neighboring buckets. As mentioned later, the impurity level becomes worse as time passes after the injection, especially in low energy machines with short Touschek lifetimes. A system that clears satellite bunches continuously is necessary for such machines.

Several methods or combinations of them have been developed to form a single bunch in a storage ring. For example, a gated RF-KO method, in which all bunches except a certain bunch are destroyed, is used in the booster of UVSOR. In this system, the RF-KO signal is turned off while the single bunch passes through the

TABLE 1. Parameters of Japanese Light Sources.

	PF	PF-AR	UVSOR	SPring-8	NewSubaru	HiSOR
Institute	KEK	KEK	IMS	JASRI	HIT	Hiroshima U
Energy (GeV)	2.5	6.5	0.75	8	1.5	0.7
Bending Radius (m)	8.66	17.825	2.2			0.87
Circumference (m)	187	377.26	53.2	1436	118.716	21.946
RF Frequency (MHz)	500	508.58	90.115	508.6	500	191.244
Harmonic Number	312	640	16	2436	198	14
α	0.061	0.00702		1.46×10^{-4}	± 0.001	
H. Tune ν_x	9.62	10.15	3.16	51.36/43.16	6.21	1.672
V. Tune ν_y	4.29	10.23	1.44	16.36/21.36	2.17	1.724
Bunch Length (mm)	10.0			35ps		
Emittance ε (nm)	36	294	140	6.9/6.2		
Initial Current (mA)	450	40 Single B	500	100	100	100

TABLE 2. Requirement for Single Bunch Purity.

	Requirement	Achieved
UVSOR(0.75GeV)	10^{-4}	10^{-4}
PF(2.5GeV)	10^{-8}	10^{-6}
PF-AR(6.5GeV)	10^{-6}	10^{-6}
Spring-8(8GeV)	10^{-9}	10^{-7}

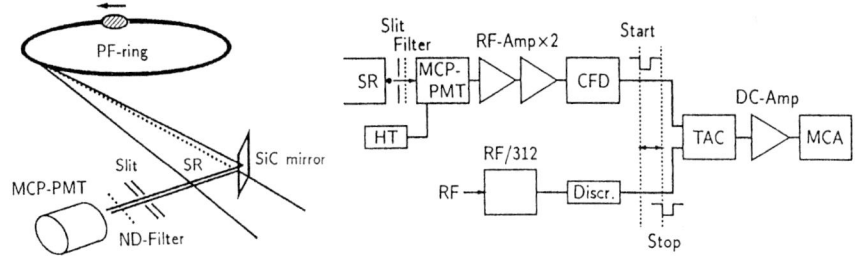

FIGURE 1. Photon counting system.

knock-out electrode for 11 ns in UVSOR. Details of this method are described in Section 4. A similar system and a short-pulse electron gun are used together in SPring-8. Although the requirement for purity in SPring-8 is better than 10^{-9} as mentioned before, the achieved purity is 10^{-7} at present. There is a plan to build a chopper after the electron gun in order to clear halos due to the field emission.

Usually, we use a photon counting system to measure the impurity. Figure 1 shows a block diagram of a photon counting system. The visible component of the synchrotron radiation is attenuated by a ND filter. A single photon is detected by a multi-channel plate-type photomultiplier (MCP-PMT). Events corresponding to the aimed bunch and other bunches are counted separately by a time-to-amplitude converter and a multi-channel analyzer. An example of a measurement is shown in Fig. 2. In this case, the impurity level was worse than 1%. The limit of measurement, determined mainly by the noise level of a MCP-PMT is approximately 10^{-6}. This limit is improved by use of an avalanche photodiode for x-ray with a low noise level, or greatly improved by a high-speed light shutter, shown in Fig. 3. When a high-voltage pulse is applied to the Pockels cell, the polarization plane rotates around the optical axis. Two polarizers, whose directions of polarization are perpendicular to each other, are set on the two sides of the Pockels cell. When the voltage is applied to the cell and rotates the polarization plane, incident light can pass through the system.

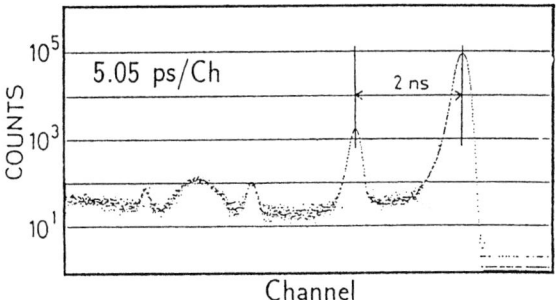

FIGURE 2. An example of an impurity measurement by the photon counting method.

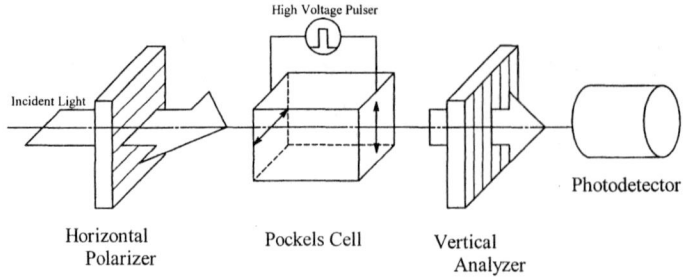

FIGURE 3. High-speed light shutter.

This shutter system is quite fast, with a rise/fall time less than 1 ns. If we close the shutter during the light pulse corresponding to arrival of the aimed bunch at the shutter and open otherwise, we can greatly improve the limit. This method has been used in SPring-8.

3 DETERIORATION OF SINGLE-BUNCH IMPURITY

Before discussing single-bunch purifiers, let us consider a phenomenon that deteriorates single-bunch purity. More than twelve years ago, I constructed a single-bunch system in UVSOR and time-resolved experiments were started. A time resolved experimenter claimed that the purity became worse during his measurement. I could not believe the phenomenon at that time. Therefore, I made a series of experiments on the phenomenon, and I found that his claim was correct. This phenomenon is explained by the following mechanism, shown in Fig. 4. Electrons thrown out of the bucket by the Touschek effect are recaptured in following buckets by radiation damping because buckets have openings due to radiation damping as seen in the figure. If scattered electrons jump into the openings, they are eventually trapped into the following buckets. Figure 5 shows this effect observed in PF. The Touschek effect is negligible in high-energy machines. Therefore this effect does not occur in SPring-8 and the 6.5-GeV PF-AR.

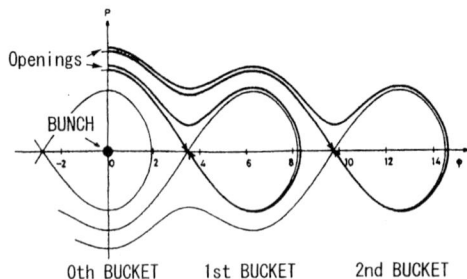

FIGURE 4. Mechanism of the impurity deterioration.

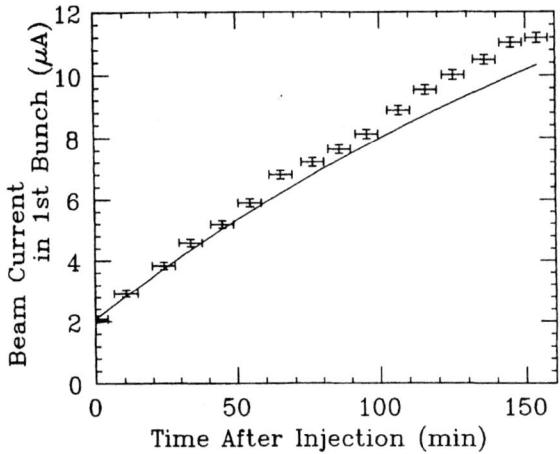

FIGURE 5. Impurity growth in PF.

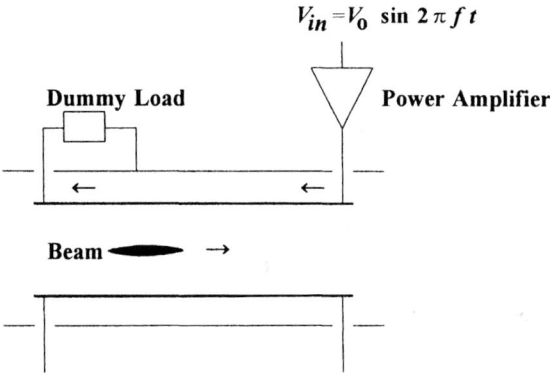

FIGURE 6. RF-KO system.

4 SINGLE-BUNCH PURIFIER

Although the effect described in the preceding section does not occur in PF-AR because the operation energy is high enough, the effect deteriorates the single-bunch purity during the injection period at 2.5 GeV. As mentioned before this effect is also observed in the 2.5-GeV PF. Furthermore, since the emittance of the PF was improved to 36 nm from 130 nm several years ago and experiments in the single-bunch mode became sophisticated, this effect affects the quality of time-resolved experiments.

Therefore, a system called the single-bunch purifier, which clears electrons in undesirable bunches, is essential for these machines.

First, let us review the principle of the RF-KO (Fig. 6). The betatron oscillation is excited if the excitation frequency f fulfills the condition

$$f = nf_{rev} \pm qf_{rev},$$

where f_{rev} is the revolution frequency, q is the decimal part of the tune and n is an integer. However, because the betatron oscillation is nonlinear, usually it is difficult to destroy beams completely with a fixed frequency. Therefore we sweep the excitation frequency in order to increase the betatron amplitude smoothly. If we can destroy only the undesirable bunches without affecting the single bunch by using the RF-KO method, we can purify the single bunch. We adopt two methods. One is to make use of the dependence of tune on the bunch current and the other is to use a gated RF-KO method. The former is used for PF-AR. Usually, the tune depends on the bunch current. If the KO frequency is set to the frequency corresponding to the tune of a bunch with low beam current, the undesirable bunches with low currents can be removed and the main bunch with a high current can survive. However, because the dependence is small and we must sweep the frequency, as mentioned above, it is difficult to find the optimum conditions for the center frequency and the sweep range. Furthermore, influence on the main bunch is unavoidable. We use a transverse feedback system that can damp the betatron oscillation together with the RF-KO system in PF-AR. Since the feedback system can stabilize the beam, detecting the oscillation of the beam, bunches with low current cannot be stabilized with this system. Therefore, if the feedback system is used together with the RF-KO system, only undesirable bunches are excited and removed. Since it is not necessary to operate the purifier after ramping up to the final energy of 6.5 GeV, as mentioned before, the KO system is turned off to avoid the influence on the main bunch after ramping up to 6.5GeV. With this system, we achieved an impurity level better than 10^{-6} in PF-AR. This level of impurity satisfies PF-AR experimenters at present.

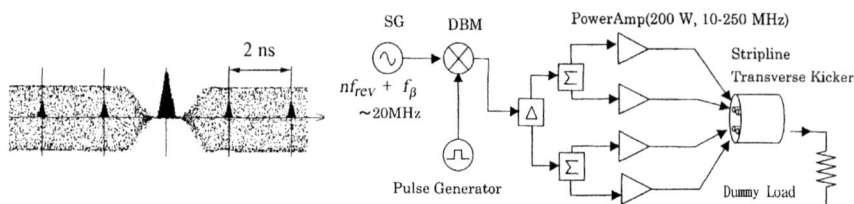

FIGURE 7. Gated RF-KO method.

Because influence on the main bunch is unavoidable with the system mentioned above, we cannot adopt the same system for the 2.5-GeV PF, in which we must continuously clear undesirable bunches during user operation. We developed a gated RF-KO Method. The principle is simple, as shown in Fig. 7. While the main bunch is passing through the RF-KO system, the KO signal is turned off. If the switching speed is high enough compared with the bunch spacing, we can clear the undesirable bunches. We can achieve an impurity level of 10^{-7} with this system. We observed the growth of the impurity level in PF due to the phenomenon mentioned above. Figure 8 shows the longitudinal bunch structure measured 1 hour after turning off the purifier. We can clearly observe the growth of undesirable bunches. Though the purifier must be continuously operated in the 2.5-GeV PF in order to avoid this deterioration, harmful influence of the purifier on beam stability has not been observed.

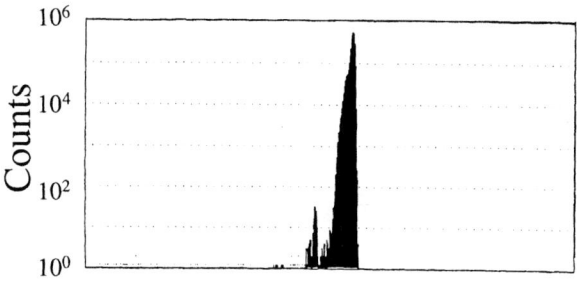

FIGURE 8. Impurity deterioration in PF.

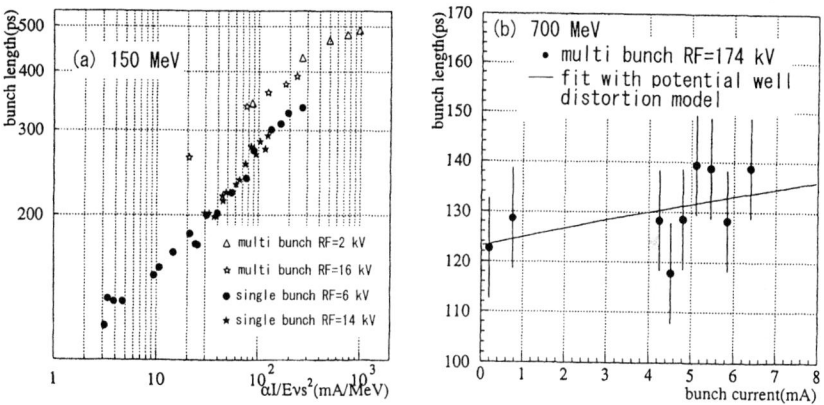

FIGURE 9. Bunch-lengthening in HiSOR at 150 and 700MeV.

5 BUNCH LENGTHENING AND LOW-ALPHA OPERATION

Gradually, longitudinal characteristics begin to show their importance for users who are interested in time-resolved experiments. Bunch lengthening is one of the issues. Phenomena of bunch lengthening are observed and measured in many electron storage rings. They are also measured in order to estimate the coupling impedances of vacuum systems. Figure 9 shows the bunch lengthening measured in HiSOR at an injection energy of 150 MeV and at the top energy of 700 MeV. The length largely depends on the bunch current. However, we have no strong requests for short bunches at present. I think control of the longitudinal emittance will be important in the near future. Operation with small alpha has been tried in several synchrotron radiation facilities. In Japan, UVSOR of the Institute for Molecular Science has also tried low-alpha configurations. An example of the bunch lengthening measured while changing the momentum compaction factor is shown in Fig. 10. New Subaru of Himeji Institute of Technology is a specially designed machine for low-alpha operation, as mentioned before. Low-alpha operation at New Subaru will be tried soon.

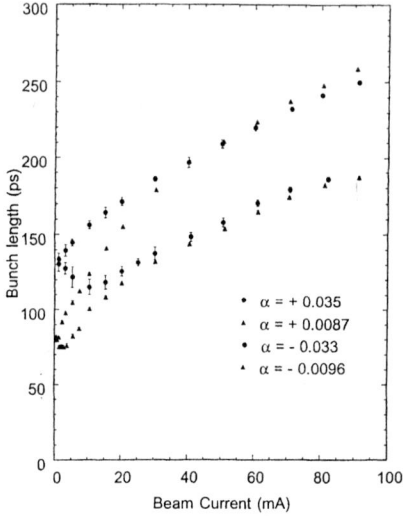

FIGURE 10. Bunch-lengthening in UVSOR.

For good or for evil, bunch lengthening does not seem to matter for most users for materials science at present. I think the longitudinal emittance will play an important role in future light sources. Also, attempts to control the bunch length will be very important in the near future.

ACKNOWLEDGMENTS

I would like to thank Dr. N. Kumagai of SPring-8, Dr. M. Kato of the Institute for Molecular Science, Dr. A. Ando of Himeji Institute of Technology, and Dr. K. Yoshida and Dr. K. Umemori of Hiroshima University, who willingly offered valuable information about their machines.

Review of Cooling

V.V. Parkhomchuk

*Budker Institute of Nuclear Physics,
Novosibirsk 630090 Russia*

I INTRODUCTION

Development of experimental investigations related to nuclear physics has resulted in a sharp rise in the quality required for particle beams (electrons, positrons, protons, antiprotons, light and heavy ions). It is especially important to obtain beams of high density and low momentum spread. The luminosity of a collider is determined by the emittance of bunches ϵ, number of particles in the bunch N, beta function at the interaction point β^*, and bunch repetition frequency f_b as

$$L = \frac{N_1 N_2}{4\pi\epsilon\beta^*} f_b. \tag{1}$$

Cooling helps decrease the beam emittance, and decreasing the momentum spread $\Delta p/p$ helps achieve stronger focusing and smaller β^*.

The easiest way to increase the particle density is to improve the focusing properties of the magneto-optical channel for particle confinement. Here, although transverse beam dimensions are reduced, the transverse beam momentum grows in accord with the condition of six-dimensional phase-space density conservation. Cooling of a beam means a decrease in the spread of particle velocities due, not to a change in focusing, but to loss of energy of the chaotic particles' motion, as a result of their interaction with the "cooler." This interaction provides for the increase in the phase-space density of the beam, and for the storage of particles, fitting them into the areas of the phase space vacated during the process of cooling.

The phase-space density of the beam cannot be increased by using any set of external electromagnetic fields, independent of the motion of specific particles of the beam (so-called Hamiltonian motion). In this case the statement (of Liouville's theorem) that the phase-space density of the beam particles is constant and determined by initial conditions, holds true. In some unique cases, it is possible to increase the phase-space density of the beam particles, purposely producing them in required areas of the phase space, as for example in the case of charge-exchange injection. To increase the phase-space density of an already existing beam of particles, it is necessary to use dissipative forces, which result in losing the energy of chaotic motion of the particles.

The mechanism of energy loss gives the cooling technique its name. Ionization cooling is based on the energy loss of particles while moving in a rather dense target. Laser cooling is based on the interaction of atoms or ions (having electrons) with laser beam. To save time, we will discuss only the cooling techniques that are widely used in accelerator practice: synchrotron radiation cooling, stochastic cooling and electron cooling.

II SYNCHROTRON RADIATION COOLING

The most developed technique used today is radiation cooling [1], based on high-power synchrotron radiation generated by the motion of light relativistic particles in magnetic fields. The energy loss rate can be written as

$$\frac{1}{E}\frac{dE}{dt} = \frac{2}{3}\frac{r_i c}{R^2}\gamma^3, \qquad (2)$$

where $r_i = e^2/(mc^2)$ is the classical radius of particles of the beam, R is the bending radius of a particle's trajectory in magnetic field B, c is the speed of light, and $\beta = v/c$, $\gamma = (1-\beta^2)^{-1/2}$ are the relativistic factors. The average energy loss is compensated by an external RF energy source, and in a proper magneto-optical system deviations from the equilibrium motion are damped. For the simplest FODO structure, where dipole magnets with uniform fields are used for bending the particles' trajectory, and the edges of the magnets are perpendicular to the beam path, both vertical and horizontal betatron oscillations can by described by an equation,

$$\frac{d^2 z}{dt^2} + \frac{1}{E}\frac{dE}{dt}\frac{dz}{dt} + \omega_z^2 z = 0, \qquad (3)$$

which gives damped oscillations with damping decrements equal to

$$\lambda_{x,z} = \frac{1}{3}\left\langle\frac{r_i c}{R^2}\right\rangle\gamma^3, \qquad (4)$$

where averaging means calculation of $1/R^2$ averaged along the beam orbit. In this magnetic structure, the energy loss increases with the energy deviation ΔE as

$$\frac{dE}{dt} \sim \frac{\gamma^4}{R^2} \sim \gamma^2 B^2 \sim \gamma_0^2 B_0^2\left(1+\frac{\Delta E}{E}\right)^2 \sim \left.\frac{dE}{dt}\right|_0\left(1+2\frac{\Delta E}{E}\right), \qquad (5)$$

where subscript 0 indicates the equilibrium value. The factor 2 results in two-fold cooling decrements in the longitudinal motion of particles, $\lambda_{||} = 2\lambda_{x,z}$.

Radiation cooling finds its widest application in the storage of intense electron and positron beams, and helps reach the high luminosity of electron-positron colliders. It is easy to see from Eq. (2) that, for the same energy E ($\gamma = E/m_i c^2$), the damping decrement is inversely proportional to the square of the mass of cooled particles. At the LHC, for protons with an energy of 8 TeV, the cooling time will be a few tens of hours.

III STOCHASTIC COOLING

The method of stochastic cooling, proposed by Van der Meer [2], is based on application of a broad bandwidth feedback system that measures a particle's position by pick-up electrodes, and corrects its motion by sending this signal (after amplification) to a kicker installed on the beam orbit. The delay time of the particle's flight from the pick-up electrodes to the kicker is adjusted to be the same as the delay in the amplification and connection lines. The peak current of each particle $Z_i eW$ ($Z_i e$ is the charge, and W is the bandwidth of the feedback system) produces a pick-up signal of voltage

$$U = Z_i eWZ \frac{x}{A_p}, \qquad (6)$$

where Z is the impedance of the pick-up line, x is the particle's displacement at the pick-up, and A_p is the pick-up aperture. For the optimal betatron phase advance $n\pi + \pi/2$ from the pick-up to the kicker ($\delta p = Z_i eU l_{kick}/(\beta c)$), when displacements at the pick-up x transfer to the velocity at the kicker corresponding to the transverse angle $\Delta p/p = x/\beta_x$ (β_x is the beta function at the kicker position), the cooling decrement is

$$\lambda = \frac{\delta p}{\Delta p} f_0 = k \frac{r_i l_{kick} \beta_x W Z}{\gamma \beta^2 A_p^2} f_0, \qquad (7)$$

where k is the gain of the feedback system, f_0 is the revolution frequency, and l_{kick} is the effective length of a kicker with the same aperture A_p as at the pick-up. Usually, the kicker is based on the traveling wave propagating in the direction opposite the beam. Limitation of the effective interaction length of the kicker by the condition $l_{kick} \approx A_p$ stimulated the use of multi-pickup-kicker systems with a high number of pickup-kickers lines $k_{kickers}$, with a relatively low gain at each line k. For the impedance Z=50 Ω that corresponds to $Z = 50/9 \cdot 10^{11} = 5/(3c)$ (in CGS units), the cooling decrement can be estimated as

$$\lambda = kk_{kickers} \frac{5 r_i \beta_x W}{3\gamma \beta^2 c A_p} f_0. \qquad (8)$$

For cooling of antiprotons, $\gamma = 8$, $\beta \approx 1$, $A_p = 10$ cm, $r_p = 1.6 \cdot 10^{-16}$ cm, $f_0 = 3 \cdot 10^6$ Hz with a cooling time $1/\lambda \approx 1$ s, the gain of the system should be $k \cdot k_{kickers} = 2.5 \cdot 10^9$. For $k_{kikers} = 100$, the gain of a single line is $k = 2.5 \cdot 10^7$, and the power of noise in this whole system is $4kTWk^2 k_{kickers} \sim 100$ kW for $W = 10^9$ s^{-1}.

This method of cooling was used for accumulating antiprotons at CERN and FNAL, and the discoveries of the W and Z bosons evidenced the great importance of this cooling method in accomplishing experiments with hadron colliding beams.

IV ELECTRON COOLING

The electron cooling method was proposed by Budker in 1965, and discussed for the first time at the Saclay Symposium on Electron-Positron Storage Rings; it was published after detailed discussion [3]. In this proposal, the friction force resulted from the motion of ions immersed in the co-moving electron beam with the same average velocity. The energy of the chaotic motion of the ions is transferred to the cold electron gas. To produce an electron beam with the same average velocity, the energy of the electron beam needs to be m_e/M_i times less; for example, to cool a 100-MeV proton beam we should accelerate the electron beam to an energy of only 50 keV. The first electron cooling experiments held at INP (Novosibirsk) in 1974 [4] demonstrated the high efficiency of this method.

At first, the theoretical estimation used a plasma model of energy exchange in an electron-ion plasma. When an ion moves past an electron with velocity V at distance ρ, the field of the ion $Z_i e/\rho^2$ kicks the electron and changes its momentum, $\Delta p_e = Z_i e^2/\rho^2 \cdot 2\rho/V$. The ion energy loss is $\Delta p_e^2/(2m_e)$, and finally the friction force can be written in the form

$$F = \frac{1}{V}\frac{dE}{dt} = \int_{\rho_{min}}^{\rho_{max}} \frac{2Z_i^2 e^4}{m_e V^2 \rho^2} n_e 2\pi\rho d\rho = \frac{4\pi Z_i^2 e^4 n_e}{m_e V^2} \ln\frac{\rho_{max}}{\rho_{min}}, \qquad (9)$$

where ρ_{max} and ρ_{min} are the maximal and minimal impact distances when it is possible to use the small-displacement (Born's) approximation for electron motion. For an electron beam with not too high a density (when the plasma frequency $\omega_e = c\sqrt{4\pi n_e r_e} < \tau^{-1} = (l_{cooling}/(c\beta\gamma))^{-1}$ is less than the inverse time of flight in the cooling section), the maximal impact distance is determined by the path of ions in the electron beam,

$$\rho_{max} = V\tau, \qquad (10)$$

and the minimal impact distances are determined by the condition that the displacement of electrons during the interaction time $Z_i \cdot e^2/(m_e \rho^2) \cdot (\rho/V)^2 \approx \rho$ becomes comparable with ρ,

$$\rho_{min} = \frac{Z_i r_e}{(V/c)^2}. \qquad (11)$$

At small ion velocities, the friction force $F \sim V^{-2}$; this helps us use an analogy in velocity space with the electric Coulomb force $F = e^2/r^2$ in normal space. If the electrons have their own chaotic motion with a distribution $f(\vec{V_e})d^3V_e$, for calculation of the friction force we need to average Eq. (9),

$$\vec{F} = \int \frac{4\pi Z_i^2 e^4 n_e L_C}{m_e|\vec{V_e}-\vec{V}|^3}(\vec{V_e}-\vec{V})f(\vec{V_e})d^3V_e, \qquad (12)$$

where $L_C = \ln(\rho_{max}/\rho_{min})$.

For example, if $f(\vec{V_e}) = const$ for $|\vec{V_e}| < V_c$, and vanishes for $|\vec{V_e}| > V_c$, the friction force grows linearly from the center to the edge of the electron velocity distribution $0 < V < V_c$, and outside it, decreases as V^{-2}. The cooling decrement can be estimated from $\lambda = F/(M_i V)$:

$$\lambda = \frac{4\pi Z_i^2 e^4 n_e L_C}{m_e M_i V_c^3} \quad \text{for} \quad |V| < V_c, \qquad (13)$$

and

$$\lambda = \frac{4\pi Z_i^2 e^4 n_e L_C}{m_e M_i |V|^3} \quad \text{for} \quad |V| > V_c. \tag{14}$$

In the first experiment at NAP-M, a 65-MeV proton beam was cooled by an electron beam with an energy of 35 keV, and the temperature was $E_{te} = 0.2$ eV, which means a 2–3 times higher energy than that of the thermal motion of electrons due to the hot cathode of the electron gun (1500 K, 0.1–0.15 eV). The velocity of thermal motion of electrons with this energy was $V_c = \sqrt{2E_{te}/m_e} = 2.3 \cdot 10^7$ cm/s, which corresponds to $\langle \lambda \rangle = \eta_e \lambda = 1/\tau_{cooling} = 1/3$ s^{-1} (η_e is the fraction of the proton orbit circumference occupied by the electron beam). In the NAP-M experiment it was discovered that the cooling time decreases with decreasing transverse velocity $V_\perp < V_c$, and in fact it turned out to be less than 0.1 s instead of 3 s. Such a dramatic increase in cooling efficiency was a result of the combined effect of two factors: first, the presence of a longitudinal magnetic field in the cooling section, and second, an extremely low spread in the longitudinal electron velocities after acceleration. The longitudinal magnetic field was used to transport the electron beam from the cathode to the proton beam cooling section, and further to the electron beam collector. The magnetic field "magnetizes" the transverse electrons motion, and as a result, the ions under cooling interact with a cool Larmor circle (with a relatively small radius $\rho_L = mV_c/eB$, where B is the magnetic field), but not with hot (and fast) free electrons [5,6]. This phenomenon resulted both in enhancement of the cooling rate, and in cooling of ions down to temperatures many times lower than the cathode temperature (1500°K). Thus, a longitudinal motion temperature of about 1°K was obtained in the proton beam at NAP-M. A useful practical equation for the magnetized cooling force results from fitting to the experimental data [7]:

$$\vec{F} = -\frac{4Z_i^2 e^4 n_e}{m_e} \frac{\vec{V}}{(V^2 + V_{\text{effe}}^2)^{3/2}} \ln \frac{\rho_{max} + \rho_{min} + \rho_L}{\rho_{min} + \rho_L}, \tag{15}$$

where $V_{\text{effe}} = \sqrt{V_{\|e}^2 + \Delta V_{\perp e}^2}$ is not the electron velocity V_c, but an effective velocity of the electron's Larmor ring, combined from the longitudinal electron velocity spread $V_{\|e}$, and the velocity component due to the transverse magnetic and electric field $\Delta V_{\perp e}$. For better cooling of the ion beam near equilibrium, we need a perfect parallelism of magnetic field lines over the cooling section, in order to achieve $\Delta V_{\perp e} = c\gamma\beta\Delta B_\perp/B$ as close to zero as possible.

V PROBLEMS IN COOLING AN INTENSE ION BEAM

Cooled beams are very sensitive to the development of instabilities, due to high density and extremely low momentum spread [8]. The limit on the longitudinal impedance is set by the Keil-Schnell criterion,

$$|Z_n/n| < \frac{A_i \beta^2 (E/e)|\eta_t|}{N e f_0} \left(\frac{\Delta p}{p}\right)^2. \tag{16}$$

After cooling, strong transverse instabilities can develop in the storage ring. A feedback system acting in a broad bandwidth can stabilize the beam. For example, at LEAR the stochastic cooling system with a bandwidth up to 500 MHz was used as a feedback system, and increased the number of stored protons from 10^8 up to $8 \cdot 10^{10}$ [8]. The instabilities are due to interaction with various elements of the storage ring. Improvement of RF cavities and vacuum chamber impedance helps increase the stored current.

But the cooling system itself can introduce some instabilities connected with the physical nature of cooling. This type of instability is the source of more fundamental limitations on achievable beam density after cooling.

A Intensity Limits for Stochastic Cooling

The stochastic cooling system has a bandwidth W, and at the kickers not only its own signal passed through amplifiers acts on the cooled ions, but also a signal from near-by ions passing the pick-up electrode during the time interval $\Delta t = 1/W$. The number of ions in this sample is $N^* = Nf_0/W$, for a coasting beam. Taking into account the second-order terms of cooling processes, we have to write the equation of cooling in the form

$$\Delta p_i^2 = (p_i - \Delta p)^2 - p_i^2, \qquad (17)$$

where Δp is the momentum correction resulting from the action of all the ions in the sample:

$$\Delta p = \delta \sum_{k=1}^{N^*} p_k, \qquad (18)$$

where $\delta = \lambda/f_0$ is the correction of ion deflection at each turn. It is easy to see from Eq. (17) that there is no correlation in the chaotic motion of ions, $\langle p_i p_k \rangle = 0$ for $i \neq k$, and

$$\Delta p_i^2 = -2\delta p_i^2 + N^* \delta^2 \langle p_k^2 \rangle. \qquad (19)$$

The optimum cooling rate corresponds to $\delta = 1/N^*$, and we have the famous Van der Meer equation for the cooling time:

$$\tau_{cool}^{-1} = \lambda = \frac{W}{N}, \qquad (20)$$

that shows reduction of the cooling rate with increasing number of ions in the beam.

In the opposite case, when a 100% correlation exists in the ion motion, and all ions move in a coherent way, the optimal value $\delta = 1/N^*$ corresponds to the coherent decrement $\delta_{coh} = \delta N^* = 1$, which means extremely fast damping of the coherent motion, just on one turn. The longitudinal motion of ions along the orbit causes mixing in the sample, and the number of turns for complete mixing is estimated as $n_{mix} = 2\pi f_0/W/(|\eta_t|\Delta p/p)$, whereas decrease of the momentum spread during the cooling process can generate bad mixing $n_{mix} \gg 1$, and thus an additional limitation on cooling of a cold beam $\delta \approx 1/(Nn_{mix})$.

B Intensity Limitation for Electron Cooling

The friction force in the electron cooling system is formed as a result of the electron beam polarization caused by the motion of each ion. This force is not localized at a point with zero dimension (as $\delta(\vec{r})$), but rather is distributed around the path of the ion in the electron beam. To demonstrate this, Fig. 1 shows the positions of electrons after passing an ion along the Z-axis, from point $-6\rho_{min}$ to $6\rho_{min}$ ($\rho_{min} = Z_i r_e/(V/c)^2$). We can see that electrons have large displacements all along the path of the ion. These displacements are real sources of the friction force acting on the ion velocity. However, the arising electric field acts not only on this particular ion (which generated the electron displacements) but also on many other ions located in the field zone. The space distribution of this field is shown in Fig. 2.

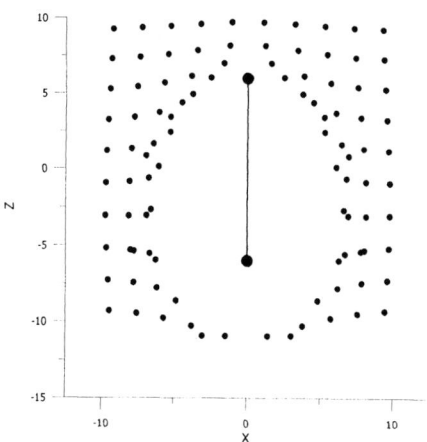

FIGURE 1. Space distribution of electrons after passing an ion from point $z = -6$ to $z = 6$ (coordinates are measured in units ρ_{min}).

At the position of the ion (point $6\rho_{min}$), the value of the friction force is

$$F = -\frac{2.5 n_e Z_i^2 e^4}{m_e V^2}, \tag{21}$$

however, far away (at $z \sim \rho_{max}$) there are also strong electric fields acting on near-by ions. The number of ions in this zone can be estimated as

$$N^* = f \frac{4\pi n_i}{3} \rho_{max}^3 \tag{22}$$

where $f \sim 1$ is a factor accounting for a more realistic shape of the perturbed zone, and ρ_{max} is an estimate of its radius.

As a result of the action of its own friction force and the action of near-by ions of the beam, we should write the change in the ion momentum over the time-of-flight of the ion through the cooling section in the form

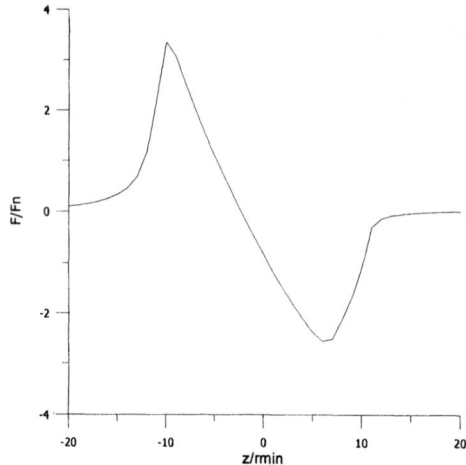

FIGURE 2. Distribution of the electric field along the ion path (coordinates are measured in units ρ_{min}, the field in units $Z_i e n_e \rho_{min}$).

$$\Delta p_i = eE_{frict}\tau = -\frac{4\pi Z_i^2 e^4 n_e L_C}{m_e V^2}\tau \frac{p_i}{M_i V} = -\delta p_i. \quad (23)$$

If all the ions in this zone have coherent motion with the same velocity V, the cooling force is increased by a factor of N^*, and the coherent decrement on a single pass of the cooling section can be written as

$$\delta_c = N^*\delta = f\frac{4\pi n_i}{3}(V\tau)^3 \frac{4\pi Z_i^2 e^4 n_e L_C}{m_e M_i V^3}\tau = \frac{f L_C}{3}\omega_i^2 \omega_e^2 \tau^4. \quad (24)$$

In stochastic cooling, the maximum value of the coherent damping can be near 1, but how large are the coherent decrements available in electron cooling?

Outside the cooling section, the equation of motion for the plasma oscillations of the electron beam can be written in the form (in the rest frame of the beams)

$$\frac{d^2 x_e}{dt^2} = -\frac{4\pi e^2 n_e}{m_e}x_e = -\omega_e^2 x_e, \quad (25)$$

and for those of the ion beam in the form

$$\frac{d^2 x_i}{dt^2} = -\frac{4\pi e^2 Z_i^2 n_i}{M_i}x_i = -\omega_i^2 x_i, \quad (26)$$

where $\omega_e = c\sqrt{4\pi n_e r_e}$ and $\omega_i = c\sqrt{4\pi n_i r_i}$ are the frequencies of plasma oscillations for the electron and ion beams.

In the cooling section, where the beams move along a common orbit, the electric fields of plasma oscillations act on both beams, and the equations can be written as

$$\frac{d^2 x_e}{dt^2} = -\frac{4\pi e^2 (n_e x_e - Z_i n_i x_i)}{m}, \tag{27}$$

$$\frac{d^2 x_i}{dt^2} = -\frac{4\pi e^2 Z_i (Z_i n_i x_i - n_e x_e)}{M_i}. \tag{28}$$

For the electric field $E = 4\pi e(n_e x_e - Z_i n_i x_i)$, it is easy to see that the oscillation equation has a simple form

$$\frac{d^2 E}{dt^2} = -(\omega_e^2 + \omega_i^2) E. \tag{29}$$

This is an equation of plasma oscillations with a frequency $\omega = \sqrt{\omega_e^2 + \omega_i^2}$.

At the entrance of the cooling section, the ion beam has some coordinates $x_i(0)$ and initial velocity $\dot{x}_i(0)$, but the "fresh" electron beam has $x_e(0) = 0$, $\dot{x}_e(0) = 0$.

The self-consistent solution of this equation can be written in the form

$$\begin{pmatrix} x_i \\ \dot{x}_i \end{pmatrix} = \begin{pmatrix} A_{11} & A_{12} \\ A_{21} & A_{22} \end{pmatrix} \cdot \begin{pmatrix} x_i \\ \dot{x}_i \end{pmatrix}_0, \tag{30}$$

where the matrix elements are calculated by integration of the equations along the cooling section, using specific initial conditions. For initial conditions $x_i(0) = 1$, $\dot{x}_i(0) = 0$, we calculate elements A_{11} and A_{21}, and for initial conditions $x_i = 0$, $\dot{x}_i = 1$ we calculate matrix elements A_{12} and A_{22}:

$$\begin{pmatrix} A_{11} \\ A_{21} \end{pmatrix} = \begin{pmatrix} x_p(\tau) \\ \dot{x}_p(\tau) \end{pmatrix}, \tag{31}$$

where τ is the interaction time in the cooling section. We obtain

$$A = \begin{pmatrix} \frac{1}{\omega^2} \left(\omega_i^2 \cos \omega \tau + \omega_e^2 \right) & \frac{1}{\omega^3} \left(\omega_i^2 \sin \omega \tau + \omega_e^2 \omega \tau \right) \\ -\frac{\omega_i^2}{\omega} \sin \omega \tau & \frac{1}{\omega^2} \left(\omega_i^2 \cos \omega \tau + \omega_e^2 \right) \end{pmatrix}. \tag{32}$$

We should remember that the determinant of this matrix $\det A \neq 1$. This system is not closed, and the electron beam can either absorb or increase the plasma oscillation energy of the ion beam. If the determinant is less than 1, it means the loss of this energy and cooling, but for $\det A > 1$ the energy increases and we have heating. The determinant of matrix A is

$$\det A = 1 + \frac{\omega_i^2 \omega_e^2}{\omega^4} \left(\omega \tau \sin \omega \tau - 2(1 - \cos \omega \tau) \right). \tag{33}$$

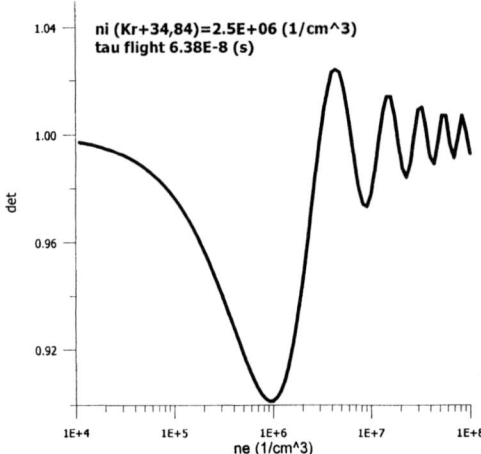

FIGURE 3. Variation of matrix determinant vs. electron beam density for Kr ion beam density $2.5 \cdot 10^6$ cm^{-3}.

For $\omega\tau \ll 1$, we have $\det A - 1 \approx -\omega_i^2 \omega_e^2 \tau^4 / 12$, and interaction with the electron beam in the cooler produces coherent cooling.

Figure 3 shows the variation of the determinant of this matrix versus the electron beam density for a Kr_{84}^{+34} beam density of $n_i = 2.5 \cdot 10^6$ cm^{-3} and $\tau = 6.38 \cdot 10^{-8}$ s. The parameters correspond to accumulation of Kr beam at the SIS synchrotron (see Fig. 5).

From Fig. 3 it follows that $\det A < 1$ for a small electron beam density, which corresponds to fast cooling of plasma oscillations in the proton beam, but for a large density $\omega_e \tau > 2\pi$ we have $\det A > 1$, and it is possible to have fast heating of coherent oscillations.

Figure 4 shows the variation of the determinant of this matrix versus the ion beam density for different electron beam densities.

From Figs. 3 and 4 it follows that the region for cooling is limited ($\det A < 1$) by conditions on the frequency of electron beam oscillations in the space-charge fields of the electron or ion beams:

$$\omega_e = c\sqrt{4\pi n_e r_e} < \frac{2\pi}{\tau}, \qquad (34)$$

$$\omega_{\text{effe}} = c\sqrt{4\pi Z_i n_i r_e} < \frac{2\pi}{\tau}. \qquad (35)$$

These conditions put limits on the maximum density of electron beam and ion beam to prevent the development of heating. If we write the maximal coherent decrement, Eq. (24), under these conditions, Eqs. (34) and (35), we find a very simple limitation:

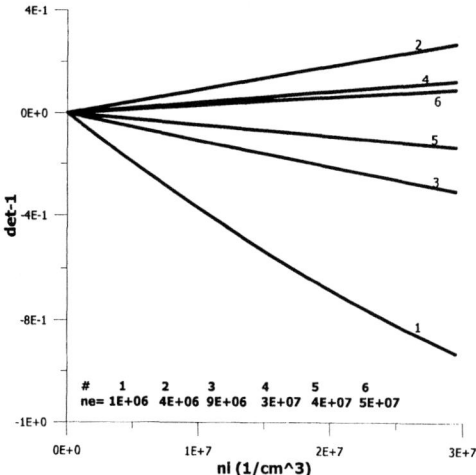

FIGURE 4. Variation of matrix determinant vs. Kr ion beam density for different densities of the electron beam.

$$\delta_c < 500 f\, L_C \frac{Z_i m_e}{M_i}. \qquad (36)$$

This means that the maximum coherent decrement corresponds to so-called inelastic interaction, where each ion captures Z_i electrons, and the momentum loss is just a ratio of masses of captured electrons to the ion mass. The question of the possibility of having a higher beam density is now open, and it may be a good problem for advanced students of this accelerator school. As you can see from Fig. 5, the development of oscillations does not mean immediate loss of the beam. The efficiency of accumulation drops slightly, but the increase in ion current continues above the instability threshold.

VI   PROJECTS USING COOLING

A good example of using electron cooling is the cooler for SIS that was designed for repeated multi-turn injection to boost SIS intensity, using a low injection current. After commissioning the cooler, beam intensities in the range of $1\text{--}5 \cdot 10^9$ ions per cycle are available. Figure 5 shows an example of accumulation of Kr beam at SIS at the time of commissioning the cooler. The cooler helped increase the intensity of the beam and decrease its emittance and momentum spread. This example of a successfully commissioned electron cooling stimulated the use of this technique in projects of heavy ion facilities, RICKEN (Japan) and CSR (China). For accumulating Pb ions, CERN will use electron cooling during accumulation of the Pb bunch in the ion-ion version of the LHC.

At high energies, the use of electron cooling for the Recycler Ring at FNAL (USA) seems to be the next step. Discussion about using cooling for the RHIC collider is

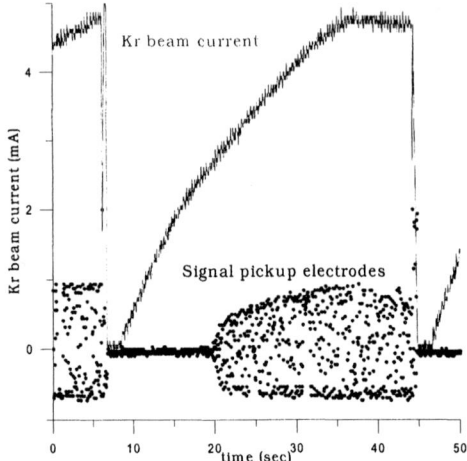

FIGURE 5. Accumulation of Kr beam during repeated multi-turn injection; after electron cooling the signal from pick-up electrodes shows development of instabilities upon repeatedly reaching the same threshold in the ion beam current.

now in progress; this will show whether it is feasible. An ion collider (ion-ion or ion-electron) with continuous cooling can achieve a higher luminosity and seems to be a good prospect. At many accelerator centers discussion of these projects with involvement of cooling techniques is under way.

ACKNOWLEDGEMENTS

The development of subjects of this article was a result of very useful discussions with my colleagues at BINP: N. Dikansky, I. Meshkov, D. Pestrikov, R. Salimov, A. Skrinsky, and B. Sukhina. I would also like to thank D. Reistad for collaboration in investigation of the "electron heating" phenomenon.

REFERENCES

1. Kolomensky A.A., Lebedev A.N., *Theory of Cyclic Accelerators*, Soviet Physics and Mathematics Publishers, Moscow, 1962.
2. Möhl D., Petrucci G., Thorndal L., Van Der Meer S., "Physics and techniques of stohastic cooling," *Phys. Rep.* **58**, 75-119 (1980).
3. Budker G.I., "Efficient method for damping of particle oscillations in proton and antiproton storage ring," *Atomnaya Energia* **22**, 346-348 (1967).
4. Budker G.I., Dikansky N.S., Kudelainen V.I., Meshkov I.N., Parkhomchuk V.V., Pestrikov D.V., Skrinsky A.N., Sukhina B.N., "First experiments on electron cooling," *IEEE Trans. Nucl. Sci.* **NS-22**, 2093-2097 (1975).

5. Derbenev Ya.S., Skrinsky A.N., "Magnetization effects in electron cooling," *Fizika Plasmy* **4**, 492-500 (1978).
6. Parkhomchuk V.V., Skrinsky A.N., "Methods of cooling of charged particles beam," *Phys. Elementary Part. and Atomic Nucl.* **12**, 557-613 (1981).
7. Parkhomchuk V.V., "New insights in the theory of electron cooling," *Nucl. Instr. Methods in Phys. Research* **A441**, 9-17 (2000).
8. Bosser J., Carli C., Chanel M., Madsen M., Maury S., Möhl D., Tranquille G., "Stabilility of cooled beams," *Nucl. Instr. Methods in Phys. Research* **A441**, 1-8 (2000).

Intrabeam Scattering

J. Struckmeier

Gesellschaft für Schwerionenforschung (GSI)
Postfach 11 05 52, 64220 Darmstadt, Germany
E-mail: j.struckmeier@gsi.de

Abstract. The Vlasov equation embodies the smooth-field approximation of the self-consistent equation of motion for charged-particle beams. This framework is fundamentally altered if we include the fluctuating forces that originate from the actual charge granularity. We thereby perform the transition from a reversible description to a statistical mechanics' description covering also the irreversible aspects of beam dynamics. It will be shown that the macroscopic effects due to the fluctuating forces scale with two distinct quantities: the magnitude of the fluctuating forces, and the system's temperature anisotropy. These analytical results are used to obtain a quantitative description for the effect of intrabeam scattering.

INTRODUCTION

Analytical approaches to the dynamics of charged-particle beams that are based on the Liouville — or equivalently on the Vlasov — equation do not include effects due to the actual charge granularity. A variety of beam phenomena are adequately described by this continuous description. As the first example, we quote the work of I.M. Kapchinskij and V.V. Vladimirskij [1] covering the description of beam transport under space-charge conditions. As a second example, we may quote the well-understood transient effects that occur if a beam is launched with a non-self-consistent phase-space density profile [2–4]. Furthermore, the various kinds of parametric resonances and instabilities that may occur in the course of beam propagation through focusing lattices and storage rings have been successfully tackled on the basis of a perturbation analysis of the Vlasov equation [5].

Despite all these achievements, there is still an important class of beam phenomena the analysis of which leads beyond the scope of the Vlasov approach. Because of the invariance of Vlasov's equation with respect to time reversal [6], we must be aware that it restricts the analysis to only reversible aspects of beam dynamics. However, a reversible, continuous description of beam dynamics no longer applies if the individual interactions of the point charges must be taken into account. Effects of elastic Coulomb scattering like the well-known phenomenon of intrabeam scattering [7] observed for intense beams that circulate in storage rings, or the process of temperature balancing within a charged particle beam — commonly referred to as beam equipartitioning — fall into this category. In order to include these irreversible effects into our analytical description of beams, the Vlasov approach must be generalized appropriately [8–11]. This will be achieved by switching from a deterministic to a statistical treatment of beam dynamics by separating the actual forces that act on the beam particles into a smooth and a fluctuating component. We will review this transition in detail in Sec. I.

If we are interested in the evolution of global beam characteristics — such as emittance and momentum spread — a direct integration of the Fokker-Planck equation is usually not worth while. One approach to simplifying the analytical description of beam optics has been presented by Lapostolle [12] and Sacherer [13], deriving equations of motion for the "root-mean-square" (rms) beam moments from the Vlasov equation. In this paper, we pursue this idea to derive a generalized set of moment equations from the Fokker-Planck equation. We thus obtain additional terms in the equations for the beam moments that allow us to describe irreversible effects within the beam not covered by the Vlasov approach. The task of determining the Fokker-Planck coefficients will be sketched in Sec. V. We hereby refer to the work of Jansen [14], who calculated these coefficients for the effect of intra-beam scattering for charged-particle beams near thermodynamic equilibrium. Including collision effects is particularly important for highly charged heavy ions and high phase-space densities. We restrict ourselves to cases where the effect of collisions is small compared with external forces or the smooth part of the self-fields. Hence we do not treat the beam's turbulent heating phase, but the long-term behavior of "relaxed" beams undergoing intra-beam scattering, which typically occurs for beams circulating in storage rings.

I LANGEVIN EQUATION

"Intrabeam scattering" denotes the physical effect of multiple small-angle Coulomb scattering, occurring within a charged particle beam that circulates in a storage ring. We start our analysis of this effect by reviewing the single-particle equation of motion for a set of charged particles interacting through Coulomb forces within the co-moving beam frame,

$$m\frac{d^2}{dt^2}\boldsymbol{x}_i - \boldsymbol{F}_{\text{ext}}(\boldsymbol{x}_i, t) - \frac{q^2}{4\pi\epsilon_o}\sum_{j\neq i}\frac{\boldsymbol{x}_i - \boldsymbol{x}_j}{|\boldsymbol{x}_i - \boldsymbol{x}_j|^3} = 0, \quad i = 1,\ldots,N. \quad (1)$$

Here m denotes the particle mass and q its charge, and $\boldsymbol{F}_{\text{ext}}$ the external force field. The sum provides the electric force generated by all particles. If a total of N particles of the same species is given, the N-body distribution function

$$\rho = \rho(\boldsymbol{x}_1, \boldsymbol{v}_1, \ldots, \boldsymbol{x}_N, \boldsymbol{v}_N, t_0) \quad (2)$$

contains the complete information on the state of the system. Equation (1) together with the knowledge of Eq. (2) defines a "reversible" system. This system is completely determined and does not contain any sources for loss of information, hence can be transformed any distance in time back and forth. In principle, all effects occurring in charged-particle beams can be derived from the time integration of Eq. (1).

Nevertheless, this picture is not adequate for the description of real N-body systems if N is very large, since the condition that the initial ρ is precisely known can never be fulfilled. In addition, the detailed knowledge of $\rho(t)$ is usually not necessary. Therefore, a statistical description of the time evolution of the particle ensemble is appropriate. This description must be consistent with exact solutions of Eq. (1) for a large number of particles N.

On the single-particle level, a statistical description means replacing the exact, fine-grained Coulomb force contained in Eq. (1) by its smoothed coarse-grained average force. The fine-grained aspect of the particle motion is then modeled by an additional fluctuating force $\boldsymbol{F}_\text{L}$ that does not depend on the instantaneous particle position in real space. As pointed out by Jowett [15], this concept constitutes "an attempt to describe the effects of the neglected microscopic degrees of freedom." In order not to introduce a systematic error into the statistical description of the N-particle ensemble, this force must vanish on the ensemble average:

$$\langle \boldsymbol{F}_\text{L} \rangle = 0.$$

In this statistical description, we must conceive $\boldsymbol{F}_\text{L}(\boldsymbol{v}, t)$ not as an ordinary vector function but as a quantity that has only statistically defined properties. Fluctuating forces of this nature are usually referred to as "Langevin forces" [16].

In performing the transition from an "exact" fine-grained description of the evolution of $\rho(t)$ according to Eq. (1) to a statistical description of this evolution, not only the fluctuating Langevin force $\boldsymbol{F}_\text{L}(\boldsymbol{v}, t)$ but also a force referred to as the "dynamical friction" force $\boldsymbol{F}_\text{fr}(\boldsymbol{v}, t)$ must be introduced. For repelling forces, the mechanism of dynamical friction is sketched in Fig. 1. We observe that the deceleration of the leftmost particle in the horizontal direction before its closest encounter with the other particles is greater than its acceleration afterwards. This means that a net deceleration, hence a friction occurs. As is easily verified, the same is true for attracting forces.

In the statistical description, the N-particle ensemble is described in terms of a smooth probability density. Accordingly, the self-field appears now as a smooth function of $\boldsymbol{x}$ and t that is equivalent to a smooth external force field. The stochastic counterpart of the deterministic single-particle equation of motion (1) can now be written as

$$m\frac{d^2}{dt^2}\boldsymbol{x} - \boldsymbol{F}_\text{ext} - q\boldsymbol{E}_\text{sc}^\text{sm} - \boldsymbol{F}_\text{fr} = \boldsymbol{F}_\text{L}, \qquad (3)$$

containing the smooth part of the Coulomb force $\boldsymbol{E}_\text{sc}^\text{sm}(\boldsymbol{x}, t)$, the dynamical friction force $\boldsymbol{F}_\text{fr}(\boldsymbol{v}, t)$, and the fluctuating Langevin force $\boldsymbol{F}_\text{L}(\boldsymbol{v}, t)$. As usual, we assumed that stochastic effects in our description are independent of the "external" force

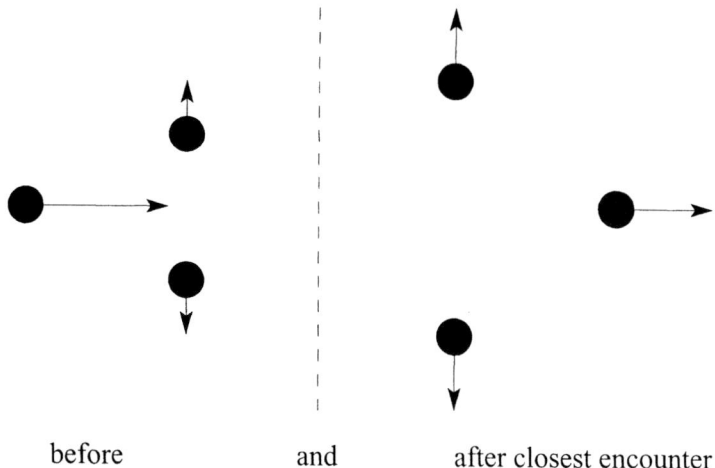

FIGURE 1. Sketch of the mechanism of dynamical friction for repelling forces between particles.

functions $\boldsymbol{F}_{\text{ext}}(\boldsymbol{x}, t)$ and $q\boldsymbol{E}_{\text{sc}}^{\text{sm}}(\boldsymbol{x}, t)$. This means that the Langevin force $\boldsymbol{F}_{\text{L}}$ as well as the friction force $\boldsymbol{F}_{\text{fr}}$ do not depend on the position $\boldsymbol{x}$ in real space.

Each particle encounters a specific realization of the Langevin force $\boldsymbol{F}_{\text{L}}(\boldsymbol{v}, t)$. These forces are defined by their statistical properties only. Therefore, direct integration of Eq. (3) is not possible. On the other hand, a deterministic equation of motion for the phase-space probability density $f(\boldsymbol{x}, \boldsymbol{v}, t)$ can be derived on the basis of Eq. (3). This topic will be reviewed in the following section.

II FOKKER-PLANCK EQUATION

We define $\boldsymbol{q} \equiv (\boldsymbol{x}, \boldsymbol{v})$ as the position vector in the 6-dimensional μ-phase-space. If the function $f(\boldsymbol{q}, t)$ represents a normalized phase-space probability density, then $f\, d\boldsymbol{q}$ provides the probability of finding a particle inside a volume $d\boldsymbol{q}$ around the phase-space point $\boldsymbol{q}$ at time t. In these terms, the generalization of Eq. (3) can be written as

$$\dot{q}_i = K_i(\boldsymbol{q}, t) + \Gamma_i(\boldsymbol{q}, t), \qquad i = 1, \ldots, 6, \tag{4}$$

with smooth functions $K_i(\boldsymbol{q}, t)$ and the random variables $\Gamma_i(\boldsymbol{q}, t)$ vanishing on the ensemble average. We now assume the random variables $\Gamma_i(\boldsymbol{q}, t)$ to be Gaussian-distributed and their time correlation to be proportional to the δ-function

$$\left\langle \Gamma_i(\boldsymbol{q}, t)\, \Gamma_j(\boldsymbol{q}, t') \right\rangle = 2 Q_{ij}(\boldsymbol{q}, t)\, \delta(t - t'). \tag{5}$$

Under these conditions, the Kramers-Moyal expansion for $\partial f(\boldsymbol{q},t)/\partial t$ terminates after the second term [17–19]. The expansion with only the first and second terms is usually called the Fokker-Planck equation

$$\frac{\partial f}{\partial t} = \boldsymbol{L}_{\text{FP}} f \qquad (6)$$

with the Fokker-Planck operator $\boldsymbol{L}_{\text{FP}}$ given by

$$\boldsymbol{L}_{\text{FP}} = -\sum_{i=1}^{6} \frac{\partial}{\partial q_i} K_i(\boldsymbol{q},t) + \sum_{i,j=1}^{6} \frac{\partial^2}{\partial q_i \partial q_j} Q_{ij}(\boldsymbol{q},t) \qquad (7)$$

We observe that the coefficients Q_{ij} are determined by the amplitude of the δ-correlated noise functions Γ_i according to Eq. (5), whereas the K_i are defined by Eq. (4). Consequently, Eq. (6) represents the deterministic equation of motion for the probability density $f(\boldsymbol{q},t)$. It is uniquely determined by the coupled set of Langevin equations (4) if Eq. (5) holds.

In terms of the special Langevin equation (3), the Fokker-Planck operator (7) reduces to

$$\boldsymbol{L}_{\text{FP}} = \sum_{i=1}^{3} \left[-\frac{\partial}{\partial x_i} v_i - \frac{1}{m}\frac{\partial}{\partial v_i} F_{\text{tot},i} + \frac{\partial^2}{\partial v_i^2} D_{ii} \right] \qquad (8)$$

with $F_{\text{tot},i}$ defined as the sum of all non-Langevin forces

$$F_{\text{tot},i}(\boldsymbol{x},\boldsymbol{v},t) = F_{\text{ext},i}(\boldsymbol{x},t) + q E^{\text{sm}}_{\text{sc},i}(\boldsymbol{x},t) + F_{\text{fr},i}(v_i,t),$$

and the diffusion coefficients D_{ii} defined by

$$\left\langle F_{\text{L},i}(v_i,t)\, F_{\text{L},j}(v_j,t') \right\rangle = 2m^2 D_{ii}(v_i,t)\delta_{ij}\,\delta(t-t'). \qquad (9)$$

The off-diagonal terms of the diffusion matrix D_{ij} vanish since the Langevin forces in Eq. (3) are not correlated for different degrees of freedom. We further note that the friction forces $F_{\text{fr},i}$ must always be decelerating. This means that $F_{\text{fr},i}$ changes sign if v_i does, and hence must be an odd function of v_i. With regard to Eq. (9), it follows that the diffusion coefficients of Eq. (8) must be even functions of the v_i

$$F_{\text{fr},i}(v_i) = -F_{\text{fr},i}(-v_i) \quad , \quad D_{ii}(v_i) = D_{ii}(-v_i). \qquad (10)$$

A Fokker-Planck equation that describes the evolution of the probability density f associated with the stochastic motion of particles in external force fields is often referred to as Kramers' equation. As will be shown in the Sec. IV, where we investigate equilibrium solutions of Eq. (6), the diffusion coefficients $D_{ii}(v_i,t)$ are uniquely determined by the friction forces $F_{\text{fr},i}(v_i,t)$.

III FOKKER-PLANCK COEFFICIENTS UNDER TIME REVERSAL

In order to gain some physical insight into the characteristics of solutions of the Fokker-Planck equation, we perform a transformation that reverses the direction of time

$$t \to -t, \qquad x_i \to x_i, \qquad v_i \to -v_i.$$

Obviously, the positions x_i and hence all quantities that depend only on the positions do not change sign under this transformation. In contrast, the velocities v_i do change sign, which means that quantities depending on the v_i may change sign under time reversal. We now separate the components of the Fokker-Planck operator (8) with respect to their behavior under time reversal,

$$\boldsymbol{L}_{\text{FP}} = \boldsymbol{L}_{\text{rev}} + \boldsymbol{L}_{\text{ir}}.$$

The "reversible" operator $\boldsymbol{L}_{\text{rev}}$ is defined as consisting of those components of Eq. (8) that change sign under time reversal

$$\boldsymbol{L}_{\text{rev}} = \sum_{i=1}^{3} \left[-\frac{\partial}{\partial x_i} v_i - \frac{1}{m} \frac{\partial}{\partial v_i} \left(F_{\text{ext},i} + q E_{\text{sc},i}^{\text{sm}} \right) \right]. \qquad (11)$$

The smooth self-field $\boldsymbol{E}_{\text{sc}}^{\text{sm}}$ is obtained from the real-space projection of the probability density $f(\boldsymbol{q}, t)$ via Poisson's equation.

The components that do not change sign constitute $\boldsymbol{L}_{\text{ir}}$:

$$\boldsymbol{L}_{\text{ir}} = \sum_{i=1}^{3} \frac{\partial}{\partial v_i} \left[-\frac{F_{\text{fr},i}(v_i, t)}{m} + \frac{\partial}{\partial v_i} D_{ii}(v_i, t) \right]. \qquad (12)$$

Here we made use of Eq. (10), which states that under time reversal $F_{\text{fr},i}$ changes sign, whereas D_{ii} does not change sign. The external forces $F_{\text{ext},i}$ have been assumed to be not velocity-dependent.

Since $\partial f / \partial t$ changes sign on time reversal, a Fokker-Planck equation with only $\boldsymbol{L}_{\text{rev}}$ of Eq. (11) remains unchanged if the direction of time is reversed. It therefore describes the reversible transformation of the probability density function $f(\boldsymbol{x}, \boldsymbol{v}, t)$. This means that earlier states are fully restored if a reversed time integration of Eq. (6) with $\boldsymbol{L}_{\text{FP}} \equiv \boldsymbol{L}_{\text{rev}}$ is carried out — just like a movie that is reversed at some instant of time t_0. Correspondingly, $\boldsymbol{L}_{\text{ir}}$ describes exactly those effects that do *not* depend on the direction of time. In other words, it describes the irreversible aspects of the particle motion. With $\boldsymbol{L}_{\text{ir}} = 0$, Eq. (6) is commonly referred to as the Vlasov equation.

The principle of irreversible effects occurring within charged particle beams can be directly observed in computer simulations. If a beam is injected in a non-equilibrium state into an ion optical system, the phase-space density f adapts itself

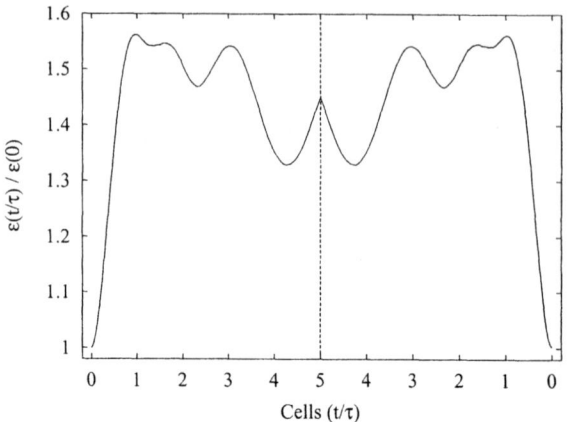

FIGURE 2. Emittance growth factors versus number of cells obtained for a non-stationary initial phase-space density at $\sigma_0 = 60°$, $\sigma = 15°$, 2500 simulation particles. The vertical dashed line marks the point of time reversal after 5 cells.

rapidly to the external force field — on the time scale of some plasma periods — until an average equilibrium is reached. This process is accompanied by a change of the rms emittances Eq. (22). In the simulation displayed in Fig. 2, a space-charge-dominated charged-particle beam is transformed along 5 focusing periods forward in time. Subsequently, the time direction of the simulation is reversed, and the beam is transformed backwards to the starting position. We observe that the initial non-equilibrium state is recovered, which means that the simulation results are compatible with the (reversible) Vlasov equation $\partial f/\partial t = \boldsymbol{L}_{\text{rev}} f$.

This is no longer true if the forward transformation exceeds a certain number of periods. Figure 3 shows the emittance variations obtained from the simulation similar to the previous case, but with the forward and the subsequent backward transformations now extending over 20 focusing periods. Obviously, the initial state is not recovered. This behavior shows that the Fokker-Planck equation for f — whose integration is realized in the numerical simulation — does contain a non-vanishing irreversible component $\boldsymbol{L}_{\text{ir}} \neq 0$. In contrast to the short-term simulation of Fig. 2, the influence of $\boldsymbol{L}_{\text{ir}}$ is now no longer negligible.

IV EQUILIBRIUM DISTRIBUTIONS IN AUTONOMOUS SYSTEMS

If the external force $\boldsymbol{F}_{\text{ext}}(\boldsymbol{x})$ of Eq. (8) is not explicitly time dependent, a stationary solution $\boldsymbol{L}_{\text{FP}} f_{\text{st}} = 0$ may exist. If it exists, it can always be written in the

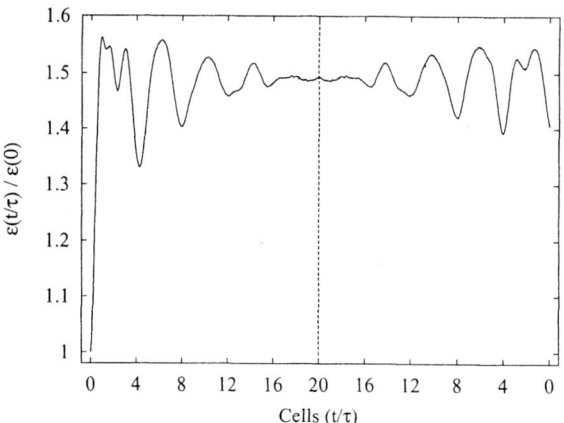

FIGURE 3. Emittance growth factors versus number of cells obtained for a non-stationary initial phase-space density at $\sigma_0 = 60°$, $\sigma = 15°$ per cell, 2500 simulation particles. The vertical dashed line marks the point of the time reversal after 20 cells.

form

$$f_{\rm st}(\boldsymbol{x},\boldsymbol{v}) = g_0^{-1}\exp\{-\phi_{\rm st}(\boldsymbol{x},\boldsymbol{v})\}\,,\qquad (13)$$

with $g_0 = \int \exp\{-\phi_{\rm st}(\boldsymbol{x},\boldsymbol{v})\}\,d\boldsymbol{x}d\boldsymbol{v}$ the normalization factor. We may define the irreversible probability current $S_{v_i}^{\rm ir}$ flowing into the v_i-direction in phase-space as

$$\boldsymbol{L}_{{\rm ir},i} f = -\frac{\partial}{\partial v_i} S_{v_i}^{\rm ir}\,.$$

Obviously, all irreversible currents $S_{v_i}^{\rm ir}$ must vanish for $f = f_{\rm st}$ to be stationary. With $\boldsymbol{L}_{{\rm ir},i}$ given by Eq. (12), this means, explicitly,

$$\frac{F_{{\rm fr},i}(v_i)}{m} = \frac{\partial D_{ii}(v_i)}{\partial v_i} - D_{ii}(v_i)\frac{\partial \phi_{\rm st}(\boldsymbol{x},\boldsymbol{v})}{\partial v_i}\,. \qquad (14)$$

Equation (14) states that for a given $\phi_{\rm st}$, the diffusion function $D_{ii}(v_i)$ is uniquely determined by the friction force function $F_{{\rm fr},i}$ — and vice versa. This mutual dependency of the diffusion effects — driving a system away from its steady state — and damping effects that cause the decay of these deviations constitutes the physical content of "fluctuation-dissipation theorems."

In agreement with Eq. (10), we express the friction force function $F_{{\rm fr},i}$ and the diffusion function D_{ii} as odd and even power series in v_i, respectively:

$$F_{{\rm fr},i}(v_i) = -m\sum_{i=0}^{\infty} a_k v_i^{2k+1}\,,\quad D_{ii}(v_i) = \sum_{i=0}^{\infty} b_k v_i^{2k}\,. \qquad (15)$$

Here we assumed that the coefficients a_k, b_k do not depend on x and the degree of freedom i — in agreement with the precondition that the stochastic effects are not influenced by the external forces.

With Eq. (15), we find that Eq. (14) can be satisfied only if ϕ_{st} is a quadratic function of the v_i. Therefore, ϕ_{st} may always be separated as

$$\phi_{st}(\boldsymbol{x}, \boldsymbol{v}) = \psi_{st}(\boldsymbol{x}) + \sum_{i=1}^{3} \frac{v_i^2}{2 \langle v_i^2 \rangle}, \tag{16}$$

the angle brackets denoting the respective averages over the phase-space density function: $\langle a \rangle = \int a f d\boldsymbol{x} d\boldsymbol{v}$. The quantity $\langle v_i^2 \rangle$ thus embodies the ensemble average of the squares of all particle velocities, also referred to as the second moment of the velocity v_i of the equilibrium distribution f_{st}. In a state of equilibrium these moments must agree for each degree of freedom; hence can be identified with the equilibrium temperature T_{eq} according to

$$kT_{eq} = m \langle v_i^2 \rangle, \; i = 1, 2, 3, \tag{17}$$

with k denoting Boltzmann's constant.

Inserting Eq. (16) into the Fokker-Planck equation (6) with (8), the generalized potential $\psi_{st}(\boldsymbol{x})$ follows from

$$\nabla \psi_{st}(\boldsymbol{x}) = -\frac{1}{kT_{eq}} \left(\boldsymbol{F}_{ext}(\boldsymbol{x}) + q\boldsymbol{E}_{sc}^{sm}(\boldsymbol{x}) \right).$$

In the final form, the equilibrium probability density of the Fokker-Planck equation (6) reads

$$f_{st} = g_0^{-1} \exp\{-\psi_{st}(\boldsymbol{x})\} \exp\left\{ -\sum_{i=1}^{3} \frac{mv_i^2}{2kT_{eq}} \right\}. \tag{18}$$

We summarize that the equilibrium distribution (18) follows directly from the assumption that the stochastic component of the particle motion is caused by Gaussian-distributed Langevin forces with a time correlation function proportional to the δ-function, regardless of the dependency of the friction and the diffusion coefficients on the v_i. For a given temperature T_{eq}, the spatial probability function following from ψ_{st} is uniquely determined by the external force $\boldsymbol{F}_{ext}$, and the stationary self-field $\boldsymbol{E}_{sc}^{sm}$. Together with the unique velocity distribution, the entire phase-space probability density function is uniquely determined, which means that no other equilibrium distribution of Eq. (6) exists — in contrast to Vlasov systems where friction as well as diffusion effects vanish. If the external force function $F_{ext}(\boldsymbol{x})$ does allow for an equilibrium, and if the friction is not negligible, arbitrary non-equilibrium density functions always settle down to a unique equilibrium. This is what we observe in long-term simulations of charged-particle beams [4,20]. Regardless of our initial phase-space filling, we always end up with a Gaussian velocity distribution if no resonance effects are involved.

V FOKKER-PLANCK COEFFICIENTS FOR INTRA-BEAM SCATTERING

The Fokker-Planck equation (6) with the operator (8) has been derived on the basis of the stochastic differential equation (3). Accordingly, the coefficients $F_{\mathrm{fr},i}$ and D_{ii} contained in (8) are related to the dynamical friction $\boldsymbol{F}_{\mathrm{fr}}$ and the Langevin force terms $\boldsymbol{F}_{\mathrm{L}}$ of Eq. (3) — which in turn models the set of single-particle equations (1). In order to learn how the Fokker-Planck coefficients for the effect of intra-beam scattering are correlated to the physical properties of the charged particle ensemble in question, it is necessary to return to the single-particle equation (1), and to analyze the process of small-angle Coulomb scattering of a pair of charged particles. Since this has been done by Chandrasekhar [8] and Jansen [14], we may restrict ourselves to reporting their results.

The method of evaluating the Fokker-Planck coefficients for intra-beam scattering effects within a charged-particle beam can be summarized as follows:

- In the first step, the velocity changes of a test particle due to scattering from a single beam particle as a function of the test particle's initial velocity and impact parameter are calculated,

- second, the expression obtained in the first step is averaged over all possible impact parameters, and

- finally, averaging over all particle velocities of the beam is performed. This means that the velocity distribution of the beam must be taken into account.

For simplicity, we assume that the *equilibrium* particle beam has an *isotropic* Maxwellian velocity distribution. If the actual beam is in a state not too far from this equilibrium state, it is justifiable to assume that friction, as well as the diffusion process, is isotropic. Then only one diffusion coefficient D in conjunction with a single friction coefficient β_f appears in our equations.

For the effect of intra-beam scattering, $F_{\mathrm{fr},i}$ can be obtained not too far from thermodynamic equilibrium as

$$F_{\mathrm{fr},i} = -m\beta_f v_i, \qquad \beta_f = \frac{16\sqrt{\pi}}{3}\frac{Z^4}{A^2} \cdot n c r_c^2 \left(\frac{mc^2}{2kT_{\mathrm{eq}}}\right)^{3/2} \cdot \ln\Lambda, \qquad (19)$$

with $r_c = e_0^2/(4\pi\epsilon_0 mc^2)$ the classical particle radius, n the particle density, T_{eq} the equilibrium temperature Eq. (17), and $\ln\Lambda$ the Coulomb logarithm.

Under these circumstances, the general form of the fluctuation-dissipation relation (14) simplifies to the well-known Einstein relation [21]

$$D \equiv D_{ii} = \beta_f \frac{k_B T_{\mathrm{eq}}}{m}. \qquad (20)$$

We will use this simple approximation in our approach.

In terms of the maximum impact parameter b_m, the Coulomb logarithm $\ln \Lambda$ is given by

$$\ln \Lambda \approx \ln \frac{b_m}{b_\perp},$$

$b_\perp$ denoting the impact parameter that corresponds to a 90° deflection. As usual for phenomena involving the long-range Coulomb forces, we must establish a reasonable upper limit for the maximum impact parameter b_m in order to keep Λ finite. In plasma physics, we usually identify $b_m \equiv \lambda_D$ with the Debye screening length. For non-neutralized systems, Jansen [14] suggested that we identify the maximum impact parameter b_m with the average distance between the particles ("ion sphere radius") rather than with the Debye screening length. The average distance can be expressed in terms of the average particle velocity v_{av} and the correlation time of individual scattering events τ as

$$b_m \equiv v_{av}\tau.$$

With $b_m = n^{-1/3}$, this means for Λ

$$\Lambda \approx \frac{3kT}{2Z^2 r_c\, mc^2\, n^{1/3}}.$$

Since β_f depends only logarithmically on Λ, any result will not depend critically on the exact value of Λ. The inaccuracy in calculating the Coulomb logarithm is a general problem, which appears in all approaches on intra-beam scattering. Of course, the assumption of isotropic diffusion and friction — embodied in a single friction coefficient β_f — could be dropped. On the other hand, this would lead to more complicated equations for the temperature relaxation processes, to be presented in the following sections.

VI   MOMENT ANALYSIS OF THE FOKKER-PLANCK EQUATION

For our purpose of estimating intra-beam scattering effects in charged-particle beams, the detailed knowledge of the evolution of the phase-space probability density f is not necessary. Therefore, a direct solution of the Fokker-Planck (Kramers) equation would not be worth while. A usual way to switch to more global physical quantities is to consider "second moments" [13] of f, similar to

$$\langle x^2 \rangle = \int x^2 f\, d\boldsymbol{x} d\boldsymbol{v}.$$

$\sqrt{\langle x^2 \rangle}$ has the physical interpretation of being proportional to the actual beam width in x.

The derivatives of the moments are calculated according to

$$\frac{d}{dt}\langle x^2 \rangle = \int x^2 \frac{\partial f}{\partial t} d\boldsymbol{x} d\boldsymbol{v},$$

with $\partial f/\partial t = \boldsymbol{L}_{\mathrm{FP}} f$ given by Eq. (6). In total, the second-order moment analysis of the Fokker-Planck equation (6) with (8) yields the following set of coupled moment equations for each phase-space plane $i = 1, 2, 3$:

$$\frac{d}{dt}\langle x_i^2 \rangle - 2\langle x_i v_i \rangle = 0, \tag{21a}$$

$$m\frac{d}{dt}\langle x_i v_i \rangle - m\langle v_i^2 \rangle - \langle x_i F_{\mathrm{ext},i} \rangle - q\langle x_i E_{\mathrm{sc},i}^{\mathrm{sm}} \rangle = \langle x_i F_{\mathrm{fr},i} \rangle, \tag{21b}$$

$$m\frac{d}{dt}\langle v_i^2 \rangle - 2\langle v_i F_{\mathrm{ext},i} \rangle - 2q\langle v_i E_{\mathrm{sc},i}^{\mathrm{sm}} \rangle = 2\langle v_i F_{\mathrm{fr},i} \rangle + 2m\langle D_{ii} \rangle. \tag{21c}$$

The rms emittance in the beam system is commonly defined as

$$\varepsilon_i^2(t) = \langle x_i^2 \rangle \langle v_i^2 \rangle - \langle x_i v_i \rangle^2. \tag{22}$$

Calculating the time derivative of Eq. (22), and inserting the moment equations (21), we find that three distinct sources for the rms emittance change can be distinguished:

$$\frac{d}{dt}\varepsilon_i^2(t) = \left.\frac{d}{dt}\varepsilon_i^2(t)\right|_{\mathrm{ext}} + \left.\frac{d}{dt}\varepsilon_i^2(t)\right|_{\mathrm{sc}} + \left.\frac{d}{dt}\varepsilon_i^2(t)\right|_{\mathrm{ir}},$$

namely the external field contribution, the contribution related to the smooth space-charge fields and the contribution due to the Langevin forces described by the irreversible part (12) of the Fokker-Planck operator.

If the external focusing forces are linear, their contribution to the change of the rms emittance vanishes:

$$\left.\frac{m}{2}\frac{d}{dt}\varepsilon_i^2(t)\right|_{\mathrm{ext}} = \langle x_i^2 \rangle \langle v_i F_{\mathrm{ext},i} \rangle - \langle x_i v_i \rangle \langle x_i F_{\mathrm{ext},i} \rangle$$

$$= 0 \quad \Longleftrightarrow \quad F_{\mathrm{ext},i} \propto x_i.$$

The contribution to the emittance change due to the smooth space-charge field $\boldsymbol{E}_{\mathrm{sc}}^{\mathrm{sm}}$ is given by

$$\left.\frac{m}{2}\frac{d}{dt}\varepsilon_i^2(t)\right|_{\mathrm{sc}} = q\left[\langle x_i^2 \rangle \langle v_i E_{\mathrm{sc},i}^{\mathrm{sm}} \rangle - \langle x_i v_i \rangle \langle x_i E_{\mathrm{sc},i}^{\mathrm{sm}} \rangle\right].$$

If we write this equation for all three spatial degrees of freedom, the electric field terms together form the physical quantity of "free field energy," i.e. the difference between the *actual* charge distribution's electrostatic field energy W and the field energy W_u of the *uniform* charge distribution having the same rms size [3,22,23]

$$\sum_{i=1}^{3} \frac{1}{\langle x_i^2 \rangle} \frac{d}{dt} \varepsilon_i^2(t) \bigg|_\mathrm{sc} + \frac{2}{mN} \frac{d}{dt}(W - W_\mathrm{u}) = 0. \qquad (23)$$

In the short-term simulation presented in Fig. 2 of Sec. III, we have demonstrated that the exchange of rms emittance and "free field energy" is indeed a reversible process.

The third contribution to the change of the rms emittance emerges from the *irreversible* Fokker-Planck operator, as given by Eq. (12)

$$\frac{m}{2} \frac{d}{dt} \varepsilon_i^2(t) \bigg|_\mathrm{ir} = \langle x_i^2 \rangle \langle v_i F_i \rangle - \langle x_i v_i \rangle \langle x_i F_i \rangle + m \langle x_i^2 \rangle \langle D_{ii} \rangle. \qquad (24)$$

Thus, the irreversible emittance growth depends on both the Fokker-Planck coefficients *and* the specific shape of the beam envelope functions. With $F_{\mathrm{fr},i}$ from Eq. (19) and linear external focusing forces

$$F_{\mathrm{ext},i} = -m\omega_i^2(t) x_i,$$

we obtain the envelope equation from the first two moment equations (21)

$$\frac{d^2}{dt^2} \sqrt{\langle x_i^2 \rangle} + \beta_f \frac{d}{dt} \sqrt{\langle x_i^2 \rangle} + \omega_i^2(t) \sqrt{\langle x_i^2 \rangle} - \frac{q}{m} \frac{\langle x_i E_{\mathrm{sc},i}^{\mathrm{sm}} \rangle}{\sqrt{\langle x_i^2 \rangle}} - \frac{\varepsilon_i^2(t)}{\sqrt{\langle x_i^2 \rangle}^3} = 0. \qquad (25)$$

With the fluctuation-dissipation theorem in the simple form of Eq. (20), Eq. (24) may be rewritten as

$$\frac{1}{\langle x_i^2 \rangle} \frac{d}{dt} \varepsilon_i^2(t) \bigg|_\mathrm{ir} = 2\beta_f \left(\frac{k_B T_\mathrm{eq}}{m} - \frac{\varepsilon_i^2(t)}{\langle x_i^2 \rangle} \right). \qquad (26)$$

Equation (26) represents a simple temperature relaxation equation. Together with the envelope equation (25), we thus find a closed set of differential equations for $\sqrt{\langle x_i^2 \rangle}$ and $\varepsilon_i^2(t)$ if we neglect the reversible emittance changes due to variations of the "free field energy" as described by Eq. (23).

VII BEAM TEMPERATURES

As usual in statistical physics, we relate the temperature of a particle ensemble to its "incoherent" motion. In general, charged-particle beams change their size while passing through an ion optical system. The total kinetic energy $m \langle v_i^2 \rangle /2$ contains a coherent part if $\langle x_i v_i \rangle \neq 0$. Therefore, the coherent part of the particles' kinetic energy must be subtracted from the total kinetic energy in order to get the incoherent part of the kinetic beam energy. We thus find for the generalized, non-equilibrium "temperature" $k_B T_i$ pertaining to the i-th degree of freedom

$$k_B T_i \equiv m \left\langle \left(v_i^{\text{inc}}\right)^2 \right\rangle, \qquad v_i^{\text{inc}} = v_i - x_i \frac{\langle x_i v_i \rangle}{\langle x_i^2 \rangle}.$$

Apart from isolated symmetry locations, the "temperatures" related to the spatial directions do not agree. Using the rms emittance, defined by Eq. (22), we can then express the non-equilibrium "temperature" $k_B T_i$ of the i-th degree of freedom as

$$k_B T_i(t) = m \frac{\varepsilon_i^2(t)}{\langle x_i^2 \rangle}. \tag{27}$$

For a coasting beam in a strong-focusing system, we have

$$T_x > T_{\text{eq}} \iff T_y < T_{\text{eq}}$$

and vice versa. With Eq. (27) and $k_B T_z = m \langle (\Delta v_z)^2 \rangle$, the longitudinal temperature in the beam frame, we may approximate the equilibrium temperature T_{eq} by

$$\frac{k_B T_{\text{eq}}}{m} = \frac{k_B}{3m} (T_x + T_y + T_z) = \frac{1}{3} \left(\frac{\varepsilon_x^2}{\langle x^2 \rangle} + \frac{\varepsilon_y^2}{\langle y^2 \rangle} + \langle (\Delta v_z)^2 \rangle \right). \tag{28}$$

VIII EQUATIONS FOR THE EMITTANCE GROWTH

Inserting the equilibrium temperature expression (28), we can rewrite Eq. (26) for the irreversible emittance growth as

$$\frac{1}{\langle x^2 \rangle} \frac{d}{dt} \varepsilon_x^2(t) \bigg|_{\text{ir}} = -\frac{2\beta_f}{3} \left(\frac{2\varepsilon_x^2(t)}{\langle x^2 \rangle} - \frac{\varepsilon_y^2(t)}{\langle y^2 \rangle} - \langle (\Delta v_z)^2 \rangle \right). \tag{29}$$

With the definition of the temperature ratios

$$r_{xy} = \frac{T_y(t)}{T_x(t)}, \qquad r_{xz} = \frac{T_z(t)}{T_x(t)}, \qquad r_{yz} = \frac{T_z(t)}{T_y(t)},$$

Eq. (29) simplifies to

$$\frac{d}{dt} \ln \varepsilon_x(t) \bigg|_{\text{ir}} = \tfrac{1}{3} \beta_f \left(r_{xy} + r_{xz} - 2 \right).$$

Obviously, the change of the emittance $\varepsilon_x(t)$ may be positive as well as negative, depending on the actual temperature ratios. Summing over all three degrees of freedom, the change of the total emittance $\varepsilon^3 = \varepsilon_x \varepsilon_z \varepsilon_z$ is found to be positive in any case

$$\frac{d}{dt} \ln \varepsilon_x \varepsilon_y \varepsilon_z \bigg|_{\text{ir}} = \tfrac{1}{3}\beta_f \left(\frac{(1-r_{xy})^2}{r_{xy}} + \frac{(1-r_{xz})^2}{r_{xz}} + \frac{(1-r_{yz})^2}{r_{yz}} \right) \geq 0. \qquad (30)$$

We thus always find a growth of ε if the coefficients of the irreversible part of the Fokker-Planck operator (12) do not vanish. This suggests relating the growth of ε due to non-vanishing friction and diffusion effects to a growth of the beam entropy [11,24].

To obtain the e-folding time τ_{ef} of the total emittance $\varepsilon = \sqrt[3]{\varepsilon_x \varepsilon_y \varepsilon_z}$, we integrate Eq. (30) along one focusing period T

$$\tau_{\text{ef}}^{-1} = \tfrac{1}{9}\beta_f \left(I_{xy} + I_{xz} + I_{yz} \right).$$

Here, I_{xy}, I_{xz}, and I_{yz} denote the "temperature imbalance integrals" similar to

$$I_{xy} = \frac{1}{T} \int_0^T \frac{[1 - r_{xy}(t)]^2}{r_{xy}(t)} dt \quad , \quad r_{xy}(t) = \frac{\varepsilon_y^2 \langle x^2 \rangle}{\langle y^2 \rangle \varepsilon_x^2}.$$

Equation (29) shows that the time evolution of the emittances depends on the detailed shape of the beam widths. Therefore, the differential equations describing the growth of $\varepsilon(t)$ can be integrated only in conjunction with the envelope equations, as given for the x-direction by Eq. (25).

IX COUPLED SET OF ENVELOPE AND EMITTANCE EQUATIONS FOR UNBUNCHED BEAMS IN STORAGE RINGS

We now switch from the beam system notation to the "trace space" notation in the laboratory frame. The respective laboratory frame quantities are marked with the subscript ℓ. Furthermore, we now use the longitudinal path length z in place of the time t as the independent variable. With γ the relativistic mass factor, the transformation relations for the particle velocities are then given by

$$v_x = c\beta\gamma x'_\ell,$$
$$v_y = c\beta\gamma y'_\ell,$$
$$\Delta v_z = c\beta\eta^{-1}(\Delta p/p)_\ell.$$

Here $\eta(z)$ denotes the longitudinal "slip factor" [4]

$$\eta(z) = \gamma^{-2} - \frac{D(z)}{\rho(z)},$$

containing the radius of curvature $\rho(z)$ of the reference orbit, and the dispersion $D(z)$ that describes the deviation of a particle in the horizontal plane x from the ring's reference orbit due to a deviation from the design momentum

$$D(z) = \frac{\Delta x}{\Delta p/p}.$$

For the sake of brevity, we use the notation $a(z) = \sqrt{\langle x^2 \rangle}$ and $b(z) = \sqrt{\langle y^2 \rangle}$ for the rms beam widths in the x- and y-directions; and $\delta(z) = \sqrt{\langle (\Delta p/p)^2 \rangle}$ for the rms momentum spread. The dispersion-induced increased beam size $A(z)$ in the bending plane evaluates to [25,26]

$$A(z) = \sqrt{a^2 + D^2 \delta^2}.$$

Neglecting the emittance changes due to variations of the "free field energy" $W - W_u$ — as described by Eq. (23) — the closed set of coupled equations for the envelopes, emittances, dispersion, and momentum spread finally reads

$$a'' + k_f a' + k_x^2(z)\, a - \frac{\tfrac{1}{2} K}{A(A+b)} a - \frac{\bar{\varepsilon}_x^2}{a^3} = 0, \tag{31a}$$

$$b'' + k_f b' + k_y^2(z)\, b - \frac{\tfrac{1}{2} K}{A+b} - \frac{\bar{\varepsilon}_y^2}{b^3} = 0, \tag{31b}$$

$$D'' + \left(k_x^2 - \rho^{-2}\right) D - \frac{\tfrac{1}{2} K}{A(A+b)} D - \frac{1}{\rho} = 0, \tag{31c}$$

$$\frac{1}{a^2}\frac{d}{dz}\bar{\varepsilon}_x^2 + \tfrac{2}{3} k_f \left(2\frac{\bar{\varepsilon}_x^2}{a^2} - \frac{\bar{\varepsilon}_y^2}{b^2} - \eta\delta^2 \right) = 0, \tag{31d}$$

$$\frac{1}{b^2}\frac{d}{dz}\bar{\varepsilon}_y^2 + \tfrac{2}{3} k_f \left(2\frac{\bar{\varepsilon}_y^2}{b^2} - \frac{\bar{\varepsilon}_x^2}{a^2} - \eta\delta^2 \right) = 0, \tag{31e}$$

$$\eta \frac{d}{dz}\delta^2 + \tfrac{2}{3} k_f \left(2\eta\delta^2 - \frac{\bar{\varepsilon}_x^2}{a^2} - \frac{\bar{\varepsilon}_y^2}{b^2} \right) = 0. \tag{31f}$$

Here, K denotes the generalized perveance, which is given for coasting beams by

$$K = \frac{2 Z e_0 I}{4\pi \epsilon_0 mc^3 \beta^3 \gamma^3}.$$

The lattice functions $k_i^2(z) = \omega_i^2(t)/c^2\beta^2\gamma$ describe the linear external focusing forces in the laboratory frame. The transverse "trace space" emittances $\bar{\varepsilon}_i$ in this frame

$$\bar{\varepsilon}_i^2 = \langle x_i^2 \rangle \langle x_i'^2 \rangle - \langle x_i x_i' \rangle^2$$

are related to the emittances (22) in beam system through $\bar{\varepsilon}_i = \varepsilon_i/c\beta\gamma$. Following from Eq. (19), the friction coefficient $k_f = \beta_f/c\beta\gamma$ can be expressed in the vicinity of equilibrium as

$$k_f = \frac{1}{\sqrt{2\pi}} \frac{Z^4}{\beta^4 \gamma^4 A^2} \cdot r_c^2 \cdot \frac{N}{\bar{\varepsilon}_x \bar{\varepsilon}_y \delta} \ln \Lambda.$$

X NUMERICAL EXAMPLE

Figure 4 shows an example of a numerical integration of the coupled set of differential equations (31). The lattice data represent the GSI experimental storage ring (ESR) [27].

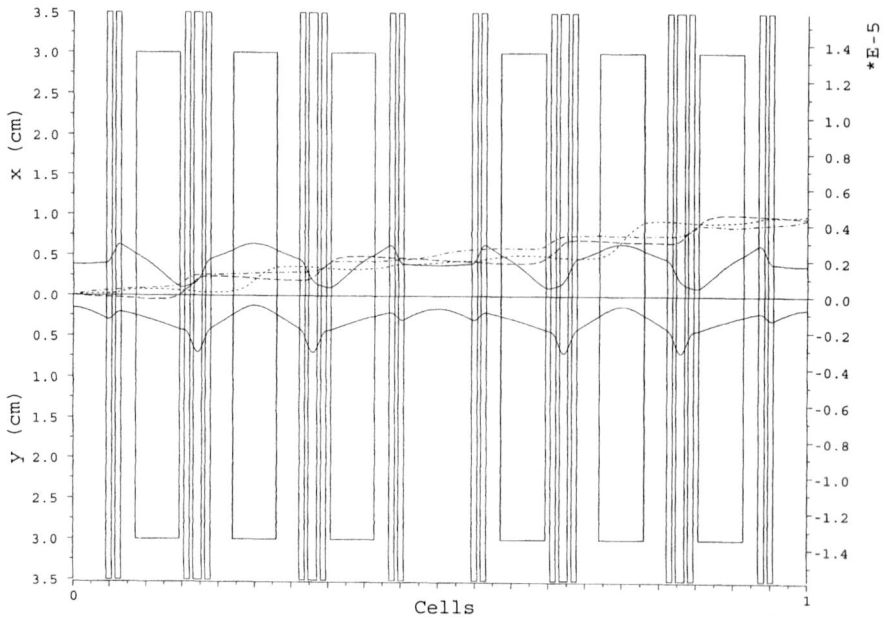

FIGURE 4. Beam envelopes (solid lines) and emittance and momentum spread growth factors (dashed lines) along one turn in the GSI Experimental Storage Ring (ESR).

The solid lines display the beam envelopes in the horizontal and vertical directions representing a coasting (unbunched) beam that passes through the lattice of quadrupoles, bending magnets, and drift spaces. The dashed lines show the emittance growth factors $(\varepsilon_x(z)/\varepsilon_x(0)) - 1$ and $(\varepsilon_y(z)/\varepsilon_y(0)) - 1$, as well as the changes of the rms momentum spread $(\delta(z)/\delta(0)) - 1$. For this particular case, the initial beam parameters in terms of emittance and momentum spread have been adjusted to yield a minimum average temperature anisotropy. According to Eqs. (31d), (31e), and (31f) this leads to the minimum overall emittance growth due to intrabeam scattering effects.

CONCLUSIONS

For the realm of charged-particle beams, we have shown that the Fokker-Planck (Kramers) equation provides a useful starting point for analytical approaches that include effects due to the actual charge granularity. The second-order moment analysis of the Fokker-Planck equation may be applied to derive a set of equations that are directly related to the beam parameters. It follows that the emittance growth rates due to intra-beam scattering effects are related to both the temperature imbalances along the focusing lattice, and a global friction coefficient β_f. The determination of the friction coefficient β_f, in particular the appropriate choice of the Coulomb logarithm from the beam parameters, needs careful examination for each particular case.

REFERENCES

1. Kapchinskij, I.M. and Vladimirskij, V.V., *Proceedings of the International Conference on High Energy Accelerators*, CERN, Geneva, 247 (1959).
2. Struckmeier, J., Klabunde, J., and Reiser, M., *Part. Accel.* **15**, 47 (1984).
3. Wangler, T.P., Crandall, K.R., Mills, R.S., and Reiser, M., *IEEE Trans. Nucl. Sci.* **32**, 2196 (1985).
4. Reiser, M., *Theory and Design of Charged Particle Beams*, New York: Wiley, 1994.
5. Hofmann, I., Laslett, L.J., Smith, L., and Haber, I., *Part. Accel.* **13**, 145 (1983).
6. Hobson, A., *Concepts in Statistical Mechanics*, New York: Gordon and Breach Science Publishers (1971).
7. Piwinski, A., *Proceedings of the 9th International Conference on High Energy Accelerators, Stanford, 1974*, SLAC, Stanford, 405 (1974).
8. Chandrasekhar, S., *Rev. Mod. Phys.* **15**, 1–89 (1943).
9. Bisognano, J. J., in *Physics of High Energy Particles, Stony Brook, 1983*, edited by M. Month, P. F. Dahl, and M. Dienes, AIP Conf. Proc. No. 127, New York: AIP, 443 (1985).
10. Struckmeier, J., *Part. Accel.* **45**, 229 (1994).
11. Struckmeier, J., *Phys. Rev. E* **44**, 830 (1995).
12. Lapostolle, P.M., *CERN report CERN-ISR/DI-70-36* (1970), *IEEE Trans. Nucl. Sci.* **NS-18**, 1055 (1971).
13. Sacherer, F.J., *IEEE Trans. Nucl. Sci.* **NS-18**, 1105 (1971).
14. Jansen, G.H., *Coulomb Interaction in Particle Beams*, New York: Academic Press, 1990.
15. Jowett, J. M., in *Physics of Particle Accelerators, SLAC, Stanford, 1985*, edited by M. Month and M. Dienes, AIP Conf. Proc. No. 153, New York: AIP, 1987, p. 864.
16. Langevin, P., *Comptes rendus* **146**, 530 (1908).
17. Kramers, H.A., *Physica* **7**, 284 (1940).
18. Moyal, J.E., *J. R. Stat. Soc. (London) B* **11**, 150 (1949).
19. Risken, H. *The Fokker-Planck Equation*, Springer, Berlin, Heidelberg, New York, 1989.
20. Lund, S.M., Barnard, J.J., and Miller, J.M., *Proceedings of the 1995 Particle Accelerator Conference, Dallas, 1995*, Piscataway: IEEE Press, 3278, (1996).
21. Einstein, A., *Ann. Physik* **17**, 549 (1905), and **19**, 371 (1906).
22. Hofmann, I. and Struckmeier, J., *Part. Accel.* **21**, 69, (1987).

23. Struckmeier, J. and Hofmann, I., *Part. Accel.* **39**, 219 (1992).
24. Struckmeier, J., *Phys. Rev. ST-AB* **3**, 034202 (2000).
25. Venturini, M. and Reiser, M., *Phys. Rev. E* **57**, 4725 (1998).
26. Holmes, J.A., Danilov, V.V., Galambos, J.D., and Olsen, D.K., *Phys. Rev. ST-AB* **2**, 114202 (1999).
27. Franzke, B., *Nucl. Instrum. & Methods* **B24/25**, 18 (1987).

EMITTANCE DILUTION IN PROTON RINGS

F. Willeke, DESY

I INTRODUCTION

The luminosity of a hadron collider like the LHC is given by

$$\mathcal{L} = \frac{\gamma}{4\pi e} \frac{N_p}{\varepsilon_N} \frac{I_p}{\beta^*}. \tag{1}$$

To obtain this expression a Gaussian transverse distribution for the two beams has been assumed; γ is the relativistic factor for the proton beams (equal energies are assumed), e is the elementary charge, N_p is the number of protons per bunch in one beam, I_p is the total beam current in the opposite beam, β^* is the amplitude function at the interaction point, and ε_N is the normalized emittance. Equal emittance and β functions for the horizontal and vertical planes and for both beams at the interaction point are assumed. The normalized emittance of the beam is an invariant. It does not change as long as all forces on the beam vary slowly in time compared to the betatron period. In order to obtain large luminosity, the emittance must be as small as possible. The factor $\frac{N_p}{\varepsilon_N}$ is also called 'beam brightness'. It is usually limited in the low energy part of the accelerator chain. Sudden changes of the fields in the accelerator or mismatches of the beam orbit and beam envelope may lead to an increase of the beam emittance and a loss of luminosity. This can happen upon transfer from one accelerator to another due to fields which are generated by the beam itself, due to fast kickers for ejection, injection or feedback or due to diagnostics devices such as wire scanners and tune kickers. In order to achieve the largest possible luminosity in large hadron colliders such as the LHC, it is essential to keep these sources of emittance dilution under control. In this lecture, the increase of emittance due to various effects will be investigated, calculated, and compared.

II SOURCES OF EMITTANCE DILUTION

In the most simple model of an ideal accelerator, the transverse beam emittance in a proton accelerator decreases linearly with the beam momentum as long as all field changes are slow compared to the betatron or synchrotron oscillation amplitudes. Therefore the product of emittance and the normalized beam momentum $\varepsilon\gamma\beta$ is an invariant. It is denoted by ε_N, the normalized emittance. Here γ is the beam energy in terms of the rest-energy and β is the ratio of particle velocity to the speed of light. The factor β can be omitted for high energy.

In reality however there are many effects which may lead to an increase of the normalized emittance. The transfer of the beam from one accelerator to the next one in an accelerator cascade is always critical.

If the beam is not injected exactly on the closed orbit in both position and angle, the beam will have a dipole moment which oscillates coherently. Since in a real accelerator, there are always nonlinear forces present, particles with different betatron amplitudes will have a slightly different betatron-tune. Therefore, the particles will diverge in betatron phase. The distribution in phase will be uniform after a while and as a result, the beam size will be increased.

The same process happens, if the phase space distribution of the incoming beam is not matched to the equilibrium shape of the circulating beam. This will lead to oscillation of higher moments of the beam distribution. Since every particle has a different tune (or betatron phase advance per revolution), the coherent oscillation will disappear after a while and we are left with increased amplitudes of incoherent oscillations. Such distortions are caused by a mismatch of the linear beam optics, by different coupling in the two accelerators or rotated quadrupoles in the beam line. Any nonlinear field in the beam line will also lead to an increase of the normalized emittance. A mismatch of the dispersion function will lead to additional dipole oscillation amplitudes for off-momentum particles.

Space charge effects may also cause emittance increase. The direct impact of collective space charge forces should not lead to emittance growth. There are many secondary effects however which are related to the mismatch induced by space charge forces. As an example we consider the detrimental interference of space charge and the slow longitudinal dynamics at transition crossing. The longitudinal space charge forces below transition are defocusing while they are focusing above transition. Near transition, the synchrotron oscillation period becomes very long and the change of space charge focusing is no longer a slow adiabatic but a sudden change. This situation is comparable with a mismatch of focusing on the the transfer from one accelerator to the next.

Proton beams may become self-excited and oscillate coherently due to various instability mechanisms. Coherent instabilities however are suppressed by a spread of the tune. A tune spread can originate from nonlinear fields or from chromaticity in conjunction with the momentum spread in the beam. The change in tune ΔQ with amplitude r due to nonlinear fields, averaged around the accelerator and over the betatron phases and denoted by D is to lowest order proportional to the square of the amplitude $\Delta Q = Dr^2$. The tune spread of a beam with an rms betatron amplitude of σ is $\delta Q = D\sigma^2$. If δQ is too large the particles cannot organize themselves to oscillate coherently. This phenomenon is called Landau Damping. Consider the case that the Landau damping for a beam at rest to too small and a coherent oscillation starts to grow. The tune spread δQ in the beam with rms amplitude σ will increase quadratically with coherent oscillation amplitude R as

$$\delta Q = D\sigma^2 \sqrt{3 + 2(R/\sigma)^2}. \qquad (2)$$

Eventually, the tune spread will become large enough that the individual particles will no longer be able to follow in phase the driving oscillation of the beam as a whole. At this point, the growth of the coherent amplitude will slow down and eventually cease completely. The individual particles will continue to oscillate with the same amplitude but with diverging phases the same way as in a free, non-driven oscillation. The dipole moment will vanish after a while, and the particles will be distributed over a larger volume in phase space than before. Instabilities are thus a prominent source of emittance increase in proton accelerators.

The effects described so far have all to do with randomizing phases after some sort of coherent excitation process has taken place. There are other effects which are of completely different nature: Scattering of beam particles among each other or with other particles such as nuclei of gas molecules, the nuclei in stripping foils, beam diagnostics tools like wire scanners of gas ionization monitors are incoherent effects which act on individual particles. These incoherent emittance growth effects can be strongly influenced by the presence of nonlinear fields, especially if a nonlinear resonance condition is fulfilled.

This lecture however will concentrate on emittance dilution due to randomizing phases in coherent oscillations.

III EMITTANCE CONSIDERATIONS

A Liouville's Law, Symplecticity, Emittance Dilution and Emittance Growths

Consider a particle beam in a circular accelerator. We neglect forces between the particles, interaction with other particles (such gas particles in the beam pipe) or any dissipative effects. All forces are 'external forces'. The equations of motion can then be formulated using the Hamiltonian formalism:

$$\dot{q}_i = \frac{\partial H}{\partial p_i} \quad \dot{p}_i = \frac{-\partial H}{\partial q_i} \qquad (3)$$

In this expression, q_i, p_i are the canonical variables and H is the Hamiltonian of the system which contains the potential from which the forces can be derived. For a motion with n degrees of freedom, this can be written in vector form by defining a $2 \times n$- dimensional phase space vector $\vec{Z}$

$$\dot{\vec{Z}} = S\nabla H; \qquad \vec{Z} = \begin{pmatrix} q_1 \\ \cdot \\ \cdot \\ q_n \\ p_1 \\ \cdot \\ \cdot \\ p_n \end{pmatrix}; \qquad S = \begin{pmatrix} 0 & E \\ -E & 0 \end{pmatrix} \qquad (4)$$

E is the $n \times n$-unity matrix and 0 is the $n \times n$ null matrix.

In order to understand the evolution of the phase space density when many particles move under the influence of the forces described by the Hamiltonian, we consider particle trajectories in the vicinity of a reference trajectory $\vec{Z}_{ref}$ which is a particular solution of the equations of motion. The Jacobian $\partial z_i(t)/\partial z_j(t = 0)$ at $\vec{Z} = \vec{Z}_{ref}$ is denoted by $\mathcal{J}(\vec{Z}_{ref})$. It describes how trajectories with an infinitesimally small initial distance from the reference trajectory propagate in time,

$$\vec{Z}(t) - \vec{Z}_{ref}(t) = \mathcal{J} \cdot (\vec{Z}(0) - \vec{Z}_{ref}(0)). \qquad (5)$$

(The argument $\vec{Z}_{ref}$ of $\mathcal{J}$ is omitted in the following.) It is useful to consider the time evolution of $\mathcal{J}$ which is obtained by using Hamilton's equations 4

$$\dot{\mathcal{J}}_{ik} = \sum_n S_{in} \cdot \frac{\partial^2 H}{\partial z_n \partial z_k(0)} = \sum_n S_{in} \cdot \sum_l \frac{\partial^2 H}{\partial z_n \partial z_l} \frac{\partial z_l}{\partial z_k(0)} = \sum_l \mathcal{J}_{lk} \cdot \sum_n S_{in} \cdot \frac{\partial^2 H}{\partial z_n \partial z_l} \qquad (6)$$

or in matrix from

$$\dot{\mathcal{J}} = S\mathcal{H}\mathcal{J}; \qquad \mathcal{H}_{ij} = \frac{\partial^2 H}{\partial z_i \partial z_j}. \qquad (7)$$

Then we find that the expression $\mathcal{J}^+ S \mathcal{J}$ is an invariant using $S^+ S = E$, $SS = -1$:

$$\frac{d}{dt}\left(\mathcal{J}^+ S \mathcal{J}\right) = \dot{\mathcal{J}}^+ S \mathcal{J} + \mathcal{J}^+ S \dot{\mathcal{J}} = \mathcal{J}^+ \mathcal{H} S^+ S \mathcal{J} + \mathcal{J}^+ SS \mathcal{H} \mathcal{J} = \mathcal{J}^+ \mathcal{H} \mathcal{J} - \mathcal{J}^+ \mathcal{H} \mathcal{J} = 0 \qquad (8)$$

Since $\mathcal{J}^+(0) S \mathcal{J}(0) = S$, the value of this invariant is S. This is the so called symplectic condition. It imposes a strong constraint on the motion of particles in Hamiltonian systems.

To investigate the consequences, let us consider a two dimensional phase space. Consider the evolution of an infinitesimal small ellipse in phase space which is generated by two orthogonal phase space vectors $\delta\vec{y}_1, \delta\vec{y}_2$ on the reference trajectory. The area of the ellipse is

$$\varepsilon_{12} = \vec{y}_1^+ S \vec{y}_2 \qquad (9)$$

We use the symplectic condition $\mathcal{J}^+ S \mathcal{J} = S$ to investigate how the area varies while the generating vectors move in time $\vec{y}_i \to \mathcal{J} \vec{y}_i$ ($i = 1, 2$). One finds

$$\varepsilon_{12} = \vec{y}_1^+ \mathcal{J}^+ S \mathcal{J} \vec{y}_2 = \vec{y}_1^+ S \vec{y}_2 = \varepsilon_{12}(0) \qquad (10)$$

This implies that an infinitesimally small phase space ellipse preserves its area during the motion. Suppose that this ellipse is initially uniformly filled with particles. Then the preservation of the phase space area means that the particle density is preserved during the motion. This is true for any reference trajectory. Any macroscopic area in phase space can be composed of many infinitely small ellipses around the whole set of initial phase space points. Since every single microscopic area is preserved during the motion, this must be true for any macroscopic area in phase space as well. Thus the initial phase space density is preserved during the motion in a Hamiltonian system. This is the manifestation of Liouville's law in the context the Hamiltonian model of the accelerator. The sketch in figure 1 illustrates these considerations.

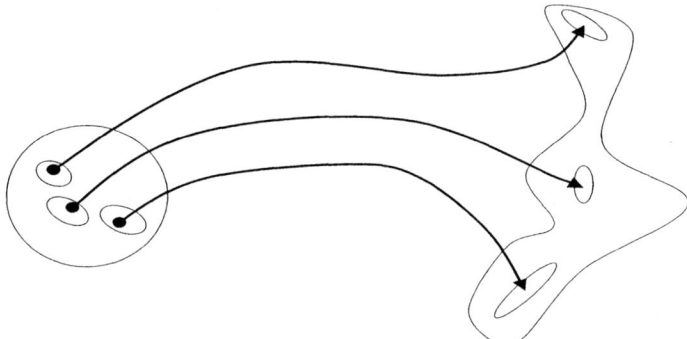

FIGURE 1. Infinitely small phase space areas defined by the area occupied by particle coordinates around any reference trajectory may change its shape but not its size during the motion.

Following these considerations, we distinguish between two different phenomena when discussing emittance effects. We talk about **dilution** if the microscopic phase space area is preserved, but the phase space area changes its shape, so that after averaging the particle density over a given (say elliptical) shape, the **average** density appears to be reduced. The dilution after a coherent oscillation of the beam is also called **filamentation**. The phase space consists of infinitesimally fine filaments which still have the original density but they enclose a certain empty phase space so that the **effective** density decreases.

There are however effects which can not be described by a single particle Hamiltonian model. If the motion of particles is coupled, the dimension of phase space needs to be increased. If a particle interacts with the gas particles in the beam pipe for example,it is clearly impractical to treat the complete coupled system. The interaction may then be better described by friction. Friction terms will lead to an increase or decrease of the emittance. Such effects, we will refer to as **emittance growth**. Emittance growth is caused by scattering of the particles with the residual gas or between the particles within the beam.

It it quite interesting to note however, that although the mechanisms for emittance dilution and emittance growth are fundamentally different, the end effect may not be different at all. Consider a beam which is receiving random small kicks. Each particle receives the same kick. Suppose, that there is time enough between kicks so that the coherent oscillation completely filaments inbetween kicks. Particle which where originally very closely together and received the same increase of amplitude, will end up at arbitrary distances in phase. A later kick therefore might have a completely different impact on the originally close particles. One particle could be further increased in amplitude by the kick, while the amplitude of another particle could be reduced. Thus, the random sequence of kicks is experienced different by each particle, the same way as the particles would have received individual random kicks. One should mention at this point, that this reasoning has been supported by rigorous treatment recently [8].

B   Action Variables

Consider a single particle traveling in the accelerator. We will assume a Hamiltonian model as discussed before. If the phase space coordinates of a particle are measured on successive turns, the phase space coordinates x, x' lie on the border of a closed curve with the form of an ellipse. The area of the ellipse $\mathcal{A}$ is π times twice the action variable or 'single particle emittance' of the particle, $\mathcal{A} = 2\pi \times J$. It is a measure of the oscillation amplitude and we can express the betatron oscillation as

$$\begin{aligned} x(s) &= \sqrt{2J \cdot \beta(s)} \cdot \cos(\Psi(s) + \phi) \\ x'(s) &= -\sqrt{2J/\beta(s)} \cdot (sin(\Psi(s) + \phi) + \alpha \cdot cos(\Psi(s) + \phi)). \end{aligned} \quad (11)$$

The area of the ellipse in units of π, the single particle emittance is

$$2 \cdot J = \gamma x^2 + 2\alpha \cdot xx' + \beta x'^2 \quad (12)$$

where β, α and Ψ are the lattice function of the linear optics model of the accelerator $\gamma = (1 + \alpha^2)/\beta$.

C   Adiabatic Invariant and Adiabatic Damping

1   Transverse Damping

If a hadron beam is accelerated, its transverse emittance will shrink with increasing beam energy. The emittance varies like one over the particle energy during the acceleration process. This will be demonstrated using a simple model. During the acceleration process, a particle picks up on each turn the amount of momentum δp in the RF resonators necessary to balance the increase in magnetic field during one revolution (a synchronous particle is considered, thus synchrotron oscillations are omitted for the sake of clarity). The transfer matrix for one revolution from cavity to cavity remains unchanged. Let this transfer matrix for the coordinates x, x' be

$$M = \begin{pmatrix} cos(2\pi Q) & \beta \cdot sin(2\pi Q) \\ -sin(2\pi Q)/\beta & cos(2\pi Q) \end{pmatrix} \tag{13}$$

Assuming perfect alignment and beam orbit, it is the longitudinal component of the beam momentum which will be increased in the cavity by an amount δp, while the transverse momentum will remain constant. Thus the slope of the trajectory x' of a particle changes by a factor $p/(p+\delta p)$. A case which can be conveniently evaluated is when the relative energy gain Δ is always the same during the whole acceleration. This corresponds to an exponential acceleration $p_n = p_0 \cdot (1 + \Delta)^n \simeq p_0 exp(n\Delta)$. The reduction of x' on each revolution is then described by the constant matrix

$$D = \begin{pmatrix} 1 & 0 \\ 0 & (1+\Delta)^{-1} \end{pmatrix} \tag{14}$$

The total revolution matrix T, the product of this matrix with the unperturbed revolution matrix M is

$$T = \begin{pmatrix} cos(2\pi Q) & \beta \cdot sin(2\pi Q)/(1+\Delta) \\ -sin(2\pi Q)/\beta & cos(2\pi Q)/(1+\Delta) \end{pmatrix}. \tag{15}$$

T has eigenvalues λ, λ^* of

$$\begin{aligned} \lambda_n &= \frac{cos(2\pi Q)(1+\frac{1}{2}\Delta) \pm i\sqrt{(1+\Delta)^2 - cos(2\pi Q) \cdot (1+\frac{1}{2}\Delta)^2}}{1+\Delta} \\ &= \frac{exp(\pm i\phi)}{\sqrt{1+\Delta}}, \quad with \end{aligned} \tag{16}$$

$$tan\phi = \frac{\sqrt{(1+\Delta)^2 - (1+\frac{1}{4}\Delta)^2 cos^2(2\pi Q)}}{(1+\frac{1}{2}\Delta)cos(2\pi Q)}.$$

The initial phase space coordinates

$$\vec{z}_0 = \begin{pmatrix} x \\ x' \end{pmatrix} \tag{17}$$

may be expanded in terms of the eigenvectors $\vec{r}, \vec{r}^*$

$$\vec{z}_0 = A\vec{r} + c.c. \tag{18}$$

By using $T\vec{r} = \lambda \vec{r}$, one obtains after n revolutions, described by n iterations of the map T, the phase space coordinates

$$\vec{z}_n = A\lambda^n \vec{r} + c.c. \tag{19}$$

We recognize in $|\lambda^n| = |\exp(i\,n\phi)|/\sqrt{(1+\Delta)^n} = 1/\sqrt{(1+\Delta)^n}$ the ratio of final to initial beam momentum $\sqrt{p_n/p_0}$. Note that the beam emittance is proportional to the square of the phase space vector $\vec{z}_n^2 = x^2 + x'^2 \propto J_n$. We thus find that the ratio $\vec{z}_n^2/\vec{z}_0^2 \propto J_n/J_0$ scales with the ratio of initial and final value of the beam momentum p_0/p_n. The quantity $J_N = J\beta\gamma$ is an invariant for each particle. Averaging the quantity over all particles in the beam, one obtains the invariant emittance or 'normalized emittance'. It is interesting to note at this point, that in order to obtain this result, no assumptions on the size of the momentum step $\delta p/p$ has been made. It has only been assumed, that it matches the magnetic field change within one revolution period. Thus transverse damping does not require that the energy changes adiabatically.

2 Longitudinal Adiabating Damping

In order to explain the phenomenon of damping during acceleration in the longitudinal phase plane, one has to assume, the parameters change adiabatically. There is a very general theorem which states that if the parameters of a non-dissipative oscillator system change slowly compared to the time in which the coordinates of a particle change significantly (e.g. compared to the oscillation period in oscillatory systems) it can be shown that the action integral $\oint p\,dq$ is an adiabatic invariant (see standard text books like [5]).

In accelerator systems, the longitudinal action $\oint \Delta E(\tau)d\tau$ is an invariant where $\Delta E(\tau) = E(\tau) - E_s$ is the energy amplitude of the synchrotron oscillation, $\tau = t - t_s$ is the time with respect to the synchronous time t_s and E_s is the energy of the synchronous particle. The linearized longitudinal equations of motion are

$$\frac{d}{dt}\Delta E = \dot{\epsilon} = \frac{eU\cos(\phi_s)\omega^2}{2\pi h}\cdot \tau$$
$$\dot{\tau} = -\frac{\alpha}{E}\epsilon \tag{20}$$

(U is the circumferential voltage, ϕ_s is the synchronous phase angle, ω is the angular rf frequency, α is the momentum compaction factor, E is the beam energy and h is the harmonic number) which may be derived from the Hamiltonian

$$H = \frac{1}{2}\frac{\alpha}{E}\epsilon^2 + \frac{1}{2}\frac{eU\cos(\phi_s)\omega^2}{2\pi h}\tau^2. \tag{21}$$

Our aim is to calculate the longitudinal action $J_s = \oint \epsilon\,d\tau$ as a function of the beam energy by making use of the property that H can be considered constant during one period. Expressing ϵ as a function of H and τ yields

$$\epsilon = \sqrt{\frac{-2HE}{\alpha} + \frac{eUE\cos(\phi_s)\omega^2 E}{2\pi h\alpha}\tau^2}. \tag{22}$$

In this expression H may be expressed in terms of the maximum τ which is $\hat{\tau}$ (note that the corresponding value of ϵ is zero)

$$\epsilon(\tau) = \sqrt{\frac{eUE\cos(\phi_s)\omega^2 E\hat{\tau}^2}{2\pi h\alpha}} \sqrt{1 - \tau^2/\hat{\tau}^2}. \tag{23}$$

We now can perform the integral over τ in the range $\pm\hat{\tau}$ which yields

$$\begin{aligned} J_s &= 2\pi\hat{\tau}^2 \sqrt{\frac{eUE\cos(\phi_s)\omega^2 E}{2\pi h\alpha}} \\ &= 2\pi\hat{\epsilon}^2 \sqrt{\frac{2\pi h\alpha}{eUE\cos(\phi_s)\omega^2 E}} \tag{24} \\ &= 2\pi\hat{\tau}\hat{\epsilon} \end{aligned}$$

Using the condition that the action J_s is an adiabatic invariant, we find that the bunch length and the energy spread scale the following way with the energy

$$\begin{aligned} \hat{\tau} &= \hat{\tau}_0 \left(\frac{U_0 E_0 \cos(\phi_{s0})}{UE\cos(\phi_s)} \right)^{1/4} \\ \hat{\epsilon} &= \hat{\epsilon}_0 \left(\frac{UE\cos(\phi_s)}{U_0 E_0 \cos(\phi_{s0})} \right)^{1/4}. \end{aligned} \tag{25}$$

Both the bunch length τ and the relative energy deviation ϵ/E scale with $1/\sqrt{E}$ which is called 'adiabatic damping'.

D Effective Emittance of Multiple Particles

Consider a beam of particles distributed over a certain area in phase space with projections z_i, p_{zi}. The multi-particle emittance of this beam can be defined by fitting ellipses around the occupied areas in the two dimensional phase sub-spaces. Consider for example the horizontal plane x, p_x. One first finds a rotated coordinate system x', p'_x which minimizes the rms value of p'_x (the index x is dropped in the following) as illustrated in figure III D.

$$\begin{pmatrix} \tilde{x}'_i \\ \tilde{p} \end{pmatrix} = R \begin{pmatrix} x_i \\ p_i \end{pmatrix} = \begin{pmatrix} x_i \cdot \cos\theta + p_i \cdot \sin\theta \\ -x_i \cdot \sin\theta + p_i \cdot \cos\theta \end{pmatrix} \tag{26}$$

We then find

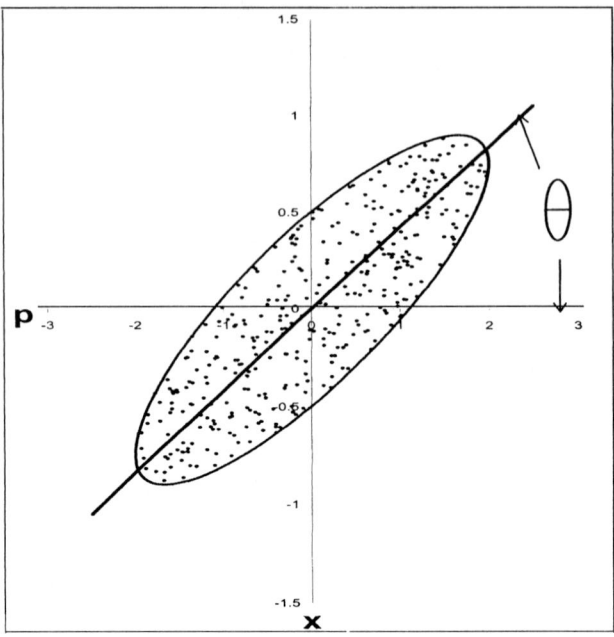

FIGURE 2. The particle distribution x, p in phase space is expressed in a coordinate system x', p' which is rotated by an angle Θ thereby minimizing the RMS-value of p'

$$\begin{aligned}(p'_i)^2 &= (-x_i \cdot \sin\theta + p_i \cdot \cos\theta)^2 \\ &= x_i^2 \cdot \sin\theta^2 + p_i^2 \cdot \cos\theta^2 - 2x_i p_i \cdot \sin\theta \cos\theta\end{aligned} \quad (27)$$

The RMS-value of p' is found by averaging over all N particles.

$$\begin{aligned}(p'_{rms})^2 &= \tfrac{1}{N}\Sigma_i(p'_i)^2 \\ &= \sin^2(\theta)\tfrac{1}{N}\Sigma_i(x_i)^2 + \cos^2(\theta)\tfrac{1}{N}\Sigma_i(p_i)^2 + 2\sin(\theta)\cos(\theta)\tfrac{1}{N}\Sigma_i x_i p_i \\ &= \tfrac{1}{N}\Sigma_i(x_i)^2 + \cos(\theta)^2 \tfrac{1}{N}\Sigma_i((p_i)^2 - (x_i)^2) - 2\sin(\theta)\cos(\theta)\tfrac{1}{N}\Sigma_i x_i p_i \\ &= \tfrac{1}{2N}\Sigma_i((x_i)^2 + (p_i)^2) - \cos(2\theta)\tfrac{1}{2N}\Sigma_i((x_i)^2 - (p_i)^2) - \sin(2\theta)\tfrac{1}{N}\Sigma_i x_i p_i\end{aligned} \quad (28)$$

The angle which minimizes p'_{rms} follows from $2p'_{rms} \cdot dp'_{rms}/d\theta = 0$

$$\tan(2\theta) = \frac{2\sum(x_i p_i)}{\sum(x_i^2 - p_i^2)} \tag{29}$$

Using this angle, $\tilde{p}'_{rms}$ is given by

$$\tilde{p}'_{rms} = \sqrt{\frac{1}{2N}\sum_i(x_i^2 + p_i^2) - \frac{1}{2N}\sqrt{(\sum_i(x_i^2 - p_i^2))^2 + 4(\sum_i p_i x_i)^2}} \tag{30}$$

The RMS-value in the orthogonal x'-direction is

$$\tilde{x}'_{rms} = \sqrt{\frac{1}{2N}\sum_i(x_i^2 + p_i^2) + \frac{1}{2N}\sqrt{(\sum_i(x_i^2 - p_i^2))^2 + 4(\sum_i p_i x_i)^2}} \tag{31}$$

The emittance (in units of π) is defined as the phase space area $\varepsilon = x'_{rms} p'_{rms}$

$$\begin{aligned} x'_{rms} p'_{rms} &= \frac{1}{2N}\sqrt{(\sum_i(x_i^2 + p_i^2))^2 - (\sum_i(x_i^2 - p_i^2))^2 - 4(\sum_i p_i x_i)^2} \\ \varepsilon &= \frac{1}{N}\sqrt{\sum_i x_i^2 \sum_i p_i^2 - (\sum_i p_i x_i)^2} \end{aligned} \tag{32}$$

This expression has first been derived by Lapostolle [1].

E Beam Emittance and Single Particle Emittance

The area of the ellipse which is obtained if the phase space coordinates of a particle are plotted against each other for successive turns is equal to $2J$, twice the value of the action. It is related to the beam emittance introduced in the previous section. If the coordinates on the periphery of the phase space ellipse are used to calculate the effective emittance ε_{eff} for a single particle using the formula of the previous section, one finds

$$x = \sqrt{2J\beta}\cos(\phi) \quad\rightarrow\quad <x^2> = J\beta$$

$$p = \beta x' = -\sqrt{2J/\beta}(\sin(\phi) + \alpha\cos(\phi)) \quad\rightarrow\quad <p^2> = \frac{J}{\beta}(1+\alpha^2) \tag{33}$$

$$<xp> = -J\beta\alpha.$$

Using these expressions, the effective emittance is

$$\varepsilon_{eff} = \sqrt{<x^2><p^2> - <xp>^2} = J \tag{34}$$

Thus the effective emittance of a single particle is half of the area of the ellipse enclosed by the coordinates. This factor of two will be recovered when calculating the beam emittance as an average over the action values for a Gaussian distribution. Integrating over the distribution function $\rho_a(J)$ of the action values J with

$$\rho(x, \tilde{x}') = 1/(2\pi\sigma^2)exp\left[-(x^2 + \tilde{x}'^2)/(2\sigma^2)\right] \tag{35}$$

($\tilde{x}' = x\alpha + x'\beta$) or equivalently

$$\rho_a(J, \phi) = 1/(4\pi J_0)exp[-J/(2J_0)] \tag{36}$$

one obtains the beam emittance

$$\varepsilon = \int d\phi \int dJ \cdot J \cdot \rho_a(J, \phi) = 2J_0. \tag{37}$$

IV EMITTANCE DILUTION

A Emittance Dilution due to Coherent Oscillation

Since nonlinear fields are unavoidable for real beams, all the particles in a beam have a (slightly) different tune. In the longitudinal phase plane, the focusing forces are inherently nonlinear (sinusoidal). In the transverse phase plane the source of non-linearities are sextupole magnets and field imperfections. As discussed before, the spread in tune and the corresponding build up of a spread in phase is the reason why the dipole moment of a coherently oscillating beam decays. The decaying observable amplitude $< x_n >$ as a function of time in units of the revolution time is described by [3,4]

$$< x_n > = \frac{exp(-\frac{(4\pi n a \Delta \cdot \sigma)^2}{2(1 + (4\pi n a \sigma^2)^2)})}{1 + (4\pi n a \sigma^2)^2} \Delta sin(\phi_n) \tag{38}$$

$$\phi_n = 2\pi Q n + \frac{2\pi a \Delta^2 n}{1 + (4\pi n a \sigma^2)^2} + 2tan^{-1}(4\pi n a \sigma^2).$$

Q is the betatron tune, σ is the RMS beam size, a is the detuning coefficient and Δ is the initial oscillation amplitude. A typical decay pattern is shown in figure 3. Eventually all the particles are uniformly distributed in phase. The dipole moment becomes zero and the beam particles are spread over a larger area in phase space. The phase space density is then diluted and the effective beam emittance is larger. This phenomenon is called filamentation. It is sketched in figure 4.

Chromaticy ξ and momentum spread $\frac{\Delta p}{p}$ in the beam also cause a tune spread. However, a chromatic tune spread for bunched beam does not immediately lead to irreversible filamentation. After a full synchrotron oscillation period, the chromatic betatron phase advance is zero for all particles (assuming all particles have the same synchrotron tune) since

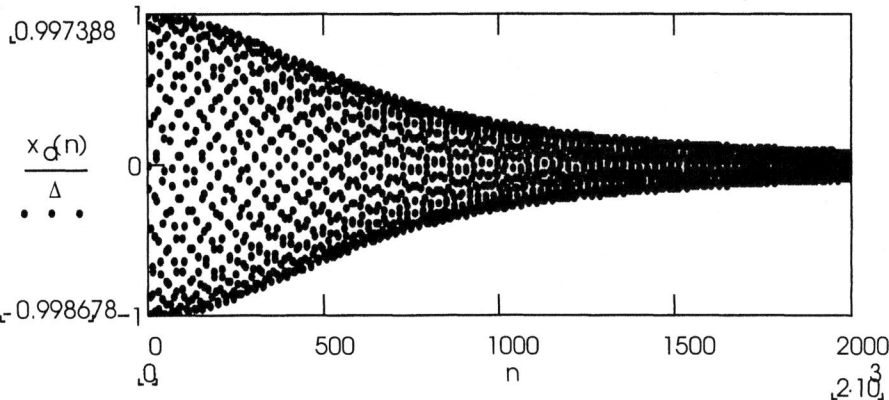

FIGURE 3. Dipole moment of a filamenting beam versus time in units of revolution time n. Δ is the initial amplitude which is $10mm$ in this example. The amplitude dependence of the tune is described by $a = 1m^{-2}$. The RMS beam size σ is $10mm$. The tune is $Q = 0.261$.

$$\Delta\phi = 2\pi Q_\beta/Q_s + \frac{\Delta p}{p} \cdot \xi \int_0^{2\pi} d\theta\, sin(Q_s\theta) \qquad (39)$$

and the dipole moment of the beam is restored. Thus, the emittance dilution is in principle recoverable. Eventually, the beam will filament nevertheless because of the nonlinearity in the longitudinal focusing force and the corresponding synchrotron tune spread.

B Filamented Distribution

The filamentation process can be described in a more formal way. Consider an initially undistorted distribution of particle coordinates $\rho(x, x')$ in a beam. If an oscillation of the beam is excited by a kick Δ, the distribution becomes $\rho(x, x'-\Delta)$. This is described by a new, non-stationary distribution $\tilde{\rho}(x, x', t)$ with $\tilde{\rho}(x, x', t = 0) = \rho(x, x' - \Delta)$. The distribution at time t is then obtained by

$$\tilde{\rho}((x, x'), t) = \tilde{\rho}(\mathcal{T}^{-1}(x, x'), t = 0) \|\mathcal{J}_\mathcal{T}\|. \qquad (40)$$

The map $\mathcal{T}(x, x')$ describes the transformation of the coordinates between $t = 0$ and t and $\|\mathcal{J}_\mathcal{T}\|$ is the Jacobian of $\mathcal{T}$. For a Hamiltonian system, its determinant is equal to one. The time dependence of the distribution is most conveniently studied by writing the distribution in terms of the action and angle variables J, ϕ. If the perturbation is a dipole kick Δ, one has

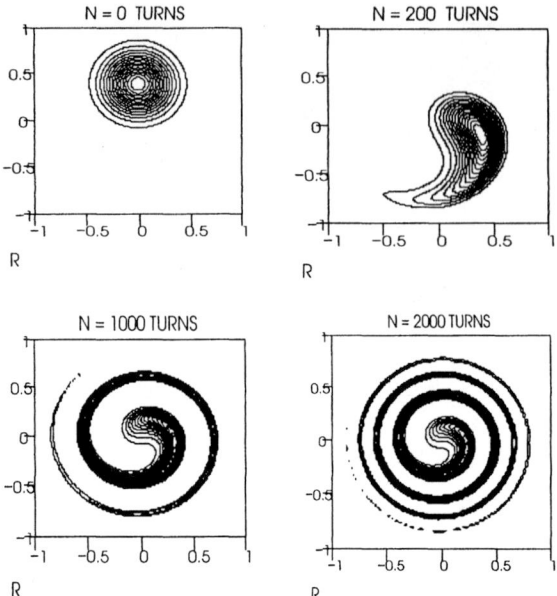

FIGURE 4. A tune spread leads to a betatron-phase spread and a decay of the dipole moment of coherent oscillations. Contour plots of an initially Gaussian distribution of phase space coordinates (1 degree of freedom) after a kick of twice the RMS-beam width of $\sigma = 10mm$ are shown after 0, 200, 1000 and 2000 turns. The tuneshift with amplitude in this example is $a = \Delta Q/x^2$, $a = 1m^{-2}$. The plot is produced by evaluating the distribution function described below.

$$\tilde{\rho}(x, x', t = 0) = \tilde{\rho}_a(J, \phi(t = 0)) = \rho_a(J(x, x' - \Delta)) \qquad (41)$$

with

$$\begin{aligned} x &= \sqrt{2J\beta}\cos(\phi) \\ x' &= -\sqrt{2J/\beta}(\sin(\phi) + \alpha\cos(\phi)) \\ J(x, x') &= \tfrac{1}{2}(\gamma x^2 + 2\alpha x x' + \beta x'^2). \end{aligned} \qquad (42)$$

In action and angle variables, the distorted distribution is

$$\begin{aligned} \tilde{\rho}_a(J, \phi(t)) &= \rho_a(\tfrac{1}{2}(\gamma x^2 + 2\alpha x(x' - \Delta) + \beta(x' - \Delta)^2)) \\ &= \rho_a(J - \sqrt{\tfrac{1}{2}J\beta}\Delta\sin(\phi) + \tfrac{1}{2}\beta\Delta^2)) \end{aligned} \qquad (43)$$

The action J is a constant of motion and because of the amplitude-dependent tune the phase ϕ is not only a linear function in time but also a function of the action J

$$\phi(t) = \phi(0) + (Q + aJ)\omega_0 \cdot t. \qquad (44)$$

In this expression, a characterizes the nonlinear tuneshift with amplitude and ω_0 is the revolution frequency. The distribution at time t is given by

$$\tilde{\rho}_a(J, \phi(t)) = \rho_a(J + \frac{1}{2}\beta\Delta^2 - \sqrt{\frac{1}{2}J\beta}\Delta \sin(\phi - \omega_0(Q + aJ))) \tag{45}$$

For large times $t \to \infty$, the phase difference for even small differences in J becomes large, which means, that the average distribution over an interval $[J, J + \delta J]$ becomes more and more uniformly distributed in ϕ. The range δJ for which the average distribution becomes essentially phase independent will shrink in time and eventually the distribution will become stationary and can be expressed as a function of the action alone $\tilde{\rho}_a(J, \phi, t \to \infty) = \hat{\rho}_a(J)$. This final distribution is thus obtained by averaging the distribution $\tilde{\rho}_a(J, \phi)$ over ϕ for each J. This can be carried out for example for an initial Gaussian distribution:

$$\hat{\rho}_a(J) = \frac{1}{2\pi}\frac{1}{2J_0}\int d\phi\ \exp(\frac{-1}{2J_0}(J - \sqrt{\frac{1}{2}J\beta}\Delta \sin(\phi) + \frac{1}{2}\beta\Delta^2))$$

$$= \frac{1}{2J_0}\exp(\frac{-1}{2J_0}(J + \frac{1}{2}\beta\Delta^2)) \cdot I_0(\frac{\sqrt{\frac{1}{2}J\beta} \cdot \Delta}{2J_0}) \tag{46}$$

Here, I_0 is the modified Bessel function. The distribution in x can also be calculated for an original Gaussian distribution $\rho(x) = 1/(\sqrt{2\pi}\sigma)\exp(-x^2/2\sigma^2)$. The projection of the oscillating distributions is expressed by a phase dependent offset in x of $\Delta\beta\sin(\phi)$.

$$\hat{\rho}(x) = \frac{1}{2\pi}\int d\phi \rho(x - \Delta\beta\sin(\phi))$$

$$= \frac{1}{\sqrt{2\pi}\sigma}\exp\{-\frac{x^2 + \frac{1}{2}\Delta^2}{2\sigma^2}\}$$

$$\frac{1}{2\pi}\int d\phi \cdot \exp\{-\frac{2 \cdot x \cdot \Delta \cdot \beta \cdot \sin(\phi) + \frac{1}{2}\Delta^2 \cdot \beta^2 \cdot \cos(2\phi)}{2\sigma^2}\} \tag{47}$$

$$\hat{\rho}(x) = \frac{1}{\sqrt{2\pi}\sigma}\exp\{-\frac{x^2 + \frac{1}{2}\Delta^2}{2\sigma^2}\}$$

$$\cdot \left(I_0(\frac{\Delta x}{\sigma^2})I_0(\frac{\Delta^2}{4\sigma^2}) + 2\sum_{k=1}^{\infty}(-1)^k I_{2k}(\frac{\Delta x}{\sigma^2})I_k(\frac{\Delta^2}{4\sigma^2})\right)$$

Such a distribution is depicted in figure 5.

C RMS-Beam Width

The horizontal rms beam width σ is calculated by

$$\sigma_x = \sqrt{\int dx x^2 \rho(x)} \tag{48}$$

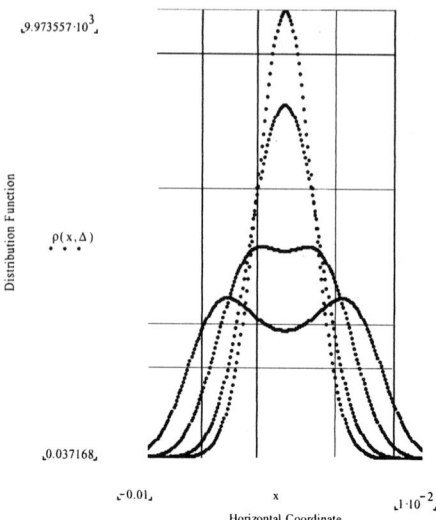

FIGURE 5. Filamented Distribution in x of an originally Gaussian distributed beam after the beam's center has been displaced by n-times its RMS width ($n = 0, 1, 2, 3$) according to equation 47

where $\rho(x)$ is the distribution in x. For evaluating emittance dilution effects, there is a convenient way to calculate the rms width of the diluted beam without calculating the distorted distribution function, which can be quite tedious, even in simple cases. If the coordinates of the particles are distorted by an excitation $x, x' \to x, \tilde{x}'$, the action of the distorted particle $\tilde{J}$ is a function of the undistorted action J, the betatron phase and the distortion Δ: $\tilde{J} = \tilde{J}(J, \phi, \Delta)$. The particles with distorted action values will distribute uniformly in phase. Consider particles with the same undistorted action J with uniformly distributed phases ϕ. After the distortion occurred, these particles will be found within a band of action values $\tilde{J}(J, \phi, \Delta)$. The average action of these particles is the average over the phase

$$<\tilde{J}> = \frac{1}{2\pi} \int d\phi \tilde{J}(J, \phi, \Delta)$$
$$= J + \Delta J(J, \Delta). \tag{49}$$

The emittance of the whole beam is obtained by averaging over the original action J using the original distribution function $\rho_a(J)$

$$\begin{aligned}\varepsilon &= \int dJ \cdot (J + \Delta J(J,\Delta)) \cdot \rho_a(J) \\ &= \varepsilon + \int dJ \cdot \Delta J(J,\Delta) \cdot \rho_a(J)\end{aligned} \qquad (50)$$

D Emittance Dilution due to Filamentation of a Dipole Oscillation

The relative increase in beam size of a filamented beam after a coherent dipole oscillation is independent of the initial distribution. This can be seen the following way: Consider a beam with a certain horizontal distribution $\rho(x)$. The rms value is defined as $\sigma_x = \sqrt{\int dx \rho(x) x^2}$. While the beam is oscillating the distribution has a variable offset and becomes $\tilde{\rho}(x,\phi) = \rho(x - \Delta\beta \cdot \sin(\phi))$ where $\phi = \omega \cdot t$ is the phase of the oscillation. Filamentation means that for each amplitude, one obtains a uniform distribution in phase, or within in each small phase interval, we obtain the original offset distribution in radius r. This is equivalent to averaging the phase dependent distribution over the phases ϕ. The new rms value of the beam width is then

$$\tilde{\sigma} = \sqrt{\frac{1}{2\pi} \int d\phi \int dx x^2 \rho(x - \beta \cdot \Delta \sin(\phi))} \qquad (51)$$

We evaluate this expression by a transformation of the integration variable from x to $u = x - \Delta \cdot \beta \sin(\phi)$.

$$\begin{aligned}\tilde{\sigma} &= \sqrt{\frac{1}{2\pi} \int d\phi \int du (u + \Delta \cdot \beta \sin(\phi))^2 \rho(u)} \\ \tilde{\sigma} &= \sqrt{\sigma^2 + (\Delta \cdot \beta)^2 \frac{1}{2\pi} \int d\phi \sin(\phi)^2} \\ &= \sqrt{\sigma^2 + \tfrac{1}{2}\Delta^2 \cdot \beta^2}\end{aligned} \qquad (52)$$

This gives the expression

$$\frac{\Delta\sigma}{\sigma} \simeq \frac{1}{4}\left(\frac{\Delta \cdot \beta}{\sigma}\right)^2. \qquad (53)$$

It is remarkable that this expression is independent of the form of the initial distribution function.

THe same result may be obtained by calculating the action $\tilde{J}$ of a particle with original J and ϕ after it has received a kick Δ

$$\tilde{J} = \tfrac{1}{2}(x^2\gamma + 2\alpha(x' + \Delta) + \beta(x' + \Delta)^2)$$

$$= J + \tfrac{1}{2}\sqrt{2J\beta}\Delta\sin(\phi) + \tfrac{1}{2}\beta\Delta^2 \qquad (54)$$

Integration over all particles which are assumed to have a Gaussian distribution in x, $\rho_a(J, \phi) = 1/(2\pi J_0) \cdot exp(-J/(2J_0))$ yields

$$\tilde{\sigma} = \sqrt{\tfrac{1}{2\pi J_0} \int d\phi \int dJ \cdot exp(-\tfrac{1}{2}J/J_0) \cdot (J + \tfrac{1}{2}\sqrt{2J\beta}\Delta\sin(\phi) + \tfrac{1}{2}\beta\Delta^2)\beta}$$

$$= \sqrt{\tfrac{1}{2\pi J_0} \int dJ exp(-\tfrac{1}{2}J/J_0) \cdot (J\beta + \tfrac{1}{2}\beta^2\Delta^2)} \qquad (55)$$

$$= \sqrt{2J_0\beta + \tfrac{1}{2}\beta^2\Delta^2}$$

$$= \sigma\sqrt{1 + \tfrac{1}{2}\left(\tfrac{\beta \cdot \Delta}{\sigma}\right)^2}$$

or

$$\frac{\Delta\sigma}{\sigma} = \frac{1}{4}\left(\frac{\Delta \cdot \beta}{\sigma}\right)^2. \qquad (56)$$

E Dispersion Mismatch

1 Distribution Function at Position with Dispersion

The dispersion D describes the closed orbit for a particle with a small momentum offset $x_\epsilon = \Delta p/p \cdot D$. Let us assume, that there is only dispersion in the horizontal plane which is the case for a flat ring with no vertical deflections. At a position with dispersion D, the coordinate of a particle with momentum deviation $\Delta p/p = \epsilon$ is a sum of the β-tron part and the momentum part

$$x = x_\beta + \epsilon \cdot D \qquad (57)$$

Thus the beam size given by the distribution of particles in x depends also on the momentum distribution and the dispersion. Let the distribution functions be Gaussian in x_β and ϵ

$$\rho(x_\beta) = \frac{1}{\sqrt{2\pi}\sigma} exp\left(-\frac{x_\beta^2}{2\sigma^2}\right) \quad \tilde{\rho}(\epsilon) = \frac{1}{\sqrt{2\pi}\sigma_\epsilon} exp\left(-\frac{\epsilon^2}{2\sigma_\epsilon^2}\right). \qquad (58)$$

At a given value of the x-coordinate, the horizontal β-tron amplitude x_β is $x - \epsilon D$. Averaging over the contributions of $\epsilon \cdot D$ by performing an integral over the momentum distribution yields the distribution function P in x as convolution of transverse and longitudinal distributions

$$P(x) = \int d\epsilon \, \rho(x - \epsilon D) \cdot \tilde{\rho}(\epsilon)$$

$$= \frac{1}{2\pi\sigma\sigma_\epsilon} \int d\epsilon \exp\left(-\frac{(x-\epsilon D)^2}{2\sigma^2}\right) \exp\left(-\frac{\epsilon^2}{2\sigma_\epsilon^2}\right)$$

$$= \frac{1}{2\pi\sigma\sigma_\epsilon} \exp\left(-\frac{x^2}{2\sigma^2}\right)$$
$$\cdot \int d\epsilon \exp\left(-\frac{1}{2\sigma_\epsilon^2}\left(\epsilon^2 \frac{\sigma^2+\sigma_\epsilon^2 D^2}{\sigma^2} - 2\epsilon\frac{x\sigma_\epsilon^2 D}{\sigma^2}\right)\right) \quad (59)$$

$$= \frac{1}{2\pi\sigma\sigma_\epsilon} \exp\left(-\frac{x^2}{2\sigma^2}\left(1 - \frac{\sigma_\epsilon^2 D^2}{\sigma^2 + \sigma_\epsilon^2 D^2}\right)\right)$$
$$\int d\epsilon \exp\left(-\frac{1}{2\sigma_\epsilon^2}\left(\epsilon\sqrt{\frac{\sigma^2+\sigma_\epsilon^2 D^2}{\sigma^2}} - \frac{x\sigma_\epsilon^2 D}{\sigma\sqrt{\sigma^2+\sigma_\epsilon^2 D^2}}\right)^2\right)$$

This gives the expected result: the beam rms-width contributions due to betatron oscillations and due to momentum oscillations add in quadrature.

$$P(x) = \frac{\exp\left(-\frac{x^2}{2(\sigma^2 + \sigma_\epsilon^2 D^2)}\right)}{\sqrt{2\pi(\sigma^2 + \sigma_\epsilon^2 D^2)}} \quad (60)$$

2 Emittance Increase due to an Dispersion Mismatch

In order to avoid emittance increase on beam transfer, the dispersion function at the end of the transfer line (out) must be matched to the periodic dispersion at the injection point (in). A mismatch of the dispersion will lead to an increase of the transverse beam emittance. Since $x_{out} = (x_\beta + \epsilon D)_{out}$ equals $x_{in} = (x_\beta + \epsilon D)_{in}$, the increase in β-tron amplitude is $\Delta x_\beta = (D_{in} - D_{out}) \cdot \epsilon = \Delta D \cdot \epsilon$. The distribution with a dispersion mismatch can be written after filamentation

$$\tilde{P}(x) = \frac{1}{2\pi\sigma\sigma_\epsilon} \frac{1}{2\pi} \int d\phi \int d\epsilon \exp\left(-\frac{(x - \epsilon D - \epsilon\Delta D \cdot \sin(\phi)^2)}{2\sigma^2}\right) \exp\left(-\frac{\epsilon^2}{2\sigma_\epsilon^2}\right)$$

$$= \frac{1}{2\pi} \int d\phi \frac{\exp\left(-\frac{x^2}{2(\sigma^2 + \sigma_\epsilon^2(D + \Delta D\sin(\phi))^2)}\right)}{\sqrt{\sigma^2 + \sigma_\epsilon^2(D + \Delta D\sin(\phi))^2}} \exp\left(-\frac{\epsilon^2}{2\sigma_\epsilon^2}\right)$$

$$(61)$$

The rms width which follows from this distribution is

$$\begin{aligned}\tilde{\sigma} &= \sqrt{\int dx\, x^2 \tilde{P}(x)} \\ &= \sqrt{\tfrac{1}{2\pi}\int d\phi\,(\sigma^2 + \sigma_\epsilon^2(D+\Delta D\sin(\phi))^2)} \\ &= \sqrt{\tfrac{1}{2\pi}\sigma^2 + \sigma_\epsilon^2(D+\tfrac{1}{2}\Delta^2 D^2)}\end{aligned} \qquad (62)$$

$$\frac{\Delta\sigma}{\sigma} = \frac{1}{4}\frac{\sigma_\epsilon^2 \Delta D^2}{\sigma^2 + \sigma_\epsilon^2 D^2}$$

The corresponding emittance dilution is $\Delta\varepsilon/\varepsilon = (1/2)\cdot(\sigma_\epsilon^2 D^2)/\sigma^2$.

The alternative way to calculate the emittance dilution is more convenient. We first sort the particles which experience a mismatch of the dispersion function $\epsilon\Delta D, \epsilon\delta D'$ according to their value of emittance ε. We then calculate the increased action due to the mismatch:

$$\tilde{J}(\phi) = \frac{1}{2}\cdot\left(\gamma(x+\Delta D\epsilon)^2 + 2\alpha(x+\Delta D\epsilon)(x'+\Delta D'\epsilon) + \beta(x'+\Delta D\epsilon)^2\right). \qquad (63)$$

The action increase depends on the betatron phase angle ϕ. The mismatched beam will filament and the particles will be distributed uniformly in phase and populating a band of action values. Each action value is labeled by the original value of the particle phase so that averaging over these phases gives the average emittance. All terms which are linear in x and x' will vanish due to the phase averaging so that we obtain

$$\tilde{J} = J + \frac{1}{2}\epsilon^2\cdot(\gamma\Delta D^2 + 2\alpha\delta D\Delta D' + \beta\Delta D^2). \qquad (64)$$

In order to obtain the new beam emittance, we have to take the appropriate average using the original distribution function of the emittance and the momentum

$$\begin{aligned}\frac{\Delta\varepsilon}{\varepsilon} &= \tfrac{1}{2}\int dJ\cdot\rho(J)\cdot\int d\epsilon\cdot\rho_\epsilon(\epsilon)\frac{\epsilon^2}{\sigma^2}\cdot\beta(\gamma\Delta D^2 + 2\alpha\delta D\Delta D' + \beta\Delta D^2) \\ &= \tfrac{1}{2}\cdot\sigma_\epsilon^2\beta(\gamma\Delta D^2 + 2\alpha\Delta D\Delta D' + \beta\Delta D^2).\end{aligned} \qquad (65)$$

Thus for the increase in rms-beam size due to a dispersion mismatch we find

$$\frac{\Delta\sigma}{\sigma} = \frac{\beta}{4}(\gamma\delta D^2 + 2\alpha\delta D\delta D' + \beta\delta D^2)\left(\frac{\sigma_\epsilon}{\sigma}\right)^2 \qquad (66)$$

F Mismatch of Lattice Function

If the form of the space space ellipse of an injected beam is mismatched to the phase space ellipse given by the lattice function at the injection point, the phenomenon of filamentation will occur as well, which leads to an increase of effective emittance.

1 Emittance Dilution due to Lattice Function Mismatch

To calculate this effect, consider a beam which is matched to lattice functions β_1, α_1 and which has a distribution of action values J as

$$\rho_a(J) = \frac{1}{2J_0} \cdot \exp\left(-\frac{J}{2J_0}\right)$$

$$x = \sqrt{2J\beta_1}\cos(\phi)$$
$$x' = -\sqrt{2J/\beta_1}(\sin(\phi) + \alpha_1 \cos(\phi)).$$

(67)

This beam is injected into a ring at a location with lattice functions β_2, α_2. The particles which had an original value J of the action, now jump to a different value of action $\tilde{J}$ in the new system. This new action value depends on the original value of the betatron phase ϕ

$$\begin{aligned}\tilde{J}(\phi) &= \gamma_2 x(\phi)^2 + 2\alpha_2 x(\phi) x'(\phi) + \beta_2 x'(\phi)^2 \\ &= J \cdot (\gamma_2\beta_1 - 2\alpha_1\alpha_2 + \alpha_1^2\beta_2/\beta_1) \cdot \cos(\phi)^2 \\ &+ J \cdot 2(\alpha_1\beta_2/\beta_1 - \alpha_2) \cdot \sin(\phi)\cos(\phi) \\ &+ J \cdot (\beta_2/\beta_1) \cdot \sin(\phi)^2.\end{aligned}$$

(68)

Particles with action values in the range $\tilde{J} + \delta\tilde{J}$ will distribute uniformly over the phase φ in the new system due to the spread in tune. Thus particles of originally the same action J will filament into a band of ellipses. This is illustrated in figure 6. Averaging over the original betatron phases ϕ yields the average action $\bar{J}$ of the band of actions as a function of the original action values J:

$$\bar{J} = J \cdot \frac{1}{2}(\gamma_1\beta_2 + \gamma_2\beta_1 - 2\alpha_1\alpha_2) = \frac{1}{2}J \cdot A.$$

(69)

Multiplying with the original action distribution function $\rho_a(J)$ and integrating over J gives the new increased value of the emittance of the whole beam.

$$\tilde{\varepsilon} = \int dJ \cdot J \cdot \rho_a(J) \cdot \frac{1}{2}A = \frac{1}{2}\varepsilon \cdot A$$

(70)

where $\varepsilon = \int dJ \, J \cdot \rho_a(J)$ is the original emittance. The new emittance is thus proportional to the original value of the beam emittance. Writing $\beta_1 = \beta$, $\beta_2 = \beta + \delta\beta$, $\alpha_1 = \alpha$ and $\alpha_2 = \alpha + \delta\alpha$ and expanding up to 2nd order in $\delta\beta$ gives

$$\frac{\Delta\varepsilon}{\varepsilon} = \frac{\tilde{\varepsilon} - \varepsilon}{\varepsilon} = \frac{1}{\beta}(\gamma\delta\beta^2 + 2\alpha\delta\alpha\delta\beta + \beta\delta\alpha^2) = \frac{\tilde{A}}{\beta}$$

(71)

with $\tilde{A} = \gamma\delta\beta^2 + 2\alpha\delta\alpha\delta\beta + \beta\delta\alpha^2$. The relative increase of the rms beam size is

$$\frac{\Delta\sigma}{\sigma} = \frac{1}{2}\frac{\tilde{A}}{\beta}.$$

(72)

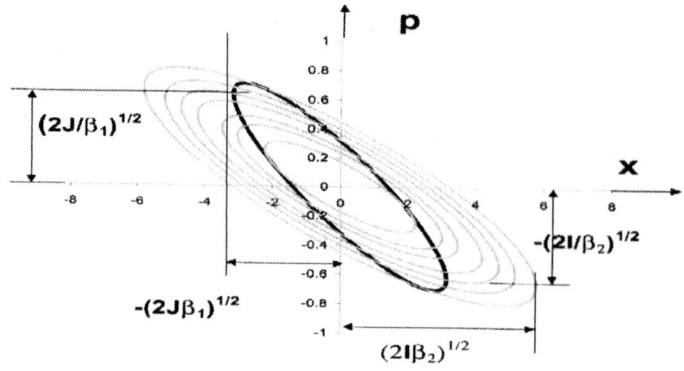

FIGURE 6. Schematic view of the filamentation process due to mismatch of the lattice functions

2 Distribution Function of a Mismatched Beam after Filamentation

The distribution function of a mismatched beam eventually becomes independent of the phase. The beam is distributed in terms of the action in the original system (for example a transfer line, denoted with index "2") as $\rho(J) = 1/(2J_0)\exp(-\frac{1}{2}J/J_0)$. It is injected into a new system (circular accelerator, denoted with index "1"). The new action $\tilde{J}$ expressed in terms of the original action variable $J(x,x')$

$$\begin{aligned}
\tilde{J}(\phi) &= \gamma_1 x^2 + 2\alpha_1 xx' + \beta_1 x'^2 \\
&= J\left(a\cos(\phi)^2 + 2b\sin(\phi)\cos(\phi) + c\sin(\phi)^2 a_1/\beta_2\right)
\end{aligned} \qquad (73)$$

$$\begin{aligned}
a &= \gamma_1\beta_2 + \gamma_2\beta_1 - 2\alpha_1\alpha_2 - \beta_1/\beta_2 \\
b &= \alpha_1 - \alpha_2\beta_1/\beta_2 \\
c &= \beta_2/\beta_1.
\end{aligned}$$

Thus we have

$$J = \tilde{J}/\left(a\cos(\phi)^2 + 2b\sin(\phi)cos(\phi) + c\sin(\phi)^2 a_1/\beta_2\right). \tag{74}$$

When the beam is circulating, filamentation will take place as discussed before, the phase distribution for particles in an arbitrarily small range $\delta\tilde{J}$ becomes uniform. In order to obtain the new distribution function $\tilde{\rho}_a(\tilde{J})$ we use the transformation of equation 74, insert it in the original distribution function $\rho_a(J)$), and average over the phase ϕ

$$\tilde{\rho}_a(\tilde{J}) = \frac{1}{2\pi}\int d\phi \; \rho(\tilde{J}/(a\cos(\phi)^2 + 2b\sin(\phi)cos(\phi) + c\sin(\phi)^2 a_1/\beta_2)) \tag{75}$$

which yields

$$\tilde{\rho}_a(\tilde{J}) = \frac{1}{2J_0}\exp\left(-\tilde{J}\cdot A/(2J_0)\right)\cdot I_0\left(\frac{\tilde{J}}{2J_0}\sqrt{A^2-1}\right). \tag{76}$$

A is $\gamma_2\beta_1 + \beta_2\gamma_1 - 2\alpha_1\alpha_2$ as previously defined and I_0 is the modified Bessel Function. The calculation is performed in reference [6]. The corresponding distribution is obtained by replacing J by $x^2 + \tilde{x}'^2$ with $\tilde{x}' = x\alpha + x'\beta$ and integrating over $\tilde{x}'$. The result, which is evaluated numerically is shown in figure 7.

G Coupling Effects

If the motion in the two transverse planes is mixed by rotated quadrupoles in the transfer line or by a rotation of the transverse coordinate system (due to interleaved horizontal and vertical deflections), the beam envelope is not matched to the lattice functions and the beam will subsequently filament. The same happens if the horizontal and vertical motions are coupled in the delivering or the receiving accelerator. These effects will be investigated in this section.

1 Emittance Blowup due to a Single Skew Quadrupole

Suppose there is perfect matching between two rings in an accelerator cascade but there is a single skew quadrupole of length l and strength k_s in the transfer line between. Then the particles receive a kick in the horizontal plane $\Delta x' = k_s l \cdot y$ (and a corresponding kick in the vertical plane). The change in emittance is for particles with action J_x and phase ϕ_x

$$\tilde{J}_x = J_x - 2k_s l\sqrt{J_x J_y \beta_x \beta_y}\cos(\phi_x)cos(\phi_y) + (k_s l)^2 J_y \beta_x \beta_y \cos(\phi_y)^2 \tag{77}$$

The beta values are taken at the position of the kick. If one averages these expression over all phases and action values, the relative change in the rms beam size σ_x becomes

$$\frac{\Delta\sigma_x}{\sigma_x} = \left(\frac{1}{2}\frac{\sigma_y}{\sigma_x}(k_s l \beta_x)\right)^2 \tag{78}$$

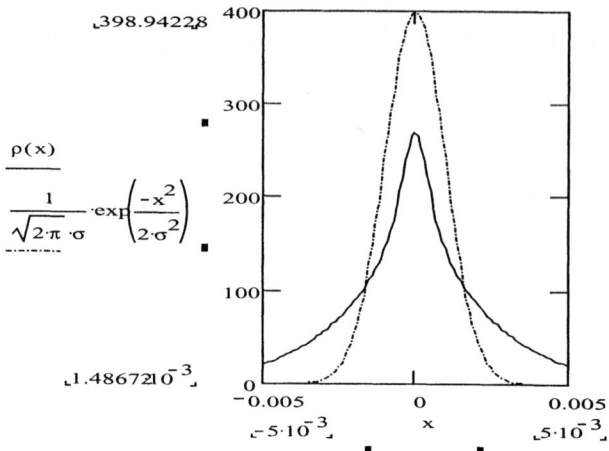

FIGURE 7. Distribution function in x as a result of filamentation after a strong lattice function mismatch ($\Delta\beta/\beta = 1$). The dashed line is the original Gaussian distribution shown for comparison. The distribution function is the result of numerical integration of equation 76 over $\tilde{x}' = x\alpha + x'\beta$.

2 Emittance Blowup due Unmatched Coupled Beam Optics

If the beam optics is coupled in one of the two accelerators involved in the transfer, the beam optics parameters are expressed as the sum of two modes [7]

$$x = \sqrt{2J_I\beta_{xI}}\cos(\Psi_I) + \sqrt{2J_{II}\beta_{xII}}\cos(\Psi_{II}) \tag{79}$$

with corresponding expressions for x', y, and y'. If the phase averaged emittance is calculated for undistorted uncoupled lattice functions $\beta_{x,y}$, $\alpha_{x,y}$ using the coupled coordinates, one finds

$$\tilde{J}_x = \sum_{i=I,II} J_{xI}(\gamma_x\beta_{xi} + \beta_x\gamma_{xi} - 2\alpha_x\alpha_{xi}). \tag{80}$$

For small coupling we may neglect the departure of the mode I-lattice functions from the undistorted horizontal ones so that we only have to consider the contribution from mode II to $\delta\sigma_x$:

$$\Delta\sigma_x/\sigma_x = \frac{1}{4}(\gamma_x\beta_{xII} + \beta_x\gamma_{xII} - 2\alpha_x\alpha_{xII})/\sigma_x^2 \tag{81}$$

H Nonlinear Effects

Nonlinear effects are not so important in emittance dilution considerations unless there are strong nonlinear fields or resonantly enhanced nonlinearities in a ring.

1 Emittance Dilution due to a Sextupole in the Transfer line

Passing through a sextupole, the particle will receive a kick

$$\Delta x' = \frac{1}{2} m \cdot l \cdot (x^2 - y^2) \tag{82}$$

where $m = e/p \cdot d^2 B/dy^2$ is the sextupole strength, l is the sextupole length and p is the beam momentum. This kick will distort the beam emittance and will lead to a mismatch at injection. The new action values for the particles are

$$\tilde{J} = J + \frac{1}{2} m \cdot l \cdot (x^2 - y^2)(\alpha x + \beta x') + \frac{1}{8}(m \cdot l)^2 (x^2 - y^2)^2 \tag{83}$$

Averaging over the betatron phases and over the distribution of action values J yields

$$\Delta \sigma_x / \sigma_x = \frac{1}{16} \left(\frac{m \cdot l \cdot \beta}{\sigma_x} \right)^2 \cdot \left(3\sigma_x^4 + 3\sigma_y^4 - 2\sigma_x^2 \sigma_y^2 \right). \tag{84}$$

2 Emittance Dilution due to Nonlinear Distortions

If there are nonlinear fields in in an accelerator, the shape of the phase space ellipse becomes distorted. If such a distorted beam is injected into an undistorted one, the distortion give rise to dilution and emittance blow up. If the tune is not too close to a resonance, the distortion can be approximately described by a canonical transformation which transforms from a Hamiltonian H with variables J, ϕ into a system $K, I\varphi$. The transformation is constructed such that the nonlinear forces are reduced to the effect of amplitude dependence of the tunes which is equivalent to requiring that K is a function of the action alone $K = K(I)$ ($\partial k/\partial \varphi = 0 \to I = $ const, $\partial K(I)/\partial I = \varphi' \to \Delta Q(I) = K(I)/\partial I$). In this transformed coordinate system, the particles are uniformly distributed in phase. Therefore we can perform the phase average to describe the filamentation process over the transformed phases by performing a phase averaging by integration. The transformation may be generated by a function $F(I, \phi)$. I and φ are the variables in the transformed system and J and ϕ the variables in the distorted system. The transformation reads

$$J = I + \partial F/\partial \phi = I + F_\phi$$
$$\varphi = \phi + \partial F/\partial I = \phi + F_I$$
$$K = H + \partial F \partial s \tag{85}$$

The increase in action is obtained by averaging over the phase variable φ

$$J = I + F_\phi - F_{\phi\phi} F_I \tag{86}$$

F can be represented by a harmonic expansion 9index n) of the distribution of the field around the ring

$$F = \sum_n f_n(I) \sin(n\phi + \phi_n)$$
$$F_{\phi\phi} = -\sum_n n^2 f_n(I) \sin(n\phi + \phi_n) \tag{87}$$
$$F_I = -\sum_n \frac{df_n}{dI} \sin(n\phi + \phi_n)$$

Inserting this in the expression for $J(\varphi)$ and integrating over φ one finds

$$\Delta I = \frac{1}{2\pi} \int d\varphi \sum_{nn'} n^2 f_n(I) \frac{df'_n(I)}{dI} \sin(n'\phi + \phi'_n) \cdot \sin(n\phi + \phi_n) = \frac{1}{2} \sum_n n^2 f_n(I) \frac{df_n(I)}{dI} \tag{88}$$

As an example for the size of the emittance distortion which can be expected from nonlinear distortions, consider a nonlinear distortion due to a single sextupole of strength $m \cdot l$. The transformation function is an infinite power series in the sextupole strength. Replacing the power series by the leading term is sufficient for our purpose in estimating the size of the effect. The first order term is obtained expressing the vector potential of the sextupole field

$$V(x, y, s) = (1/6) m(s) x^3, \tag{89}$$

(which is the remainder of the nonlinear Hamiltonian after tranformation to action and angle variables, thus $H = V$) in action and angle variables by replacing x by $\sqrt{J\beta/2} exp(\Psi(s) + \phi)$ yielding

$$V(J, \phi, s) = \sum_{n=1,3} \frac{1}{6} \binom{3}{\frac{3-n}{2}} m(s) \left(\frac{\beta}{2}\right)^{3/2} \cdot J^{3/2} e^{i \cdot n \cdot (\Psi(s) + \phi)}. \tag{90}$$

We exploit the periodicity of the sextupole structure and the lattice functions by expressing the potential as a sum of Fourier components

$$V(J, \phi, s) = \sum_{nq} v_{3nq} J^{3/2} e^{i \cdot (n\phi + (nQ + q) 2\pi s/L)}. \tag{91}$$

(L is the machine circumference, q labels the Fourier components, and Q is the linear tune). with coefficients

$$v_{3nq} = \frac{3!}{12\pi \left(\frac{3-n}{2}\right)! \left(\frac{3+n}{2}\right)! 2^{3/2}} \int dsm(s)\beta^{3/2}e^{i\cdot n \cdot (\Psi(s) - (nQ+q)2\pi s/L)} \quad (92)$$

Using the potential written in this form, the Hamilton Jacobi-Equation

$$K - H = \frac{\partial F}{\partial s} \quad (93)$$

is solved approximately for each Fourier component by preserving only the leading terms which are first order in the integrated sextupole strength ml. The solutions are proportional to $-iv_{3nq}/(nQ-q)$ where $nQ - p$ is refered to as resonance dominator. Finally we perform the inverse Fourier transformation for the term $n = 3$ which becomes important near the third integer resonance $3Q + q = 0$. This results in the desired approximate distortion function

$$f_{33} = \frac{1}{24\sqrt{2}} ml\beta^{3/2} I^{3/2} \frac{sin(3\phi)}{sin(3\pi Q)} \quad (94)$$

(a phase ϕ_{33} in the argument of the sine function has been omitted.) Using this expression in equation 88, one obtains after integrating over the phase

$$\Delta\sigma_x/\sigma_x = \frac{3}{1024}(m \cdot l \cdot \beta)^2 \sigma_x^2/(sin(3\pi Q))^2 \quad (95)$$

Comparing this emittance dilution with the one obtained after a single passage through a sextupole field in the transferline, one can see that the distortions, which are the result of averaging the nonlinear effect over many revolutions in the accelerator is much weaker than the effect of a single sextupole in the transfer line which is seen only once by the beam. However, the resonance dominator indicates that a mismatch due to nonlinear distortion can be important if the beam is extracted or injected with tunes near a low order resonance.

I   Comparison of Sources of Emittance Dilution upon Transfer using the HERA Proton Ring as an Example.

In this section, typical injection errors of HERA will be analyzed and evaluated in order to demonstrate which sources of emittance dilutions are the most important ones. The relevant parameters of HERA are:
Emittance at injection $\varepsilon = 0.125 \cdot 10^{-6} m$.
Typical β-function value in a focusing quadrupole in the regular arc $\beta = 90m$.
Dispersion function at a similar position $D = 2m$.
RMS momentum spread $\sigma_\epsilon = 10^{-3}$.

When protons are injected into HERA, the trajectory of the injected beam is carefully adjusted on the closed orbit, in both amplitude and in slope for both oscillation planes. This is performed using short bunch trains for injection, so-called pilot bunches. The precision of the BPM system allows to suppress the injection orbit errors to levels of $0.5mm$. Between injections, the injector accelerator has to cycle and the reproducibility of the injection trajectory on successive injections is also about $\simeq 0.5mm$. Furthermore there is a $\simeq 1\%$ ripple on the flat top of the injection kicker which produces trajectory errors in the order of $0.3mm$. All this causes an injection error of approximately $\Delta x \simeq 1mm$. The corresponding emittance dilution is

$$\Delta \varepsilon_D = \frac{1}{2}\frac{\delta x^2}{\beta} \qquad (96)$$

$$\Delta \varepsilon_D / \varepsilon = 0.079$$

At typical β-beat in a large accelerator is $\Delta\beta/\beta = 10\%$. In HERA the RMS gradient error of the quadrupole magnets is about $\Delta k/k = 10^{-3}$. There are $N = 100$ FODO cells with a maximum β-function of $\beta = 90m$ and the quadrupoles have a strength of $kl = 0.06m^{-1}$. Using these numbers, one can see that the arcs contribute with $\Delta\beta/\beta = \sqrt{N}kl\Delta k/k\beta = 4.8\%$ to the β-beat. If contributions from systematic gradient errors and contributions from the straight section are included, the expected β-beat is in the order of 10%. The pre-accelerator and the beam line in between can be assumed to have similar beam optics distortions which accumulates to a mismatch of $\Delta\beta/\beta = 17\%$. The corresponding emittance dilution due to mismatch is

$$\Delta \varepsilon_Q = \varepsilon \frac{\Delta\beta}{\beta} \qquad (97)$$

$$\Delta \varepsilon_Q / \varepsilon = 0.03$$

The dispersion function is distorted by the dipole fields of the corrector magnets, by closed orbit differences in the quadrupole magnets, and by optical errors. A value of $\Delta D/D = 10\%$ is typical. Contributions from the injector, the beam line and HERA are together approximately 17%. The corresponding emittance growth is

$$\Delta \varepsilon_E = \frac{(\Delta D \sigma_\epsilon)^2}{\beta} \qquad (98)$$

$$\Delta \varepsilon_E / \varepsilon = 0.005$$

Next we look at the effect of coupling. Let us consider the quadrupole tilts in the injection line as a possible source of skew quadrupole field. A pessimistic estimate is to assume that all the $N = 20$ quadrupoles in the beam line would be rotated randomly about their longitudinal axis with an RMS tilt of $\theta_{RMS} = 3mrad$. Then the corresponding emittance growth would be

$$\Delta\varepsilon_{SQ} = \frac{\varepsilon\beta^2}{4}(kl\beta\theta_{RMS})^2 N \tag{99}$$

$$\Delta\varepsilon_{SQ}/\varepsilon = 0.003$$

Beam line magnets do not have to have a field quality as high as magnets in the ring. However, if the sextupole components become too strong emittance dilution might be the result. Assume that there are $N = 20$ magnets with a bend angle of $\theta = 50mrad$ and a sextupole component of $b_3 = 1 \cdot 10^{-3}$ at a radius of $r = 20mm$, the corresponding emittance dilution due to phase space mismatch would be

$$\Delta\varepsilon_S = \frac{3}{2}(b_3\theta/r^2)^2\varepsilon^2\beta^3 N \tag{100}$$

$$\Delta\varepsilon_S/\varepsilon = 0.004$$

These contributions accumulate to a total emittance dilution of

$$\Delta\varepsilon_{total}/\varepsilon = 0.12 \tag{101}$$

which cause a significant reduction of accelerator performance.

These estimates demonstrate that the dominant source of emittance dilution is the mismatch between injection trajectory and closed orbit. The reproducibility of the injection trajectory is obviously not sufficient to preserve the emittance to the level of a few percent. Active damping of injection oscillations will be necessary. The injection damper system needs to provide a damping time of the coherent oscillation which is smaller than the decoherence time which is in the case of HERA at injection roughly 66 ms. With such a damper system, the residual emittance dilution is less than a percent.

Even if injection oscillations can be suppressed sufficiently, the expected contributions from the optics mismatch are still significant. Injection mismatch can be measured by detecting transverse shape oscillations. Quadrupoles in the beam line may then be used to empirically improve the optical matching of the incoming beam with the lattice function at the interaction point.

V EMITTANCE MEASUREMENT

In order to evaluate the emittance of the beam one can measure the transverse (or longitudinal) particle density distribution. Knowing the focusing properties of the accelerator lattice, one can calculate the emittance. Direct measurement of both x and x' is only possible using destructive methods which are better suited for low-energy beam lines rather than high energy rings. A number of devices is in use which will be described here only briefly.

A Wire Scanner

Wire scanners are widely used in proton accelerators. A thin wire (Tungsten or Carbon of a diameter of $1\mu m$) is driven through the beam with a typical speed of $(1-2)m/sec$. The particles circulate many times around the accelerator during a wire scan. The beam particle interacts with the wire. The collision rate depends on the particle density seen by the wire. In the collisions hadronic showers, secondary emitted electrons or scattered protons are produced. They can be measured downstream of the wire as a function of wire position and speed which gives the density distribution. Furthermore, the electron current which flows to replace secondary emitted electrons from the wire can also be measured to detect the density distribution. The emittance growth due to interaction of the beam with the wire can be neglected for a single scan. For example, in the HERA proton ring at 920GeV, for a $1\mu m$ Carbon wire, the emittace growth due to one scan is less than $\Delta\varepsilon/\varepsilon = 10^{-4}$. The resolution of the measurement is in the order of microns which is sufficient. A problem are that wires may brake at hight beam intensity. Repair requires a partial venting of the beam vacuum system. The principle of a wire scanner is sketched in figure 8

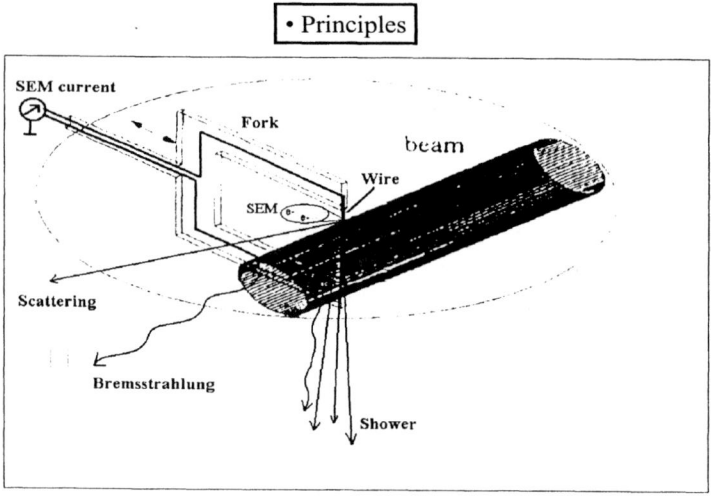

FIGURE 8. Sketch of a wires canner monitor which indicates the various detection methods (courtesy of K. Wittenburg, DESY)

B Restgas Ionization Monitor

For low beam densities, a rest gas ionization monitor provides quasi continuous monitoring of the beam emittance. This monitor uses the effect that rest gas particles get ionized by the beam. An electrical field is then applied transversely to the beam direction such that the ions get accelerated towards the cathode where they can be detected. The motion of the ion perpendicular to the electrical field is suppressed by a weak magnetic field in the same direction. Both the electric and the magnetic fields are too weak to affect the beam. The ion distribution at the cathode is a measure of the particle distribution in the beam. Rest gas ionization monitors can be made very sensitive and can detect very weak proton beam with only 10^6 protons. If the density of the beam becomes large however, the ion distribution is distorted by the space charge forces inside the beam and the measurement eventually becomes dominated by this effect. A nice feature of this monitor is that it can have a good time resolution on the order of a few tens of microseconds. This monitor is therefore suited to observe transient phenomena like shape oscillations.

C Fluorescence Monitor

A fluorescence monitor makes use of the fluorescence light which is emitted by excited atoms and molecules after interaction with the beam particles. In order to obtain a useful signal, a low gas pressure in the order of $10^{-4} mbar$ is necessary which requires a complicated vacuum system around this monitor. In order to achieve the low gas pressure, a controlled gas inlet is necessary. Strong vacuum pumps around the monitor restore the vacuum outside the monitor area so that there is only a local pressure bump. The light emitted by the gas particles in the beam produces a picture of the beam density projection in the plane of observation. A CCD-camera can be used to record the beam profile. The bandwidth of the monitor is given by the electronic camera. Therefore this monitor is well suited for continuous monitoring of the emittance and for investigation of dynamic effects. In particular, this monitor would be suited to detect shape oscillations at injection, which could be useful for empirical tuning of the optical mismatch at injection. Another advantage of the monitor is that space charge forces do not disturb the profile very much, since the photons are emitted in a time short compared to the time the gas particles need to move significantly under the influence of electrical forces.

D Emittance Measurement using Sextupoles and Beam Position Monitors

The transverse emittance of the beam can be measured also by using the beam position monitor (BPM) system. This has been proposed by Brinkmann [2]. Suppose you have two strong sextupole magnets spaced by a β-tron phase advance of $\pi/2$. The beam optical parameters should be identical at the two magnets. In the middle, at a phase of $\pi/4$, a BPM is needed. Outside the sextupole bump, there should be no effect since the kicks of the two sextupoles should cancel exactly for all the particles in the beam. Before the sextupoles are turned on, the beam orbit must be carefully centered in both sextupoles such that the orbit is unchanged by turning them on. For every particle in a beam, we find after the first sextupole (in thin lens approximation) the following changes in the coordinates:

$$\begin{aligned} x &\to x \\ x' &\to x' + \tfrac{1}{2}mlx^2 \end{aligned} \qquad (102)$$

with $m = (e/p) \cdot d^2 B_y / dx^2$ (p is the particle momentum e is the elementary charge, l is the length of the sextuple and B_y is the vertical magnetic field on the x-axis). These coordinates transform at the BPM into

$$\begin{pmatrix} 0 & \beta \\ -1/\beta & 0 \end{pmatrix} \begin{pmatrix} x \\ x' + \tfrac{1}{2}mlx^2 \end{pmatrix} = \begin{pmatrix} x'\beta + \tfrac{1}{2}ml\beta x^2 \\ -x/\beta \end{pmatrix} \qquad (103)$$

Assume, that the beam has a Gaussian distribution at the position of the first sextupole. The difference orbit for sextupoles on and sextupoles off measured by the BPM in between the sextupoles is

$$\Delta x_{BPM} = \int dx' \int dx \rho(x,x')(x'\beta + \frac{1}{2}ml\beta x^2) \qquad (104)$$

The first term of this integral is zero for a symmetric distribution. The second term is

$$\Delta x_{BPM} = \frac{1}{2}ml\beta \int xx^2 \rho(x) = \frac{1}{2}ml\beta\sigma^2 = \frac{1}{2}ml\beta^2\varepsilon \qquad (105)$$

If one inserts numbers into these expressions (for example using HERA parameters: $B_z = 1\ Tesla, l = 0.5m, r_{pole} = 30mm, \beta = 100m, pc = 40GeV, \varepsilon_N = 5 \cdot 10^{-6} mrad\ mm$), one obtains

$$\Delta x_{BPM} = 2.6mm \qquad (106)$$

which is well above the BPM resolution of $0.1mm$. Thus it is possible to measure the beam size with a precision of about 2% with this method. The measurement principle is sketched in figure 9

VI ACKNOWLEDGMENT

The author is indebted to Dr. G. Hoffstätter, DESY for enlightening discussions concerning distribution functions and adiabatic damping.

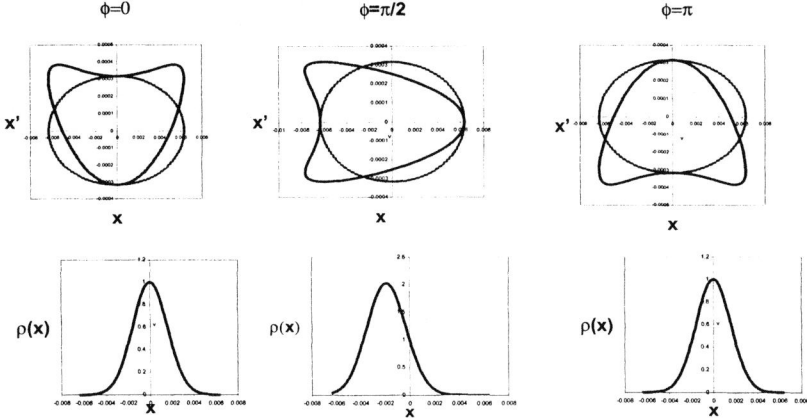

FIGURE 9. Two strong sextupoles with a betatron phase difference of 180 degree between them have a beam position monitor half way in between. Particles receive a kick proportional to the square of their distance form the axis x from the first sextupole. The so modified distribution in x' rotates in phase space and turns at the beam position monitor into a modified distribution in x with a dipole moment after 90 degree of phase rotation. The second sextupole cancels the effect of the first one.

REFERENCES

1. P.M. Lapostolle, IEEE Trans.Nucl. Sci. NS-18, No3, 1101 (1971)
2. R.Brinkmann, private communication
3. A. Chao, private communication
4. G. Hoffstaetter, private communication (2000)
5. H.Goldstein,Classical Mechanics, Addison-Wesley Publ.Co. (1959)
6. T.O. Raubenheimer, F.-J. Decker, J.T .Seeman, Beam Distribution after Filamentation, Stanford Linear Accelerator Center, Internal Report, SLAC-PUB-95-6850 (1995)
7. G. Ripken and F. Willeke, Method of Beam Optics, AIP Conf.Proc.184, VolII (1989) p 758ff
8. J. Ellison, private communication (1997)

Emittance Preservation in Linear Accelerators

M. Minty*

*DESY, Notkestrasse 85, 22306 Hamburg, Germany

Abstract.
In linear colliders preservation of the phase space density of charged particles during acceleration to high energies is essential. In practice, the electromagnetic fields which govern the beam transport may not be sufficiently well understood. This may arise, for example, from magnet and structure alignment and/or manufacturing errors, time-varying electromagnetic fields due to component vibration or imperfect regulation, or at high beam currents, from beam-induced fields. These inadequacies may be overcome using measurements of the beam response. In this report we review such methods for preserving single-bunch beam emittances with experimental results from the Stanford Linear Collider.

INTRODUCTION

Minimizing dilutions to the beam's phase space volume, or emittance, over extended periods is vital for ensuring the highest possible luminosity at colliders. The Stanford Linear Collider (SLC) is the first high-energy linear collider and much practical experience on emittance preservation comes from operating this accelerator. Since considerable understanding of beam dynamics in linear colliders was motivated, stimulated, and triggered by observations from the SLC, it may be helpful if one has been made familiar with this accelerator. For this a brief overview of the SLC is given.

The geometry of the SLC is shown in Fig. 1. The 3 km linac accelerates three bunches simultaneously - a positron followed by two electron bunches. The constant gradient structures are driven by klystrons at 2856 MHz with a 120 Hz repetition frequency. Typical beam parameters include the bunch populations of 4×10^{10} particles, the bunch lengths $\sigma_z \sim 1$ mm, the energy spreads $\delta < 0.1\%$, and normalized transverse emittances of $\gamma\epsilon_x = (4-5) \times 10^{-5}$ m-rad horizontally and $\gamma\epsilon_y = (0.5 - 1.0) \times 10^{-5}$ m-rad vertically measured at the end of the linac at 50 GeV.

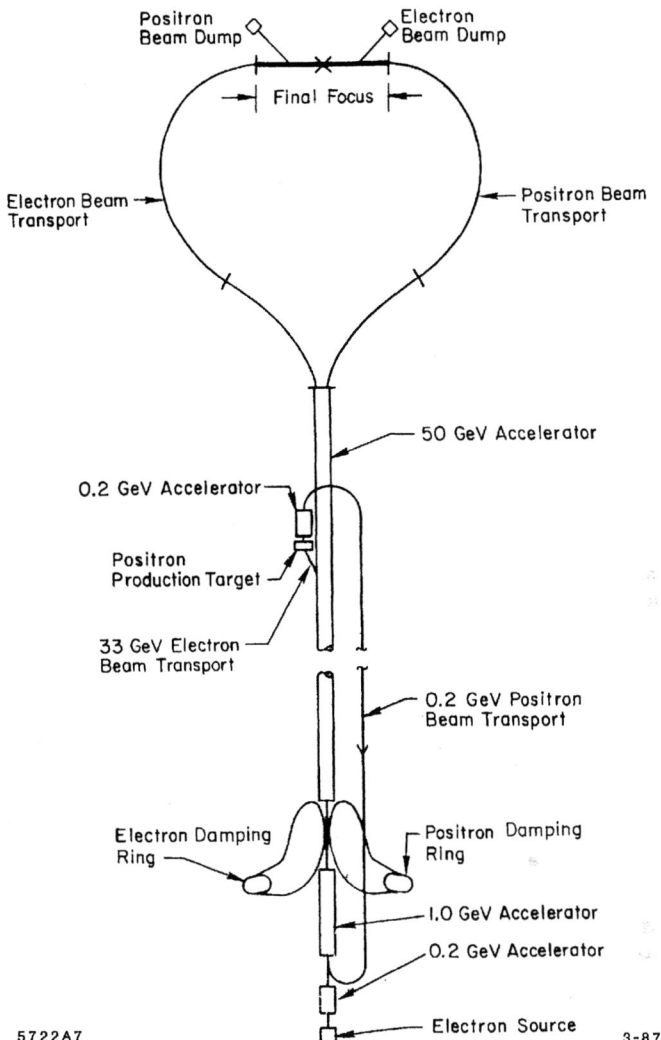

FIGURE 1. Overview of the SLC. Two electron bunches are produced at the source, accelerated to 1.2 GeV, and injected into the electron damping ring where they are damped by emission of synchrotron radiation and rf acceleration. The two bunches are then extracted, compressed, and then accelerated to high energy in the main linac. The leading electron bunch is extracted at 20 GeV and deflected onto to the positron production target. The trailing bunch is accelerated to 50 GeV and transported to the interaction point through the collider arc and final focus. The positron beam is transported at low energy through the beam transport line, accelerated to 1.2 GeV, and radiation damped in the positron damping ring. The positron bunch is then extracted, compressed, accelerated to 50 GeV (leading the 2 electron bunches), transported through the arc, and brought into collision with the electrons at the interaction point

Shown in Fig. 2 are measurements [1] of the normalized horizontal emittance made after several years of SLC operation measured at various locations along the collider as a function of bunch charge. By this time many emittance enlargement effects had been reduced significantly such as optical mismatches in the ring-to-linac transport line [2–4], dilutions arising from quadrupole and accelerator misalignments [5–9], and coupling generated in the nonplanar collider arcs [10]. In addition, both long and short term variations in the beam properties at injection [11] and in the main linac were controlled using orbit feedback [12–20] and BNS damping. At injection, fluctuations and drift of the transverse beam position and angle were regulated using launch feedback loops [13,14] while the injection phase was held nominally constant by maintaining the phase of the injected beam using analog feedback between the rf systems of the linac and damping ring [21].

From Fig. 2, emittance growth in the ring-to-linac transport line was dominated by chromatic effects [22] which were exaggerated by high-current bunch lengthening in the damping rings [23–29] and bunch compression [3,4,22]. In the main linac the emittance dilutions were governed by both wakefield and chromatic effects which became increasingly important at high beam currents [8,22,30–33]. Minimizing and stabilizing such current-dependent dilutions proved essential for achieving routine, high-luminosity operation. By the end of the SLC program, the emittance growth from the (uncoupled) damping rings to the end of the linac was routinely maintained to less than (15-20)% at bunch charges of 4×10^{10} both horizontally and vertically compared with nearly a factor of 3 increase seen in Fig. 2.

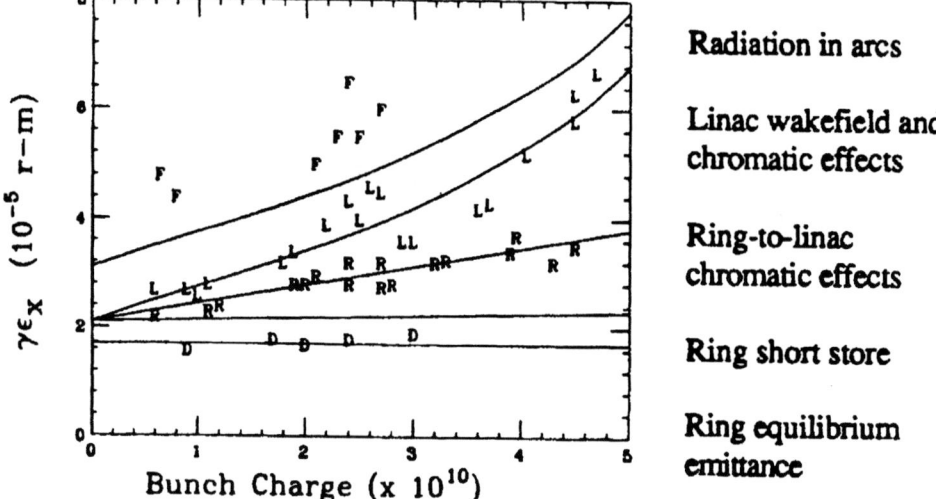

FIGURE 2. Emittance growth from the SLC linac entrance to the final focus as a function of bunch population. Measurement locations are denoted by D at the exit to the damping ring, R at the linac entrance, L at the end of the linac, and F in the final focus. Courtesy J. Seeman (2000).

In this report emittance preservation during acceleration in the linac proper will be discussed. In sections 2 and 3 basic concepts in transverse and longitudinal emittance preservation, respectively, will be reviewed. In section 4 will be described commonly applied steering algorithms. A conclusion is given in section 5. In appendix A the emittance is defined. In appendix B emittance measurement procedures are described.

TRANSVERSE EMITTANCE PRESERVATION

The transverse dynamics of a single particle in a linac is governed by three variables: the transverse wakefield $W_\perp$, acceleration $\frac{dE}{ds}$, and k the lattice focussing. In the following we will consider special cases of the general equation of motion [34,35] given by[1]

$$\frac{d}{ds}\left[E(s)\frac{dx(z,s)}{ds}\right] + E(s)k^2(s)x(z,s) = e^2 \int_z^\infty \rho(z')W_\perp(z'-z)x(z',s)dz', \quad (2)$$

where s represents the longitudinal coordinate and z gives the longitudinal coordinate relative to the bunch center. Here $\rho(z)$ is the longitudinal distribution function of the beam which, in the approximation of zero transverse dimension as in reference [34], is equal to the line density of particles in the bunch. This approximation is valid provided that the beam sizes are much less than that of the vacuum chamber so that the transverse wakefield may be taken to be uniform across the bunch's transverse dimensions. For a particle at position z the transverse wakefield is evaluated over the preceeding particles only. In the special cases to be discussed below, the lattice strength $k(s)$ will be assumed to be smoothly varying rather than consisting of discrete quadrupoles.

Case i: $W_\perp = 0$ - zero current limit
$\quad E = E_0$, the beam energy at injection, or $\frac{dE}{ds} = 0$ - no acceleration
$\quad k = k_0$ - constant gradient
In this case with initial conditions at injection $x(0) = \hat{x}$ and $x'(0) = 0$,

$$\frac{d^2x}{ds^2} + k_0^2 x = 0 \quad (3)$$

with solution $x(s) = \hat{x} \cos k_0 s$ which represents betatron oscillations of peak amplitude $\hat{x}$ about a reference trajectory $x_c + x_\eta$. In general, the deviation of the particle trajectory is given by

---

[1] the original notation of refs. [34] and [35] has been modified slightly noting

$$\gamma(s) = \frac{E(s)}{mc^2} \text{ and } k(s) = \frac{2\pi}{\lambda(s)}, \quad (1)$$

where γ is the Lorentz factor, E is the beam energy, mc^2 is the particle rest mass, $\lambda(s)$ is the instantaneous wavelength of betatron focusing, and $e^2 = r_0 mc^2$, where r_0 is the classical electron radius.

$$x = x_c + x_\beta + x_\eta$$
$$= x_c + x_\beta + \eta\delta. \tag{4}$$

For notational simplicity, the solutions for x in Eq. 3 as in the remainder of this chapter will refer to x_β (the subscript will be omitted). In Eq. 4 x_c represents the central trajectory which is defined as the mean orbit that an on-energy particle would follow through the lattice. Ideally this term is zero if the orbit is flat (for a planar linac) passing through perfectly aligned structures and magnets. Due to misalignments x_c is in practice not perfectly linear. The term $x_\eta = \eta\delta$ gives the additive contribution arising from an energy deviation of the particle, where η is the dispersion, and δ is the relative energy deviation of the particle from that of the design particle.

Case ii: $W_\perp = 0$ - zero current limit
$E = E_0(1 + G \cdot s)$ - linear acceleration with gradient G
$k = k_0$ - constant gradient
The equation of motion is

$$\frac{d^2x}{ds^2} + \frac{1}{E}\left(\frac{dE}{ds}\right)\frac{dx}{ds} + k_0^2 x = 0, \tag{5}$$

with solution for $x(0) = \hat{x}$ and $x'(0) = 0$

$$x(s) = \hat{x}\sqrt{\frac{E_0}{E(s)}} \cos k_0 s. \tag{6}$$

This result shows that the betatron oscillations damp as $\frac{1}{\sqrt{E}}$.

Damping of the beam's transverse dimensions in a linear accelerator is conceptually easy to visualize. Sketched in Fig. 3a is the particle momentum at energy E_0 decomposed vectorially into its transverse momentum $p_\perp$ and longitudinal momentum $p_\parallel$. The angle x' is

$$x' = \tan^{-1}\left(\frac{p_\perp}{p_\parallel}\right) \approx \frac{p_\perp}{p_\parallel} \text{ for } p_\perp \ll p_\parallel. \tag{7}$$

In Fig. 3b the decomposed momentum after acceleration by $p'\Delta s$ is shown. The particles' transverse momentum $p_\perp$ is unchanged by the acceleration and

$$x'(s + \Delta s) \approx \frac{p_\perp}{p_\parallel(1 + \frac{p_\parallel'}{p_\parallel}\Delta s)} \approx \frac{p_\perp}{p_\parallel}(1 - \frac{p_\parallel'}{p_\parallel}\Delta s)$$
$$\approx x'(s)(1 - \frac{p_\parallel'}{p_\parallel}\Delta s). \tag{8}$$

Differentiating gives

$$x''(s) \equiv \frac{x'(s+\Delta s) - x'(s)}{\Delta s} \approx -\frac{p_\parallel'}{p_\parallel} x'(s) \text{ or}$$

$$x'' + \frac{E'(s)}{E_0} x'(s) = 0 \qquad (9)$$

which is Eq. 5 with $k = 0$.

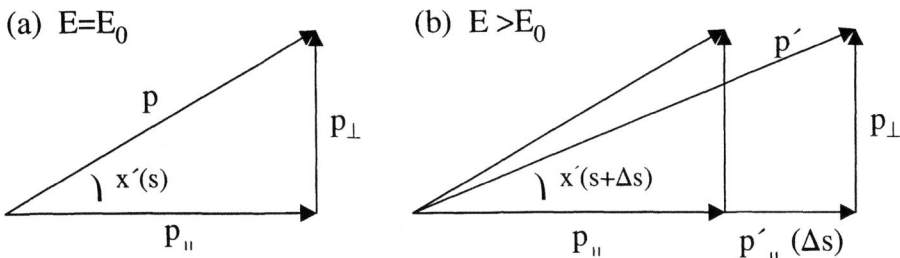

FIGURE 3. Illustration of damping of transverse oscillations in linear accelerators.

From appendix A the final emittance scales as

$$\epsilon \sim <x^2>^{\frac{1}{2}} \sim \frac{E_0}{E}\epsilon_0 = \frac{\gamma_0}{\gamma}\epsilon_0, \qquad (10)$$

where γ is again the Lorentz factor. That is, the emittance 'damps' as $\frac{1}{\gamma}$. For this reason, in practice one often expresses emittances as $\gamma\epsilon$ along the linac as it is this quantity which is conserved in the absence of dissipative forces.

So far the transverse motion of only a single particle has been considered. For a bunch consisting of multiple particles, the situation is more complicated since particles of different energy are focussed differently. The lattice focussing depends on the beam energy as

$$k^2 = \frac{ec}{E\beta}\frac{\partial B_z}{\partial x}, \qquad (11)$$

where $\beta\gamma = \sqrt{\gamma^2 - 1}$ and $\frac{\partial B_z}{\partial x}$ describes the quadrupole magnetic fields. The motion of the bunch centroid (defined as the position of the mean of the bunch charge distribution) after a net displacement of the bunch may therefore not be damped as $\frac{1}{\sqrt{E}}$. That is, a macroparticle approximation for centroid motion breaks down if the bunch has an internal energy spread.

The measured horizontal beam centroid motion is shown as a function of position along the SLC linac [1,8] in Fig. 4 for different bunch populations. At low current the centroid motion decays faster than as $\frac{1}{\sqrt{E}}$. This is due to the bunch energy spread which causes a spread in the phase advance of the particles within the bunch. At 2.5×10^{10} particles the longitudinal profile had been optimized and the centroid motion is seen to decay at least initially as $\frac{1}{\sqrt{E}}$ like that of a single particle. Towards the end of the linac and at higher bunch charge, the transverse dynamics is more complicated and $W_\perp \neq 0$ must be considered.

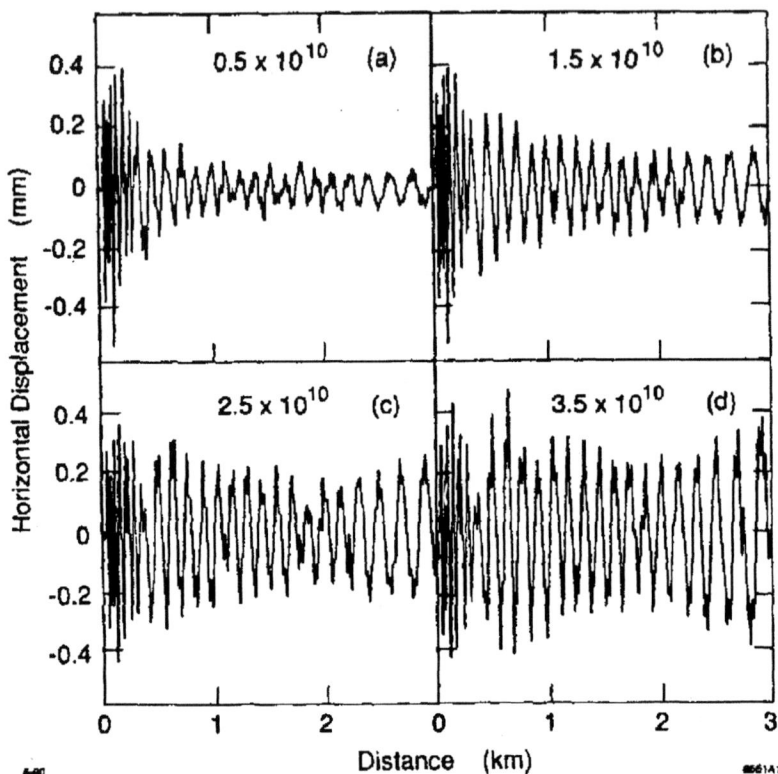

FIGURE 4. Measured horizontal trajectories versus charge at the SLC under identical injection conditions with the overall linac phase adjusted to minimize the final energy spread. Courtesy J. Seeman (2000).

Case iii: $W_\perp = W_\perp' z$ - wakefield linear along the bunch's longitudinal extent
$E = E_0$ or $\frac{dE}{ds} = 0$ - no acceleration
$k = k_0$ - constant gradient
The equation of motion is

$$x''(z,s) + k_0^2 x(z,s) = \frac{e^2}{E(s)} \int_z^\infty \rho(z') W_\perp(z'-z) x(z',s) dz'. \tag{12}$$

The essential features of the general solution [35] may be visualized using a simplified macroparticle model [1,8]. Here, the bunch is divided into three slices each having a rectangular distribution. The head (h) has charge $\frac{N}{4}$, the core (c) has charge $\frac{N}{2}$, and the tail (t) charge $\frac{N}{4}$. The core is separated from the neighboring slices by the rms bunch length σ_z. The equations of motion for each macroparticle are

$$x_h'' + k_0^2 x_h = 0 \quad \text{by causality the leading slice has no driving term}$$
$$x_c'' + k_0^2 x_c = B x_h \quad \text{the driving term is given by the effect of the head on the core}$$
$$x_t'' + k_0^2 x_t = 2B x_h + 2B x_c, \tag{13}$$

where in the last equation the first term has a factor of 2 representing the $2\sigma_z$ displacement of the head relative to the core, and the second term has a factor of 2 since the core has twice the charge of the head. The factor B is

$$B = \frac{e^2}{E}(\frac{N}{4}) W_\perp \sigma_z. \tag{14}$$

The solutions for Eq. 13 with initiappeal condition $x'(s=0) = k_0 \hat{x}$ are

$$x_h = \hat{x} \sin k_0 s$$
$$x_c = \hat{x}[(1 + \frac{B}{4k_0^2}) \sin k_0 s - \frac{B}{2k_0} s \cos k_0 s]$$
$$x_t = \hat{x}[(1 + \frac{2B}{k_0^2}) \sin k_0 s - \frac{2B}{k_0} s \cos k_0 s - \frac{B^2}{4k_0^2} s^2 \sin k_0 s]. \tag{15}$$

In the 3-slice macroparticle model the amplitudes of the head, tail, and core are all linear in the initial displacement $\hat{x}$. Each slice adds an additional power of (Bs) which is proportional to the product $NW_\perp s$ which suggests an exponential growth of the tail of the beam in the limit of many slices [8]. Shown in Fig. 5 are profile monitor measurements and trajectories for three different initial vertical displacements (settings of a vertical dipole corrector magnet) from the SLC [8]. The middle plots correspond to an optimized orbit. In the top plots the beam is kicked in one direction and in the bottom plot in the other direction. The increase in vertical amplitude towards the tail of the bunch shows the intrabunch particle displacements due to the transverse wakefields as described in Eq. 15. Also evident from this measurement is a position-energy correlation[2]. The observed decrease in energy along the bunch depends subtely on the cancellation between the rf slope and the slope of the accelerating rf. This will be discussed further in the next section.

While the emittance of a slice of the beam in Fig. 5 seems nearly preserved, for experiments the *projected emittance* (seen by projecting the distribution onto the y axis) is important and is observed here to be larger than the slice emittances.

---

[2] these measurements were obtained by deflecting the beam onto a fluorescent screen using a kicker magnet located in a dispersive region (in the collider arcs) so that the measured horizontal position indicates an energy deviation; i.e. the profile monitor shows $y(E)$ [38].

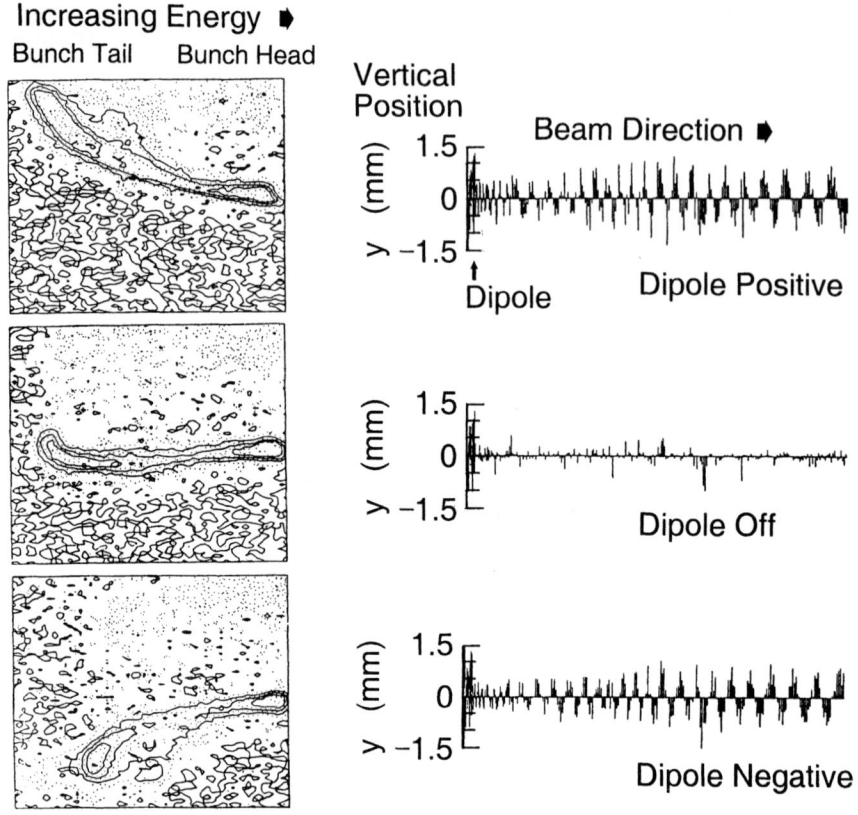

FIGURE 5. Profile monitor measurements in a region of nonzero dispersion after the end of the SLC linac and vertical centroid trajectories with a positive perturbation to the particle orbit (top), under nominal conditions (middle), and with a negative perturbation (bottom). Courtesy J. Seeman (2000).

Shown in Fig. 6 are now the transverse beam profiles measured at the end of the SLC linac for various initial beam displacements [37]. These measurements were made [60] by deflecting the beam using fast kicker magnets located within the linac so that the true transverse profile $y(x)$ is represented. Based on the above analysis we may interpret the faint tail seen with large amplitude excitations as the off-energy and off-axis tail generated by the transverse wakefields.

Case iv: $W_\perp = W_\perp' z = W_\perp z/l$
$\frac{dE}{ds} \neq 0$ or $E = E_0(1 + Gs)$

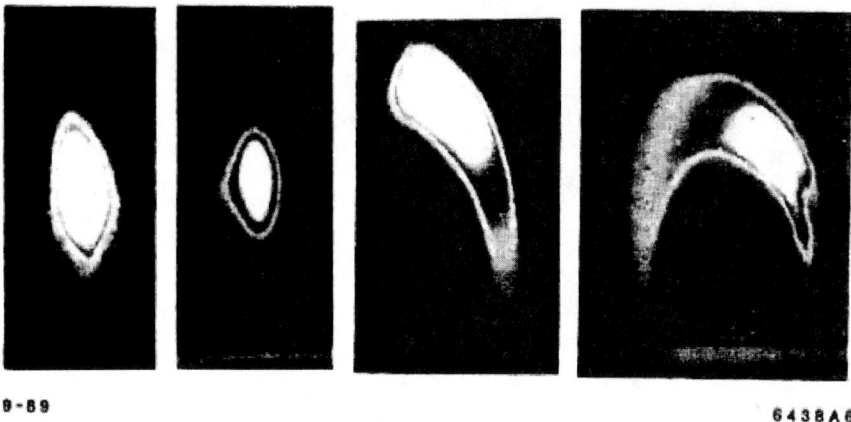

FIGURE 6. Measured beam profiles demonstrating emittance growth due to wakefields as a function of increasing oscillation amplitude. From left to right the amplitudes in the applied horizontal trajectory displacement are 0 mm, 0.2 mm, 0.5 mm, and 1.0 mm. The single-bunch charge was 2×10^{10} electrons. Courtesy J. Seeman (2000).

(a) k **adiabatic**[3]

The solution to the general equation of motion has no closed form expression. It is obtained [35] by expanding the solution $x(z,s)$ in a power series and solving recursively. In the asymptotic limit of strong wakefields ($|\eta| \gg 1$), the peak-to-initial amplitude given at the end of the linac of length L is [39]

$$\frac{x(z,L)}{x_0} = \sqrt{\frac{E_0 k_0}{E(s)k(s)}} \frac{\eta^{-\frac{1}{6}}}{\sqrt{6\pi}} e^{\frac{3\sqrt{3}}{4}\eta^{\frac{1}{3}}}, \qquad (16)$$

where

$$\eta(z) = \frac{eNW_\perp' z^2}{\sigma_z} \int_0^L \frac{ds}{E(s)k(s)} \gg 1. \qquad (17)$$

The last 2 equations show the lagging particle trajectory increasing exponentially with the transverse wakefield. Based on such observations first made at the SLC [40]-[42], this phenomenon has come to be referred to as *beam breakup* [40]-[43].

Another example of large transverse beam tails is shown [8] in Fig. 7. In this measurement the bunch was intentionally lengthened to about $2.5\sigma_z$ or 2.5 mm and made to oscillate both horizontally and vertically to sample the transverse wakefields. The profile shows $y(x)$. The head of the beam is at the lower right while the tail of the beam contributes significantly to larger projected beam emittances

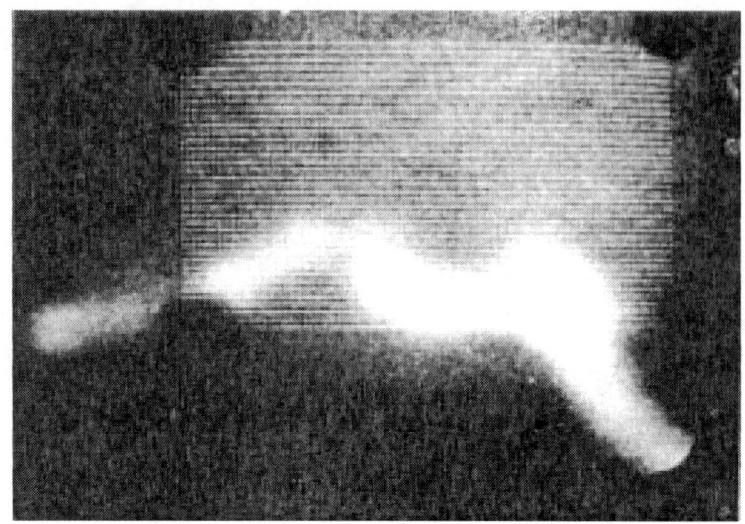

FIGURE 7. Measured transverse profile of a long bunch showing large transerverse displacements of the bunch tail. Courtesy J. Seeman (2000).

in both transverse planes.

(b) k **tailored**

The cure for the *beam breakup instability* may be motivated as the follows. With the same focussing function k experienced by all the particles within a bunch, a perturbation at the head of the bunch due to the transverse wakefield may resonantly drive particles in the tail of the bunch. By making the focussing function different across the bunch, such resonant build-up can be avoided. This is precisely the technique proposed by Balakin, Novokhatsky, and Smirnov known today as *BNS damping* [44] in their honor. Using the 3-slice model of ref. [1,8], the head of the bunch is focussed with $k = k_0 + \alpha$, the core with $k = k_0$, and the tail with $k = k_0 - \eta$. Requiring [1,44] that the head and core follow the same trajectories, the solutions for α and η are

$$\alpha = e\frac{N W_\perp \sigma_z}{4 \; E_0 k_0}, \quad \text{core follows head, and } \eta = 4\alpha. \tag{18}$$

At the SLC the variation in k was achieved by back-phasing the first part of the linac (i.e. the bunch preceeds in time the rf wave) and, to restore a small energy spread at the linac end, by forward phasing the remainder of the linac. This introduced a

---

[3] k is taken to scale with energy such that the instantaneous wavelength $\lambda(s) = 2\pi/k$ is constant; i.e. $\lambda(s) = \lambda_0$.

correlated energy spread within the bunch as shown [45] in Fig. 8 which is just as desired to achieve BNS damping on a global scale. The optimum choice of phases is current-dependent and was determined by measuring the peak amplitude response at the end of the linac following a perturbation applied after injection.

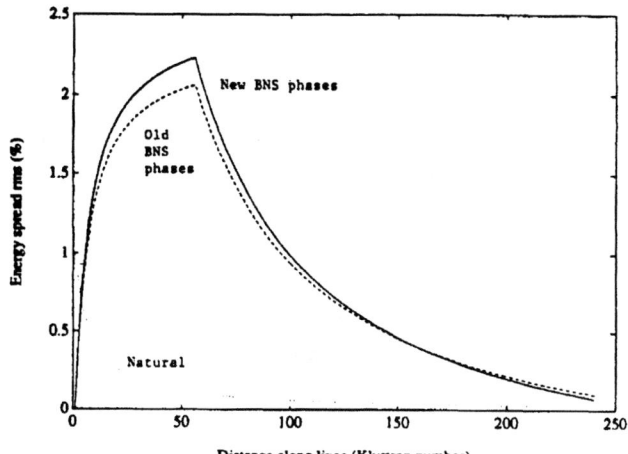

FIGURE 8. Estimated energy spread at the end of the linac (i.e. a projection along the horizontal axis of Fig. 5) for different complements of linac klystron rf phases. The two different curves are optimized for two different bunch currents. Courtesy F.J. Decker (2000).

Measured transverse centroid oscillations both with and without BNS damping invoked are shown [1] in Fig. 9 following an intentionally applied initial displacement early in the SLC linac. The measured centroid displacement normalized to initial kick amplitude was reduced by about a factor of 10 with BNS damping implemented.

(c) k exact

With BNS damping alone, the projected 6-dimensional beam emittance may assume large values along the linac where the beam energy spread is large. Subsequent emittance dilution may also occur if the dispersion is not perfectly corrected. Alternatively, a general condition on the focussing k may be determined. Substituting $x(s) = \hat{x}_0 \cos(k_0 s + \phi_0)$ in Eq. 2, leads to [44], [46]

$$k_0^2 = k^2(z,s) + \frac{r_0}{\gamma(z,s)} \int_z^\infty \rho(z') W_\perp(z'-z) x(z',s) dz'. \tag{19}$$

If $k(z,s)$ can be adjusted to exactly compensate the second term in Eq. 19 for all particles within the bunch, then the particles within the bunch follow the same trajectory and experience the same focusing. This allows the small projected 6-dimensional beam emittance to be maintained over the entire length of the linac.

The condition given in Eq. 19 is refered to as *auto-phasing* and is hard to realize in practice [8]. The adjustable parameters are the lattice focussing k, the beam

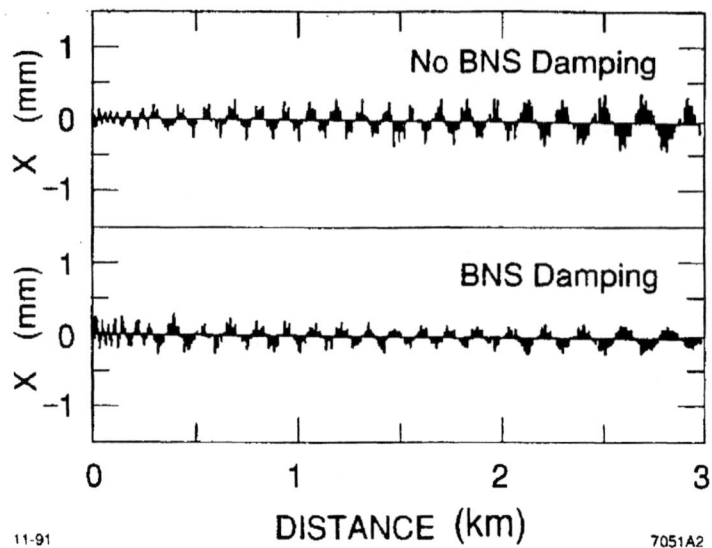

FIGURE 9. Measured horizontal trajectories obtained under nominally identical conditions without BNS damping (top) and with BNS damping (bottom) at 2×10^{10} particles. Courtesy J. Seeman (2000).

energy γ, and the longitudinal profile ρ. Of these, the easiest to control is the longitudinal profile, which is the main topic of the next section.

LONGITUDINAL EMITTANCE PRESERVATION

The energy of the particles with longitudinal density distribution $\rho(z)$ is a sum of the injected beam energy E_0, the energy gained in acceleration from each klystron ΔE_i, and the losses from longitudinal wakefields $W_\parallel$ [31,46]:

$$E(z) = E_0 + \sum_{i=1}^{N_{klys}} \left[\Delta E_i \cos(\phi_i + \phi(z)) + \Delta s_i \int_z^\infty W^i{}_\parallel(z' - z)\rho(z')dz'\right], \quad (20)$$

where ϕ_i is the klystron phase (i.e. the arrival time of the beam with respect to the crest of the rf), $\phi(z) = \frac{2\pi z}{\lambda_{rf}}$, $\lambda_{rf} = \frac{c}{f_{rf}}$ is the rf wavelength at frequency f_{rf}, and Δs gives the distance between klystrons.

The energy spread of the bunch σ_E is obtained by averaging over the particle distribution after subtracting out the mean energy $<E>$ of the bunch. Normalized to the mean energy

$$\frac{\sigma_E}{E} = \frac{1}{<E>}\left[\int_{-\infty}^\infty (E(z) - <E>)^2 \rho(z) dz\right]^{\frac{1}{2}}, \quad (21)$$

where the mean energy of the bunch is

$$<E> = \int_{-\infty}^{\infty} E(z)\rho(z)dz. \tag{22}$$

In the low-current limit ($\rho(x) \sim 0$, $W_\perp = 0$, $W_\parallel = 0$), the beam does not take away any energy from the accelerating structures and the beam is placed on the crest of the rf wave to achieve both maximum acceleration and minimum energy spread within the bunch. At higher beam currents while invoking BNS damping ($\rho(x) \neq 0$, $W_\perp \neq 0$) with $W_\parallel = 0$, the klystrons in the first part of the linac are phased to impart relatively higher energy to the head of the bunch while in the later part of the linac the energy spread is restored (recall Fig. 8).

At even higher bunch currents, ($\rho(x) \neq 0$, $W_\perp \neq 0$) with $W_\parallel \neq 0$ one must carefully balance the second and third terms in Eq. 20; that is cancel the energy variation along the bunch arising from the slope of the rf and that from the longitudinal wakefield. Shown in Fig. 10 are sketches illustrating such cancellation. The effective rf gain representing the vector sum of all accelerating stations is plotted versus time together with the projection of the charge distribution which shows the resultant energy spread of the bunch. At low current, a bunch placed on crest has minimum energy spread. Off crest, there is a position-energy correlation and the energy spread is increased. At high current, due to longitudinal wakefield, or *beam loading*, a bunch placed on crest has a large energy spread. For short, high intensity bunches, the energy spread may be minimized by placing the beam off-crest as shown. In this case the beam-induced wakefield exactly cancels the slope of the rf across the bunch.

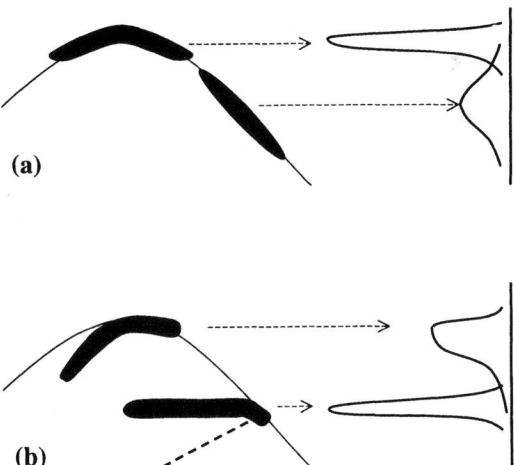

FIGURE 10. Effective energy gain and energy spread for low (a) and high (b) current bunches illustrating optimum klystron phasing for minimum energy spread.

Measurements of the beam energy spread at the end of the linac versus klystron phase may be used to determine the optimum phase settings. An example [47] is given in Fig. 11. The case of highest energy gain, case (f), corresponds to Fig. 10a for the on-crest bunch. As the rf phase was varied, an optimum condition, case (b) was attained corresponding to Fig. 10b albeit with a long low-energy tail.

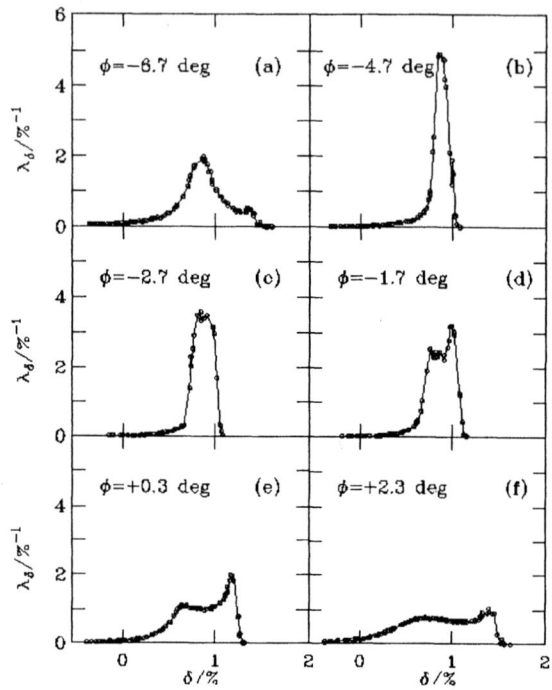

FIGURE 11. Measured energy spread at the end of the SLC linac for various relative overall linac phase. The vertical axis shows relative intensity of the energy distribution of the beam while the horizontal axis shows the relative beam energy spread. Maximum accelerating gradient, or zero absolute phase with respect to the rf crest, is seen in subplot (f). Courtesy K. Bane (2000).

Depending on the bunch length and charge the cancellation may be imperfect resulting in a nonlinear energy variation across the bunch which, as can be seen in Fig. 11, results in non-Gaussian energy distributions with energy 'tails'. Such energy tails have proven to be highly detrimental to collider performance. Highly motivated to minimize chromatic effects in the downstream final focus systems, commensurate measures are always taken to understand the longitudinal beam dynamics [48–51] and to avoid such tails in the beams' energy distribution. This may be achieved by shortening the bunch using *bunch compression*, by shaping the bunch distribution to modify the wakefield-driven term in Eq. 20 using *bunch shaping* or possibly using a combination in a scheme refered to as *over compression*.

Bunch Compression

To minimize the effects of imperfect cancellation between the accelerating rf and the longitudinal wakefields, the bunch length is made shorter before injection into the main linac using bunch compressors. A schematic view of a bunch compressor is shown in Fig. 12. It consists of a transport line with initially zero dispersion containing an acceleration structure. The bunch first passes the accelerating cavity at the zero crossing (that is, there is no net energy gain averaged over the bunch) which introduces an energy-position correlation along the bunch. The bunch is then transported through bending dipoles. The inherent energy-dependent path length of the transport line is such that particles at the front of the bunch travel a longer distance while particles in the tail travel a shorter distance which produces a compression of the bunch length.

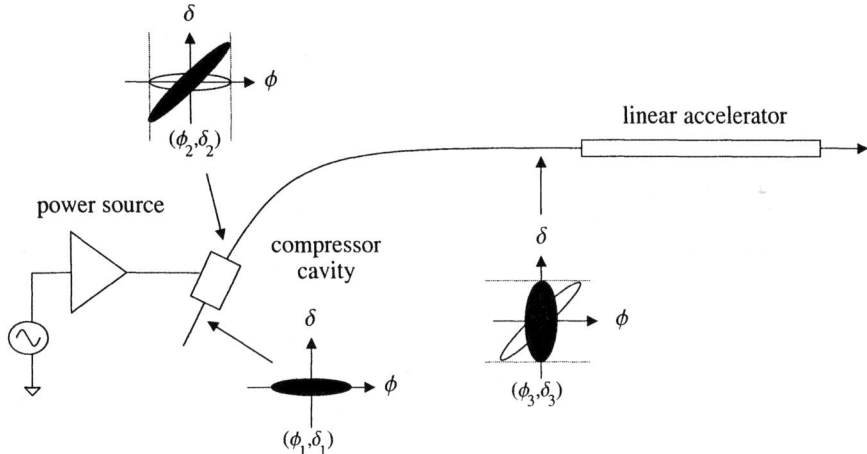

FIGURE 12. Schematic illustration of the bunch compression scheme implemented at the SLC (see also Ref. [50]).

For illustration, the bunch length at the entrance to the linac is evaluated. Initially, the bunch enters the 'compressor cavity' with particle coordinates ($\phi_1 = \frac{\omega z}{c} = \frac{\omega}{c}(\frac{\lambda_1}{2\pi}), \delta_1$). At the exit of the cavity,

$$\phi_2 = \phi_1$$
$$\delta_2 = \delta_1 + \frac{eV}{E} \sin \phi_1 \tag{23}$$

At the entrance to the linac, after passing through the dispersive region,

$$\phi_3 = \phi_2 - \frac{R_{56}\omega}{c}\delta_2$$
$$\delta_3 = \delta_2, \tag{24}$$

where $R_{56} = dz/d\delta$ represents the lattice property which translates energy into longitudinal displacement. Combining Eqs. 23-24,

$$\phi_3 = \phi_1 - R_{56}(\delta_1 + \frac{eV}{E}\sin\phi_1) \approx \phi_1(1 - \frac{R_{56}eV\omega}{Ec}) - R_{56}\delta_1. \quad (25)$$

The bunch length at the entrance to the linac is

$$\sigma_{z3} = \frac{\omega}{c}\sigma_{\phi_3} \equiv \frac{\omega}{c}<\phi_3^2>^{\frac{1}{2}}$$
$$= \frac{\omega}{c}[(1 - R_{56}\frac{eV\omega}{Ec})^2 <\phi_1^2> -2R_{56}(1 - \frac{R_{56}eV\omega}{Ec})<\phi_1\delta_1> +R_{56}^2<\delta_1^2>]^{\frac{1}{2}}$$
$$\approx \sqrt{(1 - \frac{\omega}{c}R_{56}\frac{eV}{E})^2\sigma_{z1}^2 + R_{56}^2\sigma_{\delta1}^2} \approx (1 - \frac{\omega}{c}R_{56}\frac{eV}{E})\sigma_{z1}, \quad (26)$$

assuming $<\phi_1\delta_1> = 0$ (i.e. there is no incoming $E-z$ correlation) and that the incoming energy spread is small so that $<\delta_1^2>$ is negligible. The trade-off between small bunch length and larger energy spread may be evaluated from $\sigma_\delta = <\delta_3^2>^{\frac{1}{2}}$, which is usually not of importance. The above analysis holds provided that the incoming bunch length is small ($\sin\phi_1 \sim \phi_1$) so that only the linear portion of the accelerating voltage of the compressor cavity is seen by the beam.

Bunch Shaping [52]

The delicate process of canceling the effects of the slope of the rf and the energy gradient caused by beam loading may be simplified, in principle, by adjustments to the longitudinal charge distribution $\rho(z)$ in Eq. 20. The energy gained of a particle after traversing an accelerating structure [52] is

$$E(\theta_1) = E_0\cos\theta_1 + \int_0^{(\theta_0-\theta_1)}\rho(\theta')W_{\parallel}(\theta_0 - \theta_1 - \theta')d\theta', \quad (27)$$

where θ_0 is the position of the head of the bunch and θ' varies from 0 at the head of the bunch to $\theta_0 - \theta_1$ where the net energy is evaluated. To reduce the energy spread within the bunch, $eV(\theta_1) = E(\theta_1)$ should be independent of θ_1;

$$\frac{\partial V(\theta_1)}{\partial\theta_1} = 0. \quad (28)$$

Using a realistic model of the longitudinal wakefield of the SLC structures, this equation has been solved for the charge distribution as shown [52] in Fig 13. The horizontal axis shows the phase of individual particles with respect to the head of the bunch (located at phase zero) while the different curves show different phase offsets of the bunch head relative to the crest of the rf. The points marked by 'T' indicate the extent of the bunch for a total of 5×10^{10} particles. For short bunches with large phase offsets (as needed for BNS damping), Fig. 13 shows that a charge distribution peaked towards the head of the bunch is desirable.

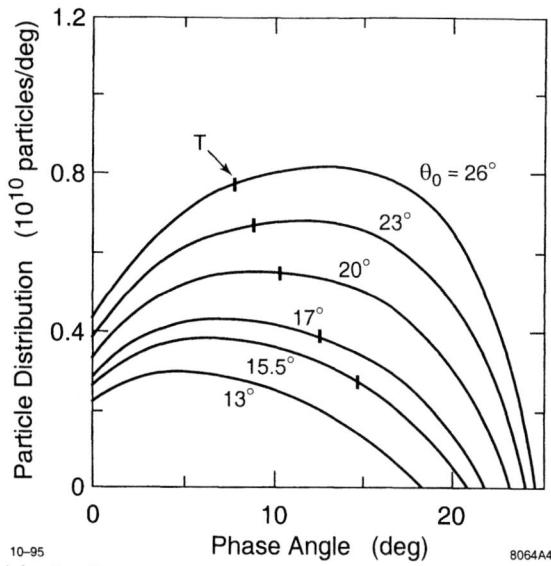

FIGURE 13. Particle distribution, or bunch shape, as a function of phase angle for minimum energy spread. Courtesy G. Loew (2000).

Overcompression

Overcompression [53,49–51] using the bunch compressor was used at the SLC for minimization of the energy tails using a combination of bunch compression and bunch shaping. It served to be particularly useful in the case that $\sin\phi_1 \neq \phi_1$ in Eq. 25; that is for the case that the nonlinear fields of the compressor cavity are important. It is hoped that future linear collider designs make note of the unfortunate consequences of bunch lengthening in the upstream damping ring systems (see [24]- [29]) and avoid having to treat such nonlinear effects.

Shown [53] in Fig. 14 are simulations of the longitudinal phase space and resultant energy spread at the entrance to the linac taking into account the nonlinear rf fields seen by a long bunch of length 1/10 that of the compressor cavity rf wavelength. The case of undercompression is shown on the left while the distributions for overcompression at higher compressor cavity voltage on the right. Overcompression yields two advantages: reduced particles in the energy tails as well as a steeper rise in the longitudinal distribution at the head of the bunch. The corresponding rf waveform including beam loading and the energy spread calculated at the end of the linac are shown [53] in Fig. 15. On the left, the under-compressed bunch has a longitudinal distribution which does not cancel the slope of the rf. The result is an extended low energy tail in the energy spectrum. With overcompression, the slope of the rf and the beam loading compensate one another nearly perfectly resulting in a small energy spread at the end of the linac.

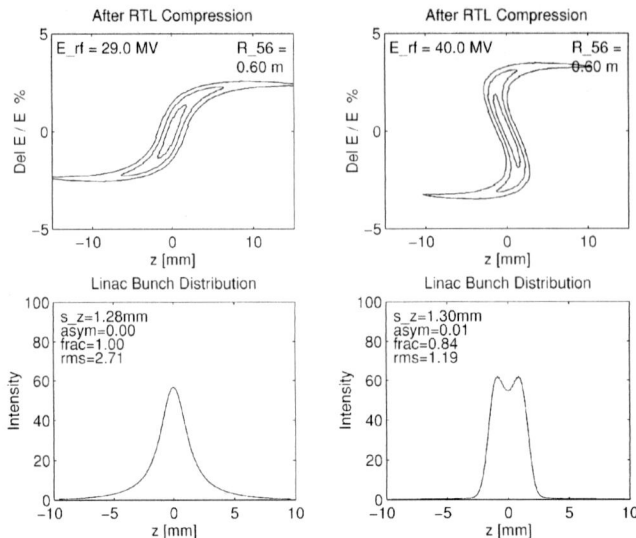

FIGURE 14. Simulation of the longitudinal phase space and beam energy distribution at injection into the linac with bunch undercompression (left) at 85% of nominal compression voltage and bunch overcompression (right) at 120% of nominal voltage. The injected bunch length was taken to be 10 mm, or approximately 1/10th of the rf wavelength. Courtesy F.J. Decker (2000).

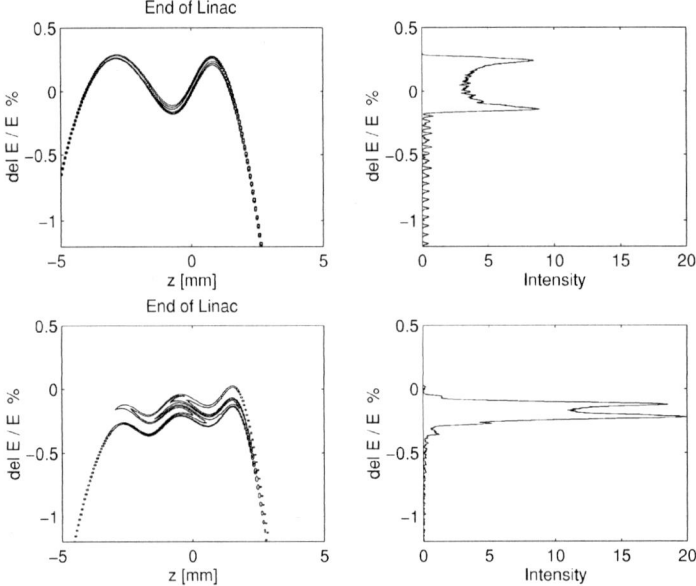

FIGURE 15. Simulation of the rf waveform and beam energy distributions at the end of the linac with bunch undercompression (top) and bunch overcompression (bottom). Courtesy F.J. Decker (2000).

Measurements of the beam at the end of the linac in a dispersive region are shown [53] in Fig. 16 with undercompression (left) and with overcompression (right). The absence of the low energy tails justified the routine use of over-compressed beams at the SLC.

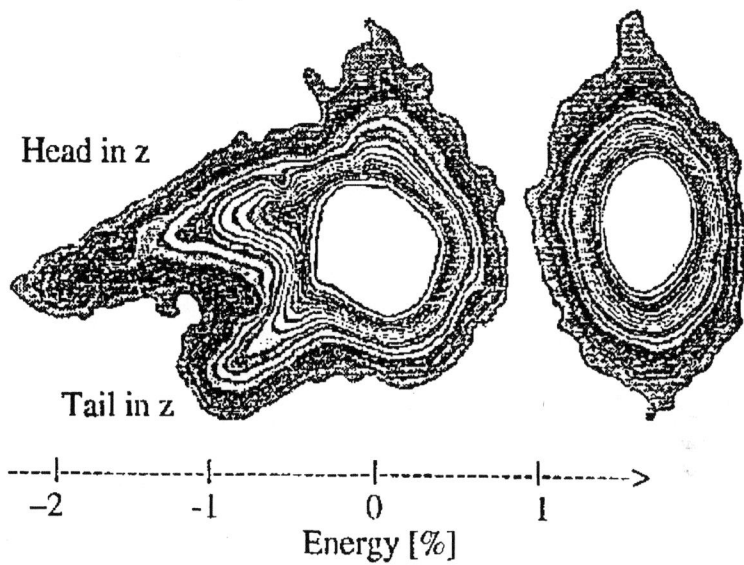

FIGURE 16. Measurements of the beam profile at a dispersive location at low compressor voltage (left) and with bunch overcompression (right). Courtesy F.J. Decker (2000).

TRAJECTORY STEERING

Motivation and perspective for this section is given with the following considerations. Recall from Appendix A that

$$\epsilon = \frac{\sigma_x^2}{\beta} \text{ with } \sigma_x^2 = <x^2>^{\frac{1}{2}}. \tag{29}$$

As discussed previously, the horizontal (x) or vertical (y) trajectory is given by a superposition of terms (c.f. Eq. 4):

$$\begin{aligned} x &= x_c + x_\beta + x_\eta \\ &= x_c + x_\beta + \eta\delta. \end{aligned} \tag{30}$$

In chapter 2 was described how to minimize contributions from the term x_β while chapter 3 focussed on minimization[4] of x_η by minimization of the beam energy

---

[4] in the absence of cross-correlations of the form $<x_c x_\eta>= 0$, for example, the dispersive term $x_\eta = \eta\delta$ contributes in quadrature

spread δ. In this section steering procedures which aim to minimize emittance dilutions arising from deviations in the central trajectory x_c and from the dispersion η will be described.

Increased beam emittances may arise from beam-to-magnet or beam-to-structure position deviations as shown conceptually in Fig. 17. The beam passing through a single misplaced quadrupole experiences the next lower-order field namely a dipole field ($<x_c x_\beta> \neq 0$). The misalignment therefore generates a betatron oscillation *and* dispersion as higher energy particles are less deflected by the dipole field. In the case of a displaced structure a betatron oscillation is also induced due to the transverse wakefield. Recall that the driving term is linear in the wakefield times the initial displacement. In either case, the ensuing orbit is such that further emittance dilutions may result downstream of the perturbation due to the initial errors.

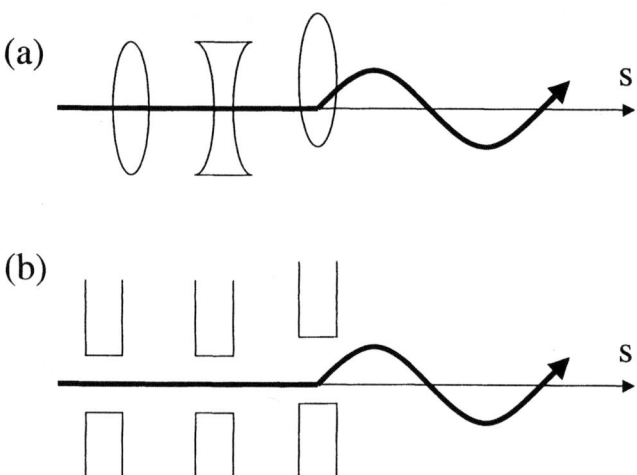

FIGURE 17. Conceptual drawing illustrating orbit perturbations due to misaligned quadrupoles (a) or structures (b).

A One-to-one steering

This algorithm aims to steer the beam so that the transverse displacements measured by beam position monitors (BPMs) are minimized. The BPMs are typically mounted near the center of quadrupoles since their sensitivity is highest at large β-function. A conceptual orbit steered one-to-one is shown in Fig. 18. The beam is successfully kicked to pass through the magnet center and, assuming that the

$$\sigma_\eta = <x_\eta^2>^{\frac{1}{2}} = \eta^2 <\delta^2>^{\frac{1}{2}}. \tag{31}$$

BPM is not offset with respect to the quadrupole, the BPM would show zero displacement. Notice however that one-to-one steering generates dispersion which contributes to emittance dilutions.

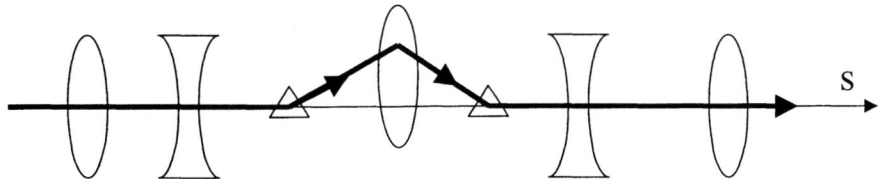

FIGURE 18. Conceptual illustration of a closed bump that would minimize the BPM reading after one-to-one steering.

In a transport line the beam centroid position measured downstream at location $s = j$ obeys

$$x_j = \sum_{i=0}^{j} \sqrt{\beta_i \beta_j} \theta_i \sin(\theta_j - \theta_i), \tag{32}$$

which has contributions from each dipole kick θ_i and depends on the β-functions at the location of the initial disturbance (i) and at the observation point (j). The corrector magnet fields to be applied to minimize the BPM readings will be solved for assuming linear transport; that is, that there are no nonlinear magnetic fields and the measurements are made at low bunch current so that nonlinear wakefield effects may be ignored.

In matrix form

$$\vec{x} = M\vec{\theta}, \tag{33}$$

where $\vec{x}$ is the set of measurements from m BPMs, $\vec{\theta}$ is the set of kick angles to be applied by n correctors, and M contains the transfer matrix elements between the correctors and the BPMs:

$$x^T = (x_0, x_1, ..., x_m) \tag{34}$$
$$\theta^T = (\theta_0, \theta_1, ..., \theta_n) \tag{35}$$
$$M_{ij} = \sqrt{\beta_i \beta_j} \sin(\phi_j - \phi_i) \tag{36}$$

Solving Eq. 37 the kick angles to be applied for minimizing the BPM readings are obtained:

$$M^T \vec{x} = M^T M \vec{\theta} \text{ or } \vec{\theta} = (M^T M)^{-1} M^T \vec{x}. \tag{37}$$

If the number of correctors equals the number of BPMs then M is a square matrix so Eq. 37 reduces to simply $\vec{\theta} = M^{-1} \vec{x}$. Otherwise the general form is taken. If

$n > m$ the matrix is overdetermined. For $n < m$ the number of unknowns exceeds the number of measurements so an independent measurement should be made after changing some parameter, for example, the beam energy. In a linear accelerator Eq. 32 must be modified [54] to include the energy scaling factor $\sqrt{\frac{E_i}{E_j}}$. This introduces m additional unknowns so additional measurements are required to constrain the solution.

As motivation for the algorithms to be used below in the discussion of beam-based alignment and dispersion-free steering, the solution is equivalently formulated in terms of a minimization procedure, which is well adapted to computational evaluation. The function to be minimized, given by Eq. 33, is

$$\sum_j \left[x_j - \sum_i M_{ij} \theta_i \right]^2, \tag{38}$$

where x_j again represents the BPM measurements and the fitting function $\sum_i M_{ij} \theta_i$ has unknowns θ_i. The minimization procedure demands

$$0 = \frac{\partial}{\partial \theta} \left[\sum_j (x_j - \sum_i M_{ij} \theta_i)^2 \right]$$

$$= 2 \sum_j \left[x_j - \sum_i M_{ij} \theta_i \right] M_{kj}, \text{ or}$$

$$\sum_j M_{kj} x_j = \sum_j \sum_i M_{ij} M_{kj} \theta_i, \tag{39}$$

which is identical to Eq. 37.

B Beam-based alignment

We define beam-based steering algorithms as ones which provide information on magnet, BPM, or structure misalignments using measurements with the beam. With this definition, one-to-one steering may also be considered a beam-based alignment algorithm since the applied kicks θ are related to the quadrupole displacements Δx by $\theta = k \Delta x$, where k is the quadrupole focussing field.

More generally, we take in this example into account that the electrical zero of the BPMs may not be coincident with the magnetic center of the quadrupoles and that the quadrupoles themselves may be displaced with respect to the reference axis. The coordinate system used is sketched in Fig. 19. Here the beam position x measured with respect to some reference axis, which is common to all magnets, is given as a sum of the quadrupole displacement x_q, the difference in location of the electrical center of the BPM and the magnetic center of the quadrupole x_{bpm}, and the measured BPM value x_m.

The beam position x_k and angle x_k' at quadrupole k are given [7] by

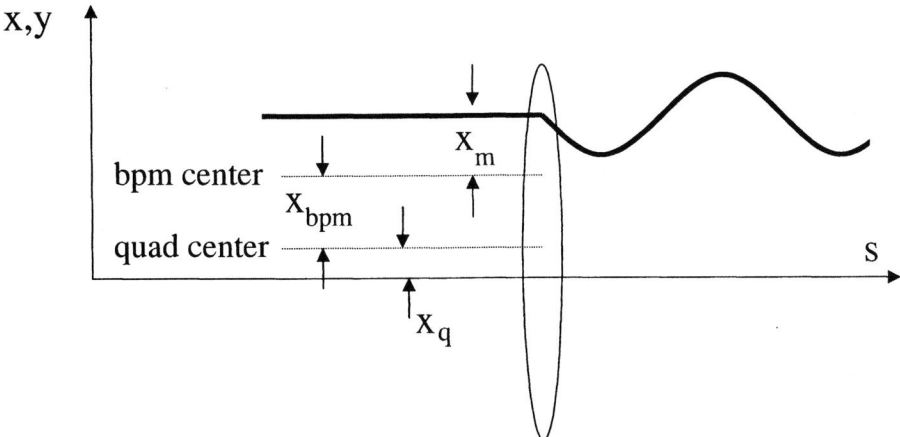

FIGURE 19. Coordinate system used in the example of beam-based alignment. (Courtesy C. Adolphsen (2000).

$$\begin{pmatrix} x_k \\ x_k' \\ 1 \end{pmatrix} = R_{j+1,k} \left\{ R_{j,j+1} \left[\begin{pmatrix} x \\ x' \\ 1 \end{pmatrix}_j + \begin{pmatrix} -x_q \\ 0 \\ 0 \end{pmatrix} \right] + \begin{pmatrix} x_q \\ 0 \\ 0 \end{pmatrix} \right\} \qquad (40)$$

where $()_j$ gives the beam position and angle with respect to the quad center, the term in $[]$ is the beam position with respect to the reference axis, $R_{j,j+1}[]$ is the beam position with respect to the reference axis transported between quad j and quad $j+1$, and the term in $\{\}$ is the beam position with respect to the quad center transported between quads j and $j+1$. Rearranging terms gives

$$\begin{pmatrix} x_k \\ x_k' \\ 1 \end{pmatrix} = R_{0,k} \begin{pmatrix} x \\ x' \\ 1 \end{pmatrix}_0 + \sum_{j=1}^{k-1} (R_{j+1,k} - R_{j,k}) \begin{pmatrix} x_{q,j} \\ 0 \\ 0 \end{pmatrix}, \qquad (41)$$

where the sum is taken over upstream quadrupoles. The function to be minimized is then

$$\sum_k \left[x_m - (x_k - x_q - x_{bpm}) \right]^2, \qquad (42)$$

where x_m are the measurements and $(x_k - x_q - x_{bpm})$ is the fitting function with unknowns x_q, x_{bpm} and the initial position and angle x_0 and x_0'.

The number of measurements is about twice the number of unknowns so the system is underconstrained. To constrain the solution, two independent measurements are required. An independent set of data may be obtained by scaling all the quadrupoles and correctors by a common factor and repeating the measurements. Multiple such scalings may be used to overdetermine the system which reduces the sensitivity of the solution to statistical errors.

C 'Wakefield bumps'

Through the early 1990's emittance dilutions were controlled by imposing tight tolerances on injection errors as precursor to BNS damping [11,34], steering using both one-to-one correction and localized beam-based alignment [7,55], and by invoking BNS damping. As the beam currents were increased, a more localized emittance preservation technique was developed in which empirically determined trajectory oscillations ('bump') were used to cancel emittance dilutions from transverse wakefields and dispersive errors. While the origins of the disturbances could not be easily localized longitudinally along the linac, the accumulated effects could be cancelled using such bumps and the emittance dilution could be reduced by a factor of almost ten [56]. The effect of the beam was determined by emittance measurements near the end of the linac (see Appendix B).

Two trajectories in both x and y are shown [56] in Fig. 20. Both trajectories produced about the same small emittance measured near the end of the linac. Notice the vertical scale which shows excursions of nearly 750 μm peak-to-peak. While wakefield bumps were used for many years, it became clear as the currents were increased that this technique was inherently unstable; small (e.g. thermal) changes in the reference line phase, for example, changed the phase advance over the bump range so that even this more localized correction scheme was not sufficiently local to be stable against realistic variations in the accelerator. Both trajectories in Fig. 20 resulted in about the same final beam emittances indicating that the procedure was not deterministic.

Physical insights were gained by simulations carried out using the program LIAR [57]. The model included representative amplitudes of the wakefield bumps which minimized the relative emittance growth at the locations of the measurements. The relative growth in normalized emittance is shown [58,59] in Fig. 21 as a function of position along the linac. There are several important conclusions to be drawn from this simulation result:

- Comparison with Fig. 8 shows that at the first emittance measurement, the optimization had been made in a location where the energy spread of the beam was large; that is, a compromise was made using wakefield bumps between correction of dispersive and wakefield-induced errors.

- Between the emittance measurement stations, there was uncontrolled emittance growth.

- Between the final emittance measurement station and extraction of the beam from the linac, there was significant emittance growth.

- Most importantly, being nonlocal in nature, small changes in the phase advance could destroy this delicate cancellation. In practice this caused significant time-dependent variations in the measured emittances [60]-[63].

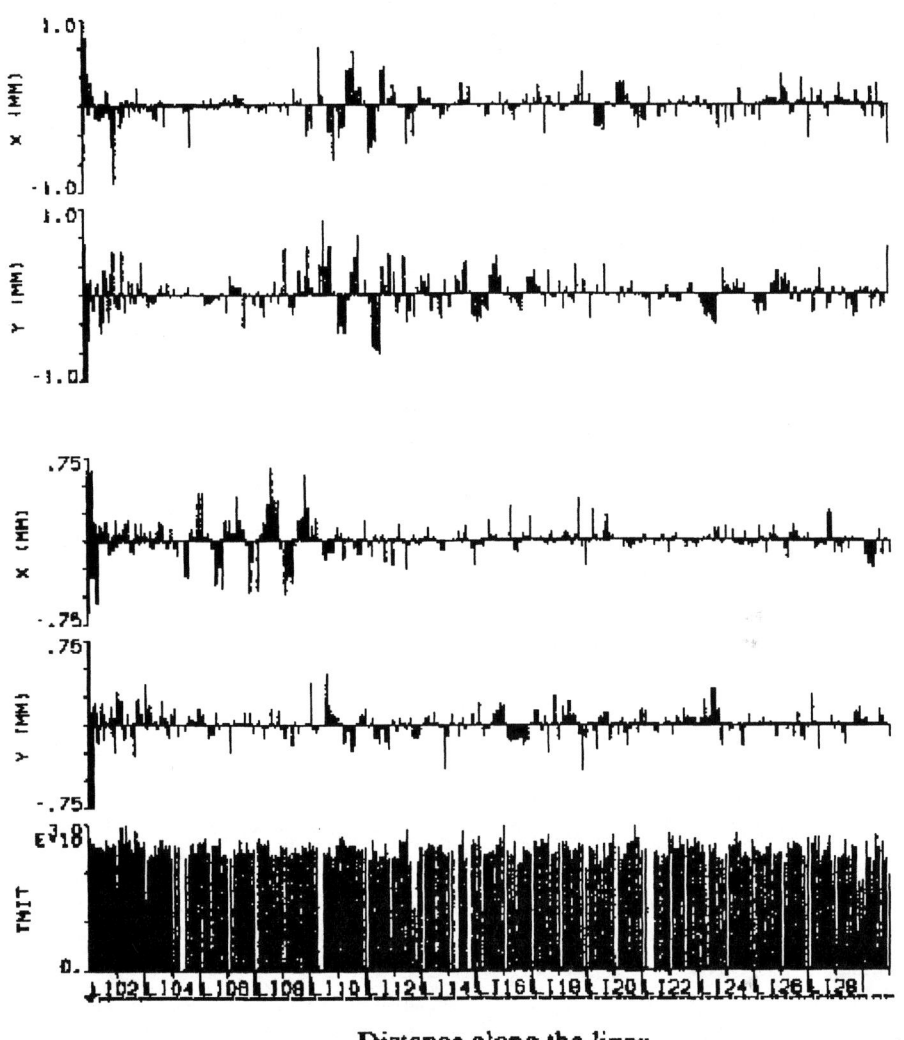

FIGURE 20. Two measured orbits with empirically determined coherent betatron oscillations used to cancel accumulated wakefield and dispesion errors. Courtesy J. Seeman (2000).

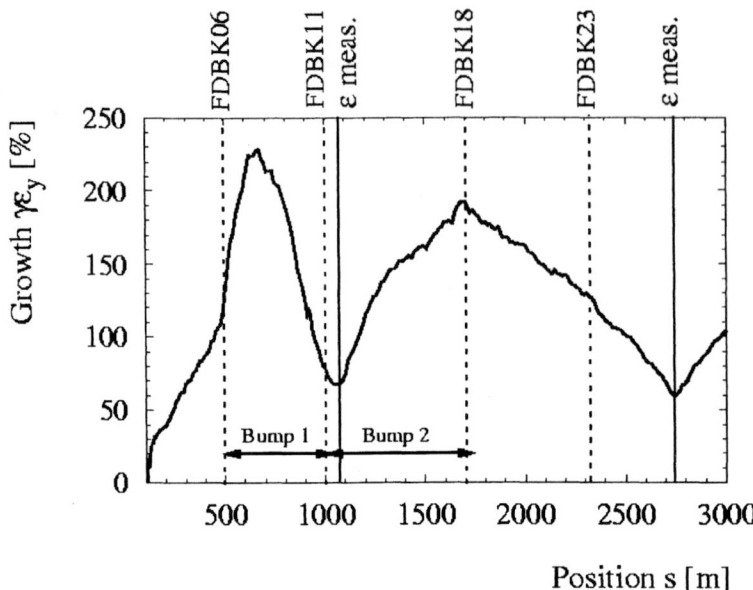

FIGURE 21. Simulated emittance growth as a function of position along the linac. The locations of the feedback loops, which controlled the amplitude of the wakefield bumps, is shown along with the locations of the emittance measurements. Courtesy R. Assmann (2000).

D Dispersion-free steering

So far we have described *one-to-one steering* which is a first step in orbit optimization but is imperfect as minimization of the BPM reading in a displaced quadrupole generates dispersive errors, *beam-based alignment* of quadrupole displacements which works beautifully at low beam currents where there is no wakefield-generated dispersion, and wakefield bumps which while more local than BNS damping is highly sensitive to small perturbations in the electromagnetic optics. With perfect implementation of either procedure, dispersive emittance dilutions may still result. As an example, consider a closed trajectory bump of the kind illustrated in Fig. 17. It has been shown [66] using LIAR and realistic optical parameters of the SLC linac that a closed 100 μm π-bump at a quadrupole located early in the linac generates nearly 0.5 mm dispersion at the end of the linac. Naively, about 6 such bumps acting independently would produce a dispersive emittance contribution equal to that typically achieved at the SLC.

Dispersion-free steering [64–66] is an algorithm which corrects even more locally dispersive errors from misaligned quadrupoles *and* dispersive errors arising from transverse wakefields. For mostly technical reasons (e.g. data acquisition and processing time) implementation was unduly delayed at the SLC. Dispersion-free steering (and rf phase stability [67–69]) proved crucial for maintaining stable linac emittances at the SLC.

The centroid trajectory[5] is given by

$$x_j = \sum_{i=1}^{j-1} \sqrt{\frac{E_i}{E_j}} \sqrt{\beta_i \beta_j} \theta_i \sin(\phi_j - \phi_i) \tag{43}$$

$$= \sum_{i=1}^{j-1} R_{12}{}^{ij} \theta_i, \tag{44}$$

where the adiabatic damping factor $\sqrt{\frac{E_i}{E_j}}$ has been included. To constrain the system, one can equivalently change the beam energy (which is in practice difficult) or as before scale the lattice. Then

$$\Delta x_j = \sum_{i=1}^{j-1} \left[R_{12}{}^{ij} - \kappa \sqrt{\frac{E_i}{E_j}} \sqrt{\beta_i \beta_j} \sin(\phi_j - \phi_i) \right] \theta_i$$

$$= \sum_{i=1}^{j-1} R_{12,\kappa}{}^{ij} \theta_i, \tag{45}$$

where the change in lattice focussing is given by

$$\kappa = \frac{\Delta K}{K} - 1, \tag{46}$$

where K is the quadrupole strength.

The function to be minimized is

$$\sum_j \left[x_j - \sum_i M_{ij} \theta_i \right]^2, \tag{47}$$

where x_j is an $M \times 1$ vector containing the difference measurements and the fitting function is given by $\sum_i M_{ij} \theta_i$ where θ_i is an $N \times 1$ vector of unknowns. The M_{ij} represents an $M \times N$ matrix containing the transfer matrix elements.

In practice it is not difficult to minimize not only the difference orbit but simultaneously the absolute orbit. In this case x_j is a $2M \times 1$ vector containing the difference measurements and the absolute orbit, M_{ij} is a $2M \times N$ matrix and θ remains an $N \times 1$ matrix. This approach was used at the SLC where in addition, to overconstrain the solution and minimize systematic errors arising from magnet hysteresis, the measurements were performed for 4-5 values of κ corresponding to energy variations of +5% to -30%. In later years, the problem of hysteresis was eliminated and the application became noninvasive as two independent measurements could be obtained *without* changing the lattice by measuring independently the orbits of the electrons and positrons which passed through the same lattice.

---

[5] a reminder: intrabunch position-energy correlations, when projected, may result in measured centroid displacements which underestimate the contributions from off-axis bunch tails

Shown in Fig. 22 are absolute ($\kappa = 0$) and difference trajectories ($\kappa = 0.9$, 0.8, and 0.7) measured after trajectory steering of the SLC linac [66] using 20-pulse BPM averaging. With an equivalent energy change of 30% ($\kappa = 0.7$), up to 1.5 mm trajectory difference was observed. Similar measurements made after iteration of dispersion-free steering [66] are given in Fig. 23. Iteration proved useful to reduce sensitivity to errors in the assumed optics even though experience showed that the first iteration yielded the largest improvements. With $\kappa = 0.3$, neglecting the errant point due possibly to a bad BPM near sector 20, the maximum orbit difference after dispersion-free steering was reduced from 1.5 mm to less than 200 μm. Notice that the rms of the measurements of the absolute orbit are actually larger following dispersion-free steering. This suggests significant BPM and/or quadrupole misalignment errors.

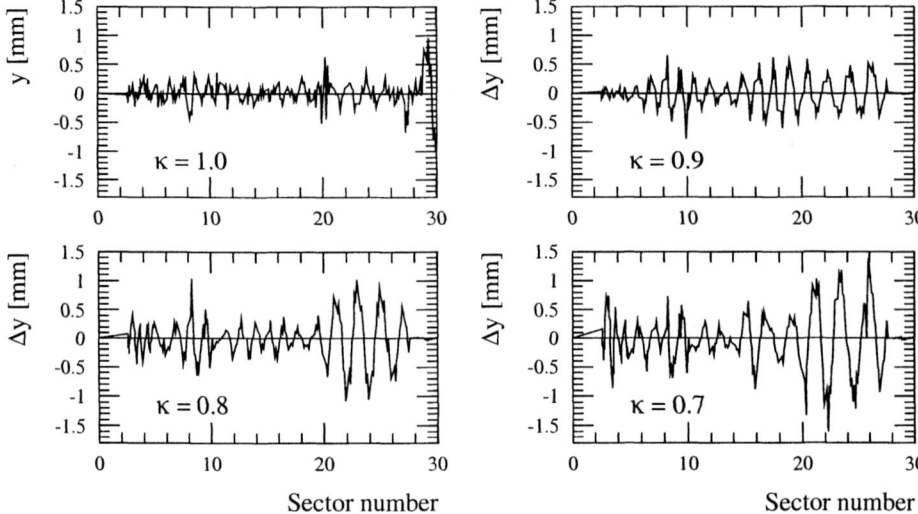

FIGURE 22. Absolute and difference vertical trajectories measured after trajectory steering before dispersion-free steering of the SLC linac. Courtesy R. Assmann (2000).

E Errors

For simplicity of expression, measurement errors have been neglected up to now. Error sources and their typical rms contributions include BPM resolution errors $\sigma(x_j) < 10$ μm, bpm misalignments $\sigma_{\text{bpm}} \sim 100$ μm, and systematic errors arising from beam jitter and/or slow drifts $\sigma_{\text{sys}} \sim 20$ μm. To propagate the measurement errors used in the minimization procedures, a weighting function may be defined as

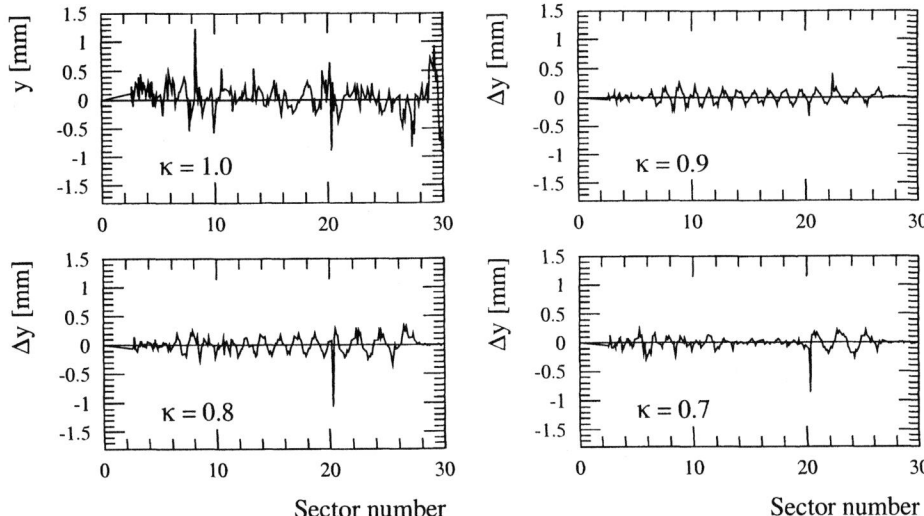

FIGURE 23. Absolute and difference vertical trajectories measured after dispersion-free steering in the SLC linac. Courtesy R. Assmann (2000).

$$w_j = \frac{1}{\sum_m \sigma_{m,j}^2}, \qquad (48)$$

where the subscripts m give the different error sources and j is a sum over the BPM measurements. The functions to be minimized then are (c.f. Eqs. 38, 42, and 47)

$$\sum_j \left[\frac{x_j - \sum_i M_{ij}\theta_i)}{\sum_m \sigma_{m,j}^2}\right]^2, \qquad \text{one-to-one}$$

$$\sum_k \left[\frac{x_m - (x_k - x_q - x_{bpm})}{\sum_m \sigma_{m,j}^2}\right]^2, \qquad \text{beam-based alignment}$$

$$\sum_j \left[\frac{x_j - \sum_i M_{ij}\theta_i}{\sum_m \sigma_{m,j}^2}\right]^2, \qquad \text{dispersion-free steering} \qquad (49)$$

A goodness of fit parameter, or χ-squared may be correspondingly constructed. In the dispersion-free steering example given above for which both the trajectory and the trajectory differences were to be simultaneously minimized,

$$\chi^2 = \sum_j \left[\frac{x_j^2}{\sigma_{bpm}^2} + \sum_\kappa \frac{\Delta x_{j,\kappa}^2}{\sigma_{sys}^2}\right], \qquad (50)$$

where the second summation over κ corresponds to the different energy scalings under which the measurements were made. The errors from BPM resolution were assumed to be negligible and the summation over errors has been simplified to reflect the dominating errors; that is, the systematic errors contribute less than the alignment errors in the measurements of the absolute trajectories while in the difference trajectory measurements the BPM misalignments cancel and are therefore set to zero.

CONCLUSION

In this report we reviewed various mechanisms causing emittance dilution in linear accelerators. We concentrated on those effects which were important for the Stanford Linear Collider. Many experimental measurement techniques were developed and used to suppress emittance growth. In particular powerful procedures were applied to overcome inadequacies in the precision to which the electromagnetic fields governing the beam transport were known. The described methods all contributed to the steady improvements [1,23,58,70–79] in the collider performance, which is summarized [79] in Fig. 24. For the next generation of linear colliders careful control of the beam emittances will be even more important as alignment tolerances are stricter at higher rf frequency. The methods described in this report should be applicable at least during initial commissioning.

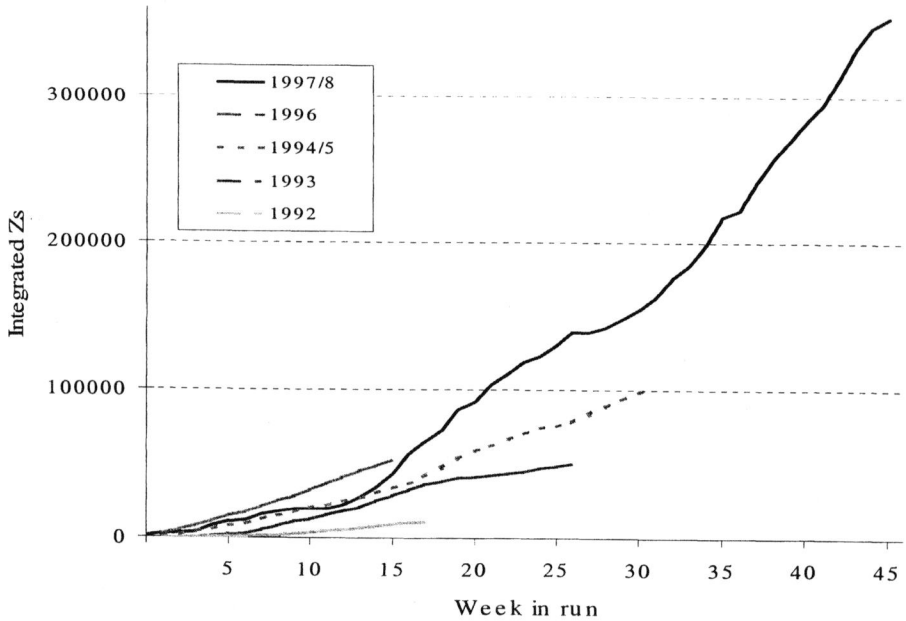

FIGURE 24. History of integrated Z^0 events recorded at the SLC from 1992 through 1998. Courtesy N. Phinney (2000).

REFERENCES

1. J.T. Seeman, 'Status of the Stanford Linear Collider', Proc. ECFA Workshop on e^+e^- Linear Colliders (LC92), Garmisch-Partenkirchen, Germany (1992) 93-120.
2. C. Adolphsen, K.L. Bane, J. Seeman, 'Effect of wake fields on first order transport in the SLC linac', Proc. of IEEE Part. Accel. Conf., San Francisco, CA (1991) 3207-3209.
3. F.J. Decker et al, 'Dispersion and betatron matching into the linac', Proc. IEEE Part. Accel. Conf., San Francisco, CA (1991) 905-907.
4. L. Merminga P.L. Morton, J. Seeman, W.L. Spence, 'Transverse phase space in the presence of dispersion', Proc. of IEEE Part. Accel. Conf., San Francisco, CA (1991) 461-463.
5. T.L. Lavine et al, 'Beam determination of quadrupole misalignments and beam position monitor biases in the SLC linac', Proc. Lin. Acc. Conf., Williamsburg, VA (1988).
6. J.T. Seeman, 'Linear collider accelerator physics issues regarding alignment', Proc. Acc. Alignment, Stanford, CA (1989) 257-262.
7. C. Adolphsen et al, 'Beam based alignment technique for the SLC linac', Part. Acc. Conf., (1989) 977-979.
8. J.T. Seeman, 'Observation of high current effects in high energy linear colliders', 1990 Joint US-CERN School on Particle Accelerators: Frontiers of Particle Beams, Intensity Limitations, Hilton Head Island, SC (1990); published in CERN US PAS (1990) 255-292.
9. J.T. Seeman et al, 'Alignment issues of the SLC linac accelerating structure', Proc. of Part. Acc. Conf., San Fransisco, CA (1991) 2949-2951.
10. T. Barklow, P. Emma, P. Krejcik, N.J. Walker, 'Review of lattice measurement techniques at the SLC', Proc. 5th ICFA Adv. Beam Dyn. Wkshp., Corpus Christi, TX (1991) 347-354.
11. T. Limberg, J. Seeman, W.L. Spence, 'Effects and tolerances of injection jitter in the SLC and future linear colliders', Proc. of 2nd Eur. Part. Acc. Conf., Nice, France (1990) 1506-1508.
12. F. Rouse et al, 'General, database-driven fast-feedback system for the Stanford Linear Collider', Proc. 1991 Part. Accel. Conf., San Fransisco, CA (1991) 1419-1421.
13. T. Himel et al, 'Use of digital control theory state space formalism for feedback at SLC', Proc. Part. Accel. Conf., San Fransisco, CA (1991) 1451-1453.
14. L. Hendrickson et al, 'Generalized fast feedback system in the SLC', Proc. of the Inter. Conf. on Accel. and Large Experimental Physics Control Systems, Tsukuba, Japan (1991) 414-419.
15. T. Himel et al, 'Adaptive cascaded beam based feedback at the SLC', Proc. 1993 Part. Acc. Conf., Washington, DC (1993) 2106-2108.
16. M.C. Ross, L. Hendrickson, T. Himel, E. Miller, 'Precise system stabilization at SLC using dither techniques', Proc. Part. Acc. Conf., Washington, DC (1993) 1972-1974.
17. L. Hendrickson et al, 'Tutorial on beam-based feedback systems for linacs', Proc. 17th Int. Linear Acc. Conf (LINAC94), Tsukuba, Japan (1994).
18. L. Hendrickson et al, 'Fast feedback for linear colliders', Proc. 16th IEEE Part. Acc.

Conf., Dallas, TX (1995) 2389-2393.
19. M.G. Minty et al, 'Feedback performance at the Stanford Linear Collider', Proc. 16th IEEE Part. Acc. Conf., Dallas, TX (1995) 662-664.
20. T. Himel, 'Feedback: Theory and accelerator applications', Ann. Rev. Nucl. Part. Sci. **47**, (1997) 157-192.
21. H. Schwarz and J.G. Judkins, 'Phase detector and phase feedback for a single bunch in a two bunch damping ring for the SLAC linear collider', Proc. 12th IEEE Part. Acc. Conf., Washington, D.C. (1987) 769-770.
22. C.E. Adolphsen, P.J. Emma, T.H. Fieguth, W.L. Spence, 'Chromatic correction in the SLC bunch length compressors', Proc. of IEEE Part. Accel. Conf., San Francisco, CA (1991) 503-505.
23. J. Seeman and J. Sheppard, 'Status of the SLC', Proc. of 'Orsay Accel. Wkshp. on New Developments in Part. Acc. Techniques', Orsay, France (1987) 122-133.
24. L.Z. Rivkin et al, 'Bunch lengthening in the SLC damping rings', Proc. 1st Eur. Part. Acc. Conf., Rome, Italy (1988).
25. K.L. Bane, 'Bunch lengthening in the SLC damping rings', Impedance and bunch instability workshop adv. phonton source, Argonne, IL (1989); K.L.F. Bane, 'Bunch Lengthening in the SLC Damping Rings', Proc. Acc. Phys. and Modeling, Upton, NY (1991), 235-289.
26. K.L. Bane and R.D. Ruth, 'Bunch lengthening calculations for the SLC damping rings', Part. Accel. Conf., Chicago IL (1989) 789-791.
27. P. Krejcik et al, 'High intensity bunch length instabilities in the SLC damping rings', Proc. Part. Acc. Conf., Washington, DC (1993) 3240-3242.
28. K. Bane et al, 'High intensity single bunch instability behavior in the new SLC damping ring vacuum chamber', Proc. Part. Acc. Conf., Dallas, TX (1995) 3109-3111.
29. K.L.F. Bane and K. Oide, 'Simulations of the longitudinal instability in the new SLC damping rings', Proc. Part. Acc. Conf., Dallas, TX (1995) 3105-3108.
30. J.T. Seeman, 'The Stanford Linear Collider', Phys. of Part. Acc.. In Jackson, J.D. (ed.) et al: Annual review of nuclear and particle science **41**, 389-428.
31. J.T. Seeman, 'Observations and cures of wakefield effects in the SLC Linac', 5th ICFA Adv. Beam Dyn. Wkshp on Effects of Errors in Accelerators, Their Diagnosis and Correction, Corpus Christi, TX (1991) 339-346.
32. C. Adolphsen, T. Slaton, 'Beam trajectory jitter in the SLC linac', Proc. of Part. Acc. Conf., Dallas, TX (1995) 3034-3036.
33. C. Adolphsen et al, 'Pulse to pulse stability issues in the SLC', Proc. of Part. Acc. Conf., Dallas, TX (1995) 645-648.
34. A.W. Chao, B. Richter, C.Y. Yao, 'Beam emittance growth caused by transverse deflecting fields in a linear accelerator', Nucl. Instrum. Meth. **178**, (1980) 1-24.
35. A.W. Chao, B. Richter, C.Y. Yao, 'Transverse wake field effects on intense bunches with application to the SLAC linear accelerator', Proc. XI Intl. Conf. on High Energy Acc., Geneva, Switzerland (1980) 597-604.
36. F.J. Decker, R. Brown, J.T. Seeman, 'Beam size measurements with noninterceptive off-axis screens', Proc. of Part. Acc. Conf., Washington, DC (1993) 2507-2509.
37. J.T. Seeman, K.L. Bane, T. Himel, W.L. Spence, 'Observation and control of emit-

tance growth in the SLC linac', Part. Accel., **30**, (1989) 97-104.
38. J.C. Sheppard *et al*, 'Implementation of nonintercepting energy spread monitors', Proc. 12th Part. Acc. Conf., Washington, D.C. (1987) 757-760.
39. A.W. Chao, (ed.) and M. Tigner, (ed.), 'Handbook of accelerator physics and engineering', Singapore, Singapore: World Scientific (1999) 650 p.
40. P.B. Wilson, 'A study of beam blow-up in electron linacs', Report No. HEPL-297 (Rev. A), High Energy Physics Laboratory, Stanford University, Stanford, California (1963).
41. R.B. Neal and W.K.H. Panofsky, Science **152**,(1966) 1353.
42. W.K.H. Panofsky and M. Bander, 'Asymptotic theory of beam break-up in linear accelerators', Rev. Sci. Instr. **39**, 206 (1968).
43. G.V. Stupakov, 'BNS damping of beam breakup instability', SLAC-AP-108 (1997).
44. V. Balakin, S. Novokhatsky, V. Smirnov, 'VLEPP: transverse beam dynamics' Proc. 12th Int. Conf. on High Energy Acc. (1983) 119-120.
45. J.T. Seeman, F.J. Decker, R.L. Holtzapple, W.L. Spence, 'Measured optimum BNS damping configuration of the SLC linac', Proc. Part. Acc. Conf., Washington, DC (1993) 3234-3236.
46. J.T. Seeman and N. Merminga, 'Mutual compensation of wakefield and chromatic effects of intense linac bunches', Proc. 1990 Linac Conf., Albuquerque, NM (1990) 387-389.
47. K.L.F. Bane *et al*, 'Measurement of the londitudinal wake field and the bunch shape in the SLAC linac', Proc. 1997 Part. Acc. Conf., Vancouver, BC, Canada (1997).
48. C. Adolphsen, K.L. Bane, J. Seeman, 'Effect of wake fields on first order transport in the SLC linac', Proc. of Part. Acc. Conf., San Fransisco, CA (1991) 3207-3209.
49. R.L. Holtzapple, 'Longitudinal dynamics at the Stanford Linear Collider', PhD thesis, (June, 1996); R.L. Holtzapple *et al*, 'Measurements of longitudinal phase space in the SLC linac', Proc. Part. Acc. Conf., Dallas, TX (1995) 3025-3027.
50. R.L Holtzapple, 'Bunch compression at the Stanford Linear Collider', Proc. of Micro Bunches: A Workshop on the Production, Measurement and Applications of Short Bunches of Electrons and Positrons in Linacs and Storage Ringe, Upton, NY (1995) 36-45.
51. F.J. Decker *et al*, 'Longitudinal phase space setup for the SLC beams', Proc. Part. Acc. Conf., Vancouver, BC, Canada (1997) 509-511.
52. G.A. Loew and J.W. Wang, 'Minimizing the energy spread within a single bunch by shaping its charge distribution', Proc. Part. Acc. Conf., Vancouver, BC Canada (1985) 3228-3230.
53. F.J. Decker, R Holtzapple, T. Raubenheimer, 'Overcompression, a method to shape the longitudinal bunch distribution for a reduced energy spread', Proc. 17th Intl. Linear Acc. Conf., Tsukuba, Japan (1994), 47-49.
54. K. Thompson *et al*, 'Operational experience with model based steering in the SLC linac', Proc. Part. Acc. Conf., Chicago, IL (1989), 1675-1677.
55. P. Emma, T.H. Fieguth, T. Lohse, 'Online monitoring of dispersion functions and transfer matrices at the SLC', Nucl. Instrum. Meth. A **288**, (1990) 313-334.
56. J.T. Seeman, F.J. Decker, I. Hsu, 'The introduction of trajectory oscillations to reduce emittance growth in the SLC linac,', Proc. 15th Intl. Conf. High Energy

Acc., Hamburg, Germany (1992) 879-881.
57. R. Assmann et al, 'LIAR - A computer program for the modeling and simulation of high performance linacs', SLAC/AP-103, 1997.
58. R. Assmann, 'Beam dynamics in the SLC', Proc. of Part. Acc. Conf., Vanouver, BC, Canada (1997).
59. R. Assmann, F.J. Decker, P. Raimondi, 'Improvements in emittance wakefield optimization for the SLAC linear collider', Proc. of 6th Eur. Part. Acc. Conf., Stockholm, Sweden (1998).
60. F.J. Decker et al, 'Diganostic beam pulses for monitoring the SLC linac', Proc. 16th Part. Acc. Conf., Dallas TX (1995) 2646-2648.
61. et al, 'Collective centroid oscillations as an emittance preservation diagnostic in linear collider linacs', Proc. 17th Part. Acc. Conf., Vancouver, Canada (1997).
62. F.J. Decker et al, 'Beam based analysis of day night performance variations at the SLC linac', Proc. 17th Part. Acc. Conf., Vancouver, Canada (1997) 506-508.
63. R.W. Assmann et al, 'Beam based monitoring of the SLC linac optics with a diagnostic pulse', Proc. 17th Part. Acc. Conf., Vancouver, Canada (1997) 497-499.
64. T.O. Raubenheimer and R.D. Ruth, 'A dispersion free trajectory correction technique for linear colliders', Nucl. Instr. and Meth. **A302**, (1991) 191-208.
65. R. Assmann et al, 'Quadrupole alignment and trajectory correction for future linear colliders: SLC tests of a dispersion-free steering algorithm', Proc. 4th Int. Workshop on Acc. Alignment, Tsukuba, Japan (1995).
66. R. Assmann, T. Chen, F.J. Decker, M. Minty, P. Raimondi, T.O. Raubenheimer, R. Siemann, 'Simultaneous trajectory and dispersion correction in the SLC linac' (1996), unpublished.
67. D. McCormick, M. Ross, T. Himel, N. Spencer, 'Thermal stabilization of low level RF distribution systems at SLAC', Proc. of Part. Acc. Conf., Washington, DC (1993) 1975-1977.
68. F.J. Decker et al, 'Effects of temperature variation on the SLC linac rf system', Proc. of Part. Acc. Conf., Dallas, TX (1995) 1821-1823.
69. J. Bogart et al, 'A fast and accurate phasing algorithm for the rf accelerating voltages of the SLAC linac', Proc. 6th Eur. Part. Acc. Conf. Stockholm, Sweden (1998) 22-26.
70. G.S. Abrams et al, 'Experimental beam dynamics and stability in the SLC linac', Proc. 14th Int. Conf. on High Energy Acc., Tsukuba, Japan (1989); Part. Accel. **30**, (1990) 91-96..
71. J.L. Turner et al, 'Vibration studies of the stanford linear accelerator', Proc. of Part. Acc. Conf., Dallas, TX (1995) 665-667.
72. F. Zimmermann et al, 'Performance of the 1994/1995 SLC final focus system', Proc. of Part. Acc. Conf., Dallas, TX (1995) 656-658.
73. P. Emma, 'The Stanford Linear Collider', Proc. of Part. Acc. Conf., Dallas TX (1995) 606-610.
74. P. Emma, L.J. Hendrickson, P. Raimondi, F. Zimmermann, 'Limitations of interaction point spot size tuning at the SLC', Proc. of Part. Acc. Conf., Vancouver, Canada (1997) 452-454.
75. R.W. Assman et al, 'Accelerator physics highlights in the 1997/98 SLC run', Proc. 1st Asian Part. Acc. Conf., Tsukuba, Japan (1998) 474-476.

76. F. Zimmermann, '1998 SLC luminosity and pinch enhancement', SLAC-CN-418 (1998) 7pp.
77. F.J. Decker, M.G. Minty, Y. Nosochkov, P. Raimondi, 'Status of the SLC linac', Proc. Eur. Part. Acc. Conf. (1998).
78. R.W. Assmann *et al*, 'Accelerator physics highlights in the 1997/98 SLC run', Proc. 1st Asian Part. Acc. Conf. (1998) Tsukuba, Japan.
79. P. Raimondi *et al*, 'Recent luminosity improvements at the SLC', Proc. 6th Eur. Part. Acc. Conf. Stockholm, Sweden (1998).
80. D.A. Edwards and M.J. Syphers, 'An introduction to the physics of high energy accelerators', Wiley series in beam physics and accelerator technology, New York, USA (1993).
81. E.D. Courant and H.S. Snyder, 'Theory of the alternating-gradient synchrotron", Annals of Physics **3**, (1958) 1-48.
82. K.L. Brown, 'A first and second-order matrix theory for the design of beam transport systems and charged particle spectrometers", SLAC-75 (1982).
83. K.L. Brown, F. Rothacker, D. Carey, C. Iselin, 'Transport. A computer program for designing charged particle beam transport systems", SLAC-91 (1977).
84. H. Wiedemann, 'Particle accelerator physics: basic principles and linear beam dynamics', Springer Verlag, Berlin, Germany (1993).
85. M.C. Ross *et al*, 'Automated emittance measurements in the SLC', Proc. of Part. Acc. Conf., Washington, DC (1989) 725-728.

Appendix A - Definition of the Beam Emittance

The reader is refered to Ref. [80]. The beam emittance ϵ describes the phase space area occupied by the beam. For a Gaussian beam with standard deviation σ, the phase space area containing a fraction F of the beam is

$$\epsilon = -\frac{2\pi\sigma^2}{\beta}\ln(1-F), \tag{51}$$

where β is the β-function at the observation point (the Twiss parameters α, β, and γ are described in numerous texts such as in Refs. [81]- [84]). Various definitions, which depend on the choice of F, are used depending on the particular application. In this report, we take $F = 15\%$ so that

$$\epsilon = \frac{\sigma^2}{\beta}. \tag{52}$$

The standard deviation σ is often taken to represent the root-mean-square (rms) of the distribution. It is given by

$$\sigma_x = \sqrt{<x^2> - <x>^2}, \tag{53}$$

where x represents either the horizontal or the vertical plane. Here $<x>$ and $<x^2>$ are the first and second moment of the beam distribution, respectively. For an intensity distribution $f(x)$,

$$<x> = \frac{\int_0^\infty x f(x) dx}{\int_0^\infty f(x) dx}$$

$$<x^2> = \frac{\int_0^\infty x^2 f(x) dx}{\int_0^\infty f(x) dx}. \tag{54}$$

Often the physical quantity of interest is given by Eq. 53 with the static position offset of the beam intensity centroid omitted so that

$$\epsilon = \frac{<x^2>^{\frac{1}{2}}}{\beta}. \tag{55}$$

Appendix B - Measurements of the Beam Emittance

The transformation between an initial beam matrix σ_0 and the beam matrix σ at a desired observation point is given by

$$\sigma = M \sigma_0 M^T, \tag{56}$$

where the beam matrix, in terms of the Twiss parameters ([81]- [84]), is

$$\sigma = \epsilon \begin{pmatrix} \beta & -\alpha \\ -\alpha & \gamma \end{pmatrix} \tag{57}$$

In an uncoupled system,

$$\sigma = \begin{pmatrix} \sigma_{11} & \sigma_{12} & 0 & 0 \\ \sigma_{21} & \sigma_{22} & 0 & 0 \\ 0 & 0 & \sigma_{33} & \sigma_{34} \\ 0 & 0 & \sigma_{43} & \sigma_{44} \end{pmatrix} \text{ and } M = \begin{pmatrix} M_{11} & M_{12} & 0 & 0 \\ M_{21} & M_{22} & 0 & 0 \\ 0 & 0 & M_{33} & M_{34} \\ 0 & 0 & M_{43} & M_{44} \end{pmatrix} \tag{58}$$

The beam matrix is symmetric with $\sigma_{12} = \sigma_{21}$, but in general $M_{12} \neq M_{21}$.

Single wire measurement of the beam emittance [85]

An (invasive) measurement of the beam emittance can be made by varying the field strength of a quadrupole located upstream of a single wire or screen. The transfer matrix is $M = SQ$, where S is the transfer matrix of the quadrupole:

$$Q = \begin{pmatrix} 1 & 0 \\ k = \pm \frac{1}{f} & 1 \end{pmatrix} \tag{59}$$

using a thin-lens approximation for which the length of the quadrupole is short compared to it's focal length f. After mutiplying matrices,

$$M = \begin{pmatrix} S_{11} + k S_{12} & S_{12} \\ S_{21} + k S_{22} & S_{22} \end{pmatrix}. \tag{60}$$

Expanding the matrix product $\sigma = (SQ)\sigma_0(SQ)^T$ and equating the (11) element on both sides, the beam size is

$$\sigma_{11} = (S_{11}^2\sigma_{110} + 2S_{11}\sigma_{120} + S_{12}^2\sigma_{220}) + (2S_{11}S_{12}\sigma_{110} + 2S_{12}^2\sigma_{120})k + S_{12}^2\sigma_{11}k^2, \quad (61)$$

which is quadratic in the field parameter k.

Procedure (for a single-wire wire scanner measurement)
1. For each value of quadrupole field strength k, scan the wire to obtain detector counts as a function of wire position.
2. For each wire scan at fixed k, fit the measured distribution to a Gaussian of the form

$$f(x) = f_0 + f_{max}e^{-\frac{(x-<x>)^2}{2<x^2>}}, \quad (62)$$

where f_0 is the basline level offset and f_{max} is the peak value of the Gaussian distribution.
3. Plot the fitted $<x^2>$ as a function of k.
4. Fit this result to a parabola. One parametrization for the fit is

$$\begin{aligned}\sigma_{11} &= A(k-B)^2 + C \\ &= Ak^2 - 2ABk + (C+AB)^2.\end{aligned} \quad (63)$$

5. Reconstruct the σ matrix by equating coefficients:

$$A = S_{12}^2\sigma_{11} \quad (64)$$
$$-2AB = 2S_{11}S_{12}\sigma_{11} + 2S_{12}^2\sigma_{12} \quad (65)$$
$$C + AB = S_{11}^2\sigma_{11} + 2S_{11}S_{12}\sigma_{12} + S_{12}^2\sigma_{22}, \quad (66)$$

and solve for $\sigma_{11}, \sigma_{12} \,(= \sigma_{21})$, and σ_{22}. The results are

$$\sigma_{11} = \frac{A}{S_{12}^2},$$
$$\sigma_{12} = -\frac{A}{S_{12}^2}(B + \frac{S_{11}}{S_{12}}),$$
$$\sigma_{22} = \frac{1}{S_{12}^2}[(AB^2 + C) + 2AB(\frac{S_{11}}{S_{12}}) + A(\frac{S_{11}}{S_{12}})^2] \quad (67)$$

6. Calculate the beam emittance from the determinant of the beam matrix $\epsilon = \sqrt{\det \sigma}$ and propagate errors:

$$\det \sigma = \sigma_{11}\sigma_{22} - \sigma_{12}^2$$
$$= \frac{AC}{S_{12}^4} \quad \text{so} \quad \epsilon = \frac{\sqrt{AC}}{S_{12}^2}. \quad (68)$$

One can also obtain the ellipse parameters $\alpha, \beta,$ and γ:

$$\beta = \frac{\sigma_{11}}{\epsilon} = \sqrt{\frac{A}{C}}$$

$$\alpha = -\frac{\sigma_{12}}{\epsilon} = \sqrt{\frac{A}{C}}(B + \frac{S_{11}}{S_{12}})$$

$$\gamma = \frac{1}{\epsilon} = \frac{S_{12}^2}{\sqrt{AC}}[(AB^2 + C) + 2AB\frac{S_{11}}{S_{12}} + A(\frac{S_{11}}{S_{12}})^2]. \quad (69)$$

As a check, the ellipse parameters should satisfy $\beta\gamma - 1 = \alpha^2$.

An example emittance measurement made in two transverse planes is shown in Fig. 25. The graphics output shows the square of the measured beam size in μm^2 as a function of the quadrupole field strength in $\frac{kG}{m}m$. The first two rows of text show the measured emittance (ϵ) and the normalized emittance ($\gamma\epsilon$).

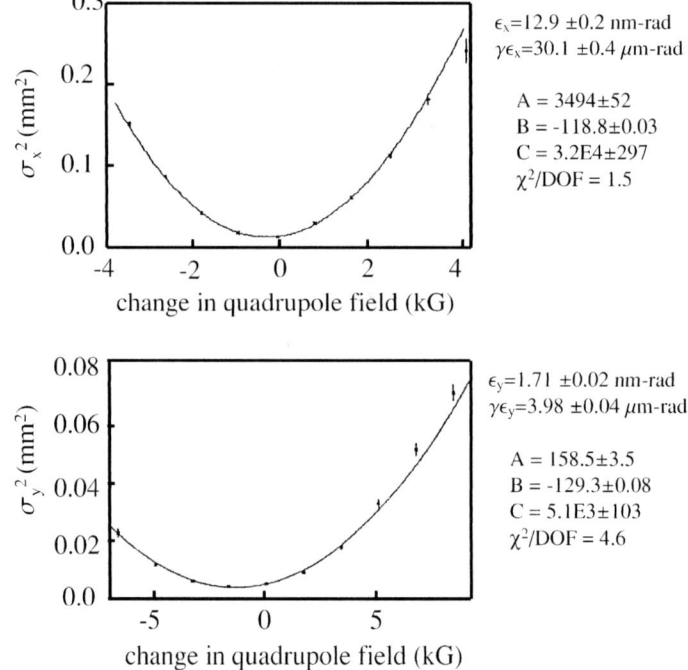

FIGURE 25. Transverse beam emittance measurements from the SLC made using a quadrupole scan and a single wire. The unit designation 'M-R' denotes mm-mrad.

Multiple wire measurement of the beam emittance [84]

The beam emittance may be measured (in many applications noninvasively) using a minimum of 3 wires if there are no coupling elements or using 4 wires with coupling. The optimum wire locations for maximum sensitivity are such that the separation between wires corresponds to a difference in betatron phase advance $\Delta\phi$ of $\frac{90°}{N_w}$, where N_w is the number of wires used in the measurement. Letting σ_i denote the measured σ_{11}'s for wire i, and considering the case of 4 wires, the matrix equation to be solved is

$$\begin{pmatrix} \sigma_1 \\ \sigma_2 \\ \sigma_3 \\ \sigma_4 \end{pmatrix} = \begin{pmatrix} c_1^2 & 2c_1 & s_1 s_1^2 \\ c_2^2 & 2c_2 & s_2 s_2^2 \\ c_3^2 & 2c_3 & s_3 s_3^2 \\ c_4^2 & 2c_4 & s_4 s_4^2 \end{pmatrix} \begin{pmatrix} \sigma_{11} \\ \sigma_{12} \\ \sigma_{22} \end{pmatrix} \quad (70)$$

Notice that M need not be a square matrix. Rewriting Eq. 70 as $A = MC$, then $M^T A = M^T M C$, or $C = (M^T M)^{-1} M^T A$; that is,

$$\begin{pmatrix} \sigma_{11} \\ \sigma_{12} \\ \sigma_{22} \end{pmatrix} = (M^T M)^{-1} M^T \begin{pmatrix} \sigma_1 \\ \sigma_2 \\ \sigma_3 \\ \sigma_4 \end{pmatrix} \quad (71)$$

which gives the beam matrix elements (σ_{ij}) in terms of the measured sigmas.

Procedure (for multiple-wire wire scanner measurement)
1. Scan each wire to obtain detector counts as a function of wire position.
2. For each wire scan, fit the distribution to a Gaussian function using Eq. 62.
3. Reconstruct the σ matrix using Eq. 62, the transfer matrix elements M_i from the model, and the σ_i from the measurements.
4. Calculate the emittance $\epsilon = \sqrt{\det \sigma}$.
5. Calculate the ellipse parameters $\alpha = -\frac{\sigma_{12}}{\epsilon}$, $\beta = \frac{\sigma_{11}}{\epsilon}$, and $\gamma = \frac{\sigma_{22}}{\epsilon}$.

Graphics
Increased tuning efficiency may be obtained from meaningful graphical representation of the experimental data. In the multiple wire emittance measurement it is useful to project the measurements to a single point along the accelerator and to plot the normalized phase space. Defined in terms of the ellipse parameters, the emittance is

$$\epsilon = \gamma x^2 + 2\alpha x x' + \beta x'^2. \quad (72)$$

Since $\beta\gamma = 1 + \alpha^2$,

$$\epsilon = \frac{1}{\beta}[x^2 + (\alpha x + \beta x')^2]$$

$$= \frac{1}{\beta}(x + p_x^2), \quad (73)$$

where $p_x = \alpha x + \beta x'$ is the canonically conjugate coordinate to x.

Procedure (for graphical representation of the emittance measurement)
1. Plot the design rms ellipse in the phase space (a circle)

$$\left(\frac{x}{\sqrt{\beta}}, \frac{\alpha x + \beta x'}{\sqrt{\beta}}\right) \tag{74}$$

at some reference point s along the trajectory. Normalize the design ellipse to unit radius.

2. In the same figure, plot the ellipse obtained from the measurements of the ellipse parameters at the reference point. Apply the same normalization as in step 1.

3. Using the model of for the lattice, for each wire project its orientation back to the reference point and add the result to the figure; that is, for each point along the wire $(x, x')_w$, do an inverse mapping to the reference point

$$\begin{pmatrix} x \\ x' \end{pmatrix}_{\text{ref pt}} = M^T \begin{pmatrix} x \\ x' \end{pmatrix}_w . \tag{75}$$

The display should summarize the measurements which might include the measured and expected beam widths at each of the wires, the measured and design beam emittances, and the beam intensity.

An example of such graphics from measurements at the SLC is shown in Fig. 26. From Fig. 26 it is immediately obvious that while the measured ellipse has roughly the same area as the design circle, the orientation of the ellipse is incorrect. From the figure, can be immediately deduced the degree of phase space coverage spanned by the wires. In the horizontal plane, for example, the wire orientations are about 0°, -45°, -22.5°, and -67.5°, which is ideal for the 4-wire measurement.

The 'measured ellipse', that is the ellipse that was reconstructed from te individual wire scans based on the measured beam widths and the model-dependent transport matrices, does not in this specially seleccted case represent the true rms distribution of the beam. The raw data used in this measurement are given in Fig. 27. For these complex particle distributions a better characterization of the rms was obtained using an 'asymmetric Gaussian' distribution fucntion for which the left and right hand sides of the measured beam profile were independently fit with two separate Gaussian functions. The fitting function used was

$$f(x) = f_0 + f_{max} e^{-\frac{(x-<x>)^2}{2<x^2>(1+\alpha[sign(x-<x>)]}} \tag{76}$$

where α represents an asymmetry factor and is zero for a perfectly Gaussian beam distribution. The σ for the left and right hand sides of the fitted distribution are $\sigma = <x^2>(1 \pm \alpha)$. For the ellipse reconstruction the average σ was used. When large tails are present in the raw data this more accurately represents the beam distribution. Based on the raw data however it is clear that even with the modified fitting algorithm, the fit only marginally represents the actual beam distributions.

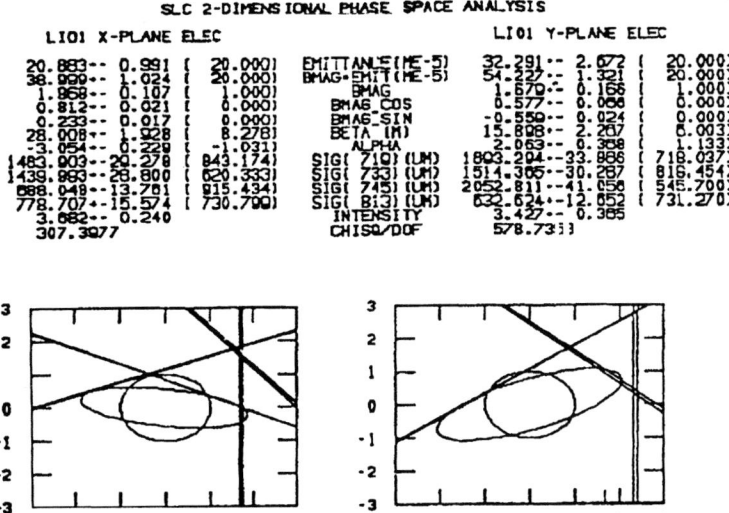

FIGURE 26. Transverse beam emittance measurements from the SLC injector made using multiple wires.

For reasonably well 'matched' beams, the graphical summary display is most useful. In this example however, the raw data are more revealing: the double-humps in the raw data are characteristic of an upstream error: a beam, if kicked transversely will filament (lose coherency due to the natural spread in betatron phase advance) resulting in such double-humps and an increased projected emittance.

If a wire is mounted at 45° with respect to x and y, then it is also possible to measure the coupling between x and y. The full σ-matrix is

$$\begin{pmatrix} \sigma_{11} & \sigma_{12} & \sigma_{13} & \sigma_{14} \\ \sigma_{21} & \sigma_{22} & \sigma_{23} & \sigma_{24} \\ \sigma_{31} & \sigma_{32} & \sigma_{33} & \sigma_{34} \\ \sigma_{41} & \sigma_{42} & \sigma_{43} & \sigma_{44} \end{pmatrix} \qquad (77)$$

where for example σ_{14} represents the coupling between x and y'. Notice that $\sigma_{14} \neq \sigma_{23}$ so that, whereas for the single plane the uncoupled beam matrix reconstruction required a minimum of 3 measurements, to fully reconstruct the coupled beam matrix a total of at least 10 measurements are needed. This includes 3 measurements in the x plane, 3 measurements in the y plane, and 4 measurements in the u plane. An example of a coupled emittance measurement is presented in Figs. 28-31. In this case the raw data are well fit using a Gaussian function. In the text of Fig. 28 the parameters ϵ_1 and ϵ_2 represent the emittance one would measure in the absence of coupling. They are in good agreement with the measured emittances ϵ_x and ϵ_y.

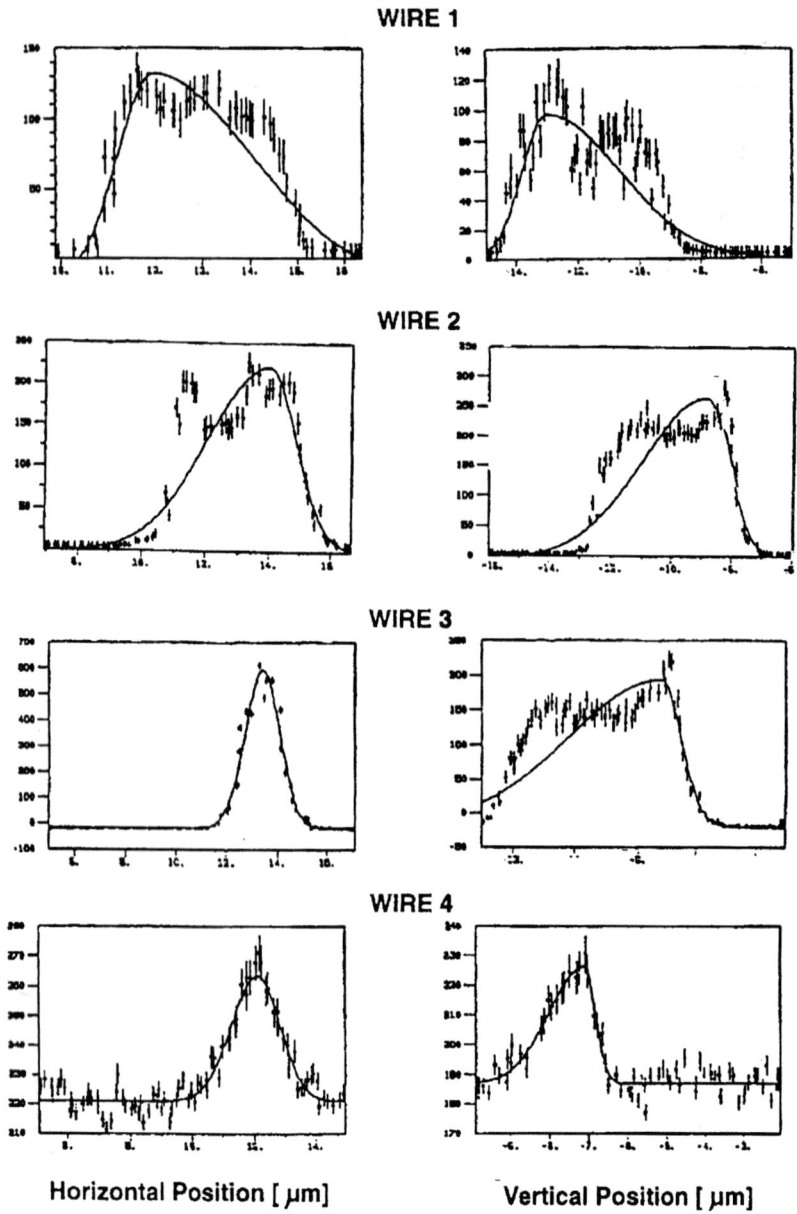

FIGURE 27. Raw data showing individual wire scans used in the emittance measurement summarized in Fig. 26 and 'asymmetric' Gaussian fits (cf. Eq.76).

```
Four Dimensional Phase Space Analysis                LI02  ELEC 2.9E10/bunch
-----------------------------------------
  epsilon_x   =   2.20        +-1.42E-02      epsilon_y   =   1.89         +-1.72E-02
epsilon-b_x   =   2.23        +-1.15E-02    epsilon-b_y   =   1.95         +-2.05E-02
      Bmag_x  =   1.01        +-*********         Bmag_y  =   1.03         +-3.03E-03
  Bmag_cos_x  =   6.355E-02   +-1.13E-02     Bmag_cos_y  =   0.248        +-1.12E-02
  Bmag_sin_x  =   0.152       +-1.50E-02     Bmag_sin_y  =   3.112E-02    +-2.90E-02
   beta_cos_x =   1.08        +-1.44E-02      beta_cos_y =   1.29         +-1.37E-02
   beta_sin_x =   0.154       +-1.41E-02      beta_sin_y =   3.214E-02    +-3.00E-02
  theta_1/deg =  -11.8        +-0.00E+00    theta_2/deg  =  -12.6         +-0.00E+00
  theta_3/deg =  -20.5        +-0.00E+00    theta_4/deg  =   11.7         +-0.00E+00
      tr(CC') =   0.245       +-0.00E+00          det C  =  -0.114        +-0.00E+00

    epsilon_1 =   2.13        +-0.00E+00      epsilon_2  =   1.71         +-0.00E+00
       -det G+=   3.265E-02   +-0.00E+00       psi+/deg  = -126.          +-0.00E+00
        det G-=   0.128       +-0.00E+00       psi-/deg  =  -89.4         +-0.00E+00
       tr(BB')=   0.358       +-0.00E+00           det B =   0.104        +-0.00E+00

      chi^2_x =   29.2                          chi^2_y  =   2.25
      chi^2_u =   7.334E-27                   chi^2/dof  =  15.7
    condition =   546.
```

FIGURE 28. Summary display for a 4-dimensional emittance measurement in the SLC linac.

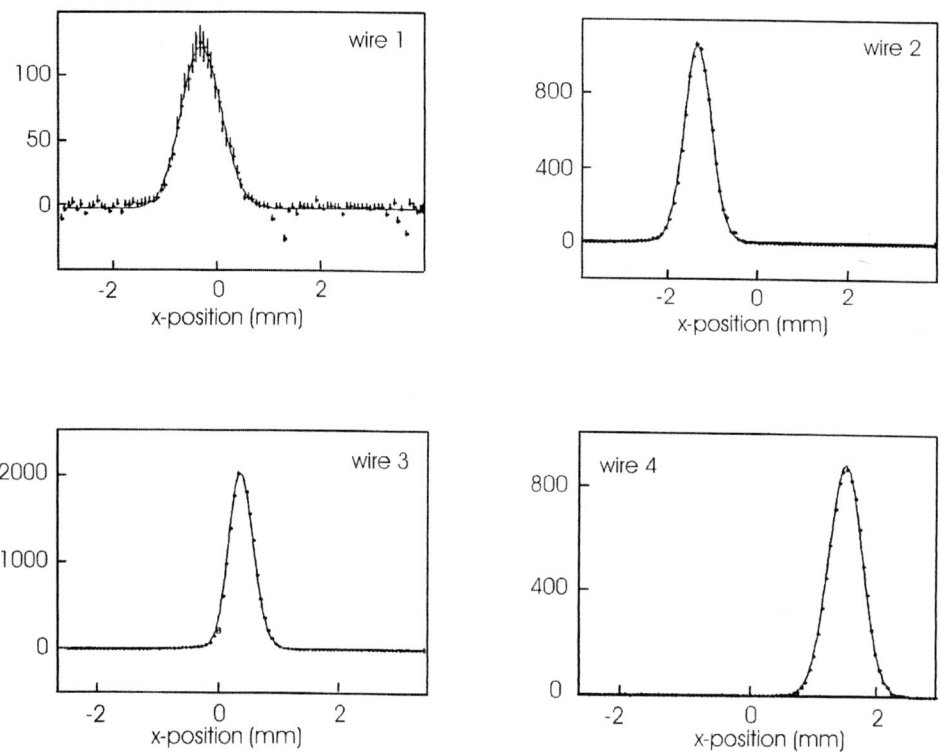

FIGURE 29. Raw x-plane data corresponding to the 4-dimensional emittance measurement summarized in Fig. 28.

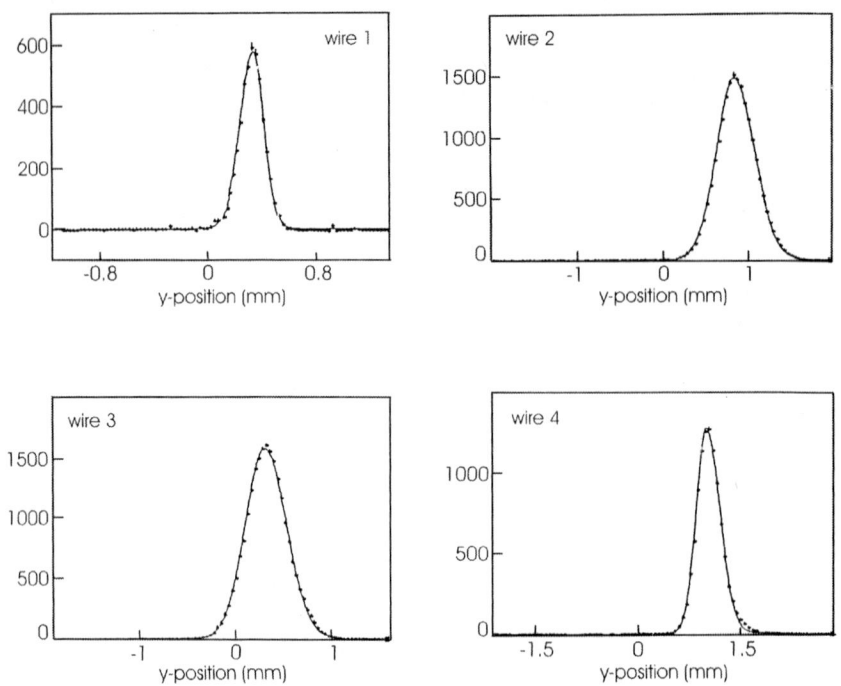

FIGURE 30. Raw y-plane data corresponding to the 4-dimensional emittance measurement summarized in Fig. 28.

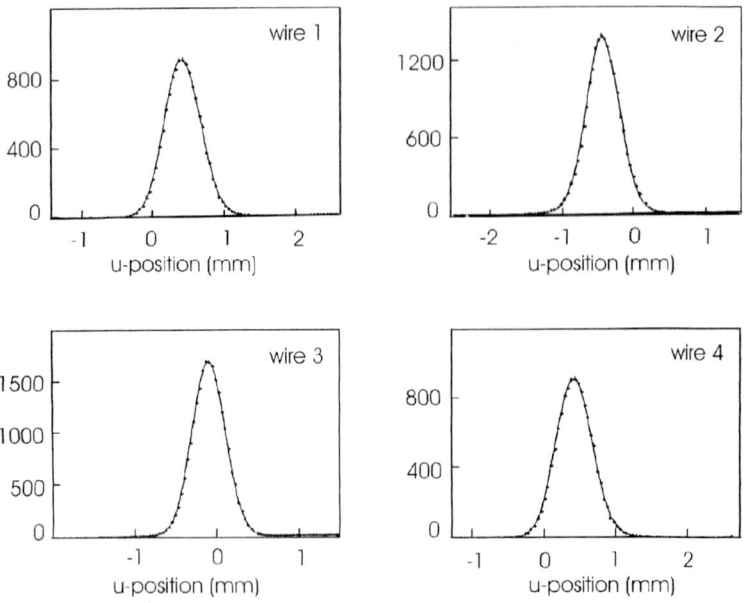

FIGURE 31. Raw u-plane data corresponding to the 4-dimensional emittance measurement summarized in Fig. 28.

Luminosity and the Beam-Beam Interaction

John T. Seeman

Stanford Linear Accelerator Center
2575 Sand Hill Road, Menlo Park, California, 94025, USA
SEEMAN@SLAC.STANFORD.EDU

Abstract. The beam-beam interaction is a strong nonlinear force and ultimately limits the luminosity performance of a colliding beam accelerator. The beam dynamics resulting from the beam-beam force has many aspects and is still under active investigation. Ultimately, the useful product of a collider, the integrated luminosity, depends strongly on the characteristics of the beam-beam interactions.

1   LUMINOSITY

Two beams are collided to provide an event rate R given by $R=\sigma_p L$ where σ_p is the cross section for the physical reaction of interest and L is the luminosity. The luminosity expresses the degree and frequency of overlap of the two beams and is the measure of the ability of the accelerator to produce events. For example in the PEP-II collider [1] operating at the Y(4S), L is about 3×10^{33} cm^{-2}s^{-1} producing B meson pairs at a rate of about 12 Hz.

A typical beam is described by a Gaussian particle density distribution function ρ which depends on the horizontal position x, vertical position y, and the longitudinal coordinate z, and the respective Gaussian rms widths σ_x, σ_y, σ_z. The density ρ is normalized to the number of beam particles N_b in each bunch. The number of bunches is k. The two beams in general have different beam sizes. The luminosity for two beams (one positive and one negative) of different sizes is

$$L = \frac{k\, N_+ N_- f_c}{2\pi \sqrt{(\sigma_{x,+}^{*2} + \sigma_{x,-}^{*2})(\sigma_{y,+}^{*2} + \sigma_{y,-}^{*2})}} \qquad (1)$$

where f_c is the collision frequency of each bunch. If both beams have the same beam sizes in the respective planes, the luminosity simplifies to

$$L = \frac{k\, N_+ N_- f_c}{4\pi \sigma_x^* \sigma_y^*}. \qquad (2)$$

In e$^+$e$^-$ colliders the beams are typically transversely very flat with the horizontal beam size much larger than the vertical size. However, for proton-(anti)proton or ion colliders the beams are typically of equal size horizontally and vertically. For equal size beams, equal beta functions β^* at the collision point, and equal emittances ε with $\sigma = (\varepsilon\beta)^{0.5}$, the luminosity is given by

$$L = \frac{k\,N_+N_-f_c}{4\pi\varepsilon\beta^*}.\tag{3}$$

2 BEAM-BEAM FORCE

The beam-beam force has been described by many authors, e.g. in Refs. 2-8. The transverse deflection [8] of, say, a positive particle passing through the collision point at position $\mathbf{r}^+$ by an opposing negative particle at $\mathbf{r}^-$ is given by

$$\Delta r' = -\frac{e^2}{2\pi\varepsilon_o cp}\frac{r^+ - r^-}{|r^+ - r^-|^2}\tag{4}$$

where e is the electron charge, c the speed of light, p the particle's momentum, and ε_o the permittivity of free space. Each $\mathbf{r}$ is a two dimensional vector and $\Delta\mathbf{r}'$ is a radial deflection between particles. This expression can cause problems in practical use because of the singularity in the denominator. Integrating over the entire opposing bunch gives the total deflection. One of the most convenient forms of this integral is the Bassetti-Erskine equation [5]. An example of the integrated beam-beam force (Figure 1) shows that the largest force (deflection) occurs when the particle's offset from the center of the opposing beam is about 1.4 σ.

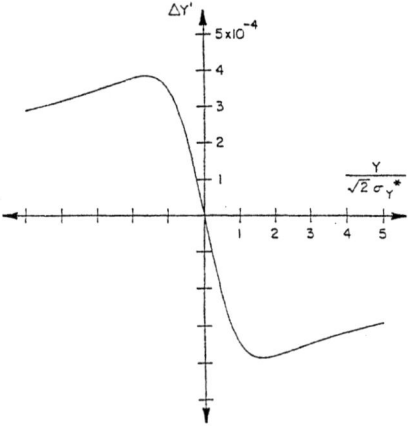

FIGURE 1. The beam-beam force depends on the separation of the test particle and the center of the opposing beam. Here is shown the vertical kick experienced by a typical electron.

3 BEAM-BEAM PARAMETER AND TUNE SHIFT

The beam-beam force is a nonlinear focusing lens for each beam. The linear focusing term can be calculated near the centroid of the opposing bunch. The resulting effective tune change is called the beam-beam parameter ξ,

$$\xi_{y,+} = \frac{r_o N_- \beta^*_{y,+}}{2\pi \gamma_+ \sigma^*_{y,-}(\sigma^*_{x,-} + \sigma^*_{y,-})} \tag{5}$$

where r_o is the classical radius of the electron. There are three additional beam-beam parameters with x and y exchanged and '+' and '-' exchanged. For round beams with equal beta functions, the beam-beam parameter simplifies to

$$\xi_+ = \frac{r_o N_-}{4\pi \gamma_+ \varepsilon_-}. \tag{6}$$

4 LUMINOSITY EXPRESSIONS

Using the beam-beam parameter equations, the expression for the luminosity can be rewritten in various forms under certain conditions, as from Furman and Zisman [7]:

Expression for $\mathcal{L}$	Conditions for validity
$\dfrac{N_+ N_- f_c}{2\pi\sqrt{(\sigma^{*2}_{x,+} + \sigma^{*2}_{x,-})(\sigma^{*2}_{y,+} + \sigma^{*2}_{y,-})}}$	general
$\dfrac{N_+ N_- f_c}{4\pi \sigma^*_x \sigma^*_y}$	$\sigma^*_{x,+} = \sigma^*_{x,-} \equiv \sigma^*_x,\ \sigma^*_{y,+} = \sigma^*_{y,-} \equiv \sigma^*_y$
$K(1+r)\xi_y \left(\dfrac{EI}{\beta^*_y}\right)_{+,-}$	$\sigma^*_{x,+} = \sigma^*_{x,-} \equiv \sigma^*_x,\ \sigma^*_{y,+} = \sigma^*_{y,-} \equiv \sigma^*_y$, $\xi_{x,+} = \xi_{x,-} \equiv \xi_x,\ \xi_{y,+} = \xi_{y,-} \equiv \xi_y$
$K(1+r)\left(\xi \dfrac{EI}{\beta^*_y}\right)_{+,-}$	$\sigma^*_{x,+} = \sigma^*_{x,-} \equiv \sigma^*_x,\ \sigma^*_{y,+} = \sigma^*_{y,-} \equiv \sigma^*_y$, $\xi_{x,+} = \xi_{y,+} \equiv \xi_+,\ \xi_{x,-} = \xi_{y,-} \equiv \xi_-$
$\dfrac{N f_c \gamma \xi}{r_0 \beta^*}$	$\sigma^*_{x,+} = \sigma^*_{x,-} = \sigma^*_{y,+} = \sigma^*_{y,-}$, $\beta^*_{x,+} = \beta^*_{x,-} = \beta^*_{y,+} = \beta^*_{y,-} \equiv \beta^*$, $N_+ = N_- \equiv N,\ E_+ = E_- \equiv E$
$\dfrac{N^2 f_c}{4\pi \epsilon \beta^*}$	$\epsilon_{x,+} = \epsilon_{x,-} = \epsilon_{y,+} = \epsilon_{y,-} \equiv \epsilon$, $\beta^*_{x,+} = \beta^*_{x,-} = \beta^*_{y,+} = \beta^*_{y,-} \equiv \beta^*$, $N_+ = N_- \equiv N$, $D^*_{x,\pm} = D^*_{y,\pm} = 0$
$\dfrac{\pi f_c \gamma^2 \epsilon_x \xi_x \xi_y (1+r)^2}{r_0^2 \beta^*_y}$	$\epsilon_{x,+} = \epsilon_{x,-} \equiv \epsilon_x,\ \sigma^*_{y,+} = \sigma^*_{y,-}$, $\beta^*_{x,+} = \beta^*_{x,-} \equiv \beta^*_x,\ \beta^*_{y,+} = \beta^*_{y,-} \equiv \beta^*_y$, $N_+ = N_- \equiv N,\ E_+ = E_- \equiv E$, $D^*_{x,\pm} = 0$

One effect of the beam-beam parameter is to shift the tune of the ring by an amount $\Delta \nu$, called the linear beam-beam tune shift or coherent beam-beam tune shift. The effect can be calculated using matrix formalism.

$$\cos 2\pi(\nu_0 + \Delta \nu)_{x,y} = \cos(2\pi\nu_0)_{x,y} - 2\pi\xi_{x,y} \sin(2\pi\nu_0)_{x,y}, \tag{7}$$

where ν_0 is the unperturbed betatron tune per crossing. The beta function β at the collision point is also changed from the nominal β_0:

$$\beta \sin 2\pi(\nu_0 + \Delta \nu)_{x,y} = \beta_0 \sin(2\pi\nu_0)_{x,y}. \tag{8}$$

This change in beta functions with beam current is called the dynamic beta effect. Consequently, the beta functions around the ring are affected and the emittance may be changed [6]. The relationship between ξ and $\Delta \nu$ depends on ν_0 as can be seen in Figure 2. To minimize the effect of the tune shift on the beam-beam effect, tunes just above the integer and half integer are often chosen.

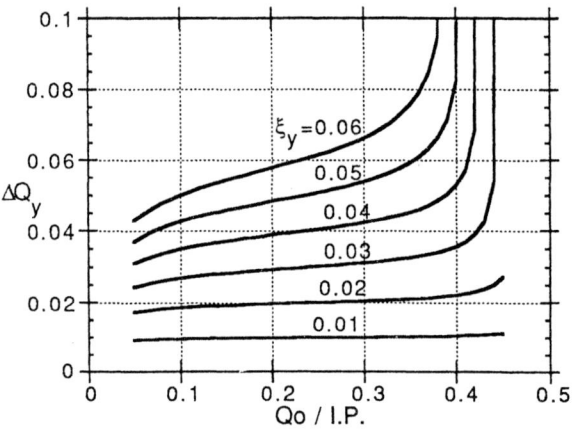

FIGURE 2. The coherent beam-beam tune shift as a function of fractional tune for different beam-beam parameters.

5 LUMINOSITY AND BEAM-BEAM OBSERVATIONS

The operational parameters of several colliders are shown in Tables 1 and 2. These accelerators include e^+e^-, pp_{bar}, and e^-p machines. The performance of these machines is limited by the beam-beam effect, total beam current, or detector backgrounds. There are constant improvement programs at these colliders to increase the luminosity with lower IP betas, more beam current, or multi-bunch beams.

TABLE 1 Examples of Luminosity and Beam-Beam Parameters of Several e^+e^- Colliders

Collider Parameter	Units	VEPP-2M	BEPC	DAΦNE	CESR	KEKB LER	KEKB HER
Energy	GeV	0.5	1.55	0.51	5.28	3.5	8.0
Beam particle		e^+e^-	e^+e^-	e^+e^-	e^+e^-	e^+	e^-
Number of rings		1	1	2	1	1 of 2	1 of 2
Circumference	m	18	240	97.7	768	3018	3018
Eff bend radius	m	1.22	10.3	--	60	16.3	104.5
Number of IRs		2	2	2	1	1	1
U_o (loss/turn)	MeV	0.0045	0.056	0.0093	1.16	0.81	3.5
$SigE/E_o$	xE3	0.39	0.26	0.4	0.59	0.7	0.7
Trans damp τ	msec	13	44	36	25	43	46
Horiz emittance	nm	30	390	1000	210	17	18
Vert emittance	nm	--	--	10	--	0.36	0.36
Crossing angle	mrad	0	0	2 x 12.5	2 x 2.3	2 x 11	2 x 11
Beta y*	cm	3	5	6	1.9	0.7	0.7
Beta x*	cm	30	120	600	120	63	63
Eta x*	cm	40	0	--	0	0	0
Q_y		3.08	6.70	5.09	9.60	44.08	42.135
Q_x		3.06	5.80	6.07	10.53	45.525	44.525
Q_s		0.010	0.021	--	0.06	--	--
Bunch length σ	mm	30	45	45	19.5	6	6
Particles/bunch	xE10	0.64	11.	8.9	13.8	2.7	2.4
Number bunches		1	1	30	36	1154	1154
Total current	mA	2 x 17	2 x 22	2 x 300	2 x 310	499	443
ξ_x		0.015	0.04	0.0	0.028	0.034	0.032
ξ_y		0.050	0.04	0.012	0.060	0.047	0.016
Luminosity L	xE30	3.0	5.0	14.	1100.	2050.	2050.
Reference(s)		9	10	11	12	13	14

As the bunch currents are raised in a collider, the beam-beam tune shifts increase until they reach their limiting values. Often, the vertical tune shift reaches its limit first. In Figure 3 are shown the luminosity and the calculated vertical beam-beam parameter as a function of current in several e^+e^- accelerators [2]. The beam-beam parameter is calculated using the fourth entry in the luminosity expression table above. Note that the measured tune shifts saturate at high current. If the current is increased further, the beam emittance grows in proportion to the current to keep the affected tune shift constant. Consequently, the luminosity grows only linearly, not quadratically with current as before. Two additional examples from LEP and BEPC are shown in Figure 4 illustrating in detail the enlargement. As can be seen from Eq. 5, lowering β^* reduces the tune shifts, and the luminosity can be increased, as shown in Figure 4 for BEPC. Lowering beta is just one method of increasing the luminosity [2, 22].

In addition to enlarging the core of the beam, the beam-beam interaction and the saturation of the tune shift with increasing current can cause enlarged beam transverse tails which extend much farther than a Gaussian profile. An enlarged transverse beam tail can be seen in the LEP example in Figure 5 [23]. These tails often reduce the beam lifetime and may cause lost particle backgrounds in the detector [24].

TABLE 2 Examples of Luminosity and Beam-Beam Parameters of Several Colliders

Collider Parameter	Units	PEP-II LER	PEP-II HER	LEP	TEVATRON	HERA proton	HERA electron
Energy	GeV	3.1	9.0	98.	900	920	27.5
Beam particle		e^+	e^-	e^+e^-	pp_{bar}	p	e^-
Number of rings		1 of 2	1 of 2	1	1	1 of 2	1 of 2
Circumference	m	2200	2200	26659	6280	6336	6336
Eff bend radius	m	13.75	165	3026	754	608	608
Number of IRs		1	1	4	2	2	2
U_o (loss/turn)	MeV	0.68	3.6	2400	--	0	86
$SigE/E_o$	xE3	0.70	0.61	1.7	0.9	0	1.0
Trans damp τ	msec	63	37	6.5	--	--	13
Horiz emittance	nm	40	49	45	3800/2200	5000/$\beta\gamma$	41
Vert emittance	nm	4	2	0.1	3800/2200	5000/$\beta\gamma$	4
Crossing angle	mrad	0	0	0	0	0	0
Beta y*	cm	1.25	1.25	5	35	50	60
Beta x*	cm	50	50	150	35	700	90
Eta x*	cm	0	0	0	0	0	0
Q_y		36.58	23.64	90.28	20.58	32.3	48.2
Q_x		38.65	24.57	76.19	20.58	32.3	49.27
Q_s		0.027	0.045	--	--	--	--
Bunch length σ	mm	13	13	13	600	191	11
Particles/bunch	xE10	9.9	5.3	43.	5.5-/23.+	7.3	3.5
Number bunches		692	692	4	6	180	189
Total current	mA	1492	800	2x3.1	--	100	50
ξ_x		0.069	0.060	--	0.019 -	0.0026	0.024
ξ_y		0.055	0.028	0.083	0.019 -	0.0007	0.061
Luminosity L	xE30	3100	3100	100.	16.	20.	20.
References		15	16	17,18	19	20	21

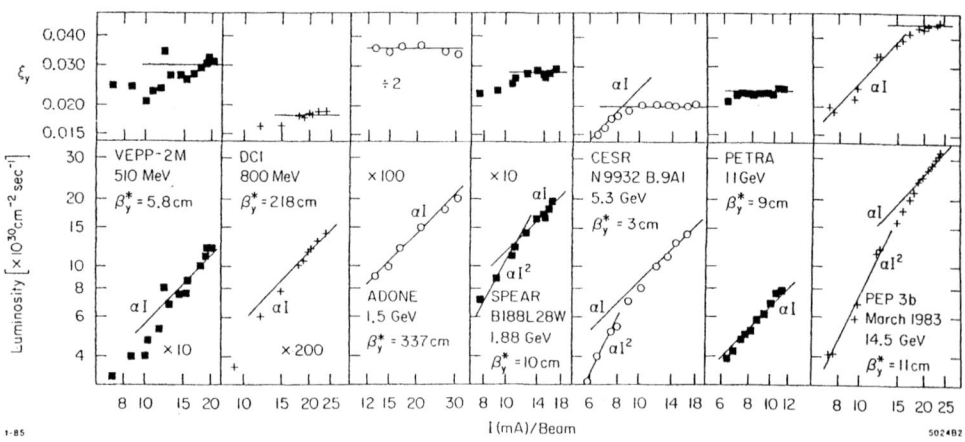

FIGURE 3. Luminosity and vertical beam-beam parameter versus current for seven e^+e^- colliders. The tune shift saturates at some current above which the luminosity grows linearly.

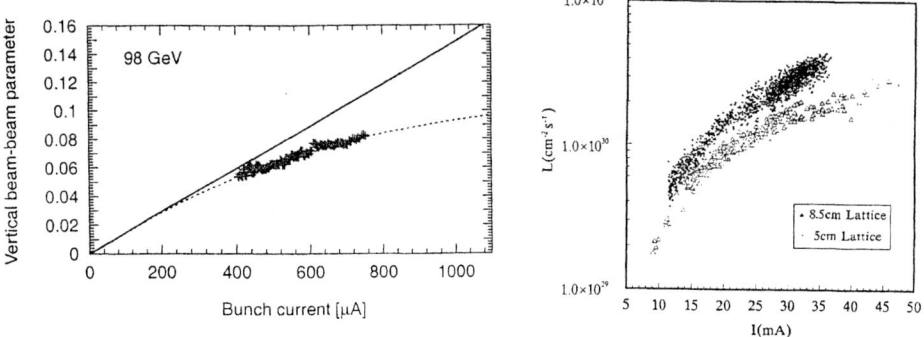

FIGURE 4. LEP (left) and BEPC (right) data showing the saturation of the beam-beam parameter.

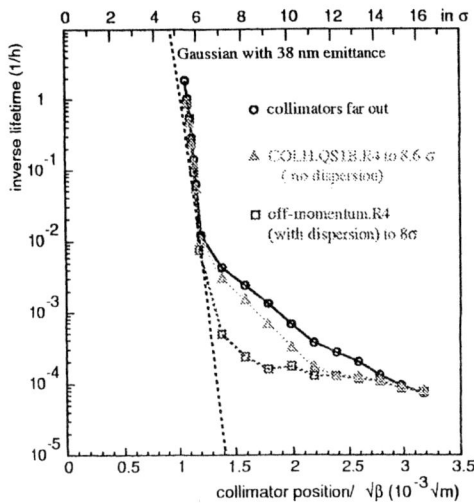

FIGURE 5. Vertical non-Gaussian beam tails as measured and simulated for LEP [23].

The beam-beam tune shifts can often be observed on a spectrum analyzer. The beam spectrum typically shows two peaks corresponding to the "σ" mode where the two bunches oscillate in phase with no beam-beam focusing and the "π" mode where the two bunches oscillate out of phase experiencing the full beam-beam force. In Figure 6 are shown measured beam-beam modes in TRISTAN. The π mode is related to the beam-beam parameter by the Yokoya factor [25], which is not unity, as real bunches that are offset during collisions are not transversely rigid ($\Delta v_y = 1.21 \xi_y$).

The above discussion applies to two beams with equal tunes. If the two tunes are not equal, as is likely in two ring colliders, then a more complicated analysis must be done [26, 27]. For widely separated tunes, both modes move and the sum gives the beam-beam parameter [27]. How the beam-beam modes change with tune separation can be seen in Figure 7, and an example of a measurement in the PEP-II HER is shown in Figure 8.

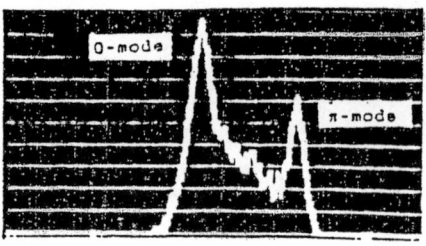

FIGURE 6. Measured beam-beam modes in TRISTAN.

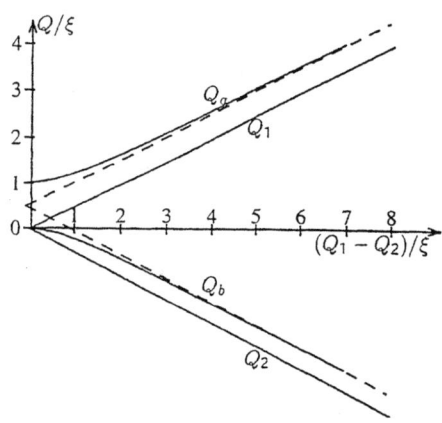

FIGURE 7. Beam-beam mode tunes Q_a, Q_b versus tune separation Q_1-Q_2. [27].

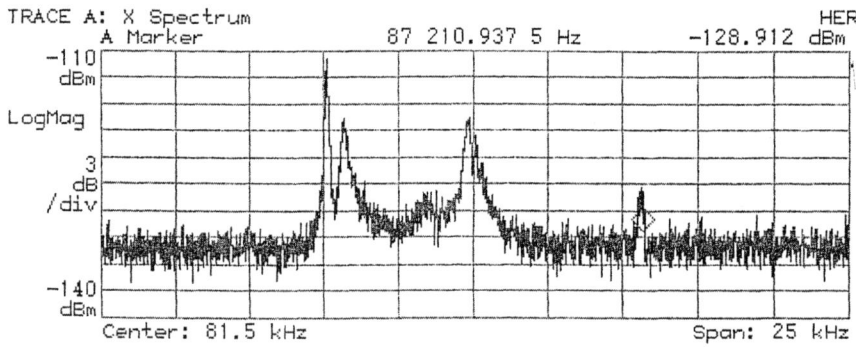

FIGURE 8. Measured beam-beam modes in PEP-II HER with different tunes for LER and HER. The narrow peak on the left is a pilot bunch in HER with no corresponding colliding bunch in LER. Note that both beam-beam modes have shifted.

The value of the saturated tune shift for a given collider should be relatively insensitive to beam energy but in practice has a strong dependence. This dependency can be seen for the PETRA collider [28] in Figure 9. The maximum value of the tune shift is more than tripled by increasing the energy from 7 GeV to 17 GeV. Many colliders see an increased tune shift roughly proportional to energy.

In electron colliders ξ often reaches values of about 0.05 to 0.08, whereas in hadron colliders perhaps it reached values of only about 0.005. This large difference is due to the strong synchrotron radiation damping in electron machines. Furthermore, in some hadron colliders it is difficult to produce enough particles to reach the tune shift limit as can be seen in Figure 10 for the antiprotons in the TEVATRON [19].

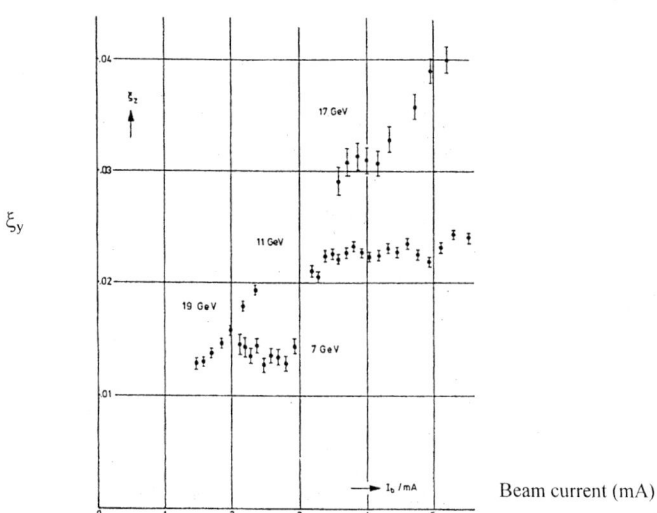

FIGURE 9. Beam-beam parameters for different PETRA energies [27].

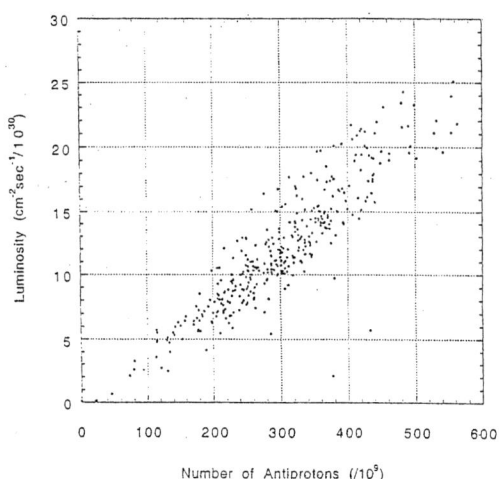

FIGURE 10. The luminosity in the TEVATRON is proportional to the number of antiprotons.

6 BEAM-BEAM RESONANCES

Beam particles can have their betatron tunes shifted onto resonances by the beam-beam interaction and be elevated to large amplitudes or be lost. Thus, the choice of tunes is very important to maximize the beam-beam tune shift and the luminosity. Every collider conducts tune plane searches for the optimal x and y tunes. Few colliders have equal tune plane "foot prints" because the synchrotron tunes are not equal and because the lattice nonlinearities that drive resonances are not equal in all machines. In Figure 11 a tune plan study in VEPP-4 is shown indicating many resonances [29]. A similar study for PEP-II is shown in Figure 12 [30].

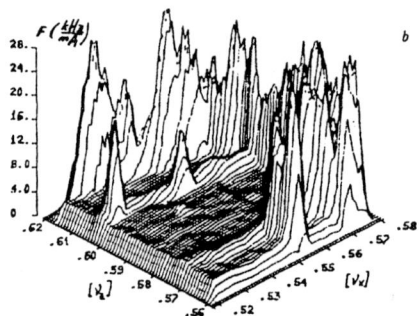

FIGURE 11. Tune plane studies in VEPP-4 [29]: e^+ loss rate versus x-y tunes.

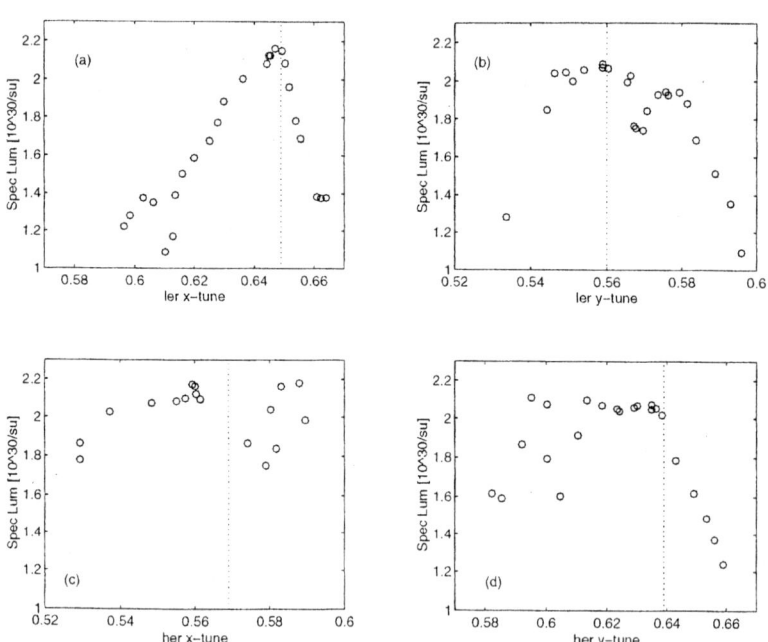

FIGURE 12. Tune plane studies in PEP-II [30]: Specific luminosity versus tunes.

7 COHERENT BEAM-BEAM MODES

Coherent beam-beam modes have theoretically set limits on collider operation. Dipole modes have been studied to explain observation of modes seen in spectrum analyzer displays [25, 31]. Although there have been predictions of other modes, only dipole modes have been seen in colliders so far. These studies have also set theoretical limits on the tune shifts but correlations with observations have been difficult. These modes are still under study.

Early in the designs of asymmetric colliders, studies indicated that coherent modes would be a problem if the two rings were to have different circumferences [32].

An observation in LEP and an associated numerical model indicate that the beam-beam interaction can enhance the head tail instability and reduce the threshold for a transverse quadrupole mode blowup ($m = -1$) for positive chromaticities [33].

Finally, it was observed in PEP-II that the beam-beam interaction strongly suppresses bunch centroid motion to the extent that the beam-beam interaction is much stronger than the bunch-by-bunch transverse feedback system [34]. This effect can been seen in Figure 13, where the feedback system gain needed to keep the beam stable is measured as a function of centering of the two colliding beams. The effect is suspected to be a form of Landau damping. For this reason, PEP-II often injects beam while in collision.

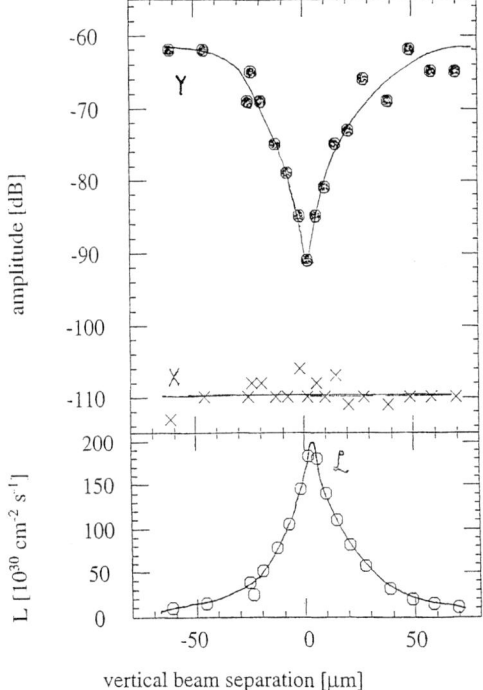

FIGURE 13. Strong dipole mode damping in PEP-II by the beam-beam interaction [34]. The feedback gain can be reduced while keeping the beam stable with more beam-beam overlap.

8 BUNCH LENGTH AND HOURGLASS EFFECTS

There are limits on how low the beta functions at the interaction point can be reduced because of the beam-beam interaction. With a finite bunch length the change in beta function with longitudinal distance from the IP will enlarge the beam size over the length of the bunch thus reducing the luminosity. Also, the enlarged beta away from the IP forces the bunch as a whole to experience a larger beam-beam tune shift.

The geometrical loss in luminosity can be calculated given the bunch length σ_z and the quadratic longitudinal dependence of the beta function at the IP, $\beta(s)=\beta_0+s^2/\beta_0$ [2, 35, 36]. The geometrical reduction in luminosity assuming $\beta_y \ll \beta_x$ and with K_0 the modified Bessel function is given by

$$L = L_0 \frac{1}{\sqrt{\pi}} \cdot \left(\frac{\beta_y^*}{\sigma_z}\right) \exp\left(\frac{\beta_y^{*2}}{2\sigma_z^2}\right) K_0\left(\frac{\beta_y^{*2}}{2\sigma_z^2}\right). \qquad (9)$$

This equation is shown graphically using PEP-II parameters [37] in Figure 14. For asymmetrical colliders with different beta functions for the two beams, the luminosity reduction expression is more complicated [35].

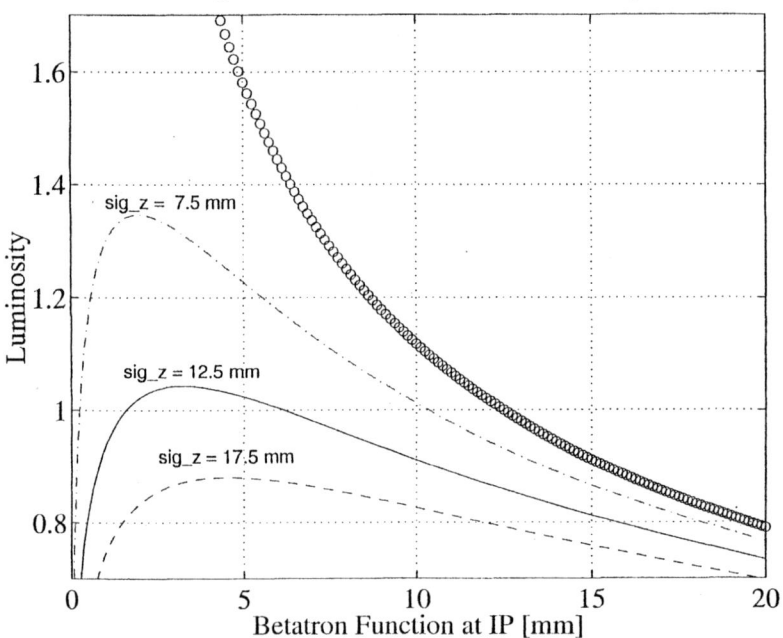

FIGURE 14. Geometrical reduction in luminosity as the β^* is lowered. The upper curve is for a an ideal zero bunch length.

Measurements of the dynamic beta and the hourglass effects have been made at the CESR collider using the CLEO detector to measure the beam sizes at the collision point [37,38]. In Figure 15 are shown the height of the luminous region that increases with the longitudinal distance from the IP and the measured β_x that decreases with the bunch current.

The longitudinal variation of the beta function over the interaction region also leads to a change in the beam-beam parameter with β^*_z/σ_z [35, 36]. The vertical beam-beam parameter for particles with a small longitudinal offset from the bunch center $z \ll \sigma_z$, with $\beta^+_y = \beta^-_y$, and using K_0 and K_1 modified Bessel functions is given by

$$\xi_y = \xi_{y0} \frac{1}{\sqrt{2\pi}} \cdot \left(\frac{\beta^*_y}{\sigma_z}\right) \exp\left(\frac{\beta^{*2}_y}{\sigma^2_z}\right) \left[K_0\left(\frac{\beta^{*2}_y}{\sigma^2_z}\right) + K_1\left(\frac{\beta^{*2}_y}{\sigma^2_z}\right) \right]. \tag{10}$$

As can be verified, ξ_y will always be equal to or larger than ξ_{y0}. The threshold for a tune shift reduction is about $\beta^*_y/\sigma_z \sim 1.5$. Measurements of the geometrical and beam-beam parameter changes with bunch length and β^*_y for several accelerators are shown in Figure 16. Because of these two hourglass effects, lowering β^*_y/σ_z significantly below unity will not be very profitable. Typical for a operating collider, any project to reduce the beta function will also include a bunch length reduction project.

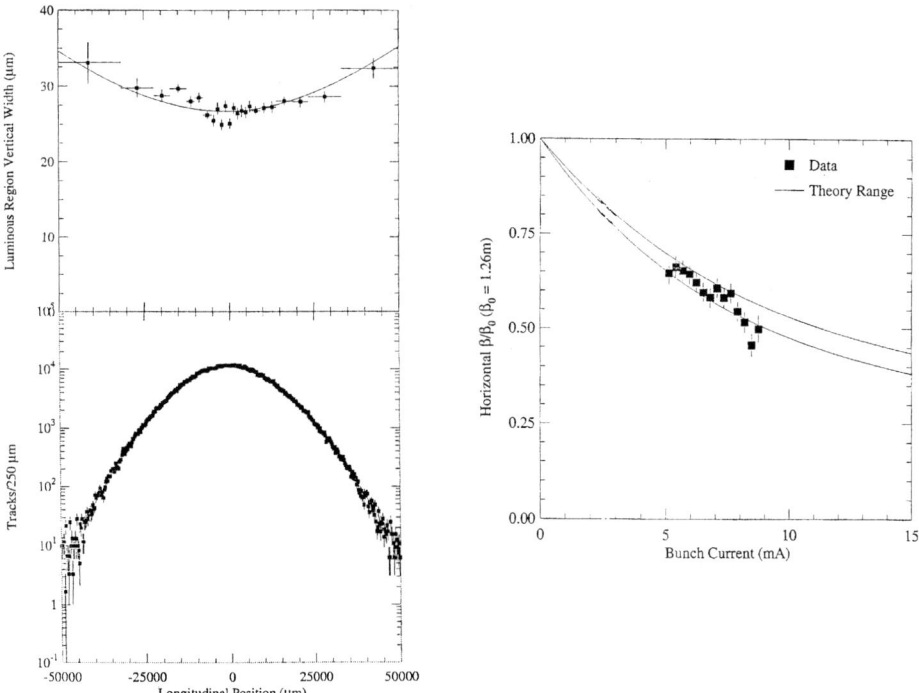

Figure 15. Measurements of the dynamic beta and hourglass effects at CESR [37,38].

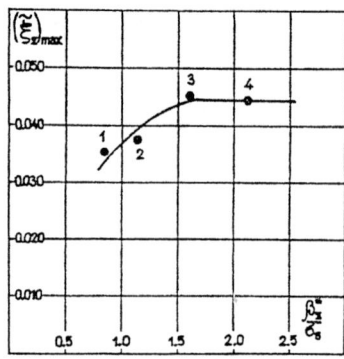

Experimental dependence of the attained space charge parameter $(\tilde{\xi}_s)_{max}$ on the β_z^*/σ_s parameter. The maximum luminosity regime. $E = 510$ MeV, $H = 7.5$ T: $\beta_z^* = 3.0$ cm (1); $\beta_z^* = 3.9$ cm (2); $\beta_z^* = 5.5$ cm (3); $\beta_z^* = 7.5$ cm (4).

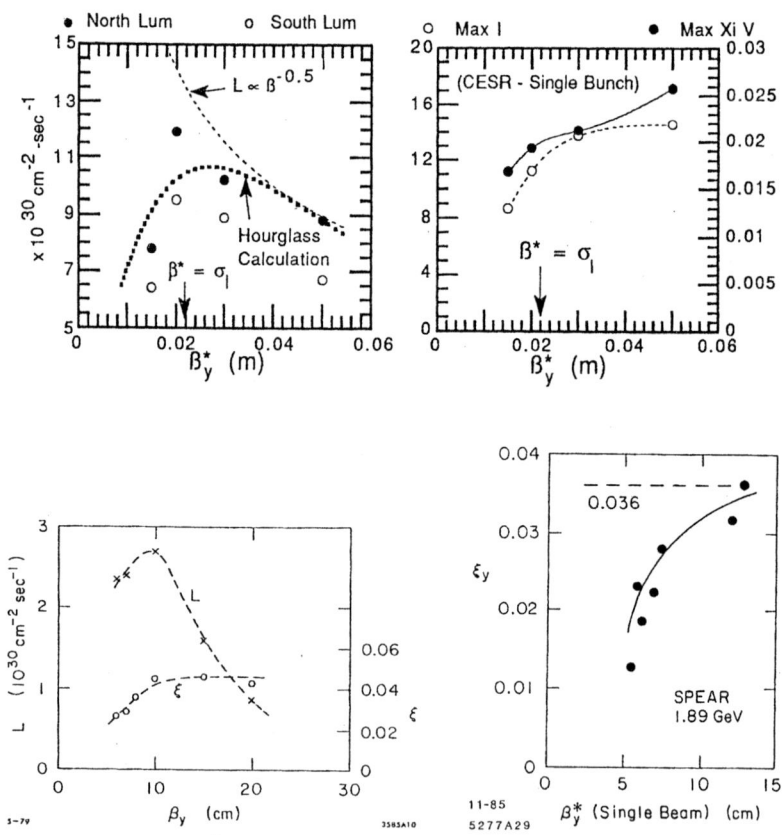

FIGURE 16. Measurements of luminosity and beam-beam parameter reduction with reduced β^*. Top: VEPP-2M is at the top [36]. Middle: CESR [5]. Bottom: SPEAR [2].

9 TWO RING AND MULTI-BUNCH EFFECTS

During the commissioning of several recent two-ring colliders (HERA, DAΦNE, KEKB, and PEP-II), several new techniques have been developed to help diagnose the beam-beam interaction. These techniques are Σ scans, beam-beam deflection scans, and longitudinal waist scans. Furthermore, several new multi-bunch effects influence the beam-beam interaction.

The first technique, the Σ scan, is to scan one beam transversely through the other beam and measure the luminosity as a function of offset [40, 41, 42]. The transverse displacements are made either by magnetic fields or conversely with the RF phase of one of the rings (used with rings with an IP crossing angle). In Figure 17 is shown the PEP-II luminosity versus the vertical position at the IP of one of the PEP-II beams with the other beam fixed. The resulting curve has a Gaussian width called Σ.

$$\Sigma_{x,y} = \sqrt{\sigma^{*2}_{x,y,+} + \sigma^{*2}_{x,y,-}} \; . \tag{11}$$

The luminosity for a collider can be written in terms of the two Σs:

$$L = \frac{k \, N_+ N_- f_c}{2\pi \Sigma_x \Sigma_y} \; . \tag{12}$$

Using the Σ scan technique all the parameters of this equation can be independently measured and a self-consistency check made. For example, in PEP-II all the input parameters were measured, but initially the overall agreement was good only to 50%. It took quite some effort and over a year of working on calibration constants and measurement errors to make this equation check, at the present, to about 5%. Much was learned and is being learned about PEP-II in the process.

The Σ scans are typically done at low bunch currents to minimize the beam-beam blow up when the beams are about one Gaussian sigma separated. Examples of beam-beam scans at high currents that exhibit distorted shapes are shown in Figure 18. Vertical scans are distorted less than the horizontal scans. In fact, the horizontal scans show local depressions at about 1 to 1.3 sigma. It is clear that centered beams are most resistant to the beam-beam effect. Transverse beam-beam feedbacks have been developed to keep the beams centered during collisions. These feedback systems work very well but can be fooled by finding the local minimum near about 2 sigma horizontally.

The Σ scans are done in the plane of the devices (magnets) that makes the position shifts. However, the natural planes of the beam may be in different planes. Thus, by scanning in several planes and analyzing the principle planes, the true Σ values are determined [41]. An example of a multi-plane scan is shown in Figure 19. Tilted beams at the interaction region can cause additional beam-beam effects [43].

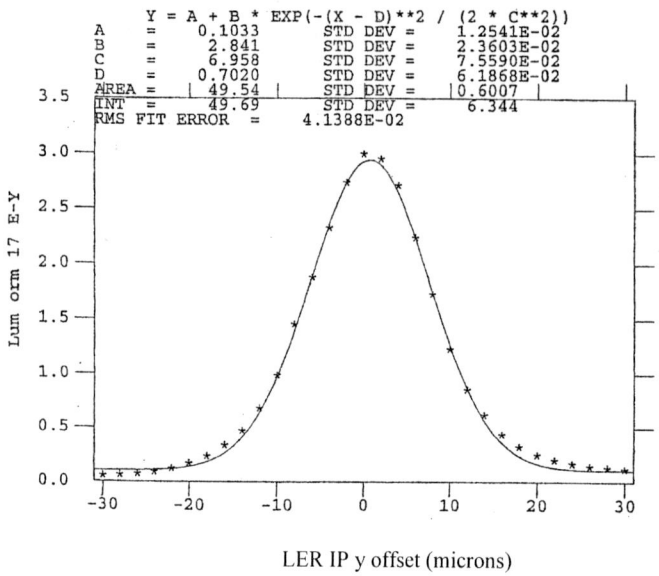

FIGURE 17. Beam-beam position scan in PEP-II to measure Σ_y. The luminosity is measured as a function of a relative vertical beam position offset. Σ_y was measured here to be 6.9 μm.

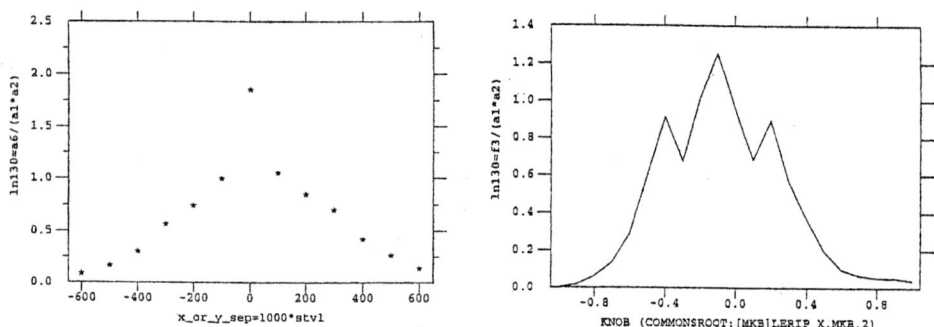

FIGURE 18. Beam-beam position scans in PEP-II at high beam currents showing shape distortions with offsets of about one sigma. Left: vertical scan. Right: horizontal scan. The vertical scale is measured luminosity.

The second scanning technique is to measure the orbit distortion of the two beams as they are scanned through each other [41,43]. The measurements are made with position monitors near the interaction region. The deflection angle at the IP is fitted. These scans have to be done at high currents so the beam-beam forces are strong. Consequently, the beams often become enlarged which causes beam losses. However, because the deflections are caused by the average beam offset, the deflection measurements are still valid. Examples of beam-beam deflection measurements from

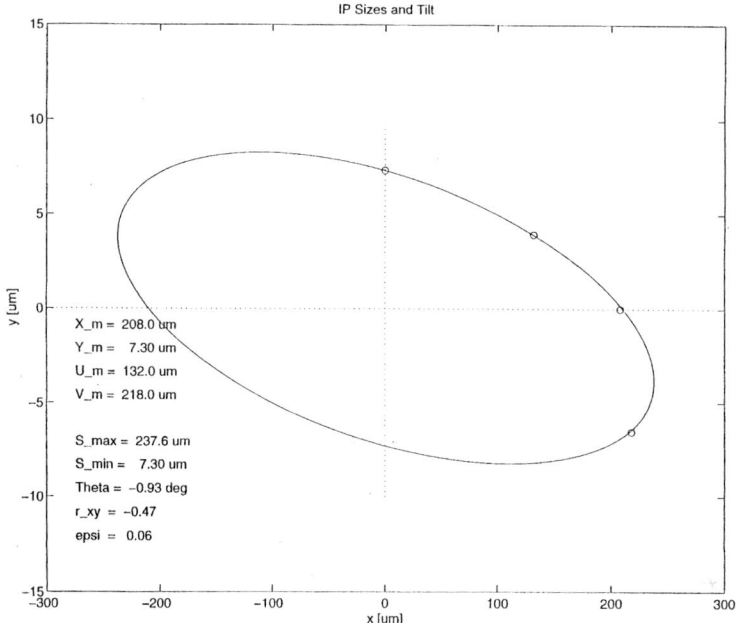

FIGURE 19. A multi-plane Σ scan can extract the principle planes of the beams [42].

KEKB [13] and PEP-II are shown in Figure 20. The measured slope S near the origin of the beam-beam deflection scan is related to the two Σs. With $R=\Sigma_y/\Sigma_x$ the slope is

$$S_{x,y}^{+/-} = \frac{2r_e N^{-/+}}{\gamma^{+/-} \Sigma_x \Sigma_{x,y}(1+R)}. \tag{13}$$

The deflections reach a maximum at a certain transverse offset. The asymptotic deflection amplitude Θ measures the horizontal beam size Σ_x knowing the energy $\gamma^{+/-}$ of the deflected bunch and the intensity $N^{-/+}$ of the deflecting bunch.

$$\Theta^{+/-} = \frac{\sqrt{2\pi}\, r_e N^{-/+}}{\gamma^{+/-} \Sigma_x \sqrt{(1+R^2)}}. \tag{14}$$

These measurements can be used to cross check the luminosity scans described above. The out of plane deflection measurements (e.g. looking in y while scanning in x) are also useful to identify beam centroid offsets [44].

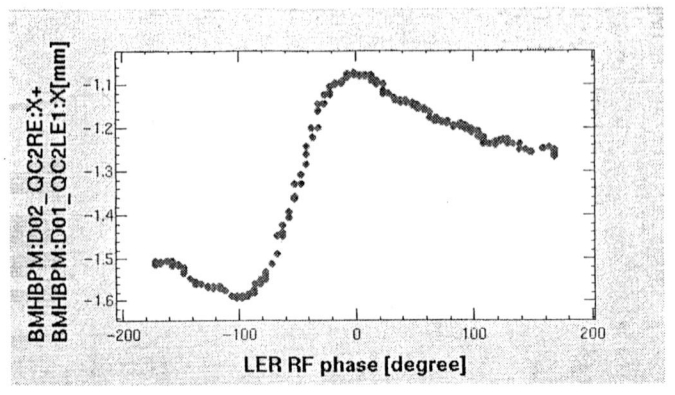

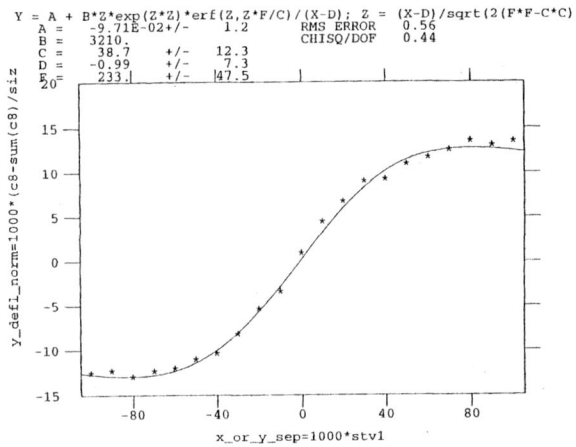

FIGURE 20. Beam-beam deflection scans. Top: Horizontal at KEKB. Bottom: Vertical at PEPII.

The third technique is to perform longitudinal waist scans to verify that the beta minima are at the collision point. In one plane there are three variables: the relative RF phase between the two beams and the two beta waist locations. In this technique the luminosity is measured as a function of the relative RF phase. The phase moves the collision point through the waist location. The IP location moves only half as far as the RF phase. Next, the magnetic waist of one beam is moved longitudinally by adjusting nearby IR quadrupoles and is then held fixed. A second longitudinal RF scan is done. The waist is moved again and a further scan is done. When enough scans are done, the optimum waist location can be determined with respect to the fixed waist of the other beam. The RF phase at the best luminosity is also the optimum. An example of this longitudinal waist scan is shown in Figure 21 for PEP-II. These measurements show that the proper vertical waist is about 2.5 mm displaced from the starting configuration. Since the vertical beta function at the IP in e^+e^- colliders is usually much smaller than the horizontal, the horizontal waist rarely moves far enough to cause problems and the waist scans are done primarily in the vertical.

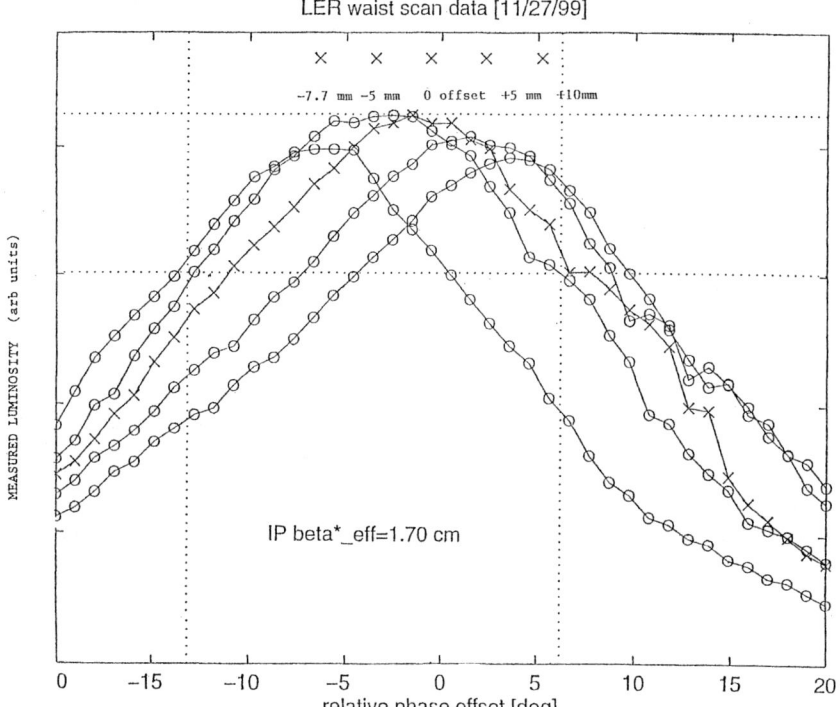

FIGURE 21. Longitudinal waist scans determine the proper location of the two beta minima and relative RF phase. The LER β_y waist was moved and the luminosity scan done.

In two-ring colliders in order to get high luminosity many bunches are used. This situation leads to multi-bunch effects which can change the beam-beam interaction.

If adjacent bunches can not be separated quickly enough at the collision point, additional long-range beam-beam interactions will occur [45, 46]. This effect is called parasitic crossings. This effect is both a dipole deflection and a (nearly) quadrupole focusing term. Remember that the beta function become very large at the locations of the parasitic crossings. There are various schemes to control this effect including crossing angles, beta function, and emittance changes. In PEP-II with bunches spaced every 2 RF buckets and a horizontal separation at the first parasitic crossing of about 10 σ_x, the luminosity for the same currents and number of bunches is reduced by about 10%.

Finally, other multi-bunch effects in the ring that can affect collisions. An example of such an effect is shown in Figure 22 where the electron cloud instability (ECI) blows up the positron beam along a train of bunches in PEP-II LER and reduces the luminosity [47, 48, 49]. A short bunch gap removes many of the offending electrons and restores the luminosity. The temporary solution is to collide many short bunches as shown in Figure 23 [50].

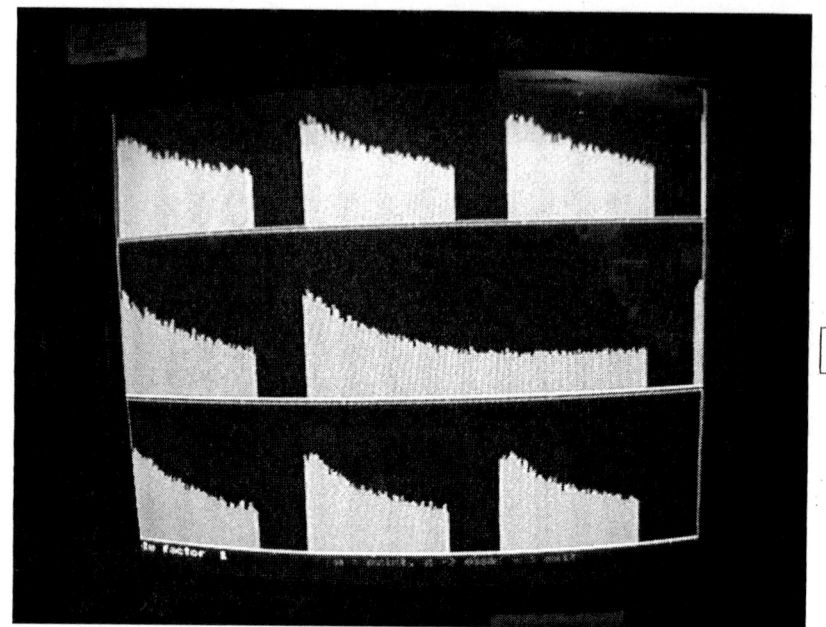

FIGURE 22. The measured luminosity for each bunch in PEP-II decreases along the bunch train because the electron cloud instability enlarges the positron beam. A long bunch gap removes the offending electrons and restores the luminosity. The measurements were made with bunches every 4 rf buckets with 8 large bunch gaps.

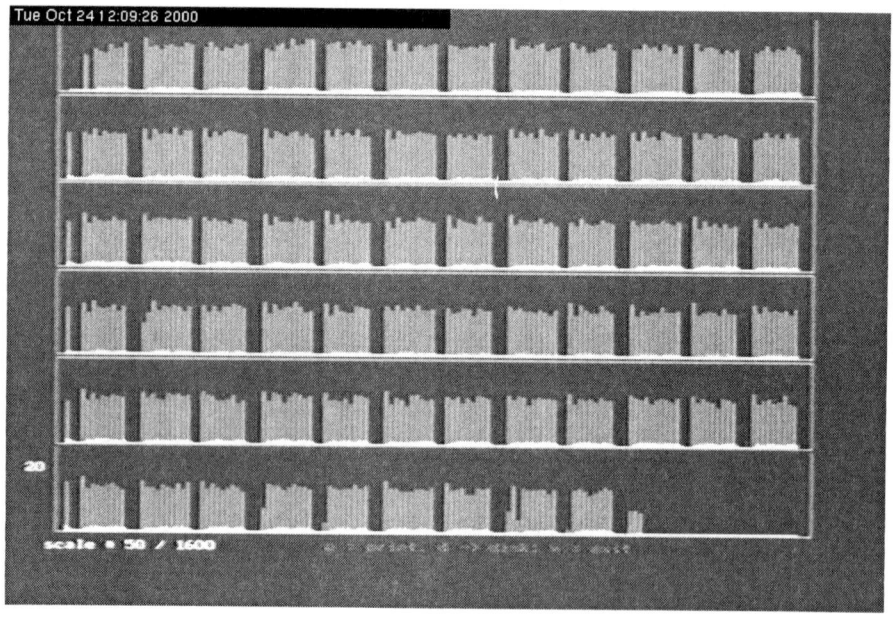

FIGURE 23. The measured luminosity for each bunch in PEP-II with short bunch gaps every 9 bunches. Bunches are spaced every 4 rf buckets using 9 bunch mini-trains followed by 3 missing bunches. The average luminosity remains much higher.

10 ACKNOWLEDGMENTS

The author wishes to thank the Joint Accelerator School (JAS) for providing this opportunity to discuss one of the most interesting accelerator physics topics of today's colliders. Many thanks go to our Russian hosts who provided the JAS School with a wonderful time with their generous hospitality, a unique educational environment, and the spectacular scenery. Discussions of the beam-beam interaction with Y. Cai, A. Chao, Y. Funikoshi, W. Kozanecki, S. Kurokawa, K. Oide, M. Placidi, D. Rice, B. Richter, D. Rubin, M. Sullivan, G. Voss, U. Wienands, and F. Willeke have been very helpful.

11 REFERENCES

1. J. Seeman, "Status of the PEP-II B-Factory," ICFA Conference on Beam Dynamics Issues for e^+e^- Factories, Frascati Physics Series Vol. 10, p. 61 (1998).
2. J. Seeman, "Observations of the Beam-Beam Interaction," Proc. of the Joint US-CERN School on Particle Accelerators, Santa Margherita di Pula, Sardiania, Janaury 1985, p. 121.
3. L. Evans, J. Gareyte, "Beam-Beam Effects," CAS CERN 87-03, p. 159 (1987).
4. J. Tennyson, "The Beam-Beam Limit in Asymmetric Colliders," Workshop on Asymmetric Colliders, LBL, AIP Conf. Proc. 214, p. 130 (1990).
5. D. Rice, "Beam-Beam Interaction: Experimental," Workshop on e^+e^- Asymmetric Colliders, LBNL, AIP Conf. Proc. 214, p. 219 (1990).
6. K. Hirata, "Beam-Beam Effects in Storage Rings," *Accelerator Physics and Engineering*, A. Chao and M. Tigner eds., p. 134 (1999).
7. M. Furman, M. Zisman, "Luminosity," *Accelerator Physics and Engineering*, A. Chao and M. Tigner eds., p. 247 (1999).
8. R. Talman, "Multiparticle Phenomena and Landau Damping," AIP Conf. Proc. 153, p. 790 (1985).
9. G. Tumaikin, A. Temnykh, 13th Intl. Conf. H.E. Acc., p. 88 (1986).
10. Z. Zhao, S. Wang, "BEPC and the Future Program at IHEP," APAC, p. 54 (1998).
11. M. Zobov *et al.*, "Status Report on DAΦNE Performance," EPAC, Vienna, Austria, p. 43 (2000).
12. S. Belomestnykh *et al.*, "Running CESR at High Luminosity and Beam Current with Superconducting RF System," EPAC, Vienna, p. 2025 (2000).
13. Y. Funikoshi, *et al.*, "KEKB Performance," EPAC, Vienna, Austria, p. 28 (2000).
14. S. Kurokawa, "Status of KEKB," Note to the KEKB Accelerator Advisory Committee, December (2000).
15. J. Seeman *et al.*, "Status Report on PEP-II Performance," EPAC, Vienna, Austria, p. 38 (2000).
16. U. Wienands *et al.*, "Status of PEP-II," Proc. of the Second Workshop on PEP-II Performance, SLAC WP-19 (2000).
17. G. Arduini *et al.*, "LEP Operation and Performance with 100 GeV Colliding Beams," EPAC, Vienna, p. 265 (2000).
18. J. Jowett, "Beam Dynamics at LEP," ICFA Conference on Beam Dynamics Issues for e^+e^- Factories, Frascati Physics Series Vol. 10, p. 15 (1998).
19. A. Sery *et al.*, "Status of R&D Activities at Fermilab," Beam-Beam Workshop, LHC99, Geneva, CERN-SL-99-039 AP, p. 33 (1999).
20. M. Bieler *et al.*, "Recent and Past Experience with Beam-Beam Effects in HERA," Beam-Beam Workshop, LHC99, Geneva, CERN-SL-99-039 AP, p. 12 (1999).
21. G. Hoffstaetter, "Future Possibilities for HERA," EPAC, Vienna, p. 13 (2000).
22. J. Seeman, M. Sullivan, "Luminosity Upgrade Possibilities for the PEP-II B-Factory,"EPAC,

Berlin, p. 421 (1998).
23. H. Burkhardt, I. Reichel, G.Roy, "Transverse Beam Tails Due to Inelastic Scattering," CERN-SL-99-068, Geneva (1999).
24. D. Shatilov, A. Zholents, "Lifetime and Tail Simulations for Beam-Beam Effects in PEP-II B-Factory," LBL –36484 and PEP-II AP Note 95.06 (1995).
25. K. Yokoya, "Tune Shift of Coherent Beam-Beam Oscillations," Part. Accel. 27, p. 181 (1990).
26. The SPEAR Group, "Beam-Beam Coupling in SPEAR," Proc. 9th Int. Conf. on High Energy Accelerators, SLAC p. 66 (1974).
27. A. Hofmann, "Beam-Beam Modes for Two Beams with Unequal Tunes," Beam-Beam Workshop, LHC99, Geneva, CERN-SL-99-039 AP, p. 56 (1999).
28. A. Piwinski, "Beam-Beam Observations at DESY," ICFA Beam Dynamics Workshop on Beam-Beam Effects in Circular Colliders," Novosibirsk, p. 12 (1989).
29. A. Temnykh, "Observation of Beam-Beam Effects on VEPP-4," ICFA Beam Dynamics Workshop on Beam-Beam Effects in Circular Colliders," Novosibirsk, p. 5 (1989).
30. M. Minty, "Tune Space Mapping Under High Luminosity Conditions at PEP-II", PEP-II Note 2/18/2000.
31. R. Meller, R. Siemann, IEEE Trans. NS-28, p. 2431 (1981).
32. K. Hirata, E. Keil, "Barycentre Motion of Beams due to Beam-Beam Interaction in Asymmetric Ring Colliders," NIM A292 p. 156 (1990).
33. K. Cornelis and M. Lamont, "Head-Tail Instabilities Enhanced by the Beam-Beam Interaction," EPAC, London, p. 1150 (1994).
34. M. Minty, "Collective Effects Associated with Ultra-High Beam Intensities in Factories," EPAC, Vienna, p. 146 (2000).
35. M. Furman, "The Hourglass Reduction Factor for Asymmetric Colliders," ABC-21/ESG-161, LBNL (1991).
36. P. Ivanov *et al.*, "Luminosity and the Beam-Beam Effects in VEPP-2M," ICFA Beam Dynamics Workshop on Beam-Beam Effects in Circular Colliders, Novosibirsk, p. 26 (1989).
37. D. Cinabro, "Observation of the Dynamic Beta Effects at CESR with CLEO," CBN 96-17, Cornell CESR Report (1996).
38. D. Cinabro, K. Korbiak, "Observation of the Hourglass Effect and Measurement of CESR Beam Parameters with CLEO," CBN 00-6, Cornell CESR Report (2000).
39. F.-J. Decker, "Multi Bunch PEP-II Operation," PEP-II Performance Workshop, SLAC-WP-19, SLAC (2000).
40. P. Bambade *et al.*, "Observation of Beam-Beam Deflections at the Interaction Point of the SLAC Linear Collider," PRL Vol. 62, No. 25, p. 2949 (1989).
41. M. Furman *et al.*, "Closed Orbit Distortion and the Beam-Beam Interaction," LBL-32435, ESG-193, DAPNIA/SPP 92-03, ABC-49 (1992).
42. M. Venturini, W. Kozanecki, "The Hourglass Effect and the Measurement of the Transverse Size of Colliding Beams by Luminosity Scans," SLAC-PUB-8699 (2000).
43. D. Sagan, "The Effect of Coupling on Luminosity Performance," CBN-95-01, Cornell (1995).
44. M. Venturini, W. Kozanecki, "Out-of-Plane Beam-Beam Deflections as a Diagnostic Tool and Application to PEP-II," SLAC-PUB-8700 (2000).
45. J. Jowett, "Parasitic Beam-Beam Effects and Separation Schemes," *Accelerator Physics and Engineering*, A. Chao and M. Tigner eds., p. 144 (1999).
46. Y. Chin, "Parasitic Crossing at an Asymmetric B Factory," LBL-30-01 (1990).
47. S. Heifets, A. Kulikov, J. Seeman, "PEP-II LER Non-Linear Vacuum Pressure Increase with Current," Intl. Workshop on e^+e^- Factories, KEK, p. 82 (1999).
48. F. Zimmermann, "Electron Cloud at the KEKB Low Energy Ring, " CERN-SL-2000-017 (AP) (2000).
49. K. Ohmi, F. Zimmermann, "Head-Tail Instability Caused by Electron Cloud in Positron Storage Ring," KEK Report, May (2000).
50. F.-J. Decker, "Luminosity and Bunch Patterns in PEP-II," Proc. PEP-II Performance Workshop, SLAC-WP-19, December 2000.

Beam-beam Interaction in Linear Collider

Kaoru Yokoya

High Energy Accelerator Research Organization (KEK), Japan

Abstract. Beam-beam interaction at the collision point of future e^+e^- linear colliders imposes strong constraints on the collider parameters. The expected phenomena include bunch deformation due to the Coulomb force, synchrotron radiation (beamstrahlung), coherent pair creation, and incoherent processes. These topics are very briefly summarized.

INTRODUCTION

The beam-beam interaction in linear colliders is basically the same Coulomb interaction as in circular colliders but has very different features in practice because

- the interaction occurs only once for each bunch (single pass) so that a very strong bunch deformation is permissible,
- quantum theoretical phenomena are important because of the high energy and the strong beam field.

The nature of the interaction has two aspects, according to the above two reasons: the classical phenomena and quantum theoretical phenomena.

The classical phenomena include bunch deformation, which is called the pinch effect or disruption effect; enhancement of the luminosity; deflection of individual particle trajectories, which causes background to the experiments; and so on.

The quantum theoretical phenomena include synchrotron radiation due to the beam field (called beamstrahlung), e^+e^- pair creation from the beamstrahlung photons in the strong beam field (called coherent pair creation), interaction between individual particles, and so on.

In this short lecture we treat only e^+e^- linear colliders because $\mu^+\mu^-$ linear colliders are not attractive, $e^\pm$-p linear colliders have not yet been studied seriously, and the beam-beam interaction in γ-γ colliders is not of primary

TABLE 1. Beam Parameters of Linear Collider Projects

			TESLA	JLC/NLC	CLIC	CLIC
CM energy	E_{cm}	TeV	0.8	1.0	1.0	5.0
Accelerating frequency	f	GHz	1.3	11.4	30	30
Bunch population	N	10^{10}	1.41	0.95	0.4	0.4
Number of bunches /pulse	N_b		4500	95	150	150
Bunch separation	t_b	nsec	189	2.8	0.67	0.67
Repetition rate	f_{rep}	Hz	3	120	150	50
Bunch length	σ_z	μm	300	120	50	25
Normalized emittance	$\gamma\epsilon_x$	μm	8	4.5	1.48	0.58
	$\gamma\epsilon_y$	μm	0.01	0.1	0.07	0.01
Beta functions	β_x^*	mm	15	12	10	6
	β_y^*	mm	0.3	0.15	0.1	0.1
Beam size	σ_x^*	nm	391	234	123	27
	σ_y^*	nm	2	3.9	2.7	0.45
Disruption parameter	D_x		0.2	0.12		0.16
	D_y		37	7.2		9.3
Upsilon	$\langle\Upsilon\rangle$			0.3	0.57	26.7
Number of photons /electron	n_γ			1.43	1.1	2.6
Beamstrahlung energy loss	δ_{BS}	%	4.7	10.3	9.2	40
Pinch enhancement	H_D		1.8	1.46	1.54	1.99
Luminosity	$\mathcal{L}$	10^{34}	5.0	1.29	1.1	14.9

concern. In e^-e^- linear colliders the disruption is important but will not be described here because of the limited time.

A few review articles on the topics are available [1–3]: [1] includes all the basic features although it is a little old; [2] is a brief summary plus a few new items such as so-called minijets; [3] is not a review of beam-beam interaction but of the beam focusing system, but it contains various practical points of the beam-beam interaction.

Table 1 briefly summarizes the parameters of the beam-beam interaction in the existing linear collider projects as of July 2000.

CLASSICAL PHENOMENA

A linear collider system is an extremely complicated facility involving a tremendous number of parameters, but the beam-beam interaction is quite simple. The important parameters are

E	Beam energy
N	Number of particles per bunch
σ_z	R.m.s. bunch length
σ_x, σ_y	Transverse r.m.s. beam size

In addition some less important ones are

β_x, β_y	Beta functions
t_b	Distance between bunches
N_b	Number of bunches
σ'_x, σ'_y	Angle spread
ϕ_c	Crossing angle
errors	Displacement $\Delta_{x,y}$, tilt, etc.

Geometric Luminosity

The geometric luminosity (luminosity calculated from the beam size and shape without taking into account the dynamical deformation) for head-on collisions of Gaussian bunches is given by

$$\mathcal{L}_{00} = \frac{f_{coll} N^2}{4\pi \sigma_x \sigma_y} \tag{1}$$

where f_{coll} is the collision frequency of bunches. In all the projects listed in Table 1 the beam pulse consists of several bunches. f_{coll} is the product of the number of bunches per pulse N_b and the pulse repetition frequency f_{rep}.

In the presence of the crossing angle ϕ_c and the effect due to the finite beta functions (hour-glass effect),

$$\mathcal{L}_0 = \mathcal{L}_{00} \times \eta(\phi_c, A_x, A_y), \quad (A_{x(y)} = \sigma_z/\beta_{x(y)}). \tag{2}$$

All the projects adopt very flat beams ($\sigma_x \gg \sigma_y$) in order to reduce the beamstrahlung, to be described in the next section. In such a case,

$$\eta(\phi_c, A_y) \approx \frac{1}{\sqrt{\pi} A_y} \exp\left[-\frac{1+c_\phi^2/4}{2 A_y^2}\right] K_0\left(\frac{1+c_\phi^2/4}{2 A_y^2}\right), \quad c_\phi \equiv \frac{\phi_c}{\sigma_x/\sigma_z} \tag{3}$$

where K_0 is the modified Bessel function.

Disruption Parameter

The classical-dynamical effects can be characterized mainly by the dimensionless parameter called the disruption parameter:

$$D_{x(y)} = \frac{2 N r_e \sigma_z}{\gamma \sigma_{x(y)} (\sigma_x + \sigma_y)}$$

where r_e is the classical electron radius and $\gamma = E/mc^2$. This is related to the tune-shift parameter ξ used in the beam-beam interaction in rings as

$$D_{x(y)} = \frac{\sigma_z}{\beta_{x(y)}} \times 4\pi \xi_{x(y)}. \qquad (4)$$

When one approximates the longitudinal distribution by a uniform distribution of length $2\sqrt{3}\sigma_z$, the equation of motion of a particle near the axis can be given approximately by

$$\frac{d^2 x}{ds^2} + \frac{D_x}{\sqrt{3}\sigma_z^2} x = 0, \qquad (-\sqrt{3}\sigma_z/2 < s < \sqrt{3}\sigma_z/2) \qquad (5)$$

(similarly for y.) Thus, the number of oscillation is $(\sqrt{3} D_{x(y)})^{1/2}/2\pi$. When $D_{x(y)}$ is small, the deflection angle of a particle with initial displacement (x_0, y_0) is given by $D_{x(y)} x(y)_0 / \sigma_z$.

Most classical phenomena can be characterized by the disruption parameters but, unless the D's are extremely small, computer simulations are indispensable to obtain useful results for application.

Luminosity Enhancement

In the presence of the pinch effects, the luminosity is

$$\mathcal{L} = \mathcal{L}_0 \times H_D(D, A) \qquad (6)$$

where H_D is called the pinch enhancement factor. (H_D is sometimes defined as $\mathcal{L} = \mathcal{L}_{00} \times H_D(D, A)$.) In $e^+ e^-$ colliders, H_D is usually larger than unity because of the Coulomb attraction. It is less than unity in $e^- e^-$ colliders.

H_D is plotted in Figure 1 for flat Gaussian beams with $A_y = 1$. The actual collision cannot be exactly head-on so that initial vertical displacement Δ_y (each beam $\pm \Delta_y/2$) is included in the figure. The same data are plotted in Figure 2 as a function of Δ_y. The general behavior of H_D as a function of D_y is

- H_D monotonically increases as D_y when $\Delta_y=0$.
- In the presence of errors ($\Delta_y \neq 0$) H_D saturates at values around 1.5 to 2 at $D_y \sim 8$ to ~ 20, and then goes down.
- H_D is insensitive to D_y in this region $8 \lesssim D_y \lesssim 20$.
- H_D is very sensitive to errors when D_y is large ($\gtrsim 50$).

When D_y is very large, the initial small displacement grows exponentially resulting in a loss of luminosity. Figure 3 illustrates the behavior of the bunch during collision. The initially displaced bunches attract each other ($t=0.2$), pass over ($t=0.4$), attract again ($t = 0.6$), and eventually the amplitudes become much larger than the initial displacement. However, because of the initial attracting process, H_D does not decrease exponentially for large Δ_y but sustains a relatively large value, as is seen in the curve for $D_y=50$ in Figure 2.

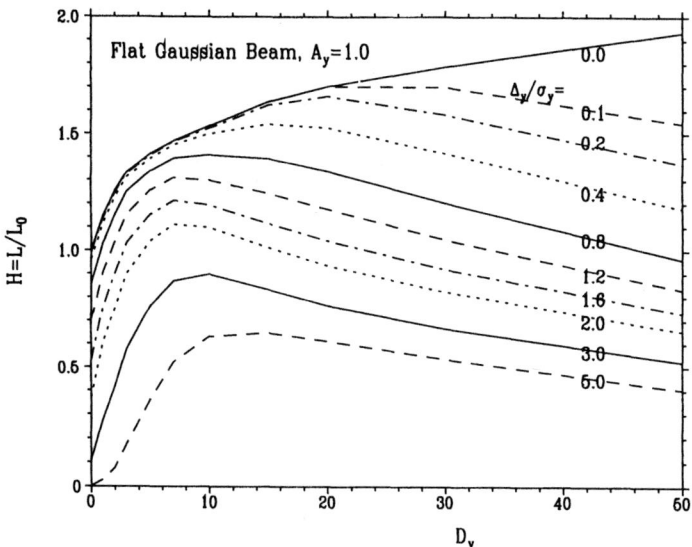

FIGURE 1. Luminosity enhancement factor for flat Gaussian beams with $A_y = 1$ as a function of D_y.

Deflection of Low Energy Particles

The typical value of the particle deflection angle is the same in the horizontal and vertical planes and is given by

$$\theta_0 = \frac{2Nr_e}{\gamma(\sigma_x + \sigma_y)}. \tag{7}$$

However, low energy particles created in the collision process (e.g., by e^+e^- pair creation) can be deflected by much larger angles. If the particle (energy ϵE, $\epsilon \ll 1$) has the same sign of charge as the on-coming beam, it is kicked out with large angles. The maximum angle is approximately given by

$$\theta_{max} \sim \left[\frac{\log(4\sqrt{3}D_x/\epsilon)}{\sqrt{3}\epsilon D_x}\right]^{1/2} \theta_0 \quad (\epsilon \ll D_x). \tag{8}$$

These particles are a potential source of background.

Flat Beams

In all the projects a very flat beam with an aspect ratio of $R \equiv \sigma_x/\sigma_y$ 50 to 200 is adopted. Flat beams are preferred (to round beams of the same

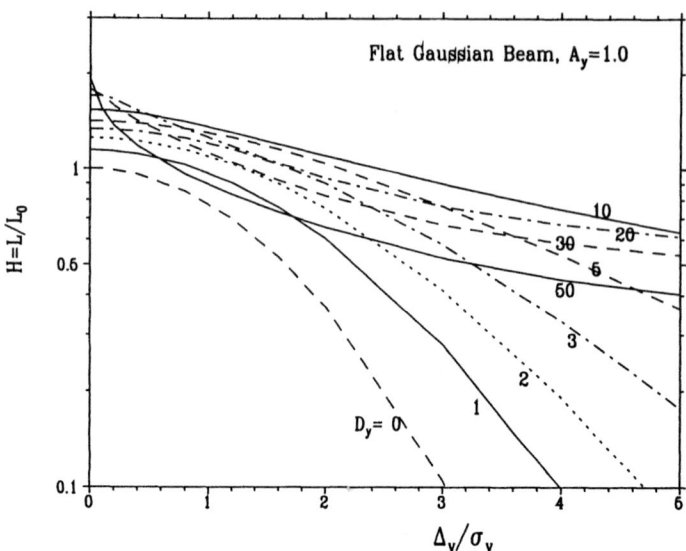

FIGURE 2. Luminosity enhancement factor for flat Gaussian beams with $A_y = 1$ as a function of Δ_y.

cross-section area $\sigma_x \times \sigma_y$, giving the same geometric luminosity) for many reasons:

- Relax beamstrahlung effects (next section).
- Chromaticity correction in the Final Focus is easier. (An accurate correction is required in the vertical plane only.)
- Large crossing angle possible without large luminosity reduction. The beams after collision can go outside the last quadrupole magnets.
- One can make use of small vertical emittance from damping rings.

The price to pay is the tight vertical orbit tolerance after the damping ring all the way down to the collision point.

Crab Crossing

Even with flat beams a crossing angle ϕ_c larger than the beam diagonal angle σ_x/σ_z may be needed for avoiding background. In such cases one can adopt the so-called crab crossing. The beams are kicked transversely by using an RF cavity at the zero phase such that the head and tail are kicked in opposite direction. If the kick angle is correct, the beams collide effectively head-on as illustrated in Figure 4. Instead of using an RF cavity one can also

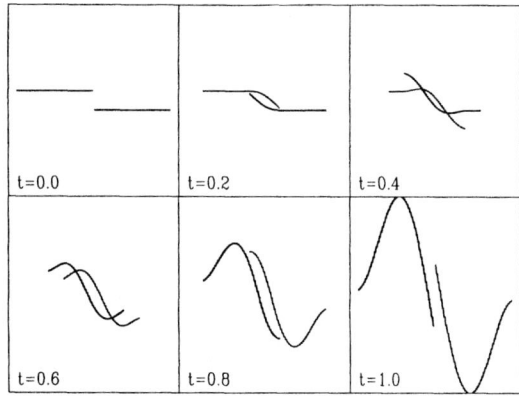

FIGURE 3. Kink instability. Side view of the flat bunches initially displaced vertically. The time and the vertical scale are arbitrary.

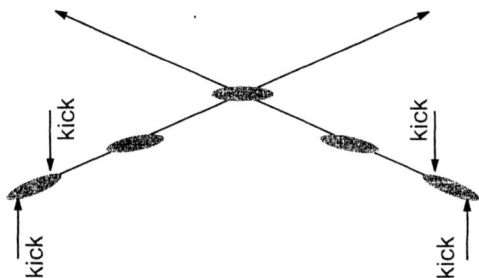

FIGURE 4. Crab crossing scheme.

use the finite dispersion function η_x^* at the collision point together with the energy slope dE/dz from head to tail of the bunch.

BEAMSTRAHLUNG

The electrons/positrons lose a considerable fraction of their energy during the collision by synchrotron radiation (called beamstrahlung) due to the strong beam field (order of several kilo-Teslas). The radiation coherence length is much shorter than the bunch length so that the uniform-field formula of radiation is valid.

The most important parameter describing the synchrotron radiation is

$$\Upsilon = \frac{2}{3}\frac{\hbar\omega_c}{E} = \frac{\lambda_e\gamma^2}{\rho} = \gamma\frac{2B}{B_s} = \frac{e}{m^3}\sqrt{|(F_{\mu\nu}p^\nu)^2|} \qquad (9)$$

where ω_c is the critical energy, λ_e the Compton wavelength, ρ the orbit curvature radius, B the beam magnetic field (the factor 2 accounts for the electric field effect), B_s Schwinger's critical field (4.4 GTesla), $F_{\mu\nu}$ the electro-magnetic field tensor, and p^ν the electron 4-momentum.

The number spectrum of radiation is given by the Sokolov-Ternov formula

$$\frac{dW_\gamma}{d\omega} = \frac{\alpha}{\sqrt{3}\pi\gamma^2}\left[\int_\xi^\infty K_{5/3}(\xi')d\xi' + \frac{y^2}{1-y}K_{2/3}(\xi)\right] \quad (10)$$

where K_ν is the modified Bessel function and

$$y = \frac{\omega}{E}, \quad \xi = \frac{2}{3\Upsilon}\frac{y}{1-y}. \quad (11)$$

The power spectrum [Eq(10) multiplied by ω] is plotted in Figure 5 for three values of Υ.

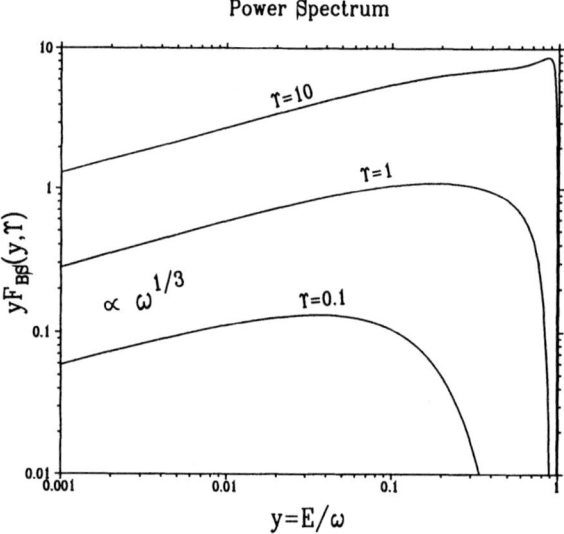

FIGURE 5. Radiation spectrum.

Various important parameters are derived from the Sokolov-Ternov formula. The number of photons per unit time is

$$\frac{dn_\gamma}{dt} = \int_0^E \frac{dW_\gamma}{d\omega}d\omega \approx \frac{5}{2\sqrt{3}}\frac{\alpha\Upsilon}{\lambda_e\gamma}\frac{1}{\sqrt{1+\Upsilon^{2/3}}}. \quad (12)$$

The average energy loss per unit time is

$$\left\langle -\frac{1}{E}\frac{dE}{dt}\right\rangle = \int_0^E \frac{\omega}{E}\frac{dW_\gamma}{d\omega}d\omega \approx \frac{2}{3}\frac{\alpha\Upsilon^2}{\lambda_e\gamma}\frac{1}{[1+(1.5\Upsilon)^{2/3}]^2}. \quad (13)$$

The average photon energy is

$$\left\langle \frac{\omega}{E}\right\rangle = \begin{cases} 0.462\Upsilon & (\Upsilon \to 0) \\ 16/63 = 0.254 & (\Upsilon \to \infty) \end{cases}. \quad (14)$$

In the limit $\Upsilon \to \infty$, the number spectrum approaches

$$\frac{dW_\gamma}{d\omega} \approx C\omega^{-2/3}, \quad (0 < \omega < E). \quad (15)$$

(This is not exact. See the small wiggle of the curve $\Upsilon=10$ near $y = 1$ in Figure 5.) This gives $\langle \omega/E \rangle = 1/4$, which is close to the exact value 16/63.

Relation to the Beam Parameters

The formulas in the previous subsection are general. They can be expressed by the beam parameters N, σ_x, σ_y, etc., for linear colliders.

The average of Υ during collision is

$$\Upsilon_{avr} \approx \frac{5}{6}\frac{Nr_e\lambda_e\gamma}{\sigma_z(\sigma_x+\sigma_y)} \quad (16)$$

and the maximum during collision is $\Upsilon_{max} \approx 2.4\Upsilon_{avr}$.
The number of photons per electron is

$$n_\gamma \approx 2.54\left[\frac{\alpha\sigma_z}{\lambda_e\gamma}\right]\frac{\Upsilon_{avr}}{\sqrt{1+\Upsilon_{avr}^{2/3}}}. \quad (17)$$

The average energy loss is

$$\delta_{BS} = \left\langle -\frac{\Delta E}{E}\right\rangle \approx 1.24\left[\frac{\alpha\sigma_z}{\lambda_e\gamma}\right]\frac{\Upsilon_{avr}^2}{[1+(1.5\Upsilon_{avr})^{2/3}]^2}. \quad (18)$$

In linear collider projects with center-of-mass energy above $E_{CM} \gtrsim 500\text{GeV}$, the change in the e^+e^- energy spectrum due to beamstrahlung is very serious. When $\Upsilon \gtrsim O(1)$, the energy spectrum is determined mainly by the average number of photons per electron. If $n_\gamma \gg 1$, the spectrum is badly destroyed because each photon carries away a considerable amount of energy. Therefore, all the projects above $E_{CM} \gtrsim 500$ GeV adopt $n_\gamma = O(1)$ (small n_γ is not adequate because of low luminosity).

In Eq(17) the factor in the brackets decreases as the energy γ goes up. (Note, the bunch length σ_z also becomes small because higher-energy projects normally adopt higher accelerating frequency and because σ_z is more or less proportional to the accelerating wavelength.) Thus, if n_γ has to be $O(1)$, large Υ is inevitable at high E_{CM}. This is obvious in Table 1. The problem associated with large Υ is described in the next section, but, so long as one keeps $n_\gamma \lesssim O(1)$, the change in the energy spectrum, in particular the high energy tip, is not very serious.

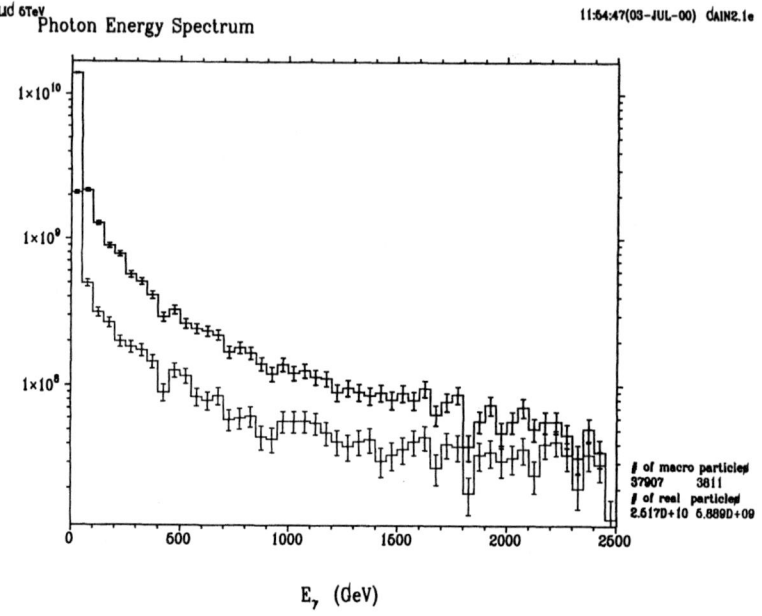

FIGURE 6. Photon spectrum for CLIC 5 TeV.

Figure 6 shows the result of a computer simulation of the beamstrahlung energy spectrum from 5 TeV CLIC. The top curve shows the first photons from each electron (i.e., full energy electrons) and the bottom curve all the photons. Figure 7 is the corresponding differential luminosity $d\mathcal{L}/E_{CM}$. About 25 (37)% of the luminosity comes from the region $0.98(0.90)E_{CM,max} < E_{CM} < E_{CM,max}$. Figure 8 is the luminosity between electron and beamstrahlung (does not included positron-beamstrahlung). Figure 9 is the luminosity between right- and left-going beamstrahlung. These luminosities are all comparable to the e^+e^- luminosity.

Beam-Beam Depolarization

The electron/positron beam can be depolarized by the beam-beam interaction by two different mechanisms, both characterized by n_γ and Υ.

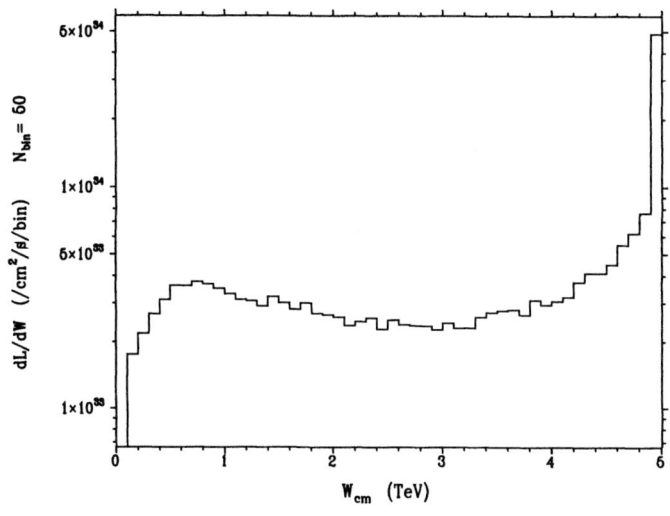

FIGURE 7. e^+e^- luminosity spectrum for CLIC 5 TeV.

- Depolarization by precession in the beam field.
  The precession angle is given by the orbit deflection angle times $(1 + \gamma a)$ where a is the coefficient of anomalous magnetic moment. The term γa is dominant at high energies. (a is a function of Υ. It is $\approx \alpha/2\pi$ at $\Upsilon = 0$ and decreases as Υ.) The depolarization is given by

$$\Delta P/P = n_\gamma^2 f_{prec}(\Upsilon), \tag{19}$$

$$f_{prec} = \frac{\log(4/3)}{(2\pi)^3} \left[\frac{a(\Upsilon)}{\alpha/2\pi}\right]^2 (1 + \Upsilon^{2/3}). \tag{20}$$

f_{prec} is plotted in Figure 10.

- Depolarization by spin-flip beamstrahlung
  The spontaneous polarization mechanism in storage rings by spin-flip synchrotron radiation works as a depolarization mechanism in the beam-beam interaction. It is estimated by

$$\Delta P/P = n_\gamma f_{flip}(\Upsilon). \tag{21}$$

f_{flip} is also plotted in Figure 10.

The depolarization by both mechanisms is not too large if $n_\gamma \lesssim O(1)$.

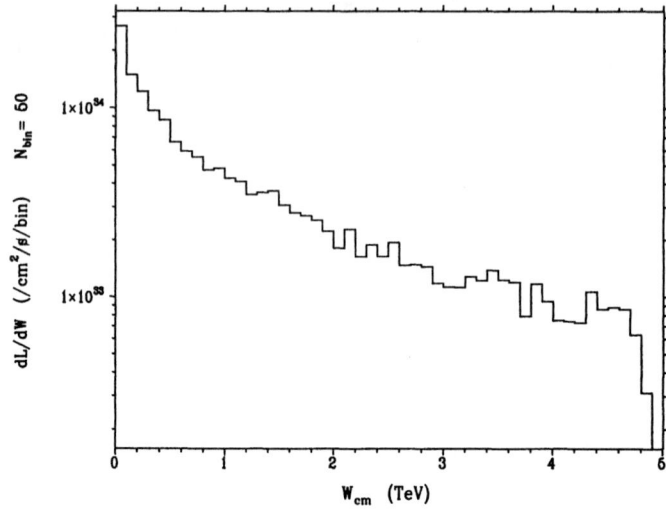

FIGURE 8. e^--γ luminosity spectrum for CLIC 5 TeV.

I COHERENT PAIR CREATION

A The Parameter χ

Usually, a photon cannot become an e^+e^- pair spontaneously because of energy-momentum conservation, but this process is possible in a strong electro-magnetic field if the photon energy is very high. The process, called coherent pair creation, is characterized by the parameter

$$\chi = \frac{\omega}{m}\frac{2B}{B_s}. \qquad (22)$$

(The factor of 2 accounts for the contribution of the electric field for the case of linear colliders.) The threshold is given by $\chi \lesssim 1$.

When this photon comes from the beamstrahlung with parameter $\Upsilon = (E/m)(2B/B_s)$, then

$$\chi = \frac{\omega}{E}\Upsilon. \qquad (23)$$

When $\Upsilon \gtrsim 1$, ω/E can be $O(1)$, and thus χ can be $\gtrsim 1$.

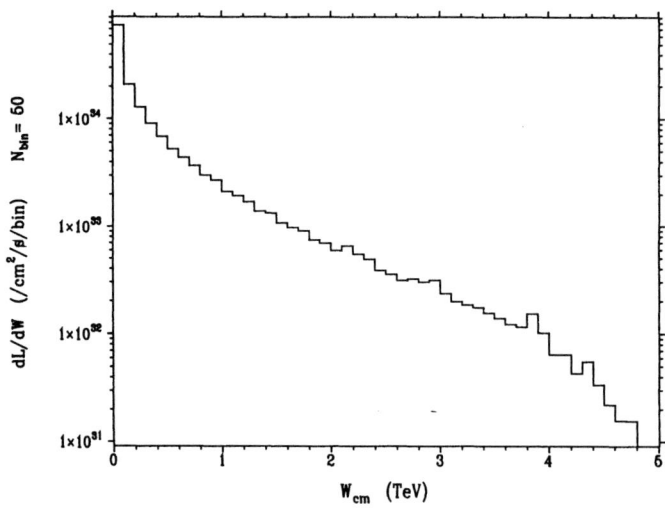

FIGURE 9. γ-γ luminosity spectrum for CLIC 5 TeV.

Spectrum of Coherent Pair

In a given field the probability of creating a pair of energies E_+ and $E_- = \omega - E_+$ in unit time is given by

$$\frac{dW_{CP}}{dE_+} = \frac{\alpha}{\sqrt{3}\pi} \frac{m^2}{\omega^2} F_{CP} \qquad (24)$$

where

$$F_{CP} = \int_\eta^\infty K_{1/3}(\eta')d\eta' + \left(\frac{E_-}{E_+} + \frac{E_+}{E_-}\right) K_{2/3}(\eta),$$

$$\eta = \frac{2}{3\chi} \frac{\omega^2}{E_+ E_-}, \qquad E_- = \omega - E_+ \qquad (25)$$

K_ν being the modified Bessel functions. The total probability per unit time is

$$W_{CP} = \int_0^\omega \frac{dW_{CP}}{dE_+} dE_+ = \begin{cases} 0.23(\alpha m^2/\omega)\chi e^{-8/3\chi} & (\chi \ll 1) \\ 0.38(\alpha m^2/\omega)\chi^{2/3} & (\chi \gg 1) \end{cases}. \qquad (26)$$

F_{CP} normalized by the total number of pairs is plotted in Figure 11. One finds

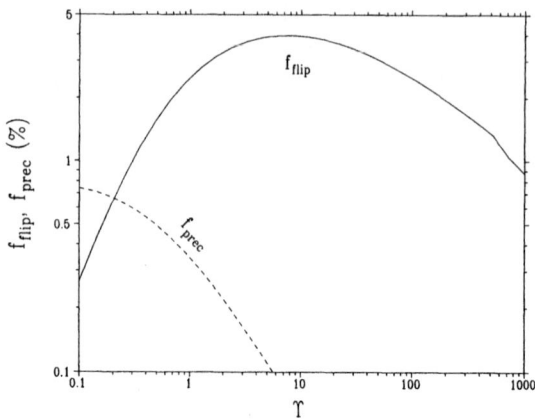

FIGURE 10. The functions f_{prec} and f_{flip} related to beam-beam depolarization.

- When $\chi \ll 1$, the pair energy is concentrated at $\sim \omega/2$, i.e., the pair particles equally share the initial photon energy.
- When $\chi \gg 1$, one of the paricle carries most of the photon energy and the other has a low energy peaked at

$$\frac{E_+}{\omega} \sim \frac{1.6}{\chi}. \tag{27}$$

From the chain $e \to$ beamstrahlung $\to$ coherent pairs, the number of pairs from one initial electron is

$$n_{coh.pair} = \left[\frac{\alpha \sigma_z}{\gamma \lambda_e} \Upsilon_{avr}\right]^2 \Xi(\Upsilon_{avr}). \tag{28}$$

Ξ is exponentially small for small Υ_{avr} and $\Xi \sim O(0.1)$ for large Υ.

$$\Xi(\Upsilon) = \begin{cases} \frac{7}{128}\exp\left(-\frac{16}{3\Upsilon}\right) & (\Upsilon \lesssim 1) \\ 0.295\Upsilon^{-2/3}(\log \Upsilon - 2.488) & (\Upsilon \gg 1) \end{cases}. \tag{29}$$

$\Xi(\Upsilon)$ is plotted in Figure 12 together with the asymptotic form in Eq(29). From Eq(17) one gets

$$n_{coh.pair} \approx \left(\frac{n_\gamma}{2.54}\right)^2 (1 + \Upsilon_{avr}^{2/3})\Xi(\Upsilon_{avr}). \tag{30}$$

For extremely large Υ_{avr},

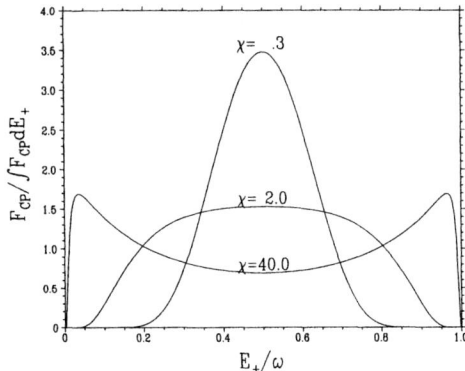

FIGURE 11. The spectrum function F_{CP} of coherent pair creation.

$$n_{coh.pair} \sim 0.046 n_\gamma^2 (\log \Upsilon_{avr} - 2.5). \tag{31}$$

The expected pair energy spectrum in the 5 TeV CLIC case is plotted in Figure 13.

The major problem when going to the very large Υ region is the low energy pair particles from coherent pair creation. They are deflected by the beam-beam force by large angles whose maximum is approximated by Eq(8).

Figure 14 shows the low energy part of the spectrum in 5 TeV CLIC (magnification of the low energy end of Figure 13). It shows the low energy exponential cut-off below Eq(27). Figure 15 shows the relation between the energy and the out-coming angle of the coherent pairs in 5 TeV CLIC.

PARTICLE-PARTICLE COLLISION

The collisions of individual particles are not of concern to accelerator physicists, but besides the high-energy processes that particle physicists are interested in, there are many processes causing only backgrounds, which accelerator physicists have to care about.

- QED processes like

$\gamma_{BS}\gamma_{BS} \to e^+e^-$	Breit-Wheeler
$e^\pm \gamma_{BS} \to e^\pm e^+e^-$	Bethe-Heitler
$e^+e^- \to e^+e^-e^+e^-$	Landau-Lifshitz

Here, γ_{BS} is the beamstrahlung photon. All these are described by the process $\gamma\gamma \to e^+e^-$ with γ being a virtual or real photon.

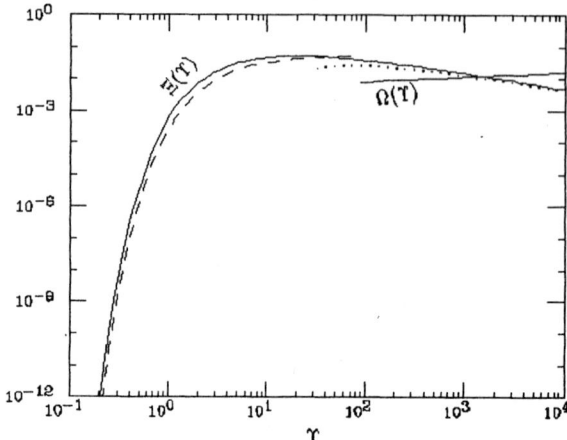

FIGURE 12. The function $\Xi(\Upsilon)$ related to the chain process beamstrahlung → coherent pair creation. The dashed and dotted curves are the asymptotic forms in Eq(29).

- QCD processes
  - Minijets $\gamma\gamma \to$ jets (γ: real or virtual)

The QED processes create pair particles (called incoherent pair creation) whose energies are much lower than those from coherent pair creation. The number of pairs from one bunch collision is about 10^6 to 10^7. This is the dominant source of low energy particles in the low Υ region. An example of incoherent pairs in 5 TeV CLIC is shown in Figure 16. Their out-coming angle is plotted in Figure 17. The top and the bottom curves are the same as in Figure 15.

Pair Monitor

For measuring the beam size at the collision point, the laser-interference monitor in FFTB at SLAC was successful for beams down to ∼50 nm. However, measurement will not be easy in next-generation linear colliders because the required precision is $\lesssim 1$ nm and because the monitor does not work during collision because of the background by beamstrahlung.

Pair creation is normally a disturbance in experiments, but its distribution carries information about the beam field. If one carefully observes the distribution, one can extract information about the beam size, beam tilt, etc. A pair monitor works for each beam separately, but only when two beams exist so that it complements the laser-interference monitor.

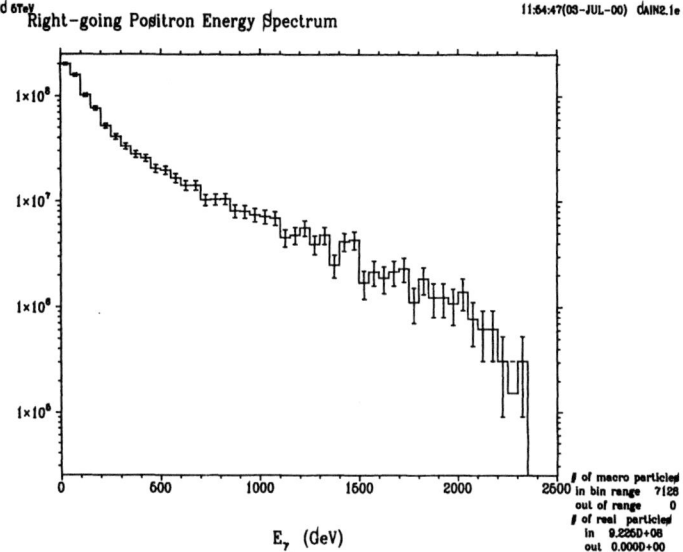

FIGURE 13. The positron energy spectrum coming from the coherent pair creation in 5 TeV CLIC.

Figure 18 shows an example of the horizontal-vertical angle distribution of incoherent pairs in 5 TeV CLIC. One finds only a small number of pairs on the horizontal plane. The ratio of the distribution in the horizontal to vertical planes contains information on σ_y/σ_x. A complication arises because the distribution is rotated by the detector solenoid. Figure 19 shows the relation between the pair energy and the radial position at 5 m downstream from the collision point (5 TeV CLIC with 4 Tesla solenoid). If one detects pairs at $r > 15$ cm, one can select a small range of energy (the rectangle in the figure). Then for these particles, the azimuthal distribution (in the x-y plane) is preserved after a well-defined rotation angle.

REFERENCES

1. K. Yokoya, P. Chen, "Beam-Beam Phenomena in Linear Colliders," Lecture Notes in Phys. 400, Springer Verlag (1992).
2. P. Chen, in *Handbook of Acelerator Physics and Engineering*, p.140, World Scientific (1999).
3. T. O. Raubenheimer, F. Zimmermann, "Final Focus Systems in Linear Colliders," Rev. Mod. Phys. **72** (2000) 95-107.

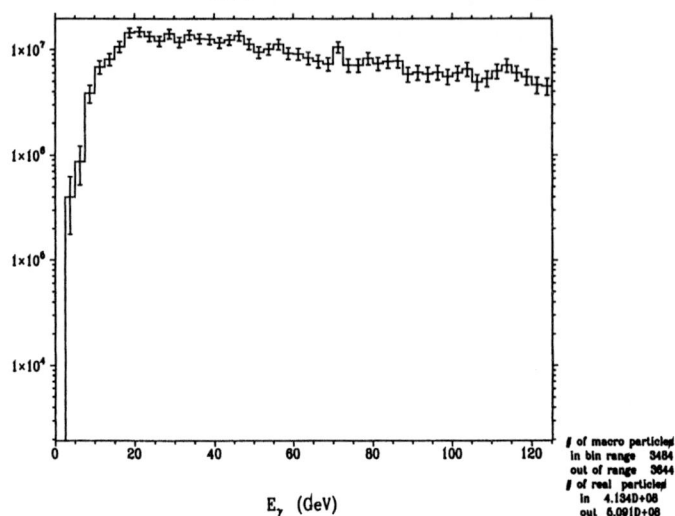

FIGURE 14. The low energy part of the positron energy spectrum (magnified view of Figure 13)

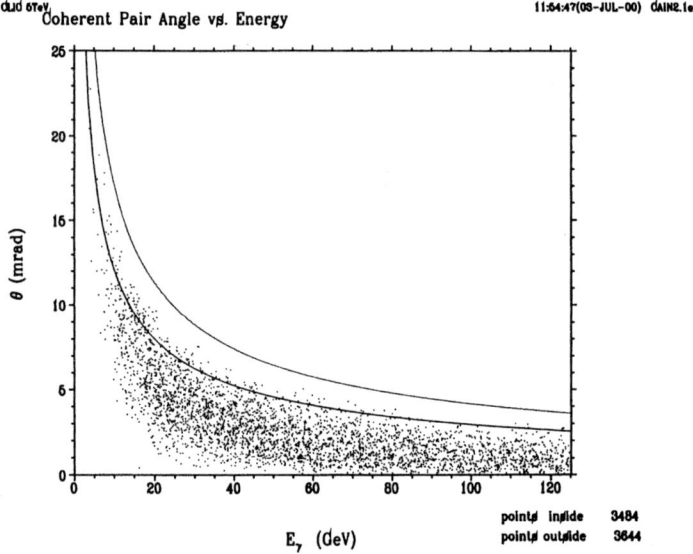

FIGURE 15. The energy-angle relation of coherent pairs in 5 TeV CLIC. The bottom curve show the relation expressed by Eq(8) and the top curve is the one multiplied by $\sqrt{2}$.

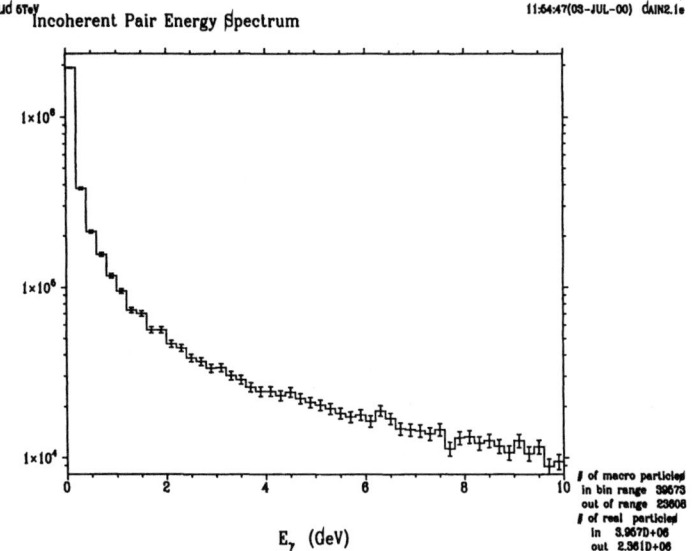

FIGURE 16. The energy spectrum of incoherent pairs in 5 TeV CLIC.

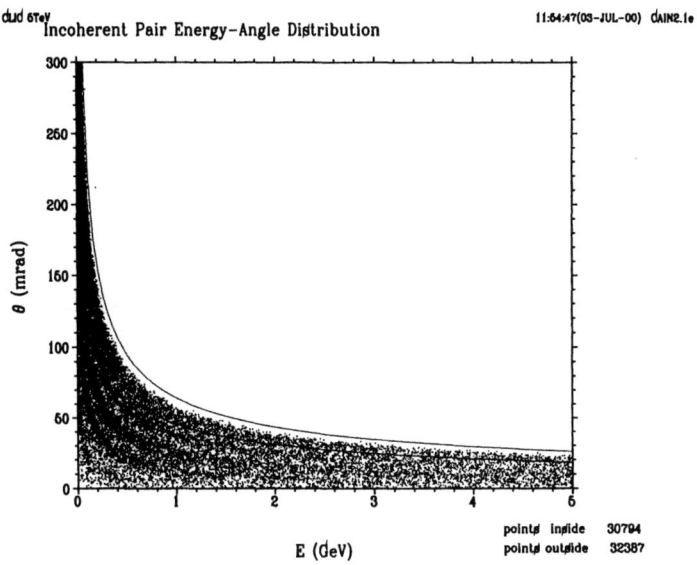

FIGURE 17. The energy spectrum of incoherent pairs in 5 TeV CLIC.

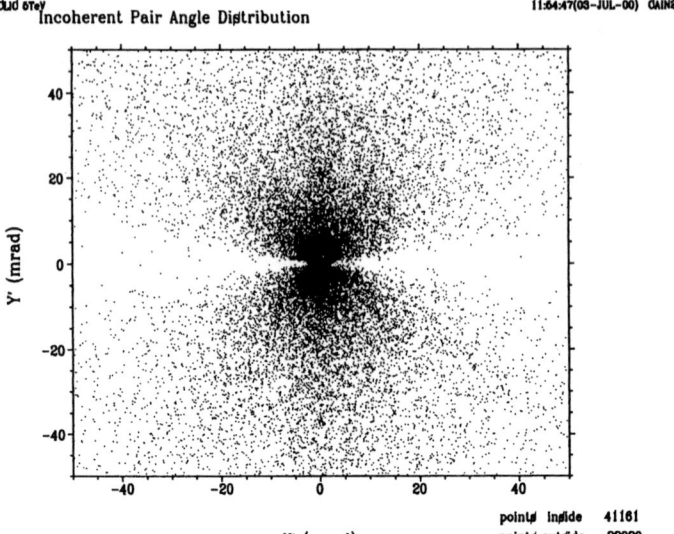

FIGURE 18. The horizontal-vertical angle distribution of incoherent pairs in 5 TeV CLIC.

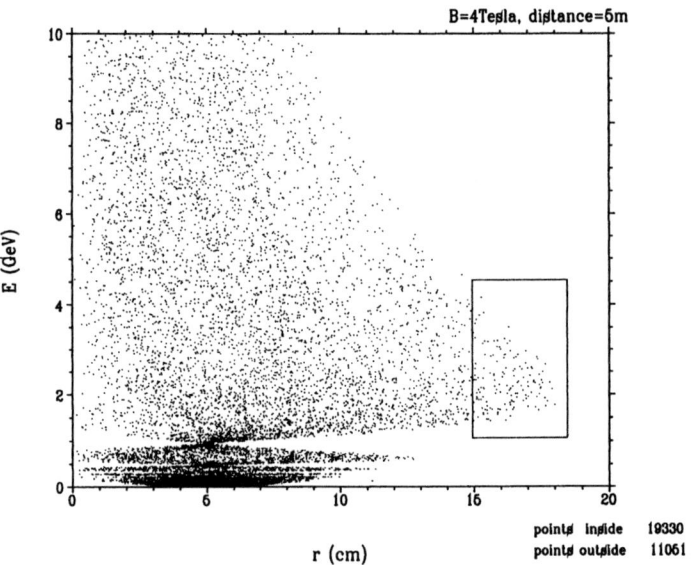

FIGURE 19. Energy and radius relation of incoherent pairs at 5 m from the collision point after rotation by the detector solenoid (4 Teslas).

Wake and Impedance

G. V. Stupakov

Stanford Linear Accelerator Center Stanford University, Stanford, CA 94309

I INTRODUCTION

In this lecture we will develop a concept of wakes and impedances for relativistic beams interacting with the surrounding environment. Among the numerous publications and reviews on this subject, we refer here to recent books [1–3], where the reader can find a more detailed treatment and further references.

We will use the CGS system of units throughout this paper.

II INTERACTION OF MOVING CHARGES IN FREE SPACE

We begin with interactions of particles that moving with constant velocity in free space. If the material walls are far from the particles, their effect in the first approximation can be neglected.

Let us consider a *leading* particle of charge q moving with velocity v, and a *trailing* particle of *unit* charge moving behind the leading one on a parallel path at a distance s with an offset x, as shown in Fig. 1. We want to find the force which the leading particle exerts on the trailing one.

We will use the following expressions for the electric and magnetic fields of a particle moving with a constant velocity (see, e.g., [4]):

$$\boldsymbol{E} = \frac{q\boldsymbol{R}}{\gamma^2 R^{*3}}, \qquad \boldsymbol{H} = \frac{1}{c}\boldsymbol{v} \times \boldsymbol{E}, \qquad (1)$$

where $\boldsymbol{R}$ is the vector drawn from point 1 to point 2, $R^{*2} = s^2 + x^2/\gamma^2$, and $\gamma = (1 - v^2/c^2)^{-1/2}$.

From Eq. (1) we find that the longitudinal force acting on the trailing charge is

$$F_l = E_z = -\frac{qs}{\gamma^2(s^2 + x^2/\gamma^2)^{3/2}}, \qquad (2)$$

and the transverse force is

$$F_t = E_x - \frac{v}{c}B_y = \frac{qx}{\gamma^4(s^2 + x^2/\gamma^2)^{3/2}}. \qquad (3)$$

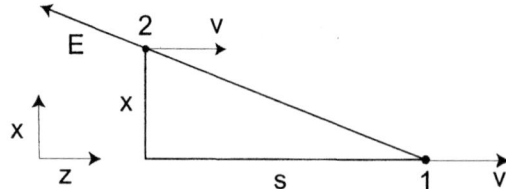

FIGURE 1. A leading particle 1 and a trailing particle 2 traveling in free space with parallel velocities v. Shown also is the coordinate system x, z.

In accelerator physics, the force $\boldsymbol{F}$ is often called *the space-charge force*.

It is easy to see that for any position given by s and x, the longitudinal force decreases with the growth of γ as γ^{-2}. For the transverse force, if $s \gg x/\gamma$, $F_t \sim \gamma^{-4}$, but for $s = 0$, $F_t \sim \gamma^{-1}$. Hence, in the limit of ultrarelativistic particles moving parallel to each other, $\gamma \to \infty$, the electromagnetic interaction in free space vanishes.

In this lecture, we will focus on the case of ultrarelativistic charges, where $v \to c$. The space-charge effects discussed above disappear in this limit, and the interaction between the particles is due only to the presence of material walls.

Note that, taking the limit $v \to c$ in Eq. (1) and recalling that $s = vt - z$, we can write the electromagnetic field of an ultrarelitivistic charge in free space as

$$\boldsymbol{E} = \frac{2q\boldsymbol{r}}{r^2}\delta(z - ct), \qquad \boldsymbol{H} = \hat{\boldsymbol{z}} \times \boldsymbol{E}, \qquad (4)$$

where $\boldsymbol{r} = \hat{\boldsymbol{x}}x + \hat{\boldsymbol{y}}y$ is a two-dimensional radius vector in a cylindrical coordinate system ($\hat{\boldsymbol{x}}$ and $\hat{\boldsymbol{y}}$ are the unit vectors in the directions of x and y, respectively).

III PARTICLES MOVING IN A PERFECTLY CONDUCTING PIPE

If particles from the above example move parallel to the axis in a perfectly conducting cylindrical pipe of arbitrary cross section, they induce image charges, on the surface of the wall, that screen the metal from the electromagnetic field of the particles. The image charges travel with the same velocity v (see Fig. 2). Since both the particles and the image charges move on parallel paths, in the limit $v = c$, according to the results in Section II, they do not interact with each other, no matter how close to the wall the particles are.

Interaction between the particles in the ultrarelativistic limit can occur if 1) the wall is not perfectly conducting, or 2) the pipe is not cylindrical (which is usually due to the presence of RF cavities, flanges, bellows, beam position monitors, slots, etc., in the vacuum chamber).

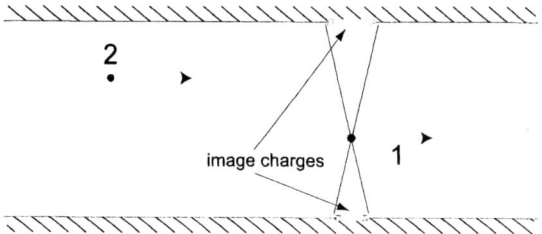

FIGURE 2. Particles traveling inside a perfectly conducting pipe of arbitrary cross section. Shown are the image charges on the wall generated by the leading charge.

IV CAUSALITY AND THE "CATCH-UP" DISTANCE

If a beam particle moves along a straight line with the speed of light, the electromagnetic field of this particle scattered off the boundary discontinuities will not overtake it and, furthermore, will not affect the charges that travel ahead of it. The field can interact only with the trailing charges in the beam that move behind it. This constitutes the principle of *causality* in the theory of wakefields, according to which the interaction of a point charge moving with the speed of light propagates only downstream and never reaches the upstream part of the beam.

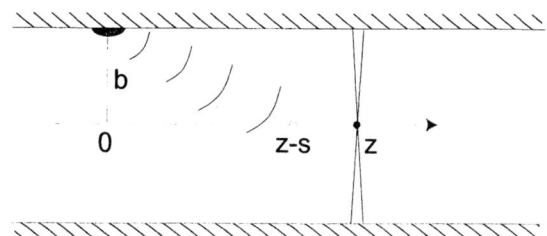

FIGURE 3. A wall discontinuity located at $z = 0$ scatters the electromagnetic field of an ultrarelativistic particle. When the particle moves to location z, the scattered field arrives to point $z - s$.

We can estimate the distance at which the electromagnetic field produced by a leading charge reaches a trailing particles traveling at a distance s behind. Let us assume that a discontinuity located at the surface of a pipe of radius b at coordinate $z = 0$ is passed by the leading particle at time $t = 0$, see Fig. 3. If the scattered field reaches point s at time t, then $ct = \sqrt{(z-s)^2 + b^2}$, where z is a coordinate of the leading particle at time t, $z = ct$. Assuming that $s \ll b$, from these two equations we find

$$z \approx \frac{b^2}{2s}. \qquad (5)$$

The distance z given by this equation is often called the *catch-up distance*. Only after the leading charge has traveled this distance away from the discontinuity, can a particle at point s behind it feel the wakefield generated by the discontinuity.

V ROUND PIPE WITH RESISTIVE WALLS

Consider a round pipe of radius b, with finite wall conductivity σ. A point charge moves along the z axis of the pipe with the speed of light, and a trailing particle follows the leading one at a distance s. Both particles are assumed to be on the axis of the pipe. Because of the symmetry of the problem, the only non-zero component of the electromagnetic field on the axis is E_z, which, depending on the sign, either accelerates or decelerates the trailing charge. Our goal now is to find the field E_z as a function of s.

If the conductivity of the pipe is large enough, we can use perturbation theory to find the effect of the wall resistivity. In the first approximation, we consider the pipe as a perfectly conducting one. In this case, the electromagnetic field of the charge is the same as in free space and is given by Eqs. (4). For what follows, we will need only the magnetic field H_θ,

$$H_\theta = \frac{2q}{r}\delta(z-ct). \qquad (6)$$

Using the mathematical identity

$$\delta(z-ct) = \frac{1}{2\pi c}\int_{-\infty}^{\infty} d\omega\, e^{-i\omega(t-z/c)}, \qquad (7)$$

we will decompose H_θ into a Fourier integral,

$$H_\theta(r,z,t) = \int_{-\infty}^{\infty} d\omega\, H_{\theta\omega}(r) e^{-i\omega t + i\omega z/c}, \qquad (8)$$

where

$$H_{\theta\omega}(r) = \frac{q}{\pi r c}. \qquad (9)$$

In the limit where the skin depth δ corresponding to the frequency ω, $\delta = c/\sqrt{2\pi\sigma\omega}$, is much smaller than the pipe radius, $\delta \ll b$, we can use the Leontovich boundary condition [5] that relates the tangential electric field $\boldsymbol{E}_t$ on the wall with the magnetic one,

$$\boldsymbol{E}_t = \zeta \boldsymbol{H} \times \boldsymbol{n}, \qquad (10)$$

where $\mathbf{n}$ is the unit vector normal to the surface and directed toward the metal, and

$$\zeta(\omega) = (1-i)\sqrt{\frac{\omega}{8\pi\sigma}}. \tag{11}$$

Combining Eqs.(10), (11) and (9), we find

$$E_{z\omega}|_{r=b} = -(1-i)\sqrt{\frac{\omega}{8\pi\sigma}}\frac{q}{\pi bc}. \tag{12}$$

Equation (12) gives us the longitudinal electric field on the wall, but we need the field on the axis of the pipe. To find the radial dependence of $E_{z\omega}$, we use Maxwell's equations, from which it follows that the electric field in a vacuum satisfies the wave equation. In the cylindrical coordinate system the wave equation for E_z is

$$\frac{1}{c^2}\frac{\partial^2 E_z(r,z,t)}{\partial^2 t} - \Delta E_z(r,z,t)$$
$$= \frac{1}{c^2}\frac{\partial^2 E_z(r,z,t)}{\partial^2 t} - \frac{\partial^2 E_z(r,z,t)}{\partial^2 z} - \frac{1}{r}\frac{\partial}{\partial r}r\frac{\partial E_z(r,z,t)}{\partial r} = 0. \tag{13}$$

Substituting the Fourier component $E_{z\omega}(r)e^{-i\omega(t-z/c)}$ into this equation, we find

$$\frac{1}{r}\frac{\partial}{\partial r}r\frac{\partial E_{z\omega}}{\partial r} = 0. \tag{14}$$

This equation has a general solution $E_{z\omega} = A + B\ln r$, where A and B are arbitrary constants. Since we do not expect E_z to have a singularity on the axis, $B = 0$. Hence the electric field does not depend on r, $E_{z\omega} = $ const, and

$$E_{z\omega}|_{r=0} = E_{z\omega}|_{r=b}, \tag{15}$$

implying that $E_{z\omega}|_{r=0}$ is given by the same Eq. (12). Note that we have shown here that in the ultrarelativistic case the longitudinal electric field inside the pipe is constant throughout the pipe cross section.

To find $E_z(z,t)$ we make the inverse Fourier transformation,

$$E_z(z,t) = \int_{-\infty}^{\infty} d\omega E_{z\omega} e^{-i\omega(t-z/c)}, \tag{16}$$

which gives

$$E_z(z,t) = (i-1)\frac{q}{\pi cb}\sqrt{\frac{1}{8\pi\sigma}}\int_{-\infty}^{\infty} d\omega\sqrt{\omega}e^{-i\omega(t-z/c)}. \tag{17}$$

The last integral can be taken analytically in the complex plane (see the Appendix), with the result

$$E_z(z,t) = \frac{q}{2\pi b}\sqrt{\frac{c}{\sigma s^3}}, \tag{18}$$

for $s > 0$. For the points where $s < 0$, located in front of the charge, $E_z = 0$ in agreement with the causality principle. The positive sign of E_z indicates that the trailing charge (if it has the same sign as q) will be accelerated in the wake.

In our derivation we assumed that the magnetic field on the wall is the same as in the case of perfect conductivity. However, the magnetic field is generated not only by the beam current, but also by a displacement current,

$$j_z^{\text{disp}} = \frac{1}{4\pi} \frac{\partial E_z}{\partial t}. \tag{19}$$

that vanishes in the limit of perfect conductivity. To be able to neglect the corrections to H_θ due to j_z^{disp}, we must require the total displacement current to be much less then the beam current. In the Fourier representation, the time derivative $\partial/\partial t$ reduces to multiplication by $-i\omega$, and the requirement is

$$\pi b^2 \frac{1}{4\pi} \omega E_{z\omega} \ll I_\omega = \frac{q}{2\pi}, \tag{20}$$

or

$$\left(\frac{\omega}{c}\right)^{3/2} \ll \sqrt{\frac{4\pi\sigma}{cb^2}}. \tag{21}$$

In the space-time domain, the inverse wavenumber c/ω corresponds to the distance s, and the condition of applicability of Eq. (18) is,

$$s \gg s_0 = \left(\frac{cb^2}{4\pi\sigma}\right)^{1/3}. \tag{22}$$

The behavior of E_z for very small values of s, $s < s_0$, can be found in Ref. 6. Here we note only that the singularity in Eq. (18) saturates at small s, and the electric field changes sign and becomes negative at $s = 0$. This field decelerates the leading charge, as expected from the energy balance consideration.

VI WAKE DEFINITION

The electromagnetic interaction of charged particles in accelerators with the surrounding environment is usually a relatively small effect that can be considered as a perturbation. In the zeroth approximation, we can assume that the beam moves with a constant velocity along a straight line. We solve Maxwell's equation, find the fields, and then take into account the effect of these fields on a particle's motion. In this approach we neglect the second-order effects because the motion along the perturbed orbit can only slightly change the fields computed in the zeroth approximation. Those corrections are usually small, especially for ultrarelativistic particles.

Another important feature of the interaction between the generated electromagnetic field and the particles is that in many cases of practical importance it is localized in a region small compared with the length of the beam orbit. It also occurs on a time scale much smaller than the characteristic oscillation times of the beam in the accelerator (such as betatron and synchrotron periods). This allows us to consider this interaction in the impulse approximation and characterize it by the amount of momentum transferred to the particle.

Taking into account the above considerations, we will introduce the notion of the *wake* in the following way. Consider a leading particle 1 of charge q moving along

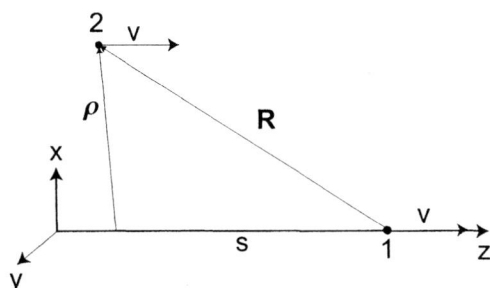

FIGURE 4. A leading particle 1 and a trailing particle 2 move parallel to each other in a vacuum chamber.

axis z with a velocity close to the speed of light, $v \approx c$, so that $z = ct$ (see Fig. 4). A trailing particle 2 of unit charge moves parallel to the leading one, with the same velocity, at a distance s with offset $\boldsymbol{\rho}$ relative to the z-axis. The vector $\boldsymbol{\rho}$ is a two-dimensional vector perpendicular to the z-axis, $\boldsymbol{\rho} = (x, y)$. Although the two particles move in a vacuum, there are material boundaries in the problem that scatter the electromagnetic field and result in interaction between the particles.

Let us assume that we solved Maxwell's equation and found the electromagnetic field generated by the first particle. We calculate the change of the momentum $\Delta \boldsymbol{p}$ of the second particle caused by this field as a function of the offset $\boldsymbol{\rho}$ and the distance s,

$$\Delta \boldsymbol{p}(\boldsymbol{\rho}, s) = \int_{-\infty}^{\infty} dt \, [\boldsymbol{E}(\boldsymbol{\rho}, z, t) + \hat{\boldsymbol{z}} \times \boldsymbol{B}(\boldsymbol{\rho}, z, t)]_{z=ct-s} \,. \tag{23}$$

Note that we integrate here along a straight line — the unperturbed orbit of the second particle. The integration limits are extended from minus to plus infinity, assuming that the integral converges.

Since the beam dynamics is different in the longitudinal and transverse directions, it is useful to separate the longitudinal momentum Δp_z from the transverse component $\Delta \boldsymbol{p}_\perp$. With the proper sign and the normalization factor c/q, these two components are called the *longitudinal* and *transverse wake functions* (or simply *wakes*),

$$w_l(\boldsymbol{\rho}, s) = -\frac{c}{q}\Delta p_z = -\frac{c}{q}\int dt E_z|_{z=ct-s},$$
$$w_t(\boldsymbol{\rho}, s) = \frac{c}{q}\Delta \boldsymbol{p}_\perp = \frac{c}{q}\int dt \left[\boldsymbol{E}_\perp + \hat{\boldsymbol{z}} \times \boldsymbol{B}\right]_{z=ct-s}. \tag{24}$$

Note the minus sign in the definition of w_l — it is introduced so that the positive longitudinal wake corresponds to the energy loss of the trailing particle (if both the leading and trailing particles have the same sign of charge). The defined wakes have dimension cm^{-1} in CGS units and V/C in SI units.[1]

Because of the causality principle, the wakefield does not propagate in front of the leading charge, hence

$$w_l(\boldsymbol{\rho}, s) \equiv 0, \qquad \boldsymbol{w}_t(\boldsymbol{\rho}, s) \equiv 0, \qquad \text{for } s < 0. \tag{25}$$

It was assumed above that the electromagnetic field is localized in space and time and the integral in Eq. (23) converges. There are cases, however, where this is not true and the source of the wake is distributed uniformly along an extended path, such as the resistive wall wake of a long pipe, considered in Section V. In this case it is more convenient to introduce the wake per unit length of the path by dropping the integration in Eq. (23):

$$w_l(\boldsymbol{\rho}, s) = -\frac{1}{q}E_z|_{z=ct-s},$$
$$\boldsymbol{w}_t(\boldsymbol{\rho}, s) = \frac{1}{q}\left[\boldsymbol{E}_\perp + \hat{\boldsymbol{z}} \times \boldsymbol{B}\right]_{z=ct-s}. \tag{26}$$

In this definition, the wakes acquire an additional dimension of inverse length, and has the dimension cm^{-2} in CGS and V/C/m in SI.

VII PANOFSKY-WENZEL THEOREM

Several general relations between longitudinal and transverse wakes can be obtained from Maxwell's equation without specifying the boundary condition for the fields.

Let us introduce the vector $\boldsymbol{R} = (\boldsymbol{\rho}, -s)$ (the negative sign in front of s is due to measuring s in the negative direction of z, see Fig. 4) and consider momentum $\Delta \boldsymbol{p}$ in Eq. (23) as a function of $\boldsymbol{R}$. Let us assume that the electric and magnetic fields are specified through the vector potential $\boldsymbol{A}(\boldsymbol{r}, t)$ and the scalar potential

---

[1] A useful relation between the units is 1 V/pC = 1.11 cm^{-1}.

$\phi(\boldsymbol{r}, t)$, and compute $\Delta \boldsymbol{p}$ for the given fields. It is convenient to use the Lagrangian formulation of the equations of motion,[2]

$$\frac{d}{dt}\frac{\partial L}{\partial \boldsymbol{v}} = \frac{\partial L}{\partial \boldsymbol{r}} \equiv \nabla L, \tag{27}$$

with the Lagrangian for the trailing *unit* charge in the electromagnetic field

$$L = -mc^2\sqrt{1 - \frac{v^2}{c^2}} + \frac{1}{c}\boldsymbol{A}\boldsymbol{v} - \phi. \tag{28}$$

Putting Eq. (28) into Eq. (27) yields ($\boldsymbol{p} = m\gamma\boldsymbol{v}$)

$$\frac{d}{dt}\left(\boldsymbol{p} + \frac{1}{c}\boldsymbol{A}\right) = \nabla\left(\frac{1}{c}\boldsymbol{A}\boldsymbol{v} - \phi\right). \tag{29}$$

Now, integrating this equation along the orbit of the trailing particle, $x = $ const, $y = $ const and $z = ct - s$, and assuming that the fields $\boldsymbol{A}$ and ϕ vanish at infinity, we find

$$\Delta \boldsymbol{p}(\boldsymbol{R}) = \int dt \nabla\left(\frac{1}{c}\boldsymbol{A}\boldsymbol{v} - \phi\right)$$
$$= \frac{q}{c}\nabla_R W(\boldsymbol{R}), \tag{30}$$

where we introduced the *wake potential* W,

$$W(\boldsymbol{R}) = \frac{c}{q}\int dt \left(\frac{1}{c}\boldsymbol{A}\boldsymbol{v} - \phi\right)$$
$$= \frac{c}{q}\int dt \left(A_z - \phi\right). \tag{31}$$

In the last equation we used $\boldsymbol{v} \approx c\hat{\boldsymbol{z}}$.

We just proved an important relation that states that all three components of the vector $\Delta \boldsymbol{p}$ can be obtained by differentiation of a single scalar function W. Recalling now the relation between the components of $\Delta \boldsymbol{p}$ and the wakes, Eq. (24), we find that

$$w_l = -\frac{\partial W}{\partial(-s)} = \frac{\partial W}{\partial s}, \qquad \boldsymbol{w}_t = \nabla_\rho W, \tag{32}$$

and hence

$$\frac{\partial \boldsymbol{w}_t}{\partial s} = \nabla_\rho w_l. \tag{33}$$

---

[2] This approach to the derivation of the Panofsky-Wenzel theorem is due to A. Chao.

This relation is usually referred to as the Panofsky-Wenzel theorem. Note that ∇_ρ is a two-dimensional gradient with respect to coordinates x and y.

One of the most important computational applications of the Panofsky-Wenzel theorem is that knowledge of the longitudinal wake function w_l allows us to find the transverse wake w_t by means of a simple integration of Eq. (33).

We now prove another important property of W: it is a harmonic function of variables x and y,

$$\Delta_\perp W \equiv \frac{\partial^2 W}{\partial x^2} + \frac{\partial^2 W}{\partial y^2} = 0. \tag{34}$$

To prove this, we will use the fact that both $\boldsymbol{A}$ and ϕ satisfy the wave equation in free space, $(\partial^2/\partial t^2 - c^2\Delta)\boldsymbol{A} = (\partial^2/\partial t^2 - c^2\Delta)\phi = 0$. Hence

$$\begin{aligned} 0 &= \frac{c}{q} \int dt \left(\frac{\partial^2}{\partial t^2} - c^2\Delta \right) (A_z - \phi) \\ &= -\frac{c}{q} \int dt \left(\frac{\partial^2}{\partial x^2} + \frac{\partial^2}{\partial y^2} \right) (A_z - \phi) + \frac{c}{q} \int dt \left(\frac{\partial^2}{\partial t^2} - c^2 \frac{\partial^2}{\partial z^2} \right) (A_z - \phi) \\ &= -\frac{\partial^2 W}{\partial x^2} - \frac{\partial^2 W}{\partial y^2} + \frac{c}{q} \int dt \left(\frac{\partial}{\partial t} + c\frac{\partial}{\partial z} \right) \left(\frac{\partial}{\partial t} - c\frac{\partial}{\partial z} \right) (A_z - \phi). \end{aligned} \tag{35}$$

The last integral in this equation vanishes because

$$\frac{\partial}{\partial t} + c\frac{\partial}{\partial z} \approx \frac{\partial}{\partial t} + \boldsymbol{v}\nabla = \frac{d}{dt} \tag{36}$$

and

$$\begin{aligned} &\int dt \left(\frac{\partial}{\partial t} + c\frac{\partial}{\partial z} \right) \left(\frac{\partial}{\partial t} - c\frac{\partial}{\partial z} \right) (A_z - \phi) \\ &= \int dt \frac{d}{dt} \left(\frac{\partial}{\partial t} - c\frac{\partial}{\partial z} \right) (A_z - \phi) \\ &= 0. \end{aligned} \tag{37}$$

It is interesting that the wake potential W turns out to be a relativistic invariant. A covariant expression for it can be written as

$$W = -\frac{1}{q} \int_{-\infty}^{\infty} A_k u^k d\tau, \tag{38}$$

where $A_k = (\phi, -\boldsymbol{A})$ is the 4-vector potential, $u^k = (c\gamma, c\gamma\boldsymbol{v})$ is the 4-vector velocity, and τ is the proper time for the particle.

VIII SYSTEMS WITH A SYMMETRY AXIS

In Section VI we defined the wake as a function of the trailing particle offset relative to the path of the leading particle. In practical applications we are also interested in how the wake depends on the trajectory of the leading particle. We will assume that the system under consideration has a symmetry axis, and choose it as the z-axis of the coordinate system (see Fig. 5). Now the leading particle 1

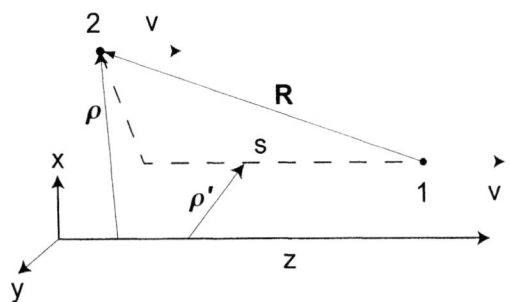

FIGURE 5. Both the leading particle 1 and the trailing particle 2 are offset relative to the axis of the chamber.

moves in the z direction with an offset given by vector $\boldsymbol{\rho}'$, and the trailing particle travels parallel to the leading one, with the same velocity, at a distance s behind the leading one, and with offset $\boldsymbol{\rho}$ relative to the axis. The vectors $\boldsymbol{\rho}'$ and $\boldsymbol{\rho}$ are the two-dimensional vectors perpendicular to the z-axis. The wake is still defined by Eq. (24), but now it will be considered as a function of $\boldsymbol{\rho}'$, $\boldsymbol{\rho}$, and s

$$w_l = w_l(\boldsymbol{\rho}, \boldsymbol{\rho}', s),$$
$$\boldsymbol{w}_t = \boldsymbol{w}_t(\boldsymbol{\rho}, \boldsymbol{\rho}', s). \tag{39}$$

Usually the vacuum chamber is designed so that the system axis serves as an ideal orbit for the beam. Deviations from it are relatively small, and both vectors $\boldsymbol{\rho}$ and $\boldsymbol{\rho}'$ are typically much smaller than the size of the vacuum chamber. We can neglect them in w_l and introduce the longitudinal wake function that depends only on s,

$$w_l(s) = w_l(0, 0, s). \tag{40}$$

If the vacuum chamber also has some symmetry elements (e.g., it has either circular, elliptical or rectangular cross section), the transverse wake on the axis, where $\boldsymbol{\rho}, \boldsymbol{\rho}' = 0$, vanishes, $\boldsymbol{w}_t(0, 0, s) = 0$. For small values of $\boldsymbol{\rho}, \boldsymbol{\rho}'$ we can expand $\boldsymbol{w}_t(\boldsymbol{\rho}, \boldsymbol{\rho}', s)$ keeping only the lowest-order linear terms. That gives a tensor relation between the transverse wake and the offsets,

$$\boldsymbol{w}_t(\boldsymbol{\rho}, \boldsymbol{\rho}', s) = \overset{\leftrightarrow}{\boldsymbol{W}}_1(s)\boldsymbol{\rho} + \overset{\leftrightarrow}{\boldsymbol{W}}_2(s)\boldsymbol{\rho}', \tag{41}$$

where $\overleftrightarrow{W}_1$ and $\overleftrightarrow{W}_2$ are the two-dimensional tensors of rank 2. An example of the wake calculation for elliptical and rectangular cross sections of the pipe can be found in Ref. 7.

IX AXISYMMETRIC SYSTEMS

In an axisymmetric system W depends only on the absolute values of ρ, ρ', and the angle θ between them. We can always chose a coordinate system such that the vector ρ' lies in the x-z plane (see Fig. 6), so that the potential function W will be a periodic even function of angle θ in a cylindrical coordinate system. Decomposing

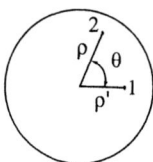

FIGURE 6. Vectors ρ and ρ' in an axisymmetric system.

W in Fourier series in θ yields

$$W(\rho,\rho',\theta,s) = \sum_{m=0}^{\infty} W_m(\rho,\rho',s)\cos m\theta. \qquad (42)$$

Putting this equation into Eq. (34) gives

$$\sum_{m=0}^{\infty}\left(\frac{1}{\rho}\frac{\partial}{\partial\rho}\rho\frac{\partial W_m}{\partial\rho} - \frac{m^2}{\rho^2}W_m\right)\cos m\theta = 0, \qquad (43)$$

from which we can find an explicit dependence of W_m of ρ,

$$W_m(\rho,\rho',s) = A_m(\rho',s)\rho^m. \qquad (44)$$

In Eq. (44) we discarded a singular solution of Eq. (43) $W_m \propto \rho^{-m}$.

It is also possible to find the dependence of W_m versus ρ' (see [8]), which turns out to be

$$A_m(\rho',s) = F_m(s)(\rho')^m. \qquad (45)$$

Using Eq. (32) we now find for the longitudinal and transverse wake functions

$$w_l = \sum w_l^{(m)}, \qquad \boldsymbol{w}_t = \sum \boldsymbol{w}_t^{(m)} \qquad (46)$$

216

where

$$w_l^{(m)} = (\rho')^m \rho^m F'_m(s) \cos m\theta,$$
$$\boldsymbol{w}_t^{(m)} = m(\rho')^m \rho^{m-1} F_m(s) \left[\hat{\boldsymbol{r}} \cos m\theta - \hat{\boldsymbol{\theta}} \sin m\theta\right], \qquad (47)$$

where $\hat{\boldsymbol{r}}$ and $\hat{\boldsymbol{\theta}}$ are the unit vectors in the radial and azimuthal directions in the cylindrical coordinate system. Remember that in this equation we assume that the leading particle is in the plane $\theta = 0$.

Equations (47) are valid for arbitrary values of ρ and ρ'. Near the axis, where the offsets are small, the higher-order terms with large values of m in these equations also become small. In this case, we can keep only the lower-order terms with $m = 0$ (*monopole*) and $m = 1$ (*dipole*) wakes. For the monopole wake we find

$$w_l \equiv w_l^{(0)} = F'_0(s), \qquad (48)$$

which shows that the longitudinal wake does not depend on the radius in an axisymmetric system. We also see that the monopole transverse wake vanishes, $w_t^{(0)} = 0$.

Since $w_l^{(0)}$ does not depend on ρ', sometimes it is more convenient to compute the monopole wake for an offset orbit, $\rho' \neq 0$, rather than on the axis.

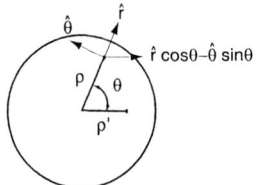

FIGURE 7. Vectors ρ and ρ' and unit vectors in cylindrical coordinate system.

For the dipole wake ($m = 1$), the vector $\hat{\boldsymbol{r}} \cos\theta - \hat{\boldsymbol{\theta}} \sin\theta$ lies in the direction of the x axis, that is in the direction of $\boldsymbol{\rho}'$, see Fig. 7. Hence,

$$\boldsymbol{w}_t^{(1)} = \boldsymbol{\rho}' F(s). \qquad (49)$$

The wake given by Eq. (49) is usually normalized by the absolute value of the offset ρ', and the scalar function $w_t^{(1)}/\rho'$ is called the transverse dipole wake w_t,

$$w_t(s) = F(s). \qquad (50)$$

Such a transverse wake has the dimension cm^{-2} or V/C/m.[3] In this definition, a positive transverse wake means a kick in the direction of the offset of the driving particle (if both particles have the same charge).

---

[3] If the original wake is defined per unit length, as in Eq. (26), then w_t will have the dimension V/C/m^2 or cm^{-3}.

X RESISTIVE WALL WAKE FUNCTIONS

We are now in a position to find the wakes generated by a particle in a circular pipe with resistive walls. Using Eq. (18) and the definition Eq. (24) gives the longitudinal wake

$$w_l(s) = -\frac{1}{2\pi b}\sqrt{\frac{c}{\sigma s^3}}. \tag{51}$$

The minus sign here means that the trailing charge is accelerated in the wakefield.

Let us now calculate the dipole transverse wake due to the resistive wall. First, we need to solve for the electromagnetic field of the leading charge q moving with an offset ρ' in a circular pipe. From the point of view of excitation of dipole modes, this charge can be considered as having a dipole moment $\boldsymbol{d} = q\boldsymbol{\rho'}$. In the zeroth approximation of perturbation theory, the electromagnetic field of a dipole moving with the speed of light in a *perfectly conducting* pipe is

$$\boldsymbol{E} = 2\delta(z-ct)\left[\frac{2(\boldsymbol{d}\cdot\boldsymbol{r})\boldsymbol{r} - d\boldsymbol{r}^2}{r^4} + \frac{\boldsymbol{d}}{b^2}\right], \qquad \boldsymbol{H} = \hat{\boldsymbol{z}} \times \boldsymbol{E}. \tag{52}$$

The first term in the expression for $\boldsymbol{E}$ is a vacuum field of a relativistic dipole, and the second one is due to the image charges, which are generated in order to satisfy the boundary condition on the metal surface.

Following the derivation in Section V, we find the magnetic field on the wall,

$$H_\theta = \frac{4d}{b^2}\cos\theta\,\delta(z-ct), \tag{53}$$

and take its Fourier transform,

$$H_{\theta\omega} = \frac{2d}{\pi c b^2}\cos\theta, \tag{54}$$

where the angle θ is measured from the direction of $\boldsymbol{d}$. Then using the Leontovich boundary condition, Eq. (12), for the electric field $E_{z\omega}$,

$$E_{z\omega}|_{r=b} = -(1-i)\sqrt{\frac{\omega}{8\pi\sigma}}\frac{2d}{\pi c b^2}\cos\theta, \tag{55}$$

and making the inverse Fourier transformation, we obtain E_z on the wall

$$E_z(z, \rho=b, \rho', t) = \frac{q\rho'}{\pi b^2}\sqrt{\frac{c}{\sigma s^3}}\cos\theta, \tag{56}$$

where $s = ct - z$. Recalling that, according to Eq. (47) the dipole wake is a linear function of ρ, we conclude that

$$E_z(z, \rho, \rho', t) = \frac{q\rho\rho'}{\pi b^3}\sqrt{\frac{c}{\sigma s^3}}\cos\theta, \tag{57}$$

and the function $F_1'(s)$ is

$$F_1'(s) = -\frac{1}{\pi b^3}\sqrt{\frac{c}{\sigma s^3}}, \tag{58}$$

which gives the following result for the transverse wake defined by Eq. (50):

$$w_t(s) = \frac{2}{\pi b^3}\sqrt{\frac{c}{\sigma s}}. \tag{59}$$

Analogous to the longitudinal wake, Eq. (18), this formula is valid only for $s \gg s_0$ (see Eq. (22)).

XI WAKEFIELD IN A BUNCH OF PARTICLES

Up to now we have studied the interaction of two point charges traveling some distance s apart. If a beam consists of N particles with the distribution function $\lambda(s)$ (defined so that $\lambda(s)ds$ gives the probability of finding a particle near point s), a given particle will interact with all other particles of the beam. To find the change of the longitudinal momentum of the particle at point s we need to sum the wakes from all other particles in the bunch,

$$\Delta p_z(s) = \frac{Ne^2}{c}\int_s^\infty ds'\lambda(s')w_l(s'-s). \tag{60}$$

Here we use the causality principle and integrate only over the part of the bunch ahead of point s. In the ultrarelativistic limit the energy change $\Delta E(s)$ caused by the wakefield is equal to $c\Delta p_z$, so Eq. (60) can also be rewritten as

$$\Delta E(s) = Ne^2\int_s^\infty ds'\lambda(s')w_l(s'-s). \tag{61}$$

Two integral characteristics of the strength of the wake are the average value of the energy loss ΔE_{av}, and the rms spread in energy generated by the wake ΔE_{rms}. These two quantities are defined by the following equations:

$$\Delta E_{\text{av}} = \int_{-\infty}^\infty ds\,\Delta E(s)\lambda(s), \tag{62}$$

and

$$\Delta E_{\text{rms}} = \left[\int_{-\infty}^\infty ds(\Delta E(s) - \Delta E_{\text{av}})^2\lambda(s)\right]^{1/2}. \tag{63}$$

As an example, let us calculate ΔE_{av} and ΔE_{rms} for the resistive wall wake given by Eq. (51) and a Gaussian distribution function,

$$\rho(s) = \frac{1}{\sqrt{2\pi}\sigma_s} \exp\left(-\frac{s^2}{2\sigma_s^2}\right), \tag{64}$$

where σ_s is the rms bunch length. Note that, since w in Eq. (51) is the wake per unit length of the pipe, we need to multiply the final answer by the pipe length L.

A direction substitution of the wake Eq. (51) into Eq. (61) gives a divergent integral when $s' \to s$.[4] To remove the divergence, we need to recall that according to Eq. (48) the longitudinal wake is equal to the derivative of the longitudinal wake potential, $w_l = F_0'(s)$ with $F_0 = (\pi b)^{-1}\sqrt{c/\sigma s}$ for $s > 0$, and $F_0 = 0$ for $s < 0$. We then rewrite Eq. (61) as

$$\begin{aligned}\Delta E(s) &= Ne^2 L \int_{-\infty}^{\infty} ds' \lambda(s') \frac{dF_0(s'-s)}{ds} \\ &= -Ne^2 L \int_{s}^{\infty} ds' \frac{d\lambda(s')}{ds} F_0(s'-s) \\ &= \frac{Ne^2 L c^{1/2}}{b\sigma_z^{3/2}\sigma^{1/2}} G\left(\frac{s}{\sigma_z}\right),\end{aligned} \tag{65}$$

where the function $G(x)$ is

$$G(x) = \frac{1}{2^{1/2}\pi^{3/2}} \int_x^{\infty} \frac{y e^{-y^2/2} dy}{\sqrt{y-x}}. \tag{66}$$

The plot of the function $G(s/\sigma_z)$ is shown in Fig. 8, where the positive values of s correspond to the head of the bunch. We see that in the resistive wake the particles lose energy in the head of the bunch and get accelerated in the tail. On the average, of course, the losses overcome the gain. For the average energy loss one can find an analytical result:

$$\Delta E_{\text{av}} = \frac{\Gamma(\frac{3}{4})}{2^{3/2}\pi^{3/2}} \frac{Ne^2 c^{1/2}}{b\sigma_z^{3/2}\sigma^{1/2}}. \tag{67}$$

Numerical integration of Eq. (63) shows that the energy spread generated by the resistive wake is approximately equal to ΔE_{av}:

$$\Delta E_{\text{rms}} = 1.06 \, \Delta E_{\text{av}}. \tag{68}$$

If the beam is traveling in the pipe with an offset y relative to the axis, it will be deflected in the direction of the offset, by the transverse wakefields. To calculate the deflection angle θ we use the relation

---

[4] The integral diverges because Eq. (51) is not valid for very small values of s, see Eq. (22).

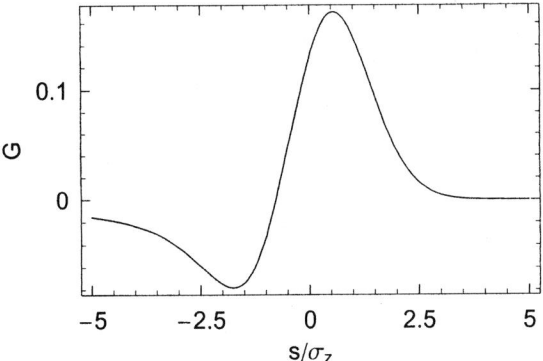

FIGURE 8. Plot of the function $G(s/\sigma_z)$.

$$\theta(s) = \frac{\Delta p_\perp(s)}{p} = yL\frac{Ne^2}{cp}\int_s^\infty ds'\lambda(s')w_t(s'-s)$$
$$= \frac{NLr_e yc^{1/2}}{\gamma b^3 \sigma_z^{1/2}\sigma^{1/2}}H\left(\frac{s}{\sigma_z}\right), \tag{69}$$

where the function $H(x)$ is

$$H(x) = \frac{2^{1/2}}{\pi^{3/2}}\int_x^\infty \frac{e^{-y^2/2}dy}{\sqrt{y-x}}. \tag{70}$$

The plot of the function $H(s/\sigma_z)$ is shown in Fig. 9.

The deflection angle averaged over the distribution function is

$$\theta_{\mathrm{av}} = \frac{\Gamma(\frac{1}{4})}{2^{1/2}\pi^{3/2}}\frac{NLr_e yc^{1/2}}{\gamma b^3 \sigma_z^{1/2}\sigma^{1/2}}, \tag{71}$$

and the rms spread is

$$\langle(\theta-\theta_{\mathrm{av}})^2\rangle^{1/2} = A^{1/2}\frac{NLr_e yc^{1/2}}{\gamma b^3 \sigma_z^{1/2}\sigma^{1/2}}, \tag{72}$$

where

$$A = \frac{2}{\pi^{5/2}}\left[K\left(\frac{3}{4}\right) - \frac{\Gamma^2(1/4)}{4\sqrt{\pi}}\right] \tag{73}$$

and K is the complete elliptic integral. The numerical value of $A^{1/2}$ is 0.186.

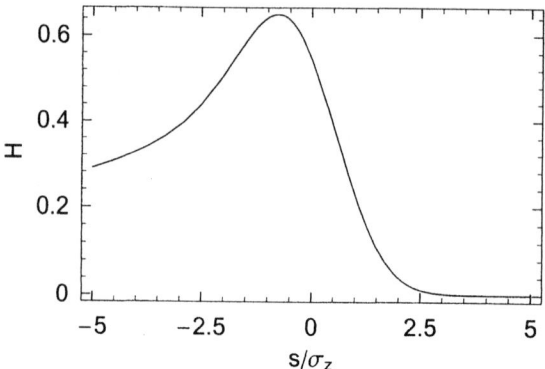

FIGURE 9. Plot of the function $H(s/\sigma_z)$.

XII DEFINITION OF IMPEDANCE AND RELATION BETWEEN IMPEDANCE AND WAKE

Knowledge of the longitudinal and transverse wake functions gives complete information about the electromagnetic interaction of the beam with its environment. However, in many cases, especially in the study of beam instabilities, it is more convenient to use the Fourier transform of the wake functions, or *impedances*. Also, it is often easier to calculate the impedance for a given geometry of the beam pipe, rather than the wake function. Recall, that in Section V we actually first computed the Fourier components of the wakes, and then, using the inverse Fourier transformation, found the wakes.

For historical reasons the longitudinal Z_l and transverse Z_t impedances are defined as Fourier transforms of wakes with different factors,

$$Z_l(\omega) = \frac{1}{c}\int_0^\infty ds\, w_l(s) e^{i\omega s/c},$$
$$Z_t(\omega) = -\frac{i}{c}\int_0^\infty ds\, w_t(s) e^{i\omega s/c}. \qquad (74)$$

Note that the integration in Eqs. (74) can actually be extended into the region of negative values of s, because w_l and w_t are equal to zero in that region.

Impedance can also be defined for complex values of ω such that $\text{Im}\,\omega > 0$ and the integrals, Eq. (74), converge. So defined, the impedance is an analytic function in the upper half-plane of the complex variable ω.

We must keep in mind that other authors sometimes introduce definitions of the impedance that differ from the one given above. In Refs. 2 and 9 the longitudinal impedance is defined as a complex conjugate to the one given by Eq. (74). Here we follow the definitions of Refs. 1 and 10.

From the definitions in Eq. (74) it follows that the impedance satisfies the following symmetry conditions:

$$\text{Re}Z_l(\omega) = \text{Re}Z_l(-\omega), \qquad \text{Im}Z_l(\omega) = -\text{Im}Z_l(-\omega),$$
$$\text{Re}Z_t(\omega) = -\text{Re}Z_t(-\omega), \qquad \text{Im}Z_t(\omega) = \text{Im}Z_t(-\omega). \tag{75}$$

The inverse Fourier transform relates the wakes to the impedances:

$$w_l(s) = \frac{1}{2\pi} \int_{-\infty}^{\infty} d\omega\, Z_l(\omega) e^{-i\omega s/c},$$
$$w_t(s) = \frac{i}{2\pi} \int_{\infty}^{\infty} d\omega\, Z_t(\omega) e^{-i\omega s/c}. \tag{76}$$

It turns out that the wakefield can actually be found if only the real part of the impedance is known. Indeed, we can rewrite Eq. (76) for w_l as

$$w_l(s) = \frac{1}{2\pi} \int_{\infty}^{\infty} d\omega \left[\text{Re}Z_l(\omega) \cos\frac{\omega s}{c} - \text{Im}Z_l(\omega) \sin\frac{\omega s}{c} \right]. \tag{77}$$

For negative values of s this formula should give $w_l = 0$, hence

$$0 = \int_{\infty}^{\infty} d\omega \left[\text{Re}Z_l(\omega) \cos\frac{\omega s}{c} + \text{Im}Z_l(\omega) \sin\frac{\omega |s|}{c} \right], \tag{78}$$

from which it follows that

$$w_l(s) = \frac{1}{\pi} \int_{\infty}^{\infty} d\omega\, \text{Re}Z_l(\omega) \cos\frac{\omega s}{c} = \frac{2}{\pi} \int_0^{\infty} d\omega\, \text{Re}Z_l(\omega) \cos\frac{\omega s}{c}. \tag{79}$$

A similar derivation for the transverse wake gives

$$w_t(s) = \frac{2}{\pi} \int_0^{\infty} d\omega\, \text{Re}Z_t(\omega) \sin\frac{\omega s}{c}. \tag{80}$$

XIII ENERGY LOSS AND $\text{Re}Z_L$

We can relate the energy loss by the bunch to the real part of the longitudinal impedance. Indeed,

$$\Delta E_{\text{av}} = N^2 e^2 \int_{-\infty}^{\infty} ds\, \lambda(s) \int_{-\infty}^{\infty} ds'\, \lambda(s') w_l(s' - s)$$
$$= N^2 e^2 \int_{-\infty}^{\infty} ds\, \lambda(s) \int_{-\infty}^{\infty} ds'\, \lambda(s') \frac{1}{2\pi} \int_{-\infty}^{\infty} d\omega\, Z_l(\omega) e^{-i\omega(s'-s)/c}$$
$$= \frac{N^2 e^2}{2\pi} \int_{-\infty}^{\infty} d\omega\, Z_l(\omega) |\hat{\lambda}(\omega)|^2, \tag{81}$$

where $\hat{\lambda}(\omega) = \int_{-\infty}^{\infty} ds \lambda(s) e^{i\omega s/c}$. Since $\hat{\lambda}(-\omega) = \hat{\lambda}^*(\omega)$, $|\hat{\lambda}(\omega)|^2$ is an even function of ω, and

$$\Delta E_{\text{av}} = \frac{Q^2}{\pi} \int_0^\infty d\omega \text{Re} Z_l(\omega) |\hat{\lambda}(\omega)|^2, \tag{82}$$

where $Q = Ne$.

For a point charge, $\lambda(s) = \delta(s)$, $\hat{\lambda}(\omega) = 1$, and the energy loss is

$$\Delta E_{\text{av}} = \frac{e^2}{\pi} \int_0^\infty d\omega \text{Re} Z_l(\omega). \tag{83}$$

XIV KRAMERS-KRONIG RELATIONS

Equations (79) and (80) relate $\text{Re} Z(\omega)$ to the wake function. Since $Z(\omega)$ is given by Fourier transformation of $w(s)$, the knowledge of $\text{Re} Z(\omega)$ allows us to find $Z(\omega)$, and hence $\text{Im} Z(\omega)$. This means that $\text{Im} Z(\omega)$ and $\text{Re} Z(\omega)$ are functionally related to each other. Mathematically this relation is manifested in the Kramers-Kronig *dispersion* relation, which can be written as

$$Z(\omega) = -\frac{i}{\pi} \text{P.V.} \int_{-\infty}^{\infty} \frac{Z(\omega')}{\omega' - \omega} d\omega', \tag{84}$$

where P.V. stands for the principal value of the integral. Taking the real and imaginary parts of this equation gives explicit relations between $\text{Re} Z$ and $\text{Im} Z$:

$$\text{Re} Z(\omega) = \frac{1}{\pi} \text{P.V.} \int_{-\infty}^{\infty} \frac{\text{Im} Z(\omega')}{\omega' - \omega} d\omega',$$

$$\text{Im} Z(\omega) = -\frac{1}{\pi} \text{P.V.} \int_{-\infty}^{\infty} \frac{\text{Re} Z(\omega')}{\omega' - \omega} d\omega'. \tag{85}$$

XV USEFUL FORMULA FOR IMPEDANCE CALCULATION

Assume that we have a solution of an electromagnetic problem corresponding to the current on the axis of a pipe with the time and space dependence given by $e^{-i\omega t + i\omega z/c}$. Specifically, we know the electric field on the axis, $E_{z\omega}(z) e^{-i\omega t}$. How can longitudinal impedance be found in terms of $E_{z\omega}(z)$?

The longitudinal wake is equal to the integrated field E_z generated by a point charge moving with the speed of light. The current corresponding to this point charge can be decomposed into a Fourier integral:

$$cq\delta(z - ct) = \int_{-\infty}^{\infty} d\omega \frac{q}{2\pi} e^{-i\omega(t - z/c)}. \tag{86}$$

Since we know the electric field generated by each harmonic, we can find the field due to the point charge as a superposition of $E_{z\omega}$:

$$E_z(z,t) = \frac{q}{2\pi} \int_{-\infty}^{\infty} d\omega\, E_{z\omega}(z) e^{-i\omega t}. \tag{87}$$

For w_l we then have

$$\begin{aligned} w_l(s) &= -\frac{1}{2\pi} \int_{-\infty}^{\infty} d\omega\, d(ct) E_{z\omega}(ct - s) e^{-i\omega t} \\ &= -\frac{1}{2\pi} \int_{-\infty}^{\infty} d\omega\, dz\, E_{z\omega}(z) e^{-i\omega(z+s)/c}. \end{aligned} \tag{88}$$

Comparing Eq. (88) with Eq. (74) we find

$$Z_l(\omega) = -\int_{-\infty}^{\infty} dz\, E_\omega(z) e^{-i\omega z/c}. \tag{89}$$

Hence the longitudinal impedance can be obtained simply by making Fourier transformation of $E_{z\omega}(z)$.

XVI SMALL PILLBOX CAVITY IN ROUND PIPE

As an example of using Eq. (89) we will derive here the longitudinal impedance $Z_l(\omega)$ for a small axisymmetric cavity (pillbox) in a round perfectly conducting pipe, see Fig. 10. We assume that the wavelength associated with the frequency ω is much larger than the dimension of the pillbox, $c/\omega \gg g, h$.

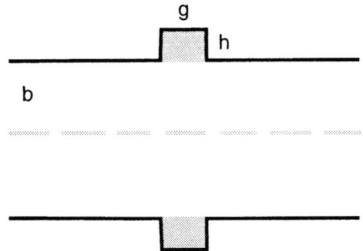

FIGURE 10. Small pillbox cavity in a round pipe. The dashed line near the wall shows the integration path in Eq. (92).

First, we need to find the solution of Maxwell's equations corresponding to a unit current $I_\omega = e^{-i\omega t + i\omega z/c}$ on the axis. Since the cavity is small, the magnitude of the magnetic field at the location of the cavity ($z = 0$) is approximately equal to H_θ on the wall of the pipe

$$H_\theta = \frac{2}{bc}. \qquad (90)$$

Because $E_{z\omega}$ does not depend on r (see Section V) we can choose the integration path in Eq. (89) close to the wall, as shown in Fig. 10, rather than the pipe axis. Along this path $E_{z\omega}$ is not equal to zero only in the cavity gap, where $|z| \sim h$, and $e^{-i\omega z} \approx 1$. We have

$$Z_l(\omega) = -\int_{-\infty}^{\infty} dz\, E_{z\omega}(z) e^{-i\omega z/c}$$

$$\approx -\int_{-\infty}^{\infty} dz\, E_{z\omega}(z) = \frac{1}{c}\frac{d\Phi}{dt} = -\frac{i\omega}{c}\Phi, \qquad (91)$$

where Φ is the magnetic flux in the cross section of the cavity, $\Phi = H_\theta h g$. As a result,

$$Z_l(\omega) = -i\omega \frac{2gh}{bc^2} = -i\omega Z_0 \frac{gh}{2\pi bc}, \qquad (92)$$

where $Z_0 = 4\pi/c = 377$ Ohm. What we obtained is a purely *inductive* longitudinal impedance, which can be rewritten as

$$Z_l(\omega) = -i\frac{\omega}{c^2}\mathcal{L}, \qquad (93)$$

where the inductance $\mathcal{L} = 2gh/b$. In CGS units the inductance has a dimension of cm, 1 cm = 1 nH.

A more detailed calculation [11] shows that our method gives only an approximate solution of the problem. In addition to the solenoidal electric field generated by the time-varying magnetic flux in the cavity, there is also a contribution due to the potential component of the electric field. This results in a different numerical coefficient in Eq. (92) which depends on the ratio g/h. For example, for $g = h$ the correction factor is 0.84.

XVII INDUCTIVE IMPEDANCE

We saw in the previous section that a small pillbox is characterized by inductive impedance if the frequency is not very large. This is a common feature of many small perturbations whose size is much smaller than the pipe radius (e.g., small holes, shallow obstacles on the wall, etc.) — for not very large frequencies their impedance is purely inductive.

The longitudinal wake corresponding to the inductive impedance can be found by using Eq. (76):[5]

$$w_l(s) = \mathcal{L}\delta'(s). \tag{94}$$

Because of the inductive wake, slices of the beam can change their energy, although the net energy loss for the bunch is zero because the real part of the impedance vanishes. We can find the energy change as a function of position within the bunch by using Eq. (61). Integration gives

$$\Delta E(s) = -e^2 N \mathcal{L} \lambda'(s). \tag{95}$$

For a Gaussian distribution function this reduces to

$$\Delta E(s) = \frac{e^2 N \mathcal{L}}{\sigma_z^2} \frac{\xi e^{-\xi^2/2}}{\sqrt{2\pi}}, \tag{96}$$

where $\xi = s/\sigma_z$. For the rms energy spread we find

$$\langle \Delta E^2 \rangle^{1/2} = 3^{-3/4}(2\pi)^{-1/2} \frac{e^2 N \mathcal{L}}{\sigma_z^2}. \tag{97}$$

XVIII CAVITY IMPEDANCE

In the more general case of a large cavity (Fig. 11) the beam excites cavity eigenmodes and the longitudinal wake in the cavity is composed of contributions

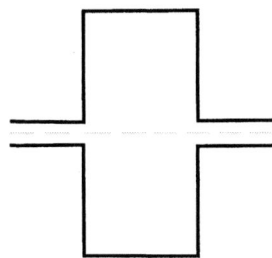

FIGURE 11. An RF cavity with beam pipes.

from single modes,

$$w_l(s) = \sum_n w^{(n)}(s). \tag{98}$$

---

[5] Since the integral involved in the calculation of $w_l(s)$ actually diverges at $\omega \to \infty$, it should be treated as a generalized function. It is easier to verify Eq. (94) by putting it into Eq. (74) and checking that the resulting impedance is given by Eq. (93).

For perfectly conducting walls, assuming that the modes do not propagate into the beam pipes,[6] the mode frequencies ω_n are real. It should be no surprise that each partial wakefield oscillates with the frequency of the mode ω_n,

$$w^{(n)}(s) = 2k_n \cos\left(\frac{\omega_n s}{c}\right), \tag{99}$$

where k_n is the *loss factor*, which depends on the geometry of the cavity and the mode number. As an example, for a cylindrical cavity with $b = l$ and TM$_{010}$ mode, $k_{010} = 4.5/l$. For a more rigorous derivation of the wake for a cavity, see Ref. 12.

The cavity impedance for this wake can be calculated by using Eqs. (74). They give

$$\mathrm{Re} Z_l = \pi k_n [\delta(\omega + \omega_n) + \delta(\omega - \omega_n)],$$
$$\mathrm{Im} Z_l = k_n \left[\frac{1}{\omega + \omega_n} + \frac{1}{\omega - \omega_n}\right]. \tag{100}$$

It is also easy to generalize the above wake for a cavity with lossy walls when the frequency of the mode has a *small* imaginary part γ_n, $\gamma_n \ll \omega_n$. The wake now decays with time as

$$w_l(s) = 2k_n e^{-\gamma s/c} \cos\left(\frac{\omega_n s}{c}\right). \tag{101}$$

Again using Eq. (74), we can calculate the impedance. It has two peaks: one in the vicinity of $\omega = \omega_n$ and the other in the vicinity of $\omega = -\omega_n$. Assuming that ω is close to ω_n, we find

$$Z_l = \frac{k_n}{\gamma - i(\omega - \omega_n)}. \tag{102}$$

REFERENCES

1. A. W. Chao, *Physics of Collective Beam Instabilities in High Energy Accelerators* (Wiley, New York, 1993).
2. B. W. Zotter and S. A. Kheifets, *Impedances and Wakes in High-Energy Particle Accelerators* (World Scientific, Singapore, 1998).
3. A. W. Chao and M. Tigner, *Handbook of Accelerator Physics and Engineering* (World Scientific, Singapore, 1999).
4. L. D. Landau and E. M. Lifshitz, *The Classical Theory of Fields*, vol. 2 of *Course of Theoretical Physics*, 4th ed. (Pergamon, London, 1979) (Translated from the Russian).
5. L. D. Landau and E. M. Lifshitz, *Electrodynamics of Continuous Media*, vol. 8 of *Course of Theoretical Physics*, 2nd ed. (Pergamon, London, 1960) (Translated from the Russian).

---

[6] This is true if the frequency of the mode is above the cutoff frequency for the pipe.

6. K. L. F. Bane and M. Sands, *The Short-Range Resistive Wall Wakefields*, Tech. Rep. SLAC-PUB-95-7074, SLAC (December 1995).
7. R. L. Gluckstern, J. van Zeijts, and B. Zotter, Phys. Rev. **E47**, 656 (1993).
8. K. L. Bane, P. B. Wilson, and T. Weiland, in *Proc. US Particle Accelerator School: Physics of Particle Accelerators, Upton, N.Y., 1983* (American Institute of Physics, New York, 1985), no. 127 in AIP Conference Proceedings, pp. 875–928.
9. P. B. Wilson, in M. Month and M. Dienes, eds., *Proc. US Particle Accelerator School: Physics of Particle Accelerators, Batavia, 1987* (American Institute of Physics, New York, 1989), no. 184 in AIP Conference Proceedings, pp. 525–564.
10. S. A. Heifets and S. A. Kheifets, Review of Modern Physics **63**, 631 (1991).
11. S. S. Kurennoy and G. V. Stupakov, Particle Accelerators **45**, 95 (1994).
12. P. Wilson, *High Energy Electron Linacs: Applications to Storage Ring RF Systems and Linear Colliders*, Tech. Rep. SLAC-AP-2884 (Rev.), SLAC (November 1991).

APPENDIX

We show here how to calculate the integral in Eq. (17),

$$I(\xi) = \int_{-\infty}^{\infty} d\omega \sqrt{\omega} e^{-i\omega\xi}, \tag{A1}$$

where $\xi = t - z/c = s/c$. First, we change the integration variable ω to $\zeta = \omega\xi$,

$$I(\xi) = \frac{1}{\xi^{3/2}} \int_{-\infty}^{\infty} d\zeta \sqrt{\zeta} e^{-i\zeta}. \tag{A2}$$

We then consider ζ as a complex variable. In order to make the integrand a single-valued function in the complex plane, we make a cut along the negative imaginary half-axis and deform the integration path from a straight line to a contour, shown in Fig. 12. The integral $I(\xi)$ can now be split into two parts, $I_1(\xi)$ and $I_2(\xi)$, corresponding to the left and right branches of the integration contour.

In the first integral $I_1(\xi)$ we change the complex integration variable ζ to the real positive variable τ, $\zeta = e^{3\pi i/2}\tau = -i\tau$, so that $\sqrt{\zeta} = e^{3\pi i/4}\sqrt{\tau} = (i-1)\sqrt{\tau}/\sqrt{2}$. This gives for I_1

$$\begin{aligned} I_1(\xi) &= \frac{1}{\xi^{3/2}} \int_{\infty}^{0} (-id\tau) \frac{(i-1)\sqrt{\tau}}{\sqrt{2}} e^{-\tau} \\ &= \frac{(i+1)}{\sqrt{2}\xi^{3/2}} \int_{\infty}^{0} d\tau \sqrt{\tau} e^{-\tau} \\ &= -\frac{\sqrt{\pi}(i+1)}{2^{3/2}\xi^{3/2}}. \end{aligned} \tag{A3}$$

For the second integral $I_2(\xi)$ we choose the real positive variable τ such that $\zeta = e^{-\pi i/2}\tau = -i\tau$, which means that $\sqrt{\zeta} = e^{-\pi i/4}\sqrt{\tau} = (1-i)\sqrt{\tau}/\sqrt{2}$ and

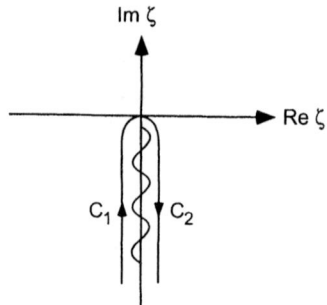

FIGURE 12. Complex plane ζ. The wavy line indicates the cut. The deformed integration contour consists of two paths, C_1 and C_2, corresponding to the integrals I_1 and I_2, respectively.

$$I_2(\xi) = \frac{1}{\xi^{3/2}} \int_0^\infty (-id\tau) \frac{(1-i)\sqrt{\tau}}{\sqrt{2}} e^{-\tau}$$
$$= I_1(\xi). \qquad (A4)$$

Hence

$$I(\xi) = I_1(\xi) + I_2(\xi) = -\frac{\sqrt{\pi}(i+1)}{2^{1/2}\xi^{3/2}}. \qquad (A5)$$

Substituting this equation into Eq. (17) gives Eq. (18).

Longitudinal Single-Bunch Instabilities

M. Migliorati and L. Palumbo

University of Rome "LA SAPIENZA" and LNF - INFN, Italy

Abstract. After introducing the concepts of longitudinal wakefield and coupling impedance, we review the theory of longitudinal single-bunch collective effects in storage rings. From the Fokker-Planck equation we first derive the stationary solution describing the natural single-bunch regime, and then treat the problem of microwave instability, showing the different approaches used for estimating the threshold current. We end the lecture with the semi-empirical laws that allow us to obtain the single-bunch behavior above threshold, and with a description of the simulation codes that are now reliable tools for investigating all these effects.

OBSERVATIONS

If we measure in a circular accelerator the RMS energy spread σ_ε and the RMS bunch length σ_z as a function of the current, parameters of great importance for the machine performance, we obtain qualitatively the behavior plotted in Figure 1.

The energy spread is almost constant up to a threshold current, called the microwave instability threshold, after which it starts to increase with the current according to a given power law (in most cases 1/3 power). The bunch length instead starts to increase from the very beginning, and, after the same threshold

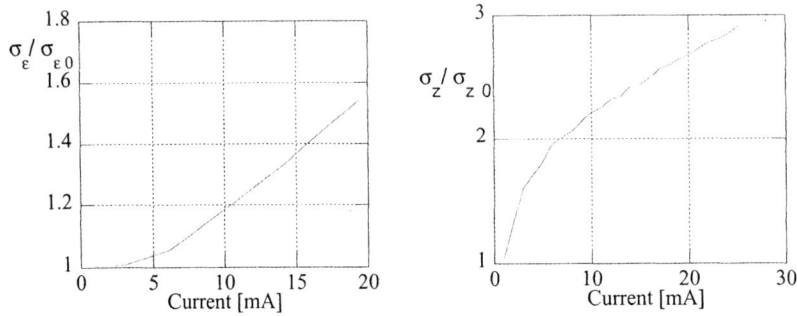

FIGURE 1. Qualitative behavior of σ_ε and σ_z as a function of current for a generic machine.

current, it grows with the same power law.

Furthermore, it may happen that above the threshold, depending on the wakefields and on the machine parameters, a sawtooth-like behavior is excited (Figure 2).

Although these effects can limit the machine performance, they make the single-bunch dynamics quite attractive from the dynamical point of view. Several physical effects are involved: RF capture, quantum fluctuations, radiation, self-fields interaction, etc. Moreover, this scenario is complicated when nonlinear effects become significant.

In the following sections we will give the basic equations and some models useful for describing the single-bunch dynamics, although a general theory able to predict single-bunch behavior in all its manifestations is still missing.

WAKEFIELD AND IMPEDANCE

Longitudinal Wake Function

The interaction of a beam with its surroundings [1] is of great importance for the study of beam dynamics since it is responsible of all the collective instabilities. The fields produced by the beam (wakefields) interact with the beam itself as in a loop system. To introduce wake fields, we consider the coordinate system of Figure 3 and call $q_1(z_1, \mathbf{r_1})$ a charge traveling with constant velocity $v = c$ along a trajectory parallel to the axis of the vacuum chamber. The longitudinal Lorentz force generated by q_1 acting on a test charge $q(z, \mathbf{r})$ following q_1 at a distance

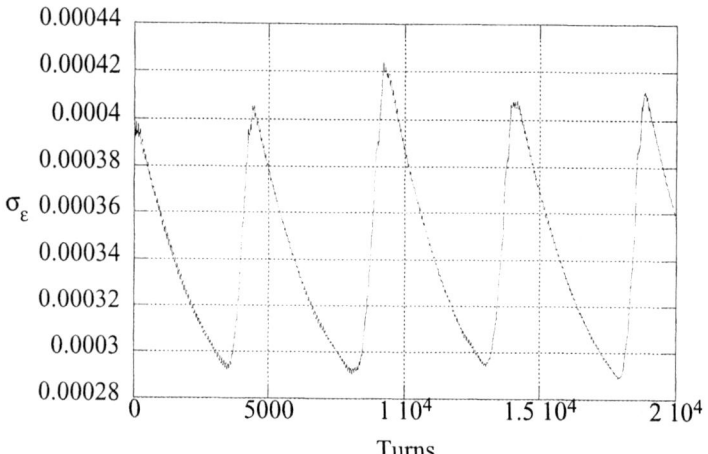

FIGURE 2. Energy spread vs number of turns for a pure inductive impedance obtained with a simulation code.

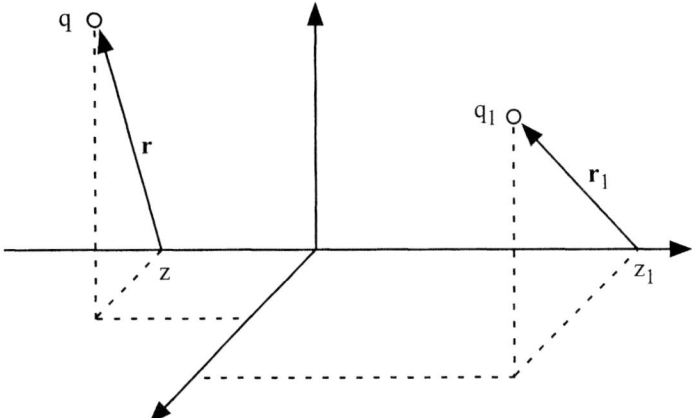

FIGURE 3. System of reference coordinate.

$\Delta z = z_1 - z$ produces on this charge an energy variation given by

$$U_\| (\mathbf{r}, \mathbf{r}_1; \Delta z) = - \int_{\text{Str}} F_\| (z, \mathbf{r}, z_1, \mathbf{r}_1; t) \, dz \qquad t = (z_1 + \Delta z)/c. \qquad (1)$$

The quantity $U_\|$ represents the energy lost (> 0) or gained (< 0) by a charge passing through a machine device, due to electromagnetic forces parallel to the particle motion. We assume that the relative energy change is so small that it does not produce any appreciable variation of the relativistic factor β.

We define the longitudinal wake function as the energy variation of a test charge q per unit charge q and q_1

$$w_\| (\mathbf{r}, \mathbf{r}_1; \Delta z) = \frac{U_\| (\mathbf{r}, \mathbf{r}_1; \Delta z)}{q q_1}. \qquad (2)$$

It can be thought of as the Green function that describes the longitudinal response of the structure to an impulsive source. It depends only on the geometrical and electromagnetic properties of the device. For $\beta = 1$, according to the causality principle, the wake function is zero for $\Delta z < 0$, i. e. for a test charge ahead of the leading one.

In general, the beam pipe is composed of structures having symmetric shapes. In the case of cylindrical symmetry it is convenient to expand the wake function in multi-polar terms. For the longitudinal case, around the vacuum chamber axis, the first monopole term is dominant and the longitudinal wakefield becomes a function only of Δz.

For a longitudinal bunch distribution $\rho(z)$ that satisfies the normalization condition

$$\int_{-\infty}^{\infty} \rho(z) \, dz = 1 \qquad (3)$$

the energy variation of a test charge at position z inside the bunch is

$$U_{\|}(z) = e^2 N_p \int_{-\infty}^{\infty} \rho(z') w_{\|}(z' - z) \, dz' \tag{4}$$

where N_p is the total number of particles in the bunch. We often call the bunch-wake-potential the energy lost $U_{\|}(z)$ normalized to $e^2 N_p$.

Coupling Impedance

We define the longitudinal coupling impedance [2] as the Fourier transform of the wake function

$$Z_{\|}(\mathbf{r}, \mathbf{r}_1; \omega) = \frac{1}{c} \int_{-\infty}^{\infty} w_{\|}(\mathbf{r}, \mathbf{r}_1; \Delta z) \exp\left[-i\omega \frac{\Delta z}{c}\right] d(\Delta z). \tag{5}$$

A real accelerator is composed of many devices connected by a vacuum chamber. For such a complicated structure, it is impossible to obtain analytical solutions of Maxwell's equations. Usually, numerical codes for the finite differences, which solve Maxwell's equations in the time domain, are used. Because of the CPU time limitations, we can analyze only a single device, or a few of them connected to an infinite pipe. Then the contributions of all the pieces are summed up to get the whole wake function. It is worth noting that this procedure might fail at high frequencies where the fields propagate in the vacuum chamber from one device to another, producing interference effects. Furthermore, numerical codes allow us to obtain only the wake potential of a distribution rather then the impulsive wake function. In Figure 4 we show as an example the wake potential of the DAΦNE machine [3] at INFN - LNF for a 2.5-mm Gaussian bunch obtained with the codes MAFIA [4] and ABCI [5].

For the study of collective effects it is convenient to distinguish between single-bunch dynamics, where the particles experience the wakefields produced by the other particles of the same bunch, and multibunch or multiturn dynamics, where the electromagnetic fields trapped in resonant structures influence other bunches or the same bunch in successive passages. Such a distinction applies also to the wakefields, called respectively short-range and long-range wakefields.

Short-Range Wakefields

Electromagnetic fields that vanish after a distance of a few bunch lengths are usually called short-range wakefields. With short-range wakefields we have a low frequency resolution of the Fourier transform, and therefore of the impedance. Even though impedance is a complicated function of frequency, with many sharp peaks, in the study of single-bunch dynamics influenced by short-range wakefields, the bunch can not resolve the details of the sharp resonances, and it rather experiences

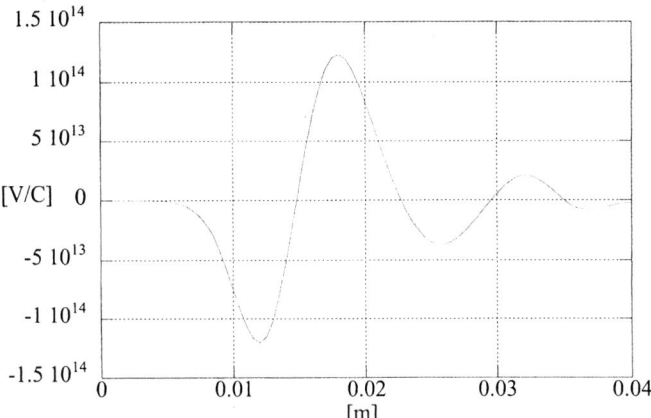

FIGURE 4. DAΦNE wake potential.

an average effect. The corresponding impedance is then smoother and broader than the actual machine impedance, and is called broad-band impedance.

For an approach to single-bunch collective effects, we can obtain the machine broad-band impedance with numerical codes. Usually the frequency behavior is simplified by using some impedance models. Such models, characterized by a small number of parameters, are useful also in the design study of the machine, when not all the devices are defined and known.

The first model historically introduced in the study of single-bunch longitudinal dynamics [6] is the so-called broad-band resonator:

$$Z_{||}(\omega) = \frac{R_s}{1 + iQ\left(\dfrac{\omega}{\omega_r} - \dfrac{\omega_r}{\omega}\right)}. \qquad (6)$$

Only three parameters are needed to determine its frequency behavior: the shunt resistance R_s, the quality factor Q, and the resonant frequency ω_r. Usually $Q \simeq 1$, ω_r is the frequency cut-off of the beam pipe, and R_s accounts for the parasitic energy loss.

The corresponding longitudinal wake function is given by the inverse Fourier transform of Eq. (5), with $Z_{||}(\omega)$ expressed by Eq. (6):

$$w_{||}(\Delta z) = \frac{\omega_r R_s}{Q} \exp\left(-\frac{\Gamma \Delta z}{c}\right)\left[\cos\left(\frac{\omega_n \Delta z}{c}\right) - \frac{\omega_r}{2Q\omega_n}\sin\left(\frac{\omega_n \Delta z}{c}\right)\right] H(\Delta z) \quad (7)$$

where

$$\Gamma = \frac{\omega_r}{2Q} \qquad \omega_n^2 = \omega_r^2 - \Gamma^2 \qquad (8)$$

and $H(\Delta z)$ is the step function.

It is worth noting that because of the low value of the quality factor, the short-range wake function vanishes rapidly with Δz.

Other impedance models have been proposed [7]. Among them we present one based on a phenomenological approach [8], which describes the impedance as an expansion in terms of $\sqrt{\omega}$ of the kind

$$Z_{\|}(\omega) = i\omega L + R + [1 + i\mathrm{sgn}(\omega)]\sqrt{|\omega|}B + \frac{1 - i\mathrm{sgn}(\omega)}{\sqrt{|\omega|}}Z_c + \dots \quad (9)$$

Every term of the expansion has a clear physical interpretation. The first term represents the inductive impedance at low frequencies typical of small discontinuities. Often it is the main contribution to the total impedance, producing a symmetric distortion of the bunch. The second term is due mainly to the RF cavities or resonant devices, and produces a shift of the bunch center of mass and a distortion of its shape. The third term represents the resistive wall impedance due to the finite conductivity of the beam pipe material, and the fourth has the same dependence on ω as the impedance of a cavity with attached tubes at high frequencies. For long enough bunches the first two terms of Eq. (9) give the main contribution to the total broad-band impedance, and are sufficient to describe the single-bunch behavior. Moreover we can obtain the values of the two parameters R and L from measurements of bunch length and synchronous phase shift versus current. The longitudinal wake function corresponding to this model with just R and L is given by

$$w_{\|}(\Delta z) = c^2 L \delta'(\Delta z) + cR\delta(\Delta z) \quad (10)$$

where δ and δ' are respectively the symbolic Dirac delta function and its derivative. Such a wake function can be easily handled analytically.

EQUATIONS OF MOTION

Single-Particle Motion

The single-particle equations of motion are

$$\dot{z} = \frac{\Delta z}{T_0} = \frac{z(t) - z(t - T_0)}{T_0} = -c\alpha_c\varepsilon, \quad (11)$$

$$\dot{\varepsilon} = \frac{\Delta \varepsilon}{T_0} = \frac{\varepsilon(t) - \varepsilon(t - T_0)}{T_0} = \frac{eV(z) - U_0}{T_0 E_0} - \frac{D}{T_0}\varepsilon - \frac{R(T_0)}{T_0 E_0} \quad (12)$$

where z is the longitudinal displacement of a particle with respect to the synchronous one ($z > 0$ means particle ahead), T_0 the revolution period, c the speed of light, α_c the momentum compaction, ε the energy variation with respect to the nominal energy E_0, $V(z)$ the voltage seen by the particle in one turn, and U_0 the energy lost per turn. A particle radiates an energy per turn equal to $U_0 + D\varepsilon + R(T_0)$ where $R(T_0)$ is a stochastic variable that accounts for the quantum fluctuations. The damping coefficient D is equal to $2T_0$ divided by the damping time τ_ε.

Since $V(z)$ is the contribution of the RF cavities and of the longitudinal wake fields, we can write Eq. (12) as

$$\dot{\varepsilon} = \frac{eV_{RF}(z) - U_0}{T_0 E_0} - \frac{e^2 N_p}{T_0 E_0} \int_{-\infty}^{\infty} \rho(z') w_{\parallel}(z'-z) dz' - \frac{D}{T_0}\varepsilon - \frac{R(T_0)}{T_0 E_0}. \tag{13}$$

Equations (11) and (13) describe the longitudinal dynamics of a single particle in a circular accelerator. The potential well in which the particle motion is confined is given by

$$\varphi(z) = \frac{\alpha_c}{L_0} \int_0^z [eV_{RF}(z') - U_0] dz' -$$
$$\frac{\alpha_c e^2 N_p}{L_0} \int_0^z dz' \int_{-\infty}^{\infty} \rho(z'') w_{\parallel}(z'' - z') dz''. \tag{14}$$

For such a motion the Hamiltonian, defined as

$$H(z, \varepsilon) = \frac{1}{2} c\alpha_c \varepsilon^2 + \frac{c}{\alpha_c E_0} \varphi(z), \tag{15}$$

satisfies the relations

$$\frac{\partial H}{\partial \varepsilon} = -\dot{z},$$
$$\frac{\partial H}{\partial z} = \dot{\varepsilon} + \frac{D}{T_0}\varepsilon + \frac{R(T_0)}{T_0 E_0} \tag{16}$$

where the last two terms in the second relation represent the non-conservative components of the system.

Since the revolution period T_0 is much smaller than the synchrotron period, we can make a linear expansion of $z(t)$ and $\varepsilon(t)$:

$$z(t) = z(t - T_0) + \dot{z}T_0,$$
$$\varepsilon(t) = \varepsilon(t - T_0) + \dot{\varepsilon}T_0, \tag{17}$$

and

$$[z + dz]_t = [z + dz]_{t-T_0} + \dot{z}|_{z+dz,\varepsilon} T_0,$$
$$[\varepsilon + d\varepsilon]_t = [\varepsilon + d\varepsilon]_{t-T_0} + \dot{\varepsilon}|_{z,\varepsilon+d\varepsilon} T_0. \tag{18}$$

By using now the relations (16) for $\dot{z}$ and $\dot{\varepsilon}$, and by noting that the partial derivative of the Hamiltonian with respect to z or ε is a function of z or ε only, we have

$$\dot{z}|_{z+dz,\varepsilon} = -\frac{\partial H}{\partial \varepsilon},$$
$$\dot{\varepsilon}|_{z,\varepsilon+d\varepsilon} = \frac{\partial H}{\partial z} - \frac{D}{T_0}\varepsilon - \frac{D}{T_0}d\varepsilon - \frac{R(T_0)}{T_0 E_0}. \quad (19)$$

If we subtract Eq. (17) from Eq. (18), using the above relation and Eq. (16), we obtain

$$dz|_t = dz|_{t-T_0},$$
$$d\varepsilon|_t = d\varepsilon|_{t-T_0}(1-D), \quad (20)$$

which give the relation between an infinitesimal area in the phase space at time t and the corresponding one at time $t - T_0$.

Before continuing our analysis, we examine the quantity $R(T_0)$. It represents the difference between the actual energy lost per turn by a particle and its average value. It is a stochastic variable describing the quantum fluctuations, and, from the definition, its average value is zero. We define the probability density $P[R(T_0)]$ such that $P[R'(T_0)]dR'(T_0)$ represents the probability that a particle, during a revolution period, radiates an energy equal to $U_0 + D\varepsilon + R(T_0)$ with $R(T_0)$ between $R'(T_0)$ and $R'(T_0) + dR'(T_0)$.

Transport and Fokker-Planck Equations

In order to study the collective single-bunch effects, we need to move from the single-particle equation of motion to an equation for an ensemble of particles. To this end we consider the longitudinal single-bunch distribution function $\Psi(z,\varepsilon;t)$ defined such that $\Psi(z,\varepsilon;t)\,dzd\varepsilon$ represents the probability of finding at time t a particle in the area $(z, z+dz, \varepsilon, \varepsilon + d\varepsilon)$ of the phase space. It satisfies the normalization condition

$$\int_{-\infty}^{\infty}\int_{-\infty}^{\infty} \Psi(z,\varepsilon;t)\,dzd\varepsilon = 1. \quad (21)$$

Its projection on the z axis gives the longitudinal bunch distribution $\rho(z)$ introduced in Eq. (4),

$$\rho(z;t) = \int_{-\infty}^{\infty} \Psi(z,\varepsilon;t)\,d\varepsilon. \quad (22)$$

We aim to derive a differential equation that describes the time evolution of the longitudinal distribution function [9].

The probability that a particle at time t has its representative point of the phase space in the area $dzd\varepsilon$ with center (z,ε) is, by definition, $\Psi(z,\varepsilon;t)\,dzd\varepsilon|_t$. The same quantity is also equal to the probability that the particle at time $t - T_0$ was in any point of the phase space such that

$$z(t - T_0) = z(t) - \dot{z}T_0 = z(t) + \frac{\partial H}{\partial \varepsilon}T_0,$$
$$\varepsilon(t - T_0) = \varepsilon(t) - \dot{\varepsilon}T_0 = \varepsilon(t) - \frac{\partial H}{\partial z}T_0 + D\varepsilon + \frac{R(T_0)}{E_0}, \quad (23)$$

and, in a revolution period, it radiated an energy equal to $U_0 + D\varepsilon + R(T_0)$. Since the probability of radiating such an energy is $P[R(T_0)] dR(T_0)$, we can write

$$\Psi[z(t), \varepsilon(t); t] \, dzd\varepsilon|_t$$
$$= \int_{-\infty}^{\infty} \Psi[z(t - T_0), \varepsilon(t - T_0); t - T_0] \, dzd\varepsilon|_{t - T_0} P[R(T_0)] dR(T_0) \quad (24)$$

where $R(T_0)$ is a stochastic variable that can assume any value from $-\infty$ to $+\infty$ with a given probability $P[R(T_0)]$.

The integral equation (24) is known as the transport equation, and it allows us to follow the time evolution of the distribution function $\Psi(z, \varepsilon; t)$ once the Hamiltonian and the probability function $P[R(T_0)]$ are known.

From the transport equation we can derive a differential equation by using the Fokker-Planck method [10]. This consists of expanding Eq. (24) in time around t keeping only linear terms. As shown in Appendix A, we obtain

$$\frac{\partial \Psi}{\partial t} = \frac{\partial \Psi}{\partial z}\frac{\partial H}{\partial \varepsilon} - \frac{\partial \Psi}{\partial \varepsilon}\frac{\partial H}{\partial z} + \frac{D}{T_0}\left(\Psi + \varepsilon\frac{\partial \Psi}{\partial \varepsilon}\right) + \frac{1}{2}\frac{\partial^2 \Psi}{\partial \varepsilon^2}\frac{\overline{R^2(T_0)}}{T_0 E_0^2} \quad (25)$$

where $\overline{R^2(T_0)}$ is the variance of the radiated energy defined as

$$\overline{R^2(T_0)} = \int_{-\infty}^{\infty} R^2(T_0) P[R(T_0)] dR(T_0). \quad (26)$$

In Eq. (25), known as the Fokker-Planck equation or diffusion equation, the first two terms in the right side represent the conservative part of the system, and the other two are related to the radiation process: damping and quantum fluctuations respectively. These effects produce an equilibrium energy distribution with an RMS, known also as the natural energy spread, equal to

$$\sigma_{\varepsilon 0} = \sqrt{\frac{\overline{R^2(T_0)}}{2DE_0^2}}. \quad (27)$$

All the fundamental elements that characterize single-bunch and multibunch collective phenomena are included in the equation. Unfortunately, there is no general solution for it; however, it gives useful information about collective effects.

BUNCH DISTORTION BELOW THRESHOLD

Stationary Solution

As a first application of the Fokker-Planck equation, we look for its stationary solution, that is, a distribution function independent of time, for which

$$\frac{\partial \Psi}{\partial t} = 0 \quad \Rightarrow \quad \Psi(z, \varepsilon; t) = \Psi_0(z, \varepsilon). \tag{28}$$

With such a condition it is possible to find a general solution of the Fokker-Planck equation of the kind

$$\Psi_0(z, \varepsilon) = \overline{\Psi} \exp\left[-\frac{H_0(z,\varepsilon)}{\overline{H}}\right] \tag{29}$$

where the subscript zero to Ψ and H means time independent, the constant $\overline{\Psi}$ is given by the normalization condition Eq. (21), and the constant $\overline{H}$ is

$$\overline{H} = \frac{c\alpha_c \overline{R^2(T_0)}}{2DE_0^2}. \tag{30}$$

Equation (29) is known as the Haissinski equation [11], and it gives the equilibrium distribution of a bunch in the presence of self-induced wakefields and external RF voltage.

A first important observation is that the distribution function can be factorized, and the energy distribution is independent of the wakefields and of the potential well where the particles are confined. In fact, if we write explicitly the Hamiltonian of Eq. (29), and use Eq. (30), we get

$$\Psi_0(z, \varepsilon) = \overline{\Psi} \exp\left[-\frac{DE_0^2 \varepsilon^2}{R^2(T_0)} - \frac{c}{\alpha_c E_0 \overline{H}} \varphi(z)\right], \tag{31}$$

from which it is easy to see that the term related to the energy always gives a Gaussian distribution with an RMS that comes out of the balance between the damping coefficient and the quantum fluctuation noise only [see Eq. (27)].

For the longitudinal distribution function, if we integrate Eq. (31) in ε, and write explicitly the potential as given by Eq. (14), we get

$$\rho_0(z) = \overline{\rho} \exp\left[-\frac{1}{L_0 E_0 \alpha_c \sigma_{\varepsilon 0}^2} \int_0^z [eV_{RF}(z') - U_0] dz' + \right.$$
$$\left. \frac{e^2 N_p}{L_0 E_0 \alpha_c \sigma_{\varepsilon 0}^2} \int_0^z dz' \int_{-\infty}^{\infty} \rho_0(z'') w_{\|}(z'' - z') dz''\right]. \tag{32}$$

The above equation, which is sometimes referred to as the Haissinski equation instead of Eq. (29), is an integral equation in the function $\rho(z)$.

Natural Regime

Let us consider first the case without wakefields, i. e. the natural regime. The longitudinal bunch distribution is given only by the RF voltage:

$$V_{RF}(z) = \hat{V} \cos\left(\phi_s - 2\pi h \frac{z}{L_0}\right) \quad (33)$$

where the synchronous phase $\phi_s = \omega_{RF} t_s$ is given by the condition

$$\cos(\phi_s) = \frac{U_0}{e\hat{V}}. \quad (34)$$

In the usual case $2\pi h z \ll L_0$, we can linearly expand $V_{RF}(z)$ around $z = 0$, obtaining

$$V_{RF}(z) = \hat{V} \cos(\phi_s) + \frac{2\pi h \hat{V} \sin(\phi_s)}{L_0} z. \quad (35)$$

If we insert Eq. (35) into Eq. (32) and use Eq. (34), we obtain

$$\rho_0(z) = \bar{\rho} \exp\left[-\frac{\omega_{s0}^2 z^2}{2\alpha_c^2 c^2 \sigma_{\varepsilon 0}^2}\right] = \bar{\rho} \exp\left[-\frac{z^2}{2\sigma_{z0}^2}\right] \quad (36)$$

where we have used the natural angular synchrotron frequency

$$\omega_{s0}^2 = \frac{c^2 \alpha_c 2\pi h e \hat{V} \sin(\phi_s)}{L_0^2 E_0}. \quad (37)$$

The longitudinal bunch distribution is then a Gaussian distribution with an RMS given by

$$\sigma_{z0} = \frac{\alpha_c c \sigma_{\varepsilon 0}}{\omega_{s0}} \quad (38)$$

and with a constant $\bar{\rho}$ obtained by the normalization condition Eq. (3),

$$\bar{\rho} = \frac{1}{\sqrt{2\pi}\sigma_{z0}}. \quad (39)$$

Wakefield Effects

Resistive Impedance

If we include the effects of the wakefield, we find that the bunch distribution is distorted. This effect can be studied analytically in the case of a pure resistive impedance for which, assuming the linear expansion of the RF voltage, Eq. (32) becomes

$$\rho_0(z) = \bar{\rho} \exp\left[-\frac{z^2}{2\sigma_{z0}^2} + \frac{e^2 N_p cR}{L_0 E_0 \alpha_c \sigma_{\varepsilon 0}^2} \int_0^z \rho_0(z') dz'\right]. \quad (40)$$

As shown in Appendix B, the analytical solution of the above equation is

$$\rho_0(z) = \frac{\exp\left(-\frac{z^2}{2\sigma_{z0}^2}\right)}{\xi_1 \sigma_{z0} \sqrt{\frac{\pi}{2}} \left[\coth\left(\frac{\xi_1}{2}\right) - \text{erf}\left(\frac{z}{\sqrt{2}\sigma_{z0}}\right)\right]} \qquad (41)$$

where the error function is

$$\text{erf}(x) = \frac{2}{\sqrt{\pi}} \int_0^z \exp(-x^2)\, dx \qquad (42)$$

and

$$\xi_1 = \frac{e^2 N_p c R}{L_0 E_0 \alpha_c \sigma_{\varepsilon 0}^2}. \qquad (43)$$

A pure resistive impedance distorts the bunch and shifts its center of mass ahead, but does not change its RMS much. In Figure 5 we show an example of longitudinal bunch distribution at different N_p.

Inductive Impedance

In the case of pure inductive impedance, the bunch distribution is given by

$$\rho_0(z) = \bar{\rho} \exp\left[-\frac{z^2}{2\sigma_{z0}^2} - \frac{e^2 N_p c^2 L}{L_0 E_0 \alpha_c \sigma_{\varepsilon 0}^2} \rho_0(z)\right] \qquad (44)$$

which can be solved with numerical tools. From Figure 6, we see that the inductive impedance does not influence the position of the bunch center of mass, but increases only the bunch length. The distribution function is no longer Gaussian, but tends to become parabolic, especially around the bunch center.

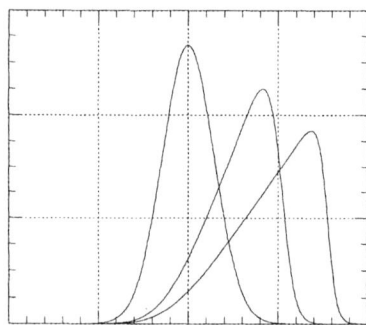

FIGURE 5. Bunch shapes at different N_p for pure resistive impedance.

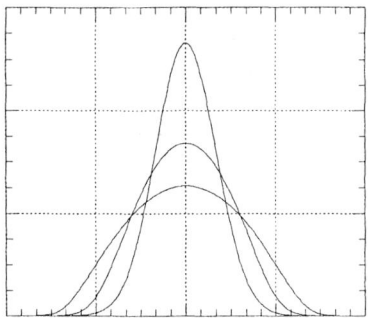

FIGURE 6. Bunch shapes at different N_p for pure inductive impedance.

Broad-Band Resonator Impedance

With the broad-band resonator model, as well as with any general wake function, it is convenient to transform Eq. (32) by introducing the wake potential of a unitary step distribution

$$S(z) = \int_0^z w_\parallel(z') \, dz' \qquad (45)$$

with which we can express Eq. (32) as

$$\rho_0(z) = \overline{\rho_1} \exp\left[-\frac{z^2}{2\sigma_{z0}^2} - \frac{e^2 N_p}{L_0 E_0 \alpha_c \sigma_{\varepsilon 0}^2} \int_{-\infty}^{\infty} \rho_0(z'+z) S(z') \, dz'\right]. \qquad (46)$$

The solution of the above equation for the resonator impedance is given in Figure 7 for different values of N_p.

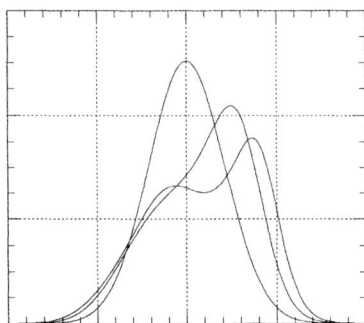

FIGURE 7. Bunch shapes at different N_p for the broad band resonator.

THRESHOLD HUNTING

Perturbation Theory

Comparing the stationary distribution with the experimental observations in a real machine, we find that they agree only at low current, where the energy spread is constant, and the longitudinal bunch distribution is distorted according to the potential well theory. However, when N_p is high, we can no longer explain the experimental observations in terms of a stationary solution. We need to explore a new dynamical regime, where the distribution function is a function of time.

In this section, we show a perturbation method generally used to obtain the behavior of $\Psi(z, \varepsilon; t)$ around the stationary solution $\Psi_0(z, \varepsilon)$ [12,13].

We start by linearizing $\Psi(z, \varepsilon; t)$:

$$\Psi(z, \varepsilon; t) = \Psi_0(z, \varepsilon) + \Psi_1(z, \varepsilon; t) \tag{47}$$

where $\Psi_1(z, \varepsilon; t)$ is a perturbation of $\Psi_0(z, \varepsilon)$. Similar expansions apply also to the single-particle potential

$$\varphi(z) = \varphi_0(z) + \varphi_1(z; t) \tag{48}$$

and to the Hamiltonian

$$H(z, \varepsilon; t) = H_0(z, \varepsilon) + H_1(z; t) \tag{49}$$

where

$$H_1(z; t) = \frac{c}{\alpha_c E_0} \varphi_1(z; t) = -\frac{c e^2 N_p}{E_0 L_0} \int_0^z dz' \int_{-\infty}^{\infty} \rho_1(z''; t) w_\|(z'' - z') \, dz'' \tag{50}$$

and

$$\rho_1(z; t) = \int_{-\infty}^{\infty} \Psi_1(z, \varepsilon; t) \, d\varepsilon. \tag{51}$$

If we substitute Eqs. (47) and (49) into the Fokker-Planck equation, and observe that $\Psi_0(z, \varepsilon)$ satisfies the stationary equation, we get

$$\frac{\partial \Psi_1}{\partial t} = \frac{\partial \Psi_1}{\partial z}\frac{\partial H_0}{\partial \varepsilon} - \frac{\partial \Psi_0}{\partial \varepsilon}\frac{\partial H_1}{\partial z} - \frac{\partial \Psi_1}{\partial \varepsilon}\frac{\partial H_0}{\partial z} - \frac{\partial \Psi_1}{\partial \varepsilon}\frac{\partial H_1}{\partial z} + \frac{D}{T_0}\left(\Psi_1 + \varepsilon\frac{\partial \Psi_1}{\partial \varepsilon}\right) + \frac{1}{2}\frac{\partial^2 \Psi_1}{\partial \varepsilon^2}\frac{\overline{R^2(T_0)}}{T_0 E_0^2}. \tag{52}$$

We now ignore the second-order terms and the effects of radiation damping and fluctuation noise on the perturbation function Ψ_1. By introducing the two action-angle variables, J and ϕ, as shown in Appendix C, we get

$$\frac{\partial \Psi_1}{\partial t} = -\omega_s(J)\frac{\partial \Psi_1}{\partial \phi} + \frac{c^2 \alpha_c e^2 N_p}{E_0 L_0 \omega_s(J)}\frac{\partial \Psi_0}{\partial J}\varepsilon \int_{-\infty}^{\infty} \rho_1(z'; t) w_\|(z' - z) \, dz'. \tag{53}$$

In order to find the solution of Eq. (53), we note that $\Psi_1(J,\phi;t)$ is periodic in ϕ, and it can then be expanded as a Fourier series, while for the time dependence, we apply the modal analysis in the frequency domain, which allows us to write

$$\Psi_1(J,\phi;t) = \exp[i\Omega t] \sum_{m=-\infty}^{\infty} R_m(J) \exp[-im\phi] \qquad (54)$$

which, introduced into Eq. (53), gives

$$i\Omega \sum_{m=-\infty}^{\infty} R_m(J) \exp[-im\phi] = i\omega_s(J) \sum_{n=-\infty}^{\infty} n R_n(J) \exp[-in\phi] +$$

$$\frac{c^2 \alpha_c e^2 N_p}{E_0 L_0 \omega_s(J)} \frac{\partial \Psi_0}{\partial J} \varepsilon \sum_{l=-\infty}^{\infty} \int_0^{2\pi} d\phi' \int_0^{\infty} R_l(J') \exp[-il\phi'] w_{\parallel}(z'-z) dJ'. \qquad (55)$$

Since the value of m in Eq. (54) specifies the angular dependence of the mth term of the Fourier expansion, m is also called the azimuthal number, and the corresponding $R_m(J)$ is called the radial function. We now exploit the symmetry of $w_{\parallel}(z'-z)$ with respect to the angle variable ϕ' and the fact that $\varepsilon(J,\phi)$ is antisymmetric in ϕ. Multiplying by $\exp[im'\phi]$ and integrating in ϕ from 0 to 2π, we obtain

$$[\Omega - m\omega_s(J)] R_m(J) = \frac{c^2 \alpha_c e^2 N_p}{2\pi E_0 L_0 \omega_s(J)} \frac{\partial \Psi_0}{\partial J} \sum_{l=-\infty}^{\infty} \int_0^{2\pi} d\phi$$

$$\times \int_0^{2\pi} d\phi' \int_0^{\infty} \varepsilon(J,\phi) R_l(J') \sin(m\phi) \cos(l\phi') w_{\parallel}(z'-z) dJ'. \qquad (56)$$

In terms of the coupling impedance this becomes

$$[\Omega - m\omega_s(J)] R_m(J) = \frac{c^2 \alpha_c e^2 N_p}{4\pi^2 E_0 L_0 \omega_s(J)} \frac{\partial \Psi_0}{\partial J} \sum_{l=-\infty}^{\infty} \int_0^{2\pi} d\phi \int_0^{2\pi} d\phi'$$

$$\times \int_{-\infty}^{\infty} d\omega \int_0^{\infty} \varepsilon(J,\phi) R_l(J') \sin(m\phi) \cos(l\phi') \exp\left[i\frac{\omega}{c}(z'-z)\right] Z_{\parallel}(\omega) dJ'. \qquad (57)$$

Equations (56) and (57) represent a generic term of the infinite set of integral homogeneous equations for $R_m(J)$. If we treat the whole set of equations as an eigenvalue problem in Ω, we get that the number of eigenvalues is the same as those of m, i. e. infinite. Each $\Omega^{(m)}$ is called a coherent frequency of the azimuthal oscillation mode m. In the limit of $N_p \to 0$, we easily obtain

$$\omega_s(J) \to \omega_{s0},$$
$$\Omega^{(m)} = m\omega_{s0}. \qquad (58)$$

In the next sections we show two methods used to solve Eq. (57).

Sacherer Equation

An attempt to evaluate the threshold of microwave instability through mode coupling is based on the simplifying hypothesis that single-particle motion is governed by a quadratic form of the Hamiltonian, of the kind

$$H(z,\varepsilon) = \frac{1}{2}c\alpha_c\varepsilon^2 + \frac{1}{2}\frac{\omega_s^2}{c\alpha_c}z^2, \tag{59}$$

which corresponds to a linear RF voltage. The hypotheses behind Eq. (59) are Ψ_0 stationary and symmetric and ω_s independent of the amplitude J.

The equations of motion of a single particle, ignoring the effects of radiation damping and quantum fluctuations, can be expressed as

$$z = \sqrt{\frac{2c\alpha_c J}{\omega_s}}\cos[\phi(t)],$$

$$\varepsilon = \sqrt{\frac{2\omega_s J}{c\alpha_c}}\sin[\phi(t)]. \tag{60}$$

If we substitute Eqs. (60) into Eq. (57) and use the relations

$$\int_0^{2\pi}\sin(\phi)\sin(m\phi)\exp[-ia\cos(\phi)]\,d\phi = i2\pi i^{-m}\frac{m}{a}J_m(a) \tag{61}$$

and

$$\int_0^{2\pi}\cos(l\phi)\exp[ia\cos(\phi)]\,d\phi = 2\pi i^l J_l(a) \tag{62}$$

where $J_m(x)$ is a Bessel function of first kind and mth order, we get

$$[\Omega - m\omega_s]R_m(J) = i\frac{mc^2e^2N_p}{E_0L_0}\frac{\partial\Psi_0}{\partial J}\sum_{l=-\infty}^{\infty}i^{l-m}$$

$$\times \int_{-\infty}^{\infty}\frac{Z_\|(\omega)}{\omega}J_m\left(\omega\sqrt{\frac{2\alpha_c J}{c\omega_s}}\right)d\omega\int_0^{\infty}R_l(J')J_l\left(\omega\sqrt{\frac{2\alpha_c J'}{c\omega_s}}\right)dJ' \tag{63}$$

where we have considered $\omega(J) = \omega_s$. In order to express the eigenvalue problem in a simple form, we introduce the two functions

$$G_{ml}(J,J') = i\frac{mc^2e^2N_p\overline{\Psi}}{\omega_s E_0 L_0}\int_{-\infty}^{\infty}\frac{Z_\|(\omega)}{\omega}J_m\left(\omega\sqrt{\frac{2\alpha_c J}{c\omega_s}}\right)J_l\left(\omega\sqrt{\frac{2\alpha_c J'}{c\omega_s}}\right)d\omega \tag{64}$$

and

$$w(J) = -\frac{1}{\overline{\Psi}}\frac{\partial\Psi_0}{\partial J} \tag{65}$$

with $\overline{\Psi}$ already defined in Eq. (29), and write Eq. (63) as

$$\left[\frac{\Omega}{\omega_s} - m\right] R_m(J) = -w(J) \sum_{l=-\infty}^{\infty} i^{l-m} \int_0^{\infty} G_{ml}(J, J') R_l(J') dJ', \qquad (66)$$

known as the Sacherer integral equation for longitudinal instabilities.

A Method of Solving the Sacherer Equation

Different methods have been proposed for solving Eq. (66) [14]. Here we show the one that uses the expansion of the radial function $R_m(J)$ in orthogonal polynomials:

$$R_m(J) = w(J) \sum_{k=0}^{\infty} a_{mk} f_{|m|k}(J). \qquad (67)$$

The absolute value of m in the polynomials $f_{|m|k}(J)$ is due to the property that the radial distributions of the terms $\pm m$ are equal. The functions $f_{|m|k}(J)$ satisfy the normalization condition

$$\int_0^{\infty} w(J) f_{|m|k}(J) f_{|m|l}(J) dJ = \delta_{kl} \qquad (68)$$

where $w(J)$ is a weight function and δ_{kl} is the Kronecher symbol.

Introducing Eq. (67) into Eq. (66), multiplying by $f_{|m|p}(J)$, and integrating in J from 0 to ∞, we obtain

$$\left[\frac{\Omega}{\omega_s} - m\right] a_{mp} = \sum_{l=-\infty}^{\infty} \sum_{k=0}^{\infty} M_{pk}^{ml} a_{lk} \qquad (69)$$

with

$$M_{pk}^{ml} = -i^{l-m} \int_0^{\infty} w(J) f_{|m|p}(J) dJ \int_0^{\infty} G_{ml}(J, J') w(J') f_{|l|k}(J') dJ'. \qquad (70)$$

Equation (69) is the generic term of a homogeneous system of equations, with m ranging from $-\infty$ to ∞ and p from 0 to ∞. If we consider the system as an eigenvalue problem, a_{mp} being the eigenvectors, the eigenvalues $\Omega^{(m,p)}$ can be evaluated once the orthogonal polynomials $f_{|m|p}(J)$ are known. These last depend only on the weight function $w(J)$, which, in turn, depends on the stationary distribution function $\Psi_0(J)$.

For the Gaussian distribution function given by Eq. (29), as shown in Appendix D, we get

$$M_{pk}^{ml} = -i\frac{c^2 e^2 N_p}{2\pi\omega_s E_0 L_0 \sigma_z \sigma_{\varepsilon 0}} i^{|l|-|m|} \frac{m}{\sqrt{p!(|m|+p)!k!(|l|+k)!}}$$

$$\times \int_{-\infty}^{\infty} \frac{Z_{\|}(\omega)}{\omega} \left(\frac{\omega \sigma_z}{\sqrt{2}c}\right)^{|m|+|l|+2p+2k} \exp\left[-\frac{\omega^2 \sigma_z^2}{c^2}\right] d\omega. \qquad (71)$$

For the RL impedance, the integral in ω can be solved, giving

$$M_{pk}^{ml} = -i \frac{c^2 e^2 N_p}{2\pi\omega_s E_0 L_0 \sigma_z \sigma_{\varepsilon 0}} \frac{i^{|l|-|m|} m 2^{-\left(\frac{|m|+|l|}{2}+p+k\right)}}{\sqrt{p!\,(|m|+p)!\,k!\,(|l|+k)!}}$$

$$\times \begin{cases} \dfrac{icL}{\sigma_z} \Gamma\left(\dfrac{|m|+|l|+2p+2k+1}{2}\right) & m+l \text{ even} \\ R\Gamma\left(\dfrac{|m|+|l|+2p+2k}{2}\right) & m+l \text{ odd} \end{cases} \quad (72)$$

where $\Gamma(x)$ is the gamma function.

Azimuthal and Radial Mode Coupling

The eigenvalues obtained with Eq. (69) are characterized by two indices, m and p, describing respectively the azimuthal and radial structure of the oscillation mode. In this case, in the limit $N_p \to 0$, the coherent frequencies are still given by Eq. (58), and all the modes with the same m but different p have the same frequency. As we increase N_p slightly, the frequencies shift away from the unperturbed values and the modes shift accordingly. The frequency shifts are initially much smaller than ω_s. Therefore, in this case, radial modes can couple only if they belong to the same azimuthal family with a given m. Therefore, if we focus our attention on the radial modes, for a given m, we can leave only the term $l = m$, thus obtaining

$$\left[\frac{\Omega}{\omega_s} - m\right] a_{mp} = \sum_{k=0}^{\infty} M_{pk}^{mm} a_{mk}. \quad (73)$$

Since the real part of $Z_\parallel(\omega)$ is an even function of ω and its imaginary part is uneven, we have that M_{pk}^{mm} is real and $M_{pk}^{mm} = M_{kp}^{mm}$. The matrix M_{pk}^{mm} is then Hermitian. As a consequence, the eigenvalues, i. e. the coherent frequencies, are always real, and radial modes do not couple. If we further increase N_p however, the frequency shifts become comparable to ω_s, so that coupling of azimuthal modes can occur. To simplify the analysis, generally only one radial mode is considered for every azimuthal family. As an example, if we retain only the radial modes with $p = 0$, assumed to be the most prominent, we get

$$\left[\frac{\Omega}{\omega_s} - m\right] a_{m0} = \sum_{l=-\infty}^{\infty} M_{00}^{ml} a_{l0} \quad (74)$$

where the matrix element M_{00}^{ml} is no longer symmetric. In this case we obtain that, above a given value of N_p, two azimuthal modes can couple producing a complex $\Omega^{(m,0)}$ and then an instability. As an example, in Figure 8 we show the coherent frequencies of the first positive modes versus N_p for a pure RL impedance. When

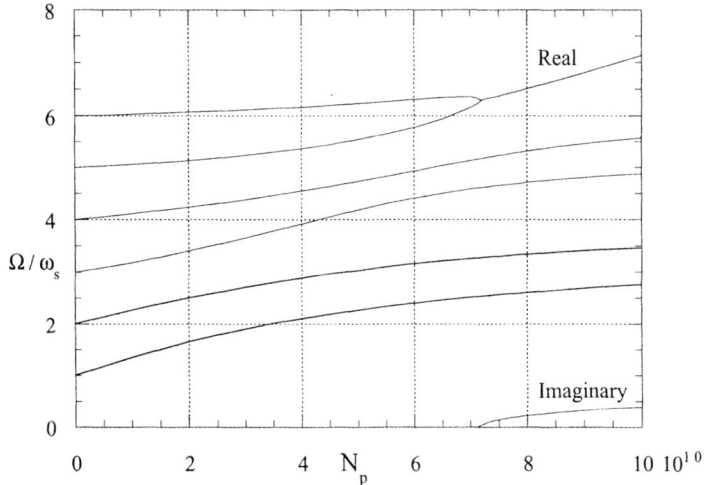

FIGURE 8. Longitudinal mode frequency Ω/ω_s versus N_p for a pure RL impedance.

N_p reaches the threshold, the modes $m = 5$ and $m = 6$ of the example couple, producing an imaginary component of Ω with values either positive and negative. According to Eq. (54), the negative term is responsible for the instability.

It is worth noting that the threshold obtained with such a method can be much higher than the measured one. The reason is that radial modes with the same azimuthal number can actually couple before a frequency shift of the order of ω_s. Equation (73) cannot predict such behavior because of the symmetry of the bunch distribution. In fact, as we will discuss in the next section, for asymmetric bunches radial mode coupling can occur even if they belong to the same azimuthal family.

Numerical Solution of the Fokker-Planck Equation

The complete problem of simultaneous accounting for azimuthal and radial modes with a perturbed stationary distribution function is very difficult to treat. There are methods that simplify the approach by using particular bunch distributions, such as the double water bag [15]. With this model we can solve analytically the eigenvalue problem of mode coupling also with an asymmetric bunch shape. For every azimuthal mode there are two radial modes, and it can be demonstrated that the distorted bunch distribution can produce radial mode coupling.

A more general method [16] divides the action variable into n intervals such that $0 = J_0 < J_1 < \cdots < J_n$, and, considers J constant, equal to its average value $\bar{J}_p$, in an interval $\Delta J_p = J_{p+1} - J_p$. Then Eq. (56) gives n equations of the kind

$(p = 0, \ldots, n-1)$

$$[\Omega - m\omega_s(\overline{J}_p)] R_m(\overline{J}_p) = \frac{c^2 \alpha_c e^2 N_p}{2\pi E_0 L_0 \omega_s(\overline{J}_p)} \left. \frac{\partial \Psi_0}{\partial J} \right|_{\overline{J}_p}$$

$$\times \sum_{l=-\infty}^{\infty} \sum_{k=0}^{n} R_l(\overline{J}_k) \Delta J_k \int_0^{2\pi} \varepsilon(\overline{J}_p, \phi) \sin(m\phi) \, d\phi$$

$$\times \int_0^{2\pi} \cos(l\phi') w_\| \left[z(\overline{J}_k, \phi') - z(\overline{J}_p, \phi) \right] d\phi'. \quad (75)$$

If we multiply by ΔJ_p we get

$$[\Omega - m\omega_s(\overline{J}_p)] R_{mp} = \sum_{l=0}^{\infty} \sum_{k=0}^{n} M_{pk}^{ml} R_{lk} \quad (76)$$

with

$$M_{pk}^{ml} = \frac{c^2 \alpha_c e^2 N_p \Delta J_p}{2\pi E_0 L_0 \omega_s(\overline{J}_p)} \left. \frac{\partial \Psi_0}{\partial J} \right|_{\overline{J}_p} \int_0^{2\pi} \varepsilon(\overline{J}_p, \phi) \sin(m\phi) \, d\phi$$

$$\times \int_0^{2\pi} \cos(l\phi') w_\| \left[z(\overline{J}_k, \phi') - z(\overline{J}_p, \phi) \right] d\phi' \quad (77)$$

Equation (76) can produce imaginary values of Ω, and thus coupling, even in the case of each azimuthal number m separately analyzed. This method can predict the threshold with better accuracy.

ABOVE THRESHOLD

Boussard Criterion

A simple method generally used to obtain a first estimate of the microwave instability threshold is known as the Boussard criterion, and it is derived from the coasting-beam theory applied to the single-bunch case [17]. To justify such an assumption, it can be observed that at high frequencies, in the microwave regime, a bunch can be thought of as a coasting beam, with an average current equal to the single-bunch peak current. With this hypothesis, the criterion fixes the instability of the bunch above the limit

$$\frac{ce^2 N_p \left| Z_\|(n)/n \right|}{(2\pi)^{3/2} E_0 \alpha_c \sigma_z \sigma_\varepsilon^2} \leq 1 \quad (78)$$

where n is a harmonic of the revolution frequency. If N_p is sufficiently high that the left side of Eq. (78) is greater than 1, then the bunch length σ_z and the energy spread σ_ε increase to restore it back to 1.

Equation (78) was derived assuming a bunch with a Gaussian longitudinal distribution. Together with Eq. (38), it allows us to obtain σ_z or σ_ε as a function of N_p above the instability threshold.

The harmonic of the revolution frequency n is generally chosen as

$$n = \frac{L_0}{2\pi\sigma_z}. \tag{79}$$

Long bunches usually interact with the vacuum chamber at low frequencies (below cut-off) where the impedance is pure inductive. In this case we have

$$\frac{Z_\parallel(n)}{n} = i\omega_0 L \tag{80}$$

which does not depend on n.

Chao-Gareyete Scaling Law

A more general scaling law for bunch lengthening above threshold was suggested by Chao and Gareyete [18]. According to their model, the bunch lengthening σ_z is a function of a single parameter ξ, which in turn depends on other machine parameters:

$$\xi = \frac{I\alpha_c}{\nu_s^2 E_0/e}. \tag{81}$$

If we assume a simple power-law behavior for the longitudinal impedance:

$$\left|\frac{Z}{n}\right| = Z_0 \omega^{a-1}, \tag{82}$$

then

$$\sigma_z \propto \left(\xi Z_0 R^3\right)^{1/(2+a)}. \tag{83}$$

For instance for the SPEAR case this results in $\sigma_0 \propto \xi^{1/1.32}$, $a = 0.68$, which means $a = -0.68$ corresponding to an impedance decreasing with frequency.

The Boussard model is a particular case of the Chao-Gareyete scaling law for $a = 1$. This corresponds to a constant longitudinal impedance, typical of storage rings with long bunches.

Numerical Simulations

Numerical simulations are a valid and reliable tool for investigating single-bunch instability. Usually a simulation code models the single bunch as an ensemble of particles obeying the turn-by-turn equations of motion. Assuming a sinusoidal time-dependent RF voltage linearized around the synchrotron particle, we have

$$z_i^n = z_i^{n-1} - L_0 \alpha_c \frac{\varepsilon_i^{n-1}}{E_0},$$

$$\varepsilon_i^n = \varepsilon_i^{n-1} + \frac{2\pi h e \hat{V} \sin(\phi_s)}{L_0} z_i^n + V_w(z_i^n) - D\varepsilon_i^{n-1} + \sigma_{\varepsilon 0} R \sqrt{2D}, \quad (84)$$

where i refers to the ith particle, and n to the nth turn, L_0 is the machine circumference, $\sigma_{\varepsilon 0}$ the natural energy spread, R a random number from a normal distribution with average 0 and variance 1, and $V_w(z_i^n)$ the voltage produced by the self-induced short-range wakefields.

Since it is impossible to simulate the motion of 10^{10} to 10^{12} particles for hundreds of thousands of turns, a smaller number of macro-particles, each one representing 10^6 or more particles, is used in the simulations. The number of macro-particles N_m must be high enough to limit the numerical noise, which scales as $1/\sqrt{N_m}$.

The voltage of the self-induced wakefields depends on the single-bunch distribution function according to the relation

$$V_w(z) = eN_p \int_{-\infty}^{\infty} w_{\parallel}(z'-z) dz' \int_{-\infty}^{\infty} \Psi(z',\varepsilon) d\varepsilon \quad (85)$$

where N_p is the number of particles per bunch. In our discrete model, we consider the N_m macro-particles distributed in N_{bin} bins, and therefore the induced voltage can be written as

$$V_w(z_i^n) = e \frac{N_p}{N_m} \sum_{\substack{k=1, N_{bin} \\ z_k > z_i^n}} N_b(z_k) w_{\parallel}(z_k - z_i^n) \quad (86)$$

with z_k the coordinate of the kth bin center, and $N_b(z_k)$ the number of macro-particles in the bin.

This method of tracking the bunch particles has been successfully used in bunch lengthening simulations [19] for the SLC [20] damping rings, SPEAR [21], PETRA, LEP [22], and DAΦNE [23]. In Figure 9 we show an example of the results obtained with the numerical simulations: two distributions of an unstable bunch and the bunch length as function of current.

REFERENCES

1. L. Palumbo, et al., CERN 95-06 (1995).
2. V. G. Vaccaro, CERN ISR-RF/66-35 (1966).
3. M. Zobov, et al., KEK Proceedings 96-6 (1996) 110.
4. R. Klatt, et al., SLAC Report 303 (1986).
5. Y. H. Chin, LBL Report 35258, UC-414 (1994).

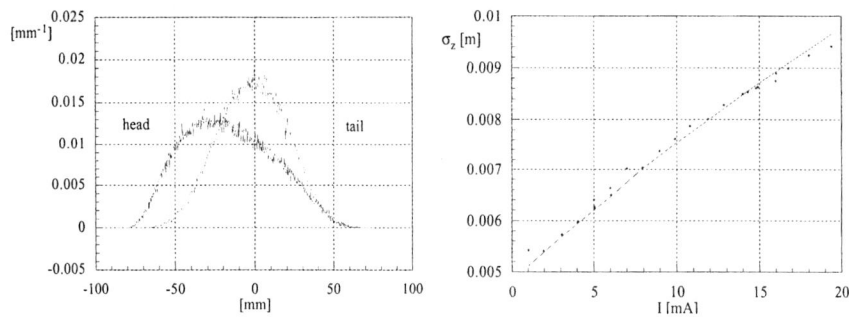

FIGURE 9. Example of numerical simulations.

6. A. Hofmann, J. Maidment, CERN LEP Note 169 (1979).
7. See, for example, S. Heifets, Frascati Physics Series, X (1998) 139.
8. K. Bane, SLAC PUB 4618 (1988).
9. A. Chao, *Physics of Collective Beam Instabilities in High Energy Accelerators*, Wiley Interscience (1993).
10. C. Bernardini, B. Touschek, LNF Internal Note 34 (1960).
11. J. Haissinski, *Il Nuovo Cimento* **18B**, (1) (1973) 72.
12. F. Sacherer, CERN/SI-BR/72-5 (1972).
13. B. Zotter, CERN-SPS/81-18,19,20 (DI) (1981).
14. J. L. Laclare, CERN 87-03, Vol I (1987).
15. A. Chao, et al., *Proc of the 1995 PAC*, Dallas (1995).
16. K. Oide and K. Yokoya, KEK Preprint 90-10 (1990).
17. D. Boussard, CERN-LABII/RF/INT75-2 (1975).
18. A. Chao and J.Gareyete, *Particle Accelerators* **25** (1990) 229.
19. A. Renieri, LNF - 75 / 11 (R) Frascati (1976).
20. K. L. F. Bane, K. Oide, *Proc of the 1993 PAC*, Washington D.C. (1993) 3339.
21. R. Sieman, NIM 203 (1982) 57.
22. T. Weiland, DESY 81-088 (1981).
23. M. Zobov, et al., DAΦNE Technical Note BM-3, Frascati (1998).

APPENDIX A

Derivation of Fokker-Planck Equation

To obtain the Fokker-Planck equation, we first substitute Eq. (23) into the distribution function $\Psi\left[z\left(t-T_0\right);\varepsilon\left(t-T_0\right);t-T_0\right]$ and then take a time linear expansion of such a function:

$$\Psi\left[z(t)+\frac{\partial H}{\partial \varepsilon}T_0,\varepsilon(t)-\frac{\partial H}{\partial z}T_0+D\varepsilon+\frac{R(T_0)}{E_0};t-T_0\right]$$
$$=\Psi\left[z(t)+\frac{\partial H}{\partial \varepsilon}T_0,\varepsilon(t)-\frac{\partial H}{\partial z}T_0+D\varepsilon+\frac{R(T_0)}{E_0};t\right]-\frac{\partial \Psi}{\partial t}T_0 \quad \text{(A1)}$$

where the time derivative of Ψ is evaluated at point $z(t),\varepsilon(t)$.
If we now expand the first term in the right side of Eq. (A1), we obtain

$$\Psi\left[z(t)+\frac{\partial H}{\partial \varepsilon}T_0,\varepsilon(t)-\frac{\partial H}{\partial z}T_0+D\varepsilon+\frac{R(T_0)}{E_0};t\right]$$
$$=\Psi\left[z(t)+\frac{\partial H}{\partial \varepsilon}T_0,\varepsilon(t)-\frac{\partial H}{\partial z}T_0+D\varepsilon;t\right]+$$
$$\frac{\partial \Psi}{\partial \varepsilon}\frac{R(T_0)}{E_0}+\frac{1}{2}\frac{\partial^2 \Psi}{\partial \varepsilon^2}\frac{R^2(T_0)}{E_0^2}+\sum_{n=3}^{\infty}\frac{1}{n!}\frac{\partial^n \Psi}{\partial \varepsilon^n}\frac{R^n(T_0)}{E_0^n}. \quad \text{(A2)}$$

Before integrating the right side of Eq. (24), we write

$$\int_{-\infty}^{\infty} R^n(T_0) P\left[R(T_0)\right] dR(T_0) = \overline{R^n(T_0)}. \quad \text{(A3)}$$

Because of the definition of probability density, we have $\overline{R^0(T_0)}=1$, and, since $R(T_0)$ has a zero average value, then $\overline{R^1(T_0)}=0$. We can now integrate the right side of Eq. (24) by using Eqs. (A1) and (A2), thus obtaining

$$\Psi\left[z(t),\varepsilon(t);t\right]dzd\varepsilon|=\left\{\Psi\left[z(t)+\frac{\partial H}{\partial \varepsilon}T_0,\varepsilon(t)-\frac{\partial H}{\partial z}T_0+D\varepsilon;t\right]-\right.$$
$$\left.\frac{\partial \Psi}{\partial t}T_0+\frac{1}{2}\frac{\partial^2 \Psi}{\partial \varepsilon^2}\frac{\overline{R^2(T_0)}}{E_0^2}+\sum_{n=3}^{\infty}\frac{1}{n!}\frac{\partial^n \Psi}{\partial \varepsilon^n}\frac{\overline{R^n(T_0)}}{E_0^n}\right\}dzd\varepsilon|_{t-T_0}. \quad \text{(A4)}$$

The next step is to expand the distribution function in the right side of Eq. (A4) around the point (z,ε), again keeping only first-order terms in time:

$$\Psi\left[z(t)+\frac{\partial H}{\partial \varepsilon}T_0,\varepsilon(t)-\frac{\partial H}{\partial z}T_0+D\varepsilon;t\right]$$
$$=\Psi\left[z(t),\varepsilon(t);t\right]+\frac{\partial \Psi}{\partial z}\frac{\partial H}{\partial \varepsilon}T_0+\frac{\partial \Psi}{\partial \varepsilon}\left(-\frac{\partial H}{\partial z}T_0+D\varepsilon\right). \quad \text{(A5)}$$

By introducing Eq. (A5) into Eq. (A4) and by using the relations (20), still ignoring second-order terms in time, we finally get

$$\frac{\partial \Psi}{\partial t}=\frac{\partial \Psi}{\partial z}\frac{\partial H}{\partial \varepsilon}-\frac{\partial \Psi}{\partial \varepsilon}\frac{\partial H}{\partial z}+\frac{D}{T_0}\left(\Psi+\varepsilon\frac{\partial \Psi}{\partial \varepsilon}\right)+\frac{1}{2}\frac{\partial^2 \Psi}{\partial \varepsilon^2}\frac{\overline{R^2(T_0)}}{T_0 E_0^2} \quad \text{(A6)}$$

where we have neglected the higher-order terms:

$$\sum_{n=3}^{\infty} \frac{1}{n!} \frac{\partial^n \Psi}{\partial \varepsilon^n} \frac{\overline{R^n(T_0)}}{T_0 E_0^n}. \tag{A7}$$

APPENDIX B

Solution of Haissinski Equation for Pure Resistive Impedance

Applying the logarithm derivative on both sides of Eq. (40), we get

$$\frac{\rho_0'(z)}{\rho_0(z)} = -\frac{z}{\sigma_{z0}^2} + \frac{e^2 N_p cR}{L_0 E_0 \alpha_c \sigma_{\varepsilon 0}^2} \rho_0(z) \tag{B1}$$

which is a Bernoulli differential equation, the general solution of which is

$$\rho_0(z) = \frac{\exp\left[-U_0(z)\right]}{k - \xi_1 \int_0^z \exp\left[-U_0(x)\right] dx} \tag{B2}$$

with

$$U_0(z) = \int_0^z \frac{x}{\sigma_{z0}^2} dx = \frac{z^2}{2\sigma_{z0}^2} \tag{B3}$$

and

$$\xi_1 = \frac{e^2 N_p cR}{L_0 E_0 \alpha_c \sigma_{\varepsilon 0}^2}. \tag{B4}$$

The constant k is obtained by the normalization condition (3). If we use an auxiliary function $f(z)$ such that

$$f(z) = k - \xi_1 \int_0^z \exp\left[-U_0(x)\right] dx \tag{B5}$$

then

$$\exp\left[-U_0(z)\right] = -\frac{f'(z)}{\xi_1}. \tag{B6}$$

The normalization condition with such a function gives

$$\int_{-\infty}^{\infty} \frac{f'(z)}{f(z)} dz = -\xi_1 \tag{B7}$$

which we can integrate, obtaining

$$\ln\left[f(z)\right]\Big|_{-\infty}^{\infty} = \ln\left[\frac{k - \xi_1 \int_0^{\infty} \exp\left[-U_0(x)\right] dx}{k - \xi_1 \int_0^{-\infty} \exp\left[-U_0(x)\right] dx}\right] = -\xi_1. \tag{B8}$$

By solving the above equation with respect to k we get

$$k = \sqrt{\frac{\pi}{2}\xi_1\sigma_{z0}\frac{1+e^{-\xi_1}}{1-e^{-\xi_1}}} = \sqrt{\frac{\pi}{2}\xi_1\sigma_{z0}\coth\left(\frac{\xi_1}{2}\right)} \tag{B9}$$

and then we can write the distribution function as

$$\rho_0(z) = \frac{\exp\left(-\frac{z^2}{2\sigma_{z0}^2}\right)}{\xi_1\sigma_{z0}\sqrt{\frac{\pi}{2}}\left[\coth\left(\frac{\xi_1}{2}\right) - \mathrm{erf}\left(\frac{z}{\sqrt{2}\sigma_{z0}}\right)\right]} \tag{B10}$$

for which we have used the definition of error function

$$\mathrm{erf}(x) = \frac{2}{\sqrt{\pi}}\int_0^z \exp\left(-x^2\right) dx. \tag{B11}$$

APPENDIX C

Derivation of Linearized Vlasov Equation

If we ignore the second-order infinitesimal

$$\frac{\partial \Psi_1}{\partial \varepsilon}\frac{\partial H_1}{\partial z} \cong 0 \tag{C1}$$

and the effects of radiation damping and fluctuation noise on the perturbation function Ψ_1, we get

$$\frac{\partial \Psi_1}{\partial t} = \frac{\partial \Psi_1}{\partial z}\frac{\partial H_0}{\partial \varepsilon} - \frac{\partial \Psi_0}{\partial \varepsilon}\frac{\partial H_1}{\partial z} - \frac{\partial \Psi_1}{\partial \varepsilon}\frac{\partial H_0}{\partial z}. \tag{C2}$$

It is convenient, at this point, to introduce the two action angle variables, J and ϕ, defined as

$$J(H) = \frac{1}{2\pi}\oint_{-\text{motion}} \varepsilon\, dz = \frac{1}{\pi}\int_{z_{\min}}^{z_{\max}}\left(\frac{2H}{c\alpha_c} - \frac{2\varphi(z)}{\alpha_c^2 E_0}\right)^{\frac{1}{2}} dz,$$

$$\dot{\phi} = \omega_s(J) = \frac{\partial H}{\partial J}, \tag{C3}$$

where the line integral is extended to a whole period of motion, and is opposite to the direction described by a particle in the phase space, in order to have a positive value of the action variable J. The values of $z_{\min}$ and $z_{\max}$ depend on the Hamiltonian H. One property of J is that it is a function only of the Hamiltonian, and it is proportional to the region, in the phase space, enclosed by the trajectory.

It is therefore a constant of the motion for a conservative system. Furthermore, the stationary distribution Ψ_0 depends only on J and not on ϕ.

With the introduction of the generating functions related to the Legendre transformations, it is possible to demonstrate that

$$\frac{\partial J}{\partial z} = \frac{\partial \varepsilon}{\partial \phi},$$
$$\frac{\partial J}{\partial \varepsilon} = -\frac{\partial z}{\partial \phi}. \tag{C4}$$

The above definitions allow us to write Eq. (C2) in a more compact form. In fact we have

$$\frac{\partial H_0}{\partial \varepsilon} = \frac{\partial H_0}{\partial J}\frac{\partial J}{\partial \varepsilon} = -\omega_s(J)\frac{\partial z}{\partial \phi},$$
$$\frac{\partial H_0}{\partial z} = \frac{\partial H_0}{\partial J}\frac{\partial J}{\partial z} = \omega_s(J)\frac{\partial \varepsilon}{\partial \phi}, \tag{C5}$$

and then Eq. (C2) becomes

$$\frac{\partial \Psi_1}{\partial t} = -\frac{\partial \Psi_0}{\partial \varepsilon}\frac{\partial H_1}{\partial z} - \omega_s(J)\left(\frac{\partial \Psi_1}{\partial z}\frac{\partial z}{\partial \phi} + \frac{\partial \Psi_1}{\partial \varepsilon}\frac{\partial \varepsilon}{\partial \phi}\right). \tag{C6}$$

The term in parentheses is just the derivative of Ψ_1 with respect to ϕ, and the first term on the right side can be written as

$$\frac{\partial \Psi_0}{\partial \varepsilon}\frac{\partial H_1}{\partial z} = \frac{\partial \Psi_0}{\partial J}\frac{\partial J}{\partial \varepsilon}\frac{\partial H_1}{\partial z} = \frac{1}{\omega_s(J)}\frac{\partial H_0}{\partial \varepsilon}\frac{\partial \Psi_0}{\partial J}\frac{\partial H_1}{\partial z} \tag{C7}$$

since Ψ_0 is independent of ϕ. If we now use the definition of Hamiltonian (15) and Eq. (50), we have

$$\frac{\partial \Psi_0}{\partial \varepsilon}\frac{\partial H_1}{\partial z} = -\frac{c^2\alpha_c e^2 N_p \varepsilon}{E_0 L_0 \omega_s(J)}\frac{\partial \Psi_0}{\partial J}\int_{-\infty}^{\infty}\rho_1(z';t)w_\parallel(z'-z)\,dz' \tag{C8}$$

which, introduced in Eq. (C6), gives finally

$$\frac{\partial \Psi_1}{\partial t} = -\omega_s(J)\frac{\partial \Psi_1}{\partial \phi} + \frac{c^2\alpha_c e^2 N_p}{E_0 L_0 \omega_s(J)}\frac{\partial \Psi_0}{\partial J}\varepsilon\int_{-\infty}^{\infty}\rho_1(z';t)w_\parallel(z'-z)\,dz'. \tag{C9}$$

APPENDIX D

Matrix Elements for Gaussian and Parabolic Distributions

In terms of action-angle variables, the stationary Gaussian distribution function Eq. (29) can be written as

$$\Psi_0(J) = \overline{\Psi}\exp\left[-\frac{\omega_s J}{\overline{H}}\right]. \tag{D1}$$

The constant $\overline{\Psi}$, obtained with the normalization condition (21), is

$$\overline{\Psi} = \frac{\omega_s}{2\pi \overline{H}} \tag{D2}$$

and then
$$\Psi_0(J) = \overline{\Psi} \exp\left[-2\pi \overline{\Psi} J\right] \tag{D3}$$
with which the weight function becomes
$$w(J) = 2\pi \overline{\Psi} \exp\left[-2\pi \overline{\Psi} J\right]. \tag{D4}$$

The normalization condition on the polynomials is therefore

$$2\pi \overline{\Psi} \int_0^\infty \exp\left[-2\pi \overline{\Psi} J\right] f_{|m|k}(J) f_{|m|l}(J) \, dJ = \delta_{kl}. \tag{D5}$$

The polynomials that satisfy such a relation are the generalized Laguerre polynomials $L_l^{(|m|)}$ for which we have

$$\sqrt{\frac{l! k!}{(|m|+l)! (|m|+k)!}} \int_0^\infty x^{|m|} L_l^{(|m|)}(x) L_k^{(|m|)}(x) e^{-x} dx = \delta_{kl}, \tag{D6}$$

which, compared with Eq. (D5), allows us to conclude that

$$f_{|m|l}(J) = \sqrt{\frac{l!}{(|m|+l)!}} \left(2\pi \overline{\Psi} J\right)^{\frac{|m|}{2}} L_l^{(|m|)}\left(2\pi \overline{\Psi} J\right) \tag{D7}$$

and, therefore, the matrix elements (70) become

$$M_{pk}^{ml} = -i \frac{mc^2 e^2 N_p \overline{\Psi}}{\omega_s E_0 L_0} i^{l-m} \sqrt{\frac{p! k!}{(|m|+p)! (|l|+k)!}} \int_{-\infty}^\infty \frac{Z_\|(\omega)}{\omega} d\omega$$
$$\times \int_0^\infty x^{\frac{|m|}{2}} e^{-x} L_p^{(|m|)}(x) J_m\left(\omega \sqrt{\frac{\alpha_c x}{\pi \overline{\Psi} c \omega_s}}\right) dx$$
$$\times \int_0^\infty y^{\frac{|l|}{2}} e^{-y} L_k^{(|l|)}(y) J_l\left(\omega \sqrt{\frac{\alpha_c y}{\pi \overline{\Psi} c \omega_s}}\right) dy. \tag{D8}$$

The last two integrals can be solved; in fact we have

$$\int_0^\infty x^{\frac{|m|}{2}} e^{-x} L_p^{(|m|)}(x) J_m\left(a\sqrt{x}\right) dx = S(m) \frac{1}{p!} \left(\frac{a}{2}\right)^{|m|+2p} \exp\left[-\frac{a^2}{4}\right] \tag{D9}$$

where
$$S(m) = \begin{cases} (-1)^m & \text{if } m < 0 \\ 1 & \text{if } m \geq 0 \end{cases} \tag{D10}$$

and then

$$M_{pk}^{ml} = -i\frac{c^2 e^2 N_p \overline{\Psi}}{\omega_s E_0 L_0} i^{|l|-|m|} \frac{m}{\sqrt{p!\,(|m|+p)!k!\,(|l|+k)!}}$$
$$\times \int_{-\infty}^{\infty} \frac{Z_{\parallel}(\omega)}{\omega} \left(\frac{\omega}{2}\sqrt{\frac{\alpha_c}{\pi\overline{\Psi} c \omega_s}}\right)^{|m|+|l|+2p+2k} \exp\left[-\frac{\omega^2 \alpha_c}{2\pi\overline{\Psi} c \omega_s}\right] d\omega, \quad (D11)$$

which, in terms of bunch length σ_z and natural energy spread $\sigma_{\varepsilon 0}$, due to the relation

$$2\pi\overline{\Psi} = \frac{1}{\sigma_z \sigma_{\varepsilon 0}}, \quad (D12)$$

can also be written as

$$M_{pk}^{ml} = -i\frac{c^2 e^2 N_p}{2\pi\omega_s E_0 L_0 \sigma_z \sigma_{\varepsilon 0}} i^{|l|-|m|} \frac{m}{\sqrt{p!\,(|m|+p)!k!\,(|l|+k)!}}$$
$$\times \int_{-\infty}^{\infty} \frac{Z_{\parallel}(\omega)}{\omega} \left(\frac{\omega \sigma_z}{\sqrt{2}c}\right)^{|m|+|l|+2p+2k} \exp\left[-\frac{\omega^2 \sigma_z^2}{c^2}\right] d\omega. \quad (D13)$$

The same procedure for the case of a parabolic bunch distribution leads to

$$f_{|m|l}(J) = \sqrt{\frac{2(m+2l+1/2)\,l!\,\Gamma(m+l+1/2)}{(m+l)!\,\Gamma(l+1/2)}}$$
$$\times \left(\frac{J}{J_{max}}\right)^{m/2} P_l^{m,-1/2}\left(1 - 2\frac{J}{J_{max}}\right) \quad (D14)$$

with $P_l^{m,-1/2}$ the Jacobi polynomials and

$$M_{pk}^{ml} = -i\frac{3c^2 e^2 N_p}{2\pi\omega_s E_0 L_0 z_{max}^2} i^{|l|-|m|} m \sqrt{\frac{(m+2p+1/2)\,\Gamma(m+p+1/2)\,\Gamma(p+1/2)}{p!\,(m+p)!}}$$
$$\times \sqrt{\frac{(l+2k+1/2)\,\Gamma(l+k+1/2)\,\Gamma(k+1/2)}{k!\,(l+k)!}}$$
$$\times \int_{-\infty}^{\infty} \frac{Z_{\parallel}(\omega)}{\omega} \frac{J_{m+2p+1/2}\left(\frac{\omega}{c}z_{max}\right) J_{l+2k+1/2}\left(\frac{\omega}{c}z_{max}\right)}{\frac{\omega}{c}z_{max}} d\omega. \quad (D15)$$

Transverse Mode Coupling Instabilities

Jacques Gareyte

CERN SL Division, 1211 Geneva 23, Switzerland

Abstract. Transverse mode coupling instabilities (TMCI) emerged between 1974 and 1980 as the main limitation of dense bunches in electron synchrotrons and storage rings. A two-particle model allows one to calculate the beam break-up (BBU) instability in linacs. Extending this to synchrotrons shows that the BBU instability is suppressed below a threshold intensity by synchrotron oscillations. The classical theory of head–tail modes, together with the general properties of coupling impedances, is used to show how single bunches become unstable when head–tail modes couple together: this is the TMCI threshold. Above threshold, observations in both proton and electron synchrotrons can be described by BBU theory.

INTRODUCTION

The concept of mode coupling was invented by Sacherer in 1977 to explain the longitudinal turbulence observed in dense particle bunches [1]. However, for reasons that took us almost 20 years to understand, it did not help much in dealing with longitudinal instabilities, whereas it proved extremely fruitful when applied to transverse collective modes. This report addresses exclusively the problem of transverse mode coupling.

Transverse mode coupling is the ultimate transverse instability, that which remains when all other possible mechanisms have been eliminated. It is observed for single bunches (thereby excluding coupled-bunch instabilities) and for a value of the chromaticity equal to zero (in the absence of head–tail instability).

A first record of what was probably a manifestation of transverse mode coupling appeared in 1974: at the e^+e^- collider SPEAR I (Standford Linear Accelerator Center) a beating pattern was observed in the coherent betatron signal for large values of the bunch current [2]. A year later SPEAR II was commissioned. This machine had a denser bunch than SPEAR I, and fast losses of the bunch were observed above a threshold current. A vertical instability was suspected, although beam position monitors showed only a weak coherent betatron signal, and the phenomenon persisted for zero chromaticity [3]. This remained a puzzle.

In 1979 the larger e^+e^- storage ring PETRA started operation at DESY (Hamburg). Again, fast losses occurred even for zero chromaticity, again a vertical instability was suspected but no large coherent signals were observed. Whereas

FIGURE 1. Single-bunch beam break-up.

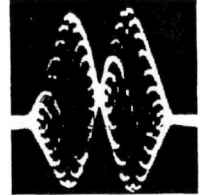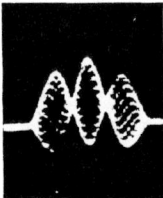

FIGURE 2. Head–tail modes $m = 0, 1, 2$.

at SPEAR II the phenomenon did not really deteriorate the performance of the machine, in PETRA it occurred at a bunch current well below that corresponding to the beam–beam limit, and was therefore a serious limitation.

Although, as already mentioned, Sacherer had drawn the attention of the community to mode coupling in 1977, and the extension of this concept to the transverse degrees of freedom was rather straightforward, a suitable theory was not developed immediately. The untimely death of Sacherer in a mountain accident in 1978 can probably explain this delay. In 1980 Kohaupt proposed the transverse mode coupling mechanism to explain the phenomenon observed in PETRA [4]. It was immediately understood that this was going to be the major limitation of electron storage rings. Kohaupt's paper came just in time to allow a thorough redesign of the injection system of LEP, taking this new phenomenon into account.

In fact, transverse mode coupling is just the manifestation in synchrotrons of a phenomenon that had been known in linacs since the 1960s: beam break-up instability (BBU) [5]. When a dense bunch of particles travels along a linac, the head of the bunch induces electromagnetic wakefields in the linac structure, which tend to deflect the bunch tail. After a while the bunch tail oscillates about the unperturbed trajectory of the head, as shown in Fig. 1, and can be lost on the aperture. In synchrotrons, the RF longitudinal focusing imposes special patterns of oscillation for the collective transverse motion: the head–tail modes [6]. Examples of head–tail modes are shown in Fig. 2. For low enough bunch intensity, the head–tail modes keep their identity, the wakefields merely change their frequency in proportion to the beam current.

Above a certain current, the wakefields are strong enough compared to RF focusing to destroy the head–tail modes: this is the threshold for mode coupling. For currents well above this threshold, the wakefields dominate and the RF focusing can be ignored: we observe the equivalent of the BBU instability in linacs. In the following we will first use a simple two-particle model to derive most of the prop-

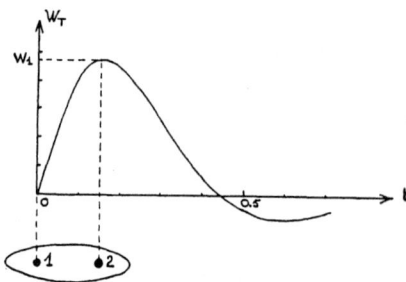

FIGURE 3. Example of transverse wakefield ($Q = 1$ resonator) generated by head particle 1, deflecting tail particle 2.

erties of BBU and mode coupling. Then we will outline a general theory of mode coupling based on the standard model of bunched-beam instabilities. Finally we will show that BBU is observed in electron as well as in proton synchrotrons, and that it can be well predicted.

FROM BBU TO MODE COUPLING: A TWO-PARTICLE MODEL

This subject is treated in detail by Chao [7]. We start with a model for the wakefield. The beam enclosure is made up of a number of elements of various shapes and electrical properties, each one producing its own wakefield. We retain only the strong, short-range wakefields, since we are interested in dense, isolated bunches. We assume that the cumulative effect of all the elements can be represented by the response to a shock excitation of a single resonator with a low quality factor Q (we will often take $Q = 1$), as shown in Fig. 3. This model is very convenient and its validity has been well tested experimentally in many existing machines. As a further simplification, the bunch charge is concentrated in two macroparticles, particle 1 at the head and particle 2 at the tail. Particle 1 excites the wakefield, which evolves in time according to

$$W_T = \omega_r \frac{R_T}{Q} e^{-\epsilon t} \sin S\omega_r t \qquad (1)$$

and deflects particle 2, which in the worst case is situated at the maximum W_1 of the wake.

In Eq. (1) ω_r is the resonant frequency, R_T the maximum resistive component of the corresponding coupling impedance, $S = (1 - 1/4Q^2)^{1/2}$ and $\epsilon = \omega_r/2Q$.

The Case of a Linac

Suppose particle 1 is displaced at time $t = 0$ by $y_1(0)$. It will oscillate according to

$$y_1 = y_1(0) \cos \omega_\beta t \qquad (2)$$

where ω_β is its betatron frequency. The wake force exerted on particle 2 is then (by definition of the wakefield) $f = e(Ne/2)y_1 W_1$, so that

$$\ddot{y}_2 + \omega_\beta^2 y_2 = \alpha y_1 \qquad (3)$$

with $\alpha = (Ne^2/2)(W_1/m\gamma)$, where Ne is the bunch charge, m the particle mass, γ the energy devided by mc^2, and W_1 the wakefield per meter of structure.

The solution of Eq. (3) is

$$y_2(t) = y_2(0) \cos \omega_\beta t + y_1(0) \frac{\alpha}{2\omega_\beta} t \sin \omega_\beta t . \qquad (4)$$

Because particle 2 is driven by particle 1, its amplitude increases linearly with time and at the end of a linac of length L reaches

$$\hat{y}_2 = \hat{y}_1 \alpha \frac{L}{2\omega_\beta c} . \qquad (5)$$

The Case of a Synchrotron

General Derivation

The fundamental difference between a linac and a synchrotron is that in a synchrotron particles 1 and 2 exchange places after half a synchrotron oscillation, $T_s/2$. We can still treat the problem in a simple way if we assume that the wakefield is constant between particles 1 and 2 and decays only after the passage of particle 2. We must furthermore keep track of all possible initial conditions, and we do this by considering the vector

$$\begin{vmatrix} y_1 \\ \dot{y}_1/\omega_\beta \\ y_2 \\ \dot{y}_2/\omega_\beta \end{vmatrix} . \qquad (6)$$

The complete solution of Eq. (3) is Eq. (4) with the addition of the terms

$$\frac{1}{\omega_\beta} \dot{y}_2(0) \sin \omega_\beta t + \frac{\alpha}{\omega_\beta} \dot{y}_1(0) \left[\frac{\sin \omega_\beta t}{2\omega_\beta^2} - t \frac{\cos \omega_\beta t}{\omega_\beta} \right] . \qquad (7)$$

We find the solution after half a synchrotron period by putting $t = T_s/2 = \pi/\omega s$:

$$\begin{vmatrix} y_1(T_s/2) \\ \frac{1}{\omega_\beta}\dot{y}_1(T_s/2) \\ y_2(T_s/2) \\ \frac{1}{\omega_\beta}\dot{y}_2(T_s/2) \end{vmatrix} = \begin{vmatrix} A(T_s/2) & 0 \\ B(T_s/2) & A(T_s/2) \end{vmatrix} \begin{vmatrix} y_1(0) \\ \frac{1}{\omega_\beta}\dot{y}_1(0) \\ y_2(0) \\ \frac{1}{\omega_\beta}\dot{y}_2(0) \end{vmatrix} \tag{8}$$

with

$$A(T_s/2) = \begin{vmatrix} \cos\mu/2 & \sin\mu/2 \\ -\sin\mu/2 & \cos\mu/2 \end{vmatrix}, \tag{9}$$

$$B(T_s/2) = \Upsilon \begin{vmatrix} \sin\mu/2 & \frac{2}{\mu}\sin\mu/2 - \cos\mu/2 \\ \frac{2}{\mu}\sin\mu/2 + \cos\mu/2 & \sin\mu/2 \end{vmatrix} \tag{10}$$

where $\mu = 2\pi\omega_\beta/\omega_s$ and $\Upsilon = (\pi Ne^2 W_1/4m\gamma\omega_\beta\omega_s)$.

By exchanging particle 1 and particle 2 we find the solution for the second half of the synchrotron period, so that the transformation matrix for one complete period is

$$T = \begin{vmatrix} A & B \\ O & A \end{vmatrix} \begin{vmatrix} A & O \\ B & A \end{vmatrix} = \begin{vmatrix} A^2 + B^2 & BA \\ AB & A^2 \end{vmatrix}. \tag{11}$$

The characteristic equation of T is complicated. However, knowing that, as in the case of symplectic matrices, the eigenvalues of T can be grouped in reciprocal pairs, we can write it:

$$\left(\lambda + \frac{1}{\lambda}\right)^2 + 2\left(\lambda + \frac{1}{\lambda}\right)\left[\frac{2\Upsilon^2}{\mu^2}(\cos\mu - 1) + \Upsilon^2\cos\mu - 2\cos\mu\right] +$$

$$\left\{4\Upsilon^2(\cos\mu - 1) + \left[\Upsilon^2 + \frac{2\Upsilon^2}{\mu^2}(\cos\mu - 1) - 2\cos\mu\right]^2\right\} = 0. \tag{12}$$

Simplified Treatment

The equation for the eigenvalues λ can be simplified if we note that μ is a large number. For instance in LEP ω_β/ω_s is of the order of 1000. Therefore terms divided by μ^2 can be dropped to yield

$$\lambda + \frac{1}{\lambda} = -(\Upsilon^2 - 2)\cos\mu \pm \Upsilon\sin\mu\sqrt{4 - \Upsilon^2}. \tag{13}$$

From Eq. (13) it appears that if $\Upsilon > 2$, the motion becomes unstable. By writing $\lambda = \exp(i\mu + i\phi)$ it is easy to show that a solution of Eq. (13) is

$$\sin\phi/2 = \Upsilon/2. \tag{14}$$

From this we conclude the following:

- At small intensity $\phi = \Upsilon$. The effect of the interaction is a phase shift.
- As Υ approaches 2, ϕ approaches π. For $\Upsilon > 2$, Eq. (14) can be satisfied only with ϕ imaginary. Then one of the eigenvalues grows exponentially: this is the transverse mode coupling instability, or TMCI. Writing $\Upsilon = 2 + \epsilon$, with $\epsilon \ll 1$, Eq. (14) gives $\cos\phi = 1 - \Upsilon^2/2 = -(1 + 2\epsilon)$ and with $\phi = \pi + \Delta\phi$ we obtain $\Delta\phi = 2\sqrt{\epsilon}\,i$. The amplitude grows as $\exp(\Delta\phi t/T_s)$, that is, with a growth rate $1/\tau = 2\sqrt{\epsilon}/T_s$. The square root dependence shows that above threshold but very close to it, the growth rate reaches a value of the order of $1/T_s$.
- Looking at Eq. (5) and the definition of Υ, we see that for $\Upsilon = 2$, the amplitude of the tail particle grows to twice the amplitude of the leading particle in half a synchrotron period.

We interpret all this in the following way: by periodically exchanging particle 1 and particle 2 the synchrotron oscillation prevents the growth of both particles from accumulating, and therefore stabilizes the beam. However, when the trailing particle grows during half a synchrotron period to twice the amplitude of the leading particle, this stabilizing mechanism ceases to be effective, and the growth does accumulate over many synchrotron periods. For larger values of the synchrotron frequency, the trailing particle has less time to grow before the particles are interchanged, and the threshold of the instability increases.

Synchrobetatron Resonances

It is instructive to examine the effects that have been neglected by making the approximation $\mu \gg 1$, or $\omega_\beta \gg \omega_s$. This will introduce us to a phenomenon that is very important in large machines, coherent synchrobetatron resonance.

We have to solve Eq. (12). The beam will be stable if all solutions for $\lambda + 1/\lambda$ are real and between -2 and 2. Figure 4 shows in grey the unstable areas. We find the well-known instability limit for $\Upsilon > 2$, but in addition there are thin areas of instability at smaller beam intensity originating at integer values of ω_β/ω_s: these are coherent synchrobetatron resonances. We observe that the width of these additional unstable regions shrinks as ω_β/ω_s increases. In the case of LEP, where $\omega_\beta/\omega_s \approx 1000$, we can imagine that they become extremely thin and that our previous approximation is well justified.

However, all the treatment done so far uses a smooth approximation and assumes that the sources of wakefields (the coupling impedances) are smoothly distributed around the machine. This is of course not true, especially in LEP, where a large fraction of the coupling impedance is located in the accelerating cavities, which are gathered in a few long straight sections. This introduces azimuthal harmonics of the force. As a consequence, a pattern similar to that of Fig. 4 repeats itself at every integer value of the betatron tune. LEP, with $Q_s \simeq 0.1$ and $Q_\beta \simeq N + 0.25$, is operated in the vicinity of synchrobetatron resonances of order 2 and 3. It must be

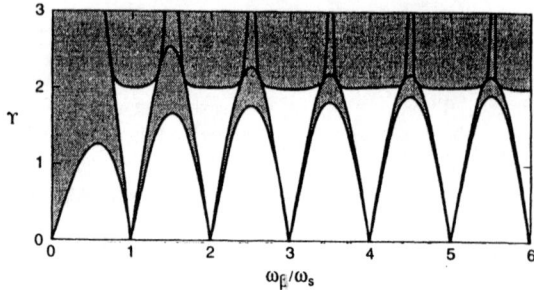

FIGURE 4. Regions of stability for a two-particle beam (shaded is unstable).

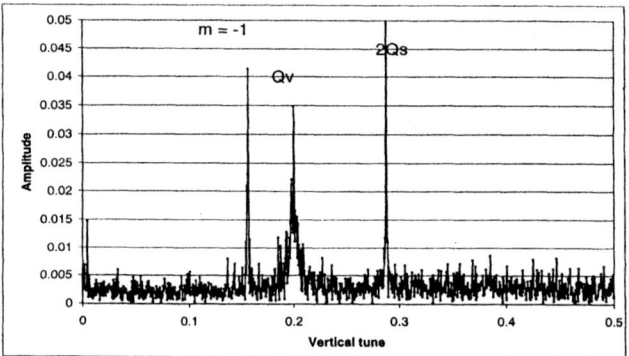

FIGURE 5. Vertical betatron spectrum at LEP, close to TMCI threshold, showing mode $m = -1$ (through its coupling to mode $m = 0$) and the strong signature of the coherent synchrobetatron resonance $2Q_s$ (here Q_s is 0.135).

tuned very carefully during beam accumulation and energy ramping to avoid these resonances. Figure 5 shows a vertical betatron spectrum observed for a bunch current about equal to half the TMCI threshold: the strong line at $2Q_s$ is the signature of the coherent synchrobetatron resonance $Q_\beta = N + 2Q_s$. If the tune wanders too close to this line, the beam is lost well below the TMCI threshold [8].

HEAD–TAIL MODES AND MODE COUPLING

A bunch of two particles can oscillate in two different modes: the particles are either in phase (this is mode $m = 0$) or out of phase (this is mode $m = 1$). The existence of two modes is just enough to demonstrate the concept of mode coupling. A real bunch exhibits an infinite number of "head–tail modes." However,

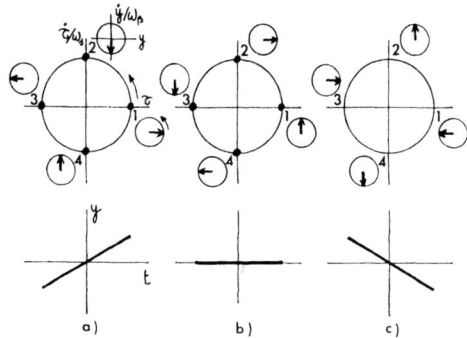

FIGURE 6. Head–tail mode $m = -1$.

the low-order modes $m = 0$ and $m = 1$ are the easiest to drive and are most often encountered.

In the following we illustrate mode $m = 1$ by considering four particles distributed with appropriate betatron phase shift ($\pi/2$ between successive particles) along a synchrotron orbit. In Fig. 6 the large circles represent the synchrotron orbit and the small ones represent the betatron orbits of the four particles. Particle 1 has first maximum positive elongation and particle 3 maximum negative elongation, while particles 2 and 4 are centred. As time proceeds, in the centre picture particles 1 and 3 cross the axis while particles 2 and 4 get maximum but opposite elongation, so that they do not contribute to the coherent signal: the bunch seems to oscillate as a whole around its fixed centre. While this fast betatron motion takes place, the particles showly migrate along the synchrotron orbit, so that after $t = T_s$ particle 1 is again at the bunch head. For an external observer, the bunch has made one more oscillation than the individual particles: the frequency of mode $m = 1$ is $\omega_1 = \omega_\beta + \omega_s$.

With a different initial distribution of betatron phases we can generate mode $m = -1$ with frequency $\omega_\beta - \omega_s$. In general, mode m has $|m|$ nodes and oscillates at $\omega_\beta + m\omega_s$.

In our two-particle model it can be shown [7] that the frequencies of the two modes 0 and -1 approach each other as the intensity is increased, until the modes merge for $\Upsilon = 2$ and become unstable: hence the name transverse mode coupling instability (Fig. 7).

A nice illustration of mode coupling is given (Fig. 8) by observations made at the storage ring PEP (SLAC): after the beam has been kicked transversely, one observes a beating pattern with diminishing frequency as modes 0 and -1 approach each other close to TMCI threshold [9].

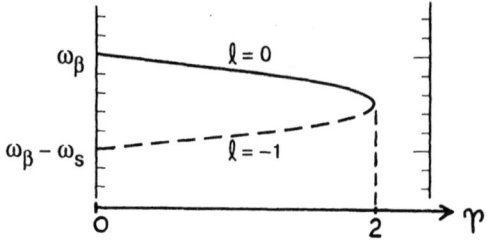

FIGURE 7. Merging of the two coherent modes in the two-particle model.

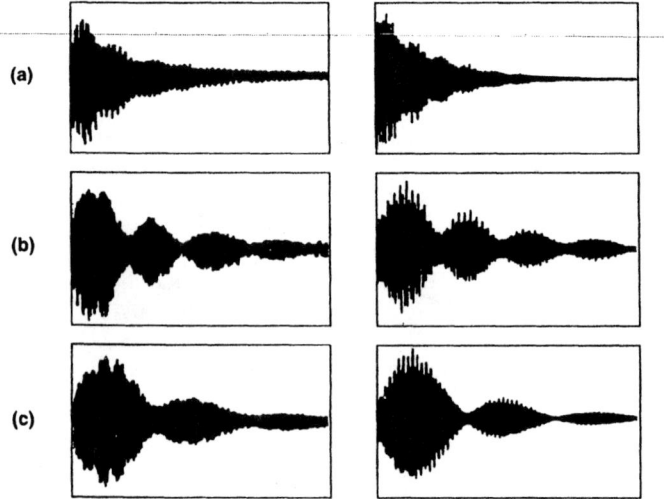

FIGURE 8. Observation (left) and simulation (right) of a beam position monitor signal after the beam has been kicked in PEP. The beam intensity varies between 0.86 (a) and 0.99 (c) of the TMCI threshold. Mode coupling with a decreasing beating frequency is observed.

OUTLINE OF A GENERAL THEORY OF TMCI

The principle as well as many fundamental aspects of TMCI can be demonstrated with the two-particle model. To treat the case of a real bunch in a self-consistent way is difficult and beyond the scope of this paper. However, it is relatively easy, using the standard theory of bunched beam stability and invoking basic principles, to demonstrate very interesting and useful properties of mode coupling.

Basics of the Bunched-Beam Stability Theory

At low intensity, particle bunches in a synchrotron can oscillate transversely according to "head–tail modes" (see Fig. 2).

A head–tail mode m has m nodes and generates a coherent signal on the synchrotron satellite $m\omega_s$ of the betatron lines. Modes are described by a set of orthogonal functions $g_m(\omega)$.

In the presence of wakefields each mode is coupled to all the others. This is described by the interaction matrix

$$M_n^m = Ci \sum_{p=-\infty}^{+\infty} i^{m-n} g_m(p) Z_T(p) g_n(p) \tag{15}$$

where $p = \omega/\omega_0$, the frequency divided by the revolution frequency, Z_T is the transverse coupling impedance, and C a real constant. At low-intensity modes m and n are separated in frequency by $(m-n)\omega_s$ and therefore cannot influence each other significantly. In this situation we can often neglect the off-diagonal terms in M. This is no longer true close to the TMCI threshold when mode frequencies may come close to each other.

Head–tail modes are standing-wave patterns: each point on the mode signal is either in phase or in antiphase with the others. As a consequence their spectra are either symmetric (m even) or antisymmetric (m odd) with respect to the origin of frequencies, as shown in Fig. 9.

We consider ultrarelativistic beams, in which a particle can affect only those coming behind. Therefore the wakefields obey causality, which means that their Fourier transforms, the coupling impedances, are symmetric (for the reactive part) or antisymmetric (the resistive part) with respect to the origin of frequencies, as shown in Fig. 9 for the $Q = 1$ resonator model. Using these general properties and considering formula (15) we see that:

- The effect of a mode m on itself comes only through the reactive part of the transverse coupling impedance Z_T. It gives a real M_m^m, which means a real frequency shift. There is no possibility of instability if we neglect the off-diagonal terms.

- For two modes $m \neq n$ we have

$$M_n^m = (-1)^{m-n} M_m^n . \tag{16}$$

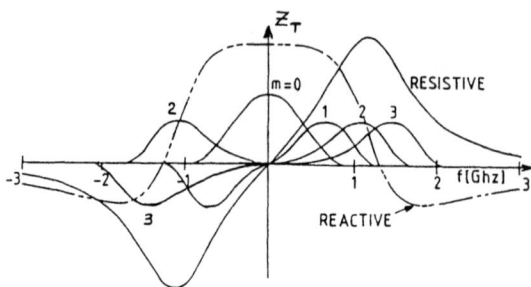

FIGURE 9. Head–tail mode spectra superimposed on the broad-band coupling impedance ($Q = 1$ resonator model).

Influence of Mode Parity

Often the interaction brings the frequency of two modes close to each other. In this case the dynamics is dominated by the coupling of these two modes. Therefore it is instructive to look at a simplified model in which only two modes interact [10]. Stability is determined in this case by the eigenvalues of a 2×2 matrix:

$$\begin{vmatrix} \lambda - m - M_m^m & -M_n^m \\ -M_m^n & \lambda - n - M_n^n \end{vmatrix} = 0 , \tag{17}$$

$$\lambda = \frac{1}{2}(m + M_m^m + n + M_n^n) \pm \sqrt{(m + M_m^m - n - M_n^n)^2 + 4M_n^m M_m^n} . \tag{18}$$

At low intensity the eigenfrequencies are well separated and the first term under the square root dominates.

In Fig. 10 we describe a situation, which is usual, where this term diminishes as the intensity increases, thus reducing the separation of the eigenmodes. The cross terms then dominate the square root, and two different situations are possible:

- m and n have the same parity. According to Eq. (16) $M_n^m = M_m^n$ and the effect of the cross terms is to push apart the eigenmodes. Modes of the same parity repel each other as in Fig. 10(b).

- m and n have different parities. The second term under the square root is negative. As intensity is increased, modes attract each other. For $I = I_{th}$ they merge and acquire an imaginary component: this is TMCI [Fig. 10(a)].

Influence of Bunch Length

TMCI is usually observed in electron storage rings, machines that have dense short bunches. Using Fig. 9 it is easy to understand why short and long bunches have very different behaviors.

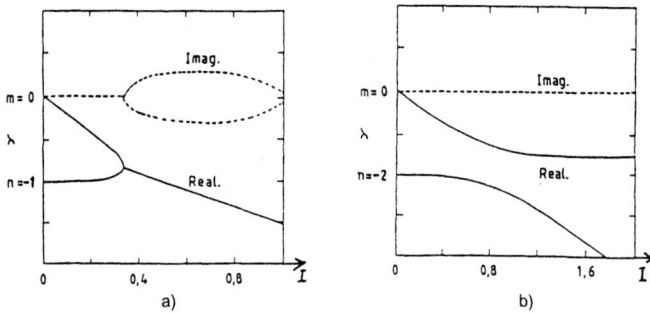

FIGURE 10. Eigenvalues as a function of bunch intensity for the cases $(m - n)$ odd (a) and $(m - n)$ even (b).

A short bunch has a spectrum that extends to high frequency. A typical case is shown in Fig. 11. Mode $m = 0$ interacts strongly with the inductive part of the broad-band coupling impedance and as a result its frequency shifts down as intensity increases. On the contrary, mode $m = -1$ interacts with the capacitive part at higher frequency, and is shifted up towards mode 0. In addition both modes interact strongly with the resistive part of the coupling impedance, and this results in important cross terms. Therefore all the ingredients are present to generate a strong TMCI. The most unstable case occurs when the bunch length $\sigma_s = c/\omega_r$ (ω_r being the resonant frequency of the broad-band resonator model, or in general the frequency corresponding to the peak of the resistive part of the coupling impedance).

On the contrary, for a long bunch the spectra of modes $m = 0$ and $m = -1$ peak at low frequency. Both modes couple to the inductive part of the coupling impedance, and therefore are shifted in the same direction. Moreover, their coupling to the resistive part of the coupling impedance is weak. As a consequence, when the two modes merge, they cannot develop a strong instability and are pulled apart as intensity increases. This behavior is shown in Fig. 12. Modes of higher order can couple in this case to produce TMCI, but high-order modes are more difficult to drive than low-order ones.

THRESHOLD FORMULA AND CURES

In the case of short bunches, it is possible to calculate the TMCI threshold in an approximate but very useful way. We neglect the frequency shift of mode $m = -1$ (see Fig. 11) and declare that instability occurs when the tune shift of mode $m = 0$ equals Q_s, the initial tune separation between modes 0 and -1. We have

$$\Delta Q_{(m=0)} = i \frac{I}{2\pi E/ef_0} \frac{1}{2\pi} \int \beta_T Z_T(\omega) h(\omega, \sigma_s) d\omega \tag{19}$$

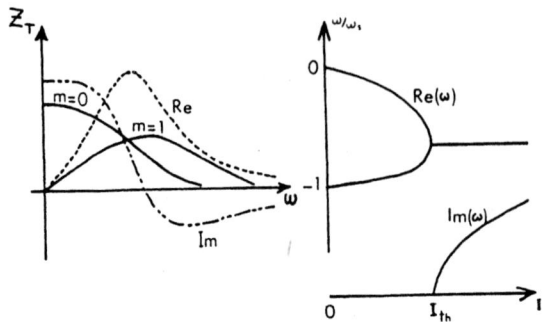

FIGURE 11. Mode coupling for short bunches.

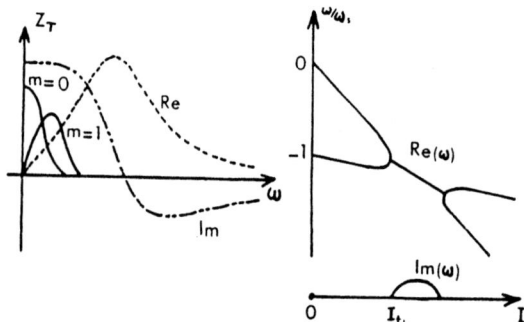

FIGURE 12. Mode coupling for long bunches.

where σ_s the bunch length, $h = gg^*$ is the bunch power spectrum, β_T the betatron function at the location of the coupling impedance Z_T, I the bunch current, E the energy, and f_0 the revolution frequency.

Often the "kick factor"

$$k_T(\sigma_s) = \frac{1}{2\pi} \int_{-\infty}^{+\infty} d\omega Z_T(\omega) h(\omega, \sigma_s) \tag{20}$$

is used instead of the impedance. This gives the formula for the threshold current

$$I_{th} = \frac{2\pi Q_s E/e f_0}{\sum \beta_T k_T(\sigma_s)} . \tag{21}$$

Remark: the threshold found with the two-particle model was

$$\Upsilon = 2 = \frac{\pi N e^2 W_1}{4 m \gamma \omega_\beta \omega_s} .$$

Using $\omega_\beta = c/\beta_T$, $I = Nef_0$, it can be put into the form

$$I_{th} = \frac{16 Q_s E/e f_0}{\beta_T (2\pi R W_1)} . \tag{22}$$

For a $Q = 1$ resonator model, it can be checked that $(2\pi R W_1)$, the wakefield around the machine, corresponds to 2.6 k_T for $\omega_r \sigma_s = 1$. Therefore, the two formulae are about the same.

Using Eq. (21) the actions that can be taken to increase the threshold are obvious. In LEP they were all useful and concurred to raise the stored bunch current close to the beam–beam limit. They are:

– increase injection energy from 20 GeV to 22 GeV,

– increase Q_s up to 0.13 by increasing the RF voltage,

– decrease β_T at the location of RF cavities,

– decrease Z_T by replacing copper cavities by superconducting cavities,

– increase σ_s by using wigglers.

Another way to counteract TMCI is to use feedback damping, as for any other instability. However, TMCI is a strong instability. As we have seen, just above threshold the growth rate reaches values of the order of $1/T_s$, which for LEP means an e-folding time of seven turns. Therefore a very powerful feedback system is required, with all the problems that such a system entails.

Moreover the cross term, which couples two modes, is dominated by the resistive component of the coupling impedance, which peaks at high frequency (see Fig. 9). To be fully effective a feedback should act in the same frequency range; this is very difficult to implement.

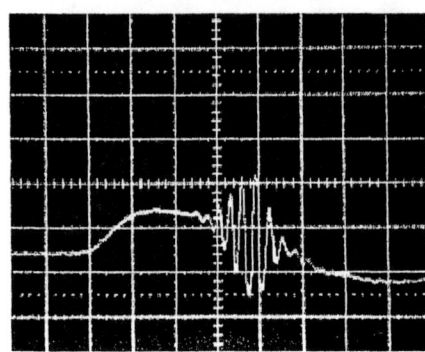

FIGURE 13. Beam break-up on a long proton bunch close to transition in the CERN PS. Horizontal scale: 5 ns/div.

To get around this problem the use of a "reactive" feedback was proposed. As we have seen, for short enough bunches the rigid mode $m = 0$ is involved. This mode can be influenced at low frequency, and the idea was to counteract its negative tune shift with intensity in an attempt to push the point of crossing with mode $m = -1$ to higher intensity. This worked in PEP (SLAC) and VEPP-4M (Novosibirsk), but for modest values of Q_s in the range of 0.02 [11]. Moreover, in PEP at least, it was noted that a usual, resistive feedback had a better performance. A more sophisticated approach was used for LEP [12]. It gave good results in simulation but had a small effect in practice.

ABOVE TMCI THRESHOLD

We will now illustrate with two examples the dynamics of the instability well above threshold.

For machines with a large value of Q_s, like LEP, the beam is lost extremely fast and it is difficult to observe anything. In the case of medium or low Q_s, we can observe the growth of the coherent oscillation.

The Case of Long Bunches: the CERN PS

Figure 13 show a single-turn signal from a vertical wide-band position monitor [13]. A high-frequency wiggle (at about 700 MHz) develops towards the tail of the 7-m-long proton bunch. This happens close to transition energy, therefore with a very small Q_s. In fact, the signal grows much faster than the synchrotron period. The dominating frequency, 700 MHz, corresponds to the frequency cut-off of the beam pipe, that is, to the peak of the resistive component of the coupling impedance in the classical broad-band model.

In order to interpret this observation in terms of TMCI theory, we have to invoke mode numbers $m > 30$! Instead, we propose in this case to use a BBU approach. We consider the synchrotron as a long linac structure of length $L = 2\pi Rn$, where n is the number of turns. We neglect synchrotron motion for the time being. We will see later on how we can reintroduce the notion of threshold, even in the absence of RF voltage.

The theory of BBU developed previously with two macroparticles is not suitable for long bunches. A long train of macroparticles, or bunchlets, is necessary in this case. A suitable theory has been developed for linacs by Yokoya [14] and can be extended to synchrotrons. Using our low-Q resonator model, we calculate the amplitude y of the tail (the last bunchlet) to be after n turns [15]

$$\frac{y}{\delta} = \frac{1}{2\sqrt{2\pi}} \frac{1}{\omega_r \tau} \cdot [\Omega\tau]^{1/4} \cdot \exp[-\epsilon\tau + \sqrt{\Omega\tau}] \tag{23}$$

where

$$\tau = \frac{4\sigma_s}{c}, \quad \Omega\tau = \frac{Ne}{E/e}\beta_T \omega_r \frac{R_T}{Q} \cdot n,$$

and δ is the initial amplitude of the bunch. For an instability that develops right after injection into a synchrotron, δ can be the injection error. When the instability develops at a certain moment in the acceleration cycle, for instance in the PS close to transition when the synchrotron motion is almost frozen, δ is the average closed-orbit deviation in the different objects that generate the transverse coupling impedance.

In old machines like the CERN PS or SPS, we have a fair estimate of the coupling impedance after many observations and cross-checks involving different phenomena. Using this estimate and Eq. (23) we can predict the growth rate observed rather well.

The Case of Medium-Length Bunches: the CERN SPS

We have already considered two extreme cases of bunch length. We have described TMCI in LEP with $\sigma_s = 1$ cm, and BBU in the PS with $\sigma_s = 1.8$ m. Now we will examine the case of the SPS used as a positron and electron accelerator in its role as LEP injector [16]. Bunches are injected into this machine with $\sigma_s = 16$ cm. Regardless of whether the RF is active or not, for N of the order of 4×10^{10} particles we observe the loss of a fraction of the beam after 11 turns, as shown in Fig. 14(b). A wide-band vertical monitor shows a signal growing exponentially from injection up to turn $n = 10$, as shown in Fig. 14(a). The signal shown is the output of a broad-band filter centred around 1.5 GHz. There is no signal at the bunch frequencies (< 500 MHz). Again we observe an instability that grows fast compared to the synchrotron frequency (T_s is 70 turns here), and makes the bunch wiggle at the cut-off frequency of the beam pipe (in the SPS, around 1.3 GHz).

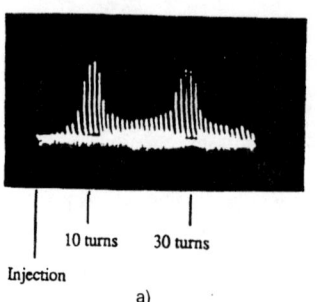 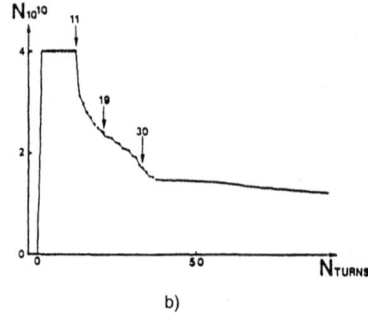

FIGURE 14. Beam position monitor signal filtered around 1.5 GHz (left) and beam intensity (right) as a function of number of turns after injection in the SPS of a positron bunch at 3.5 GeV/c.

Applying formula (23) with the well-proven broad-band impedance model of the SPS and $\delta = 1$ mm, we calculate that the beam tail should be scraped against the beam pipe 15 turns after injection (whereas we observe 11 turns).

Threshold of BBU in Synchrotrons

In the first part of this report we have shown, by using the two-particle model, how the synchrotron motion suppresses the BBU instability below a threshold intensity, by swapping the head and the tail continuously. In fact, in the case of the SPS instability described above, we observe a clear threshold regardless of whether the RF is active or not. In Fig. 14 $N = 4 \times 10^{10}$, but for $N \leq 2 \times 10^{10}$ there is no instability.

This can be explained as follows. In a synchrotron, particles of different energies turn at different angular velocities. After a time T, particles with a momentum difference ΔP are separated by a time delay ΔT given by

$$\frac{\Delta T}{T} = \eta \frac{\Delta P}{P} \qquad (24)$$

with $\eta = (1/\gamma_{tr}^2) - (1/\gamma^2)$, γ_{tr} being the γ at transition energy.

This differential streaming of particles prevents any build-up of a coherent wave along the bunch if it is too large.

As a rule of thumb let us conjecture that the threshold is reached when the ΔT accumulated over an e-folding time $\tau = T$ is equal to the period of the wave $1/f_W$.

In the SPS we have $\eta = 1.85 \times 10^{-3}, \Delta P/P = 2 \times 10^{-3}$, and $f_W = 1.5$ GHz. The revolution period is 23 μs. Therefore the instability can be observed only if $\tau < 8$ turns. This is consistent with the observations described above.

CONCLUSIONS

The beam break-up instability, which is a strong limitation in linacs, is suppressed in synchrotrons below a threshold intensity by the differential streaming of particles in the presence of energy spread. In proton accelerators this effect disappears close to transition energy, and BBU can be observed there.

Below threshold, particle bunches in a synchrotron can oscillate coherently according to well-defined standing-wave patterns, the head–tail modes. For zero chromaticity these modes are stable for single bunches.

However, as the bunch current increases, the wakefields couple head–tail modes together until by the merging of two modes of different parities, a travelling-wave pattern is created along the bunch. This transition is called transverse mode coupling instability. Above threshold, the travelling-wave pattern grows in a way similar to BBU in linacs.

REFERENCES

1. Sacherer, F.J., "Bunch Lengthening and Microwave Instability", *IEEE Trans. Nucl. Sci.* **NS-24** (3), 1393 (1977).
2. The SPEAR Group, "Fast Damping of Transverse Coherent Dipole Oscillations in SPEAR", in *9th Int. Conf. on High Energy Accelerators*, Stanford, 1974, Conf 740522, AEC, Washington DC, 1974, p. 338.
3. The SPEAR Group, "SPEAR II Performance", in *Particle Accelerator Conference*, Washington DC, 1975, *IEEE Trans. Nucl. Sci.* **NS-22** (3), 1366 (1975).
4. Kohaupt, R.D., "Transverse Instabilities in PETRA", in *11th Int. Conf. on High Energy Accelerators*, CERN, Geneva, 1980, edited by W.S. Newman, Birkhaüser, Basle, 1980, p. 526.
5. Helm, R., and Loew, G., "Beam Break Up", in *Linear Accelerators*, edited by P.M. Lapostolle and A.L. Septier, North-Holland, Amsterdam, 1970, Chapter B.
6. Gareyte, J., and Sacherer, F., "Head–Tail Type Instabilities in the CERN PS and Booster", in *9th Int. Conf. on High Energy Accelerators*, Stanford, 1974, Conf 740522, AEC, Washington DC, 1974, p. 341.
7. Chao, A.W., *Physics of Collective Beam Instabilities in High Energy Accelerators*, Wiley, New York, 1993.
8. Cornelis, K., "Resonant Behaviour of Head–Tail Modes", in *Particle Accelerator Conference*, New York, 1999, edited by A. Luccio and W.W. MacKay, IEEE Computer Soc. Press, Piscataway, 1999. Also CERN SL-99-029 OP.
9. The PEP Group, "Comparison Between Experimental and Theoretical Results for the Fast Head–Tail Instability in PEP", in *12th Int. Conf. on High Energy Accelerators*, Fermilab, 1983, edited by F.T. Cole and R. Donaldson, Fermilab, Batavia, 1984, p. 209.
10. Chin, Y.H., Transverse Mode Coupling Instabilities in the SPS, CERN/SPS/85-2 (DI-MST).
11. Myers, S., "Stabilization of the Fast Head–Tail Instability by Feedback", in *Particle Accelerator Conference*, Washington, 1987, *IEEE Trans. Nucl. Sci.* **NS-35**, 503 (1987).
12. Danilov, V., and Perevedentrev, E., Feedback System for Elimination of the Transverse Mode Coupling Instability, CERN SL/93-38 (AP).
13. Cappi, R., Métral, E., and Métral, G., "Beam Break Up Instability in the CERN PS Near Transition", in *7th European Particle Accelerator Conference*, Vienna, 2000. Also CERN/PS 2000-017 (AE).
14. Yokoya, K., Cumulative Beam Break-Up in Large Scale Linacs, DESY 86-084.

15. Brandt, D., Gareyte, J., "Fast Instability of Positron Bunches in the SPS", in *1st European Particle Accelerator Conference*, Rome, 1988, World Scientific, Singapore, 1989, p. 690.
16. Gareyte, J., "Observation and Correction of Instabilites in Circular Accelerators", in *Frontiers of Particle Beams: Intensity Limitations*, Hilton Head Island, South Carolina, 1990, US–CERN School on Particle Accelerators, edited by M. Dienes, M. Month and S. Turner, Lecture Notes in Physics 400, Springer, Berlin, 1992.

Polarized Beams in Accelerators and Storage Rings

Yu.M. Shatunov

*Budker Institute of Nuclear Physics,
Novosibirsk 630090, Russia*

Abstract. A general approach to the analysis of spin motion in realistic accelerator fields is presented. A method of finding the "spin closed orbit" and spin tune in arbitrary field configurations is given. It is shown that spin-orbital resonances perturb the spin motion and can cause depolarization. Methods of safe resonance crossing for protons are presented, including schemes with a few Siberian snakes and spin rotators in the rings. The effects of radiation on the spin of electrons are considered and illustrated by experiences in electron storage rings in the energy range from 500 MeV to 60 GeV.

INTRODUCTION

Spin was discovered in atomic experiments and is now an important part of investigations in nuclear and high energy physics. Spin belongs to the fundamental properties of particles, as well as their mass and electric charge. Cross sections of particle interactions depend on spin variables. During the last two decades physicists have shown an increasing interest in experiments with polarized particles in the whole energy range. Some old accelerators have been upgraded for polarization techniques. Most new generation machines are designed and constructed with consideration of beam polarization.

The object of my lecture is to outline spin problems that are of interest and concern to accelerator physicists, and possible ways of overcoming these problems.

To consider spin behavior in accelerator fields, we have to write Dirac's equation and solve it. Unfortunately, this absolutely exact approach encounters technical problems, and examples of its successful application are few. But there is another way to describe spin motion, which is based on the dynamical properties of spin. As we know from non-relativistic quantum mechanics, spin motion in any magnetic field is a precession due to interaction of a magnetic moment, associated with spin, with this field. Generalization of this approach to the relativistic case leads us to the so-called BMT equation describing spin motion in arbitrary electric and magnetic fields.

I GENERAL SPIN DYNAMICS

A Spin and Magnetic Moment

We begin our consideration with regular quantum mechanics, remembering that it describes spin by a two-component wave function (spinor) [1],

$$\Psi = \begin{pmatrix} f \\ g \end{pmatrix}, \qquad \Psi^\dagger = \begin{pmatrix} f^* & g^* \end{pmatrix},$$

that satisfies the Schrödinger equation[1]:

$$i\dot{\Psi} = \hat{H}\Psi.$$

In the important case of spin 1/2, the Hamiltonian operator $\hat{H}$ for spin in the magnetic field $\vec{B}$ is

$$\hat{H} = \frac{1}{2}q\left(\hat{\vec{\sigma}} \cdot \vec{B}\right),$$

where $\hat{\vec{\sigma}}$ is the spin operator with components (Pauli's matrices)

$$\sigma_x = \begin{pmatrix} 0 & 1 \\ 1 & 0 \end{pmatrix} \qquad \sigma_y = \begin{pmatrix} 0 & -i \\ i & 0 \end{pmatrix} \qquad \sigma_z = \begin{pmatrix} 1 & 0 \\ 0 & -1 \end{pmatrix}, \qquad (1)$$

and q is the so-called *gyromagnetic ratio*.

It is well known that the solution is a precession of the spin vector $\vec{S} = \langle\Psi^\dagger|\hat{\vec{\sigma}}|\Psi\rangle$ (the quantum-mechanical average of the spin operator $\hat{\vec{\sigma}}$), around the magnetic field direction ($\vec{B}\|z$) with frequency $\vec{\Omega} = -q\vec{B}$, ($S_z = |f|^2 - |g|^2 = const$).

This precession is caused by interaction of the magnetic moment associated with spin, $\vec{\mu} = q \cdot \vec{S} = (q_0 + q') \cdot \vec{S}$, with the magnetic field $\vec{B}$. In classical mechanics the same precession is described by the equation

$$\dot{\vec{S}} = \left[\vec{\mu} \times \vec{B}\right] = \left[\vec{\Omega} \times \vec{S}\right]. \qquad (2)$$

The *normal* part $q_0 = e/m_0$ of the gyromagnetic ratio $q = q_0 + q'$ reflects the spinning of the electric charge e of the particle (m_0 being its rest mass). The *anomalous* part q' appears due to the particle's interaction with the physical vacuum. The quantity $a = q'/q_0 = (g-2)/2$ is called the magnetic moment anomaly.[2] It is a fundamental property of each type of particle as well as its mass. Measurement of these anomalies is a special chapter of modern spin physics. Table 1 presents experimental results [2] for some "important" (for accelerators) particles.

---

[1] Here and later we will use the units $\hbar = c = 1$.
[2] We used the Lande factor $g = \mu/\mu_0$; here $\mu_0 = \frac{1}{2}q_0$ is Bohr's magneton.

TABLE 1. Magnetic moment anomalies

particle	anomaly $a = q'/q_0$	accuracy
$e^{\pm}$, electron/positron	$1.159652193 \times 10^{-3}$	$\pm 1. \times 10^{-11}$
$\mu^{\pm}$, muons	1.1659230×10^{-3}	$\pm 8.4 \times 10^{-6}$
p, proton	1.79284739	$\pm 6.3 \times 10^{-8}$
d, deuteron	-0.1429878	$\pm 5 \times 10^{-7}$

B BMT Equation

There are a few ways to generalize Eq. (2) to the relativistic case (see for example [3]). We will try to give here a more visual derivation based only on Lorentz transformations. Following Thomas [4], we consider a relativistic particle moving with velocity $\vec{V}$ in the magnetic and electric fields $\vec{B}, \vec{E}$ along a trajectory which is determined by the equation of motion:

$$\dot{\vec{V}} = -\frac{q_0}{\gamma}\left(\vec{E} + \left[\vec{V} \times \vec{B}\right]\right) . \qquad (3)$$

We take in the laboratory frame ("L") at time $t = 0$ two other frames: the rest frame of the particle ("C"-frame) and an inertial frame "I," moving with the particle velocity $\vec{V}(t = 0)$ and coinciding at $t = 0$ with "C," see Fig. 1.

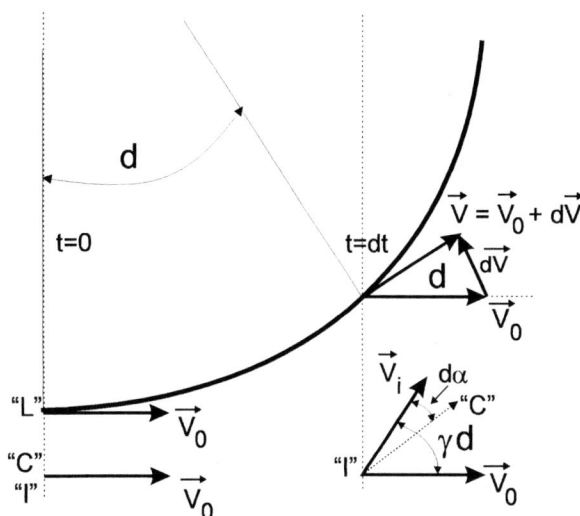

FIGURE 1. Relativistic transformation of the rest frame "C."

At time $t = dt$ the particle will rotate by an angle

$$d\vec{\alpha} = \frac{\left[d\vec{V} \times \vec{V}\right]}{V^2} = -\frac{q_0}{\gamma}\left(\vec{B}_{\perp} + \frac{\left[\vec{E} \times \vec{V}\right]}{V^2}\right) \cdot dt .$$

After Lorentz transformation with the relativistic factor γ into the moving frame we find the magnetic field $\vec{B}_C = \gamma \left(\vec{B}_\perp + [\vec{E} \times \vec{V}] \right) + \vec{B}_\parallel$, where $\vec{B}_\perp$, $\vec{B}_\parallel$ are the components of the laboratory magnetic field $\vec{B}$, transverse and parallel to $\vec{V}$. The spin changes during the time $d\tau = dt/\gamma$, according to Eq. (2), by

$$d\vec{S}_I = -q \left[\vec{B}_C \times \vec{S} \right] \cdot d\tau .$$

To find the spin shift in the rest frame, we have to take into account that the "C" frame itself rotates relatively to the inertial frame by an angle $d\vec{\phi}$. Subtracting this rotation we get

$$d\vec{S} = (d\vec{S})_I - \left[d\vec{\phi} \times \vec{S} \right] .$$

The angle $d\vec{\phi}$ can be found from the following simple arguments. First, it is obvious that at time dt the "old" rest frame has the angle $-d\vec{\alpha}$ versus the "new" velocity $\vec{V} + d\vec{V}$. Second, in the moving frame both these directions are rotating (same as the "new" velocity) γ times faster than in the laboratory frame. Thus $d\vec{\phi} = \gamma d\vec{\alpha} - d\vec{\alpha} = (\gamma - 1) d\vec{\alpha}$ and we come to

$$d\vec{S} = \left(-\frac{q}{\gamma} \left[\vec{B}_C \times \vec{S} \right] - \frac{\gamma - 1}{V^2} \left[[\dot{\vec{V}} \times \vec{V}] \times \vec{S} \right] \right) \cdot dt . \qquad (4)$$

This expression contains two terms: the first one is responsible for the direct spin rotation by the fields; the second one is due to relativistic kinematics (the so-called Thomas precession).

Substitution of Eq. (3) and $\vec{B}_C$ into Eq. (4) after simple operations yields finally the BMT equation [5]:

$$\frac{d\vec{S}}{dt} = \dot{\vec{S}} = \left[\vec{W} \times \vec{S} \right]$$

$$-\vec{W} = \left(\frac{q_0}{\gamma} + q' \right) \vec{B}_\perp + \frac{q}{\gamma} \vec{B}_\parallel + \left(\frac{q_0}{\gamma + 1} + q' \right) \left[\vec{E} \times \vec{V} \right] . \qquad (5)$$

Let us do a brief analysis of this equation.

1. First of all it is necessary to emphasize that the time and fields in this equation are taken in the laboratory frame, meanwhile the spin vector $\vec{S}$ belongs to the particle rest frame (only there does spin make sense, because it is an internal degree of freedom of the particle). Vector $\vec{W}$ has the meaning of spin precession frequency. As we see, the contributions of the normal and anomalous parts of the magnetic moment to the precession are quite different. Comparison of the vector $\vec{W}$ with the particle revolution frequency (the Larmor frequency can be easily obtained from Eq. (3)),

$$-\dot{\vec{\Omega}}_L = \frac{q_0}{\gamma} \vec{B} + \frac{\gamma q_0}{\gamma^2 - 1} \left[\vec{E} \times \vec{V} \right] , \qquad (6)$$

shows that the existence of Thomas precession makes the spin motion and our possibilities to control it more versatile.

2. In the particular case without an electric field ($\vec{E} = 0$), the spin of a particle without anomaly ($q' = 0$) will move similarly to the particle velocity (the angle $\widehat{\vec{S}\vec{V}} = const$). The situation closest to this appears for low energy electrons ($\gamma \simeq 1$), due to the smallness of the anomaly ($a = q'/q_0 \sim 10^{-3}$). To control the spin vector in polarized electron sources (PES), there is only one way: to use combinations of transverse electric and longitudinal magnetic fields.

    As an example, Fig. 2 shows a schematic of the Z-spin manipulator for 100-keV electrons used in the Amsterdam PES [6]. Originally longitudinal polarization of electrons is rotated by $\pm 90°$ in the horizontal plane by two electrostatic bends ($\alpha = \pm 107.7°$), and by a set of solenoids after each bend, to an arbitrary direction, as required.

3. A nice example of analyzing the BMT equation for practical applications is provided by the muon ($g - 2$) experiments [7,8]. When the muons move in an ultimately homogeneous transverse magnetic field, there is a "magic" value of γ at which the electrostatic focusing applied for muon motion stability does not affect the spin precession. Subtracting $\vec{\Omega}_L$ from $\vec{W}$, we get the relative spin precession:

$$\vec{w} = q'\vec{B} + q_0 \left(a_\mu - \frac{1}{\gamma^2 - 1} \right) \left[\vec{E} \times \vec{V} \right].$$

With a_μ from Table 1, at the "magic" $\gamma \simeq 29.304$, the term with the electric field is canceled, and spin will precess relatively to the velocity only because of the magnetic moment anomaly.

Recently it was proposed [9] to apply a radial dipole electric field together with the guiding magnetic field to fully compensate the anomalous precession ($\vec{w} = 0$) at any muon energy. In principle, in this particular case we can measure a spin deflection caused by the muon electric dipole moment, if it exists.

4. For intermediate electron energies ($\gamma \sim 10^3$ to 10^4), the application of electric fields is restricted by practical reasons, and combinations of longitudinal and transverse magnetic fields become more suitable for controlling the direction of polarization.

    Figure 3 shows a spin rotator designed for the VEPP-4M storage ring to obtain longitudinal polarization at the interaction point (IP) after 90° rotations: first from the vertical direction to the radial one by the solenoids, then in the horizontal plane by a bending magnet ($\alpha = 90°/\gamma a$). The vertical polarization has to be restored after the IP by reverse-order manipulations.

5. Spin manipulations for high energy electrons ($\gamma a \gg 1$) are much easier to perform by a sequence of magnets with transverse fields. For example, Fig. 4 shows schematically the side and top views of the "mini spin rotator" for the electron-proton collider HERA.

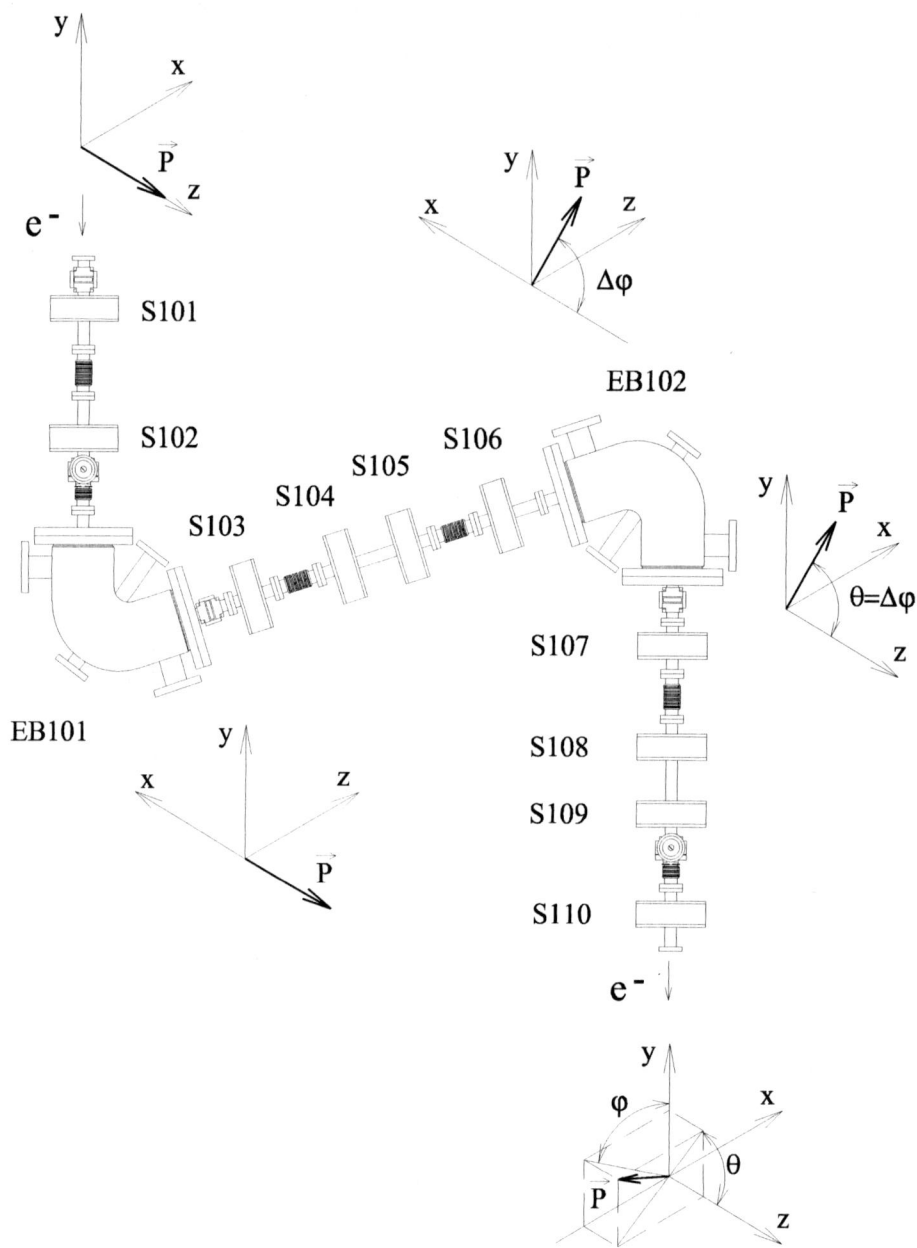

FIGURE 2. Spin manipulator for 100-keV polarized electrons

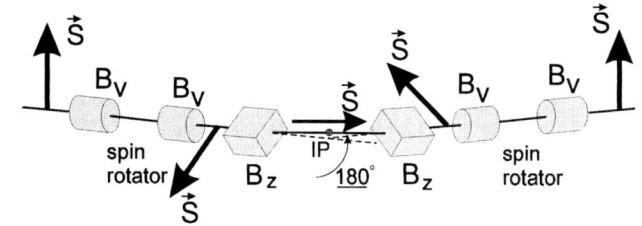

FIGURE 3. Spin rotator for VEPP-4M ($E = 4.7$ GeV).

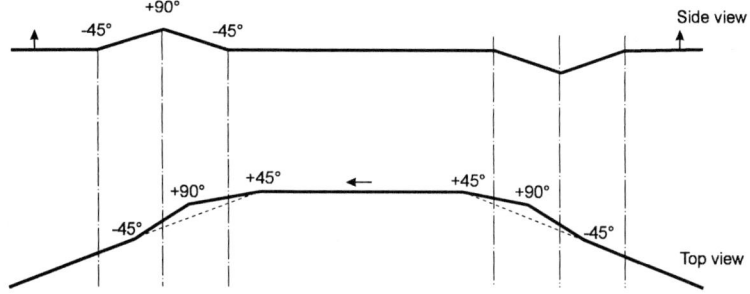

FIGURE 4. Scheme of spin rotator for HERA electron ring.

On both sides of the IP, the electron ring of HERA has chains of mechanically movable magnets with horizontal and vertical fields, which provide the "right" spin orientation: longitudinal in the IP and vertical in the arcs [10]. We shall discuss later, why spin has to be vertical in arcs.

Since the proton's anomaly is not small ($a_p \simeq 1.79$), control of proton spin can be practically accomplished only by transverse fields in the whole energy range.

II "SPIN'S CLOSED ORBIT" IN ACCELERATORS

Before considering spin behavior in circular machines, let us recall the main features of particle orbital motion. The first is the existence of the periodic closed orbit (CO), $\vec{R}_s(\theta + 2\pi) = \vec{R}_s(\theta)$. Small deviations from the CO,

$$\vec{r} = \vec{R} - \vec{R}_s = x\vec{e}_x + z\vec{e}_z, \qquad |x|, |z| \ll |\vec{R}_s|,$$

are solutions of the corresponding equations of betatron and synchrotron oscillations with characteristic frequencies ν_x, ν_z, ν_s in an accelerator orthonormal frame: $\vec{e}_x$, $\vec{e}_y = \vec{V}/V$, $\vec{e}_z$ [11].

Following the orbital motion, we shall look at spin vector behavior, not in time but along the generalized azimuth θ of the closed orbit. Moreover, we can split the spin precession frequency, Eq. (5), into two parts: $\vec{W} = \vec{W}_0 + \vec{w}$, where $\vec{W}_0$ results from fields on the CO, $\vec{W}_0(\theta + 2\pi) = \vec{W}_0(\theta)$; meanwhile fields appearing in $\vec{w}$ are due only to particles' deviations from $\vec{R}_s$, ($|\vec{w}| \ll |\vec{W}_0|$) [15],

$$\frac{d\vec{S}}{d\theta} \equiv \vec{S}' = \left[\left(\vec{W}_0 + \vec{w}\right) \times \vec{S}\right]. \tag{7}$$

After that, it is not difficult to write components of the periodic part of the precession frequency, Eq. (5), in the accelerator frame[3]:

$$\begin{align}
(W_0)_x &= \nu_0 K_x, & K_x &= \frac{B_x}{B_0}, \\
(W_0)_y &= (1+a) K_y, & K_y &= \frac{B_y}{B_0}, \\
(W_0)_z &= \nu_0 K_z, & K_z &= \frac{B_z}{B_0},
\end{align} \tag{8}$$

where we introduced the notation $\nu_0 = \gamma q'/q_0 = \gamma a$.

Any accelerator is a piece-wise sequence of N arbitrary local fields. As we know, the spin transformation through each piece is a rotation by an angle ϕ_i around the axis $\vec{n}_i = \vec{B}_i/B_i$, which can be carried out by the orthogonal O(3) or the unitary SU(2) matrices. The latter formalism is more compact. A 2×2 unitary transport matrix has the form

$$M = \begin{pmatrix} m_{11} & m_{12} \\ -m_{12}^* & m_{11}^* \end{pmatrix}. \tag{9}$$

The matrix elements are expressed in the general case via the angle ϕ_i and axis $\vec{n}_i$, of rotation:

$$m_{11} = \cos\frac{\phi_i}{2} - i(\vec{n}_i)_z \sin\frac{\phi_i}{2}, \qquad m_{12} = \left((\vec{n}_i)_x - i(\vec{n}_i)_y\right) \sin\frac{\phi_i}{2}. \tag{10}$$

After one particle turn a spin map (from θ_i to $\theta_i + 2\pi$) is the product of the local spin rotations that can be presented in a form similar to Eq. (9):

$$T = M_N \cdot M_{N-1} \cdot \cdots \cdot M_2 \cdot M_1 = M_i = I \cdot \cos\frac{\phi}{2} - i(\vec{n}_s \cdot \vec{\sigma}) \cdot \sin\frac{\phi}{2}, \tag{11}$$

where $\vec{\sigma}$ has Pauli matrices, Eq. (1), as its components, and I is the unit matrix.

This means (see Fig. 5) that we have replaced the sum of all local spin rotations by one rotation through an angle ϕ around an axis $\vec{n}_s$. At the next particle turn this procedure will proceed, because spin moves in the same fields, and so on. Thus, vector $\vec{n}_s$ has the sense of a periodic axis of spin rotations, $\vec{n}_s(\theta_i + 2\pi) = \vec{n}_s(\theta_i)$; and $\nu = \phi/2\pi$ gives the

---

[3] Here and later we use dimensionless units: magnetic fields are normalized to the average bending field $B_0 = 1/2\pi \oint B_z d\theta$; lengths are measured in units of the mean radius R_0; the time unit is the particle revolution time $\Delta t = 2\pi R_0/V$.

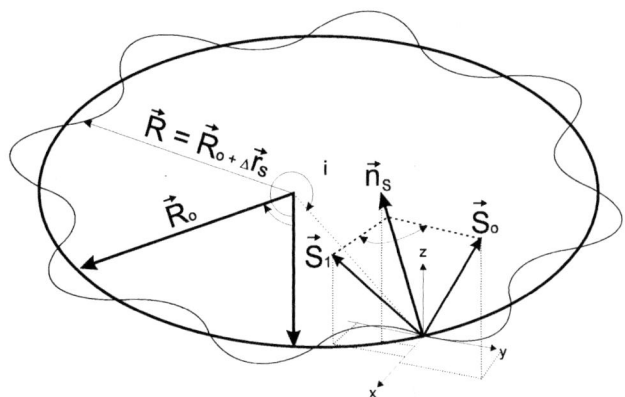

FIGURE 5. Spin rotation at $\theta = \theta_i$ after one particle turn.

number of spin rotations per particle's turn, or the spin tune. On the other hand, now it is clear that the vector $\vec{n}_s(\theta)$ is a periodic solution of the BMT equation, Eq. (5), on the closed orbit $\vec{R}_s(\theta)$: $\vec{n}'_s = [\vec{W}_0 \times \vec{n}_s]$. Thus we can call $\vec{n}_s$ the spin's closed orbit.

From Eq. (11) it is easy to get methods of finding $\vec{n}_s$ and the spin tune ν:

$$\vec{n}_s = \frac{i}{2\sin\pi\nu} \operatorname{tr}(\vec{\sigma} \cdot T); \qquad \cos\pi\nu = \frac{1}{2}\operatorname{tr}T. \qquad (12)$$

Two other orthogonal eigen-solutions of the BMT equation, η_1 and η_2, rotate with frequency ν in the plane perpendicular to the $\vec{n}_s$-vector and satisfy the periodicity conditions:

$$\vec{\eta}_{1,2}(\theta + 2\pi) = e^{2\pi i \nu}\vec{\eta}_{1,2}(\theta). \qquad (13)$$

Defining a complex vector $\vec{\eta} = \vec{\eta}_1 - i\vec{\eta}_2$, we can decompose any spin via the eigenvectors:

$$\vec{S} = S_n\vec{n}_s + \operatorname{Re}(iS_\perp\vec{\eta}^*), \qquad (14)$$

where $S_n = (\vec{S} \cdot \vec{n}_s) = const$ is an integral of motion, and $S_\perp^2 = 1 - S_n^2$.

Sometimes, instead of the rotating frame $\vec{n}_s, \vec{\eta}_1, \vec{\eta}_2$, it is more convenient to use periodic vectors $\vec{n}_s, \vec{e}_1, \vec{e}_2$, satisfying the relation $\vec{e} = \vec{e}_1 - i\vec{e}_2 = \vec{\eta}e^{-i\nu\theta}$. Spin precession in this frame looks utterly simple: $\vec{S}' = \nu[\vec{n}_s \times \vec{S}]$.

1. To illustrate the technique described above, let us give a few examples of finding the "spin vectors" $\vec{n}_s, \vec{\eta}, \vec{e}$, and the spin tune, starting with an ideal accelerator.

   In an ideal flat machine (see Fig. 6), a guiding field is everywhere perpendicular to the plane of the closed orbit:

   $$\vec{B} = K_z B_0 \cdot \vec{e}_z.$$

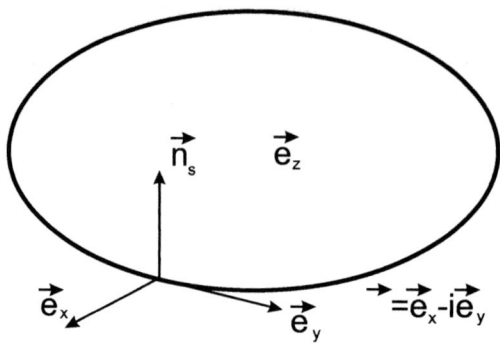

FIGURE 6. Precession axis $\vec{n}_s$ in an ideal accelerator.

We assume also that there is no coupling of the x and z oscillations. Hence, the periodic precession is

$$\vec{W}_0 = \nu_0 K_z \cdot \vec{e}_z,$$

and the spin mapping matrix is

$$M_z = V(2\pi\nu_0) = \begin{pmatrix} e^{-i\pi\nu_0} & 0 \\ 0 & e^{i\pi\nu_0} \end{pmatrix}.$$

From here, using Eqs. (1) and (12), we easily find

$$\vec{n}_s = \vec{e}_z; \qquad \nu = \nu_0 = \gamma a. \qquad (15)$$

The perpendicular solution $\vec{\eta}$ is evident:

$$\vec{\eta} = (\vec{e}_x - i\vec{e}_y)e^{i\nu_0\tilde{\theta}},$$

where we denote the bend of the velocity vector $\vec{V}$ on the way from 0 to azimuth θ by $\tilde{\theta} = \int_0^\theta K_z d\theta$. Similarly, for the periodic vector,

$$\vec{e} = (\vec{e}_x - i\vec{e}_y)e^{i\nu_0(\tilde{\theta}-\theta)}.$$

We would like to emphasize in Eq. (15) the remarkable ratio between the spin tune and the particle energy in a flat machine. This means that experimental measurement of the spin tune gives a calibration of the absolute particle energy, because the magnetic anomalies and particle's masses are known with extremely high accuracy (see Table 1). This method was developed at the VEPP-2M storage ring [12,13] and today it is applied widely in accelerator practice. A number of precise experiments have been done around the world based on this method of energy calibration [14,42].

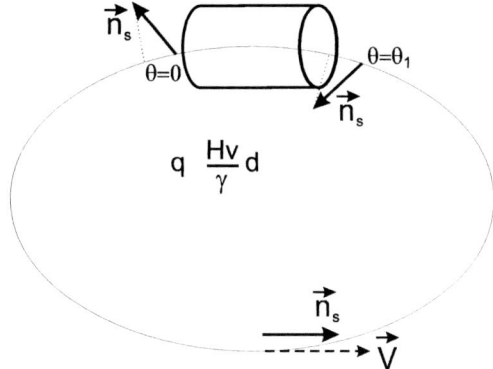

FIGURE 7. Precession axis $\vec{n}_s$ in a flat machine with one solenoid.

2. Now we insert, into any straight section $\tilde{\theta}_1 < \tilde{\theta} < 2\pi$ of an ideal accelerator, a solenoid with longitudinal field B_y that rotates the spin by an angle $\phi = (q/\gamma) \int B_y d\theta$, and we suppress the induced coupling by some compensation scheme.

In this case a full spin map for a point $\tilde{\theta}_1$ (see Fig. 7) will be

$$T_{\tilde{\theta}_1} = V(2\pi\nu_0) \cdot M_y(\phi) = \begin{pmatrix} e^{-i\pi\nu_0} \cdot \cos\phi/2 & e^{-i\pi\nu_0} \cdot \sin\phi/2 \\ -e^{i\pi\nu_0} \cdot \sin\phi/2 & e^{i\pi\nu_0} \cdot \cos\phi/2 \end{pmatrix}, \quad (16)$$

from which we get the components of vector $\vec{n}_s$ at point $\tilde{\theta}_1$:

$$n_x = \frac{\sin\phi/2 \cdot \sin\pi\nu_0}{\sin\pi\nu}; \quad n_y = \frac{\sin\phi/2 \cdot \cos\pi\nu_0}{\sin\pi\nu}; \quad n_z = \frac{\cos\phi/2 \cdot \sin\pi\nu_0}{\sin\pi\nu};$$

and the spin tune

$$\cos\pi\nu = \cos\phi/2 \cdot \cos\pi\nu_0. \quad (17)$$

We can see that the solenoid shifts the spin tune and tilts $\vec{n}_s$ from the vertical direction. The most interesting situation, shown in Fig. 7, arises when the rotation angle $\phi = \pi$. In this particular case ($\cos\pi\nu = 0$, $n_y = 1$), the periodic solution n_s always lies in the CO plane; it is longitudinal along the straight section opposite the solenoid insertion; the spin tune $\nu = 1/2$, independently of the particle energy.

An orthogonal solution $\vec{\eta}$ outside the solenoid ($0 < \tilde{\theta} < \tilde{\theta}_1$) is composed of

$$\vec{\eta}_x = -\cos\left(\pi - \tilde{\theta}\right)\nu_0, \quad \vec{\eta}_y = \sin\left(\pi - \tilde{\theta}\right)\nu_0, \quad \vec{\eta}_z = -i, \quad (18)$$

and $\vec{\eta}(\theta + 2\pi) = -\vec{\eta}(\theta)$.

A similar situation will arise, if the spin is flipped by a rotator around an axis at an angle α to the velocity. A unitary matrix of this transformation is given by

$$S_\alpha = \begin{pmatrix} 0 & -ie^{-i\alpha} \\ ie^{i\alpha} & 0 \end{pmatrix}.$$

Such rotators have been given a special name "Siberian snakes" in honor of the place of their invention [17].

3. It is interesting to consider a scheme with two Siberian snakes in a ring, see Fig. 8. Let us add a second Siberian snake at the azimuth $\tilde{\theta} = \tilde{\theta}_1 + \pi$, with an axis perpendicular to that of the first snake, $\alpha_2 = \alpha_1 \pm \pi/2$.

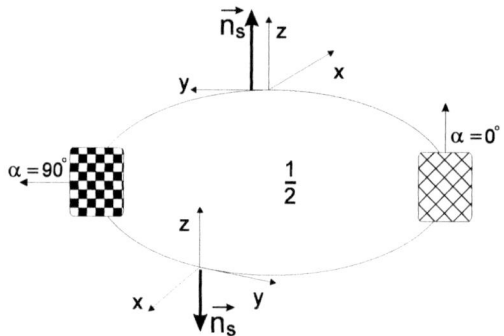

FIGURE 8. Two snakes in an ideal accelerator.

The total spin transport matrix for point θ_1 (before the first snake) is

$$T_{\theta_1} = V(\pi\nu_0) \cdot S_{\alpha_2} \cdot V(\pi\nu_0) \cdot S_{\alpha_1} = \begin{pmatrix} e^{\pm i\pi/2} & 0 \\ 0 & e^{\mp i\pi/2} \end{pmatrix}. \quad (19)$$

From Eq. (12) it is easy to get $\cos \pi\nu = 0$; $(\vec{n}_s)_z = 1$. As in the previous case, here the spin tune is 1/2, but $\vec{n}_s$ is vertical in both arcs and changes its sign while passing each snake. From a general point of view it is clear that this configuration is more stable against spin distortions.

4. The matrix technique is able to generalize the insertion of an arbitrary even number of Siberian snakes in a flat machine (flat except snake areas). We install $2N$ snakes, see Fig. 9, on azimuths $\tilde{\theta}_1, \tilde{\theta}_2, \ldots, \tilde{\theta}_{2N} = 2\pi$ and denote the axis direction of the ith snake by α_i.

As always

$$T_{\tilde{\theta}=0} = S(\alpha_{2N})V(\nu_0\Delta\tilde{\theta}_{2N})\ldots S(\alpha_2)V(\nu_0\Delta\tilde{\theta}_2)S(\alpha_1)V(\nu_0\Delta\tilde{\theta}_1), \quad (20)$$

where $\Delta\tilde{\theta}_i = \tilde{\theta}_i - \tilde{\theta}_{i-1}$ is the velocity rotation between the ith and $(i-1)$th snakes ($\Delta\tilde{\theta}_1 = \tilde{\theta}_1$).

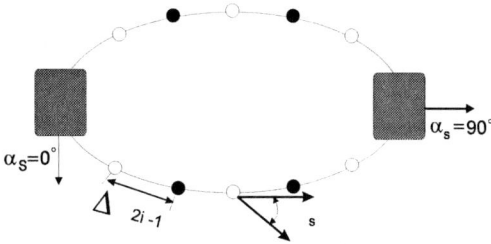

FIGURE 9. $2N$ snakes in an ideal accelerator.

If we note that a spin flip by any snake $S(\alpha_i)$ is equivalent to a combination of a rotation around the vertical axis by $2\alpha_i$, and around the velocity by π, ($S(\alpha_i) = S(0)V(2\alpha_i)$); $M \cdot M = -I$ and $V(\phi)S(0) = S(0)V(-\alpha)$, then after some algebra we derive

$$T_{\tilde{\theta}=0} = (-1)^N V(-2\alpha_{2N})V(-\nu_0\Delta\tilde{\theta}_{2N})...V(-2\alpha_2)V(-\nu_0\Delta\tilde{\theta}_2)V(2\alpha_1)V(\nu_0\Delta\tilde{\theta}_1).$$

Therefore, a one-turn spin map simplifies to rotations only around the vertical axis, and finally $T_{\tilde{\theta}=0} = V(\mu)$, where

$$\mu = 2\sum_{i=1}^{N}(\alpha_{2i-1} - \alpha_{2i}) + \nu_0 \sum_{i=1}^{N}(\Delta\tilde{\theta}_{2i-1} - \Delta\tilde{\theta}_{2i}) \tag{21}$$

is the spin phase advance.

From Eq. (21) we can conclude that uniform snake positioning

$$\sum_{i=1}^{N}(\Delta\tilde{\theta}_{2i-1} - \Delta\tilde{\theta}_{2i}) = 0$$

provides the spin tune independency of the particle energy. Moreover, the spin tune itself can be controlled by a choice of the first sum in Eq. (21). Particularly, when

$$\sum_{i=1}^{N}(\alpha_{2i-1} - \alpha_{2i}) = \frac{\pi}{2},$$

we again have $\nu = \frac{1}{2}$.

III SPIN RESONANCES

A Spin Perturbation

As we discussed in the previous section, spin precessions of off-closed orbit particles can be split into two parts, $\vec{W} = \vec{W}_0 + \vec{w}$. Since $\vec{w}$ is induced by small particle deviations from the CO, ($|\vec{w}| \ll |\vec{W}_0|$), we can consider this part of the precession only as a perturbation and apply appropriate methods [16].

To do that, first of all, we should derive correct formulas for components of the perturbation. As seen from Eq. (5), the spin precession has a simpler form in the frame of orthogonal unit vectors, related to the total velocity $\vec{V} = \vec{R}'_s + x'\vec{e}_x + z'\vec{e}_z$:

$$\vec{a}_1 = \frac{[\vec{V} \times \vec{e}_z]}{|[\vec{V} \times \vec{e}_z]|} \simeq \vec{e}_x - x'\vec{e}_y,$$

$$\vec{a}_2 = \frac{\vec{V}}{|\vec{V}|} \simeq \vec{e}_y + x'\vec{e}_x + z'\vec{e}_z,$$

$$\vec{a}_3 = [\vec{a}_1 \times \vec{a}_2] \simeq \vec{e}_z - z'\vec{e}_y, \qquad (22)$$

which is slightly different from that in the accelerator frame $\vec{e}_x$, $\vec{e}_y$, $\vec{e}_z$.[4] Common consideration of the spin and orbital equations of motion in the linear approximation yields expressions for the components of spin perturbation:

$$w_1 = \vec{w} \cdot \vec{a}_1 = \nu_0 z'' + (\nu_0 + \tfrac{a}{\gamma})K_x\tfrac{\Delta\gamma}{\gamma} + (1+a)K_y x',$$

$$w_2 = \vec{w} \cdot \vec{a}_2 = (1+a)(K'_x x + K'_z z + \Delta K_y - K_y \frac{\Delta\gamma}{\gamma}) + a(K_x x' + K_z z'),$$

$$w_3 = \vec{w} \cdot \vec{a}_3 = -\nu_0 x'' + (\nu_0 + \tfrac{a}{\gamma})K_z\tfrac{\Delta\gamma}{\gamma} + (1+a)K_y z', \qquad (23)$$

where x, z (and their derivatives) are solutions of the orbital equations that already include fields and their gradients on the closed orbit, field errors $\Delta K_{x,z}$, and the linear coupling. From these expressions it is clear that, for high energies ($\nu_0 \gg 1$), a contribution of w_2 can be neglected, compared with w_1 and w_3.

From a general point of view, we can suppose that, similarly to one periodic $\vec{n}_s$, for an off-orbit particle there is a solution of the BMT equation which depends on a particle's phase-space coordinates and is periodic with all phases [18]:

$$\vec{n}\left(x, x', z, z', \frac{\Delta\gamma}{\gamma}, \theta + 2\pi\right) = \vec{n}\left(x, x', z, z', \frac{\Delta\gamma}{\gamma}, \theta\right). \qquad (24)$$

A spread of $\vec{n}$ around $\vec{n}_s$ results in a decrease of the visible polarization $\langle \vec{n} \cdot \vec{n}_s \rangle(\theta)$, caused by the beam dimensions.

Using Eq. (14) we should look for a vector $\vec{n}$ in the form

$$\vec{n} = \sqrt{1 - |C|^2}\, \vec{n}_s + \operatorname{Re}(iC\vec{\eta}^*), \qquad (25)$$

where C is a complex constant. Substitution of Eq. (25) into Eq. (5) leads to an equation for C:

$$C' = \sqrt{1 - |C|^2}\, w_\perp - i w_\| C; \quad w_\perp = (\vec{w} \cdot \vec{\eta}); \quad w_\| = (\vec{w} \cdot \vec{n}_s), \qquad (26)$$

and, with the assumption $|C| \ll 1$, in the first approximation this is truncated[5] to

$$C' = w_\perp. \qquad (27)$$

---

[4] In the new frame, spin precesses with a frequency $\vec{W}_a = \vec{W} - \vec{\Omega}_a$, where $\vec{\Omega}_a = 1/2 \sum [\vec{a}_i \times \vec{a}_i\,']$ is the frequency of the new frame rotations with respect to the laboratory frame.
[5] Obviously, in the first approximation $w_\|$ only modulates the spin tune and does not change $\vec{n}$ itself.

B Single-Resonance Model

In the case where the perturbation contains only one frequency ($w_\perp = w_k e^{i\delta\theta}$), Eq. (27) has an exact forced solution,

$$C = -i\frac{w_k}{\sqrt{\delta^2 + |w_k|^2}} e^{i\delta\theta}, \qquad (28)$$

where δ is the imbalance between the spin tune and the perturbation frequency. Substitution of $\hat{C} = Ce^{-i\delta\theta}$ performs the transition to the rotating frame, where spin precesses in a time-independent "field" $\vec{h}$ with frequency $h = \sqrt{\delta^2 + |w_k|^2}$, see Fig. 10a.

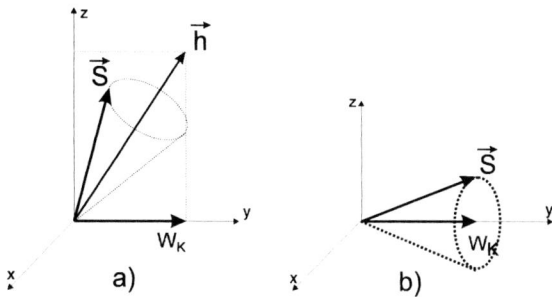

FIGURE 10. Spin precession in the resonant frame.

By a coincidence of the frequencies we have a spin resonance. On the resonance top $\delta = 0$, $|\hat{C}| = |C| = 1$, the vector $\vec{n}$ is perpendicular to $\vec{n}_s$ and the spin rotates around $\vec{n}$, with frequency w_k (Fig. 10b). This quantity is the Fourier harmonic of the perturbation:

$$w_k = \frac{1}{2\pi}\int_0^{2\pi}(\vec{w}\times\vec{\eta})e^{-i\delta\theta}\,d\theta, \qquad (29)$$

and is called the resonance strength.

Analysis in terms of a single resonance is a very visual and also a powerful method of studying spin behavior. It is enough to mention nuclear magnetic resonance, which today has wide practical applications. Moreover, the single-resonance approach works well in the case of more complicated spin perturbations. Indeed, in the vicinity of one Fourier harmonic w_k of the perturbation spectrum we can neglect the action of all other fast variables compared with the slowest one $\delta_k = \nu - \nu_k \leq |w_k|$. Thus, we can consider this resonance as isolated.

C Linear Resonances

In the general case with a perturbation $\vec{w}$, using Eq. (23) we can obtain from Eq. (27) that the deviation $\Delta\vec{n} = \vec{n} - \vec{n}_s$ can be presented in the form[6]

---

[6] Integration of Eq. (27) is carried out by a turn-by-turn summation with a small imaginary part added to frequencies of the integrand, which satisfies the periodicity condition similar to Eq. (13): $f(\theta + 2\pi) = e^{2\pi i \nu_i} f(\theta)$.

$$\Delta \vec{n} = \text{Im}\left(\vec{\eta}^{*} \int_{-\infty}^{\theta} w_{\perp}\, d\theta\right) = \text{Im}\left[\vec{\eta}^{*}\left(\sum_{i} \frac{A_{\pm\nu_i} I_{\pm\nu_i}(\theta)}{e^{2\pi i(\nu \pm \nu_k)} - 1} + \frac{I_0(\theta)}{e^{2\pi i \nu} - 1}\right)\right], \qquad (30)$$

where $A_{\pm\nu_i}$ and ν_i are amplitudes and frequencies of the orbital motion; $\nu_k = k \pm \nu_x \pm \nu_z \pm \nu_s$.
$I_{\pm\nu_i}$ and I_0 are the so-called spin-orbital integrals. The quantities $I_{\pm\nu_i}$ give contributions of the betatron and synchrotron oscillations to deviations $\Delta \vec{n} = \vec{n} - \vec{n}_s$; meanwhile I_0 denotes a correction of the periodic spin solution on the actual closed orbit, which is slightly different from the design one.

We see that deviation $\Delta \vec{n}$ increases nearby spin resonances[7]:

$$\nu = \nu_k = k \pm \nu_x \pm \nu_z \pm \nu_s \qquad (k \text{ is an integer}). \qquad (31)$$

For an ideal flat machine at high energy, where $\vec{n}_s$ is vertical, the spin perturbation

$$w_{\perp} = (w_1 - iw_2)e^{i\nu_0 \tilde{\theta}} = (\nu_0 z'' - i[(1+a)K_z' z + aK_z z'])e^{i\nu_0 \tilde{\theta}} \simeq \nu_0 z'' e^{i\nu_0 \tilde{\theta}} \qquad (32)$$

is driven only by "free" vertical oscillations

$$z = A_z f_z + A_z^{*} f_z^{*},$$

where f_z, f_z^{*} are the Floquet solutions of the homogeneous equation for the vertical deviations [11]. Here, only vertical betatron resonances $\nu = k \pm \nu_z$ can occur. The strengths of these intrinsic resonances can be found in the corresponding resonance frames, similar to Eq. (29):

$$w_k(+\nu_z) = A_z \frac{\nu_0}{2\pi} \oint g_z f_z e^{-i(k+\nu_z)\theta} d\theta; \qquad w_k(-\nu_z) = A_z^{*} \frac{\nu_0}{2\pi} \oint g_z f_z^{*} e^{-i(k-\nu_z)\theta} d\theta. \qquad (33)$$

As we see, the resonant harmonics are proportional to the amplitude of the betatron oscillations A_z, and grow with increasing energy. It is evident also that, if p is the periodicity of the focusing lattice g_z, then resonances with $k = l \cdot p$ are the strongest.

Most modern accelerators are approximately flat, but different kinds of imperfections excite other types of spin resonances. Let us give a brief analysis of possible resonances in an approximation of weak distortions for high energy machines.

Integer spin resonances $\nu = k$ (also called imperfection resonances) occur due to the existence of longitudinal and radial fields on the closed orbit. As seen from Eqs. (23) and (29), the resonant harmonics w_k of the spin resonances caused by the longitudinal field differ from the corresponding Fourier harmonics in the spectrum of K_y only by the factor q/q_0,

$$w_k = \frac{q}{2\pi q_0} \oint K_y e^{-ik\theta} d\theta. \qquad (34)$$

---

[7] We restrict our consideration to linear resonances. The effect of higher-order spin resonances $\nu_k = k + k_x \nu_x + k_z \nu_z$ on spin is much weaker, but requires more complicated mathematics.

The action of radial fields ($K_x \neq 0$) on spin is more complicated, because in addition to dipole kick by K_x, spin is affected by radial fields in quadrupole magnets, where the particle travels with forced deviations Z_s, caused by the same K_x,

$$Z_s = \frac{1}{2i}\left(f_z \int_{-\infty}^{\theta} K_x f_z^* d\theta - c.c.\right).$$

The strengths of the imperfection resonances $\nu = k$ from radial fields can be expressed in two ways: either via harmonics of Z_s'', after substitution of Z_s instead of z_s into Eq. (32):

$$w_k = \frac{\nu_0}{2\pi} \oint Z_s'' e^{-ik\theta} d\theta, \tag{35}$$

or, if we know K_x, directly via radial fields:

$$w_k = \frac{\nu_0}{2\pi} \oint K_x F_{\nu=k}(\theta) e^{-ik\theta} d\theta, \tag{36}$$

using the so-called spin response function [19]

$$F_\nu(\theta) = \nu_0 e^{i\nu_0\tilde{\theta}} \left(f_z^* \int_{-\infty}^{\theta} K_z f_z' e^{-i\nu_0\tilde{\theta}} d\theta - f_z \int_{-\infty}^{\theta} K_z (f_z^*)' e^{-i\nu_0\tilde{\theta}} d\theta \right).$$

Function $F_\nu(\theta)$ is a characteristic periodic function of an accelerator which describes the sensitivity of the $\vec{n}$ axis to vertical kicks: $\nu_o F_\nu(\theta) = |d\vec{n}/dz'|$. $|F_\nu(\theta)| \sim 1$, except in narrow resonance regions $\nu \approx k \pm \nu_z$, where it can grow strongly.

Other linear machine imperfections are skew gradients $g_{zx} = g_{zx} = g = dB_x/dx$ that are responsible for the coupling of vertical and radial motions. In the case of weak coupling $g \ll g_z$, we can express vertical deviations as a sum of free z and forced z_f oscillations:

$$Z = A_z f_z + A_x \Phi f_x + \frac{\Delta\gamma}{\gamma}\Psi_z + c.c., \tag{37}$$

where Φ and Ψ_z are periodic functions of the azimuth, which can be easy obtained from the inhomogeneous equation of vertical oscillations [11]:

$$\Phi = \frac{e^{i\nu_x\theta}}{2i}\left(f_z \int_{-\infty}^{\theta} f_z^* g \cdot f_x d\theta - f_z^* \int_{-\infty}^{\theta} f_z g \cdot f_x d\theta\right); \quad \Psi_z = \text{Im}\left(f_z \int_{-\infty}^{\theta} f_z^* g \psi_x d\theta\right).$$

Putting Z from Eq. (37) into Eq. (32), we will find that, besides $\nu = k \pm \nu_z$, the skew gradient coupling causes all other types of linear spin resonances. For example, the second term in Eq. (37) determines the strengths of radial betatron resonances $\nu = k \pm \nu_x$:

$$w_k(+\nu_x) = A_x \frac{\nu_0}{2\pi} \oint g f_x e^{-i(k+\nu_x)\theta} d\theta; \quad w_k(-\nu_x) = A_x^* \frac{\nu_0}{2\pi} \oint g f_x^* e^{-i(k-\nu_x)\theta} d\theta. \tag{38}$$

The vertical dispersion Ψ_z complicates the picture of integer resonances because, in the presence of a synchrotron oscillation of energy $\gamma = \gamma_0(1 + \Delta\gamma \cos \nu_s \theta)$, it generates two lines $\nu = k \pm \nu_s$ of equal strengths:

$$w_k(\pm\nu_s) = \frac{\Delta\gamma}{\gamma} \frac{\nu_0}{2\pi} \oint g\psi_x e^{-ik\theta} d\theta. \tag{39}$$

D Spin Resonance Crossing

The isolated resonance model allows us to obtain a correct answer to what happens with the spin while crossing the resonance. This question is crucial for the acceleration of polarized particles, when the spin tune is changed many times ($\nu_0 = \gamma q'/q_0$).

Froissart and Stora [20] found that, as a result of one resonance crossing with a detuning rate $\delta' = const$, the residual projection $S_n = (\vec{S} \cdot \vec{n})$ averaged over the spin phases is described by the formula

$$S_n = S_n(0)(2e^{-\pi|w_k|^2/(2\delta')} - 1). \tag{40}$$

The final result depends on the spin phase advance near the resonance top:

$$\Delta\phi = \int h(\theta)d\theta \simeq |w_k|^2/\delta'.$$

The polarization is preserved by adiabatic changing of parameters ($\Delta\phi \gg 1$), and the spin flips down together with the $\vec{n}$-axis. In the opposite case of fast crossing ($\Delta\phi \ll 1$), the spin tilts only slightly from an initial direction with a depolarization $\Delta S_n \simeq |w_k|^2/\delta'$. Polarization may be fully lost in an intermediate situation $\Delta\phi \simeq 1$.

The practically important case of periodic resonance crossing can be treated also in the single-resonance approximation [16]. Such a situation occurs, for instance, due to synchrotron oscillation of particle energy. Since synchrotron oscillations are relatively slow ($\nu_s \ll 1$), we can consider them as only a modulation of the spin tune.

Thus we have the spin perturbation

$$w_\perp = w_k e^{i\delta\theta}; \qquad w_\| = \nu_0 \frac{\Delta\gamma}{\gamma} \cos(\nu_s \theta) = \Delta_0 \cos(\nu_s \theta).$$

Substitution of $\hat{C} = C e^{i \sin(\nu_s \theta)\Delta_0/\nu_s}$ into Eq. (26) performs a transformation into a frame rotating with frequency $w_\|$ around $\vec{n}_0$. Fourier series expansion of the spin perturbation in the new frame yields

$$\hat{w}_\perp = w_k e^{i\delta\theta} e^{i \sin(\nu_s \theta)\Delta_0/\nu_s} = w_k e^{i\delta\theta} \sum_{m=-\infty}^{+\infty} J_m\left(\frac{\Delta_0}{\nu_s}\right) e^{im\nu_s \theta}, \tag{41}$$

where J_m are the Bessel functions. We can conclude from this expression that, instead of one resonance ν_k, there is a set of so-called modulation resonances with frequencies $\nu_k \pm m\nu_s$.

In the general case, the spectrum of the side-band resonances is complicated enough. Their strengths are proportional to the "parent" resonance and depend on the modulation index $\chi = \Delta_0/\nu_s$:

$$|w_k^m| = |w_k| J_m(\chi).$$

But in the case of small modulation indexes $\chi \ll 1$, the spectrum simplifies (see Fig. 11), and drops rapidly with the number m[8]:

$$|w_k^m| \simeq |w_k| \frac{\chi^m}{m!}.$$

If $|w_k^m| \ll \nu_s$, every modulation resonance can be considered separately from others, and the result of its crossing is given by Eq. (40).

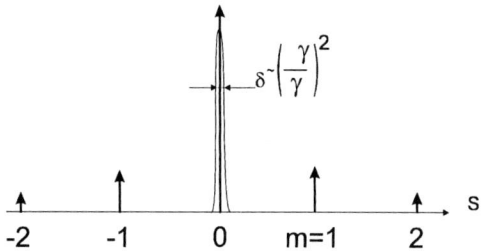

FIGURE 11. Synchrotron modulation resonances.

During acceleration, crossing of the whole area of the modulation resonances is unavoidable. If the fast crossing condition is satisfied on each side-band, then, taking account of the "conservation law" $\sum |w_k^m|^2 = |w_k^2|$, the total depolarization does not exceed $\Delta S_n \simeq |w_k|^2/\delta'$, as happens without any modulation.

Less predictable are polarization losses due to slow crossing when $|w_k| \sim \nu_s$. It is practically impossible to arrange values of δ', w_k, ν_s, to provide the adiabaticity condition for each line of the side-band spectrum. But fortunately there is a combined approach that helps maintain the polarization in this case too. Actually, positioning of the side-bands is determined by the formula

$$m\nu_s = \sqrt{\delta^2 + |w_k|^2}.$$

At $|w_k| \sim \nu_s$, lines with $m < |w_k|/\nu_s$ are absent, and we can optimize parameters in such a way as to cross the central resonance ($m = 0$) adiabatically, but meet the fast crossing condition for all possible side-bands.

The ideas of fast and adiabatic resonance crossing were successfully proved in 1975 at the electron-positron storage ring VEPP-2M [21]. It was shown that weak resonances like $\nu = \nu_z - 2$, with estimated strengths $w_k \sim 10^{-5}$, can be crossed without visible

---

[8] The modulation index χ grows with particle energy and achieves values $\chi \sim 1$ only in high energy electron storage rings ($E = 50$ to 100 GeV).

depolarization by a regular energy change ($\delta' \simeq 10^{-8}$). However, the polarization was lost on the resonances $\nu = 1$ and $\nu = 4 - \nu_z$, with resonant harmonics $w_k \geq 10^{-4}$ ($p = 4$). To overcome the imperfection resonance $\nu = 1$ adiabatically, a special 0.2 T×m solenoid was turned on near the resonance, then the resonant harmonic increased to $w_1 = 0.025$. Since at VEPP-2M $\nu_s \simeq 0.007$, the first possible side-band ($m = 5$) did not contribute to the depolarization ($w_k^5 < 10^{-6}$). As a result, on the resonance top a longitudinal polarization was obtained with a lifetime of about 200 seconds,[9] and full spin flip was measured after a ten-second crossing.[10]

IV ACCELERATION OF POLARIZED PROTONS

Next we shall survey an experimental situation for obtaining high energy polarized protons. First let us summarize the preceding section, and "draw" a picture of the resonances. For protons, the imperfection resonances repeat with the energy at a period of $\Delta E = 523.35$ MeV. On both sides of each imperfection resonance $\nu = k$ there is a pair of the betatron "satellites" $\nu = k \pm \nu_{z,x}$. Additionally, each mentioned resonance is surrounded by synchrotron side-bands. The highest "trees" in this resonance "forest" correspond to $k = l \cdot p$. According to Eqs. (33), (35), and (38), the strengths of the intrinsic and imperfection resonances grow as $\sqrt{\gamma}$, because the normalized emittance of the proton beam is conserved by acceleration. It is clear that without special efforts "safe" acceleration is a hopeless thing.

Historically, the first attempts to go through the resonance "forest" with polarized protons were undertaken in the middle 70s at ZGS, Argonne, where visible levels of beam polarization were achieved at energies up to 12 GeV [22]. Afterwards, polarized protons were accelerated successfully at Saclay [23], in Tsukuba [24] and in Jülich [25].

A AGS Results

A polarized proton facility (Fig. 12) is now in operation at Brookhaven National Laboratory [27].

First the polarization arises from a laser optically pumping rubidium vapors, then it passes by charge and spin exchange to 5-keV H$^-$ ions [26]. After that, the H$^-$ ions are delivered with a degree of polarization of $\simeq 80\%$ to the synchrotron AGS via the 2-MeV RFQ, 200-MeV linac, and 2.5-GeV booster ring. Then electrons are stripped by injection through a thin carbon foil, and protons are accelerated up to the AGS top energy of 25 GeV. During 2.5 seconds of the AGS acceleration cycle, the spin has to overcome more than 250 linear resonances of different strengths.

In the beginning of this activity a strategy of fast resonance crossing was exploited for both imperfection and intrinsic resonances. Using data of beam polarimetry as feedback, the resonance strengths of imperfection resonances were suppressed by compensation

---

[9] Radiative effects essential for electrons will be discussed later.
[10] This method was later named "partial Siberian snake."

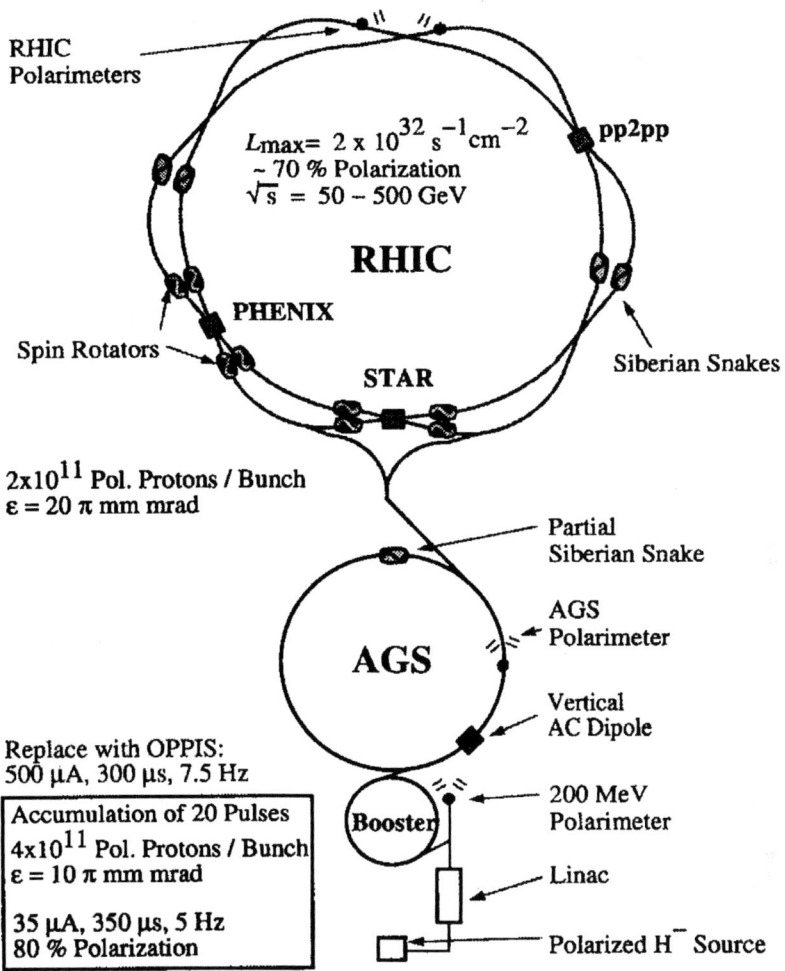

FIGURE 12. Schematic of polarized proton facility at BNL.

of the corresponding harmonics of vertical closed orbit distortions, using combinations of 95 dipole steering coils. Fast jumps of the betatron tunes, induced by twelve pulsed quadrupoles, resulted in fast passage over the more dangerous intrinsic resonances ($0 + \nu_z$, $24 - \nu_z$, $12 + \nu_z$, and so on). Figure 13 presents the data of polarization measurements [28].

In spite of all measures for fast crossings, already at 10 GeV residual polarization did not exceed 15 to 20%. Application of a solenoidal partial Siberian snake ($w_k = 0.05$) against the imperfection resonances (similar to VEPP-2M) significantly improved the results in 1994: 30% polarization was maintained up to 22 GeV.

The next improvement was creation of the adiabaticity condition for spin while passing through the intrinsic resonances. This was done by coherent excitation of the vertical betatron oscillations by a radial RF magnetic field [29]. Because of closeness to the resonance $\nu = k \pm \nu_z$, the spin response function (see Eq. (36)) increases the corresponding resonant harmonic many times, thus at the vertical amplitude of $A_z \simeq 2$ cm it reaches a value of $w_k \approx 0.1$. An advantage of this procedure is proton orbital motion adiabaticity as well, so that the emittance enhancement is negligible after the RF-dipole is turned off.

Further gains are expected from a modification of the machine optics to minimize the corresponding spin-orbital integrals in Eqs. (30) and (33), and from coupling suppression. The final goal of the AGS team (top curve on Fig. 14) is to achieve 70% polarization at the RHIC injection energy of 25 GeV.

B RHIC and Helical Snakes

The experience described above and simple numerical analysis show that, for higher energies, only the application of Siberian snakes, which in principle eliminate all spin resonances, can safeguard polarized beams. Evidently, only snakes with transverse fields are suitable for such high energies (e.g. 250 GeV, the RHIC top energy for protons). The only problem is to minimize inevitable orbit excursions down to a suitable level in the whole machine energy range. The best solution was found with special spiral dipoles (also called helical magnets) [30].

In the paraxial approximation, magnetic fields in a helical magnet with period λ can be presented in the form

$$B_x = -h \sin \kappa y, \qquad B_y = 0, \qquad B_z = h \cos \kappa y, \qquad (42)$$

where $|\kappa| = 2\pi/\lambda$, and h is the field magnitude. The field at the magnet entrance (at $y = 0$) is chosen vertical. The sign of κ indicates the spiral's helicity, which we will characterize by a parameter $s = \kappa/|\kappa|$.

A particle trajectory in such a field is also a helix:

$$x = -\frac{\rho}{\gamma|\kappa|}(1 - \cos \kappa y) + x'_0 y + x_0, \qquad z = \frac{\rho}{\gamma|\kappa|} \sin \kappa y + (z'_0 - \frac{\rho s}{\gamma})y + z_0, \qquad (43)$$

where $\rho = q_0 h/|\kappa|$, ($c = 1$), and x_0, z_0, x'_0, z'_0 are initial coordinates and slopes. It is easy to see that the dimensionless parameter ρ is the trajectory pitch in a co-moving frame.

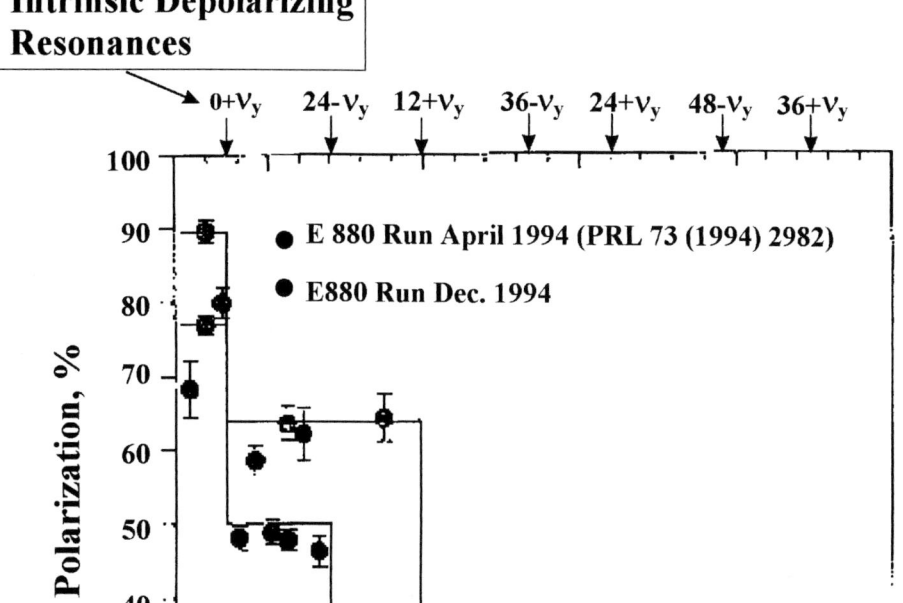

FIGURE 13. Polarized proton acceleration in AGS (first stage).

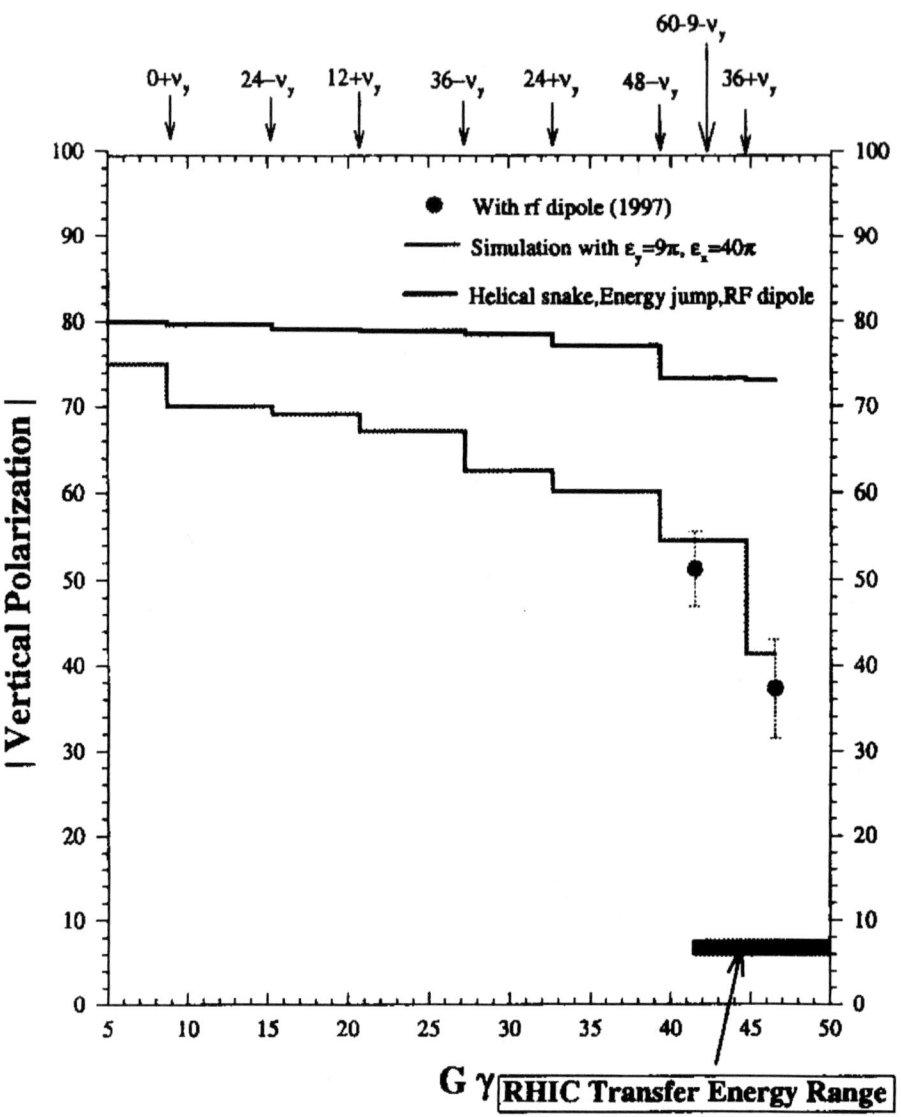

FIGURE 14. Polarized proton acceleration in AGS (second stage).

Spin motion in a helical magnet is conveniently considered in a rotating reference frame:

$$\vec{e}_1 = \vec{e}_x \cos \kappa y + \vec{e}_z \sin \kappa y, \qquad \vec{e}_2 = \vec{e}_y, \qquad \vec{e}_3 = \vec{e}_z \cos \kappa y - \vec{e}_x \sin \kappa y. \qquad (44)$$

Subtracting from the BMT precession, Eq. (5), the angular velocity of the new frame ($\vec{\Omega}_e = \kappa \vec{e}_y$), we find that spin precesses ($d\vec{S}/dy = \vec{W}_e \times \vec{S}$), relatively to $\vec{e}_1, \vec{e}_2, \vec{e}_3$, at a constant frequency:

$$W_1 = 0, \qquad W_2 = -\kappa, \qquad W_3 = -(1 - \nu_0)\frac{\rho \kappa}{\gamma}. \qquad (45)$$

We note that the vectors $\vec{e}_1, \vec{e}_2, \vec{e}_3$ are periodic and we can use here the same approach as for circular accelerators. The periodic solution $\vec{n}$ and the spin tune ν in this case are obvious:

$$\vec{n} = -\frac{1}{\sqrt{(1 + A^2 \rho^2)}} (A\rho \vec{e}_3 - s\vec{e}_2), \qquad \nu = \sqrt{(1 + A^2 \rho^2)}. \qquad (46)$$

Dependence on energy enters in Eq. (46) only through a quantity $A = (a + 1/\gamma)$, which for ultra-relativistic particles is constant and does not differ from the magnet anomaly a.

Note that $\vec{e}_1, \vec{e}_2, \vec{e}_3$ coincide with $\vec{e}_x, \vec{e}_y, \vec{e}_z$ after each period of the helix ($y = m\lambda$); we can conclude that vector $\vec{n}$ and the spin tune ν from Eq. (46) determine in the laboratory frame the 2×2 matrix M (see Eqs. (9) and (10)) of spin transformation at one helix period. For a more general case (the field direction at the entrance forms an angle ϕ to $\vec{e}_z$), the matrix elements of M are

$$m_{11} = \cos \pi \nu - in_2 \sin \pi \nu, \qquad m_{12} = ie^{-i\phi} n_3 \sin \pi \nu, \qquad (47)$$

where $n_{2,3}$ are the components of vector $\vec{n}$ from Eq. (46).

Equations (42), (46) and (47) provide a basis for designing spin rotators that can consist of several magnets with an integer number of periods and with different helicities s_i and pitch parameters ρ_i. As we know from Section II, it is desirable to have a universal scheme of Siberian snakes that can provide an arbitrary snake axis in the horizontal plane ($0 \leq \alpha_s \leq \pi$). It is also clear that any scheme has to satisfy certain symmetry conditions to restore the particle orbit. An analysis of possible solutions shows that the simplest scheme of a Siberian snake based on four one-period helical magnets has to meet the symmetry requirements shown in Fig. 15.

Table 2 gives examples of four versions of this scheme for different α_s. The orbit deviation is calculated at the RHIC injection energy of 25 GeV ($\phi = 0$).

The values of the pitch parameters in Table 2 are chosen for $\lambda = 2.4$ m. Such a helical magnet design has been developed for the RHIC. The maximum magnetic field is about 4.5 T. Two snakes of four magnets, $\alpha_1 = \alpha_2 \pm \pi/2$, will be inserted into each ring of the RHIC, hence the polarization is to be vertical in the arcs (see Fig. 12).

To deliver longitudinal spin in the interaction points, spin rotators will be installed on both sides of the IP. Moreover, since between a location of the rotator and a collision

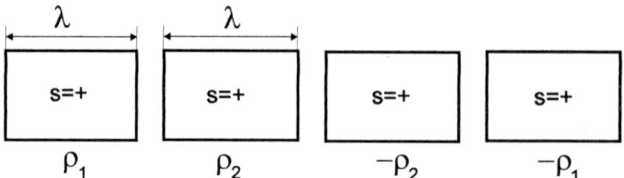

FIGURE 15. Scheme of a Siberian snake for the RHIC formed by four helical magnets.

point, there are horizontal bending magnets giving additional spin rotation around the vertical axis, the rotator has to bring spin not to an exactly longitudinal, but to a required direction in the horizontal plane, which depends on the proton energy. Analysis shows that combinations from the same helical magnets ($\lambda = 2.4$ m) of both helicities can satisfy all constrains for spin and orbit [30].

TABLE 2. Versions of helical snakes

α_s	ρ_1	ρ_2	Field integral (T× m)	Z_{max} (cm)
0°	0.267	−0.355	24.47	2.1
−45°	0.435	−0.198	24.88	3.5
45°	0.154	−0.493	25.45	2.7
90°	0.581	−0.163	29.25	4.6

At present one Siberian snake is already installed in the ring, and the first tests of proton acceleration up to 100 GeV will be performed this fall.

V POLARIZED ELECTRONS

Since electrons are light particles, they radiate energy when subjected to acceleration, Eq. (3), directed to the orbit center. This synchrotron radiation provides radiative damping of the particle oscillation, but on the other hand, quantum fluctuations of the photon emission heat the orbital motion. As a result, the amplitudes of the betatron oscillation scale linearly with γ, and the acceleration of polarized electrons becomes a very difficult operation.

The application of Siberian snakes is not a cure for electrons, because these quantum fluctuations, strongly increasing with energy, act at random upon spin causing depolarization.

A Spin Diffusion

Let us consider the effect of quantum fluctuations on spin in the vicinity of an isolated spin resonance $\delta = \nu - \nu_k \sim |w_k|$ [31].

As shown in Fig. 16, from a single quantum emission $\omega = \delta\gamma$, a precession axis $\vec{n}$ jumps to $\vec{n}_1 = \vec{n} + \delta\vec{n}$, because of a change of detuning $\Delta\delta = \nu\delta\gamma$, and so do the

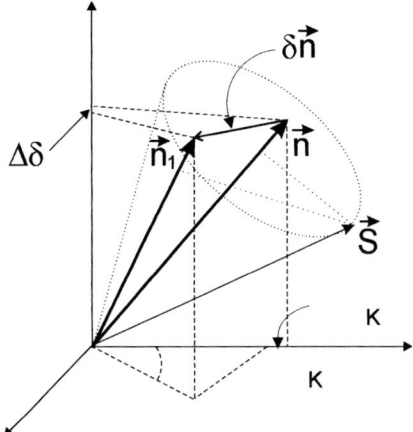

FIGURE 16. A quantum emission near a spin resonance.

resonant harmonics depending on parameters of a new orbit after the radiation: ($\vec{n} = \vec{n}(X, \Delta\gamma/\gamma)$).[11] From Eq. (14) it is easy to find the change of the spin projection

$$\delta S_n = (\vec{S} \cdot \delta\vec{n}) = S_n(\vec{n} \cdot \delta\vec{n}) + S_\perp \mathrm{Im}(\vec{\eta}^* \cdot \delta\vec{n}) \approx -\frac{1}{2}(\delta\vec{n})^2 S_n.$$

Since the radiation does not depend on the spin phase, the term with $S_\perp$ averaged over time gives zero, but the remaining term results in spin diffusion. The mean polarization of the beam $S_n = \langle S_n \rangle$ will decay because of quantum emission at a rate

$$S_n' = -\frac{1}{2} \cdot S_n \left\langle \vec{d}^2 \cdot \frac{d(\delta\gamma/\gamma)^2}{dt} \right\rangle = -\alpha_+ \cdot S_n, \qquad (48)$$

where we introduced the vector of spin-orbit coupling $\vec{d} = \gamma\, d\vec{n}/d\gamma$, and the mean rate of energy diffusion

$$\frac{d(\delta\gamma/\gamma)^2}{dt} = \frac{55}{24\sqrt{3}} q_0^5 \gamma^2 \langle |\vec{B}|^3 \rangle_\theta, \qquad (49)$$

which is well-known from the theory of synchrotron radiation [32]. Thus, we see that in the electron case spin diffusion leads to an exponential decrease of the polarization with characteristic time $\tau_d = \alpha_+^{-1}$.

Considering the resonance crossing, in addition to the Froissart-Stora conditions, Eq. (40), we have to take into account a limitation from spin diffusion on the crossing time, $\Delta t \sim w_k/\delta' \ll \tau_d$. These limitations are very serious, because, as is evident from Eq. (30), the spin-orbit coupling is sensitive to closeness to the spin resonances and,

---
[11]) We do not discuss here weak processes of direct action of radiation fields on spin.

according to estimations, it constrains possibilities for polarized electron acceleration to a range of a few GeV.

First of all, the energy dependence of deviations $\Delta \vec{n}$ is related to the spin tune $\nu_0 = \gamma q'/q_0$. In particular, presenting a contribution of the imperfection resonances to $\Delta \vec{n}$ as a series,

$$\Delta \vec{n} \simeq \text{Im} \left(\vec{\eta}^* \sum_k \frac{w_k}{\nu_0 - k} e^{ik\theta} \right), \tag{50}$$

and taking its derivative over energy, we find that the spin-orbit coupling amplifies the action of the integer resonances upon spin:

$$\vec{d}'^2 \simeq \nu_0^2 \sum_k \frac{|w_k|^2}{(\nu_0 - k)^4}. \tag{51}$$

Thus, in electron machines the imperfection resonances are even more dangerous. We recall that the resonance harmonics of imperfection resonances w_k, Eqs. (35) and (36), originate mainly from vertical distortions of the CO caused by radial fields.

The intrinsic resonances in spin-orbit coupling arise from random fluctuations of the orbital amplitudes $A_{\nu i}$:

$$\vec{d}'^2 \simeq \left| \gamma \frac{dC}{d\gamma} \right|^2 = \nu_0^2 \left| \gamma \frac{d}{d\gamma} \int_{-\infty}^{\theta} Z'' e^{-i\nu_0 \tilde{\theta}} \right|^2. \tag{52}$$

Direct calculation of $\vec{d}'^2$ is quite cumbersome even in the case of weak X-Z coupling, Eq. (37). But we can state that the diffusion process approximately equalizes the depolarization on the vertical $\nu = k \pm \nu_z$ and radial $\nu = k \pm \nu_x$ intrinsic resonances, Eqs. (33) and (38). It is clear that in an almost ideal machine only instant jumps of the vertical betatron amplitude are responsible for spin diffusion. But at the moment of emission $\Delta Z = \Delta Z' = 0$. This means that the amplitude jumps of the free and forced oscillations are equal to each other: $\delta A_z \sim g \cdot \delta A_x$. Hence, enhancement of the vertical beam size near the coupling resonance $\nu_x - \nu z = k$ does not increase the depolarization, because it develops adiabatically to spin.

B Radiative Polarization

Fortunately aggravation of electron spin by radiative effects is more than compensated for by electron self-polarization at high energies. That is especially important for positrons, for which a high-intensity polarized source is unavailable.

To explain this phenomenon, we shall consider the influence of classical synchrotron radiation on spin (including radiation by the magnetic moment). In the second approximation of the BMT equation we can write

$$S'_n = (\vec{S} \cdot \vec{n})' = (\vec{S}' \cdot \vec{n}) + (\vec{S} \cdot \vec{n}'). \tag{53}$$

The first term in this expression describes how spin moves in the magnetic field of the classical radiation of the rotating electron magnetic moment $\vec{\mu} = q\vec{S}$.

In the electron rest frame ("C"-frame), vector $\vec{\mu}$ rotates with frequency $\vec{\Omega}_C = -q\vec{B}_C$. A radiative magnetic field induced by this rotation[12]

$$\vec{h}_\mu \approx \frac{2}{3}\dddot{\vec{\mu}} = -\frac{2}{3}q\Omega_C^2 \left[\vec{\Omega}_C \times \vec{S}\right] = q^4 B_C^3 \left[\vec{b} \times \vec{S}\right], \qquad \vec{b} \equiv \frac{\vec{B}}{B} = \frac{\vec{B}_C}{B_C},$$

is moving together with spin (Fig. 17) and, in spite of its smallness, slowly deflects spin, $d\vec{S}/d\tau = -q[\vec{h}_\mu \times \vec{S}] \sim q^4 B_C^3 \left(\vec{b}(\vec{S}\cdot\vec{S}) - \vec{S}(\vec{b}\cdot\vec{S})\right)$.

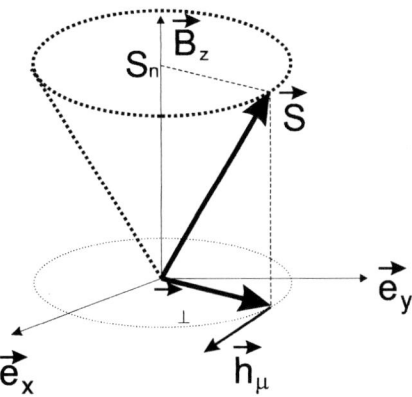

FIGURE 17. Sokolov-Ternov polarization.

After relativistic transformation to the laboratory frame, averaging over fast precession phases, and integration around CO, we come to an equation[13]:

$$(S_n')_\mu = (\vec{S}' \cdot \vec{n}) \sim q^5 \gamma^2 \langle |\vec{B}|^3 (\vec{b}\cdot\vec{n}_s)\rangle_\theta \cdot S_\perp^2, \qquad S_\perp^2 = 1 - S_n^2. \tag{54}$$

A solution of this equation

$$S_n = \frac{1 - e^{-t/\tau}}{1 + e^{-t/\tau}} \tag{55}$$

is a build-up of spins along $\vec{n}_s$, with a decrement $\sim \tau^{-1} = q^5 \gamma^2 \langle |\vec{B}|^3 (\vec{b}\cdot\vec{n}_s)\rangle_\theta$, as shown in Fig. 18.

At first glance, the above description of radiative polarization appears very questionable, because the power of the magnetic moment radiation is many orders of magnitude

---

[12] It can be derived by the retarded time development in the near zone, similarly to what is done for a radiating charge [33]. For classical radiation of an orbiting charged rotator see Ref. [43].
[13] We neglect in this case the action of $\vec{h}_\mu$ on the particle motion.

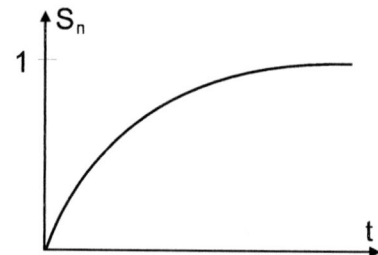

FIGURE 18. Polarization build-up versus time.

less than that of the charge radiation. Nevertheless, a strict treatment by Sokolov and Ternov [34] of the Dirac equation in a homogeneous magnetic field, including the quantum fluctuations and interaction with the electrodynamical vacuum, gives a solution,

$$S_n = \frac{8}{5\sqrt{3}}(1 - e^{-t/\tau_p}), \tag{56}$$

that is quantitatively slightly different from Eq. (55): the final polarization level is 0.924 instead of 1 and the polarization time $\tau_p = (8/5\sqrt{3})\tau$.

The second term in Eq. (53) is the impact of the "magnetic" part [43] of the full radiative friction force $\vec{f}_r$ on the $\vec{n}$-axis. From the quasiclassical form of the synchrotron radiation intensity in the non-spin-flip case [32] we find

$$\vec{f}_r = -\vec{V}W_{cl}\left[1 - \frac{3\gamma^2}{2m}|\dot{\vec{V}}|\left(\frac{55\sqrt{3}}{24} - \frac{e}{|e|}(\vec{S}\cdot\vec{b})\right)\right], \quad W_{cl} = \frac{2}{3}e^2|\dot{\vec{V}}|^2\gamma^4.$$

Since the recoil caused by this friction is longitudinal, we can substitute into Eq. (53)

$$\vec{n}' = \left(\dot{\vec{p}}\cdot\frac{\partial}{\partial\vec{p}}\right)\vec{n} \simeq \left(\vec{f}_r\cdot\frac{\partial\gamma}{\partial\vec{p}}\right)\frac{d\vec{n}}{d\gamma} = (\vec{f}_r\cdot\vec{V})\frac{\vec{d}}{\gamma m}.$$

Deflection $\Delta\vec{n}$ induced by the friction force depends on the projection $\vec{S}\cdot\vec{b}$, and it is not symmetric with respect to $\vec{n}_s$, as we can see from Fig. 19.

As a result of many rotations of spin around synchronously swinging $\vec{n}$, they both damp to $\vec{n}_s$. Similarly to Eq. (54), averaging over fast precession phases and integrating around CO, we obtain

$$(S_n')_f = (\vec{S}\cdot\vec{n}') \simeq -q_0^5\gamma^2\left\langle|\vec{B}|^3(\vec{b}\cdot\vec{d})\right\rangle_\theta S_\perp^2. \tag{57}$$

To get the full influence of radiation on spin, we combine all three radiative effects, Eqs. (48), (54) and (57):

$$S_n' = \alpha_- S_\perp^2 - \alpha_+ S_n. \tag{58}$$

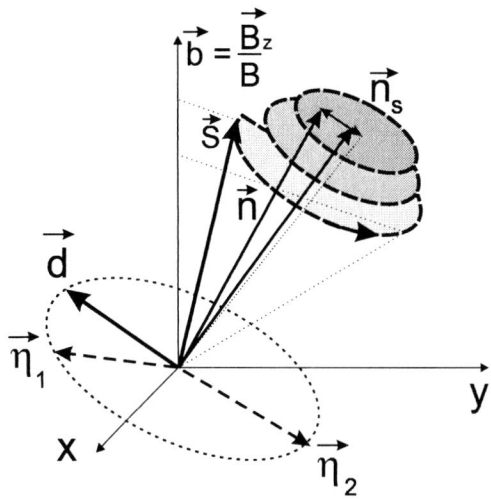

FIGURE 19. Kinetic self-polarization.

This equation reflects the competition between radiative polarization and spin diffusion. The result is that spins build up similarly to Eq. (56), with a characteristic time $\tau = \alpha_+^{-1}$, to an equilibrium polarization $S_{max} = \alpha_-/\alpha_+ < 1$.

The latter result is a manifestation of quantum nature of the spin-radiation system. Equation (58) is actually written for quantum-mechanical average of the spin-operator components, thus for spin $\frac{1}{2}$ Eq. (58) implies $\langle \hat{S}_\perp^2 \rangle = \frac{1}{2}\frac{3}{2} - \frac{1}{4} = \frac{1}{2}$ and $\langle \hat{S}_n \rangle = \frac{1}{2}S_n$, in terms of the unit spin vector $\vec{S}$.

Although the classical model of radiative polarization is adequate for large anomalies and even gives a satisfactory agreement with exact results for $g = 2$ particles [18], it should not be overused [44]. For instance, contrary to the classical model, at $0 < g < 1.2$ spins build up in the "wrong" direction because of strong spin-orbit coupling effects [18].

In more correct studies of radiative polarization in an inhomogeneous magnetic field, Derbenev and Kondratenko [18] derived expressions for α_- and α_+:

$$\alpha_- = q_0^5 \gamma^2 \left\langle |\vec{B}|^3 \vec{b} \cdot (\vec{n}_s - \vec{d}) \right\rangle_\theta,$$

$$\alpha_+ = \frac{5\sqrt{3}}{8} q_0^5 \gamma^2 \left\langle |\vec{B}|^3 \left[1 - \frac{2}{9}(\vec{n}_s \cdot \vec{V}) + \frac{11}{18}\vec{d}^2 \right] \right\rangle_\theta, \quad (59)$$

that coincide in their main terms with the estimated coefficients of Eq. (58).

Analysis of the Derbenev-Kondratenko formulae shows that, in an inhomogeneous magnetic field, because of spin-orbit coupling, radiative polarization can reach a higher degree than the 0.924 predicted for the homogeneous case. Moreover, in the presence of spin-orbit coupling, radiative polarization can occur even when vector $\vec{n}_s$ is perpendicular to the guiding field $\vec{B}$. Such a situation happens in electron storage rings with one Siberian snake [35].

The polarization rise-time τ_p in a given machine depends on the energy as γ^{-5}. We can see from Eq. (59) that it can be reduced significantly by dipoles with higher fields or by a special polarization wiggler.

C Experimental Observations

Experimental radiative polarization has a long history. It has been observed in all electron (positron) storage rings at energies from 500 MeV to 60 GeV (see for example [36]). Figure 20 demonstrates one of the first observations of polarization build-up in the VEPP-2M storage ring ($E = 650$ MeV) [12].

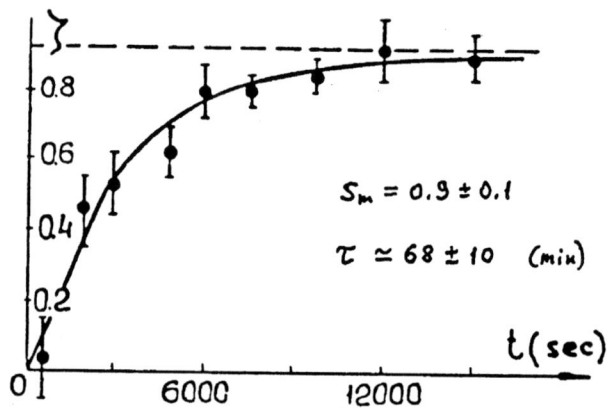

FIGURE 20. The first evidence of spin build-up versus time in VEPP-2M.

Pioneer studies of the influence of spin resonances on radiative polarization were done at VEPP-2M [21] and at SPEAR [37]. Figure 21 presents a scan of the measured values of S_{max} versus the energy in the tune gap $\nu = 8$ to 8.5 at the SPEAR storage ring.

In spite of the relatively low energy, the SPEAR data have indicated a lot of resonances, including synchrotron side-bands. These results have provided evidence of the effect of spin diffusion on resonances and underlined the seriousness of the problem of getting radiative polarization at high energy machines.

A number of measures were developed later to minimize depolarization. We hope that the items listed below will be understandable without additional explanations after our introduction to spin theory.

First of all it is necessary to suppress the resonant harmonics of the imperfection resonances.

1. Since as a rule the radiative polarization is required at a fixed energy, it is desirable to choose the spin tune near $\nu = k + 1/2$ between two imperfection resonances. Then in practice it is enough to suppress only these two nearest resonant harmonics w_k and w_{k+1} (spin harmonic matching).

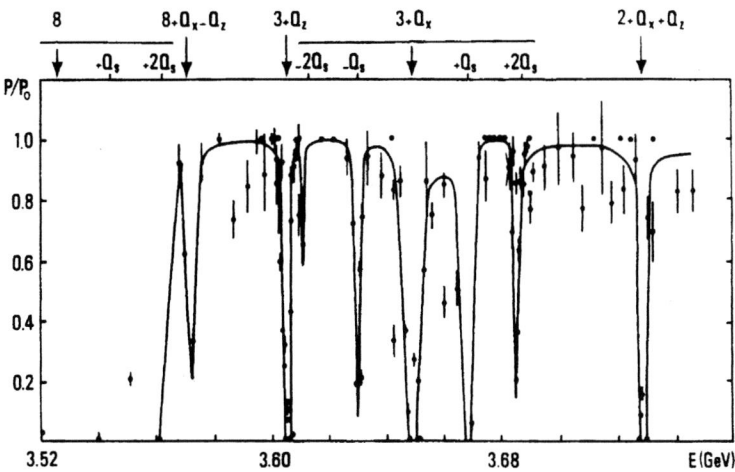

FIGURE 21. Energy scan of radiative polarization at SPEAR.

2. Similarly to the proton case, before spin harmonic matching, the alignment of all the focusing elements in the machine has to be made as good as possible.

3. The number and quality of beam position monitors and steering coils must be sufficient to suppress the higher-order modes of the vertical closed orbit distortions ($\vec{d}^2 \approx \nu_0^4 (K_x)_k$).

4. It is evident that in addition to spin harmonics matching it is useful to have locally controlled orbit bumps around the final focus quadrupoles, where the machine optical functions are maximal.

5. The machine lattice has to be stable against optics distortions.

To minimize the contribution of the intrinsic resonances it is important:

1. to choose the tunes as far away as possible from the resonant conditions $\nu = k \pm \nu_{x,z,s}$;

2. to minimize the parasitic vertical dispersion Ψ_z and betatron coupling in the arcs by preceding alignment;

3. to decrease the residual coupling by a set of skew-quadrupoles around the machine, especially near the final focus system;

4. to modify, if possible, the machine lattice to cancel the spin-orbital integrals (see Eq. (30)).

And of course, a precise polarimeter is the key to success, because at high energies small corrections of the CO or the coupling are not visible to the BPM and beam profile monitors, but meanwhile they considerably affect the depolarization.

Separate problems are created by insertion devices, such as detector fields, spin rotators and wigglers. It is relatively easy to compensate for $\vec{n}_s$-axis tilts caused by detectors' solenoids. A more complicated problem appears with the installation of spin rotators, which are usually designed to deflect the spin axis by an angle

$$\phi = \int_0^l |\vec{W}_0| d\theta$$

and to cause some closed-orbit deviations. To provide the same spin rotation for off-orbit particles, we have to provide focusing inside the rotator, which should satisfy the so-called spin transparency conditions, which can be written in a general form

$$\int_0^l w_\perp d\theta = 0,$$

where $w_\perp$ is given by Eqs. (23) and (26).

Equipped with all the instrumentation discussed above for minimizing spin diffusion, attempts to obtain polarized beams succeeded at HERA [38] at an energy of about 30 GeV, and at LEP [39] at 45 GeV. In HERA a longitudinal polarization of $\simeq 70\%$ was obtained with the spin rotators on (see Fig. 4) in two interaction regions.

Despite all these measures, the equilibrium polarization level drops with further increases in energy. Derbenev, Kondratenko and Skrinsky [19] estimated the achievable S_{max} in the linear approximation with realistic assumptions about tolerances of the alignment.

Figure 22 shows the results of radiative polarization facilities all around the world. At all high energy machines, "improved" with the above-mentioned measures, the experimental points go practically along the linear prediction curve, although the preceding measurements had been much worse.

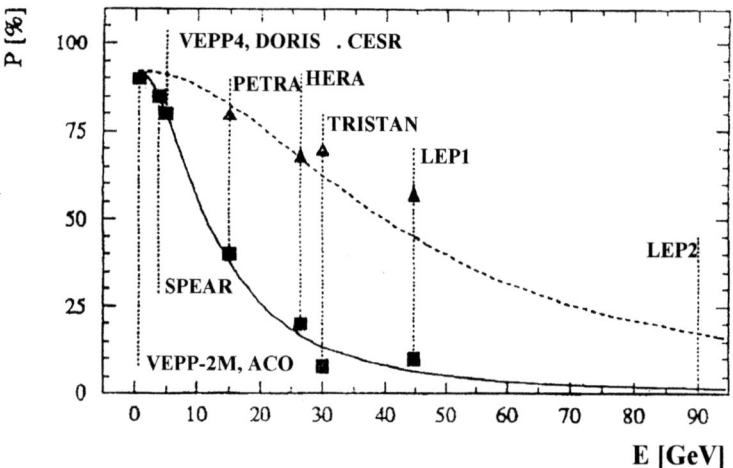

FIGURE 22. Measurements of radiative polarization all around the world.

In contradiction to the general picture are the latest measurements at LEP at energies above 45 GeV, see Fig. 23.

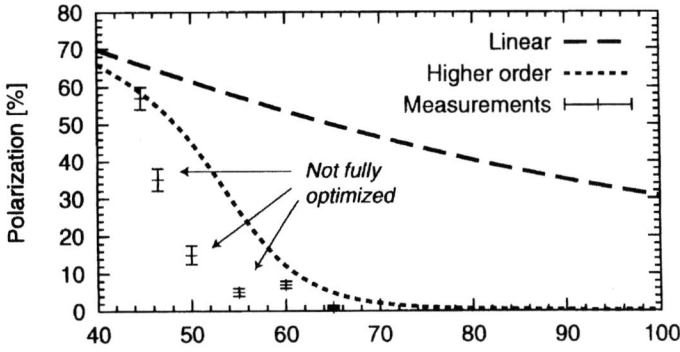

FIGURE 23. Radiative polarization measurements at LEP vs. energy, GeV.

This discrepancy was explained [40] by the fact that the machine was not fully optimized at that time, and by a wrong choice of the synchrotron tune. This probably corresponds to reality, but an estimation shows another possible reason for this phenomenon, related to enhancement of the stochasticity in the spin motion at such high energies [19].

D Spin Tune Spread

In considering synchrotron modulation (see Eq. (41)), we assumed that the distribution of $\Delta\gamma/\gamma$ does not change with time (see Fig. 11). In this case the spin tune spread at the central line of the spectrum is determined by the second approximation of the particle dynamics.[14] The dynamical tune spread depends on the beam size σ_x, the momentum compaction factor α, and the chromaticity of the radial betatron oscillations [41]:

$$(\sigma_\nu)_d = \langle \Delta\nu \rangle \simeq \frac{\nu_0 \sigma_x^2}{\alpha} \gamma \frac{d\nu_x}{d\gamma}.$$

Numerical estimations of the "dynamical" tune spread give $\sigma_\nu \leq 10^{-5}$. But this "natural" value can be controlled by sextupole corrections. Zero chromaticity of the ring significantly reduces this dynamical spin tune spread [42].

Because of quantum fluctuations, the spin tune diffuses within the equilibrium distribution (after a few damping times τ_0), which is widened by the same random effects. Let us estimate a "stochastic" width of this distribution, considering the "top view" of the spin cone in the resonant frame, see Fig. 24.

---

[14] At $\nu_s \gg w_k$ the widths of the side-bands are proportional to their number m and determined dynamically by the nonlinearity of the longitudinal potential well, which is usually much larger than the transverse nonlinearity.

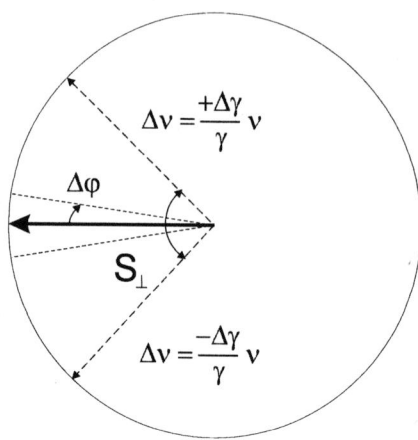

FIGURE 24. "Top view" of the spin motion in the resonant frame.

The transverse component of the spin on the closed orbit is a reference for all others that are swept around by the synchrotron oscillations. A small energy deviation $\delta\gamma/\gamma$ results in a spin phase advance during a half-period of the synchrotron oscillation:

$$\delta\phi = \nu_0 \int_0^{\pi/\nu_s} \frac{\delta\gamma}{\gamma} \sin(\nu_s \theta) d\theta \simeq \frac{\nu_0}{\nu_s} \frac{\delta\gamma}{\gamma}.$$

The mean-square angle randomly accumulates because of energy diffusion during the same time, $\delta\theta \sim \pi/\nu_s$. This quantity arises again and again in spite of averaging over the next oscillations. That is why it can be treated as an estimate of the stochastical spin tune spread:

$$(\sigma_\nu)_{st} \sim \frac{1}{2\pi} \int_0^{\pi/\nu_s} (\delta\phi)^2 d\theta \sim \frac{\nu_0^2}{\nu_s^3} \frac{d}{dt}(\delta\gamma/\gamma)^2 \simeq \frac{\nu_0^2}{\nu_s^3} \frac{\sigma_\gamma^2}{T_0}, \qquad (60)$$

where we used the formula for the root-mean-square energy spread,

$$\sigma_\gamma = \sqrt{\frac{1}{2} \frac{d(\delta\gamma/\gamma)^2}{dt} T_0}.$$

Here $(\sigma_\nu)_{st}$ scales as the seventh power of the energy. It is evident that talk of synchrotron modulation makes no sense when $(\sigma_\nu)_{st} \sim 1$. Numerical estimations give $(\sigma_\nu)_{st} \simeq 10^{-7}$ for the low energy machine VEPP-2M, for example. Meanwhile for LEP the spin tune spread reaches the "critical" value in the energy range 60–100 GeV. If so, this means that predictions of the linear theory are no longer valid for these energies, because of diffusive overlapping of resonances that could reduce the equilibrium polarization.

REFERENCES

1. L.D. Landau, E.M. Lifshitz, *Quantum Mechanics*, Moscow (1963).
2. Particle Data Group, *Phys. Rev.* **D54**, 21 (1996).
3. V.B. Berestetsky, E.M. Lifshitz, L.P. Pitaevsky, *Relativistic Quantum Theory*, v. 1, Moscow (1968).
4. L.H. Thomas, *Nature*, **117**, 514 (1926).
5. V. Bargman, L. Michel, V.L. Telegdi, *Phys. Rev. Lett.* **2**, 435 (1959).
6. Y.B. Bolkhovitinov et al., in *Proc. 12th Int. Symp. on High Energy Spin Physics*, 730 (1996).
7. J. Bailey et al., *Nuclear Physics* **B150**, 1 (1979).
8. V. Hughes et al., in *Proc. 10th Int. Symp. on High Energy Spin Physics*, 717 (1992).
9. Ya.K. Semertzidis, in *Proc. Workshop on Nuclear Electric Dipole Moment* (1999).
10. J. Buon, K. Steffen, *NIM* **A245**, 248 (1986).
11. A.A. Kolomensky, A.N. Lebedev, *Theory of Circular Accelerators*, Moscow (1962).
12. L.M. Kurdadze et al., in *Proc. 5th Int. Symp. on High Energy Physics*, 148 (1975).
13. Ya.S. Derbenev et al., *Part. Accel.* **10**, 177 (1980).
14. Yu.M. Shatunov, A.N. Skrinsky, *Sov. Phys. Uspekhi* **32**(6), 548 (1989).
15. Ya.S. Derbenev, A.M. Kondratenko, A.N. Skrinsky, *Sov. Phys. Doklady* **192**, 1255 (1970).
16. Ya.S. Derbenev, A.M. Kondratenko, A.N. Skrinsky, *JETP* **60**, 1216 (1971).
17. Ya.S. Derbenev, A.M. Kondratenko, *Part. Accel.* **8**, 115 (1978).
18. Ya.S. Derbenev, A.M. Kondratenko, *Sov. Phys. JETP* **37**, 968 (1973).
19. Ya.S. Derbenev, A.M. Kondratenko, A.N. Skrinsky, Preprint INP 77-60, Novosibirsk, (1977).
20. M. Froissart, R. Stora, *NIM* **7**, 297 (1960).
21. Ya.S. Derbenev et al., in *Proc. 10th Int. Conf. on High Energy Accelerators*, v. 2, 272 (1977).
22. T. Khoe et al., *Part. Accel.* **6**, 213 (1975).
23. T. Aniel et al., *Colloq. Phys.* (France), **C-2**, 499 (1985).
24. N. Horikawa et al., in *Proc. 9th Int. Symp. on High Energy Spin Physics*, 171 (1990).
25. A. Lehrach, Ph.D. Thesis (Germany) (1997).
26. A.N. Zelensky, in *Proc. 13th Int. Symp. on High Energy Spin Physics*, 618 (1998).
27. T. Roser, in *Proc. 13th Int. Symp. on High Energy Spin Physics*, 182 (1998).
28. H. Huang et al., in *Proc. 12th Int. Symp. on High Energy Spin Physics*, 529 (1996).
29. H. Huang et al., in *Proc. 13th Int. Symposium on High Energy Spin Physics*, 492 (1998).
30. V.I. Ptitsyn, Yu.M. Shatunov, *NIM* **A398**, 126 (1997).
31. Ya.S. Derbenev, A.M. Kondratenko, *JETP* **62**, 430 (1972).
32. A.A. Sokolov, I.M. Ternov, *Synchrotron Radiation*, Moscow (1961).
33. J.D. Jackson, *Classical Electrodynamics* (1975).
34. A.A. Sokolov, I.M. Ternov, *Sov. Phys. Doklady* **8**, 1203 (1964).
35. V.I. Ptitsyn, Yu.M. Shatunov, in *Proc. 12th Int. Symp. on High Energy Spin Physics*, 555 (1996).
36. D.P. Barber, in *Proc. 12th Int. Symp. on High Energy Spin Physics*, 99 (1996).
37. J.R. Johnson et al., *NIM* **204**, 261 (1983).
38. D.P. Barber at al., *Phys. Lett.* **B343**, 436 (1995).
39. R. Assmann et al., CERN Report SL 94-08 (1994).
40. R. Assmann, in *Proc. Workshop* (Chamonix) (1999).

41. I.A. Koop et al., in *Proc. 8th Int. Symp. on High Energy Spin Physics*, 1023 (1988).
42. I.B. Vasserman et al., *Phys. Lett.* **B198**, 302 (1987).
43. V.L. Ginzburg, Theoretical Physics and Astrophysics, Moscow 1975.
44. J.D. Jackson, *Revs. Modern Phys.*, **48** (3), 417 (1976).

Vlasov Equation and Landau Damping

D.V. Pestrikov

Budker Institute of Nuclear Physics
630090 Novosibirsk, Russian Federation

Abstract. Among different approaches the Vlasov equations present the most comprehensive way to study the coherent oscillations of beams in cases, where coherent oscillations develop faster than the rates of beam relaxation due to stochastic processes. The differences in the frequencies of the beam particle motions result in decoherence of the beam and in Landau damping of coherent modes. This is a special damping mechanism which appears due to differences in the interactions of various beam particles with excited coherent oscillations.

I INTRODUCTION

One of the most important limitations on the operation performance of particle accelerators and storage rings is due to coherent oscillations of bunches, or due to instabilities of these oscillations. These phenomena may result in increases in the beam (or bunch) sizes, in increases of their emittances, or in particle losses from the beam. Most of these harmful effects occur in cases where coherent oscillations of the beam become unstable or approach the threshold of instability. For these reasons, study of the stability of coherent oscillations of beams is of great practical importance.

Although useful predictions concerning stability conditions of coherent oscillations can be obtained by using particular simplifying models, it is desirable to have a universal tool that enables analytic or numerical studies of arbitrary problems related to coherent oscillations of beams. In the cases where amplitudes of coherent oscillations vary faster than the relaxation of the beam because of stochastic processes, such a tool is provided by the Vlasov equation. It describes variations of the single-particle distribution function of the beam, taking into account self-consistent fields of the beam and neglecting effects of collisions or other stochastic processes.

For these reasons, Vlasov equations never describe beam heatings. However, solutions to these equations can be used for calculations of collision integrals, which take into account energy exchange between the beam particles and coherent fluctuations of the beam (see, e.g. [1]). This energy exchange results in a special damping mechanism of coherent oscillations – Landau damping. In the simplest case of one-dimensional coherent and incoherent oscillations, this damping is due to a difference in the energy flow for particles moving slower and faster than the phase velocity of a coherent mode. Such a difference is possible only in beams where the particles have different oscillation or revolution frequencies, i.e. in beams with frequency spreads. The operation conditions in particle accelerators and storage rings may demand the study of more complicated cases, where induced fields, or the lattice nonlinearity, couple the oscillations. The Vlasov equation technique provides a universal tool capable of treating all these problems.

Because of the self-consistency of the beam fields contributing to the Vlasov equation generally, it describes nonlinear variations of the beam distribution functions. Thus, except for special cases, calculation of exact solutions to this equation often is possible only numerically. Fortunately, a wide class of important problems related to beam current limitations can be solved by studying the development of small coherent oscillations of beams, when the Vlasov equation can be linearized in the deviation of the distribution function from its stationary value. The comprehensive theory of linear coherent oscillations in particle accelerators and storage rings, based on averaging the Vlasov equation, was developed by many people, beginning with pioneering papers [2], [3]. The theory now enables evaluation of the stability criteria and the time evolution of small coherent oscillations for arbitrary configurations of the beams in circular and linear machines. Numerous examples of applications of this theory can be found (see, e.g. [4], [5]). Below, we focus on those aspects of the theory which are specific for storage rings and circular colliders.

II COHERENT OSCILLATIONS AND FLUCTUATIONS

The Vlasov equation can be obtained by reducing the Liouville equation for N-particle distribution function to the equation, describing the evolution of the one-particle distribution function (see, e.g. [6]). Such a one-particle (or simple) distribution function defines the average number of particles in a given phase-space volume. We denote that function using $Nf(\mathbf{r}, \mathbf{p}, t)$, where N is the number of particles in the beam, and $\mathbf{r}$ and $\mathbf{p}$ are coordinates of particles in the phase-space. Integrating Nf over the total phase-space volume of the beam, we find $\int d^3r d^3p N f(\mathbf{r}, \mathbf{p}, t) = N$, so that

$$\int d^3r d^3p f(\mathbf{r}, \mathbf{p}, t) = 1. \qquad (1)$$

To find the equation describing the variations of f, we use the technique developed by Klimontovich [7]). Generally, the state of a beam is given if we know the phase-space trajectories of its particles: $\mathbf{r}_a = \mathbf{r}_a(t)$ and $\mathbf{p}_a = \mathbf{p}_a(t)$. In this case, we also know the microscopic beam density:

$$\Phi(\mathbf{p},\mathbf{r},t) = \sum_{a=1}^{N} \delta[\mathbf{r}-\mathbf{r}_a(t)]\delta[\mathbf{p}-\mathbf{p}_a(t)]. \qquad (2)$$

Using

$$\Delta N(\Gamma) = \int_{\Delta\Gamma} d^3r d^3p \Phi(\mathbf{p},\mathbf{r},t), \qquad (3)$$

this function defines the instantaneous number of particles in a given phase-space volume $\Delta\Gamma$. According to the equations of motion, the function Φ obeys the continuity equation

$$\frac{\partial \Phi}{\partial t} + \frac{\partial}{\partial r_\alpha}[v_\alpha \Phi] + \frac{\partial}{\partial p_\alpha}[F_\alpha \Phi] = 0. \qquad (4)$$

Here, $\mathbf{v} = \{v_x, v_y, v_z\}$ is the particle velocity, $\mathbf{F} = \{F_x, F_y, F_z\}$ is the force acting on a particle and we use a convention that

$$A_\alpha B_\alpha \equiv \sum_{\alpha=x,y,z} A_\alpha B_\alpha.$$

In cases of practical interest, initial conditions of the particles are known with some probability, while the forces $\mathbf{F}$ may contain both systematic and random parts. For these reasons, the phase-space trajectories of particles also are random functions of time. Hence, the function Φ can be split into systematic and random parts,

$$\Phi = Nf(\mathbf{p},\mathbf{r},t) + \delta\Phi, \quad \langle \delta\Phi \rangle = 0, \qquad (5)$$

where the brackets $\langle \ldots \rangle$ indicate statistical averaging over the fluctuations. With this definition the average part of Φ presents the one-particle distribution function. This is a nonsingular function in phase-space. The random function $\delta\Phi$ describes (coherent) fluctuations of the number of particles in the beam near a given point in the phase-space: Schottky noise and others. Many-particle distribution functions are defined as statistical averages of powers of Φ (see, e.g. [7] or [4]).

For a beam in a particle accelerator or a storage ring, the force $\mathbf{F}$ comprises the part describing effects of the guiding fields, the particle focusing near the closed orbit, the part describing effects of random forces due to particle collisions or to other fluctuations, and probably the part describing beam cooling. Separating $\mathbf{F}$ into its random and non-Hamiltonian parts,

$$\mathbf{F} = -\frac{\partial H}{\partial \mathbf{r}} + \Delta\mathbf{F}, \quad \Delta\mathbf{F} = \mathbf{F}_c + \delta\mathbf{F}, \qquad (6)$$

we obtain from the continuity equation that its regular part describes the variations of the distribution function,

$$\frac{\partial f}{\partial t} + [H; f] = -\frac{\partial [(F_\alpha)_c f]}{\partial p_\alpha} - \frac{1}{N}\left\langle \frac{\partial}{\partial p_\alpha}(\delta F_\alpha \delta \Phi)\right\rangle, \qquad (7)$$

while the beam fluctuations are described by the random part of Eq.(4):

$$\frac{\partial \delta \Phi}{\partial t} + [H; \delta \Phi] + \frac{\partial [\delta F_\alpha N f]}{\partial p_\alpha} = -\delta\left\{\frac{\partial}{\partial p_\alpha}(\Delta F_\alpha \delta \Phi)\right\}. \qquad (8)$$

Here

$$[H; f] = \frac{\partial H}{\partial p_\alpha}\frac{\partial f}{\partial r_\alpha} - \frac{\partial H}{\partial r_\alpha}\frac{\partial f}{\partial p_\alpha}$$

denotes the Poisson bracket. The random force δF_α may contain a Hamiltonian part

$$\delta F_\alpha = -\frac{\partial \delta H}{\partial r_\alpha} + \delta(F_\alpha)_c.$$

In this case, Eq.(8) can be rewritten in a form very similar to Eq.(7):

$$\frac{\partial \delta \Phi}{\partial t} + [H; \delta \Phi] + [N\delta H; f] = -\frac{\partial [\delta(F_\alpha)_c N f]}{\partial p_\alpha} - \delta\left\{\frac{\partial}{\partial p_\alpha}(\Delta F_\alpha \delta \Phi)\right\}. \qquad (9)$$

Equations (7) and (8) show that the variations of the beam distribution function and of the beam fluctuations are coupled. The variations of the distribution function change the beam fluctuations and vice versa. Generally, the right-hand side (rhs) in Eq.(7) is expressed in terms of multiparticle distribution functions and therefore Eqs (7) and (8) are as complex as the initial equation, Eq.(4). The statistical description of beams is simplified for the problems, satisfying the time hierarchy condition [8]. Usually, it means that time variations of dynamic variables occur during time intervals (τ_{col}) much shorter than the time intervals that are specific for the variations of the distribution function f (the relaxation time τ_R). Provided that

$$\tau_{col} \ll \tau_R, \qquad (10)$$

the rhs in Eq.(7) can be expressed in terms of the single-particle distribution function only. Thus we arrive at the kinetic equation, which usually is written in the form

$$\frac{\partial f}{\partial t} + [H; f] = \text{St}[f] \propto -\frac{f}{\tau_R}. \qquad (11)$$

Here, St[f] is a collision integral. Although calculation of collision integrals is beyond the scope of this lecture, we note that in many cases of practical interest for beam physics the rhs in Eq.(11) can be written as the Fokker-Planck collision integral, [1]

$$\text{St}[f] = -\frac{\partial}{\partial p_\alpha}\left(\langle\Delta p_\alpha\rangle f + \frac{1}{2}\frac{\partial}{\partial p_\beta}[\langle\Delta p_\alpha \Delta p_\beta\rangle f]\right). \qquad (12)$$

Generally, the rhs in Eq.(11) describes processes that result in variation of the phase-space volume of the beam, while the left-hand side, in particular, describes coherent oscillations of the beam. If the specific time intervals describing the variations of f due to these oscillations (τ_{coh}) are substantially shorter than τ_R,

$$\tau_{coh} \ll \tau_R, \qquad (13)$$

we may neglect in Eq.(11) the variations of f due to collisions (St[f] $\to$ 0) and thus arrive at the Vlasov equation

$$\frac{df}{dt} = \frac{\partial f}{\partial t} + [H; f] = 0. \qquad (14)$$

This equation describes processes where the distribution function is one of the integrals of motion. For this reason, general solutions to Eq.(14) are

$$f(\mathbf{r},\mathbf{p},t) = f_{in}(\mathbf{r}_{in}[\mathbf{r},\mathbf{p},t], \mathbf{p}_{in}[\mathbf{r},\mathbf{p},t]), \qquad (15)$$

where the subscripts in indicate initial distribution functions of the beam and initial positions of the beam particles in the phase-space. Unfortunately, except for very rare cases, Eq.(15) cannot be used directly.

III   UNPERTURBED COHERENT OSCILLATIONS

For the sake of simplicity, we assume that without other perturbations particles move in a storage ring along the closed orbit of the perimeter $\Pi = 2\pi R_0$, performing betatron and possibly synchrotron oscillations. If y and x are the vertical and horizontal offsets of a particle, using conventional notations of particle accelerator physics (see, e.g. [9]), we write

$$x = \sqrt{J\beta_x}\cos(\psi_x + \phi_x) + \eta\frac{\Delta p}{p}, \quad p_x = \frac{p}{R_0}\frac{dx}{d\theta},$$

$$y = \sqrt{J\beta_y}\cos(\psi_y + \phi_y), \quad p_y = \frac{p}{R_0}\frac{dy}{d\theta}, \quad I_{x,y} = \frac{pJ_{x,y}}{2},$$

$$\theta = \omega_0 t + \phi_s, \quad \frac{d\phi}{dt} = \omega_0\alpha_p\frac{\Delta p}{p}, \quad \alpha_p = \frac{1}{\gamma^2} - \alpha. \qquad (16)$$

---

[1] For example, such collision integrals describe the variations of f due to synchrotron radiation of particles, due to the electron cooling, due to stochastic cooling, and in many other cases.

These equations generate the canonical transformation to the action-phase variables of the unperturbed oscillations of particles ($\{x_\alpha, p_\alpha\} \to \{I_\alpha, \psi_\alpha\}$). In these variables the Hamiltonian and equations of motion read:

$$H = \omega_\alpha I_\alpha, \quad \dot{I}_\alpha = 0, \quad \dot{\psi}_\alpha = \omega_\alpha, \tag{17}$$

so that solutions to Eq.(14) can be written using Eq.(15):

$$f(I, \psi, t) = Q(I) \prod_\alpha \chi_\alpha(\psi_\alpha - \omega_\alpha t), \tag{18}$$

where the functions $Q(I)$ ($I = \{I_x, I_y, I_s\}$) and $\chi_\alpha(\psi_\alpha + 2\pi) = \chi_\alpha(\psi_\alpha)$ are calculated using the initial distribution function. As seen from Eq.(18) and from Fig. 1, only the zero harmonic of the distribution function remains unaltered with time. All other harmonics oscillate with frequencies ω_α. This means that coherent oscillations of a beam are described by the Fourier harmonics of f in the oscillation phases,

$$f = \sum_{\mathbf{m}} f_{\mathbf{m}}(I, t) \exp(im_\alpha \psi_\alpha). \tag{19}$$

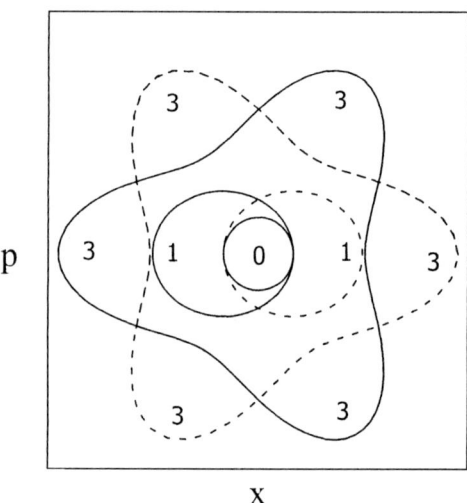

FIGURE 1. Schematic contours of several harmonics of the distribution function in phase-space. The numbers show the multipole numbers of modes. Dashed curves show plots after a half oscillation period.

Without interaction between particles these harmonics are independent and present the oscillation modes of the beam. Near the origin ($I \to 0$) we can write $f_{m_\alpha} \propto I^{|m_\alpha|/2}$. Since the combinations

$$D_m = \int d\Gamma (\sqrt{I})^{|m_\alpha|} e^{im_\alpha \psi_\alpha} f$$

define the multipoles of the distribution function in phase-space, the sets of harmonic numbers m_α define the multipole number of a coherent oscillation mode. For example, the set $\sum_\alpha |m_\alpha| = 1$ describes dipole oscillations, the set $\sum_\alpha |m_\alpha| = 2$ presents quadrupole modes, and so on.

Dipole modes describe the oscillations of the beam centroids:

$$\langle y \rangle = \int d^3 I \sqrt{2I_y} \int_0^{2\pi} \frac{d^3\psi}{(2\pi)^3} \cos(\psi_y) f. \tag{20}$$

In this case and if ω_0 is the revolution frequency, the beam current contains the harmonics

$$\omega_{mn} = m_y \omega_y + n\omega_0, \quad m_y = \pm 1.$$

Without frequency spreads the beam centroids (as well as other multipole momenta of f) oscillate with constant amplitudes, which are determined by initial deflections of the beam. Effects of the frequency spreads ($\omega_\alpha(I)$) result in decays of the amplitudes of multipole modes (Fig. 2). This means that, although particle equations of motion are Hamiltonian, multipole momenta may obey non-Hamiltonian equations of motion.

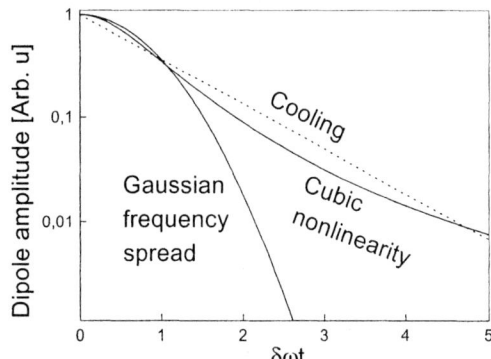

FIGURE 2. Dependence of the amplitude of dipole oscillations on time. Solid lines – variations due to frequency spreads, dotted line – due to beam cooling.

For smooth frequency distribution functions coherent oscillations decay until new kicks occur. Final widths of the beam frequency spectra may result in echoing dependencies of (dipole) momenta on time. In real cases, such spectra occur because of the finite number of particles in the beam. This means that, after initial decay, coherent oscillations in such a beam can appear again spontaneously (Fig. 3). The amplitudes and phases of these spontaneous coherent fluctuations are random functions of time. The interaction of the beam particles with the fields induced by

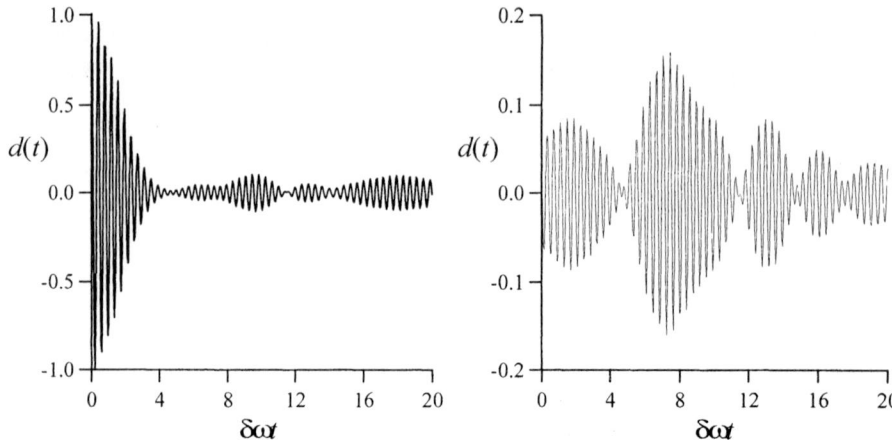

FIGURE 3. Dipole momentum vs. time. The beam contains 200 particles with equal initial amplitudes. Same frequency spread. In left– initial phases are equal, in right – initially uniform distribution in phases.

these fluctuations may result in beam heating (see, e.g. [4] for more detail). That is the main difference between coherent oscillations and coherent fluctuations of the beam.

IV COHERENT OSCILLATIONS AND WAKES

Coherent oscillations of intense beams can become unstable because of their interaction with the fields, that the beam induces in the surrounding media. These can be electrodes in the vacuum chamber, secondary particles that the beam produces inside the vacuum chamber, or other beams, which can be encountered by the beam moving along the closed orbit. In such cases the Hamiltonian describing the particle motions is

$$H = \omega_\alpha I_\alpha - L, \quad L = \frac{e}{c}\mathbf{v}\mathbf{A}(\mathbf{r},t) - eA_0(\mathbf{r},t), \qquad (21)$$

where $\mathbf{A}(\mathbf{r},t)$ and $A_0(\mathbf{r},t)$ are the vector and scalar potentials of electromagnetic fields induced by the beam, e is the charge of the particle, and c is the speed of light. These potentials and the fields perturbing the particle oscillations,

$$\mathbf{E} = -\frac{1}{c}\frac{\partial \mathbf{A}}{\partial t} - \mathrm{grad}\, A_0, \quad \mathbf{H} = \mathrm{rot}\,\mathbf{A},$$

are calculated by solving the Maxwell equations. Assuming that these equations are already solved, we can write the Lagrangian describing the perturbations due to induced fields L in the form

$$L = Ne^2 \int d^3 I' \int_0^{2\pi} \frac{d^3 \psi'}{(2\pi)^3} \int_0^\infty d\tau K(\mathbf{r}_\perp, \theta[t], \mathbf{r}'_\perp, \theta'[t-\tau], \tau) f(\mathbf{r}', \mathbf{p}', t-\tau), \quad (22)$$

where the kernel K is calculated by using the Green functions of the corresponding electromagnetic problem. For example, the kernel describing the interaction with the radiation fields can be written as

$$K = \frac{v_\alpha v'_\beta}{c^2} G_{\alpha,\beta}(\mathbf{r}_\perp, \theta[t], \mathbf{r}'_\perp, \theta'[t-\tau], \tau), \quad \text{div}\mathbf{A} = 0.$$

Because of the causality condition, the field radiated by a particle passing an electrode located at a distance $l_\perp$ from the closed orbit will affect the motions of the particles that pass this electrode after a time interval of at least $\Delta t = 2l_\perp/c$. Thus particles in the beam interact via retarding fields. For this reason, such induced fields are now commonly called wakefields (or simply wakes) of the beam.

The Vlasov equation describing the oscillations of the intense beam,

$$\frac{\partial f}{\partial t} + \omega_\alpha \frac{\partial f}{\partial \psi_\alpha} = [L; f], \quad (23)$$

is a very complicated integro-differential nonlinear equation. Even numerical calculation of the solutions to this equation is not easy. Fortunately, we may point out at least two large classes of problems of substantial practical importance where Eq.(23) can be solved by using the perturbation theory. These are (1) predictions concerning the stability of a given beam distribution relative to the excitations in the beam of small coherent oscillations, and (2) predictions concerning the time dependence of small coherent oscillations after an initial kick. In both cases, we may assume that perturbations due to induced fields of the particle motions during the oscillation periods or during the revolution periods in the ring are small ($|\Delta f| \ll f_0$). Correspondingly, we write

$$f = f_0(I) + \Delta f(I, \psi, t), \quad L = L_0[f_0] + \Delta L[\Delta f], \quad (24)$$

and linearize the Vlasov equation with respect to Δf:

$$\frac{\partial \Delta f}{\partial t} + \omega_\alpha \frac{\partial \Delta f}{\partial \psi_\alpha} - [L_0; \Delta f] = [\Delta L; f_0]. \quad (25)$$

Here, $f_0(I)$ is the stationary distribution function describing a beam without coherent oscillations. In the most general case, such a function obeys the Vlasov equation,

$$\frac{\partial f_0}{\partial t} + \omega_\alpha \frac{\partial f_0}{\partial \psi_\alpha} = [L_0; f_0], \quad (26)$$

and, since the Lagrangian is a periodic function of θ and of the oscillation phases:

$$L(\theta + 2\pi) = L(\theta), \quad L(\psi + 2\pi) = L(\psi),$$

the geometry of the phase-space of an intense beam can differ very strongly from that of an unperturbed beam (see, e.g. [10], [11]). In particular, Eq.(26) can describe various distortions of beam focusing and excitations of nonlinear resonances. If these distortions are strong, the action-phase variables of unperturbed oscillations (I, ψ) become inconvenient variables to describe stationary distributions of the beam. Moreover, if the perturbation L is a nonlinear function of the particle coordinates, effects of nonlinear resonances may define the buckets in the beam phase-space, and therefore the phase-space may contain multiple-connected regions. Thus, prior to study of the stability of coherent oscillations we have to calculate possible solutions to Eq.(26). Usually, that demands the use of numerical methods (see, e.g. [12]).

Fortunately, in many cases where we have to study coherent oscillations of beams, most such phenomena are not very important because of weak strengths and weak nonlinearity of the induced fields. Provided that the working points in the oscillation frequencies are far from nonlinear resonances, the main effect of the stationary induced fields is to shift of the particle oscillation frequencies by an amount

$$\Delta \omega_\alpha(I) = -\int_0^{2\pi} \frac{d\theta d^3\psi}{(2\pi)^4} \left(\frac{\partial L_0}{\partial I}\right). \tag{27}$$

Assuming that such frequency shifts are already included in ω_α, we rewrite Eq.(25) in the form

$$\frac{\partial \Delta f}{\partial t} + \omega_\alpha \frac{\partial \Delta f}{\partial \psi_\alpha} = -\frac{\partial \Delta L}{\partial \psi_\alpha}\frac{\partial f_0}{\partial I_\alpha} + \delta[L_0, \Delta f]. \tag{28}$$

General eigensolutions to this equation with periodic coefficients read

$$\Delta f = \exp(-i\omega t) X_\omega(I, \psi, t), \quad X_\omega\left(t + \frac{2\pi}{\omega_0}\right) = X_\omega(t). \tag{29}$$

Therefore, both eigenfunctions X_ω and eigenfrequencies ω can be obtained by using a single-turn map for Δf. As seen from Eq.(28), the eigenfrequencies ω generally are shifted relative to the frequencies of unperturbed oscillations ($\omega_m = m_\alpha \omega_\alpha$) by the coherent frequency shift ($\Delta \omega_m = \omega - \omega_m$). In many cases the values of coherent frequency shifts for the most important multipole modes are small compared to the distance from the working point in oscillation frequencies to the lines of the ring resonances:

$$|\omega - m_\alpha \omega_\alpha| \ll \min |l_\alpha \omega_\alpha + n\omega_0|, \tag{30}$$

where l_α and n are integers. Then, systematic variations of solutions to Eq.(28) are calculated by averaging this equation over the revolution period of particles in the ring. Simple calculations result in (see, e.g. [4] for more detail)

$$(\omega - m_\alpha \omega_\alpha[I])f_{\mathbf{m}} = Ne^2 \left(\mathbf{m} \frac{\partial f_0}{\partial \mathbf{I}} \right)$$

$$\times \sum_{\mathbf{m}'} \sum_{n=-\infty}^{\infty} \int d^3 I K_{\mathbf{m},\mathbf{m}'}^{n,n}(I, I'; \omega + n\omega_0) f_{\mathbf{m}'}(I'). \tag{31}$$

Here, $\mathbf{m}' - \mathbf{m}$ should satisfy the condition in Eq.(30), $\theta_s = \omega_0 t$, and

$$\Delta f = e^{-i\omega t} \sum_{\mathbf{m}} f_{\mathbf{m}}(I) \exp(im_\alpha \psi_\alpha), \quad f_{\mathbf{m}}(I) = \int_0^{2\pi} \frac{d\theta_s}{2\pi} f_{\mathbf{m},\omega}(I, \theta_s), \quad \text{Im}\,\omega > 0.$$

An error in the calculations of eigenfrequencies ω and eigenfunctions $f_{\mathbf{m}}$ by using Eq.(31) is of the order of

$$\frac{|\omega - m_\alpha \omega_\alpha|}{\min |l_\alpha \omega_\alpha + n\omega_0|} \ll 1. \tag{32}$$

Typically, the amplitudes of coherent oscillations vary during the time intervals, which substantially exceed the periods of betatron oscillations of particles. Then, with an accuracy of the order of $|\Delta \omega_m|/\omega_b \ll 1$ (where ω_b is ω_x or ω_y) we can replace Eq.(31) by the equation that describes the coupling of synchrotron modes only:

$$(\omega - m_\alpha \omega_\alpha[I])f_{\mathbf{m}} = Ne^2 \left(\mathbf{m} \frac{\partial f_0}{\partial \mathbf{I}} \right) \sum_{m_s', n=-\infty}^{\infty} \int d^3 I K_{\mathbf{m},\mathbf{m}'}^{n,n}(I, I'; \omega + n\omega_0) f_{\mathbf{m}'}(I'). \tag{33}$$

Here, $\mathbf{m} = \{m_x, m_y, m_s\}$. Finally, if the amplitudes of coherent oscillations vary for many periods of the synchrotron incoherent oscillations, we may neglect in Eq.(33) the coupling of the synchrotron and the synchrobetatron modes to arrive at an even more reduced integral equation:

$$(\omega - m_\alpha \omega_\alpha[I])f_{\mathbf{m}} = Ne^2 \left(\mathbf{m} \frac{\partial f_0}{\partial \mathbf{I}} \right) \sum_{n=-\infty}^{\infty} \int d^3 I K_{\mathbf{m},\mathbf{m}}^{n,n}(I, I'; \omega + n\omega_0) f_{\mathbf{m}}(I'). \tag{34}$$

In this case, the value of the coherent frequency shift is assumed to be substantially small compared to the distance in the unperturbed spectrum of coherent oscillations. For this reason, the multipole modes in Eq.(34) are uncoupled. Except for very special cases, equations obtained for f_m present homogeneous integral equations of the Fredholm type. They definitely have the spectra of eigenvalues. This means that Eqs.(31), (33) and (34) have non-zero solutions only in the case, where $\Delta\omega_m$ coincides with one of the eigenvalues of these equations. These eigensolutions present complete sets of the eigenfunctions. If the eigenfrequencies do not degenerate, any solution to these equation reads:

$$f_{\mathbf{m}}(I,t) = \sum_{a=0}^{\infty} C_a(f_{\mathbf{m}}[I])_a \exp[-i(\Delta\omega_{\mathbf{m}})_a t]. \qquad (35)$$

The coefficients C_a are calculated using an initial condition $f_{\mathbf{m}}(I, t = 0) = g_{\mathbf{m}}(I)$ and the orthogonality conditions for the functions $(f_{\mathbf{m}})_a$. Existence of the spectra of eigenfrequencies of coherent oscillations is a specific feature of the spectra coherent oscillations in storage rings. It is due to periodic motion of the beam particles along the closed orbit, when the perturbations due to the fields induced by any part of the beam (maybe in a long time) finally perturb the oscillations of the source particles. In such a case, induced fields close the feedback between oscillations of all beam particles. However, if such perturbations propagate along the beam during too long time intervals, the feedback will be looped in a time when the beam parameters change in such a way that the assumptions used in the calculations of our general integral equations no longer hold. During the time intervals when the beam feedback is not closed, the excitations of oscillations of various parts of the beam do not affect the source particles. Correspondingly, coherent oscillations lose the spectra of eigensolutions and develop similarly to the beam breakup oscillations observed in linacs. This is described by the possibility of transforming general integral equations into homogeneous integral equations of the Volterra type, which are known to have no eigensolutions at all (for single-bunch transverse instabilities such a transformation was studied e.g. in [13]).

V LONGITUDINAL OSCILLATIONS OF THE COASTING BEAM

The general integral equation, Eq.(34), becomes especially simple in the cases where we study coherent oscillations of coasting beams. For example, to describe the longitudinal coherent oscillations of such a beam ($m_x = m_y = 0$) we may often consider two-dimensional phase-space:

$$\Delta p = p - p_0, \quad I_s = R_0 \Delta p,$$
$$\psi_s = \theta - \theta_0, \quad \omega_s = \omega_0' \Delta p = \frac{\omega_0 \alpha_p}{p} \Delta p. \tag{36}$$

Here, $\Pi = 2\pi R_0$ is the perimeter of the closed orbit. For longitudinal oscillations only longitudinal components of the Green function are important:

$$K = \frac{v^2}{c^2} G_{s,s}(\mathbf{r}, \mathbf{r}', \theta, \theta', \tau). \tag{37}$$

If the beam interacts with surrounding electrodes, the Green function is a non-singular function of the particle coordinates and hence can be presented as a series in powers of x and y:

$$G_{s,s}(\mathbf{r}, \mathbf{r}', \theta, \theta', \tau) \simeq G_{s,s}(0, 0, \theta, \theta', \tau) + x \left. \frac{\partial G_{s,s}}{\partial x} \right|_0 + x \left. \frac{\partial G_{s,s}}{\partial x} \right|_0 + \dots.$$

Calculating the Fourier harmonics of $G_{s,s}(\mathbf{r}, \mathbf{r}', \theta, \theta', \tau)$ in betatron phases, we find that only even powers of transverse coordinates contribute to $K^{n,n}_{m,m}$. Now, if $l_\perp$ is the vacuum chamber radius, and estimating the derivatives $(\partial G_{ss}/\partial x)_0 \simeq G_{ss}(0, 0, \theta, \theta', \tau)/l_\perp$, we obtain

$$K^{n,n}_{m,m} = \left(\frac{v^2}{c^2}\right) [G_{ss}(0, 0, \theta, \theta', \tau)]^{n,n}_{m_s,m_s} + O\left(\frac{a^2}{l_\perp^2}\right), \tag{38}$$

where a is the beam radius. For most important cases, when $a \ll l_\perp$ we can omit the second term in Eq.(38) to write

$$K^{n,n}_{m,m} = \left(\frac{v^2}{c^2}\right) [G_{ss}(\theta, \theta', \tau)]^{n,n}_{m_s,m_s}, \quad G_{ss}(\theta, \theta', \tau) \equiv G_{ss}(0, 0, \theta, \theta', \tau). \tag{39}$$

Now, using

$$\overline{G_{ss}(\theta, \theta', \tau)} = \sum_n [G_{ss}(\theta, \theta', \tau)]^{n,n} e^{in(\theta-\theta')}$$

and

$$f = f_0(\Delta p) + e^{-i\omega t} \sum_{m_s} f_{m_s} e^{im_s \psi_s},$$

we conclude that for longitudinal oscillations of a coasting beam $m_s = n$ and

$$K_{m,m}^{n,n}(\omega + n\omega_0) = \delta_{m_s,n}\left(\frac{v^2}{c^2}\right)[G_{ss}]_{n,n}(\omega + n\omega_0),$$

while the integral equation, Eq.(34), reads

$$(\omega - n\omega_0'\Delta p)f_n = \frac{Ne^2n}{R_0}\frac{\partial f_0}{\partial \Delta p}G_n(\omega + n\omega_0)\int_{-\infty}^{\infty} d\Delta p' f_n(\Delta p'). \tag{40}$$

Here we define $G_n(\omega) = [G_{ss}]_{n,n}(\omega)$ and assumed $v/c \simeq 1$. If the frequency width of $G_n(\omega)$ substantially exceeds the value of the coherent frequency shift ($|\omega - n\omega_0'\Delta p|$), we can replace in the rhs of Eq.(40) an exact value of $G_n(\omega + n\omega_0)$ by its approximation $G_n(n\omega_0)$. That results in

$$(\omega - n\omega_0'\Delta p)f_n = \frac{Ne^2n}{R_0}\frac{\partial f_0}{\partial \Delta p}G_n(n\omega_0)\int_{-\infty}^{\infty} d\Delta p' f_n(\Delta p'). \tag{41}$$

A similar equation can be obtained if we express the longitudinal perturbation in the coasting beam in terms of harmonics of the induced electric field (see, e.g. [4]):
[2]

$$(\omega - n\omega_0[\Delta p])f_n = -ie\overline{E_n(n\omega_0)}\frac{\partial f_0}{\partial \Delta p}. \tag{42}$$

The value $\overline{E_n(n\omega_0)}$ is usually expressed in terms of the longitudinal coupling impedance

$$\overline{eE_n(\omega)} = -\frac{Ne^2\omega_0 Z_n(\omega)}{2\pi R_0}\int d\Delta p' f_n(\Delta p').$$

Now, comparing Eqs.(41) and (42) we express the coupling impedance in terms of the Fourier harmonic of the Green function,

$$\frac{iZ_n(\omega)}{\omega} = \frac{2\pi}{\omega_0^2}G_n(\omega). \tag{43}$$

Taking as solutions to Eq.(41) the functions $f_{n,\omega}(\Delta p) = C[\partial f_0/\partial \Delta p]/(\omega - n\omega_0'\Delta p)$, we obtain the dispersion equation defining the eigenfrequencies of the problem,

---

[2] The reader can easily prove that consistent transformation of the relevant Vlasov equation contains the same averaging steps and assumptions that we used in obtaining the general integral equation.

$$1 = -\frac{\Omega_n^2}{n\omega_0'} \int_{-\infty}^{\infty} d\Delta p \frac{\partial f_0/\partial \Delta p}{\omega - n\omega_0(\Delta p)}, \quad \text{Im}\omega > 0. \quad (44)$$

Like all previous calculations this equation holds only in the region $\text{Im}\omega > 0$, where solutions describe unstable oscillations. A dispersion equation describing stable solutions $\text{Im}\omega < 0$ should be obtained by using analytic continuation of the rhs of Eq.(44). Both the eigenfunctions $f_{n,\omega}(\Delta p)$ and the integrand in Eq.(44) have a pole singularity in ω, or in Δp, when ω tends in its complex plane to its real axes. For this reason, the rhs in Eq.(44) as a function of the complex variable ω has a cut along the axes $\text{Im}(\omega) = 0$. According to the Landau rule, the analytic continuation of this function is obtained deforming the integration contour in the plane of a complex variable Δp to keep the integrand singularity above the integration contour (see Fig. 4). The resulting equation reads

$$1 = -\frac{\Omega_n^2}{n\omega_0'} \int_{-\infty}^{\infty} d\Delta p \frac{\partial f_0/\partial \Delta p}{\omega - n\omega_0(\Delta p)} + 2\pi i \frac{\Omega_n^2}{(n\omega_0')^2} \frac{\partial f_0}{\partial \Delta p}\bigg|_{n\omega_0(\Delta p)=\omega}, \quad \text{Im}\omega < 0. \quad (45)$$

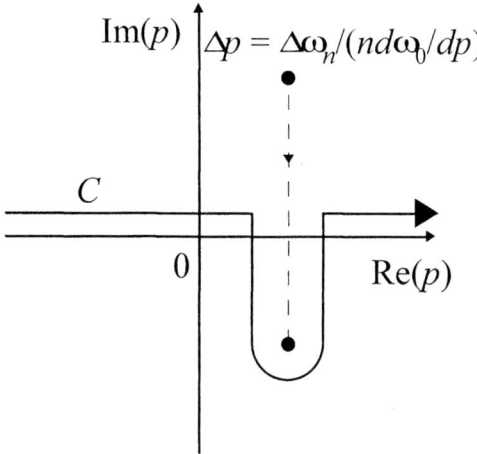

FIGURE 4. The integration contour in the dispersion equation for decaying solutions.

In all these equations we defined the value

$$\Omega_n^2 = n^2 \frac{Ne^2 \omega_0 \omega_0'}{2\pi R_0} \left[\frac{-iZ_n(n\omega_0)}{n} \right], \quad (46)$$

which means the square of the (complex) coherent frequency shift of the monochromatic coasting beam:

$$f_0(\Delta p) = \delta(\Delta p), \quad \omega = \pm\sqrt{\Omega_n^2}. \tag{47}$$

These solutions describe unstable oscillations, if $\Omega_n^2 < 0$ (the negative mass instability), or if the beam interacts with dissipative electrodes $\mathrm{Re}(Z_n) \neq 0$.

VI LANDAU DAMPING

These instabilities can be suppressed because of the revolution frequency spread of the beam which provides a special damping mechanism for coherent oscillations – Landau damping. For example, substituting in Eq.(44)

$$f_0 = \frac{\Delta}{\pi[\Delta p^2 + \Delta^2]},$$

we obtain:

$$1 = \frac{\Omega_n^2}{(\omega + i|n|\delta\omega)^2}, \quad \omega = -i|n|\delta\omega \pm \sqrt{\Omega_n^2}, \quad \delta\omega = |\omega_0'|\Delta. \tag{48}$$

According to these equations the oscillations are stable if the parameters of the beam obey the stability condition $|n|\delta\omega \geq \mathrm{Im}\sqrt{\Omega_n^2}$. This stability condition holds beforehand provided that Z_n/n satisfies the Z/n (or Keil-Schnell [14]) criterion

$$|Z_n/n| \leq \frac{pv|\alpha_p|}{eI_b}\frac{\Delta^2}{p^2}. \tag{49}$$

Here, $I_b = Ne\omega_0/(2\pi)$ is the beam current. This criterion predicts quite realistic thresholds in the values of the coupling impedance and in the beam parameters for particle energies above the transition energy of the ring. Below transition energy the values of thresholds may substantially exceed that predicted by the Z/n criterion (for example, in experiments on electron cooling in NAP-M the threshold currents exceeded Z/n thresholds by about a factor of 200 [15]).

Deeper insight into the nature of Landau damping can be obtained if we inspect the case where $\Omega_n^2 > 0$ and $\Omega_n \gg |n|\delta\omega$, while the roots of the dispersion equation are [3]

$$\omega \simeq \pm\Omega_n + i\pi\frac{\Omega_n^3}{2(n\omega_0')^2}\left.\frac{\partial f_0}{\partial \Delta p}\right|_{n\omega_0'\Delta p = \pm\Omega_n}. \tag{50}$$

---

[3] Closer inspection (see, e.g. [16]) shows that apart from the eigensolutions discussed here, the Vlasov equation may have an alternative set of eigensolutions without any damping. This set of the eigensolutions is complete. The oscillations in real beams are presented by combinations of these undamped eigensolutions and therefore decay with the Landau damping decrements, which we discussed above.

This formula shows that oscillations are damped by interaction of the beam particles with the excited wave and only in the case where the number of beam particles obtaining energy from the wave $(n\omega_0'\Delta p < \Omega_n)$ exceeds the number of particles exciting the wave $(n\omega_0'\Delta p > \Omega_n)$. In our simple model, it holds if $(\partial f_0/\partial \Delta p)_{\Delta p=\pm\Omega_n/[n\omega_0']} < 0$. In an inverse case, when $(\partial f_0/\partial \Delta p)_{\Delta p=\pm\Omega_n/[n\omega_0']} > 0$ such a mechanism results in Landau anti-damping. We see that an adequate interpretation of Landau damping demands a beam description beyond the self-consistent field approximation. That is not an accidental fact. It indicates a deep mutual coupling of collective and individual-particle effects in the beams of charged particles. In particular, the beam Schottky noise reaches its equilibrium level by Landau damping of coherent fluctuations. Typically this occurs during time intervals that are substantially shorter than the momentum relaxation time of the beam. However, in very cold beams (or in beams with parameters close to instability thresholds) we can encounter the parameter regions where the Landau damping times are comparable to the beam relaxation times. In such cases, the mutual relaxation of the beam particles and of the beam coherent fluctuations results in new equilibria, where the beam phase-space volumes and noise levels relax simultaneously and their equilibrium values, generally, depend on the beam intensity (see, e.g. [4]).

Returning to studies of coherent oscillation stability we note that, for engineering purposes, instead of solving the dispersion equations it is more important to predict the regions of coupling impedances or of beam parameters where coherent oscillations do not lose their stability. For a given beam distribution function, these stability diagrams can be calculated, if in the dispersion equation ω tends toward its real axes $(\omega \to \omega + i0)$. For example, if we define in Eq.(44) a new dimensionless variable $\zeta = \Omega_n^2/n^2\delta\omega^2$, the equation for the stability border (see also Fig. 5) is

$$\frac{1}{\zeta(\omega)} = n\omega_0'\Delta^2 \int_{-\infty}^{\infty} d\Delta p \frac{\partial f_0/\partial \Delta p}{\omega - n\omega_0(\Delta p) + i0}, \quad -\infty < \omega < \infty. \quad (51)$$

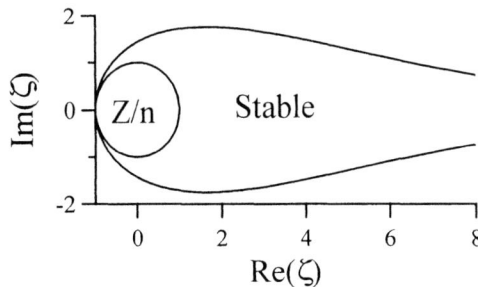

FIGURE 5. Stability diagram for longitudinal coherent oscillations in a coasting beam with a Gaussian momentum distribution function. The circle shows the Z/n region.

VII TRANSVERSE OSCILLATIONS OF A COASTING BEAM

Another simple application of the described above theory is in the case of transverse coherent oscillations in coasting beams. For simplicity, we assume dipole (vertical) modes ($\mathbf{m} = \{0, m_y = \pm 1, m_s\}$) and the smoothed focusing approximation for betatron oscillations. In addition to Eq.(36), we write

$$y = \sqrt{\frac{2I_y}{\gamma M \omega_y}} \cos\psi_y, \quad p_y = \frac{p}{R_0}\frac{dy}{d\theta}, \quad \dot\psi_y = \omega_y(I_y, \Delta p). \tag{52}$$

Because of the azimuthal symmetry of f_0 we have $K^{n,n}_{\mathbf{m},\mathbf{m}} \propto \delta_{m_s,n}$. Hence, the longitudinal mode number coincides with n. The distribution functions describing coherent oscillations depend on two phase variables:

$$f = f_0(I_y, \Delta p) + e^{-i\omega t}\sum_{m_y,n} f_{m_y,n} e^{im_y\psi_y + in\psi_s}, \tag{53}$$

while the amplitudes $f_{m_y,n}$ obey the general integral equation:

$$(\omega - \omega_{mn})f_{\mathbf{m}} = Ne^2\left(m_y\frac{\partial f_0}{\partial I_y} + \frac{n}{R_0}\frac{\partial f_0}{\partial \Delta p}\right)\int d^3 I\, K^{n,n}_{\mathbf{m},\mathbf{m}}(I, I'; \omega + n\omega_0)f_{\mathbf{m}}(I'). \tag{54}$$

Here, $\omega_{mn} = m_y\omega_y - n\omega'_0\Delta p$. Before we calculate the kernel $K^{n,n}_{\mathbf{m},\mathbf{m}}(I, I'; \omega + n\omega_0)$, we note that for smooth distribution functions the derivatives in Eq.(54) are estimated by using

$$\frac{\partial f_0}{\partial I_y} \sim -\frac{f_0}{p\epsilon_y}, \quad \frac{\partial f_0}{\partial \Delta p} \sim -\frac{f_0}{p\sigma_\delta},$$

where ϵ_y is the vertical emittance of the beam and σ_δ is the relative beam momentum spread. Then, the ratio of the first to the second term in the large parentheses in this equation is estimated by using

$$m_y\frac{\partial f_0}{\partial I_y} + \frac{n}{R_0}\frac{\partial f_0}{\partial \Delta p} \sim -\frac{f_0}{p\epsilon_y}\left[m_y + n\frac{\epsilon_y}{R_0\sigma_\delta}\right].$$

Thus for oscillations with harmonic numbers $1 \ll |n| \ll R_0\sigma_\delta/\epsilon_y \sim \sigma_\delta/(\nu_y\theta_b^2)$ ($\theta^2 \simeq \epsilon_y\nu_y/R_0$) we may neglect in Eq.(54) the term proportional to $\partial f_0/\partial\Delta p$ and may rewrite this equation in the form

$$(\omega - \omega_{mn})f_{\mathbf{m}} = Ne^2 m_y\frac{\partial f_0}{\partial I_y}\int d^3 I\, K^{n,n}_{\mathbf{m},\mathbf{m}}(I, I'; \omega + n\omega_0)f_{\mathbf{m}}(I'). \tag{55}$$

Now, we calculate the kernel $K_{m,m}^{n,n}$ assuming no wake excitations by the beam transverse currents. Then, only the component $G_{s,s}$ contributes in the kernel:

$$K_{m,m}^{n,n} = \int_0^{2\pi} \frac{d\psi_y d\psi_y'}{(2\pi)^2} [G_{s,s}^{n,n}(y,y',\omega+n\omega_0)] e^{-im_y[\psi_y-\psi_y']}$$

$$= \frac{(I_y I_y')^{|m_y|/2}}{2\gamma M \omega_y} \left. \frac{\partial^{2|m_y|} G_{s,s}^{n,n}(y,y',\omega)}{\partial y^{|m_y|} \partial (y')^{|m_y|}} \right|_0 [1+O(a^2/l_\perp^2)], \quad m_y = \pm 1.$$

Here, a is the beam radius, and $l_\perp$ is the distance from the orbit to the electrodes. Neglecting the terms of the order of $(a^2/l_\perp^2) \ll 1$, we rewrite Eq.(55) in the form

$$(\omega - \omega_{mn}) f_{mn} = \frac{\partial f_0}{\partial I_y} \left(\frac{Ne^2 m_y}{2\gamma M \omega_y} \right) \left(\frac{\partial^2 G_{s,s}^{n,n}(y,y',\omega)}{\partial y \partial z y} \right)_{y,y'=0} \int dI_y' d\Delta p' (I_y I_y')^{|m_y|/2} f_{mn}. \quad (56)$$

If we define the deflecting force for dipole coherent oscillations using

$$\overline{(F_y)_{m_y,n}(\omega)} = -\frac{Ne^2 \omega_0 Z_n^\perp(\omega)}{\Pi} \int dI_y' d\Delta p' (y)_{m_y} f_{mn}, \quad (y)_{m_y} = \sqrt{\frac{R_0 I_y}{2p\nu_y}}, \quad (57)$$

where $Z_n^\perp(\omega)$ is the transverse coupling impedance, then the rhs in the integral equation is expressed in terms of this deflecting force provided that

$$Z_n^\perp(\omega) = \frac{2\pi R_0}{\omega_0} \left(\frac{\partial^2 G_{s,s}^{n,n}(y,y',\omega)}{\partial y \partial z y} \right)_{y,y'=0}. \quad (58)$$

Comparing this formula and Eq.(43), we obtain the relationship between the longitudinal and transverse coupling impedances in the following form ($v \simeq c$):

$$Z_n^\perp(\omega) = \frac{ic}{\omega} \left(\frac{\partial^2}{\partial y \partial y'} Z_n^\parallel(y,y',\omega) \right)_{y,y'=0}. \quad (59)$$

For pipes with cross sections close to round this formula can be simplified even more:

$$Z_n^\perp(\omega) \simeq \frac{c}{l_\perp^2} \frac{iZ^\parallel(\omega)}{\omega}, \quad (60)$$

where $l_\perp$ is a pipe radius. Taking as solutions to Eq.(56) the functions

$$f_{m,n} = C \frac{(I_y)^{|m_y|/2} (\partial f_0/\partial I_y)}{\omega - \omega_{mn}}, \quad \text{Im}\,\omega > 0, \quad (61)$$

335

we arrive at the dispersion equation defining the frequencies of coherent oscillations:

$$1 = \frac{\Omega_{m,n}}{A_m} \int dI_y d\Delta p \frac{(I_y)^{|m_y|} \partial f_0/\partial I_y}{\omega - \omega_{mn}[I_y, \Delta p]}. \quad (62)$$

Here, $A_m = \int dI_y (I_y)^{|m_y|} \partial f_0/\partial I_y$ is the normalization constant, while the value

$$\Omega_{mn} = \frac{Ne^2 m_y}{2\gamma M \omega_y} \left. \frac{\partial^{2|m_y|} G_{s,s}^{n,n}(y, y', \omega + n\omega_0)}{\partial y^{|m_y|} \partial (y')^{|m_y|}} \right|_0 A_m \quad (63)$$

determines the coherent frequency shift of the monochromatic coasting beam. For a non-resonant interaction we can replace in Eq.(63) an exact value of ω by its unperturbed value ω_{mn}. Since $\text{Im}[G(\omega)]$ is an odd function of its argument, the stability condition for monochromatic beam can be written in a very general form [3]:

$$-\text{Im}\Omega_{mn} > 0, \quad m_y(m_y \omega_y + n\omega_0) > 0. \quad (64)$$

This condition definitely violates for azimuthal harmonics with $|n| > |m_y|\nu_y$. Note also that, according to Eq.(63), for a beam interacting with surrounding electrodes the values of Ω_{mn} exponentially decrease with an increase in the multipole number of betatron oscillations (m_y or m_x). In such cases, we can focus our studies on calculations for dipole oscillations. [4]

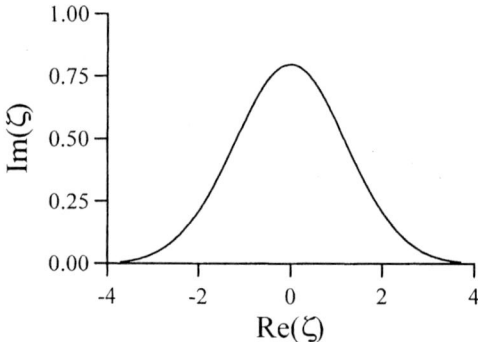

FIGURE 6. Stability diagram for transverse coherent oscillations in a coasting beam with a Gaussian momentum distribution function. The frequency spread is due to ring chromaticity. Oscillations are stable below the curve.

Unstable modes can be stabilized by Landau damping. For transverse oscillations in the coasting beam this damping can be produced by a (cubic) nonlinearity of

---

[4] Such a simplification cannot be used in describing coherent instabilities in the beams, where the Coulomb interactions between particles dominate (e.g. the beam-beam instability, the instability of an ion beam due to its interaction with a cooling electron beam and some others).

the guiding fields, or by chromatic dependence of the oscillation frequencies on the particle energy. For the latter case, the stability diagram is shown in Fig. 6. Since the frequency spread is due to differences in particle momenta, no Landau anti-damping for transverse oscillations is possible in this case.

ACKNOWLEDGMENTS

I thank Professors N. Dikansky, A. Lebedev, V. Parkhomchuk and L. Palumbo for their valuable discussions.

REFERENCES

1. Dikansky N.S., Pestrikov D.V., The Influence of the Ordering Effects on Relaxation of Coasting, Cold Beam in Storage Rings. in *Proceedings of ECOOL 1984*, p. 275, Kernforschungszentrum Karlsruhe, 1984; see also *Sov. Journ. of Tech. Phys.*, **56**, p. 289, 1986.
2. Lebedev A.N., in *Proceedings of the 6th Intern. Conf. on High Energy Accel.*, Cambrige (Mass.), p. 284, 1967.
3. Derbenev Ya.S., Dikansky N.S., in *Proceedings of the 7th Intern. Conf. on High Energy Accel.*, v. 2, p. 294, Yerevan 1970; see also Preprint INP 315. Novosibirsk 1969, SLAC-TRANS-0106, 1969; Preprint INP 318, Novosibirsk 1969, SLAC-TRANS-0114, 1969.
4. Dikansky N.S., Pestrikov D.V., *Physics of Intense Beams and Storage Rings.* AIP Press, New York, 1994.
5. Chao A.W., *Physics of Collective Beam Instabilities in High Energy Accelerators*, Wiley, New York, 1993.
6. Chattopadhyay S., Some Fundamental Aspects of Fluctuations and Coherence in Charged Particle Beams in Storage Rings, CERN 84-11, CERN, 1984.
7. Klimontovich Yu.L., *Statictical Theory of Nonequilibrium Processes in Plasma*, Moscow Univ. 1964 (in Russian).
8. Bogolubov N.N., *Problems of Dynamical Theory in Statistical Physics*, Gostekhizdat, Moscow, 1946 (in Russian).
9. Kolomensky A.A., Lebedev A.N., *Theory of Cyclic Particle Accelerators.* Wiley, New York, 1962.
10. Lebedev A.N., On the Bunch Lengthening Effect in Storage Rings, in *Physics with Intersecting Storage Rings*, p. 184, Academic Press, New York, 1971.
11. Haissinski J., *Il Nuovo Cimento*, **18B**, (1), 72, 1973.
12. Oide K., Effects of the Potential-Well Distortion on the Longitudinal Single-Bunch Instability, in *Proceedings of the Fourth Advanced ICFA Beam Dynamics Workshop on Collective Effects in Short Bunches*, p. 64, KEK Report 90-21, 1991; Oide K. and Yokoya K., KEK-Preprint-90-10, 1990.
13. Pestrikov D.V., *Particle Accelerators*, **41**, 13, 1993.
14. Keil E., Schnell W., CERN-ISR-TH-RF-69-48, CERN 1969.

15. Parkhomchuk V.V., Pestrikov D.V., *Sov. Journ. of Tech. Phys.*, **50**, 1411, 1980; Dementiev E.N., Dikansky N.S., et al., ibid, **50**, 1717, 1980.
16. Balescu R., *Statistical Mechanics of Charged Particles*, Wiley-Interscience, New York, 1963.

Transverse Instabilities

D.V. Pestrikov

Budker Institute of Nuclear Physics
630090 Novosibirsk, Russian Federation

Abstract. In this lecture, we discuss transverse coherent oscillations for cases where the fields induced by the beam have a long memory providing multibunch or resonant instabilities.

I INTRODUCTION

The name "transverse instabilities" covers a wide class of collective effects including transverse coherent oscillations of beams. In the preceding lecture [1], we discussed specific features of the transverse instabilities of coasting beams. Here we focus on the instabilities of bunched beams containing one or many bunches. In this case, a description of the incoherent oscillations of particles should include synchrotron oscillations so that transformation to the unperturbed action-phase variables of, for example, vertical betatron and synchrotron oscillations is obtained using

$$y = \sqrt{\frac{2I_y}{\gamma M \omega_y}} \cos \phi_y, \quad p_y = \frac{p}{R_0} \frac{dy}{d\theta}, \quad \frac{d\phi_y}{dt} = \omega_y + \frac{d\omega_y}{dp} \Delta p,$$

$$\theta = \omega_0 t + \phi, \quad \phi = a_s \cos(\psi_s), \quad \frac{d\phi}{dt} = \frac{d\omega_0}{dp} \Delta p,$$

$$\phi_y = \int dt' \omega_y = \psi_y + \frac{d\omega_y/dp}{d\omega_0/dp} \int dt' \frac{d\phi}{dt'} = \psi_y + \frac{d\omega_y}{d\omega_0} \phi. \tag{1}$$

We see that, in the bunched beam, dependence of the frequencies of betatron oscillations on particle energy results in sinusoidal modulation of the phases of the betatron oscillations by the synchrotron oscillation of particles. The depth of this modulation is determined by the ring chromaticity:

$$\frac{d\omega_y}{d\omega_0} = \nu_y + \frac{d\nu_y}{d\ln \omega_0} = \nu_y + \frac{1}{\alpha_p} \frac{d\nu_y}{d\ln p}, \quad \alpha_p = \frac{1}{\gamma^2} - \alpha, \tag{2}$$

where α is the ring momentum compaction factor. Note that above the transition energy of the ring ($\gamma_{tr} = 1/\sqrt{\alpha}$) the value of $d\nu_y/d\ln\omega_0$ is negative, while below the transition energy it is positive.

II INTEGRAL EQUATION

Without synchrobetatron mode coupling, the spectra of transverse (vertical) coherent oscillations are calculated by solving the following general integral equation (see, e.g. [1]):

$$(\omega - m_a\omega_a[I])f_\mathbf{m} = Ne^2 m_y \frac{\partial f_0}{\partial I_y} \sum_{n=-\infty}^{\infty} \int d^3 I K_{\mathbf{m},\mathbf{m}}^{n,n}(I, I'; \omega + n\omega_0) f_\mathbf{m}(I'), \quad (3)$$

where the kernel is calculated by using the Green function of the relevant electromagnetic problem. For the cases where the beam interacts with surrounding electrodes, the dipole betatron modes $m_y = \pm 1$ are most important. Then, $K_{\mathbf{m},\mathbf{m}}^{n,n}(I, I'; \omega + n\omega_0) \propto \sqrt{I_y I_y'}$, while transverse coherent oscillations of bunches are described by the following distribution functions:

$$f = f_0(I_y)\rho(a_s) + e^{-i\omega t}\sqrt{I_y}\frac{\partial f_0}{\partial I_y} \sum_{m_y=\pm 1}\sum_{m_s} f_\mathbf{m} e^{im_y(\psi_y)_s + im_s\psi_s}. \quad (4)$$

Since y and p_y in the kernel of Eq.(3) depend on the betatron phase variable in the combination ϕ_y, calculation of the Fourier amplitudes $K_{\mathbf{m},\mathbf{m}}^{n,n}$ results in the substitution of $e^{in\phi}$ for

$$\exp\left(i\left[n + m_y\frac{d\omega_y}{d\omega_0}\right]\phi\right) = e^{in_1\phi}.$$

Then, calculating the Fourier harmonics in the phases of the synchrotron oscillations, we obtain

$$\{e^{in_1\phi}\}_{m_s} = \int_0^{2\pi} \frac{d\psi_s}{2\pi} e^{in_1\phi - im_s\psi_s} = J_{m_s}(n_1 a_s),$$

where $J_m(x)$ is the Bessel function. Thus, for transverse coherent oscillations the amplitudes f_m obey the following integral equation:

$$f_m(a_s) = -\rho(a_s)\sum_{n=-\infty}^{\infty} \Omega_{mn}(\omega + n\omega_0) J_{m_s}(n_1 a_s)$$

$$\times \int_0^\infty da_s' a_s' J_{m_s}(n_1 a_s') f_m(a_s') \int_0^\infty dI_y \frac{I_y \partial f_0/\partial I_y}{\omega - m_s\omega_s - m_y\omega_y(I_y)}. \quad (5)$$

Here, the function $\Omega_{mn}(\omega + n\omega_0)$ means the coherent frequency shift of the monochromatic coasting beam interacting with the same electrodes. According to this equation, coherent modes in the bunched beam are constructed from coupled azimuthal modes of the coasting beam. If we can neglect excitation of wakes by transverse beam currents, the value $\Omega_{mn}(\omega + n\omega_0)$ can be expressed in terms of the transverse coupling impedance:

$$\Omega_{mn}(\omega) = -m_y \frac{Ne^2\omega_0}{4\pi p\nu_y} Z_n^\perp(\omega). \tag{6}$$

Generally, Eq.(5) can be solved numerically. Analytic studies of particular problems by using this equation, typically, demand more simplified assumptions.

III RESONANT INSTABILITIES

Equation (5) enables substantial simplifications in the case of the resonant instability. Such an instability was first observed at VEPP-2 (BINP, Novosibirsk [2]) and was understood to be due to interaction of the beam with a high-Q parasitic cavity [3]. To describe such an instability we take

$$Z_n^\perp(\omega) = \frac{c}{l_\perp^2}\left(\frac{iZ(\omega)}{\omega}\right), \quad Z(\omega) = \frac{Z_0}{1 + iQ\left(\frac{\omega_k}{\omega} - \frac{\omega}{\omega_k}\right)} \tag{7}$$

and assume that the width of the resonance ($\omega = \pm\omega_k$) is much smaller than the revolution frequency of the particles:

$$\lambda_k = \frac{\omega_k}{2Q} \ll \omega_0.$$

Then, near the resonance,

$$\omega_k \simeq n\omega_0 + m_y\omega_y, \tag{8}$$

in Eq.(5) in the sum over n we may neglect the contributions of all harmonics of the beam current except the resonant one. The resulting integral equation is reduced to the following:

$$f_m = -\rho(a_s)\frac{Ne^2\omega_0^2 m_y}{4\pi p}\left(\frac{\overline{\beta_y}}{l_\perp^2}\right)\frac{Z_0/(2Q)J_{m_s}(n_1 a_s)}{\Delta\omega_m + i\lambda_k - \varepsilon}$$
$$\times \int_0^\infty da_s' a_s' J_{m_s}(n_1 a_s') f_m(a_s') \int_0^\infty dI_y \frac{I_y \partial f_0/\partial I_y}{\omega - m_s\omega_s - m_y\omega_y(I_y)}. \tag{9}$$

Here, we used that, near the resonance in Eq.(8), the value of the coupling impedance can be replaced by its approximation:

$$\frac{iZ(\omega)}{\omega} \simeq -\frac{Z_0/(2Q)}{\omega + i\lambda_k - \omega_k}, \qquad (10)$$

and defined the detuning from the resonance $\varepsilon = \omega_k - m_y\omega_y - n\omega_0$ as well as the frequency shifts

$$\Delta\omega_m = \omega - m_y\omega_y(0), \quad \Delta\omega_y(I_y) = \omega_y(0) - \omega_y(I_y).$$

Although Eq.(9) can be solved in a very general case, below we simplify the calculations, assuming a short bunch ($a_s \to 0$). Then, only betatron oscillations ($m_s = 0$) can be unstable, and simple calculations result in the following dispersion equation:

$$1 = \frac{-m_y\Omega_y^2}{\Delta\omega_m + i\lambda_k - \varepsilon}\int_0^\infty dI_y \frac{I_y \partial f_0/\partial I_y}{\Delta\omega_m - m_y\Delta\omega_y(I_y)}. \qquad (11)$$

Here,

$$\Omega_y^2 = \frac{Ne^2\omega_0^2(Z_0/Q)}{8\pi p}\left(\frac{\overline{\beta_y}}{l_\perp^2}\right). \qquad (12)$$

As seen from Eq.(11), the resonant interaction of a beam with an RF-cavity can result in instability of the coherent oscillations due either to dissipation in the cavity (dissipative-type instability), or to interaction of the beam oscillation mode with the RF field in a sum-type resonance. Following Derbenev and Dikansky [3], we call the latter a dynamical-type instability.

A Dynamical-Type Instability

To study dynamical-type resonant instability, in Eq.(11) we take $\lambda_k = 0$ (no dissipation) and neglect the frequency spread of betatron oscillations. The resulting dispersion equation is

$$1 = \frac{m_y\Omega_y^2}{[\Delta\omega_m - \varepsilon]\Delta\omega_m}. \qquad (13)$$

Its roots

$$\Delta\omega_m = \frac{\varepsilon \pm \sqrt{\varepsilon^2 + 4m_y\Omega_y^2}}{2} \qquad (14)$$

may describe unstable oscillations ($\Delta\omega_m^2 < 0$) if the beam oscillation mode is in sum-type resonance ($m_y < 0$) with the excited RF-field. The instability occurs if the beam current exceeds the threshold value

$$I_{th} = \frac{N_{th} e \omega_0}{2\pi} = \frac{4p}{e\omega_0(Z_0/Q)} \left(\frac{l_\perp^2}{\overline{\beta_y}}\right) \left(\frac{\varepsilon}{2}\right)^2. \quad (15)$$

Here, $\overline{\beta_y}$ is the value of the vertical β-function averaged over the cavity length along the closed orbit. In the exact resonance ($\varepsilon = 0$) the rise-time of this instability varies $\propto 1/\sqrt{N}$,

$$\frac{1}{\tau_m} = \sqrt{\frac{Ne^2\omega_0^2(Z_0/Q)}{8\pi p} \left(\frac{\overline{\beta_y}}{l_\perp^2}\right)}, \quad m_y = -1. \quad (16)$$

As seen from Fig. 1, the oscillations become unstable when coherent frequency shifts of the fast mode ($\Delta\omega_m = [\varepsilon + \sqrt{\varepsilon^2 + 4m_y\Omega_y^2}]/2$) and of the slow mode ($\Delta\omega_m = [\varepsilon - \sqrt{\varepsilon^2 + 4m_y\Omega_y^2}]/2$) merge. This is one of the specific features of dynamical-type instabilities. Similar properties were found for instabilities due to coupling of the bunch synchrobetatron modes [4].

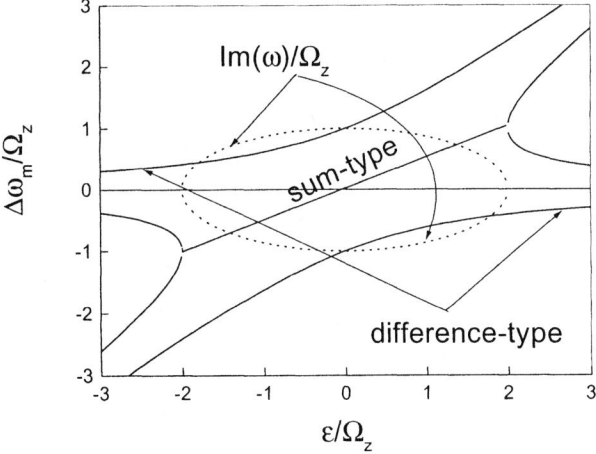

FIGURE 1. Dependence of the real (solid) and imaginary (dotted) parts of the coherent frequency shift on detuning from the resonance. No dissipation in the cavity.

B Dissipative-Type Instabilities

Dissipation in the RF-cavity can result in dissipative-type instability of coherent oscillations of the beam. In this case, and neglecting Landau damping, the dispersion equation is

$$1 = \frac{m_y \Omega_y^2}{[\Delta\omega_m + i\lambda_k - \varepsilon]\Delta\omega_m}. \quad (17)$$

The roots of this equation,

$$\Delta\omega_m = \frac{1}{2}\left[\varepsilon - i\lambda_k \pm \sqrt{(\varepsilon - i\lambda_k)^2 + 4m_y\Omega_y^2}\right], \quad (18)$$

are complex numbers. The oscillations are stable ($-\text{Im}\Delta\omega_m > 0$) if $m_y\Omega_y > 0$. This condition holds for any beam current. Thus the oscillations in a sum-type resonance are always unstable, whereas the oscillations in a difference-type resonance always decay. The decay is due to redistribution of the decrement of the RF-mode between the coupled oscillations of the cavity field and the coherent oscillation mode. The decrement being transferred to coherent oscillation increases when the beam oscillation frequency is tuned to the resonance and reaches its maximum value $\lambda_k/2$ in the exact resonance (Fig. 2). Hence, separate unstable coherent modes can be damped by using a passive RF cavity tuned to the difference-type resonance.

The difference between the stability conditions of coherent oscillations interacting in sum- and difference-type resonances can be summarized in the following stability criterion: coherent oscillations of a bunch interacting with a passive cavity will be stable if contributions of harmonics in difference-type resonances dominate.

The oscillations in the sum-type resonances can be damped only in the case where both oscillators have damping. If λ_k is the damping decrement for the cavity field oscillation, there should be some effect providing damping of coherent oscillations of the beam without interaction with the cavity. This can be the damping due to the beam cooling (for example, due to the synchrotron radiation of particles), due to feedback systems, or due to Landau damping. For simplicity, we assume that such damping is described by the decrement λ_m. Then the dispersion equation is

$$1 = -\frac{\Omega_y^2}{[\Delta\omega_m + i\lambda_k - \varepsilon][\Delta\omega_m + i\lambda_m]}. \quad (19)$$

Calculating the roots of this equation and analyzing their imaginary parts, we find the stability condition:

$$\Omega_y^2 \leq \lambda_k\lambda_m\left(1 + \frac{\varepsilon^2}{[\lambda_k + \lambda_m]^2}\right), \quad m_y = -1. \quad (20)$$

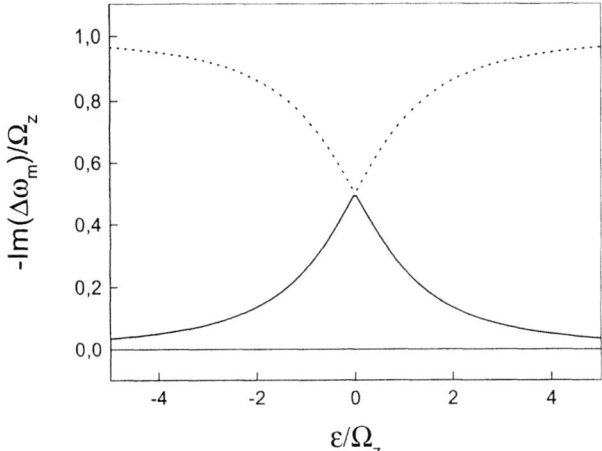

FIGURE 2. Dependence of the decrements of the slow (beam) oscillation mode (solid line) and of the fast (cavity) mode (dotted line) on the detuning from the resonance. Difference-type resonance.

This stability condition is especially simple in the case where $\lambda_k = \lambda_m = \lambda$, when it reads

$$\Omega_y^2 \leq \lambda^2 + \frac{\varepsilon^2}{4}. \tag{21}$$

Using the above-mentioned analogy of the resonant and synchrobetatron mode coupling instabilities, we conclude that stabilization of the synchrobetatron mode coupling instabilities demands the damping of both coupled modes (see, e.g. [5] for more detail).

Since resonant instabilities are due to wakes, which decay during many revolution periods of particles, the increments of these instabilities are determined by the total beam current. This holds both for a single bunch in the ring and for beams containing many bunches. Specific features of resonant instabilities in more general cases can be studied by using the dispersion equation describing spectra of both betatron ($m_s = 0$) and synchrobetatron coherent oscillations:

$$1 = -\frac{m_y \Omega_y^2}{\Delta \omega_m + i\lambda_k - \varepsilon} \int_0^\infty da_s a_s dI_y \frac{\rho(a_s) J_{m_s}^2(n_1 a_s) I_y \partial f_0/\partial I_y}{\omega - m_s \omega_s(a_s) - m_y \omega_y(I_y)}. \tag{22}$$

IV MULTIBUNCH INSTABILITIES

If the beam contains more than one bunch and wakes are longer than the bunch-to bunch distance, coherent oscillations propagate along the beam. This increases the family of coherent oscillation modes of the beam and of instabilities due to multibunch effects. In many cases, these instabilities indicate manifold collective phenomena in the beam, which can limit its total current. Studies of multibunch instabilities were started long ago (see, e.g. [6]). During the last decade these became especially important for the design and construction of high-current storage rings such as electron-positron factories (see, e.g. [7], [8]).

If the harmonic number of the RF system of the ring is h, the beam can contain h bunches separated along the closed orbit by a distance $\Delta s = \Pi/h$. Ideally, the bunches can have the same parameters and distribution functions. Also, if the bunch to bunch distances are equal, the filling pattern of the beam is uniform; otherwise, it is nonuniform. In this lecture we discuss the simplest cases of nonuniform filling patterns.

Concerning the wakefields, in contrast to the previous Section, we assume that the transverse coupling impedance is not a sharp function of the frequency, thus excluding conditions, which are required for the resonant instabilities. For example, this occurs if the fields, induced by particular bunches in the beam, decay during some small number of turns in the ring.

A Uniform filling patterns

In a beam with a uniform filling pattern, identical M_b bunches ($h = kM_b$) are separated longitudinally by a distance $\Delta s = \Pi/M_b$, so that the azimuthal positions of particles in a bunch a are

$$\theta_a = \frac{2\pi a}{M_b} + \omega_0 t + \phi, \quad a = 1, 2, \ldots, M_b. \tag{23}$$

The current of the beam is calculated as the sum of the bunch currents:

$$\mathbf{j}(\mathbf{r},t) = \sum_{a=1}^{M_b} \int d^3p \mathbf{v} f_a.$$

If the bunches are identical and the revolution period is T_0, the spectrum of the beam current contains the following harmonics:

$$\omega_{mn} = m_y \omega_y + n\omega_0 \to m_y \omega_y + (M_b l + q)\omega_0, \tag{24}$$
$$l = 0, \pm 1, \pm 2, \ldots, \quad q = 0, 1, \ldots, M_b - 1.$$

Correspondingly, the amplitudes of $f_{m,a}$ obey the following integral equation (compare to Eq.(5)):

$$f_{m,a}(a_s) = -\rho(a_s) \sum_{l=-\infty}^{\infty} \sum_{q=1}^{M_b} \Omega_{m,lM_b+q}(\omega + [lM_b + q]\omega_0) J_{m_s}(n_1 a_s)$$

$$\times \exp\left(\frac{2\pi i q a}{M_b}\right) \sum_{b=1}^{M_b} \exp\left(-\frac{2\pi i q b}{M_b}\right) \int_0^\infty da'_s a'_s J_{m_s}(n_1 a'_s) f_{m,b}(a'_s) \quad (25)$$

$$\times \int_0^\infty dI_y \frac{I_y \partial f_0/\partial I_y}{\omega - m_s \omega_s - m_y \omega_y(I_y)}.$$

Here, the value n_1 is calculated by using

$$n_1 = lM_b + q + m_y \frac{d\omega_y}{d\omega_0}. \quad (26)$$

The most significant features of the spectra of multibunch instabilities are found by using the simplified model, where we assume that the beam bunches are short $[\rho(a_s) = 2\delta(a_s^2)]$ so that only betatron modes $m_s = 0$ are important. In this case, the eigensolutions to Eq.(25) are

$$f_{m,a}(a_s) = 2\delta(a_s^2) C_a, \quad C_a = \exp\left(\frac{2\pi i q a}{M_b}\right) C_q(a), \quad (27)$$

and the equation is reduced to the following dispersion equations:

$$\Lambda_q = M_b \sum_{l=-\infty}^{\infty} \Omega_{m,lM_b+q}(\omega + [lM_b + q]\omega_0), \quad (28)$$

$$1 = -\Lambda_q \int_0^\infty dI_y \frac{I_y \partial f_0/\partial I_y}{\omega - m_y \omega_y(I_y)}. \quad (29)$$

Equation (29) describes the Landau damping of multibunch modes. The spectra of multibunch instabilities of a monochromatic beam with a uniform filling pattern are determined by Eq.(28). Assuming the localized transverse coupling impedance (see Eq.(6)), we write

$$\Delta\omega_{m,q} = \frac{-M_b N e^2 m_y \omega_0}{4\pi p \nu_y} \sum_{l=-\infty}^{\infty} Z^\perp(m_y \omega_y + q\omega_0 + M_b l \omega_0). \quad (30)$$

Generally, this equation describes both single-bunch and multibunch effects. The contributions of these effects to $\Delta\omega_{m,q}$ can be separated by using the Poisson formula

$$\sum_{l=-\infty}^{\infty} A_l = \sum_{k=-\infty}^{\infty} \int_{-\infty}^{\infty} dl\, A(l) \exp(2\pi ilk). \tag{31}$$

The reader can easily find that, because of the symmetry of $Z^\perp(\omega)$, the single-bunch part of $\Delta\omega_{m,q}$,

$$(\Delta\omega_{m,q})_{sb} = \frac{-M_b N e^2 m_y \omega_0}{4\pi p \nu_y} \int_{-\infty}^{\infty} dl\, Z^\perp(m_y \omega_y + q\omega_0 + M_b l \omega_0)$$

contributes only to the frequency shift. The instabilities of a monochromatic beam are due to the multibunch contributions:

$$(\Delta\omega_{m,q})_{mb} = \frac{-M_b N e^2 m_y \omega_0}{4\pi p \nu_y} \sum_{k\neq 0} \int_{-\infty}^{\infty} dl\, Z^\perp(m_y \omega_y + q\omega_0 + M_b l \omega_0) \exp(2\pi ilk).$$

Simple calculations result in ($T_b = T_0/M_b$)

$$(\Delta\omega_{m,q})_{mb} = -\frac{N e^2 m_y}{2 p \nu_y} \sum_{k=0}^{\infty} \exp(2\pi ik[m_y \nu_y + q])$$
$$\times \sum_{a=1}^{M_b} \exp\left(\frac{2\pi ia[m_y \nu_y + q]}{M_b}\right) Z^\perp(kT_0 + aT_b). \tag{32}$$

Here,

$$Z^\perp(t) = \int_{-\infty}^{\infty} \frac{d\omega}{2\pi} Z^\perp(\omega) e^{-i\omega t}. \tag{33}$$

According to these equations the increments of the multibunch modes ($[1/\tau] = \text{Im}(\Delta\omega_{m,q})_{mb}$) depend on the total number of bunches in the beam only in the case where the wakes last longer that the revolution period in the ring. This is the case of multiturn instabilities (see, e.g. [9]). In the case of short-range wakes, when the bunches in the beam interact during a single turn in the ring, the increments of the multibunch modes are determined by the number of interacting bunches. If we define this number using $M < M_b$, the increments are

$$\frac{1}{\tau_{mb}} = -\frac{N e^2 m_y}{2 p \nu_y} \sum_{a=1}^{M} \sin\left(\frac{2\pi ia[m_y \nu_y + q]}{M_b}\right) Z^\perp(aT_b). \tag{34}$$

In particular, if only neighbor bunches interact ($M = 1$), the increments of the multibunch modes are determined by the current of a single bunch.

According to these equations approximately half of the multibunch modes are unstable. As seen from Figs. 3 and 4 in the case of short-range interactions in the beam, the stability condition does not depend very strongly on the values of the betatron tunes.

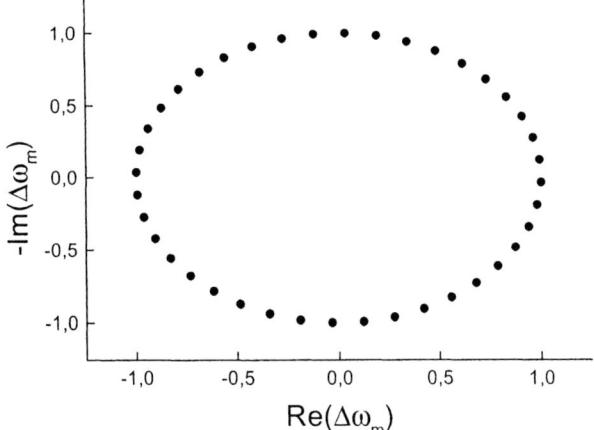

FIGURE 3. Mode map of complex coherent frequency shifts. A stepwize wake couples oscillations of neighbor bunches. Uniform filling pattern, 40 bunches in the beam, $\nu_y = 0.23$.

However, as the number of interacting bunches increases, the mode map becomes less and less symmetric. As seen from Fig. 5, if the number of interacting bunches approaches the number of bunches in the beam, the leading unstable mode appears in the spectrum. The value of the increment of this mode as well as its distance from the neighbor modes substantially depend on the position of the betatron tune relative to a half-integer number.

More generally, the sum of the decrements of the multibunch modes is equal to the multiturn parts of the sum of the decrements of separate bunches interacting with the same electrodes. Hence, the maximum increments of multibunch modes will be smaller than the maximum decrements, if betatron tunes of particles are chosen to provide the stability of coherent oscillations of single bunches. If coherent oscillations are damped by Landau damping or feedback systems, the larger beam currents can be obtained by suitable tuning the working point of the ring.

According to Eq.(27) in a beam with the uniform filling pattern, the amplitudes of the eigenvectors of the multibunch modes are constant along the beam. For this

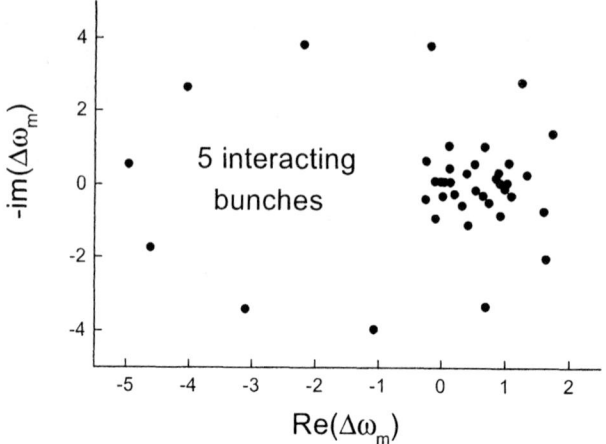

FIGURE 4. Same as Fig. 3, but for the case where the wakes couple oscillations of 5 neighbor bunches in the beam.

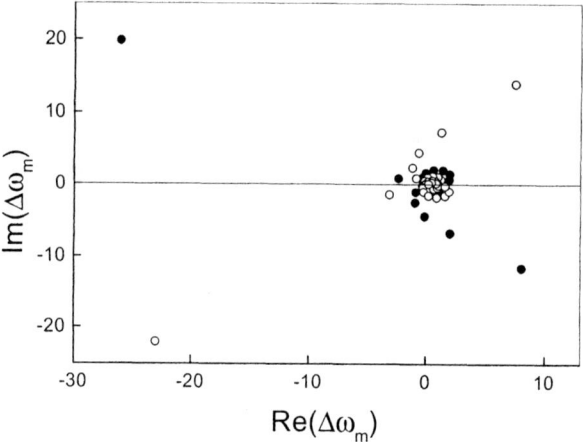

FIGURE 5. Same as Fig. 3, but for the case where the wakes couple oscillations of 35 neighbor bunches in the beam. Full circles: $\nu_y = 0.23$, open circles: $\nu_y = 0.73$.

reason, the oscillations of bunches propagate along the beam without amplification:

$$f_{m,a}(t) = \sum_{q=0}^{M_b-1} \exp\left(-i\omega_{m,q}t + \frac{2\pi i q a}{M_b}\right) C_q(a). \tag{35}$$

The coefficients $C_q(a)$ are determined from the initial conditions. These instabilities can be cured by conventional single-bunch feedback systems.

B Beam With a Gap

The simplest nonuniform filling patterns occur when the beam forms a train, or several trains, with uniform filling of the first M bunches, while remaining buckets of bunches are empty. Such gaps in the beams can be necessary for some technical reasons, or to break the coupling of the bunch oscillations along the beam. In the latter case, the beam can contain several trains separated by gaps.

For simplicity we discuss the case, where the beam contains only one train followed by a gap empty of bunches. The main features of the multibunch instabilities in such a beam are determined by the fact that the excitation of oscillations of head-on bunches in the train by the wakes induced by the tail-on bunches can be weaker, or even substantially weaker, than that inside the train. For this reason, a smaller amount of the energy of coherent oscillations of the last bunches can be transferred to coherent oscillations of the head-on bunches. Thus the energy of coherent oscillations will be accumulated at the end of the train, while the amplitudes of these oscillations will amplify from the head to the tail of the train. The values of the amplification coefficients depend on the strengths of the bunch wakes, on the time dependences of the wakes, and on the number of bunches interacting in the train.

The character of the spectra of the multibunch oscillations is essentially different for the cases of short-range and of long-range interactions between bunches. If the wakes decay during a time interval longer than the duration of the gap, the oscillations of the last bunches in the train are coupled with the oscillations of the head-on bunches. In this case, the eigenvalue problem for coherent oscillations of bunches has solutions describing the multibunch modes of the beam. For example, these modes can be calculated by solving Eq.(25) numerically with the condition $f_{m,b} = 0$, if $b > M$. That will result in a set of eigenfrequencies ($\Delta\omega_{m,q}$; $q = 0, 1, \ldots M - 1$) and eigenvectors ($V_{m,q}(a)$, $a = 1, 2, \ldots, M$), which can be used to calculate both the stability conditions of the modes and the time dependencies of oscillations of particular bunches in the train:

$$f_{m,a}(t) = \sum_{q=0}^{M_b-1} \exp\left(-i\omega_{m,q}t\right) V_{m,q}(a) f_{m,a}(0). \tag{36}$$

Examples of the mode spectra and of the distributions of the eigenvectors along the train are shown in Figs. 6 and 7. Except for the values of the maximum decrements (increments), the frequency maps look very similar as the gap length increases (Fig. 6). On the contrary, the amplifications in the eigenvector values along the train become sharper for shorter trains.

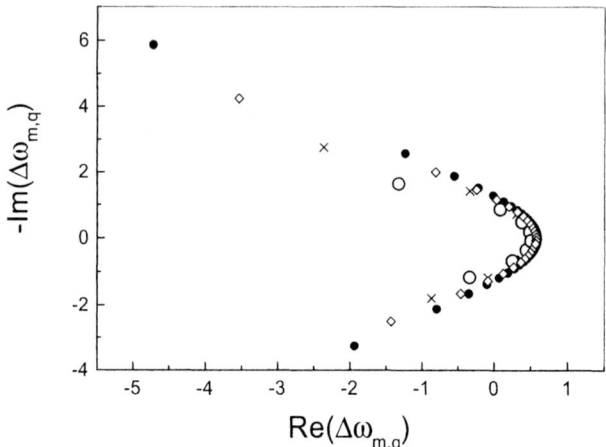

FIGURE 6. Mode maps for several filling patterns. Full circles: $M = 36$, open diamonds: $M = 25$, crosses: $M = 15$, open circles: $M = 8$. Resistive wall wakes, $M_b = 40$, $\nu_y = 0.2$.

Since the eigensolutions depend exponentially on time, the modes can be stabilized by Landau damping or by a single-bunch feedback system. However, the amplification of eigenvectors along the train may result in larger oscillation amplitudes of tail bunches. If the eigenfrequency spectrum has no leading unstable mode, that can limit the operational performance of the beam, especially near the thresholds of instabilities.

A special type of transverse instability occurs when the wake duration is shorter than the length of the gap. In this case, the oscillations of the head-on bunches are not coupled with the oscillations of the tail-on ones, so that the feedback along the beam is broken. After some simple calculations, this can be seen from Eq.(25) directly. The resulting eigenvalue problem is then reduced to a single-bunch one, while the bunch-to-bunch interaction in the train is described by the excitations of the oscillations of subsequent bunches driven by the foregoing bunches of the train. Such a phenomenon is called as the beam breakup instability. It was observed and, at least initially, understood long ago in linacs (for example, in Ref. [10]). Recently, interest in multibunch beam breakup instabilities in storage rings has risen because of expected limitations on the operational performance in B-factories due to fast ion instability [11] and to photo-electron instability [12]. The main features of these multibunch instabilities can be summarized as follows.

- Resonant excitation of the oscillations of subsequent bunches, no eigenmodes.

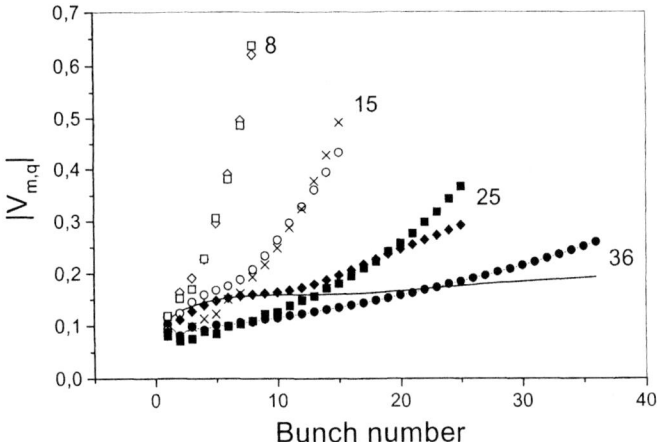

FIGURE 7. Dependence of the amplitudes of eigenvectors on the bunch number in the beam. Number of bunches in the train is shown near each curve. One curve in a group corresponds to $q = 3$, another to $q = M - 3$. Resistive wall wakes, $M_b = 40$.

- Typically, non-exponential time dependencies of the oscillation amplitudes.
- Time dependencies of the oscillation amplitudes are sensitive to the distributions of initial deflections of bunches along the train.
- Except for special cases, substantial amplification of the oscillation amplitudes along the train. Conventional feedback damping systems do not eliminate the amplification.

Since beam breakup instabilities are due to resonance excitations, the only way to eliminate them is to provide counter-phase resonant excitation of oscillations. With any conventional damping mechanisms that suppress the oscillation using forces proportional to the bunch velocity, the amplitudes of oscillations of bunches initially grow during the damping time. Thus the instability and amplification of the oscillation amplitudes is not eliminated. For the same reason, the time dependencies of the oscillation amplitudes are sensitive to the distributions of betatron tunes along the train producing BNS-damping [13]. For transverse oscillations both conventional and BNS-damping limit the values of the oscillation amplitudes after the time when the amplitudes pass their maximum values (for fast-ion instability see, e.g. [14]).

In storage rings, even in the case where the bunches in the train interact via strong short-range wakefields, this interaction is inevitably accompanied by that due to weak but long-range wakes. Without parasitic RF cavities, such long-range wakefields can be provided by the finite resistivity of the walls of the vacuum chamber. These weak and long-range wakes close the feedback between oscillations of the tail-on and head-on bunches in the train and again provide the multibunch mode spectrum in such a beam. If we denote the strength of the strong wake by W and the strength of the weak long-range wake by w, and the amplitude of the first bunch in the train by $A_1(t)$, the eigenfrequencies of multibunch modes are calculated by using the dispersion equation, written in the first approximation in w:

$$\frac{A_1(t)}{A_1(t+T_0)} = \frac{w}{\Delta\omega_m} P_{M-1}\left(\frac{W}{\Delta\omega_m}\right) = 1. \qquad (37)$$

Here, $P_M(x)$ is a polynomial function of its argument of the order of M. For example, if a strong wake is produced by the interaction of the neighbor bunches in the train, $P_M = (W/\Delta\omega_m)^M$. If the number of bunches is large ($M \gg 1$), dependencies of the roots in Eq.(37) on the strength of the weak wake are determined by the factor $(w/W)^{1/M}$. Thus the values of the frequency shifts and the oscillation decrements will be determined mainly by the strength of the strong wake, while the effect of the weak wake in these values will be hidden. Because of the exponential dependencies of the eigensolutions on time, the instability can be eliminated by using conventional feedback damping systems. Nevertheless, amplification along the train can be strong. Special feedback is necessary to suppress amplification of the oscillation amplitudes along the train.

V CONCLUSIONS

Long memory in the induced fields results in numerous instabilities of transverse coherent oscillations of beams in storage rings. Interactions with fields that decay during time intervals substantially exceeding the revolution period in the ring can result in resonant instabilities of coherent oscillations. Such instabilities are very strong, but occur when the particle oscillation frequencies are found within the stopband depending on the beam intensity. Except for some form-factors, resonant instabilities have similar properties for coasting and bunched beams. In the latter case, the family of possible resonances becomes wider because of possible synchrobetatron ones. Besides the standard damping techniques, the instability can be cured by changing the particle oscillation frequency, or by using special passive RF cavity tuned to form the difference-type resonance with an unstable mode.

Another important class of transverse instabilities in modern colliders and storage rings is due to interactions of bunches in a multibunch beam. The coupling of oscillations of the bunches in the beam results in special multibunch coherent oscillation modes. The number of multibunch modes coincides with the number of bunches in the beam. Typically, about a half of these modes are unstable. If the number of bunches in the beam is large, the global stability criteria of the beam oscillations depend very weakly on the frequencies of particle oscillations. The increments of these instabilities depend on the number of interacting bunches.

Thus the instabilities can limit the multibunch operational performance of the ring. Distributions of the eigenvectors of multibunch modes along the closed orbit depend on the beam filling pattern. In beams containing gaps empty of bunches, the amplitudes of eigenvectors can have substantial amplification towards the tail of the bunch train. This can result in additional limitations on the multibunch performance of the beam, especially near thresholds of instabilities.

ACKNOWLEDGMENTS

I thank Professors N. Dikansky, H. Fukuma, K. Oide, K.Yokoya for their valuable discussions.

REFERENCES

1. Pestrikov D.V., Vlasov Equation and Landau Damping, *These Proceedings*.
2. Auslender V.L., Dikansky N.S., et al., *Atomnaya Energia*, **22**, 198, 1967.
3. Derbenev Ya.S., Dikansky N.S., Transverse coherent resonance effects in storage devices, in *Proceedings. of All Union Part. Acc. Conf.*; Moscow, v.2, p. 391, 1970, CERN-Trans-69-15. - May 1969.
4. Kohaupt R.D., DESY Report 80/22, 1980.
5. Pestrikov D.V., *Nuclear Instruments and Methods in Physics Research*, **A**-373, 179, 1996.
6. Kolomensky A.A., Lebedev A.N., Collective Effects in Storage Rings, in *Proceedings. of All Union Part. Acc. Conf.*; Moscow, v.2, p. 261, 1970.
7. Kurokawa S., Review Luminosity Limitations, *These Proceedings*; see also Akai K., Akasaka N., Enomoto A., et al., Commissioning of the KEKB B-FACTORY, in *Proceedings of PAC-99*, New York, March 29–April 2, 1999, v.1, p. 288.
8. Fukuma H., Residual Gas and Ion Effects, *These Proceedings*.
9. Dikansky N., Ph. D. Thesis, Novosibirsk, 1969; see also Dikansky N.S., Pestrikov D.V., *Physics of Intense Beams and Storage Rings*, AIP Press, New York, 1994.
10. Panofski W.K.H., Bander M., *Rev. Sci. Instrum.*, **39**, 206, 1968.
11. Raubenheimer T.O., Zimmermann F., *Phys. Rev.*, **E 52**, 5487, 1995; Stupakov G.V., Raubenheimer T.O., Zimmermann F., *Phys. Rev.*, **E 52**, 5487, 1995; see also Pestrikov D.V., *Nuclear Instruments and Methods in Physics Research*, **A**-412, 294, 1998.
12. Ohmi K., *Phys. Rev. Lett.*, **75** (8), 1526, 1995.
13. Balakin V.E., Novokhatski A.V., Smirnov V.P., in *Proc. of the 12th Intern. Conf. on High Energy Accel.*, p. 119, Fermilab, 1983.
14. Pestrikov D., *Phys. Rev. ST Accel. Beams*, **2**, 044403, 1999.

Head-Tail Instabilities

D.V. Pestrikov

Budker Institute of Nuclear Physics
630090 Novosibirsk, Russian Federation

Abstract. After its discovery in 1969 by C. Pellegrini and M. Sands, the head-tail instability was recognized around the world as one of the most important effects limiting single-bunch currents in storage rings.

I INTRODUCTION

Numerous observations indicate that an important class of instabilities of coherent oscillations of bunches is due to wakefields that decay completely during the period of bunches in the beam. In these cases, coherent oscillations are made unstable by the interaction of coherent oscillations of various parts of a single bunch, while coherent oscillations of different bunches in the beam are independent of each other. Correspondingly, such wakefields result in single-bunch instabilities. The increments of these instabilities are independent of the particle oscillation frequencies.

In 1969 Pellegrini [1] and Sands [2] announced a new mechanism of instabilities of transverse coherent oscillations due to broad-band interactions of the bunch particles and due to chromatic shift of the betatron frequencies of interacting particles along the bunch. The increments of these instabilities did not depend on the betatron tunes, but were proportional to the ring chromaticity ($d\nu_b/d\ln p$, where $p = \gamma M v$ is the particle momentum). They called the head-tail effect or the head-tail instability. Subsequent studies showed that this instability occurs in cases where the wakefields are excited mainly by the longitudinal current of the bunch and where the instability risetime substantially exceeds the periods of synchrotron oscillations of the bunch particles. These studies started the era of the head-tail instabilities in the particle accelerator physics. Single-bunch instabilities in many electron or positron storage rings and colliders (see, e.g. [3]) as well as in hadron storage rings and synchrotrons (see, e.g. [4]) were identified with head-tail instabilities.

The instability mechanism can be explained by using the following simple picture. Let us consider a bunch consisting of two particles. Initially particle 1 having the synchrotron phase $\phi_1 = a_s \cos\psi_s$ is the head-on, while particle 2 ($\phi_2 = -a_s \cos\psi_s$) is the tail-on. We assume that each particle induces in the surrounding electrodes a wakefield proportional to its instantaneous offset, so that the force perturbing the oscillations of the tail-on particle is $F_{1,2} \propto z_1 w(s_1 - s_2)$. Because of the retardation of the induced fields, the oscillations of the tail-on particle do not perturb the oscillations of the head-on particle (see Fig. 1). In a half period of synchrotron oscillation the particles change their positions in the bunch, so that particle 2 becomes head-on and, correspondingly, excites the oscillations of particle 1. The

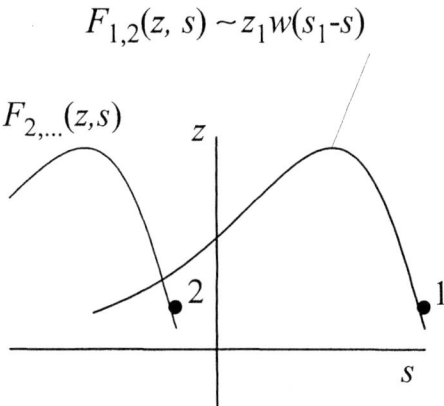

FIGURE 1. Schematic dependence of the wakefields on the distance inside the bunch.

total energy of (dipole) coherent oscillations in this case consists of the energy of the betatron and synchrobetatron modes,

$$E = \sum_{m_b=\pm 1} \sum_{m_s=-\infty}^{\infty} E_{m_b,m_s}.$$

In the case where the oscillation amplitudes grow during many periods of synchrotron oscillations, while $a_s d\nu_b/d\ln\omega_0 \ll 1$, the systematic variation of the energy E is (see, e.g. [5])

$$\overline{\frac{dE}{dt}} = \sum_{m_b=\pm 1} \overline{\frac{dE_{m_b,0}}{dt}} + 2 \sum_{m_b=\pm 1} \sum_{m_s=1}^{\infty} \overline{\frac{dE_{m_b,m_s}}{dt}} \quad (1)$$

$$\propto \left(\nu_b a_b^2\right) a_s \frac{d\ln\nu_b}{d\ln\omega_0} \left[4 - 2\sum_{m_s=1}^{\infty} \frac{1}{m_s^2 - 1/4}\right].$$

Here, a_b is the initial betatron oscillation amplitude, ν_b is the betatron tune, and a line over a value indicates time averaging. If we define the decrements of the betatron and synchrobetatron modes using

$$\delta_{m_b,0} = -\frac{1}{E_{in}} \overline{\left(\frac{dE_{m_b,0}}{dt}\right)}, \quad \delta_{m_b,m_s} = -\frac{1}{E_{in}} \overline{\left(\frac{dE_{m_b,m_s}}{dt}\right)}, \quad E_{in} = \frac{\nu_b a_b^2}{2}, \quad (2)$$

Eq.(1) yields

$$\overline{\frac{dE}{dt}} = -E_{in}\left[\delta_{1,0} + \delta_{-1,0} + \sum_{m_b=\pm 1}\sum_{m_s\neq 0}\delta_{m_b,m_s}\right].$$

Since

$$\sum_{m_s=1}^{\infty}\frac{1}{m_s^2 - 1/4} = 2,$$

Eq.(1) describes the process without systematic variations of the total energy of coherent oscillations ($\overline{dE/dt} = 0$). However, the energies of the betatron and synchrobetatron modes vary systematically in opposite directions. For particle energies above transition energy of the ring ($\gamma_{tr} = 1\sqrt{\alpha}$) we have

$$\frac{d\nu_b}{d\ln\omega_0} \simeq -\frac{1}{\alpha}\frac{d\nu_b}{d\ln p}, \quad \gamma\alpha \gg 1. \tag{3}$$

In this case, the energies of the betatron modes decay, while the energies of all synchrobetatron modes grow. Thus the synchrobetatron modes are unstable. This shows that the head-tail instability is due to the coupling of betatron and synchrotron oscillations of particles, which provides the lattice chromaticity. According to Eq.(1) the sum of decrements of all modes affected by the head-tail effect is zero. Another instability occurs when the interaction of the bunch particles is so strong that it couples the betatron and synchrobetatron coherent modes. These mode-coupling instabilities will be discussed by Gareyte [6].

II INTEGRAL EQUATION

The simplest way to study the head-tail instability provides the Vlasov equation technique. As in my preceding lecture [7] we assume that unperturbed oscillations of particles are described by the following equations:

$$y = \sqrt{\frac{2I_y}{\gamma M \omega_y}}\cos\phi_y, \quad p_y = \frac{p}{R_0}\frac{dy}{d\theta}, \quad \frac{d\phi_y}{dt} = \omega_y + \frac{d\omega_y}{dp}\Delta p,$$

$$\theta = \omega_0 t + \phi, \quad \phi = a_s\cos(\psi_s), \quad \frac{d\phi}{dt} = \frac{d\omega_0}{dp}\Delta p,$$

$$\phi_y = \int dt'\omega_y = \psi_y + \frac{d\omega_y/dp}{d\omega_0/dp}\int dt'\frac{d\phi}{dt'} = \psi_y + \frac{d\omega_y}{d\omega_0}\phi. \tag{4}$$

Assuming dipole betatron oscillations ($m_z = \pm 1$) and linear synchrotron oscillations, we write the distribution function in the linearized Vlasov equation in the form

$$f = f_0(I_y)\rho(a_s) + \frac{e^{-i\omega t}\sqrt{I_y}\partial f_0/\partial I_y}{\omega - m_s\omega_s - m_y\omega_y(I_x, I_y)} \sum_{m_y, m_s} f_m e^{im_y(\psi_y)_s + im_s\psi_s}. \quad (5)$$

Assuming that the wakefields can be described in terms of the transverse coupling impedance and a single-bunch interaction in the beam as well as linear synchrotron oscillations, we find that coherent betatron ($m_s = 0$) and synchrobetatron ($m_s \neq 0$) modes obey the following integral equation:

$$\Delta\omega_m f_m = \rho(a_s) \int_{-\infty}^{\infty} dn \Omega_{mn}(\omega_{mn}) J_{m_s}(n_1 a_s) \int_0^{\infty} da'_s a'_s J_{m_s}(n_1 a'_s) f_m(a'_s). \quad (6)$$

Here, $\omega_{mn} = m_y\omega_y + n\omega_0$, $n_1 = n + m_y d\omega_y/d\omega_0$, and

$$\Omega_{mn}(\omega_{mn}) = -m_y \frac{Ne^2\omega_0}{4\pi p \nu_y} Z_n^\perp(\omega_{mn}). \quad (7)$$

The eigenvalues $\Delta\omega_m$ define coherent frequency shifts of the monochromatic bunch. Landau damping of these modes is described by the following dispersion equation

$$1 = -\Delta\omega_m \int_0^{\infty} \frac{dI_x dI_y I_y \partial f_0/\partial I_y}{\omega - m_s\omega_s - m_y\omega_y(I_x, I_y)}. \quad (8)$$

III SIMPLE MODEL

Equation (6) is too complicated to be solved in a general form for arbitrary wakefields. However, useful predictions concerning the stability of coherent oscillations and typical values of the instability increments can be obtained by using the simplified model. We assume local wakefields [1] so that $Z_n^\perp(\omega) \to Z^\perp(\omega)$ and assume that all particles in the bunch have equal amplitudes of synchrotron oscillations:

$$\rho = \delta(a_s^2 - a_0^2). \quad (9)$$

In this case, solutions to Eq.(6) exist only on the bunch surface in longitudinal phase space $[f_m \propto \delta(a_s^2 - a_0^2)]$. Then, simple calculations give the eigenvalues

$$\Delta\omega_m = -\frac{m_y Ne^2\omega_0}{4\pi p}\left(\frac{\overline{\beta_y}}{R_0}\right)\int_{-\infty}^{\infty} dn Z^\perp\left(n - m_y \frac{d\nu_y}{d\ln\omega_0}\right) J_{m_s}^2(na_0). \quad (10)$$

---

[1] In the local wakefield approximation we assume that the azimuthal size of the electrodes ($\Delta\theta$) providing the particle interactions is so small that typical harmonic numbers of the wake signal hold $|n|\Delta\theta \ll 1$.

The mode decrements are determined by the imaginary part of the transverse coupling impedance:

$$\frac{1}{\tau_m} = -\mathrm{Im}\Delta\omega_m = \frac{m_y N e^2 \omega_0}{4\pi p} \left(\frac{\overline{\beta_y}}{R_0}\right) \int_{-\infty}^{\infty} dn \mathrm{Im}\left\{Z^\perp\left(n - m_y \frac{d\nu_y}{d\ln\omega_0}\right)\right\} J^2_{m_s}(na_0). \tag{11}$$

We note that, since $\mathrm{Im} Z^\perp$ is an odd function of the frequency and since $\sum_{m_s=-\infty}^{\infty} J^2_{m_s}(n) = 1$, then in agreement with Eq.(1) the sum of the decrements of the betatron and of the synchrobetatron modes is zero:

$$-\sum_{m_s=-\infty}^{\infty} \mathrm{Im}\Delta\omega_m = \frac{m_y N e^2 \omega_0}{4\pi p}\left(\frac{\overline{\beta_y}}{R_0}\right) \int_{-\infty}^{\infty} dn \mathrm{Im}\left\{Z^\perp\left(n - m_y \frac{d\nu_y}{d\ln\omega_0}\right)\right\} = 0.$$

This indicates that the head-tail instability is a global instability of a bunch when at least some betatron or synchrobetatron modes are unstable. For the same reasons, from Eq.(11) we conclude that for local wakes the mode decrements vanish proportional to $d\nu_y/d\ln\omega_0$. Thus for dipole betatron oscillations we can rewrite Eq.(11) in the form

$$\frac{1}{\tau_m} = -\mathrm{Im}\Delta\omega_m = \frac{N e^2 \omega_0}{4\pi p} \left(\frac{\overline{\beta_y}}{R_0}\right) \int_{-\infty}^{\infty} dn \mathrm{Im}\left\{Z^\perp\left(n - \frac{d\nu_y}{d\ln\omega_0}\right)\right\} J^2_{m_s}(na_0). \tag{12}$$

Estimations of particular values for the mode decrements and frequency shifts demand knowledge of the function $Z^\perp(\omega)$. Without particular calculations of induced fields, simple and realistic expressions for decrements can be obtained by using a model where

$$Z^\perp(\omega) = \frac{iR_0}{l_\perp^2} \frac{Z_0 \theta_0}{\theta_0[\omega/\omega_0] + i}. \tag{13}$$

Here, the parameter $R_0\theta_0$ determines the interaction radius in the bunch, and $l_\perp$ is a typical transverse distance from the vacuum chamber wall to the bunch center. Substituting $Z^\perp$ from Eq.(13) into Eq.(10), we find

$$\Delta\omega_m = \frac{-i m_y N e^2 \omega_0 Z_0}{4\pi p}\left(\frac{\overline{\beta_y}}{l_\perp^2}\right)\int_{-\infty}^{\infty} \frac{dn J^2_{m_s}(n)}{n - m_y\xi + ib}, \quad b = \frac{a_0}{\theta_0}. \tag{14}$$

Here, we defined ξ as the chromatic betatron phase advance over the bunch length:

$$\xi = a_0 \frac{d\nu_y}{d\ln\omega_0}. \tag{15}$$

Correspondingly, the mode decrements are determined by using the formula

$$\frac{1}{\tau_m} = \frac{N e^2 \omega_0 Z_0}{4\pi p}\left(\frac{\overline{\beta_y}}{l_\perp^2}\right) \int_{-\infty}^{\infty} \frac{dx(x-\xi) J^2_{m_s}(x)}{(x-\xi)^2 + b^2}. \tag{16}$$

A Small Chromaticity

Simple analytic expressions for the oscillation decrements can be obtained for the case where ξ is small ($|\xi| \ll \min\{b, |m_s|\}$). In this parameter region we expand the integral in Eq.(16) in a power series of ξ. The leading contribution reads

$$\frac{1}{\tau_m} = \frac{Ne^2\omega_0 Z_0}{4\pi p}\left(\frac{\overline{\beta_y}}{l_\perp^2}\right) F, \quad F = \int_{-\infty}^\infty \frac{dx(x-\xi)J_{m_s}^2(x)}{(x-\xi)^2 + b^2} \simeq 4\xi H_{m_s}, \qquad (17)$$

$$H_{m_s} = \int_0^\infty \frac{dx\, x}{x^2 + b^2}\frac{d}{dx}J_{m_s}^2(x). \qquad (18)$$

Calculation of the factor H is simplified in the asymptotic regions. If the bunch length is short compared to the particle interaction radius ($b \ll |m_s| + 1$), the integral in Eq.(18) is calculated taking $b = 0$: [2]

$$H_{m_s} = \int_0^\infty \frac{dx}{x}\frac{d}{dx}J_{m_s}^2(x) = \frac{1}{\pi[m^2 - 1/4]}. \qquad (19)$$

Substituting this expression into Eq.(17), we obtain

$$\frac{1}{\tau_m} \simeq \frac{Ne^2\omega_0 Z_0}{2\pi pc}\left(\frac{\overline{\beta_y} R_0}{l_\perp^2}\frac{2}{\pi}\right)\frac{\xi}{m_s^2 - 1/4}. \qquad (20)$$

Since above the transition energy we have $\xi \propto -d\nu_y/d\ln p$, for small positive chromaticity the head-tail effect results in damping of the betatron modes and in antidamping of all synchrobetatron coherent modes of a bunch.

---

[2] For synchrobetatron modes ($m_s \neq 0$), the integral in Eq.(19) is calculated by integrating the right-hand side by parts,

$$\int_0^\infty \frac{dx}{x}\frac{d}{dx}J_{m_s}^2(x) = \int_0^\infty \frac{dx}{x^2}J_{m_s}^2(x),$$

and using the formula [8]

$$\int_0^\infty J_{m_s}^2(qx)x^{-\rho}dx = \frac{(q/2)^{\rho-1}\Gamma(\rho)\Gamma(m_s - \frac{\rho-1}{2})}{2\Gamma^2\left(\frac{1+\rho}{2}\right)\Gamma\left(m_s + \frac{1+\rho}{2}\right)}, \quad 2m_s + 1 > \rho > 0,$$

which results in Eq.(19). For the betatron modes ($m_s = 0$) we write

$$H_0 = -2\int_0^\infty \frac{J_0(x)J_1(x)}{x}dx.$$

Simple calculations using

$$\pi J_0(x) = 2\int_0^{\pi/2} d\alpha\, \cos(x\sin\alpha),$$

result in $H_0 = -4/\pi$.

If the bunch length is long ($b \gg |m_s| + 1$), the main contribution to H_{m_s} gives the asymptotic region of the Bessel function ($x > |m_s| + 1$):

$$H_{m_s} = \int_0^\infty dx\, J_{m_s}^2(x) \frac{x^2 - b^2}{(x^2 + b^2)^2} \simeq \frac{1}{\pi} \int_{|m_s|+1}^\infty \frac{dx}{x} \frac{x^2 - b^2}{(x^2 + b^2)^2}.$$

$$= -\frac{1}{\pi b^2} \left[\ln \frac{\sqrt{b^2 + (|m_s| + 1)^2}}{|m_s| + 1} - \frac{b^2}{b^2 + (|m_s| + 1)^2} \right].$$

Substituting this expression into Eq.(17), we obtain

$$\frac{1}{\tau_m} = -\frac{Ne^2 \omega_0 Z_0}{2\pi p} \left(\frac{\bar{\beta}_y}{l_\perp^2} \right) \frac{2\xi}{\pi b^2} \left[\ln \frac{\sqrt{b^2 + (|m_s| + 1)^2}}{|m_s| + 1} - \frac{b^2}{b^2 + (|m_s| + 1)^2} \right]. \quad (21)$$

For betatron modes ($m_s = 0$) the accuracy of this formula is less than 20% if $b > 15$. For synchrobetatron modes, for example, for $|m_s| = 3$, the accuracy in Eq.(21) becomes less than 10% if $b > 26$. Note that, for positive ring chromaticity, all modes obeying $b \gg |m_s| + 1$ decay.

B Large Chromaticity

If the lattice chromaticity is large ($|\xi| \gg \min\{b, |m_s| + 1\}$) the mode decrements generally decrease with an increase in $|\xi|$. For example, for short bunches ($b \ll |m_s| + 1$) the main contribution to decrements comes from the region where $|x| \gg m_s$. Evaluation of the integral in Eq.(16) with logarithmic accuracy, we obtain

$$\frac{1}{\tau_m} \simeq -\frac{Ne^2 \omega_0 Z_0}{4\pi p \xi} \left(\frac{\bar{\beta}_y}{l_\perp^2} \right) \frac{2}{\pi} \ln \left(\frac{|\xi|}{|m_s| + 1} \right). \quad (22)$$

Wider dependences of mode decrements on ξ are shown in Figs. 2 and 3.

IV SMOOTH LONGITUDINAL DISTRIBUTIONS

For non-singular distribution functions in the amplitudes of synchrotron oscillations $\rho(a_s)$, Eq.(6) generally has an infinite spectrum of eigenvalues near each mode frequency $m_y \omega_y + m_s \omega_s$. These eigensolutions are numbered by using the eigenmode number k describing the number of oscillations of the eigenfunctions in the space of the amplitudes of the synchrotron oscillations (the so-called radial mode number). Generally, this radial mode number varies between zero (no oscillations between the origin and infinite amplitude) and infinity ($k = 0, 1, \ldots \infty$). If the radial modes are well defined (so that $|\text{Re}\,\omega_k| \gg |\text{Im}\,\omega_k|$), they can be observed in the bunch frequency spectra (concerning such observations in the spectra of longitudinal coherent oscillations see, e.g. [9]).

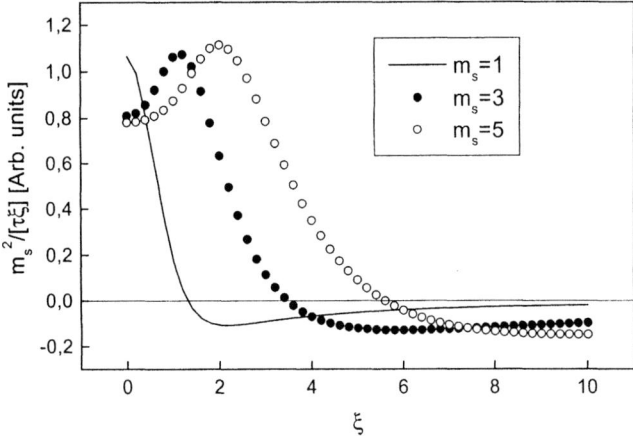

FIGURE 2. Dependences of decrements on the betatron chromatic phase advance ξ. Short bunch $a_0 \ll (|m_s| + 1)\theta_0$.

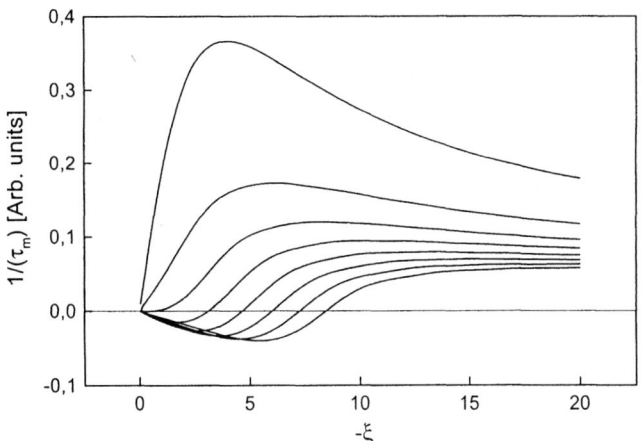

FIGURE 3. Dependences of decrements on the betatron chromatic phase advance ξ. The bunch length corresponds to $a_0 = 3b$, from top to bottom m_s varies from $m_s = 0$ to $m_s = 7$.

The sum of these eigenvalues is determined by the trace of the kernel in Eq.(6):

$$\sum_{k=0}^{\infty} \Delta\omega_{m,k} = -m_y \frac{Ne^2\omega_0}{4\pi p} \left(\frac{\overline{\beta_y}}{R_0}\right) \int_0^{\infty} da_s a_s \rho(a_s) \int_{-\infty}^{\infty} dn Z_n^{\perp}(\omega_{mn}) J_{m_s}^2(n_1 a_s). \quad (23)$$

Comparing this expression and Eq.(10), we see that the sum of eigenfrequencies of the radial modes is obtained by averaging the eigenfrequencies calculated in the previous Section with the given distribution function $\rho(a_s)$.

An important property of Eq.(6) is that it can be transformed into an integral equation with a symmetrical kernel. Taking in Eq.(6) $f_m = w(a_s)\sqrt{\rho(a_s)/a_s}$, we transform this equation to the form

$$\Delta\omega_m w = \int_0^{\infty} da'_s K(a_s, a'_s) w(a'_s), \quad (24)$$

where

$$K(a_s, a'_s) = \int_0^{\infty} dn \Phi(n) P(n, a_s) P(n, a'_s), \quad P(n, a_s) = \sqrt{a_s \rho(a_s)} J_{m_s}(n a_s), \quad (25)$$

and (for local wakefields)

$$\Phi(n) = \Omega_m\left(n - m_z \frac{d\nu_z}{d\ln\omega_0}\right) + \Omega_m\left(-n - m_z \frac{d\nu_z}{d\ln\omega_0}\right). \quad (26)$$

Because of the symmetry of the kernel $K(a_s, a'_s) = K(a'_s, a)$, the eigenvalues of Eq.(24) can be written in the form [3]

$$\Delta\omega_{m,k} = \int_0^{\infty} dn\Phi(n) \left|\int_0^{\infty} da_s P(n, a_s) w_k\right|^2. \quad (27)$$

Using this equation, we conclude that the mode is stable if

$$-\text{Im} \int_0^{\infty} dn\Phi(n) |P_k|^2 > 0, \quad P_k = \int_0^{\infty} da_s P(n, a_s) w_k$$

---

[3] Because of the symmetry of the kernel, the solutions of Eq.(24) and of its adjacent equation

$$\Delta\omega_m w(a'_s) = \int_0^{\infty} da K(a, a') w(a)$$

coincide. For this reason, the total set of eigenfunctions present the functions $w_k(a)$ and $w_k^*(a)$, which enable

$$\sum_{k=0}^{\infty} w_k(a) w_k^*(a') = \delta(a - a'), \quad \int_0^{\infty} da w_k(a) w_{k'}^*(a) = \delta_{k,k'}.$$

Multiplying both sides of Eq.(24), written for the eigenfunction w_k, by w_k^* and integrating this equation over a_s, we arrive at Eq.(27).

or

$$\int_0^\infty dn \text{Im}\left\{Z^\perp\left(n - m_z \frac{d\nu_z}{d\ln\omega_0}\right) - Z^\perp\left(n + m_z \frac{d\nu_z}{d\ln\omega_0}\right)\right\}|P_k|^2 > 0. \quad (28)$$

The reader can easily see that this stability criterion demands the requirement mentioned in our previous lecture [7] that, in the contributions to the mode decrements, the bunch frequencies in the difference resonances with wakes should dominate.

Since the right-hand side in Eq.(23) is a finite value, the eigenfrequencies $\Delta\omega_{m,k}$ decay asymptotically ($k \to \infty$) at least as $\Delta\omega_{m,k} \propto 1/k^2$. Typically, dependences of the mode frequency shifts on the radial mode number depend on the spectrum of the wakefields. The above-mentioned decay $\Delta\omega_{m,k} \propto 1/(k^2+c)$, where c is a constant, occurs when the integrals in the right-hand side of Eq.(23) converge well. In such cases, the eigenvalue spectrum contains a well-defined ground state with a frequency shift $\Delta\omega_{m,0}$ well separated from the higher-order modes ($k = 1, \ldots$). If the sums of eigenvalues diverge logarithmically (see, for example, Eqs.(21) and (22)), which corresponds to leading contributions from the regions where $Z^\perp(\omega) \propto 1/\omega$, the spectra of the radial modes contain the regions where $\Delta\omega_{m,k} \propto 1/(k+c')$. Thus the distance between the eigenvalues of the ground-state mode and of its first satellites is not so dramatic. Except for special cases, the spectra of the radial modes should be calculated numerically. Several examples enabling analytic calculations of these spectra can be found in the book [5]. In particular, for local wakefields the maximum increments can be estimated by extrapolation of the coherent spectrum from its high-frequency region. If the bunch length is a_0, the resulting expression for the complex coherent frequency shift reads (see, e.g. [5] for more detail)

$$\Delta\omega_m \simeq \rho(0)\frac{\Phi[(|m_s|+1)/a_0]}{|m_s|+1}. \quad (29)$$

A Mixing of Radial Modes

The simultaneous effect of the single-bunch and multibunch interactions generally mixes the radial modes. In a bunch where all particles have equal amplitudes of synchrotron oscillations, the effect of this mixture is simple. The total decrement of the mode is equal to the sum of the single-bunch and multibunch interactions:

$$\frac{1}{\tau_m} = (\delta_m)_{sb} + (\delta_m)_{mb}.$$

However, for beams with smooth distributions over the amplitudes of synchrotron oscillations, this effect can be more complicated. For simplicity, we discuss the simplest case, where the long memory of the wakefields is due to the resonant interaction that we described in a lecture [7]. We also assume a single bunch in the beam and neglect Landau damping of the oscillations.

For local single-turn wakes the total integral equation describing both interactions reads

$$\Delta\omega_m f_m = \rho(a_s) \int_0^\infty dn \Phi(n) J_{m_s}(na_s) \int_0^\infty da_s' a_s' J_{m_s}(na_s') f_m(a_s')$$
$$+ \frac{m_y \Omega_y^2}{\Delta\omega_m + i\lambda_k - \varepsilon} \rho(a_s) J_{m_s}(n_1 a_s) \int_0^\infty da_s' a_s' J_{m_s}(n_1 a_s') f_m(a_s'). \quad (30)$$

Here, ε is the detuning from the resonance, λ_k is the damping decrement of the cavity resonant mode,

$$n_1 = n_0 + m_y \frac{d\omega_y}{d\omega_0}, \quad (31)$$

n_0 is the resonant azimuthal harmonic number, and

$$\Omega_y^2 = \frac{Ne^2 \omega_0^2 (Z_0/Q)}{8\pi p} \left(\frac{\overline{\beta_y}}{l_\perp^2}\right) \quad (32)$$

is the square of the coherent frequency shift due to the resonant interaction. Taking $f_m = w\sqrt{\rho/a_s}$ and $g(a) = \sqrt{\rho a} J_{m_s}(n_1 a_s)$, we transform Eq.(30) into the integral equation with a symmetrical kernel:

$$\Delta\omega_m w = \int_0^\infty da_s' K(a_s, a_s') w(a_s') + \frac{m_y \Omega_y^2}{\Delta\omega_m + i\lambda_k - \varepsilon} g(a_s) \int_0^\infty da' g(a_s') w. \quad (33)$$

Here, the kernel $K(a_s, a_s')$ is determined in Eq.(25). We find solutions to Eq.(33) by using the series

$$w(a_s) = \sum_{q=0}^\infty C_q w_q(a_s), \quad C_q = \int_0^\infty da_s w(a_s) w_q^*(a_s), \quad (34)$$

where $w_q(a_s)$ are the eigensolutions of Eq.(24):

$$\Lambda_q w_q = \int_0^\infty da_s' K(a_s, a_s') w_q(a_s'), \quad \int_0^\infty da_s w_q(a_s) w_{q'}^*(a_s) = \delta_{q,q'}.$$

After simple calculations we transform Eq.(33) into the system of the algebraic equations:

$$[\Delta\omega_m - \Lambda_p] C_p = \frac{m_y \Omega_y^2}{\Delta\omega_m + i\lambda_k - \varepsilon} G_p \sum_{q=0}^\infty G_q^* C_q, \quad (35)$$

where

$$G_q = \int_0^\infty da \sqrt{\rho a} J_{m_s}(n_1 a) w_q.$$

Taking $C_p = G_p/[\Delta\omega_m - \Lambda_p]$, we obtain the dispersion equation of the problem:

$$1 = \frac{m_y \Omega_y^2}{\Delta\omega_m + i\lambda_k - \varepsilon} \sum_{q=0}^{\infty} \frac{|G_q|^2}{\Delta\omega_m - \Lambda_q}. \tag{36}$$

If the resonant interaction is weak, the roots of this equation are close to the single-bunch spectrum:

$$\Delta\omega_{m,q} \simeq \Lambda_q + \frac{m_y \Omega_y^2 |G_q|^2}{\Lambda_q + i\lambda_k - \varepsilon}, \quad |\Lambda_q| \gg \frac{\Omega_y^2 |G_q|^2}{[\text{Re}(\Lambda_q) - \varepsilon]^2 + (\text{Im}\Lambda_q + \lambda_k)^2}. \tag{37}$$

It hardly holds if $\lambda_k \simeq -\text{Im}\Lambda_q$ and $\text{Re}\Lambda_q \simeq \varepsilon$. On the other hand, the values Λ_q generally decrease with an increase in q. We may expect that the resonant interaction will result in strong mixing of the radial modes with $\Omega_y^2 |G_q|^2 \gg |\Lambda_q|$.

If the resonant interaction is strong, since the eigenfrequencies Λ_q have real and imaginary parts, different radial modes may have different positions relative to the resonance. In particular, if only one mode dominates, it may stabilize coherent oscillations because of their detuning from the resonance or damping of the single-bunch oscillation. Similarly, the resonant interaction can stabilize the leading unstable single-bunch mode if the cavity is tuned to the difference resonance to that mode.

Without the leading modes, and provided that the single-bunch modes are well defined so that $|\text{Re}\Lambda_q| \gg |\text{Im}\Lambda_q|$, the interference of the radial modes can result in "effective Landau damping" of the instability.

V DAMPING

The modes that are unstable because of head-tail instability, can be stabilized by Landau damping, by broad-band damping systems or by the bunch cooling. For electron or positron storage rings, the latter is synchrotron radiation damping, while for ion rings it can be, for example, electron cooling (see, e.g. [10]). Beam cooling will stabilize the modes if the cooling decrements exceed the increments of unstable modes. Since generally the mode increments decrease with increase in the synchrotron mode number (m_s), while the damping decrements due to cooling increase according to

$$\lambda_m = \lambda_y + |m_s|\lambda_s,$$

the higher synchrobetatron modes can more likely be damped by beam cooling. Here, $\lambda_{y,s}$ are the cooling decrements of the vertical and the synchrotron oscillations of particles.

A Landau Damping

Landau damping of synchrobetatron modes may be due to both betatron and synchrotron frequency spreads of the bunch. For simplicity we discuss the case where Landau damping of synchrobetatron modes is caused by cubic nonlinearity of the lattice focusing:

$$m_y \omega_y(I_x, I_y) = m_y \omega_y(0) + aI_y - bI_x, \tag{38}$$

so that the dispersion equation, Eq.(8), reads

$$1 = -\Delta\omega_m \int_0^\infty \frac{dI_x dI_y I_y \partial f_0/\partial I_y}{z_m - aI_y + bI_x}, \quad z_m = \omega - m_s\omega_s - m_y\omega_y(0). \tag{39}$$

As stated in a lecture [11], we can use this equation to plot the stability diagrams. Taking, for example, as f_0, a Gaussian distribution function

$$f_0 = \frac{1}{p^2 \epsilon_x \epsilon_y} \exp\left[-\frac{I_x}{p\epsilon_x} - \frac{I_y}{p\epsilon_y}\right],$$

where $\epsilon_{x,y}$ are the horizontal and vertical bunch emittances, and defining

$$\delta\omega = \frac{p\epsilon_x|b| + p\epsilon_y|a|}{2}, \quad \zeta = \frac{\Delta\omega_m}{\delta\omega}, \quad \nu = \frac{z_m}{\delta\omega},$$

we rewrite Eq.(39) in the form

$$1 = \zeta \int_0^\infty \frac{dw\, g(w)}{\nu - w}, \quad \mathrm{Im}\,\nu > 0, \tag{40}$$

where we also defined, as

$$g(w) = \int_0^\infty dx\, dy\, y \exp(-x - y)\delta[w - Ay + Bx], \quad A = \frac{p\epsilon_y a}{\delta\omega}, \quad B = \frac{p\epsilon_x b}{\delta\omega}, \tag{41}$$

an effective betatron frequency distribution function in the bunch. This equation shows that, except for a special case where $A = B$, the function $g(w)$ vanishes depending on the signs of A and/or B, when the variable w is either positive or negative. Correspondingly, the modes having coherent frequency shifts in these regions have no Landau damping, while the stability diagrams in these regions coincide with the axes $\mathrm{Im}\,\zeta = 0$ (see Figs. 4 and 5).

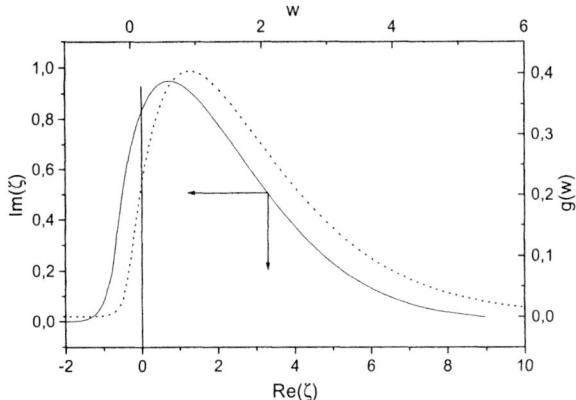

FIGURE 4. Stability diagram (solid line) for a bunch with a Gaussian betatron distribution function. Oscillations are stable below the solid line. Parameter A is positive, $A \gg |B|$. Dotted line shows the effective betatron distribution function.

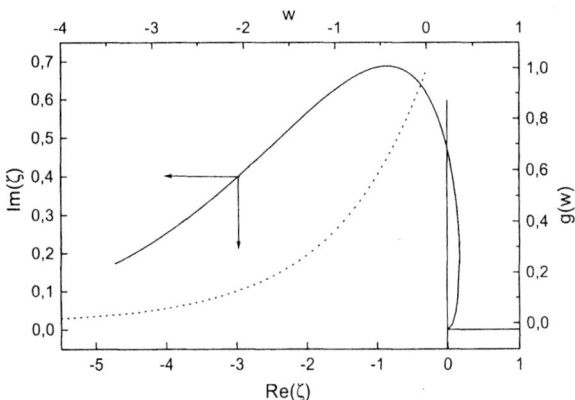

FIGURE 5. Same as Fig. 4, but $B \gg |A|$, $B > 0$. Oscillations are stable inside the region bounded by the solid line.

In particular, this means that changing the sign of the cubic nonlinearity of the lattice (the sign of a or b) changes the values of the Landau damping decrements and hence the stability conditions for synchrobetatron modes. Such a dependence of the Landau damping decrements on the sign of the cubic nonlinearity of the lattice was observed and reported, for example, in Refs. [3] and [12]. Because of the decrease in mode increments with an increase in synchrotron mode number (m_s) Landau damping more easily suppresses the synchrobetatron modes with higher m_s.

B Artificial Damping

The head-tail unstable modes, which are not damped by beam cooling or by Landau damping, should be stabilized by using damping systems providing artificial damping of coherent oscillations. Since the instability is due to wakefields varying during the time intervals corresponding to the bunch length or shorter, the frequency band-width of such a damping system should be large. Traditional single-bunch feedback systems are not necessarily the optimal devices for that purpose (see, e.g. [13]).

In this lecture we discuss in more detail artificial simultaneous damping of head-tail modes of a short bunch by using the effect of the fast damping of coherent oscillations. Initially, this phenomenon was discovered during commissioning of the collider VEPP-2 in Novosibirsk (see, e.g. [14]) and was understood as an increase of the damping of dipole coherent oscillations due to their interactions with a piece of the vacuum chamber containing a plate terminated by its characteristic impedance (Fig. 6).

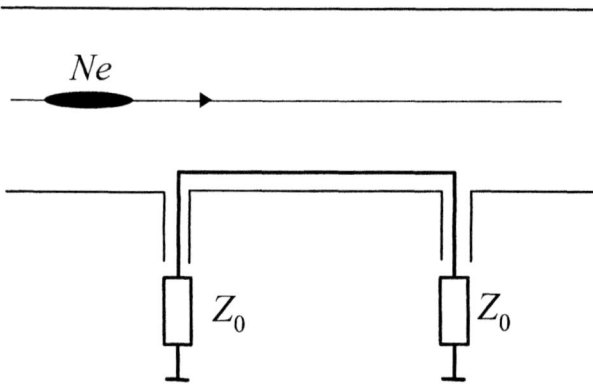

FIGURE 6. A matched plate insertion in a straight section of the storage ring.

Electromagnetically, such a piece of the vacuum chamber represents a piece of a double-connected waveguide, which enables propagation, along the piece, of TEM waves with a continuous spectrum of eigenfrequeces ($\omega = kc$, c is the speed of light) in the range

$$0 < \omega < \frac{c}{l_\perp}, \qquad (42)$$

where $c/l_\perp$ is the cutoff frequency of the duct. If the bunch length is larger than the transverse dimensions of the duct, this frequency range is most typical for coherent oscillations of the bunch. Outside the edge regions of the plate, the electrical field of the excited TEM wave is directed perpendicular to the plate and opposite to the transverse bunch current. If the plate is terminated by its characteristic impedance, the energy of coherent oscillations is dissipated in the plate output load. The interaction of coherent oscillations with such an insertion results in the damping of betatron coherent oscillations of the bunch. If the bunch length is smaller than the length of the plate ($l_\perp < \sigma_s \ll 2l$), the damping decrements due to interaction with such an insertion reads [15]

$$\frac{1}{\tau_0} \simeq \frac{Nr_0 c(Z_0 c)}{4\gamma l_\perp^2} \frac{l}{\Pi}. \qquad (43)$$

Here, Z_0 is the characteristic impedance of the plate. The head-tail instability due to interaction of a bunch with the matched plate was studied by Derbenev et al. [16], [17] (see also [5]). For short bunches ($\xi \ll |m_s| + 1$) the decrements of dipole betatron oscillations are

$$\frac{1}{\tau_{m_s}} = \frac{Nr_0 c(Z_0 c)}{4\gamma l_\perp^2} \left\{ \frac{l}{\Pi} J_{m_s}^2 \left(\frac{\sigma_s}{2} \frac{d\nu_z}{dR_0} \right) + \frac{\sigma_s/R_0}{\pi^3 [m_s^2 - (1/4)]} \left[1 - \frac{d\nu_z}{d\ln R_0} \left(\frac{\beta_0}{R_0} \right) \right] \right\}. \qquad (44)$$

Here, β_0 is the lattice β-function in the edge region of the plate. This expression shows that simultaneous damping of the betatron and synchrobetatron modes is possible, if the ring chromaticity is negative and if

$$l > \frac{8\sigma_s \beta_0}{3\pi^2 R_0} \left| \frac{d\nu_z}{d\ln R_0} \right|. \qquad (45)$$

Note that the sum of decrements in Eq.(44) over all synchrobetatron modes (m_s) is equal to the decrement of a bunch having zero length [the right-hand side in Eq.(43)] and hence is always positive. That is a specific feature of the sums of decrements due to interaction of coherent oscillations with a damping system.

In the multibunch beam with bunch period T_b, the interaction with such an insertion will not yield multibunch interactions if the length of the plate is less than $T_b c/2$ ($\gamma \gg 1$).

VI CONCLUSIONS

The head-tail effect results in a slow global instability of the transverse single-bunch coherent oscillations. If the chromatic betatron phase advance over the bunch length is small, for any sign of the lattice chromaticity, either betatron or synchrobetatron coherent modes are simultaneously unstable.

The instability is due to wakefields induced by the longitudinal current of the bunch and to the redistribution of coherent decrements between synchrobetatron modes. The coupling required for such redistribution is provided by the lattice chromaticity of the ring. However, an asymptotic increase in the chromatic betatron phase advance over the bunch length results in the same sign of decrements for the betatron and of initial (in m_s) synchrobetatron coherent modes. If higher synchrobetatron modes are damped by beam cooling or Landau damping, this can be used to provide simultaneous damping of the betatron and initial synchrobetatron modes of the bunch.

For short relativistic bunches the betatron and initial synchrobetatron coherent modes can be damped simultaneously by a broad-band damping system that provides sufficiently fast decay of the betatron modes, and choosing a proper sign of the lattice chromaticity.

ACKNOWLEDGMENTS

I thank Professors Ya. Derbenev, N. Dikansky, H. Fukuma and K. Satoh for their valuable discussions and help.

REFERENCES

1. Pellegrini C., *Il Nuovo Cimento*, **64A**, 447, 1969.
2. Sands M., The Head-Tail Effect: an Instability Mechanism in Storage Rings, SLAC TN-69-8, Stanford, 1969.
3. Hofmann A., Muelhaupt G., in *Proceedings of the 8th International Conference on High Energy Accelerators*, p. 306, CERN, 1971.
4. Kimura Y., Miyahara Y., Sassaki H., et al, in *Proceedings of the 10th International Conference on High Energy Accelerators*, v. 2, p. 30, Serpukhov, 1977.
5. Dikansky N.S., Pestrikov D.V., *Physics of Intense Beams and Storage Rings*. AIP Press, New York, 1994.
6. Gareyte J., Mode Coupling Instabilities, in *these Proceedings*.
7. Pestrikov D. V., Transverse Instabilities, in *these Proceedings*.
8. Gradshteyn I.S. , Ryzhik I.M., *Tables of Integrals, Sums and Products*, Academic Press, New York, 1965.
9. Nakajima K., Ebihara K., Ogata A. and Satoh K., in *Proceedings of the 5th Symposium on Accelerator Science and Technology*, p. 297, KEK, 1984; see also Zotter B., in *Proccedings of the Fourth Advanced ICFA Beam Dynamics Workshop on Collective Effects in Short Bunches*, p. 1, KEK, 1991.
10. Parkhomchuk V.V., Review of cooling, in *these Proceedings*.
11. Pestrikov D. V., Vlasov Equation and Landau damping, in *these Proceedings*.

12. Kobayashi Y., Ohmi K., Measurement of the Beam Decoherence Due to the Octupole Magnetic Fields at the Photon Phactory Storage Ring. KEK Preprint 98-112, 1998.
13. Kikutani E., *Particle Accelerators*, **52**, 251, 1996.
14. Auslender V.L., Dikansky N.S., et al., *Atomnaya Energia*, **22**, 198, 1967.
15. Derbenev Ya.S., Dikansky N.S., in *Proceedings of the 7th Intern. Conf. on High Energy Accel.*, v. 2, p. 294, Yerevan 1970; see also Preprint INP 315. Novosibirsk 1969, SLAC-TRANS-0106, 1969; Preprint INP 318, Novosibirsk 1969, SLAC-TRANS-0114, 1969.
16. Derbenev Ya.S., Dikansky N.S., Pestrikov D.V., in *Proceedings of the 8th International Conference on High Energy Accelerators*, p. 370, Geneva 1971.
17. Derbenev Ya.S., Dikansky N.S., Pestrikov D.V., Stability of a bunched beam interacting with matched lines, Preprint INP 7-72, Novosibirsk 1972; CERN-Trans-72-16, CERN, May 1972; *Particle Accelerators*, **8**, 129, 1978.

Beam Feedback Systems

Makoto Tobiyama

Accelerator Laboratory,
High Energy Accelerator Research Organization (KEK),
1-1 Oho, Tsukuba 305-0801, Japan

Abstract. Outlines of bunch-by-bunch feedback systems for suppressing multibunch instabilities in electron/positron storage rings are shown. The design principles and functions of the feedback components are reviewed. The application of the feedback system as a tool to analyze instabilities using transient-domain techniques is also shown.

I INTRODUCTION

In recent high-beam-current, multibunch storage rings, especially particle factories such as B-factories, Φ-factories or synchrotron radiation facilities, beam instabilities are very strong because of many impedance sources such as RF cavities, bellows, radiation protection structures or special vacuum structures around interaction regions or insertion devices, and because of rather high beam currents. Moreover, the instability may have many unstable oscillation modes which are not easy to investigate or to avoid. Once instability occurs, it usually limits the maximum storable currents, reduces beam lifetimes, or enlarges the effective emittance in both the transverse and longitudinal planes, and spoils qualities of the beams such as luminosity or brilliance.

To overcome the instabilities, we must first reduce the sources of impedance. Higher-order modes (HOMs) coming from the RF cavities should be suppressed with a damped structure. Installing HOM absorbers near the high-Q impedance source will be effective in reducing the impedance. We should pay great attention to making the structures of vacuum components smooth. This is also effective in preventing breakdown due to heating, sparking, multipactoring and so on. The second method is to introduce a tune-splitting mechanism, which may be passive or active, to enhance the Landau damping effect. The third method is to employ beam feedback systems, and this is the main theme of this article.

In general, a beam feedback system consists of three main parts: a bunch oscillation detection system, a signal processing system that shifts the phase of the oscillation, cuts unnecessary signals and ensures synchronization between the feedback signals and the beam, and high-power amplifiers and kickers to kick the beam.

With a beam feedback system, we can first damp the coherent dipole instabilities whose growth rate exceeds radiation damping, Landau damping or head-tail damping. Also, we can damp injection oscillations much more quickly than with a radiation damping mechanism. Combining feedback systems and bunch oscillation memory systems, we can make a transient-domain analysis of the instability, which is a powerful tool for analyzing instabilities.

II SIMPLE EXAMPLE OF FEEDBACK SYSTEMS

As a simple example of a feedback system, we first consider an automobile cruise control system [1] (Fig. 1). If the driver keeps the gas pedal angle constant without

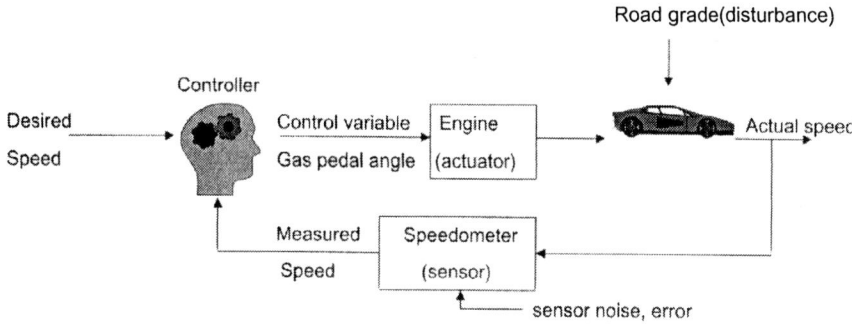

FIGURE 1. Component block diagram of automobile cruise control.

watching the speedometer, the speed will change with disturbances such as road grade or wind. This is called "open-loop control." Normally, the driver will watch the speedometer and tune the gas pedal angle to keep the speed constant. This is called "closed-loop control." Since he feeds back the speed information to the car, the effects of the disturbances will be greatly reduced. However, in some cases, for example if the driver is unskilled, the system can be unstable, and the speed may increase or decrease very rapidly, or introduce uncomfortable oscillations.

To see the effect of the closed-loop control system quantitatively, we need a set of mathematical relations among the variables of the system. For this example, we will ignore the dynamic response of the car and consider only the static behavior, though the dynamic response is also surely important in the system. We assume all the relations may be approximated as linear. For the vehicle, we measure speed on a flat road at 100 km/h and find that a unit change in the gas pedal angle causes a 10 km/h change in speed. For a grade change of 1%, the speed is reduced by 10 km/h. With these relations, we can draw the block diagram shown in Fig. 2(a).

In the open-loop case, Fig. 2(b), the output speed is given by

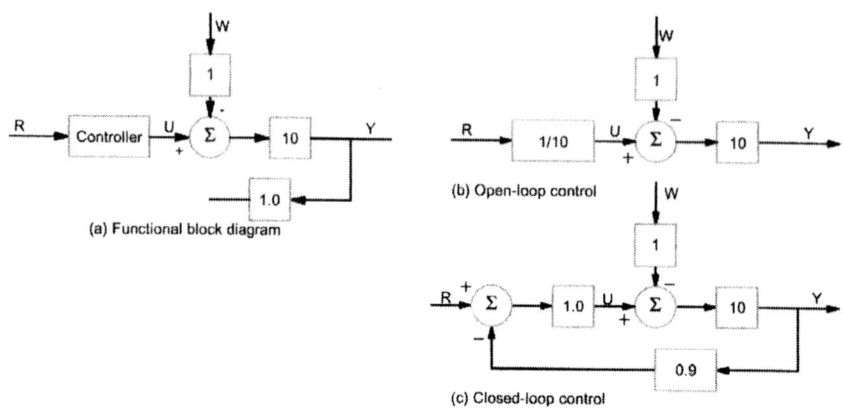

FIGURE 2. Block diagram of cruise control.

$$Y_{ol} = 10(U - W)$$
$$= R - 10W.$$

If $W = 0$, on a flat road, and $R = 100$, then the speed will be 100 km/h and there are no errors. However, if $W = 1$, on a 1% grade, then the speed will be 90 km/h, and we have 10 km/h (10%) error. Let us compare this with the feedback scheme shown in Fig. 2(c). In this case the output speed is

$$Y_{cl} = 10(U - W),$$
$$U = R - 0.9Y_{cl},$$
$$Y_{cl} = R - W.$$

In this case, if the input set-point R equals 100 km/h and the grade W is 1%, the speed will be 99 km/h, only 1 km/h (1%) error— the feedback works to reduce the error by a factor of 10.

We will not go further into the details of feedback theory [1] here, though a precise understanding of it is surely important. We note here two important points to consider. The first is the *positive* feedback situation. In the above example, we added the feedback speed information *negatively* to the controller. If we add the information *positively*, the error is not reduced but amplified. In this situation, it is in general not easy to control the system without a self-limiting (none-linear) element in the system. A frequency oscillator is an example of this application. The second point is feedback gain. Increasing the feedback gain, reduces the steady-state errors, and it provides faster response of the system with the gain. However, it also reduces the stability of the feedback system under realistic conditions. The system can easily begin to oscillate or go out of control. Recall the screeching of a sound system (usually called howling) when one raises the gain of the power amplifier too much. We must keep a gain margin and a phase margin in the feedback system to avoid unstable situations.

III BEAM FEEDBACK SCHEME

Figure 3 shows the phase-space view of a bunch-by-bunch feedback scheme. In

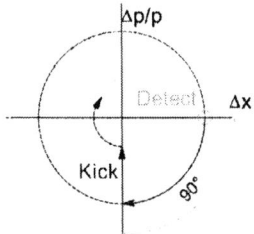

FIGURE 3. Phase-space view of the bunch-by-bunch feedback scheme.

the beam feedback system, we first detect the bunch position, then make a 90° phase shift to the beam, rejecting the unnecessary static component in the signal processing unit. Here the static component in the feedback loop may be the residual closed-orbit distortion (COD) for a transverse plane, or the change in equilibrium phase in the longitudinal plane. Next, we wait for the bunch re-arrival at the kicker section and kick the bunch to change the angular divergence in the transverse plane, or momentum in the longitudinal plane, by the feedback kicker.

Suppose our ring contains equally spaced bunches. The feedback system must be able to respond to an arbitrary pattern of oscillations of the bunches. If all bunches require a kick of the same amplitude and sign, this is the same as a DC correction. In reality, the minimum frequency is not 0 Hz but the base-band betatron frequency or synchrotron frequency, depending on the feedback plane. The highest possible frequency we will need comes from the case where all the subsequent bunches have equal magnitude with completely opposite sign. Thus, the upper end of the bandwidth must exceed half the bunch frequency, $f_b/2$. The center frequency of this bandwidth is somewhat arbitrary: when there are no bunches passing through the kicker, it does not matter what kick is applied.

Figure 4(a) shows a typical transverse feedback diagram. We detect the bunch positions by two beam position monitors (BPMs), which are arranged so that the betatron phase advance is around 90°. The two signals are summed vectorially to make a suitable phase shift to the feedback kicker, around 90°. With the signal-processing part we reject the DC component and adjust the delay, around one revolution period minus the signal delay in the circuit and cables.

Figure 4(b) shows a typical longitudinal bunch-by-bunch feedback system. We detect the longitudinal position of a bunch, which is the phase of the bunch. Since

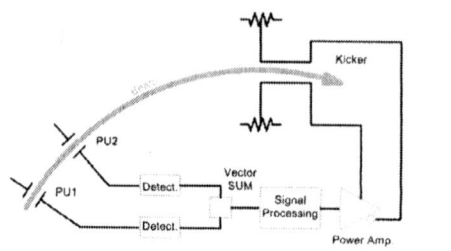
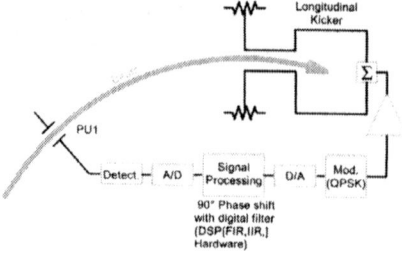

(a) Typical transverse beam feedback system (b) Typical longitudinal bunch-by-bunch feedback system

FIGURE 4. Typical bunch-by-bunch feedback diagram.

the synchrotron frequency is much lower than the revolution frequency, we must make a 90° phase shift by using a digital filter. As in the transverse case, we also provide a one-turn delay here. In some cases, if we need to use a higher-frequency kicker, we need a modulation circuit before the amplifiers.

IV FEEDBACK SYSTEM HARDWARE

A Pickup electrodes

In the design of pickup electrodes for bunch feedback systems, we have several requirements, as follows:

- Vacuum safety structure.
  All the components should be strong enough to withstand typical mechanical shocks, such as welding, vacuum baking, small impact or cable related shocks. Additionally, it should be tough enough to withstand the extracted beam power or high peak-voltage. Trapped modes in the structure should be damped as much as possible.

- Sufficient, but not excessive output power.
  Though we need enough output from the beam for a good signal-to-noise ratio, unnecessary or excessive power is troublesome because it can burn out attenuators, RF components or cables.

- Wideband frequency response and clear impulse response.
  For the separation of signals from two successive bunches, we need wideband and clear response.

Figure 5 shows a button-type electrode used in the KEKB feedback systems and its simplified equivalent circuit. It forms a high-pass filter with a time constant of $\tau = 1/CR$. Now, there are good simulation codes such as HFSS [2] and MAFIA [3] for calculating the RF characteristics with good accuracy without the need to manufactured test specimens.

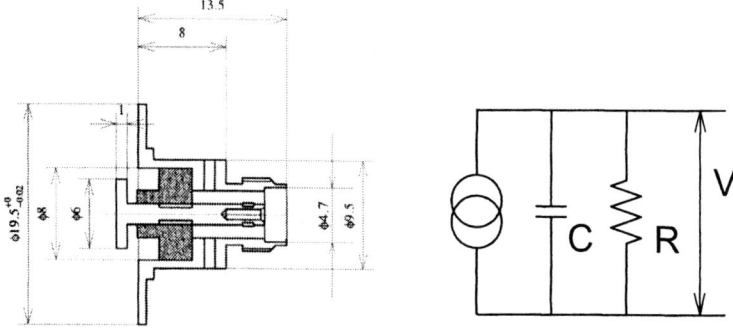

FIGURE 5. KEKB button electrode for feedback systems and its equivalent circuit.

B Bunch position detection electronics

The requirements for bunch position detection electronics are as follows:

- Quick output.
  Since we must determine the proper feedback kick voltage before the bunch arrives at the kicker section, the processing time of the detection circuit should be much shorter than the revolution period. Consequently, the detection circuit should be a simple hardware system without any software corrections.

- Wideband response.
  To distinguish the signals from two adjacent bunches, the circuit should have a high enough bandwidth. We must remember to check the bandwidth and the step response of the RF components of the circuit.

- Sufficient signal-to-noise ratio.
  The signal level and the noise-figure of the amplifiers in the circuit should be well tuned.

To achieve these features, we must accept some undesirable features such as poor linearity or saturation at high amplitude. The absolute positions and gains are usually not important. And we can accept some bunch-current dependence of the output of the feedback if the dependence is not severe.

Figure 6 shows a longitudinal bunch position detection circuit. It consists of a band-pass filter (BPF) to make a sinusoidal burst-like signal from the BPM signal, a double-balanced mixer (DBM) that works as the multiplier between the two arbitrary inputs, a low pass filter (LPF), a frequency multiplier and a phase shifter.

We first extract the $n\omega_{RF}$ component, which is the detection frequency, from the beam signal with the BPF. If the beam is oscillating longitudinally with a

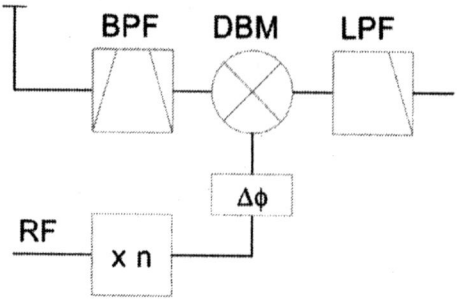

FIGURE 6. Longitudinal bunch-by-bunch position detection circuit.

synchrotron frequency of ω_s and an amplitude of Φ, the signal after the BPF will be expressed as

$$I_b \cos(n\omega_{RF}t + \Phi \sin \omega_s t).$$

By multiplying the signal with the n-th hermonic of RF synchronized signal 90° shifted from the beam phase by the phase shifter ($\sin(n\omega_{RF}t)$) using the DBM, we can split the signal into lower and higher frequency components as

$$I_b \cos(n\omega_{RF}t + \Phi \sin \omega_s t) \times \sin(n\omega_{RF}t)$$
$$= \frac{1}{2}I_b \left(\sin(2n\omega_{RF}t + \Phi \sin \omega_s t) - \sin(\Phi \sin \omega_s t)\right).$$

By rejecting the higher frequency component with the LPF, and assuming the amplitude of the oscillation to be small enough, the final output is proportional to the longitudinal displacement of the bunch as

$$\propto I_b \sin(\Phi \sin \omega_s t) \sim \Phi \sin \omega_s t.$$

Note that the sensitivity is roughly proportional to the detection frequency and the output is proportional to the bunch current. The total step response of the system is in many cases dominated by the bandwidth and the response function of the first bandpass filter since the bandwidths and step responses of other components have sufficient margins compared to the required specifications.

A transverse bunch detection circuit is shown in Fig. 7. The difference between the beam-induced signals of the two facing electrodes, which is proportional to the transverse displacement of a bunch, is made by using a hybrid circuit. Contrary to the longitudinal case we need a phase-independent component, so we multiply the $n\omega_{RF}$ signal with nf_{RF} in-phase using the DBM:

$$I_b \cos(n\omega_{RF}t + \Phi \sin \omega_s t) \times \cos(n\omega_{RF}t)$$
$$= \frac{1}{2}I_b \left(\cos(2n\omega_{RF}t + \Phi \sin \omega_s t) + \cos(\Phi \sin \omega_s t)\right)$$
$$\rightarrow \frac{1}{2}I_b.$$

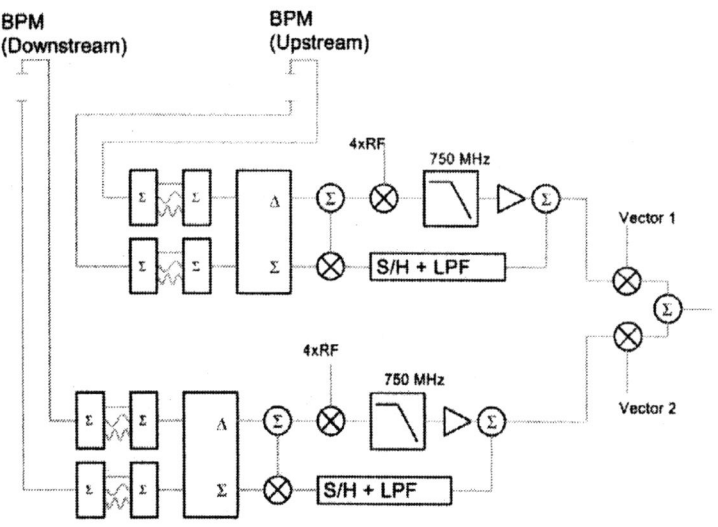

FIGURE 7. Transverse bunch by bunch position detection circuit.

C Signal processing

Since the speed of signals in the cables, detection electronics or amplifiers is much slower than that of beam, and as we want to concentrate the feedback components in one section of the ring, we need to wait for the beam to come back to the feedback section after one turn. In smaller rings, typically of circumference less than 300 m, we can use cable delay. For large rings, such as the KEKB, with a circumference of 3 km, the use of a cable delay of 10 μs is hopeless. In that case we can use a digital delay which converts the analog signal to a digital signal with a very fast ADC, records the digital signal in memory, re-reads it after some time and re-converts it to an analog signal with a very fast DAC.

The DC component in the feedback signal, which will be the residual COD in the transverse feedback, the change in the equilibrium phase or beam-loading effect in the longitudinal feedback, should be rejected as much as possible, because the feedback system tries to correct those errors, in vain. The DC signal wastes very "expensive" feedback power and causes undesirable rapid saturation at the ADC or the power amplifier section. To reject the DC component in the loop, there are several useful methods. For a small ring using long cable delay for one-turn delay, we can form a notch filter to reject the nf_{rev} component combining the cable delay and analog subtracter. Note, however, that this type of cable, with a very long cable and tuning capacitors, is sensitive to temperature changes of the surroundings so careful handling is necessary. For a large ring with digital delay, we need to take great care to reject the offset before the ADC, because the dynamic range of a fast ADC is very limited, in many cases only 8-bit. For automatic correction, we can add offset suppressor feedback loops with very slow local feedback loops before the ADC. Within the dynamic range of the ADC, we can form a DC rejection filter in the digital signal processing logic.

There are two approaches to making a digital filter. The first is to use a software-based system using many digital signal processors (DSPs) [4]. Taking advantage of the great progress in digital circuit technology, using DSPs of the highest execution speed enables us to design a complicated floating-point digital filter with considerable speed. The strong points of a software-based signal process system are as follows:

- Completely programmable. It is easy to change the type of filter by replacing the DSP code.

- Good filtering characteristics. We can form multi-tap filters easily with good accuracy.

- Flexibility. It is applicable to both small rings and large rings by changing the number of DSPs on boards.

However, it also has such weak points as:

- Large-scale and complicated design. As the speed of a single DSP is much slower than the given time limit of a calculation, we require many parallel processing cards. For a fixed number of coefficients in the digital filter, the number of DSP cards increases with the number of bunches times the synchrotron frequency. The difficulty of tuning and maintenance correspondingly increases.

- It is in general necessary to use the down-sampling technique to avoid this complexity [4]. In this case we cannot employ the same system as for the transverse feedback.

Because of its great flexibility and established usability, many rings such as ALS (LBNL), PEP-II, SPEAR (SLAC), PLS (Pohang), DAΦNE (Frascatti) and BESSY-II (BESSY) are using the same system for longitudinal feedback. In PEP-II, 40 DSP's are used for one ring, which provide an aggregate multiple-accumulation rate of 1.6×10^9 operations/s.

The second approach, which we have employed for the KEKB rings, is to make the simplest digital filter that satisfies our minimum requirements with fast hardware logic circuits without the down-sampling technique [5]. The structure of the filter is the hardware two-tap FIR proposed F. Pedersen [6]. As shown in Fig. 8, it has only two taps, A(+1) at $-90°$ and B(-1), at $-270°$ of the oscillation. The frequency response has very wide peaks at f_s, $3 \times f_s$, $\cdots$ and has zeros at 0 (DC), $2 \times f_s$, $\cdots$, where f_s is the synchrotron frequency in the longitudinal case. Strong points of the hardware two-tap FIR filter are:

- Simple and applicable to both longitudinal and transverse feedback systems. The phase shift and time delay are tunable by selection of the tap positions while preserving the time differences between the two taps.

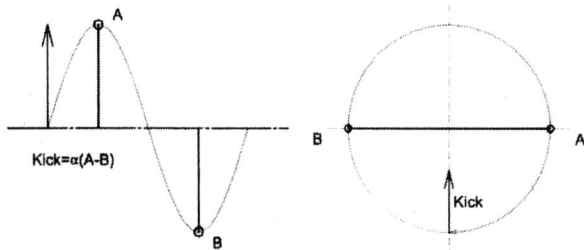

FIGURE 8. Tap positions for the 2-tap FIR filter.

With this simplification, the function of the filter is fairly limited and the following weak points arise:

- Very limited flexibility. To reduce the complexity, the structure of the filter needs to be strongly geared towards a particular ring. Application to other rings is in general very difficult.

- No sharp-filtering effect around the center frequency. This is, in practice, not a serious problem. Experiments with beams show that the measured S/N of the detection signal is good enough, typically better than 40 dB.

- Complication of the high-frequency digital circuits around the ADC and the DAC. At KEKB, these difficulties have been overcome by developing custom LSIs to multiplex and demultiplex the fast digital data.

Figures 9 and 10 show a block diagram and a photo of the two-tap FIR filter.

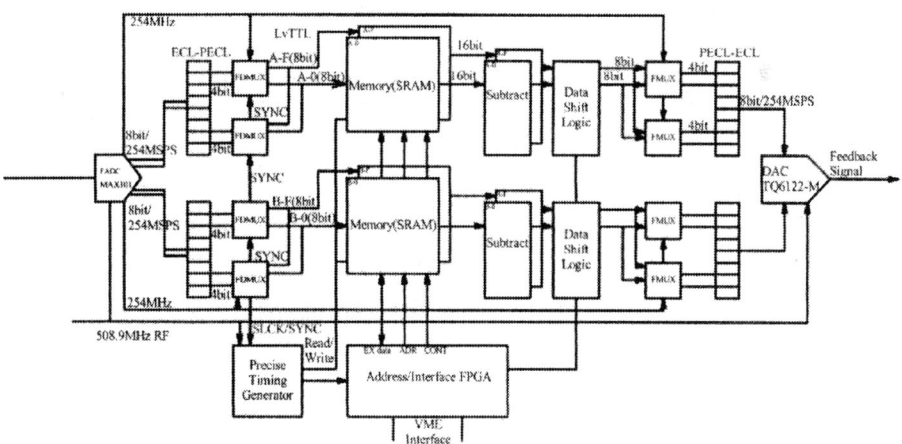

FIGURE 9. Block diagram of the 2-tap FIR filter board developed at KEK.

FIGURE 10. Photo of the 2-tap FIR filter board for KEKB.

D Kickers and power amplifiers

We first show rough formulae for estimating the necessary feedback voltage for a desired damping time. For the longitudinal case, to damp a synchrotron oscillation of maximum energy deviation ΔE with a feedback damping time of τ_ϵ, we need

$$V_{feedback} = 2\frac{1}{\tau_\epsilon}T_0(\Delta E/e)$$

where T_0 is the revolution time. For example, assuming E=3.5 GeV, $(\Delta E/e)/E = 0.1\%$, $T_0 = 10$ μs and $\tau_\epsilon = 10$ ms, we need 7 kV/turn, which is a fairly tough value to achieve.

For the transverse case, we have

$$V_{feedback} = 2\frac{1}{\tau_x}T_0(E/e)\frac{1}{\sqrt{\beta_m\beta_k}}x_{max}$$

where τ_x is the feedback damping time, x_{max} is the maximum amplitude, β_m and β_k are the betatron functions at the monitor and the kicker, respectively, assuming the betatron phase advance between the kicker and the monitor is 90°. Again, to damp $x_{max} = 1$ mm oscillation with $\tau_x = 1$ ms in the case of $E = 3.5$ GeV, $\beta_m = \beta_k = 10$ m and $T_0 = 10\mu$s, we need $V_{feedback} = 7$ kV/turn. Though it depends on the mode of oscillation, this kick voltage is not an unreasonable value.

To act as wideband kickers, stripline-type kickers are widely used. The structure(Fig. 11) consists of a rod or a plate as inner conductor and the wall as outer conductor. The characteristic impedance of the kicker is normally designed to be

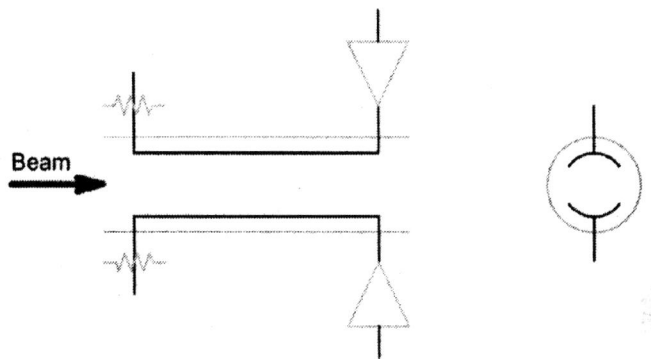

FIGURE 11. Sketch of a strip-line type kicker.

50 Ω. We know from the Panofsky-Wenzel theorem [7] that we need a change in the longitudinal electric field to kick the beam, even in the transverse direction. Therefore, the deflecting power must be supplied from the downstream port. If we supply the power in phase, we can kick the beam longitudinally. If the power is supplied with opposite phase, the beam will be deflected transversely. The shunt impedance of the kicker is calculated based on the direction of the kick. The longitudinal shunt impedance $R_\parallel$ is given [8] by

$$R_\parallel T^2 = 2Z_L g_\parallel^2 \sin^2 k\ell,$$

where T^2 is the transit-time factor, Z_L is the characteristic impedance of the kicker, $g_\parallel$ is the longitudinal geometric factor, k is the wave number, and ℓ is the length of the stripline. The transverse shunt impedance $R_\perp$ is given by

$$R_\perp T^2 = 2Z_L \left(g_\perp \frac{2\ell}{h} \frac{\sin k\ell}{k\ell} \right)^2,$$

where $g_\perp$ is a transverse geometric factor and h is the distance between the facing electrodes. From the formulas above, we know that:

- The longitudinal shunt impedance is very low, around 100 Ω. Since the shunt impedance is a periodic function of the frequency, we can use higher-frequency components which helps in the design of wideband power amplifiers.

- The transverse shunt impedance is proportional to $\sin^2(k\ell)/(k\ell)^2$ so we can use only the base-band region. A longer stripline has larger shunt impedance but the bandwidth decreases rapidly. Typical shunt impedance is around a few kΩ.

For longitudinal kickers, ALS and PEP-II have adopted a series-drift-tube type kicker [9], in principle a stripline kicker but combined with two electrodes to make good use of feedback power. The measured shunt impedance for one unit is about 320 Ω. At DAΦNE, an over-coupled cavity (namely very wideband) has been developed for the longitudinal kicker [10]. A quality factor less than 5 is achieved around the center frequency of 1 GHz with a shunt impedance greater than 600 Ω. This kind of longitudinal kicker has been adopted at DAΦNE, KEKB-LER, SPEAR, PLS and BESSY-II. Since this kind of kicker has no directivity, we need wideband circulators to protect the amplifiers from the beam-induced power.

The maximum power required for the feedback system is easily evaluated as

$$P_{MAX} = V_{MAX}^2 / 2R_{sh}$$

where V_{MAX} is the maximum feedback voltage. Actually, however, the maximum power of available wideband amplifiers is limited for technical and the economic reasons. The cost of power for a wideband amplifier is in general extremely high. We usually restrain ourselves to selecting smaller and cheaper amplifiers.

If the power of the amplifiers is sufficient, not saturating at the needed feedback voltage, the oscillation should be damped exponentially, with the designed feedback damping time. However, when the amplifiers are not powerful enough to supply the necessary power, the oscillation will be damped linearly, not exponentially. We call this regime "bang-bang damping." The effective damping time in this region is much longer than the feedback damping time. In some cases which the growth rate is huge, the feedback system fails to damp the oscillation after the oscillation exceeds some threshold amplitude (cannot re-capture the oscillation).

V TRANSIENT-DOMAIN ANALYSIS

The transient behavior of the beam just after closing or opening of the feedback loop reveals many important characteristics of the coupled-bunch motions as well as the performance of the feedback systems. This powerful method of analyzing instabilities is known as transient-domain analysis [4,11]. By preparing a large-scale memory board that can accumulate bunch positions for every bunch with many turns, we can record and analyze the time evolution of oscillation for every bunch at the transient of feedback on/off. Figure 12(a) shows the time evolution of the betatron-oscillation components for each bunch. By taking the Fourier transform of the bunch amplitude over time, taking into account the betatron phase advances, we calculate the evolution of the modes of the instability over time, as shown in Fig. 12(b).

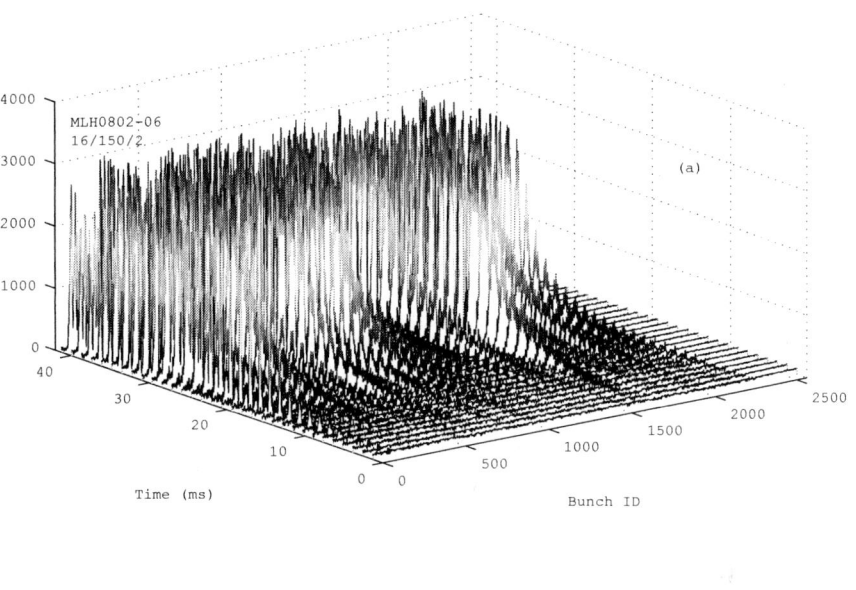

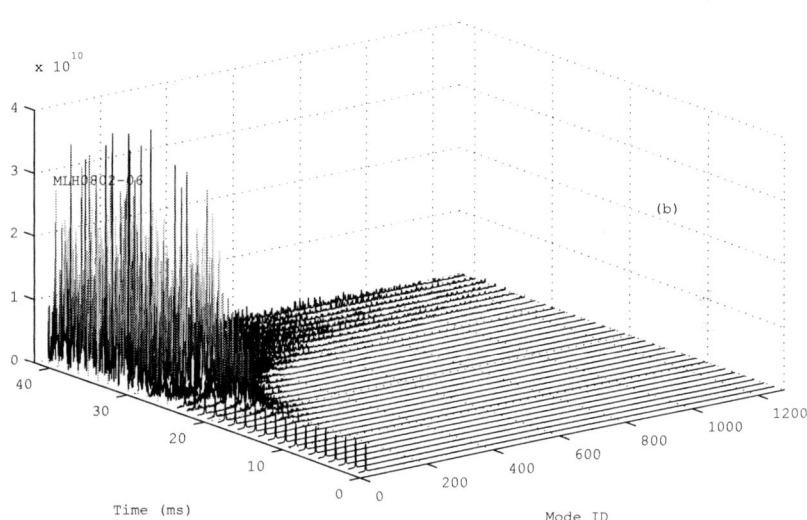

FIGURE 12. Example of a growing transient due to horizontal instability in the KEKB LER. (a) Time evolution of the betatron amplitude for each bunch; (b) Time evolution of the modes of the instability.

Comparing this method with the ordinary method of analyzing instabilities, we can get the full spectrum of bunch motion in a single transient. And the motion can be studied in the small-oscillation situation where the oscillation may be regarded as linear.

SUMMARY

We have seen the outline of the bunch-by-bunch feedback systems and the application of the feedback systems to analyze the instabilities. However, these complicated systems are also fairly expensive and require skills to operate them and keep them in good condition. You may ask, "Is such a costly and complicated system surely necessary?" I believe, that making all possible efforts to reduce impedance sources in the ring is surely valuable, even if the work is costly and boring. The power of bunch feedback systems is limited and they can damp only the dipole oscillation within some limited amplitude. However, we must also consider the total cost performance and the possibility of unknown broad-band impedance sources. With a beam feedback system we can suppress the instabilities coming from uncertain or unknown broad-band impedance sources. Also we can diagnose the impedance source by transient-domain analysis, helping to locate the source. Beam feedback systems can greatly contribute to the high quality operation of a ring.

ACKNOWLEDGMENTS

The author would like to express his sincere appreciation to Professor S. Kurokawa for his continued help and useful suggestions. He also wishes to express his gratitude to Dr. E. Kikutani, Dr. J. W. Flanagan, Professor S. Hiramatsu, Professor T. Kasuga, Dr. T. Obina, and Dr. Y. Ohnishi for numerous thoughtful discussions. Discussions with the people who are developing the feedback systems at SLAC, DAΦNE, and Pohang have been very fruitful. He thanks the commissioning group of KEKB for their great patience with the feedback systems, especially in the early stage of commissioning.

REFERENCES

1. G. F. Franklin et al., *Feedback Control of Dynamic Systems*, Addison Wesley, 1991.
2. Ansoft HFSS (High-Frequency Structure Simulator) V7, Ansoft Corporation, 1999.
3. Maxwells equations by the Finite Integration Algorithm (MAFIA), Computer Simulation Technology (CST) Co., Germany, 1999.
4. J. D. Fox, R. Larsen, S. Prabhakar, D. Teytelman, A. Young, A. Drago, M. Serio, W. Barry and G. Stover, in Proc. 1999 Part. Accel. Conf., New York, 1999, p. 636.
5. M. Tobiyama and E. Kikutani, Phys. Rev. ST Accel. Beams **3**, 012801 (2000).
6. F. Pedersen, "Multibunch Feedback— Transeverse, Longitudinal and RF Cavity Feedback," presented at the Workshop "Factories with e^+e^- Rings," Benalmadena, Spain, Nov. 1992.

7. W. K. H. Panofsky and W. A. Wenzel, Rev. Sci. Instr. **27**, 967 (1956)
8. G. R. Lambertson, "Dynamic Devices—Pickups and Kickers," in *Physics of Accelerators,* eds. M. Month and M. Dienes, AIP Conf. Proc. **153** 1414 (1987).
9. J. N. Corlett, J. Johnson, G. Lambertson and F. Voelker, in Proc. 1994 European Part. Accel. Conf., London, 1994, p. 1626.
10. R. Boni, A. Gallo, A. Ghigo, F. Marcellini, M. Serio and M. Zobov, Part. Accel. **52**, 95 (1996).
11. S. Prabbhaker, J. D. Fox, D. Teytelman, and A. Young, Phys. Rev. ST Accel. Beams **2**, 084401 (1999).

Space Charge

N. A. Vinokurov

Budker Institute of Nuclear Physics
Novosibirsk, Russia

Abstract. The space-charge effects in a straight beamline are considered.

INTRODUCTION

In these short notes a few simple models are described. They may be useful for obtaining rough estimates of space-charge effects, and for testing the numerical results provided by computer codes.

SPACE-CHARGE FORCES

Coasting Cylindrical Beam

1. Consider first a round coasting (i. e. unbunched, so the beam parameters do not depend on time) cylindrical (i.e. the beam parameters do not depend on the coordinate along the beam axis z) beam having charge density $\rho(r)$, where $r = \sqrt{x^2 + y^2}$ is the distance from the beam axis z. Suppose that all the particles are moving along the z axis with the same velocity v. Then the electric and magnetic fields have only one component in the cylindrical system of coordinates (r, α, z), and we can use Gauss's and Stokes's theorems to find the fields, respectively:

$$E_r = \frac{4\pi}{r} \int_0^r \rho(r')r'dr', \qquad (1)$$

$$B_\alpha = \frac{v}{c} \frac{4\pi}{r} \int_0^r \rho(r')r'dr', \qquad (2)$$

where c is the velocity of light. According to the derivation method, Eqs. (1) and (2) are also valid in the presence of a coaxial round cylindrical vacuum chamber. The Lorentz force has only a radial component[1]:

$$F_r = eE_r - e\beta B_z = \frac{1}{\gamma^2} eE_r, \qquad (3)$$

---

[1] For the vacuum chamber with finite conductivity there is also the z component of electric field.

where $\beta = v/c$, and $\gamma = (1 - \beta^2)^{-1/2}$ is the relativistic factor (the ratio of the particle energy to its rest-frame energy mc^2). According to Eq. (3), the magnetic part of the Lorentz force is subtracted from the electric one, and therefore the net value of the Lorentz force decreases significantly with particle energy growth.

When a particle of a beam passes a length l, the radial component of the momentum grows as

$$\Delta p_r = F_r \frac{l}{v} = \frac{el}{v\gamma^2} \frac{4\pi}{r} \int_0^r \rho(r') r' dr'. \tag{4}$$

Since Δp_r is invariant with respect to the boost along the z axis, it is interesting to calculate it in the beam rest frame. Here only the electric field induced by the charge density ρ/γ,

$$E'_r = \frac{4\pi}{r} \int_0^r \frac{\rho(r')}{\gamma} r' dr', \tag{5}$$

exists. Because of Lorentz contraction, the force acts during a time interval $l/\gamma v$, and we again get the result of Eq. (4).

For the simplest case of homogeneous charge distribution with beam radius a,

$$\rho = \begin{cases} \dfrac{I}{v\pi a^2} & r \leq a \\ 0 & r > a \end{cases}, \tag{6}$$

where I is the beam current, we have

$$E_r = \begin{cases} \dfrac{2I}{va^2} r & r \leq a \\ \dfrac{2I}{vr} & r > a \end{cases}. \tag{7}$$

If the beam is propagating inside a conducting round pipe of radius b, we can choose the potential to be zero on the inner pipe surface. Then according to Eq. (1) the potential inside the beam pipe is

$$\varphi(r) = \int_r^b E_r(r') dr' = 4\pi \int_r^b \int_0^{r'} \rho(r'') r'' dr'' \frac{dr'}{r'}. \tag{8}$$

For the homogeneous charge distribution given by Eq. (6), we get

$$\varphi(r) = \begin{cases} \dfrac{2I}{v} \ln\dfrac{b}{a} + \dfrac{I}{v}\left(1 - \dfrac{r^2}{a^2}\right) & r \leq a \\ \dfrac{2I}{v} \ln\dfrac{b}{r} & r > a \end{cases}. \tag{9}$$

2. For an elliptical beam cross-section,

$$\rho(x,y) = \frac{I}{v\pi ab} \vartheta\left(1 - \frac{x^2}{a^2} - \frac{y^2}{b^2}\right), \tag{10}$$

where

$$\vartheta(t) = \begin{cases} 0 & t<0 \\ 1 & t\geq 0 \end{cases} \tag{11}$$

is the step function, the electric field inside the beam ($x^2/a^2 + y^2/b^2 \leq 1$) is also a linear function of the coordinates [1]:

$$E_x(x,y) = \frac{4I}{v}\frac{x}{a(a+b)},$$
$$E_y(x,y) = \frac{4I}{v}\frac{y}{b(a+b)}. \tag{12}$$

Using Eq. (10) in the rest frame, and coming back to the laboratory frame, we can easily get the magnetic field $B_x = -\beta E_y$, $B_y = \beta E_x$ and the Lorentz force

$$F_x(x,y) = \frac{4eI}{v\gamma^2}\frac{x}{a(a+b)},$$
$$F_y(x,y) = \frac{4eI}{v\gamma^2}\frac{y}{b(a+b)}. \tag{13}$$

Longitudinal Electric Field

1. Let the beam pipe have different radii b_- at $z < -L_t$ and b_+ at $z > L_t$ ($2L_t$ is the length of the transition region). Then there is a longitudinal electric field $E_z(r,z)$ in the transition region, and

$$U = -\int_{-\infty}^{\infty} E_z(r,z)dz = \varphi_+(r) - \varphi_-(r) = \frac{2I}{v}\ln\frac{b_+}{b_-}. \tag{14}$$

Note that the voltage U does not depend on r. For $b_- < b_+$ particles are decelerated, and the beam "loses" corresponding power UI. This power is deposited to the energy of the fields. Moving from left to right, the beam induces the fields in additional space, contained between the cylinders $r = b_-$ and $r = b_+$. The additional power passing through the right part of the beam pipe is

$$P = \int_{b_-}^{b_+} \frac{cE_r B_\alpha}{4\pi} 2\pi r dr = \frac{2I^2}{v}\ln\frac{b_+}{b_-}, \tag{15}$$

which is just IU. To prove energy conservation in the general case, we need to take into account the variation of the beam velocity. Then

$$U(r) = \varphi_+(r) - \varphi_-(r) = \frac{2I}{v_+}\ln\frac{b_+}{a} - \frac{2I}{v_-}\ln\frac{b_-}{a} + I\left(1-\frac{r^2}{a^2}\right)\left(\frac{1}{v_+} - \frac{1}{v_-}\right). \tag{16}$$

Averaging $U(r)$ over the beam cross-section ($\langle r^2/a^2 \rangle = \frac{1}{2}$), we get the power

$$IU = \frac{2I^2}{v_+}\left(\ln\frac{b_+}{a}+\frac{1}{4}\right)-\frac{2I^2}{v_-}\left(\ln\frac{b_-}{a}+\frac{1}{4}\right). \tag{17}$$

It can also be derived from the total energy conservation:

$$\gamma_+\frac{mc^2}{e}I+\frac{2I^2}{v_+}\left(\ln\frac{b_+}{a}+\frac{1}{4}\right)=\gamma_-\frac{mc^2}{e}I+\frac{2I^2}{v_-}\left(\ln\frac{b_-}{a}+\frac{1}{4}\right). \tag{18}$$

2. Now let the beam current depend on time as $I(t-z/v)$. Since

$$t-\frac{z}{v}=\gamma\left(t'+v\frac{z'}{c^2}\right)-\frac{\gamma}{v}(z'+vt')=-\frac{z'}{\gamma v}, \tag{19}$$

for the ideal-conducting round beam pipe the field in the rest frame is purely electrostatic. If

$$\left|\frac{d}{dz'}\ln I\left(-\frac{z'}{\gamma v}\right)\right|=\frac{1}{\gamma v}\left|\frac{d}{dt}\ln I\left(t-\frac{z}{v}\right)\right|<<\frac{1}{b}, \tag{20}$$

i. e. the bunch in the rest frame is much longer than the pipe radius, Eqs. (8) and (9) are valid approximately. Then the corresponding longitudinal electric field in the beam ($r \leq a$) is

$$E_z = -\frac{\partial}{\partial z'}\varphi'(r,z')=-\left(2\ln\frac{b}{a}+1-\frac{r^2}{a^2}\right)\frac{d}{dz'}\frac{I}{\gamma v}=\left(2\ln\frac{b}{a}+1-\frac{r^2}{a^2}\right)\frac{1}{\gamma^2 v^2}\frac{\partial}{\partial t}I\left(t-\frac{z}{v}\right). \tag{21}$$

Note that E_z is Lorentz invariant, and therefore Eq. (21) is valid in the laboratory frame also. For a thin beam ($a << b$), the dependence of E_z on the transverse coordinate r inside the beam is rather weak, and we can replace r^2/a^2 by its average value ½. Another simplification of Eq. (21) can be obtained for the parabolic longitudinal distribution

$$I(t)=\begin{cases}\dfrac{3}{4}\dfrac{Q}{\tau}\left(1-\dfrac{t^2}{\tau^2}\right) & |t|\leq\tau \\ 0 & |t|>\tau\end{cases}, \tag{22}$$

where Q and τ are the charge and the duration of the bunch. Then E_z is a linear function of $t-z/v$ inside the bunch ($|t-z/v|\leq\tau$),

$$E_z = -\left(3\ln\frac{b}{a}+\frac{3}{4}\right)\frac{Q}{\gamma^2 v^2\tau^3}\left(t-\frac{z}{v}\right). \tag{23}$$

3. Let the beam pipe have a small azimuthal groove with depth h and width w, i. e. the pipe surface equation is

$$r(z)=\begin{cases}b+h & |z|\leq\dfrac{w}{2} \\ b & |z|>\dfrac{w}{2}\end{cases}. \tag{24}$$

For a long bunch ($\frac{w}{v}\left|\frac{d}{dt}\ln I\left(t-\frac{z}{v}\right)\right| \ll 1$), the magnetic field inside the groove can be approximated by Eq. (2). Therefore the magnetic flux in the groove is

$$\Phi = \frac{2Iw}{c}\ln\frac{b+h}{b} \approx \frac{2Iwh}{cb}. \tag{25}$$

Another way to obtain this flux is to calculate the magnetic field in the groove from the surface density of the image current, $i = -I/(2\pi b)$: $B_a = -4\pi i/c$, $\Phi = B_a wh$. According to Eq. (25), the groove inductance is $L = 2wh/b$.

As the flux is time-dependent, the corresponding voltage between the groove banks, U, is

$$-\int_{-\infty}^{\infty} E_z dz = U = \frac{1}{c}\frac{d\Phi}{dt} \approx 2\frac{wh}{c^2 b}\frac{dI(t)}{dt}. \tag{26}$$

The longitudinal impedance is defined as $Z(\omega) = U(\omega)/I(\omega)$. In our case

$$Z = -\frac{1}{c^2}i\omega L = -2i\frac{wh\omega}{c^2 b}. \tag{27}$$

Comparing Eqs. (26) and (21) we can say that the Coulomb longitudinal forces are equivalent to the effective (negative) inductance per unit length:

$$\left(\frac{L}{l}\right)_c = -\frac{2}{\gamma^2 \beta^2}\left(\ln\frac{b}{a}+\frac{1}{4}\right). \tag{28}$$

TRANSVERSE MOTION

Laminar Beam

As it was shown earlier, the components of the Lorentz force inside the uniformly charged beam are linear functions of the transverse coordinates (see Eq. (13)). Let's consider a round beam. The equation of transverse motion of a particle inside such a beam has the form[2]:

$$\gamma m \frac{d^2 r_1}{dt^2} = \frac{2eI}{\gamma^2 va^2} r_1. \tag{29}$$

If the initial particle velocities depend linearly on the initial coordinates $\frac{dr_1}{dt}(0) = Ar_1(0)$, then the linear dependence (with a time-dependent coefficient A) will be conserved during the motion, because of the linearity of Eq. (29). This means that particle trajectories do not cross. Such a beam is frequently referred to as a laminar

---

[2] Here and below we consider the paraxial approximation $|dr_1/dt| \ll v$.

beam. In the phase space (r, r'), all particles lie in the interval bounded by the points $(0, 0)$ and $(a, da/dt)$. Equation (29) is also valid for the boundary particles $r_1 = a$. Then, changing the independent variable $t = z/v$, we obtain

$$\frac{d^2 a}{dz^2} = \frac{2I}{(\beta\gamma)^3 I_0} \frac{1}{a}, \qquad (30)$$

where $I_0 = mc^3/e \approx 17$ kA (for electrons) is the characteristic current. The first integral of Eq. (30) is

$$\frac{a'^2}{2} + \frac{2I}{(\beta\gamma)^3 I_0} \ln a = \frac{2I}{(\beta\gamma)^3 I_0} \ln a_0, \qquad (31)$$

where $a' = da/dz$, and a_0 is the minimum beam radius (where $a' = 0$). Using Eq. (31) we can get the expression for the angular divergence,

$$a' = \pm 2\sqrt{\frac{I}{(\beta\gamma)^3 I_0} \ln \frac{a}{a_0}}. \qquad (32)$$

Non-relativistic beams are frequently characterized by the perveance $P = IU^{-3/2} \propto I/\beta^3$ (eU is the particle kinetic energy), then the dimensionless constant in Eq. (30) can be expressed as $P\sqrt{m/(2e)}$.

Linear external focusing can be easily taken into account by adding the corresponding force to Eq. (29). Then Eq. (30) will be generalized as follows,

$$\frac{d^2 a}{dz^2} + K(z)a - \frac{2I}{(\beta\gamma)^3 I_0} \frac{1}{a} = 0, \qquad (33)$$

where $K(z)$ is the focusing rigidity. For example, for a thin focusing lens installed at $z = 0$, $K(z) = \delta(z)/F$, F is the focal length, and $\delta(z)$ is Dirac's delta-function.

To generalize further our consideration of transverse motion, taking account of the space-charge force, we will discuss a beam with finite emittances.

Kapchinsky-Vladimirsky Equation

1. A particle beam can be described in more detail by using the distribution function $f(x, x', y, y')$ in the four-dimensional phase space of the transverse coordinates and angles. For a uniform elliptical beam (Eq. (10)) we can choose a distribution function of the form

$$f = \frac{1}{\pi^2 \varepsilon_x \varepsilon_y} \delta\left(\frac{\beta_x x'^2 + 2\alpha_x x x' + \gamma_x x^2}{\varepsilon_x} + \frac{\beta_y y'^2 + 2\alpha_y y y' + \gamma_y y^2}{\varepsilon_y} - 1\right), \qquad (34)$$

where $\alpha_x, \beta_x, \gamma_x, \alpha_y, \beta_y, \gamma_y$ are z-dependent Twiss parameters, and ε_x and ε_y are constants called emittances. Distribution in Eq. (34) is referred to as the Kapchinsky-Vladimirsky (KV) distribution and corresponds to uniform particle distribution over the surface of a four-dimensional ellipsoid. Integration of the distribution function over the angles x' and y' leads to the charge density, Eq. (10), with $a^2 = \varepsilon_x \beta_x$ and $b^2 = \varepsilon_y \beta_y$. We can write the particle trajectory equations using the Lorentz force, Eq. (13),

$$x'' = -K_x(z)x + \frac{4I}{(\beta\gamma)^3 I_0} \frac{x}{a(a+b)},$$
$$y'' = -K_y(z)y + \frac{4I}{(\beta\gamma)^3 I_0} \frac{y}{b(a+b)}, \quad (35)$$

where K_x and K_y are the rigidities of the external focusing. Equation (35) is the set of two Hill's equations with total rigidities

$$K_x^{tot} = K_x(z) - \frac{4I}{(\beta\gamma)^3 I_0} \frac{1}{a(a+b)},$$
$$K_y^{tot} = K_y(z) - \frac{4I}{(\beta\gamma)^3 I_0} \frac{1}{b(a+b)}. \quad (36)$$

Then the envelope equations with these rigidities are

$$a'' + K_x(z)a - \frac{4I}{(\beta\gamma)^3 I_0} \frac{1}{a+b} - \frac{\varepsilon_x^2}{a^3} = 0,$$
$$b'' + K_y(z)b - \frac{4I}{(\beta\gamma)^3 I_0} \frac{1}{a+b} - \frac{\varepsilon_y^2}{b^3} = 0. \quad (37)$$

Equations (37) are referred to as the Kapchinsky-Vladimirsky equations. For the special distribution they reduced the problem of the evolution of the distribution function to the problem of the evolution of two transverse beam sizes a and b. It is worth noting that these are exact equations.

2. Frequently people use the so-called root-mean-square emittance

$$\varepsilon_x^{rms} = \sqrt{\langle x^2 \rangle \langle x'^2 \rangle - \langle x x' \rangle^2}. \quad (38)$$

To understand the origin of this combination we can write down the "symplectic distance" between two points (particles) on the phase plane

$$S = (x_1 \; x_1') \begin{pmatrix} 0 & 1 \\ -1 & 0 \end{pmatrix} \begin{pmatrix} x_2 \\ x_2' \end{pmatrix} \quad (39)$$

which is the area of the parallelogram constructed on vectors $(x_1 \; x_1')$ and $(x_2 \; x_2')$. As $\langle\langle S \rangle\rangle_{1/2} = 0$, we can calculate

$$\langle\langle S^2 \rangle\rangle_{1/2} = \langle\langle (x_1 x_2')^2 \rangle\rangle_{1/2} - 2\langle\langle x_1 x_2' x_2 x_1' \rangle\rangle_{1/2} + \langle\langle (x_2 x_1')^2 \rangle\rangle_{1/2}$$
$$= 2\langle x^2 \rangle \langle x'^2 \rangle - 2\langle x x' \rangle^2. \quad (40)$$

For the Kapchinsky-Vladimirsky distribution $\varepsilon_x^{rms} = \varepsilon_x/4$.

Comparison of the two nonlinear terms in Eq. (37) for a round beam, $a = b$, $\varepsilon_x = \varepsilon_y = \varepsilon$, shows that the laminar beam approximation, Eq. (33), is valid when

$$\frac{\varepsilon}{\beta_x} \ll \frac{2I}{(\beta\gamma)^3 I_0} . \qquad (41)$$

The left side of Eq. (41) is the square of the local angular spread, and the right side is the dimensionless perveance, which is the square of the "characteristic angular divergence" (see Eq. (32)). In the opposite case, we can neglect the space-charge force term in Eq. (37) and return to the single-particle approximation. It is worth noting that the condition (41) depends on the beta-function. Therefore, it may be different in different places along the beamline. In particular, the places where the beta-function (and the beam size) are maximal, are the most sensitive to the influence of the space-charge forces.

Emittance Degradation

For a coasting beam and Kapchinsky-Vladimirsky transverse distribution, the space-charge force is linear and does not change the emittances. But for a bunched beam the current I depends on time. Correspondingly, at different parts of the bunch the corrections to the focusing rigidities (see Eq. (36)) are different. Therefore the Twiss parameters of the beam also become different and, averaged over the whole bunch, the transverse emittance (the so-called projection emittance) increases.

Consider a beam passing through free space of length L. According to Eq. (36), the horizontal optical strength is

$$\frac{1}{F_x} = K_x^{tot} L = -\frac{4I(s)}{(\beta\gamma)^3 I_0} \frac{L}{a(a+b)} , \qquad (42)$$

where s is the coordinate along the bunch. The Twiss matrix transformation

$$\mathbf{J} = \mathbf{T}\mathbf{J}_0\mathbf{T}^{-1} = \begin{pmatrix} 1 & 0 \\ -\frac{1}{F_x} & 1 \end{pmatrix} \begin{pmatrix} \alpha_{x0} & \beta_{x0} \\ -\gamma_{x0} & -\alpha_{x0} \end{pmatrix} \begin{pmatrix} 1 & 0 \\ \frac{1}{F_x} & 1 \end{pmatrix}$$

$$= \begin{pmatrix} \alpha_{x0} + \frac{\beta_{x0}}{F_x} & \beta_{x0} \\ -\gamma_{x0} - \frac{2\alpha_{x0}}{F_x} - \frac{\beta_{x0}}{F_x^2} & -\alpha_{x0} - \frac{\beta_{x0}}{F_x} \end{pmatrix} , \qquad (43)$$

and averaging over both the transverse distribution function and the bunch length gives the rms projection emittance,

$$\varepsilon_x^2 = \langle x^2\rangle\langle x'^2\rangle - \langle xx'\rangle^2 = \langle \varepsilon_{x0}\beta_x\rangle\langle\varepsilon_{x0}\gamma_x\rangle - \langle\varepsilon_{x0}\alpha_x\rangle^2$$

$$= \varepsilon_{x0}^2\left(\langle\beta_x\rangle\langle\gamma_x\rangle - \langle\alpha_x\rangle^2\right) = \varepsilon_{x0}^2\left[1 + \beta_{x0}^2\left\langle\frac{1}{F_x^2}\right\rangle - \beta_{x0}^2\left\langle\frac{1}{F_x}\right\rangle^2\right], \quad (44)$$

and the emittance increase,

$$\Delta(\varepsilon_x^2) = \varepsilon_x^2 - \varepsilon_{x0}^2 = (\varepsilon_{x0}\beta_{x0})^2\left(\left\langle\frac{1}{F_x^2}\right\rangle - \left\langle\frac{1}{F_x}\right\rangle^2\right). \quad (45)$$

For a Gaussian bunch $I = I_{max}\exp\left(-\frac{s^2}{2\sigma^2}\right)$, $\left\langle\frac{1}{F_x^2}\right\rangle - \left\langle\frac{1}{F_x}\right\rangle^2 = \left(\frac{1}{F_x}\right)_{max}^2\left(\frac{1}{\sqrt{3}} - \frac{1}{2}\right)$.

Assuming for simplicity a round beam $a^2 = b^2 = 4\varepsilon_{x0}\beta_{x0}$ and a small initial emittance $\varepsilon_{x0} \ll \varepsilon_x$, we can easily find from Eqs. (42) and (45):

$$\Delta\varepsilon_x \approx \frac{0.14 I_{max} L}{(\beta\gamma)^3 I_0}. \quad (46)$$

The other reason for emittance degradation is the nonlinearity of the Lorentz force for real transverse distributions (which differ from the Kapchinsky-Vladimirsky distribution). The contribution from nonlinearity increases the numerical coefficient in Eq. (46) from 0.14 to approximately 0.2. The emittance degradation described above is not a truly irreversible process, therefore it can be partly com-pensated by proper downstream optics.

LONGITUDINAL MOTION

Debunching of a Long Bunch

The equation of longitudinal motion,

$$\frac{d}{dt}\left(\gamma m\frac{dz}{dt}\right) = eE_z, \quad (47)$$

can be rewritten approximately as

$$\gamma^3 m\frac{d^2 s}{dt^2} = eE_z, \quad (48)$$

where $s = z - v_0 t$, v_0 is the average velocity and $\gamma = (1 - v_0^2/c^2)^{-1/2}$. For parabolic charge distribution, Eq. (22), the longitudinal field E_z depends on s linearly (see Eq. (23)), and we can repeat the study done above for the transverse motion of a laminar beam. Substituting Eq. (23) into Eq. (48) gives

$$\frac{d^2s}{dt^2} = 3\left(\ln\frac{b}{a} + \frac{1}{4}\right)\frac{eQ}{\gamma^5 ml^3} s, \tag{49}$$

where $l = v_0 \tau$. An equation for the bunch length $l = s_{max}$ follows from Eq. (49):

$$\frac{d^2l}{dt^2} = 3\left(\ln\frac{b}{a} + \frac{1}{4}\right)\frac{eQ}{\gamma^5 m} \frac{1}{l^2}. \tag{50}$$

The most significant difference between Eq. (50) and Eq. (30) is the inverse square dependence of the right side. The first integral of Eq. (50) is

$$\frac{1}{2}\left(\frac{dl}{dt}\right)^2 + 3\left(\ln\frac{b}{a} + \frac{1}{4}\right)\frac{eQ}{\gamma^5 m}\frac{1}{l} = 3\left(\ln\frac{b}{a} + \frac{1}{4}\right)\frac{eQ}{\gamma^5 m}\frac{1}{l_0} = \frac{1}{2}\left(\frac{dl}{dt}\right)^2_{\infty}, \tag{51}$$

where l_0 is the minimum bunch length and $(dl/dt)_\infty$ is the maximum velocity deviation of the bunch end. The corresponding maximum energy deviation ($\delta\gamma = \gamma^3 \beta \frac{1}{c}\frac{dl}{dt}$) is

$$(\delta\gamma)_{max} = \gamma^3 \beta \frac{1}{c}\sqrt{6\left(\ln\frac{b}{a} + \frac{1}{4}\right)\frac{eQ}{\gamma^5 m}\frac{1}{l_0}} = \sqrt{8\left(\ln\frac{b}{a} + \frac{1}{4}\right)\beta\gamma\frac{I_{max}}{I_0}}. \tag{52}$$

Short-Wavelength Density Modulation

Now we'll consider the opposite limiting case of short-wavelength longitudinal density modulation. It is important for some electronic devices (klystrons, traveling-wave tubes, free electron lasers, etc.).

In the beam rest frame, space charge induces only an electric field. Therefore, the equations in hydrodynamic approximation have the form

$$\frac{\partial \rho}{\partial t} + div(\rho\mathbf{v}) = 0,$$

$$m\frac{d\mathbf{v}}{dt} = e\mathbf{E}, \tag{53}$$

$$div\mathbf{E} = 4\pi\rho.$$

Linearization

$$\rho = \rho_0 + \rho_1,$$

$$\frac{d\mathbf{v}}{dt} = \frac{\partial \mathbf{v}}{\partial t} + (\mathbf{v}\nabla)\mathbf{v} \approx \frac{\partial \mathbf{v}}{\partial t}, \tag{54}$$

leads to

$$\frac{\partial \rho_1}{\partial t} + \rho_0 div\mathbf{v} = 0,$$

$$m\frac{\partial \mathbf{v}}{\partial t} = e\mathbf{E}, \tag{55}$$

$$div\mathbf{E} = 4\pi\rho_1,$$

and hence to the equation for plasma oscillation,

$$\frac{\partial^2 \rho_1}{\partial t^2} + \omega_p^2 \rho_1 = 0, \quad (56)$$

where $\omega_p = \sqrt{4\pi\rho_0 e/m}$ is the rest-frame plasma frequency. The general solution of Eq. (56) is

$$\rho_1 = A(\mathbf{r})\cos(\omega_p t) + B(\mathbf{r})\sin(\omega_p t). \quad (57)$$

Returning to the laboratory frame, we can find the space period of the plasma oscillations,

$$L_p = \frac{2\pi\gamma\beta c}{\omega_p} = \sqrt{\pi(\beta\gamma)^3 I_0/j}, \quad (58)$$

where j is the current density. For uniform charge distribution, Eq. (6),

$$L_p = \pi a \sqrt{(\beta\gamma)^3 I_0/I}. \quad (59)$$

An interesting application of the above theory is suppression of the shot noise in the current of an electron beam from a cathode. Electrons leave the cathode at uncorrelated moments of time with approximately zero velocity spread at the output of the electron gun. After they pass length $L_p/4$, the initial density fluctuations disappear, and by accelerating the beam at this point we can "freeze" the charge distribution (L_p increases significantly after acceleration). This is how a low-noise traveling-wave tube works.

CURRENT LIMITATIONS

Energy Spread and Maximum Current

Frequently the source of an electron beam is a cathode at potential U. Then the electron total energy is eU and the kinetic energy T is (see Eq. (9))

$$T = e(U - \varphi) = e\left(U - \frac{2I}{v}\ln\frac{b}{a} + \frac{I}{v}\left(1 - \frac{r^2}{a^2}\right)\right). \quad (60)$$

The kinetic energy spread (i. e., the energy difference for particles at $r = 0$ and $r = a$) is

$$\Delta\gamma = \frac{\Delta T}{mc^2} = e(U - \varphi) = \frac{I}{\beta I_0}. \quad (61)$$

Since the kinetic energy must be positive at any r, the current is limited:

$$I < \frac{U\beta c}{2\ln\frac{b}{a}+1}. \tag{62}$$

To obtain higher currents, sheet beams are used. Then for a uniformly charged rectangular beam $|y| < a$, $|x| < w/2$, $w \gg a$,

$$\varphi(y) = \frac{2\pi i}{v}\left(b - \frac{a}{2} - \frac{y^2}{2a}\right), \tag{63}$$

where $2b$ is the vertical aperture of the conducting beam pipe (very wide in the x direction), and $i = I/w$ is the linear current density. Then instead of Eq. (61) and Eq. (62) we get

$$\Delta\gamma = \frac{\pi i a}{\beta I_0} \tag{64}$$

and

$$I < U\beta c \frac{w}{2b-a}. \tag{65}$$

Maximum Equilibrium Current Density

The equilibrium solution of Eq. (33) for constant focusing, $K = const$,

$$Ka - \frac{2I}{(\beta\gamma)^3 I_0}\frac{1}{a} = 0, \tag{66}$$

leads to a current density

$$j = \frac{I}{\pi a^2} = \frac{KI_0(\beta\gamma)^3}{2\pi}. \tag{67}$$

But there is no way to organize equal constant focusing in both transverse directions.[3] Therefore, a longitudinal magnetic field is used to keep the beam radius constant. As a uniform longitudinal magnetic field $B_z = B$ does not provide true focusing, we need to derive the equilibrium condition from the beginning.

At the beam rest frame the equilibrium condition is

$$eE_r + \frac{e}{c}v_\alpha B_z = -\frac{mv_\alpha^2}{r} \tag{68}$$

(the cylindrical coordinates r, α, z are used, and $v_\alpha \ll c$ is assumed). We will neglect the beam-induced contribution to the magnetic field B_z. According to Eq. (7) the radial electric field is linear in r. Therefore, choosing the velocities as $v_\alpha = -\omega r$, we can satisfy the equilibrium equation Eq. (68) for all r [4]:

---

[3] We can use a bending magnet with field index $n = 0.5$, but then the equilibrium trajectory is curved.
[4] In this case the beam is rotated as a rigid body.

$$\omega^2 - \omega_c\omega + \frac{\omega_p^2}{2} = 0, \tag{69}$$

where $\omega_c = \frac{eB}{mc}$ is the cyclotron frequency. The roots of Eq. (69) are

$$\omega = \frac{\omega_c}{2} \pm \sqrt{\frac{\omega_c^2}{4} - \frac{\omega_p^2}{2}}. \tag{70}$$

The maximum charge density takes place for the zero discriminant of Eq. (70), i. e. at $\omega_c = \omega_p\sqrt{2}$,

$$j_{max} = \gamma\rho_0\beta c = \beta\gamma \frac{e}{mc} \frac{B_0^2}{8\pi}, \tag{71}$$

and is referred to as the Brillouin current density. For example, if $B_0 = 10$ kG and $\beta\gamma = 1$, then $j_{max} \approx 23$ kA/cm^2.

ELECTRON GUNS

The conditions of electron guns differ from those discussed above, because near the cathode particle velocities are low, and we can not use the paraxial approximation successfuly.

For an infinite-plane cathode at $x = 0$, the set of equations for a beam propagating in the positive x direction,

$$\frac{d^2\varphi_1}{dx^2} = -4\pi\frac{j}{v},$$
$$\frac{mv^2}{2} - e\varphi_1 = 0, \tag{72}$$

with initial conditions $\varphi_1(0) = 0$ and $d\varphi_1/dx(0) = 0$, has the solution

$$\varphi_1(x) = \left(-9\pi j\sqrt{\frac{m}{2e}}\right)^{2/3} x^{4/3} \tag{73}$$

(the current density j is negative). If the voltage at the gap d is $U = \varphi_1(d)$, the well-known Child's law

$$-j = \frac{1}{9\pi d^2}\sqrt{\frac{2e}{m}} U^{3/2}, \tag{74}$$

and the planar diode perveance

$$P = \frac{-jS}{U^{3/2}} = \frac{S}{9\pi d^2}\sqrt{\frac{2e}{m}}, \tag{75}$$

where S is the cathode area, can be easily obtained. In the general case we can not apply Eq. (74) for the finite transverse beam size, but now we'll find a special gun configuration, called Pierce's gun, for which Eqs. (74) and (75) are valid. Consider the two-dimensional problem with a beam in quadrant $x > 0$, $y < 0$, propagating as before in the positive x direction (see Fig. 1).

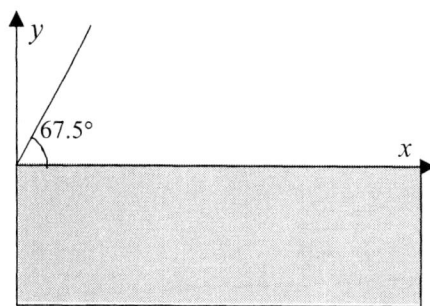

FIGURE 1. Geometry of the simplest Pierce's gun. The beam (gray) occupies the lower quadrant. The focusing zero-potential electrode (zero equipotential) is shown in the upper quadrant.

Let the potential in this quadrant be given by Eq. (73), so that the beam particles are moving as in the infinite diode case. The solution of the Laplace equation in the upper quadrant $x > 0$, $y > 0$, for the boundary conditions $\varphi(0,y) = 0$, $\varphi(x,0) = \varphi_1(x)$, can be obtained by the analytic continuation $\varphi(x,y) = \text{Re}\varphi_1(x+iy)$. In particular, the zero equipotential equation

$$\text{Re}(x+iy)^{4/3} = r^{4/3}\cos\left(\frac{4}{3}\alpha\right) = 0, \tag{76}$$

where $x + iy = re^{i\alpha}$, gives $\alpha = 3\pi/8 = 67.5°$. The zero-potential (connected with the cathode at $x = y = 0$) electrode installed at this position provides the necessary potential inside the beam. The same electrode can limit the beam from the bottom also, and thus create a sheet beam. To construct the gun, we solved the inverse problem, finding the proper boundary conditions from the given potential. Solving the same problem in cylindrical coordinates, we can construct a parallel cylindrical beam and find the corresponding shape of the focusing electrodes. Using the solution for the spherical diode, we can construct a converging (conical) round beam.

For an arbitrary electrode configuration, the electron beam parameters can be calculated numerically.

In conclusion let us write down the set of equations for a laminar electron beam:

$$m\mathbf{v} = \nabla S,$$
$$\frac{m\mathbf{v}^2}{2} + e\varphi = 0,$$
$$\nabla(\rho\mathbf{v}) = 0, \tag{77}$$
$$\Delta\varphi = -4\pi\rho.$$

Expressing all variables through the action S, we can obtain Spangenberg's equation,

$$\nabla\left\{\nabla S \Delta\left[(\nabla S)^2\right]\right\} = 0. \tag{78}$$

PROBLEMS

1. A hollow beam has inner radius a_1 and outer radius a_2. Find the space-charge-induced kinetic energy "spread" and the maximum beam current for gun voltage U. Consider also the limiting cases $a_1 \to 0$ and $a_1 \to a_2$.
2. A cylindrical beam collimator has opening radius b and length L. For the KV distribution find the maximum emittance of a round uncharged beam that can pass through the collimator.
3. Derive the equation for the variation of the vertical size of a laminar sheet beam (i. e. an infinitely wide beam). Compare the solution with the round beam case.
4. Find the maximum perveance of a round laminar beam for the collimator of Problem 2.
5. A coasting beam passed through a straight drift section of length L. Estimate the emittance degradation due to the nonlinearity of the transverse space-charge field. Compare this with the similar result for a bunched beam with a linear transverse space-charge field.
6. Consider the equilibrium condition and the maximum current density for a sheet (infinitely wide) beam in a longitudinal magnetic field.
7. A Brillouin beam leaves the solenoid passing through its short end-field region. Show that the beam rotation will be stopped outside the solenoid, i. e. the particles' angular momenta with respect to the beam axis will vanish.
8. For a low current density, a beam has two equilibrium angular velocities in the longitudinal magnetic field. What is their meaning?

ACKNOWLEDGMENTS

The author thanks E.A. Perevedentsev for numerous stimulating discussions and advice, and also for help in preparation of these lectures.

REFERENCES

1. Lawson, J.D., The Physics of Charged-Particle Beams, Clarendon, Oxford, 1977.

Space-Charge Effects in Circular Accelerators

Shinji Machida

KEK
Oho 1-1, Tsukuba-shi, Ibaraki-ken
305-0801 Japan
e-mail: shinji.machida@kek.jp

Abstract. Space-charge effects in a circular accelerator are discussed, with emphasis on self-consistent treatment. After we review some basic aspects of space-charge effects, we describe observations, cures, and simulation methods.

INTRODUCTION

In an accelerator, a bunch of particles is confined in a small space with focusing elements. Modern accelerator utilizes the alternating-gradient principle (or strong focusing) in the transverse direction and rf focusing in the longitudinal. Although nonlinearities are inevitably caused by fabrication errors and mis-alignment of magnets, the focusing force is essentially linear with respect to the transverse coordinates. In the longitudinal direction, the sinusoidal rf wave is approximately linear near the stable fixed point. Therefore, the particles constituting a beam should have simple harmonic oscillations in both the transverse and longitudinal directions, and the oscillation frequency or its value normalized to the revolution frequency (which is called tune) is equal for all the particles in the beam independent of their oscillation amplitudes.

Once interactions among particles are included, the particle dynamics becomes complicated. For the time being, let us assume that there is no longitudinal focusing and the particles are distributed uniformly We consider only the transverse motions. The linear external force and its quadratic potential are distorted by defocusing forces due to Coulomb interactions among particles. As an illustration, a sum of external and space-charge forces and its potential are shown in Fig. 1 when the charge distribution is Gaussian. Four cases are depicted depending on beam intensity.

In fact, the situation becomes even more complicated. Once the Coulomb potential is added to an external focusing force, the overall potential is no longer the same as before. Then the particle distribution adjusts itself to the overall potential.

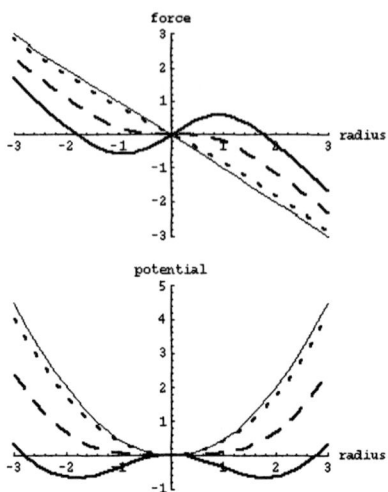

FIGURE 1. Simple sum of external linear focusing force and space-charge force due to Gaussian particle distribution (top), and potential (bottom). Four cases are depicted depending on beam intensity: (thin solid) zero intensity, (dotted) linear defocusing force near the center is 0.2 times the external force, (dashed) linear defocusing force has the same amplitude as the external force, and (thick solid) linear defocusing force is stronger than the external force.

This process continues until self-consistency is satisfied. As we will see later, a self-consistent treatment is essential in studying space-charge effects. In the following sections, we will discuss some basic aspects of space-charge effects, observations of these effects in accelerators, some methods to cure them, and ways to simulate them.

NONNEUTRAL PLASMA ASPECT OF A BEAM

Before going into a specific discussion of space-charge effects in circular accelerators, let us look at some nonneutral plasma aspects of a beam. One of the underlying characteristics of plasma is Debye shielding. That is, in neutral plasma, a charged particle attracts oppositely charged particles around it and its Coulomb forces are terminated in a relatively short range. If we observe the particle from outside its Debye length, or Debye sphere, it is electrically neutral. A concept similar to Debye shielding can be applied to a charged-particle beam, although it contains no oppositely charged particles. In the case of a charged-particle beam, which we may call nonneutral plasma, an external electro-magnetic focusing force is shielded.

Let us consider again the situation we mentioned in the Introduction. Even if the external focusing force is linear, the potential of a charged-particle beam modifies the overall focusing function. Then, the particle distribution adjusts itself to the modified potential. What will happen if the potential of a beam gets very strong? Is the external focusing force overcome so that all the particles disappear to infinity? Maybe so, when the beam intensity is extraordinarily strong. However, until the intensity gets so high, the following phenomena can be expected. When the intensity is increased, especially near the center of the beam, the defocusing force exceeds the external focusing one. The beam size gets larger and the charge density becomes thinner until the defocusing force becomes comparable to the focusing one. Now let us look at the edge region of a beam. Suppose a resulting particle distribution is more or less uniform up to the edge so that the defocusing force is linear, then the focusing force is cancelled everywhere in a beam. The focusing force suddenly resumes outside the beam. In a realistic beam, such a rigid distribution is not a final state, and a soft edge is formed, but two aspects may still be valid. Namely, in the core region, the external focusing force is cancelled. The beam retrieves a focusing force near the edge. We could consider that the core region is Debye shielded and the characteristic distance of the edge region is the Debye length.

For a numerical example, let us calculate some characteristic parameters of a beam at injection in the KEK PS booster. In fact, some parameters have meaning only when the system reaches equilibrium. A hadron beam never attains equilibrium, but we can quote values with such an approximation in mind.

The particle density is

$$n = \frac{n_t}{(\pi a^2)(2\pi R)} = 6.9 \times 10^{14} \ [1/\text{m}^3], \tag{1}$$

where n_t is the total number of particles (3×10^{13}), R is the average radius of the machine (6 [m]), and a is the average beam radius (6.1 [mm]).

The transverse rms velocity is

$$v_{rms} = \frac{c\beta \epsilon_{rms}^{un.}}{x_{rms}} = 9.5 \times 10^4 \ [\text{m/s}], \tag{2}$$

where c is the speed of light, β is the Lorentz factor (0.283 at 40 [MeV]), $\epsilon_{rms}^{un.}$ is the un-normalized rms emittance (13.6 [πmm-mrad]), and x_{rms} is the transverse rms beam size, which is assumed to be $a/4$ (1.5 [mm]).

The beam temperature is

$$k_B T = \gamma m v_{rms}^2 = 97.6 \ [\text{eV}], \tag{3}$$

where k_B is the Boltzmann factor, γ is another Lorentz factor (1.043), and m is the rest mass of a proton.

The plasma frequency is

$$\omega_p = \sqrt{\frac{q^2 n}{\epsilon_0 \gamma^3 m}} = 32.4 \times 10^6 / 2\pi \ [\text{Hz}], \tag{4}$$

where q is the unit charge, and ϵ_0 is the vacuum permittivity.

Finally, the Debye length is

$$\lambda_D = \frac{v_{rms}}{\omega_p} = 2.9 \text{ [mm]}. \tag{5}$$

The plasma parameter, defined as the number of particles inside the Debye sphere, is much larger than unity and therefore the collision frequency is much less than the plasma frequency. As a result, the beam can be characterized as a nonneutral plasma in the following sense. First, collective behavior is dominant compared with collisional effects. That justifies use of the mean-field approximation. Second, the characteristic length of the system, here taken as the rms beam size, is just a little larger than the Debye length. This means that a beam shows collective behavior while a long range Coulomb force still exists.

INCOHERENT TUNE SHIFT

Tune shift and tune spread

Transverse space-charge effects in a synchrotron are estimated from the transverse tune shifts of individual particles. The formula for evaluating the tune shift was first derived by Laslett and included contributions from image charge and current on the beam pipe and surrounding material, such as magnet poles [1]. The so-called Laslett tune shift formula is written as [1]

$$\Delta\nu_y = -\frac{r_p n_t F}{\pi \beta^2 \gamma^3 \epsilon_y (1 + \sqrt{\epsilon_x/\epsilon_y}) B_f},$$

$$F = 1 + \frac{b(a+b)}{h^2}[\epsilon_1(1 + B_f(\gamma^2 - 1)) + \epsilon_2 B_f(\gamma^2 - 1)\frac{h^2}{v^2}], \tag{6}$$

where, r_p is the classical proton radius, n_t is the total number of particles in a ring, β and γ are the Lorentz factors, $\epsilon_{x,y}$ is the horizontal and vertical emittance, B_f is a bunching factor, which is the ratio of average to peak line density and therefore is between 0 and 1, a is horizontal beam size, b is vertical beam size, $2h$ is the vertical aperture of the vacuum chamber, $2v$ is the height of the magnet gap, and ϵ_1 and ϵ_2 are the image force coefficients given in Ref. 2.

In fact, the tune shift calculated above indicates only the "average" value. Unless the particle distribution is uniform in the transverse direction and therefore the

---

[1] The definition of Laslett tune shift now has several variations. Some people, especially those dealing with energy-frontier machines, consider only that with image charge and current to be a Laslett tune shift and distinguish it from direct space-charge tune shift in free space. On the other hand, Eq. (6) without contribution from images is also called a Laslett tune shift. In the latter case, normally the machine under discussion has low enough energy so that the direct space charge is dominant.

space-charge force is linear with respect to transverse coordinates, each individual particle has a different tune shift. The shifted tune is an area instead of a point in the tune diagram, as shown in Fig. 2. The area describing tune spread due to certain beam parameters is called a necktie diagram [2].

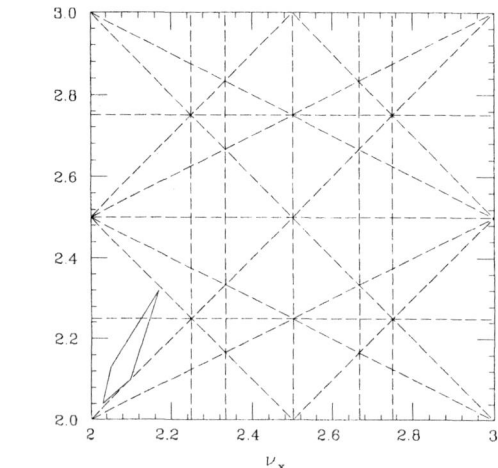

FIGURE 2. Space-charge tune spread on tune diagram. The KEK PS booster is taken as an example. Incoherent tune is scattered from the bare tune of (2.17, 2.32) down to integer tune (and even below) depending on its amplitude.

Let us estimate the magnitude of tune spread due to octupole space-charge fields, which are the lowest nonlinear terms [3]. Suppose a beam has a Gaussian distribution in the transverse direction and is uniform longitudinally. The space-charge potential up to octupole (4th order) is thus

$$V = \frac{r_p \lambda}{\beta^2 \gamma^3}[-\frac{1}{\sigma_x(\sigma_x+\sigma_y)}x^2 - \frac{1}{\sigma_y(\sigma_x+\sigma_y)}y^2 + \frac{2\sigma_x+\sigma_y}{12\sigma_x^3(\sigma_x+\sigma_y)^2}x^4 + \frac{1}{2\sigma_x\sigma_y(\sigma_x+\sigma_y)^2}x^2y^2 + \frac{2\sigma_x+\sigma_y}{12\sigma_y^3(\sigma_x+\sigma_y)^2}y^4 + ...], \quad (7)$$

where λ is line density, and $\sigma_{x,y}$ is rms beam size. The first two terms correspond to the linear space charge, which gives the same tune shift for all the particles. Rewrite Eq. (7) with action $I_{x,y}$ and angle $\psi_{x,y}$ variables as

---

[2] There are at least two necktie diagrams in accelerator literature. Another one is the diagram showing the stability region in terms of horizontal and vertical focusing strength.
[3] Usually it is a reasonable assumption that a particle distribution has symmetry in configuration space.

$$x = \sqrt{2\beta_x I_x} \cos \psi_x, \qquad (8)$$
$$y = \sqrt{2\beta_y I_y} \cos \psi_y, \qquad (9)$$

where $\beta_{x,y}$ is amplitude function. Then the octupole terms (x^4, x^2y^2, y^4) are

$$V_o = \frac{r_p \lambda}{\beta^2 \gamma_3}[\frac{2\sigma_x + \sigma_y}{8\sigma_x^3(\sigma_x + \sigma_y)^2}\beta_x^2 I_x^2 + \frac{1}{2\sigma_x \sigma_y(\sigma_x + \sigma_y)^2}\beta_x \beta_y I_x I_y + \frac{\sigma_x + 2\sigma_y}{8\sigma_y^3(\sigma_x + \sigma_y)^2}\beta_y^2 I_y^2], \qquad (10)$$

where we have used

$$\langle \cos^4 \psi_{x,y} \rangle = \frac{3}{8}, \qquad (11)$$
$$\langle \cos^2 \psi_x \cos^2 \psi_y \rangle = \frac{1}{4}. \qquad (12)$$

Since the tune is defined as an average over one turn,

$$\nu_{x,y} = \frac{1}{2\pi} \int_{s=0}^{s=C} \frac{d\psi_{x,y}}{ds} ds = \frac{1}{2\pi} \int \frac{\partial V}{\partial I} ds, \qquad (13)$$

the amplitude-dependent tune shifts due to octupole terms are

$$\Delta \nu_x = \frac{r_p \lambda}{2\pi \beta^2 \gamma^3}[\frac{2\sigma_x + \sigma_y}{4\sigma_x^3(\sigma + x + \sigma_y)^2}\beta_x I_x + \frac{1}{2\sigma_x \sigma_y(\sigma_x + \sigma_y)^2}\beta_x \beta_y I_y], \qquad (14)$$

$$\Delta \nu_y = \frac{r_p \lambda}{2\pi \beta^2 \gamma^3}[\frac{\sigma_x + 2\sigma_y}{4\sigma_y^3(\sigma + x + \sigma_y)^2}\beta_y I_y + \frac{1}{2\sigma_x \sigma_y(\sigma_x + \sigma_y)^2}\beta_x \beta_y I_x]. \qquad (15)$$

On the tune diagram, four particles with different amplitudes of $(I_x = I_y = 0), (I_x = 0, I_y = I_{y0}), (I_x = I_{x0}, I_y = 0), (I_x = I_{x0}, I_y = I_{y0})$ are as shown in Fig. 3. In a real beam, higher-order terms above octupole also exist.

In a bunched beam, synchrotron oscillations make the line density time dependent. In the Laslett formula, the variation of the line density in the longitudinal direction is taken into account by using the bunching factor. That is, only the place where the peak line density exists is considered. In reality, the beam is spread very long in the longitudinal direction, a few meters to a hundred meters. A particle with synchrotron oscillations feels line density oscillations depending on the synchrotron oscillation phase. Usually synchrotron oscillations are much slower than betatron ones, and transverse space-charge tune shift is considered to vary with synchrotron frequency.

A study of space-charge effects in circular accelerators often included the evaluation of incoherent tune shift with various boundary conditions, as in Ref. 3. The evaluation of incoherent tune, however, does not necessarily mean the estimation of "space-charge effects." Naively speaking, when tune is shifted and a particle

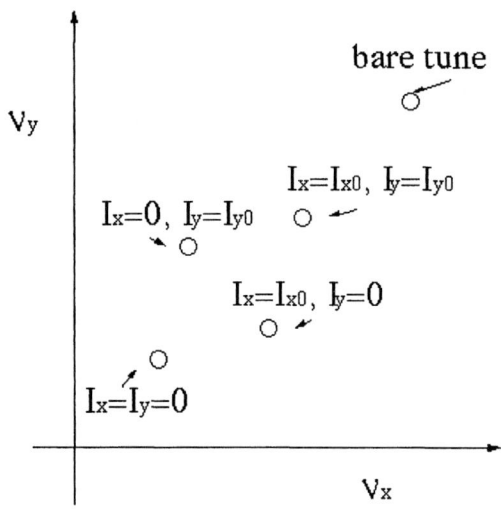

FIGURE 3. Bare tune and amplitude-dependent tune shift. We consider only octupole terms of a Gaussian distribution. Four extreme cases are indicated where the amplitude of both horizontal planes is zero or maximum, and one of the planes has maximum amplitude while the other has zero.

crosses one of the lattice resonances, its amplitude is exponentially growing. If there are a bunch of such particles in a beam, the emittance of the beam becomes large [4]. However, a resonance of incoherent tune is not so obvious. In the next section, we will clarify that the naive picture does not work [4] [5].

Resonance of incoherent tune?

Integer resonance

First, we will consider integer resonance crossing of a space-charge-dominated beam. Here the equation of motion is simplified as follows. The single-particle motion with a driving term of integer resonance but without space charge is

$$\frac{d^2 x}{ds^2} + k_q x = F(s). \qquad (16)$$

The driving term of integer resonance does not depend on amplitude x. When (linear) space charge force is introduced, the equation becomes

---

[4] We limit our discussion to circular accelerators since the periodic nature of particle dynamics is assumed as a source of resonances.
[5] Although the self-consistent view was originally formulated by Sacherer, the review article by Baartman [4] explains it nicely. We will follow his notation.

$$\frac{d^2 x}{ds^2} + k_q x = k_{sc}(x - x_c) + F(s). \tag{17}$$

Remember that the space-charge force is proportional with respect to the gravitational center x_c of a beam, not to the machine coordinate x. Take the average of Eq. (17),

$$\frac{d^2 x_c}{ds^2} + k_q x_c = F(s). \tag{18}$$

Namely, the motion of the gravitational center x_c is not affected by the space-charge force. By taking the difference of Eqs. (17) and (18), we get

$$\frac{d^2(x - x_c)}{ds^2} + k_q(x - x_c) = k_{sc}(x - x_c). \tag{19}$$

This shows that an individual particle motion whose focusing force is modified by space-charge is not affected by the driving term of integer resonance. Simulation results based on a more realistic model agree with the above prediction [5].

Half-integer resonance

Secondly, let us examine the effects of a half-integer resonance: in the case the driving term is proportional to coordinate x. The equation of motion without space charge becomes

$$\frac{d^2 x}{ds^2} + k_q x = x F(s). \tag{20}$$

As before, we introduce the space-charge force as

$$\frac{d^2 x}{ds^2} + k_q x = k_{sc}(x - x_c) + x F(s). \tag{21}$$

Taking the average, we get

$$\frac{d^2 x_c}{ds^2} + k_q x_c = x_c F(s). \tag{22}$$

The motion of the gravitational center is not affected by the space-charge force but is subject to half-integer resonance. Then consider individual motion

$$\frac{d^2(x - x_c)}{ds^2} + k_q(x - x_c) = k_{sc}(x - x_c) + (x - x_c) F(s). \tag{23}$$

The equation of motion suggests that the incoherent tune calculated with k_q and k_{sc} determines the condition of half-integer resonances. However, this is not correct since the k_{sc} is a function of beam size and k_{sc} becomes smaller when the tune approaches the resonance and the beam size becomes larger. Self-consistent treatment is needed.

SELF-CONSISTENT TREATMENT

Envelope equations

In order to construct a self-consistent model with space charge, envelope equations were introduced and analyzed by Smith and Sacherer in the 1960s [6]. In the 1-D case, the envelope equation is

$$\frac{d^2 x}{ds^2} + k(s)x - \frac{\epsilon^2}{x^3} - \frac{2\pi r_p \lambda}{\beta^2 \gamma^3} = 0. \tag{24}$$

We then define

$$\tilde{x} = x/x_p, \tag{25}$$

where $x_p = \sqrt{\epsilon \beta}$. Changing the variable,

$$\phi = \int \frac{ds}{\nu \beta}, \tag{26}$$

and introducing n−th harmonics of a half-integer error component, we get

$$\frac{d^2 \tilde{x}}{d\phi^2} + (\nu^2 + 2\nu \Delta \nu_s \cos n\phi)\tilde{x} - \frac{\nu^2}{\tilde{x}^3} - 2\nu \Delta \nu_{sc} = 0, \tag{27}$$

where $\Delta \nu_s$ is the resonance width of the half-integer error and

$$\Delta \nu_{sc} = \frac{1}{2\nu} \frac{4\pi r_p R^3}{\beta^2 \gamma^3} \frac{\lambda}{2a}, \tag{28}$$

$$a = \sqrt{\epsilon R/\nu}. \tag{29}$$

When $\Delta \nu_s = 0$, the solution is

$$\tilde{x} = 1 + \frac{\Delta \nu_{sc}}{2\nu}. \tag{30}$$

If we introduce perturbation δ into the above solution, the equation of δ becomes

$$\frac{d^2 \delta}{d\phi^2} + (4\nu^2 - 6\nu \cdot \Delta \nu_{sc})\delta = -2\nu \cdot \Delta \nu_s \cos n\phi. \tag{31}$$

The resonance occurs when

$$\frac{3}{4}\Delta \nu_{sc} = \nu - \frac{n}{2}. \tag{32}$$

Namely, the tune shift can be larger than the distance between the bare tune and the nearest resonance line.

Similarly, in the 2-D case, the envelope equations are

$$\frac{d^2x}{ds^2} + k_x(s)x - \frac{\epsilon_x^2}{x^3} - \frac{4r_p\lambda}{\beta^2\gamma^3}\frac{1}{x+y} = 0, \tag{33}$$

$$\frac{d^2y}{ds^2} + k_y(s)y - \frac{\epsilon_y^2}{y^3} - \frac{4r_p\lambda}{\beta^2\gamma^3}\frac{1}{x+y} = 0. \tag{34}$$

Define the dimensionless variables as

$$\tilde{x} = x/x_p, \tag{35}$$
$$\tilde{y} = y/y_p, \tag{36}$$

where $x_p = \sqrt{\epsilon_x\beta_x}$ and $y_p = \sqrt{\epsilon_y\beta_y}$. By changing the variables,

$$\phi_x = \int \frac{ds}{\nu_x\beta_x}, \tag{37}$$

$$\phi_y = \int \frac{ds}{\nu_y\beta_y}, \tag{38}$$

and introducing the n-th harmonics of a half-integer error component, the equations become

$$\frac{d^2\tilde{x}}{d\phi^2} + (\nu_x^2 + 2\nu_x\Delta\nu_{sx}\cos n\phi)\tilde{x} - \frac{\nu_x^2}{\tilde{x}^3} - \frac{b\omega_p^2}{a\tilde{x}+b\tilde{y}} = 0, \tag{39}$$

$$\frac{d^2\tilde{y}}{d\phi^2} + (\nu_y^2 + 2\nu_y\Delta\nu_{sy}\cos n\phi)\tilde{y} - \frac{\nu_y^2}{\tilde{y}^3} - \frac{a\omega_p^2}{a\tilde{x}+b\tilde{y}} = 0, \tag{40}$$

where $\Delta\nu_{sx}$ and $\Delta\nu_{sy}$ are the resonance widths of the half-integer resonance,

$$\omega_p^2 = \frac{2\lambda}{\pi}\frac{r_p R}{ab}\frac{1}{\beta^2\gamma^3}, \tag{41}$$

and

$$a = \sqrt{\epsilon_x R/\nu_x}, \tag{42}$$
$$b = \sqrt{\epsilon_y R/\nu_y}. \tag{43}$$

The space-charge-induced frequency shifts for a beam with a constant envelope $\tilde{x} = 1$, and $\tilde{y} = 1$ are

$$\Delta\nu_{scx} = \frac{b}{a+b}\frac{\omega_p^2}{2\nu_x}, \tag{44}$$

$$\Delta\nu_{scy} = \frac{a}{a+b}\frac{\omega_p^2}{2\nu_y}. \tag{45}$$

In the case of equal tune and emittance,

$$\nu_x = \nu_y = \nu, \tag{46}$$

$$\Delta\nu_{scx} = \Delta\nu_{scy} = \Delta\nu_{sc} = \frac{\omega_p^2}{4\nu}. \tag{47}$$

Then,

$$\frac{d^2\tilde{x}}{d\phi^2} + (\nu^2 + 2\nu\Delta\nu_{sx}\cos n\phi)\tilde{x} - \frac{\nu^2}{\tilde{x}^3} - \frac{4\nu\Delta\nu_{sc}}{\tilde{x}+\tilde{y}} = 0, \tag{48}$$

$$\frac{d^2\tilde{y}}{d\phi^2} + (\nu^2 + 2\nu\Delta\nu_{sy}\cos n\phi)\tilde{y} - \frac{\nu^2}{\tilde{y}^3} - \frac{4\nu\Delta\nu_{sc}}{\tilde{x}+\tilde{y}} = 0. \tag{49}$$

We consider δ_x and δ_y about the constant solution of

$$\tilde{x} = \tilde{y} = 1 + \frac{\Delta\nu_{sc}}{2\nu}. \tag{50}$$

Then, we find there are two solutions, one is symmetric mode ($\delta_x = \delta_y$) with a frequency of $2(\nu - \frac{1}{2}\Delta\nu_{sc})$, and the other is antisymmetric mode ($\delta_x = -\delta_y$)) with a frequency of $2(\nu - \frac{3}{4}\Delta\nu_{sc})$. Therefore, a resonance occurs at

$$\frac{1}{2}\Delta\nu_{sc} = \nu - \frac{n}{2} \tag{51}$$

for symmetric mode and

$$\frac{3}{4}\Delta\nu_{sc} = \nu - \frac{n}{2} \tag{52}$$

for antisymmetric mode. In the case of unequal tune, the motions are decoupled in each plane, and a resonance occurs at

$$\frac{5}{8}\Delta\nu_{sc} = \nu - \frac{n}{2}. \tag{53}$$

When we introduced the envelope equations, we did not state what beam size or emittance really means. Can it be 100% beam size, 95%, rms, or any definition? In fact, when the envelope equations were introduced, it was assumed that the particle distribution is uniform both in 1-D and 2-D so that the space-charge force is linear. The beam size or emittance was 100% [6].

Sacherer later found that envelope equations of the same form hold for rms beam size, and they barely depend on the distribution inside provided that the emittance

---

[6] The distribution that gives uniformity in the 2-D configuration space has a peculiar form in phase space such that each particle has a constant total emittance surface of 4-D phase space and the particles are distributed uniformly. That distribution is called the KV distribution after Kapchinskij and Vladimirskij [7].

evolution is know *a priori* and the distribution has elliptical symmetry [8]. That is also true for a 3-D bunched beam. That makes the envelope equation more useful since what we need to know is rms beam size, not the detailed distribution. In other words, as long as their rms beam sizes are equal, different beams behave in the same way as far as space charge is concerned. The concept of an "equivalent beam" was based on these considerations. Namely, an equivalent beam is a set of beams with the same rms emittance.

Envelope equations with dispersion

You may think the envelope equations do not include the dispersion function that changes the envelope size in a circular machine. The extended envelope equations for a circular machine with dispersion were found by Lee and Okamoto [9] and by Venturini and Reiser [10]. Essentially, the envelope equation for the dispersion function is added:

$$\frac{d^2 D_x}{ds^2} + k_x(s)D_x - \frac{1}{\rho} - \frac{4r_0 \lambda}{\beta^2 \gamma^3} \frac{D_x}{X(X+Y)} = 0, \quad (54)$$

$$\frac{d^2 x}{ds^2} + k_x(s)x - \frac{\epsilon_x^2}{x^3} - \frac{4r_0 \lambda}{\beta^2 \gamma^3} \frac{x}{X(X+Y)} = 0, \quad (55)$$

$$\frac{d^2 y}{ds^2} + k_y(s)y - \frac{\epsilon_y^2}{y^3} - \frac{4r_0 \lambda}{\beta^2 \gamma^3} \frac{y}{Y(X+Y)} = 0, \quad (56)$$

where D_x is the linear dispersion, and X and Y are the beam sizes including both betatron oscillations and dispersion with momentum spread. Usually, there is no dispersion in a vertical plane so that $Y = y$. One reasonable (not unique) definition of X is

$$X^2 = x^2 + D_x \langle (dp/p)^2 \rangle, \quad (57)$$

where we take the rms value as x and the momentum spread as (dp/p).

OBSERVATIONS

HIMAC experiment

Incoherent tune shift cannot be observed because of the collisionless plasma nature. Coulomb forces among individual particles are averaged and the pickup monitor outside the beam cannot see individual particle motions. That nature actually assures the use of the mean field approximation in order to perform macro-particle simulation, discussed later.

Even though the incoherent tune is not visible, the coherent motion is easy to observe. The lowest coherent mode is dipole. In fact, it is worth mentioning that the oscillation we excite to measure the machine tune is the tune of a coherent dipole mode with space-charge effects, not the incoherent one. As we have already seen, the tune of a dipole coherent mode is not affected by direct space charge so that we can infer incoherent tune from it. The effect of space charge through image charge and current on dipole mode was also estimated by Laslett and it is called the Laslett coherent tune shift [7].

The experimental setup for measuring a dipole mode is to have two electrodes for each transverse plane. The difference of the induced charge or current on these electrodes is assumed to be proportional to the transverse beam displacement.

In a similar way, a coherent quadrupole mode can be observed. In this case, instead of taking the difference of two electrodes, two signals are summed. In order to subtract the DC bias, we can take the difference of the sum of two horizontal plates and the sum of two vertical plates. Figure 4 show such a device installed in HIMAC at NIRS. The observed frequency shown in Fig. 5 is proportional to the beam intensity, as expected in Ref. 11.

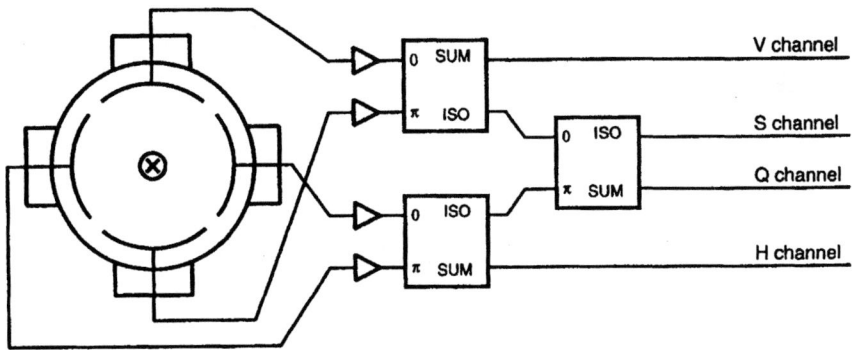

FIGURE 4. Quadrupole pick-up monitor installed in HIMAC of NIRS. Sum signal of two electrodes, instead of difference, is Fourier analyzed to obtain coherent quadrupole frequency. Other signals, like horizontal and vertical dipole frequency, and intensity sum, are also monitored.

CERN PS experiment

CERN PS gives clear results which confirm that the incoherent tune does not deteriorate a beam, but the coherent one does [12]. In Fig. 6, beam emittance is depicted as a function of beam intensity. It indicates that nothing happens in the emittance when the incoherent tune becomes as large as the distance between

---

[7] When we discuss a coherent mode of a beam, people tend to think only of Laslett coherent dipole tune shift. The coherent mode we discuss here is not only a dipole mode, but includes all other higher modes.

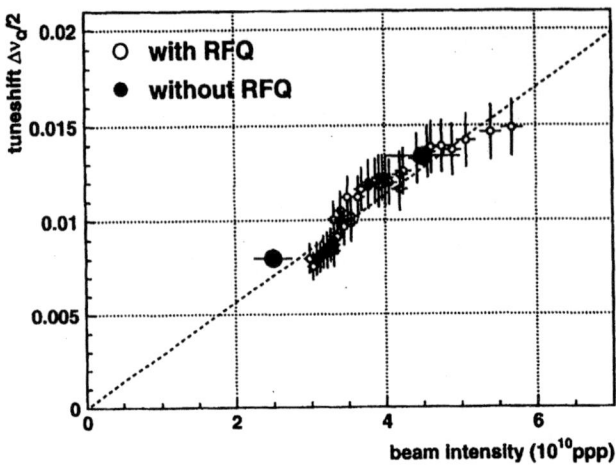

FIGURE 5. Coherent quadrupole frequency as a function of beam intensity. For some beam intensities, RFQ was used to initiate quadrupole oscillations. Cited from Ref. 11.

the bare tune and the nearest resonance, which is $\nu_y = 6$. When the intensity is increased further and incoherent tune shift is considerably larger than the distance, the emittance starts growing.

KEK PS booster experiment

So far we have discussed the 1-D phenomena of space-charge effects. Space charge also excites coupling between two planes. The most familiar one is a $2\nu_x - 2\nu_y = 0$ resonance. Sometimes it is called a Montague resonance after his first analysis in the 1960s [13].

If we look at the potential of Eq. (7), the lowest coupling term is $x^2 y^2$. Other terms like xy, xy^2, and $x^2 y$, are wiped out because of a symmetrical particle distribution in configuration space. Without machine errors, harmonics of h, where $2\nu_x - 2\nu_y = h$ and h is a multiple of superperiodicity, can be a driving term by space-charge force. However, $h = 0$ is the strongest since the space-charge force is always defocusing around the machine.

We may think this is not harmful because it is a differential coupling resonance and the emittance in each plane is bound. This may be so in a collider type of accelerator complex where horizontal and vertical emittances are similar. That is not the case in a high intensity accelerator or a small accelerator for medical and industrial usage. In particular, a synchrotron following an injector such as a linac has a multi-turn injection scheme to increase circulating beam intensity. After multi-turn injection, the emittance of one transverse plane, usually the horizontal,

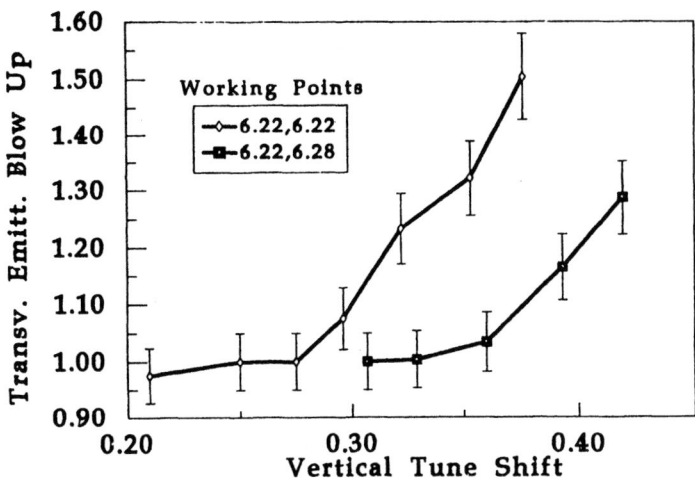

FIGURE 6. Beam emittance as a function of incoherent tune shift. No intensity effects are observed when incoherent tune becomes equal to the distance between bare tune and resonance. Emittance blow-up starts when tune shift is more than 0.28 while the distance to the resonance is 0.22 in one case, and when tune shift is more than 0.35 while the distance is 0.28 in the other. (From Ref. 12, © 1993 IEEE.)

becomes much larger than that of the vertical. What is worse, in the design stage of such a synchrotron, the vertical aperture, which is mainly limited by the gap of a bending magnet, is squeezed as much as possible, whereas the horizontal one is widened to accommodate the large emittance. In these synchrotrons, once the intensity is increased and the bare tune is slightly above the coupling resonance, the asymmetry of emittance is broken and horizontal emittance flows into vertical. Smaller vertical acceptance cannot take the large emittance originating from the horizontal and beam loss occurs in the vertical plane.

Experimentally, such a phenomenon was observed in the KEK PS booster [14]. The bare tune is (2.17,2.32) just above the coupling resonance of $2\nu_x - 2\nu_y = 0$. In fact, the KEK PS booster does not have to have asymmetric emittance at injection because of the H^- injection scheme, although it was first started with a multiturn injection scheme with H^+ particles long ago. Since it used to use multi-turn injection and about four time larger horizontal acceptance than vertical, larger horizontal emittance was formed intentionally at the beginning to decrease spacecharge effects and increase beam intensity. When the beam intensity is moderate, asymmetry of the transverse emittance is preserved up to the extraction channel. However, as the intensity increases, the horizontal emittance shrinks and the vertical one expands, as shown in Figs. 7, 8, and 9. Unfortunately, the operating point of the KEK PS booster is fixed without any knob, and there is no way to avoid the

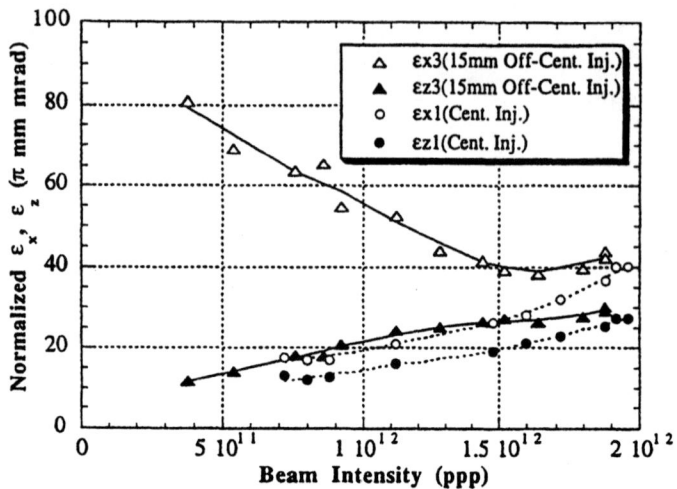

FIGURE 7. Beam emittance as a function of intensity. Initially the horizontal emittance is larger than the vertical, and this is preserved when the intensity is low (4×10^{11} ppp). When beam intensity is increased, coupling between two transverse emittance occurs. The large horizontal emittance shrinks and the small vertical one blows up. Cited from Ref. 14.

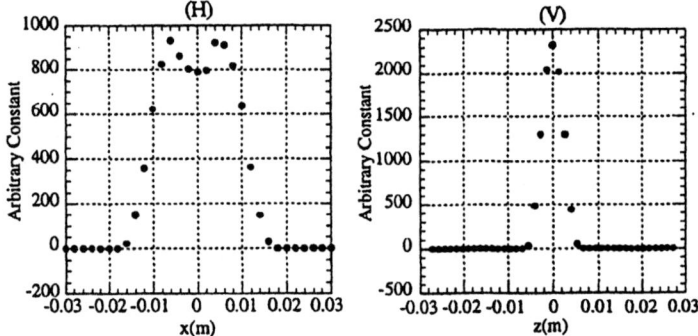

FIGURE 8. Simulation results of KEK PS booster with low intensity. Beam profile at 128 turns after injection. The bare tune is (2.17,2.32). When beam intensity is low, the initial asymmetric emittance (the horizontal one is four times larger than the vertical) is preserved.

coupling at present.

Asymmetric emittance is created not only at multi-turn injection but during the painting process we will discuss later. In order to obtain the desired particle distribution at the end of painting, understanding of the coupling mechanism due to space-charge potential is essential.

BEAM PROFILE MEASUREMENTS

Space-charge effects become important for a low energy hadron beam. Since the beam is low energy and hadron, we cannot expect any synchrotron radiation coming from it, which is quite an useful signal for measuring the beam profile of an electron beam. The only process we can use to measure beam profile is electric interactions, namely direct or indirect measurement of beam charge. Here we survey typical measurement methods and discuss their problems, especially when they are applied to measure space-charge dominated beams.

Ionization profile monitor

When a beam hits surrounding residual gas in a vacuum chamber, the gas is ionized. Ionized gas or scattered electrons are collected on two parallel electrodes when high voltage (a few to 10 kV) is applied between them (Fig. 10) [15]. If, and only if, the electric field lines are perpendicular to the electrodes in the beam region, the distribution of collected ions or electrons should reflect that of particles in the beam. One good thing about this monitor is that the measurement does not destroy the beam at all. Although we may need to make the vacuum pressure worse in order to get more ions and electrons in the monitor region, in principle, interaction between a beam and residual gas is always there. This is a typical non-destructive beam profile monitor.

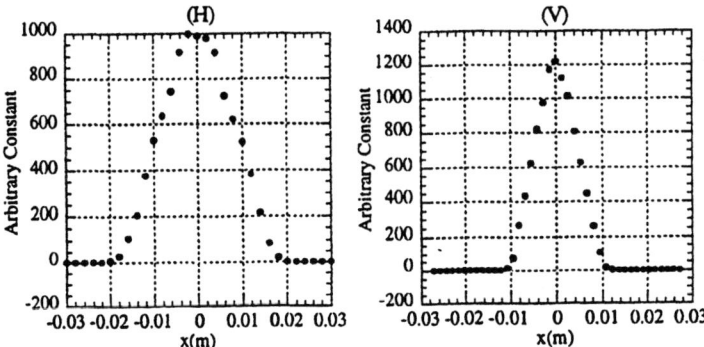

FIGURE 9. Simulation results of KEK PS booster with high intensity. Beam profile at 128 turns after injection. The bare tune is (2.17,2.32). When beam intensity is high, the initial asymmetric emittance is not preserved because of coupling between horizontal and vertical.

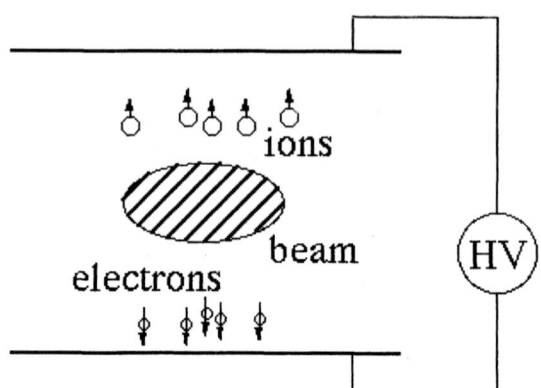

FIGURE 10. Schematic view of ionization profile monitor. When a beam hits residual gas, the gas is ionized and electrons are also produced. Voltage applied in the transverse direction collects ions and electrons. A simple electrode or MCP detects the charge distribution of either kind. If the space-charge field due to the beam can be ignored, the charge distribution reflects the particle distribution of the beam.

This is the case when beam intensity is not high. However, a high intensity beam itself creates electric potential around it of the same order as that from an externally applied voltage, and the electric field lines are distorted. For example, the field flux is expanded in the beam area and the ions and electrons are diminished when the beam reaches the electrodes. Therefore this type of profile monitor gives no reliable signal for high intensity beams. Note that this is the space-charge force created by a beam and seen by ions and electrons moving in the transverse direction so that there is no cancellation of beam current by magnetic fields. The space-charge force is merely determined by line density in the laboratory frame.

Flying wire

Although we cannot have a wire in a beam to measure beam profile all the time, we can make a wire fly across a beam. When a wire of 10 or a few 10 micrometers flies across a beam with a speed of a few 10 m/s, secondary particles are produced by scattering. If we know speed of the wire and measure the amount of secondary particles with good time resolution, we can reconstruct the beam profile at the place where the wire crossed. This method obviously is not fully non-destructive. A few percent of particles are scattered by the wire and lost (Fig. 11) [16].

The flying wire monitor may be a better choice compared with the two others, but it has some problems. First, it takes a few milliseconds to cross a beam of typical size. Although a few milliseconds may seem short enough, a profile change faster than that time scale is not detected. Second, when a wire crosses a high intensity beam, even if the beam hitting rate is small in fractional terms, the total energy deposited on the wire becomes large. If the wire is heated beyond a certain threshold, it breaks. Third, when beam energy is low, the scattering angle becomes large and more distortion of the circulating beam is expected. Again, this technique works well for low intensity beams, but problems emerge when the intensity gets high.

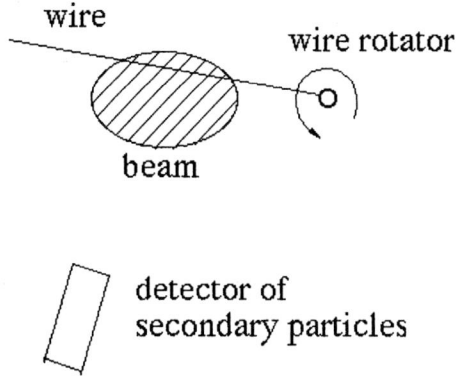

FIGURE 11. Schematic view of flying wire profile monitor. When a wire crosses a beam, it produces secondary particles. If we know the speed of the wire and the time structure of the detected secondary particles, the beam profile can be reconstructed.

Beam scraping

This technique was invented at CERN and called BEAMSCOPE [17]. Suppose there is a local aperture minimum, which can be set by a scraper. Now excite the bump orbit at the scraper's location so that large-amplitude particles are lost. As the bump excitation becomes larger, smaller-amplitude particles are getting lost, and finally all the particles are gone. If we measure the circulating beam intensity as a function of the bump excitation, which gives beam displacement, the beam profile can be reconstructed as the differential of circulating beam intensity as a function of displacement (Fig. 12).

One difference from the other methods is that the particle distribution is obtained as a function of a betatron amplitude, not of a real-space position, because the scraper peels a beam in phase space. The method is totally destructive, and it takes time to get a full profile. If we use it to measure the profile of a space-charge-dominated beam, the main problem is its varying intensity during the measurement. For example, the lattice functions depend on beam intensity and there is no longer a simple relation between beam size and emittance. Furthermore, this method cannot detect a fast change of beam profile.

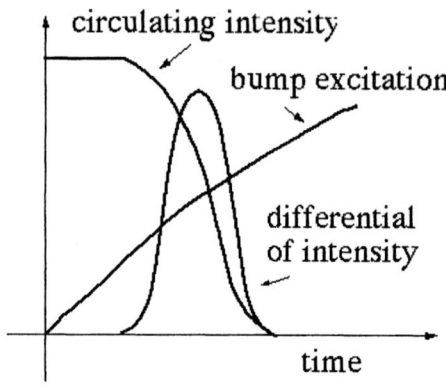

FIGURE 12. Conceptual signal of BEAMSCOPE profile monitor. When a bump is excited and the machine aperture is reduced locally, beam loss starts with large-amplitude particles. Since circulating intensity reflects the population of particles inside a certain aperture, the differential of the intensity is proportional to an initial particle distribution.

Other methods

There are several other ways to measure the beam profile, most of them in the development stage. One uses a shaped electron beam that is scanned perpendicular to the beam. The reflected electron rays contains information on the particle distribution of a beam.

Another novel idea uses a molecular sheet beam. When a beam hits the molecular gas sheet, the gas is ionized. By collecting ions with MCP and looking at secondary electrons by CCD, the beam profile is reconstructed. Generation of an intense molecular beam is essential for this monitor [18]. As an example Hashimoto et. al. Ref. [18] use O_2 gas. The idea of detecting the luminescence of gas excited by the passage of a beam was also been tested. Nitrogen gas has the advantage of quick response time [19].

In any case, reliable measurement of the beam profile in a high intensity beam is expected to become possible.

CURES

Since the space-charge force is increased in dense beams, the most straightforward and probably the only way to lower its effects is to make particles evenly distributed in a large configuration space. There are mainly two ways to realize such a cure. One is to shape the potential of the external focusing force and the other is to re-arrange the distribution. Particle distribution in either case ends up being more uniform in real space.

Longitudinal

Before discussing the above cures, let us review the ordinary focusing scheme. In longitudinal phase space, the focusing force comes from sinusoidal rf voltage. In the center region, force is quite linear and the resulting potential is quadratic with respect to the distance from the center. At large amplitude, the nonlinear force becomes significant and there is a boundary that separates bound and unbound motions. The situation is like that of a pendulum with a fixed-length arm. The bounded region in phase space is as shown in Fig. 13 (a) and it is called a separatrix. Suppose that the particle density inside the separatrix is constant, then the projection in phase, namely configuration space, becomes Fig. 13 (b), which has a dense core in the center region. In reality, the particle density is greater at the center in phase space, and the density in the core is even more increased.

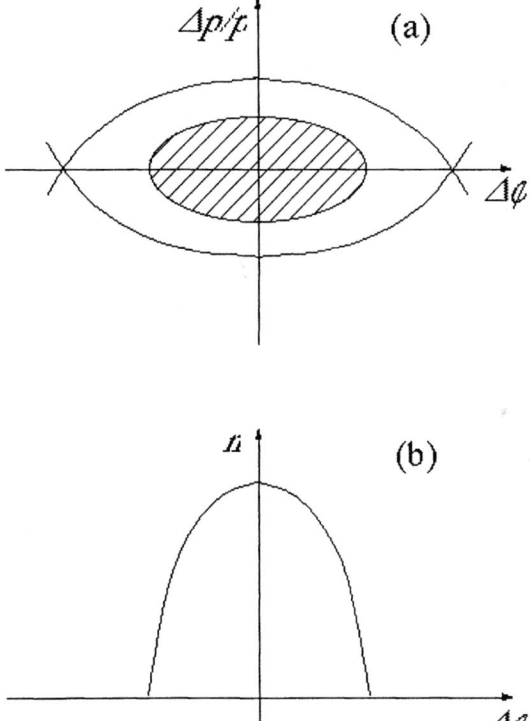

FIGURE 13. (a) Separatrix and particle distribution in longitudinal phase space and (b) its projection to real space.

Let us first discuss the way to shape the potential. If we add higher harmonic waves of the fundamental rf frequency, normally the second or third one, the focusing potential has a flat bottom near the center and a steeper wall near the edge. That makes the separatrix more rectangular in shape, and the particles are distributed more evenly in the longitudinal direction. If we really want to make the separatrix rectangular, higher and higher harmonics should be included.

Alternatively, we can excite a single rf wave both at the beginning and at the end of a beam so that a particle in between does not feel any focusing force. Once the particle reaches the place where rf is applied, a strong restoring force keeps it inside. The bucket defined by two single rf waves is called a barrier bucket [20]. Using a barrier bucket, we can make an almost uniform beam in the longitudinal direction. This helps cure space-charge effects during an injection and stacking process (Fig. 14). It may be difficult to accelerate a beam captured by a barrier bucket.

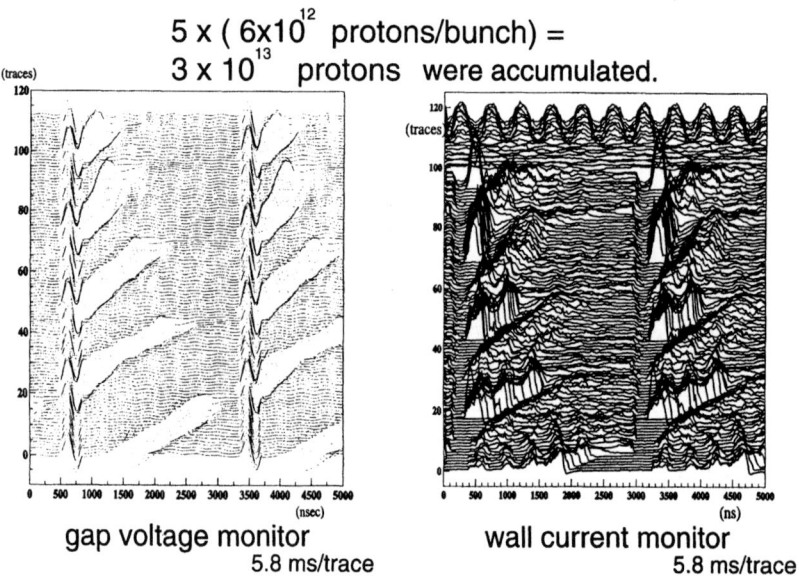

FIGURE 14. RF pulse (left) and beam current (right) of barrier bucket scheme. In this example, barrier rf voltage moves five times to accumulate five incoming beams. Although line density increases when the distance between two barriers is shortened, it is almost uniform and no structure exists, unlike Fig. 13 (b). Cited from Ref. 20.

The other way to make a distribution even is to populate more large-amplitude particles and fewer small-amplitude ones. The distribution can be re-arranged by some diffusion excited with an external focusing force or, in a more direct way, by injecting more particles at large amplitude. The latter is called a longitudinal painting technique. We will explain the meaning of painting below in connection with transverse painting.

The former scheme enforces a proper shaping of focusing potential throughout an accelerator complex. For instance, if two machines are connected in cascade and both have space-charge effects at injection, we need to shape a potential in both machines. To the contrary, the latter scheme assures less space-charge effects afterwards if it is once applied in the first machine.

Transverse

In transverse phase space, the external focusing force is kept as linear as possible. The introduction of nonlinear potential may make the particle distribution more uniform in configuration space, which is good from the space-charge point of view, but at the same time, nonlinear driving terms excite the resonance and the dynamic aperture is reduced. Therefore, re-arranging the particle distribution in phase space is the only way to make uniform particle distribution.

The most effective way to re-arrange distribution can be performed at H^- injection. Beams are injected turn by turn until a necessary number of particles is stacked. If we change the injection point in phase space turn by turn, more area than the emittance of a single beam is filled after many turns. This is called transverse painting injection. Usually, in both the horizontal and the vertical planes, painting is performed at the same time. Assuming the emittance of an injected beam is relatively small compared with the phase-space area to be filled, the following change of injection point gives uniform distribution in real space at the end:

$$x_{cod} = x_m\sqrt{1 - \frac{t}{T}} + x_{off}, \tag{58}$$

$$y_{cod} = y_m\sqrt{\frac{t}{T}} + y_{off}, \tag{59}$$

where T is the injection period and t is time. Since its injection scheme produces a large-amplitude particle in one plane with a small-amplitude one in the other at the beginning, and moves in the opposite direction, it is called anti-correlated painting. Typical beam shape after painting is shown in Fig. 15. The other way is also possible, namely injecting a large- or small-amplitude particle in both planes at the beginning and moving in the same direction, which is called correlated painting (Fig. 16.) Correlated painting produces a rectangular shape in real space at the end [21] and the distribution in real space is not uniform.

In principle, anti-correlated painting gives an ideal uniform distribution and seems a better choice than the correlated one. This may be the case, but the some care has to be taken on the horizontal and vertical coupling. In particular, there is a large asymmetry of the emittance at the beginning, and the painting

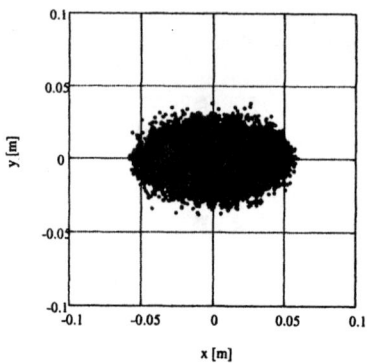

FIGURE 15. Real-space particle distribution after anti-correlated painting. If injector has zero emittance, the distribution becomes uniform (KV) in real space. (Courtesy of Noda at JAERI.)

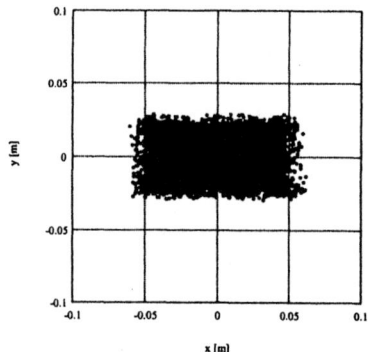

FIGURE 16. Real-space particle distribution after correlated painting. (Courtesy of Noda at JAERI.)

scheme works only when two planes do not interact. Optimization of the operating point to avoid coupling due to space-charge potential, which we have already shown taking the KEK PS booster as an example, is necessary.

SIMULATION

In this section, we show a simulating method for space-charge effects. As an example, we describe the code Simpsons [22].

Field calculation

A calculation of space-charge fields is based on a particle-in-cell (PIC) method [23]. As a coordinate system, a cylindrical one (r, j, z) is taken for a 3-D beam and a polar one (r, j) for a 2-D beam. The fractional charge of each macro particle is allocated to grid points nearby according to area weighting. The poisson equation is then solved with a boundary condition of circular cross-sectional beam pipe. The contribution from the magnetic field is simply taken into account with a $1/\gamma^2$ factor.

Under the Lorentz gauge, a scalar potential satisfies

$$\Delta\phi - \epsilon_0\mu_0\frac{\partial^2\phi}{\partial t^2} = -\frac{\rho}{\epsilon_0}. \tag{60}$$

We introduce the ordering,

$$\frac{\partial^2\phi}{\partial z^2} \sim \epsilon_0\mu_0\frac{\partial^2\phi}{\partial t^2} \ll \frac{1}{r}\frac{\partial}{\partial r}(r\frac{\partial\phi}{\partial r}) + \frac{1}{r^2}\frac{\partial^2\phi}{\partial\varphi^2}. \tag{61}$$

Then,

$$\frac{1}{r}\frac{\partial}{\partial r}(r\frac{\partial\phi}{\partial r}) + \frac{1}{r^2}\frac{\partial^2\phi}{\partial\varphi^2} = -\frac{\rho}{\epsilon_0}. \tag{62}$$

Once a charge distribution at grid points is obtained, it is Fourier transformed in the azimuthal direction:

$$\phi = \sum_m \phi_m \exp(im\varphi), \tag{63}$$

$$\frac{\rho}{\epsilon_0}(\equiv n) = \sum_m n_m \exp(im\varphi), \tag{64}$$

with the inverse transform,

$$\phi_m = \frac{1}{2\pi}\int_0^{2\pi} \phi(r, z, \varphi)\exp(-im\varphi)d\varphi, \tag{65}$$

$$n_m = \frac{1}{2\pi}\int_0^{2\pi} n(r, z, \varphi)\exp(-im\varphi)d\varphi. \tag{66}$$

Note that $n_{-m} = n_m^*$. Then,

$$n_m(r, z) = \frac{1}{r}\frac{\partial}{\partial r}(r\frac{\partial}{\partial r}\phi_m(r, z)) - \frac{m^2}{r^2}\phi_m(r, z). \tag{67}$$

The general solution to the equation ($m \geq 0$) is

$$\phi_m = \alpha(\frac{r}{b})^m + W(r) - (\frac{r}{b})^m W(b). \tag{68}$$

On the boundary condition, $\phi_m = 0$ at $r = b$ (b is the beam pipe radius), it becomes

$$\phi_m = W(r) - (\frac{r}{b})^m W(b), \qquad (69)$$

where

$$W(r, z) = \int_0^r \ln\frac{r}{r'} n_m(r', z) r' dr' \qquad (70)$$

for $m = 0$, and

$$W(r, z) = \frac{r^m}{2m}\int_0^r r'^{(1-m)} n_m(r', z) dr' - \frac{r^{-m}}{2m}\int_0^r r'^{(1+m)} n_m(r', z) dr' \qquad (71)$$

for $m \neq 0$, where $\phi_{-m} = \phi_m^*$ is used to compute ϕ_m for $m < 0$. We replace the integral with summation at grid points. The electric fields are

$$E_r = \sum_m \frac{\partial}{\partial r}\phi_m \exp(im\varphi), \qquad (72)$$

$$E_\varphi = \sum_m \frac{1}{r}\frac{\partial}{\partial \varphi}\phi_m \exp(im\varphi), \qquad (73)$$

$$E_z = \sum_m \frac{\partial}{\partial z}\phi_m \exp(im\varphi). \qquad (74)$$

The differentiation is done analytically beforehand except with z.

Tracking procedures

The independent variable is *time* so that a snapshot of a beam in configuration space is obtained. At each time step, space-charge fields are calculated and applied to each macro particle as an impulse kick. Then, the particle position is advanced with the newly calculated momentum. Those two computations are repeated for as many turns as specified initially.

The size of grids is recalculated at each time step so that all the macro particles are included with the minimum required grid size. Once a particle amplitude becomes larger than the pipe radius, it is regarded as a lost particle. The number of grid points is optimized with some tests. For example, radial grid points are fixed as 40 in order to reproduce a smooth field curve.

The number of longitudinal grids is determined in such a way that a half FODO cell is divided into at least 5 grids. Since a beam stretches over several transverse focusing periods and transverse beam size is accordingly modulated [8], it is necessary to divide a beam in the longitudinal direction such that harmonic components

---

[8] This is true for a spallation source type of synchrotron, where the rf harmonic number is a few, whereas the number of focusing periods is a few tens.

driven by transverse space charge are included. Once the longitudinal grid size is determined, a time step is calculated as

$$dt \sim \frac{dz}{\beta c}. \tag{75}$$

The number of azimuthal modes is usually taken up to four, which is enough to model octupolar shape in configuration space.

The space-charge defocusing force modifies the lattice functions and an optical matching condition. If the force is linear, all the particles feel the same focusing force and therefore the modified matching condition with space charge applies to the whole beam. In reality, however, the distribution is not uniform and nonlinear space-charge force cannot make the whole beam matched. In the simulation, the rms envelope equations including space-charge force are solved to find the matched optical condition for the rms emittance. In the 3-D simulation, a peak current is assumed to solve the envelope equations so that only the center of a bunch is matched.

As an initial particle distribution, we can specify several kinds, such as KV, waterbag, parabolic, and Gaussian for transverse planes, and uniform, parabolic, and Gaussian for longitudinal ones. We adopt a concept of equivalent beam so that the rms emittance is fixed when we compare two or more distributions.

Presentation of simulation results

Simulation results are presented in several ways. Plotting of macro particles in phase space every n turns (n should be reasonably large unless the particle distribution changes very quickly) is the most common one. With its projection to position and gradient axes, it shows the particle distribution and its evolution in the most primitive way.

Some statistical quantities, such as second- or higher-order moments, can be derived from the particle ensemble. An example is the rms emittance defined by the second moments and their cross term. When a machine has its physical acceptance, we can define the fraction of particles that fit in the acceptance. The fraction of particles outside the acceptance can be regarded as particle loss. Then, we can trace particle loss as a function of time or turn.

Instead of dealing with all the macro particles, we also see a single-particle trajectory, the so-called Poincaré map. Several particle trajectories with different initial coordinates reveal phase-space structure. In a synchrotron, it is common to look at the single-particle coordinates at one fixed point in a ring. As a tracking study in a linac, however, it is also possible to plot coordinates with other periodicity such as the one introduced by initial mismatch or coherent-mode oscillations.

A single-particle trajectory gives the tune of an individual particle. It is defined as a frequency at which the following quantity has a peak, as shown in Fig. 17 (left) [24],

$$m_j^{klm}(\nu) = \frac{1}{n_{max}} \sum_n^{n_{max}} x_j^k(n) y_j^l(n) z_j^m(n) \exp(-2\pi i \nu n). \tag{76}$$

Each individual particle has its own incoherence tune. This results in an incoherent tune spread in a tune diagram.

The characteristic frequency of the whole beam is also calculated. One example is a coherent frequency defined in the following. First, a moment of the particle coordinates is calculated every turn:

$$M_{coh.}^{klm}(n) = \frac{1}{N} \sum_j^N x_j^k(n) y_j^l(n) z_j^m(n). \tag{77}$$

Then, the frequency spectrum of the moment is

$$M_{coh.}^{klm}(\nu) = \frac{1}{n_{max}} \sum_n^{n_{max}} M_{coh.}^{klm}(n) \exp(-2\pi i \nu n). \tag{78}$$

The frequency at which the above quantity has a peak is the coherent tune, as shown in Fig. 17 (right). The sum of $k + l + m$ defines the order of the coherent modes.

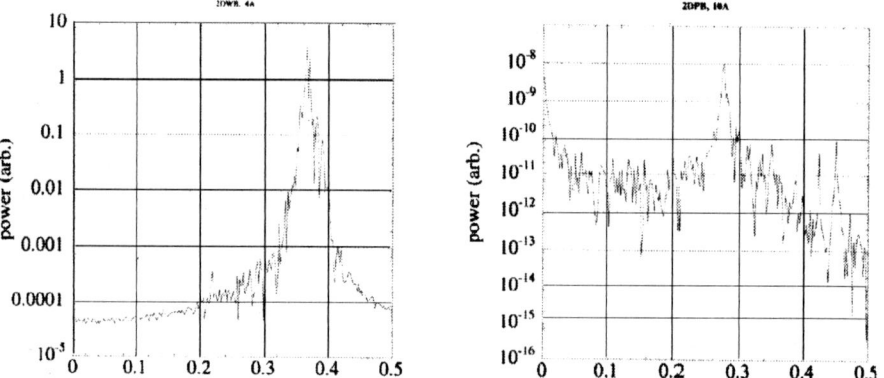

FIGURE 17. Fourier spectrum of single-particle motion (left) and of second moment (right).

SUMMARY

We have described some properties of a space-charge-dominated beam in circular accelerators. The keyword is self-consistency. The collective effects are dominant and any approximations based on single-particle dynamics plus an external force as a perturbation do not give the right answer.

There are some cures that help reduce space-charge effects. Essentially, making particles in a beam uniformly distributed in a 3-D configuration space is a way we can try. This can be realized by changing the shape of the external potential or re-distributing in phase space. Both are effective for the longitudinal direction and only the latter is possible in the transverse direction.

Some experimental observations exist which support the analysis. However, the emittance (beam profile) measurement of a high intensity beam is difficult, although not impossible. More accurate and conclusive experimental data and the setup of measurements are expected to become available soon.

Since the experimental study is limited, we rely heavily on simulation studies to check the theoretical analysis. As an example, we show the simulation method and its numerical example using the code Simpsons. Simulation study has become realistic thanks to the rapid increase of computational power. However, phenomena we study in a circular machine last a few 10 ms to seconds, and it is still not possible to track a fully 3-D beam within a reasonable CPU time. In addition, the validation of a simulation code is a crucial issue because not all the parameters the simulation predicts can be observed experimentally. Further study in that aspect is also expected.

ACKNOWLEDGMENT

We would like to acknowledge Prof. Shin-ichi Kurokawa, the school director of our local region, for giving us the great opportunity for teaching and learning (in fact) the subject. We also appreciate Prof. Yoshiharu Mori for his continuous encouragement on the study of space-charge problems. Finally, we would like to thank the school organizing committee members for a wonderful school and stay on the boat.

REFERENCES

1. Laslett, L. J., *Proc. of 1963 Summer Study on Storage Rings*, BNL-7534, (1963) p. 324.
2. For example, Bovet, C., Gouiran, R., Gumowski, I., and Reich, K. H., *A Selection of Formulae and Data Useful for the Design of A. G. Synchrotrons*, CERN/MPS-SI/Int., 1970.
3. Zotter, B., CERN 85-19, (1985) p. 253.
4. Baartman, R., *AIP Conf. Proc.* **448** (1998) 56.
5. Hofmann, I. and Beckert, K., *IEEE Trans. Nucl. Sci.*, **NS-32, No.5**, (1985) 2264.
6. Smith, L., *Int. Conf. on High Energy Acc.* Dubna, 1963, p. 1232. Also, F. Sacherer, Ph.D. Thesis, University of California (UCRL-18454), 1968.
7. Kapchinskij, I. M. and Vladimirskij, V. V., *Proc. Int. Conf. on High Energy Accelerators*, CERN, Geneva, 1959, p. 274.
8. Sacherer, F. J., *IEEE Trans. Nucl. Sci.*, **NS-18, No.3**, (1971) 1105.
9. Lee, S. Y. and Okamoto, H., *Phys. Rev. Lett.*, **80** (1998) 5133.
10. Venturini, M. and Reiser, M., *Phys. Rev. Lett.*, **81** (1998) 96.

11. Uesugi, T., Machida, S., and Mori, Y., *Proc. of 7th European Particle Accelerator Conference*, 2000, p. 1333.
12. Cappi, R., Garoby, R., Hancock, S., Martini, M., and Riunaud, J. P., *Proc. of the 1993 Particle Accelerator Conference*, 1993, p. 3570.
13. Montague, B. W., CERN 68-38, 1968.
14. Sakai, I., Adachi, T., Arakida, Y., Irie, Y., Kitakawa, K., Machida, S., Mori, Y., Shimosaki, Y., Someya, H., and Yoshimoto, M., *Proc. of 7th European Particle Accelerator Conference*, 2000, p. 2261.
15. Kawakubo, T., et. al., *Nuclear Instruments and Methods* **A302** (1991) 397.
16. Elmfors, P., et. al., *Nuclear Instruments and Methods* **A396** (1997) 13.
17. Schonauer, H., *IEEE Trans. on Nucl. Sci.*, **NS-26, No.3**, (1979) 3294, and CERN/PS/BR 79-8, 1979.
18. Hashimoto, Y., et. al., *Proc. of 7th European Particle Accelerator Conference*, 2000, p. 1729.
19. Burtin, G., et. al., *Proc. of 7th European Particle Accelerator Conference*, 2000, p. 256.
20. Fujieda, M., Iwashita, Y., Noda, A., Mori, Y., Ohmori, C., Sato. Y., Yoshii, M., Blaskiewicz, M., Brennan, J. M., Roser, T., Smith, K. S., Spitz, R., and Zaltsmann, A., *Phys. Rev. ST Accel.* **Beams 2**, 122001 (1999).
21. Beebe-Wang, J., Fedotov, A. V., Wei, J., and Machida, S., *Proc. of 7th European Particle Accelerator Conference*, 2000, p. 1286
22. Machida, S. and Ikegami, M., *AIP Conf. Proc.* **448** (1998) 73.
23. For example, Birdsall, C. K. and Langdon, A. B., *Plasma Physics via Computer Simulation*, New York: MacGraw-Hill (1985).
24. Orlov, Y. and Soffer, A., CLNS 92/1178.

Insertion and Crossing Region Design

U. Wienands, P. Beloshitsky

SLAC, *Stanford, CA 94309, USA* and
CERN, *Geneva, Switzerland*

Abstract. This article is the summary of the 5-afternoon tutorial on insertions for circular machines. Roughly half the course (Part 1) was spent discussing interaction regions, We start by recapitulating basic beam optics including building blocks. This provides the tools to analyze the basic structure of interaction regions and explore the parameter space. This simple example is then successively refined and made more realistic. Examples of realized interaction regions for both hadron and electron machines are shown and their salient features and differences explained. A brief discussion of solenoid-decoupling brings Part 1 to a close. In Part 2 we discussed various utility sections. Dispersion suppressors are presented in detail discussing the principles as well as the practical implementation of flexible suppressors using LEP as an example. Injection schemes, both single-turn and multi-turn stacking, are presented in depth. The matching of wiggler and undulator insertions and a discussion of the impact of these devices on beam parameters closes out Part 2.

Part 1

I MATRIX OPTICS

The solution of the equation of motion of a particle in a beam-guidance system can be found in the literature and we will not repeat here what has already been described with much clarity [1–3]. But to clarify our notation and spare the reader constant referencing to other papers we will briefly state the most salient relations of matrix optics.

In matrix optics, a particle's coordinates in a curvilinear coordinate system centered on the reference axis and following the beam line are described by a 6-vector,

$$X = \begin{pmatrix} x \\ x' \\ y \\ y' \\ l \\ dp \end{pmatrix}. \qquad (1)$$

In linear approximation, the effect of the elements of a beam guidance system on this particle is described by 6×6 matrices R, and regular matrix multiplication rules apply. For "back-of-the-envelope" estimates, thin-lens approximation is often used, where the magnetic elements are described by zero-length elements and field-free drift spaces.

A   Drift Space

The simplest "element" is a drift space of length L. The beam coordinates at the end are

$$x_2 = x_1 + L \times x'_1,$$
$$x'_2 = x'_1, \qquad (2)$$

and identical in the y plane. The matrix R is then

$$R = \begin{pmatrix} 1 & L & 0 & 0 & 0 & 0 \\ 0 & 1 & 0 & 0 & 0 & 0 \\ 0 & 0 & 1 & L & 0 & 0 \\ 0 & 0 & 0 & 1 & 0 & 0 \\ 0 & 0 & 0 & 0 & 1 & 0 \\ 0 & 0 & 0 & 0 & 0 & 1 \end{pmatrix}. \qquad (3)$$

B   Quadrupole Magnet

A quadrupole magnet behaves as a lens with equal focal lengths in either plane but of opposite sign. For a thin quadrupole, the beam coordinates transform as

$$x_2 = x_1,$$
$$x'_2 = x'_1 - \frac{x_1}{f} \qquad (4)$$

and similar in y, and its R matrix is

$$R = \begin{pmatrix} 1 & 0 & 0 & 0 & 0 & 0 \\ -1/f & 1 & 0 & 0 & 0 & 0 \\ 0 & 0 & 1 & 0 & 0 & 0 \\ 0 & 0 & 1/f & 1 & 0 & 0 \\ 0 & 0 & 0 & 0 & 1 & 0 \\ 0 & 0 & 0 & 0 & 0 & 1 \end{pmatrix}. \qquad (5)$$

Note the minus sign in the focusing plane.

C Dipole Magnet

A dipole (bending) magnet changes the direction of the reference line by its bending angle α. It introduces dispersive terms and exhibits edge focusing depending on the angle of its pole faces. For a wedge dipole with bending radius ρ we have

$$x_2 = x_1,$$
$$x_2' = x_1' - x_1 \frac{\sin \alpha}{\rho} + \delta p \sin \alpha, \qquad (6)$$

and its R matrix is

$$R = \begin{pmatrix} 1 & 0 & 0 & 0 & 0 & 0 \\ -\frac{\sin \alpha}{\rho} & 1 & 0 & 0 & 0 & \sin \alpha \\ 0 & 0 & 1 & 0 & 0 & 0 \\ 0 & 0 & 0 & 1 & 0 & 0 \\ 0 & 0 & 0 & 0 & 1 & 0 \\ 0 & 0 & 0 & 0 & 0 & 1 \end{pmatrix}. \qquad (7)$$

The sector dipole exhibits only horizontal edge focusing. Edge focusing is often ignored in thin-lens approximation as it is usually small compared to the effect of quadrupole lenses.

D Composite Systems

To describe optical modules, from quadrupole doublets to entire beam lines or rings, the matrices are multiplied following ordinary matrix algebra.

Quadrupole doublet and triplet

Let us consider a quadrupole doublet consisting of a defocusing and a focusing quadrupole separated by a drift space of length L_D. The magnets have focal lengths f_F and f_D. The matrix is then

$$R_d = R_{D-quad} \circ R_{drift} \circ R_{F-quad},$$

$$= \begin{pmatrix} 1 + \frac{L_D}{f_D} & L_D & 0 & 0 & 0 & 0 \\ \frac{1}{f_D} + \frac{1}{f_F} + \frac{L_D}{f_D f_F} & 1 + \frac{L_D}{f_F} & 0 & 0 & 0 & 0 \\ 0 & 0 & 1 + \frac{L_D}{f_F} & L_D & 0 & 0 \\ 0 & 0 & \frac{1}{f_D} + \frac{1}{f_F} + \frac{L_D}{f_D f_F} & 1 + \frac{L_D}{f_F} & 0 & 0 \\ 0 & 0 & 0 & 0 & 1 & 0 \\ 0 & 0 & 0 & 0 & 0 & 1 \end{pmatrix}. \qquad (8)$$

Suppose the quadrupoles have equal focal lengths but of opposite sign, i.e. $f_F = -f_D = f$. In this case,

$$R_d = \begin{pmatrix} 1 + \frac{L_D}{f} & L_D & 0 & 0 & 0 & 0 \\ -\frac{L_D}{f^2} & 1 - \frac{L_D}{f} & 0 & 0 & 0 & 0 \\ 0 & 0 & 1 - \frac{L_D}{f} & L_D & 0 & 0 \\ 0 & 0 & -\frac{L_D}{f^2} & 1 + \frac{L_D}{f} & 0 & 0 \\ 0 & 0 & 0 & 0 & 1 & 0 \\ 0 & 0 & 0 & 0 & 0 & 1 \end{pmatrix} \quad (9)$$

and, since $R_{21} < 0$ and $R_{43} < 0$, this doublet focuses in either plane. The effective focusing power decreases with decreasing L_D: moved too close the two quadrupoles "fight each other." The fact that the symmetric thin-lens doublet has the same focusing power in either plane ($R_{21} = R_{43}$) does *not* result in anastigmatic behavior, however, as the principal planes for the x plane and the y plane ($H_{1,x}, H_{2,y}$) are at quite different locations. Fig. 1 shows this in a graphical form.

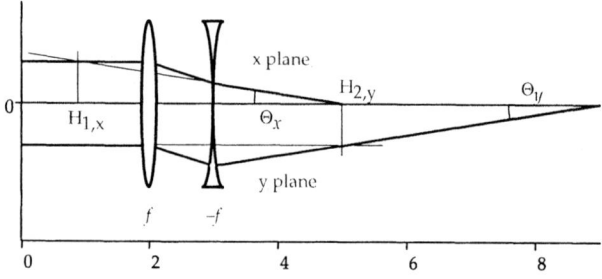

FIGURE 1. Imaging in a doublet

Simple FODO array

For a circular machine the FODO array uses the quadrupole strength most efficiently. Letting it begin at the center of a quadrupole, we find its R matrix:

$$R = \begin{pmatrix} 1 - \frac{L^2}{2f^2} & 2L\left(1 \pm \frac{L}{2f}\right) \\ -\frac{L}{2f^2}\left(1 \mp \frac{L}{2f}\right) & 1 - \frac{L^2}{2f^2} \end{pmatrix} \quad (10)$$

where L is the spacing between the quadrupoles and f is the focal length. Here we have restricted the form to the 2×2 matrix for one plane only; the upper sign applies to the horizontal plane and the lower, to the vertical plane.

For matched lattice functions, we can show that a phase angle μ exists such that

$$R = \begin{pmatrix} \cos\mu + \alpha\sin\mu & \beta\sin\mu \\ -\gamma\sin\mu & \cos\mu - \alpha\sin\mu \end{pmatrix}, \qquad (11)$$

where $\gamma = (1+\alpha^2)/\beta$. It follows that

$$\cos\mu = 1 - \frac{L^2}{2f^2}, \quad \sin\frac{\mu}{2} = \frac{L}{2f}, \quad \beta = 2L\frac{1 \pm \sin\frac{\mu}{2}}{\sin\mu} \qquad (12)$$

and

$$\alpha = 0. \qquad (13)$$

II MATCHING

Insertion of a section into an otherwise regular accelerator lattice should not disturb the beam parameters in the remainder of the machine. This can be achieved in two ways:

- By making the insert transparent to the beam or
- By ensuring that the lattice (Twiss) parameters at the end of the insert have exactly the same value as at the beginning.

Even though the former is a special case of the latter, there are important differences between these two approaches.

A Transparent Section

In the description of a ring (or beam line) by a product of matrices we can insert a unit matrix at any place without changing the properties of the line. If we can design an optical insertion so that its overall matrix becomes the unit matrix, we can insert such a section anywhere into the ring or beam line without affecting the optical properties of the machine (to first order).

From Eq. (11) it follows that any beam line with an integer phase advance $\mu/(2\pi)$ is a unit (I) section. If the line is symmetric, each half will be a $-I$ section with 180° phase advance.

B Twiss-Parameter Matching

The Twiss parameters of a ring describe a "machine ellipse"; an ellipse in phase space (here understood to mean $x' - x$ space) that will be occupied by a matched beam of a certain emittance ϵ. A matched beam in a ring maintains its density distributions turn after turn. Figure 2 shows such an ellipse and its relation to the Twiss parameters (β, α, γ). Since this ellipse completely describes the beam properties at each location of a ring (again to first-order) for matching an insertion

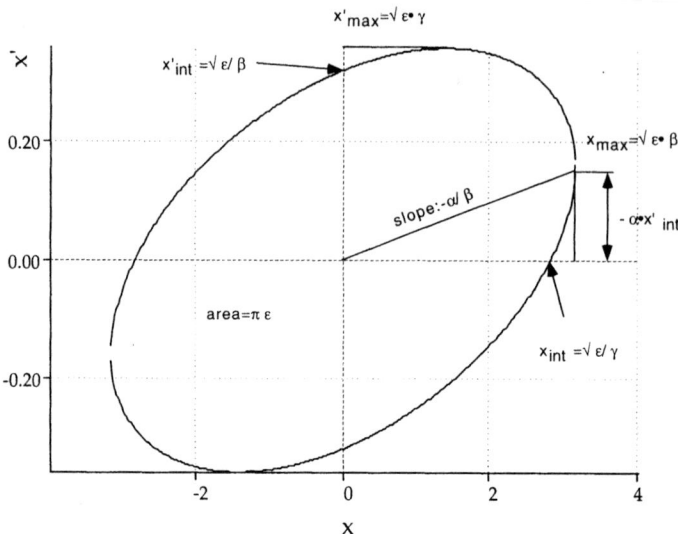

FIGURE 2. Machine ellipse

into a ring it is sufficient to demand that the machine ellipse have the same shape at the exit from the insertion as it has at the entrance. Liouville's theorem already ensures that the area stays the same; this leaves two parameters to be matched: e.g. the tilt angle and the length of the major axis, or the lattice functions β and α. Compared to the unit section this is one parameter (the phase advance) less—two if we consider the two-dimensional problem in x and y—and the resulting increase in flexibility is significant. Equation (11) provides the transformation between the matched Twiss parameters and the R matrix describing the optics of the insertion.

We can extend this prescription to a section matching points in a lattice with different Twiss parameters. For this we need to know how to propagate Twiss functions through a lattice. Following Ref. [3], we find this transformation by transporting a matched beam through the insertion optics. The matched beam distribution (in one plane) is described by

$$\gamma_1 x_1^2 + 2\alpha_1 x_1 x_1' + \beta_1 x_1'^2 = \epsilon \tag{14}$$

and this can be written in matrix form as

$$X_1^t B_1^{-1} X_1 = \epsilon \tag{15}$$

where

$$B_1 = \begin{pmatrix} \beta_1 & -\alpha_1 \\ -\alpha_1 & \gamma_1 \end{pmatrix} \text{and} X_1 = \begin{pmatrix} x_1 \\ x_1' \end{pmatrix}. \tag{16}$$

Transported to location 2 through a beam line this takes on the form

$$X_2^t B_2^{-1} X_2 = \epsilon \qquad (17)$$

where $X_2 = RX_1, X_2^t = X_1^t R^t$, and therefore

$$X_1^t R^t B_2^{-1} R X_1 = \epsilon, \qquad (18)$$

and

$$B_1^{-1} = R^{-1} B_2^{-1} (R^t)^{-1}, \qquad (19)$$

or

$$B_2 = R B_1 R^t, \qquad (20)$$

which we can then write in explicit form as

$$\begin{pmatrix} \beta_2 \\ \alpha_2 \\ \gamma_2 \end{pmatrix} = \begin{pmatrix} R_{11}^2 & -2R_{11}R_{12} & R_{12}^2 \\ -R_{11}R_{21} & 1+2R_{12}R_{21} & -R_{12}R_{22} \\ R_{21}^2 & -2R_{21}R_{22} & R_{22}^2 \end{pmatrix} \begin{pmatrix} \beta_1 \\ \alpha_1 \\ \gamma_1 \end{pmatrix}, \qquad (21)$$

It remains to express α, β, γ and $\sin \mu$ in terms of the magnet strengths and distances between magnets. It is here that the analytic approach quickly becomes impractical and computer programs like MAD, DIMAD or others are called for to find the solution numerically.

C General Matching Conditions

In the above we have ignored dispersion completely. A straight insertion without dipoles will maintain zero dispersion and therefore can be matched at a point of zero dispersion without explicit matching of the dispersion. If non-zero dispersion needs to be matched, this results in additional conditions on the phase advance.

The total number of variables to be matched is then eight: four Twiss parameters and four dispersive terms (two of each at each end of the insert). If the insert is to prepare certain beam conditions, e.g. a low-β value at an interaction point, the number of parameters further increases. Symmetries can reduce the number of parameters to be matched. In general there need to be at least as many optical elements as there are parameters to match; this then sets the minimum number of independent magnets to be used.

III INTERACTION REGION DESIGN CONSIDERATIONS

A β Function at the Interaction Point

Particle colliders bring two counterrotating beams together at one or more interaction points (IP) where the beams collide, producing the physics events to be detected. The rate of these events is called luminosity, and following Chao [4] the luminosity of a collider is given by

$$\mathcal{L} = \frac{N^2 f}{4\pi\sigma^2} \; [\text{cm}^{-2} \; \text{s}^{-1}], \tag{22}$$

where N is the number of particles per bunch and f, the collision frequency. Round beams of size σ and head-on collision are assumed.

The electromagnetic fields associated with the two beams exert a mutual force on each other, which turns out to be a (nonlinear) focusing force. The linear part is termed the *beam-beam tune shift* and is given by

$$\xi = \frac{N r_0 \beta^*}{4\pi\sigma^2 \gamma}, \tag{23}$$

where r_0 is the classical radius of the particle type, γ, the relativistic factor and β^* the β function (assumed to be equal in x and y) at the interaction point.

Putting this together we find the luminosity

$$\mathcal{L} = \frac{N f \gamma \xi}{r_0 \beta^*} \tag{24}$$

to be directly proportional to the beam-beam tune shift and inversely proportional to β^*. Under these assumptions and for a limit on the tune shift (which depends on whether or not there is radiation damping) lowering β^* is the *only* way to increase luminosity once the number of bunches and particles per bunch (*i.e.* the beam current) in the ring is set. It is for this reason that lower and lower β functions are desired at the IP.

The above assumption of round beams does not hold for electron machines, where $\epsilon_y \ll \epsilon_x$ because of quantum excitation. For electron machines, Eq. (23) becomes

$$\xi_{x,y} = \frac{N r_0 \beta^*_{x,y}}{2\pi \sigma^*_{x,y}(\sigma^*_x + \sigma^*_y)\gamma}, \tag{25}$$

from which it follows that

$$\frac{\xi_y}{\xi_x} = \sqrt{\frac{\beta_y/\beta_x}{\epsilon_y/\epsilon_x}}, \tag{26}$$

and, since usually $\xi_y \approx \xi_x$, the ratio β_y/β_x should be about equal to the emittance ratio and $\beta_y \ll \beta_x$. Beam aspect ratios as small as 0.02 are not uncommon.

B Crossing Angle

The colliding beams have to be pulled apart at some point to avoid collisions in other regions of the machine. One way to do this is by colliding at a certain, non-zero crossing angle, see Fig. 3

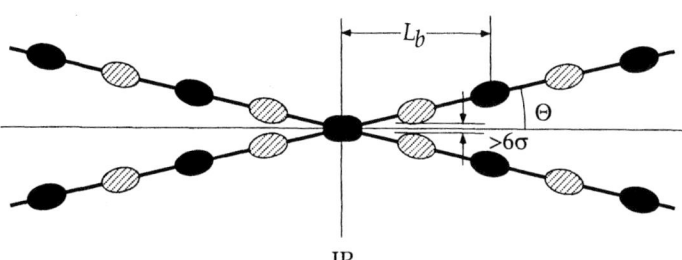

FIGURE 3. Crossing-angle geometry

To avoid significant tune shift from the first parasitic crossing, one aims to have the beams separated by more than 6σ. It then follows immediately that the crossing angle should be larger than ± 3 times the beam divergence at the IP, which is $\sqrt{\epsilon/\beta^*}$.

Often, head-on collisions are desired because the synchro-betatron coupling generated by a crossing angle can cause destructive beam-beam interaction. In this case, the separation of the beams is achieved by electrostatic or—in the case of two-ring colliders with different beam energies or like particles—magnetic beam separators. In this case, the required bending angle (per beam) will be larger than $3\sqrt{\epsilon/\beta^*}$, by a considerable amount if the beam separation goes only by the energy difference (*e.g.* for the *B*-Factories).

IV A SIMPLE INTERACTION REGION

We are now ready to investigate basic interaction region optics.

To reiterate: The goal of an interaction region system is to focus the beam down to as small a waist as is practical, *i.e.* create a low-β section. In addition, some free drift space is needed for the detector to be placed around the IP. We then expect to place a quadrupole at a distance L_{IP} from the interaction point. Symmetry arguments demand that at the minimum-β point (IP) α be 0 and we can without undue restriction set $\alpha = 0$ at the other end as well; we will place a quadrupole there to achieve this. The whole arrangement is then reflected about the IP to make up the complete interaction region insertion, Fig 4.

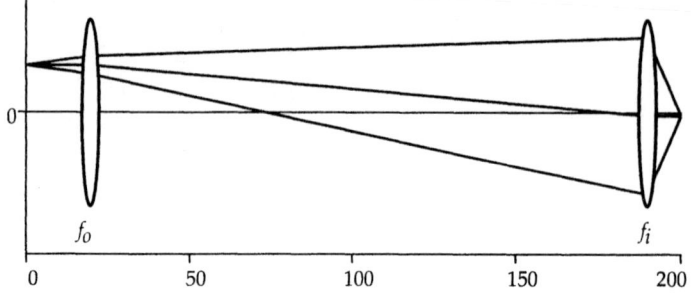

FIGURE 4. Imaging in a simple interaction region telescope

Without knowing more than β^* at the IP and L_{IP} we can directly estimate the β function in the QI quadrupole by using Eq. 21:

$$\beta_2 = \beta^* + L^2/\beta^*. \tag{27}$$

The β functions rises quadratically with distance from the IP.

A Transparent Match

It is instructive to first match the region as a unit section akin to section II A. In this case we have four matrix elements to match, and the variables to achieve this are the two quadrupole strengths as well as the drift lengths.

The half-insertion region up to the IP is described by

$$R = \begin{pmatrix} 1 & 0 \\ \frac{-1}{f_o} & 1 \end{pmatrix} \circ \begin{pmatrix} 1 & L_D \\ 0 & 1 \end{pmatrix} \circ \begin{pmatrix} 1 & 0 \\ \frac{-1}{f_i} & 1 \end{pmatrix} \circ \begin{pmatrix} 1 & L_{IP} \\ 0 & 1 \end{pmatrix}. \tag{28}$$

For the phase advance through this section to be $\mu = 0.5$, R_{12} and R_{21} have to be zero. Solving for f_i and f_o, the focal lengths of the quadrupoles, yields the conditions

$$f_i = \frac{L_D L_{IP}}{L_D + L_{IP}}, \quad f_o = \frac{L_D^2}{L_D + L_{IP}} \tag{29}$$

and, with Eq. (28),

$$R = \begin{pmatrix} \frac{-L_{IP}}{L_D} & 0 \\ 0 & \frac{-L_D}{L_{IP}} \end{pmatrix}. \tag{30}$$

To see how the Twiss parameters transform we use Eq. (21) to find

$$\begin{pmatrix} \beta_2 \\ \alpha_2 \\ \gamma_2 \end{pmatrix} = \begin{pmatrix} \frac{L_{IP}^2}{L_D^2} & 0 & 0 \\ 0 & 1 & 0 \\ 0 & 0 & \frac{L_D^2}{L_{IP}^2} \end{pmatrix} \circ \begin{pmatrix} \beta_1 \\ \alpha_1 \\ \gamma_1 \end{pmatrix} \tag{31}$$

We observe:

1. $\frac{1}{f_i}$ and $\frac{1}{f_o}$ are > 0, i.e. both lenses are focusing.

2. The "magnification" of β is $\frac{L_{IP}^2}{L_D^2}$, i.e. the placement of the lenses determines the value of β at the IP; to get low β at the IP we need $L_D \gg L_{IP}$.

3. α is left unchanged for any value of β; this is a consequence of $\mu = 0.5$.

Item 1 indicates that we should be able to replace each lens by a doublet to get this to work in both planes. Item 2 shows that significant space is needed along the beam axis for low β^*. Using Eq. (29) we further see that the focal length f_i approaches L_{IP} for large L_D, while f_o increases to approach L_D.

As an aside, let us have a brief look at the parameters for the inner quadrupole, QI. The gradient required for QI is

$$G = kB\rho = \frac{B\rho}{fL_{QI}} \tag{32}$$

and its excitation current is approximately

$$I = \frac{1}{0.8\pi} GR^2 = \frac{1}{0.8\pi} \frac{B\rho}{L_{QI}} \frac{L_{IP}\epsilon}{\beta^*} \tag{33}$$

where we used Eq. (27) and a general formula for quadrupole excitations [5]. The magnet power is then

$$P \propto I^2 L = \frac{1}{(0.8\pi)^2} \frac{(B\rho)^2}{L_{QI}} \frac{\epsilon^2}{\beta^{*2}} L_{IP}^2, \tag{34}$$

since the electrical resistivity of the coil will be proportional to the conductor length, thus the magnet length L. The power increases quadratically with distance of the inner focusing magnet from the IP.

B Twiss-Parameter Match

The way the interaction region above was matched is restrictive: once the magnets are placed the β function at the IP is given. This is usually not desirable; most machines need the flexibility to adjust and optimize the IP beam parameters as operating experience is gained. To match the Twiss parameters, only two constraints have to be met: $\alpha = 0$ at the IP and β^* being the desired value. Two variable parameters should be sufficient to meet these, and we want these to be the magnet settings. In this way, we gain the flexibility to adjust β^* once the machine is built.

To match the section we will use f_i to make $\alpha = 0$ at the IP. The strength of the outer quad f_o can then be adjusted to achieve the desired β^*. The matrix to the IP is

$$R = \begin{pmatrix} 1 + \frac{L_{IP}}{f_i} + \frac{L_{IP} + L_D(1 + L_{IP}/f_i)}{f_o} & L_{IP} + L_D(1 + L_{IP}/f_i) \\ 1/f_i + \frac{1 + L_D/f_i}{f_o} & 1 + L_D/f_i \end{pmatrix}. \tag{35}$$

Using these matrix elements in Eq. 21 and solving for $\alpha_2 = 0$ is tedious but can be accomplished rather easily using *Mathematica* or *Maple*. We end up with a rather complex expression for f_i, which is not very insightful, and with an expression for β^*,

$$\beta^* = \frac{\Delta}{2\beta_i} - \frac{\Delta}{2\beta_i}\sqrt{1 - \frac{L_{IP}^2 4\beta_i^2}{\Delta^2}}, \tag{36}$$

with

$$\Delta = \frac{L_D^2 + \beta_1^2(1 + L_D/f_o)^2}{2\beta_1} - L_{IP}. \tag{37}$$

We conclude that for the solution to exist, L_{IP} has to be smaller than $\Delta/2\beta_i$ (else Eq. (36) has no real solution). Observation 2 from above still holds in the sense that as L_D increases β^* becomes smaller, although this is not trivially obvious from Eq. (36) in this form. Furthermore, the value of β^* is primarily driven by f_o.

Numeric example

As an example we design a 10-cm β^* insertion for a 2000-m ring. Assume that β_i at the beginning of the insertion is 30 m. We begin with a 2π insertion.

- Using Eq. (27) we find that, for 10 m free drift to the insertion quadrupole, $\beta \approx 950$ m, which is large but acceptable.

- We need a magnification L_D/L_{IP} of $\sqrt{30/0.1}$ which is 17, therefore the second quad will be 170 away from the insertion quad and the whole IR will be 2×180 m long.

- From Eq. 29 we find the focal lengths of the quadrupoles to be $f_i \approx 9.5$ m and $f_o \approx 160$ m.

Using MAD or DIMAD we can easily verify the optics of this simple example.

In order to optimize this design we will now relax the phase-advance condition. To reduce the large β functions at QI we *have* to move QI closer to the IP. To reduce the length of the drift between the quads it is instructive to look at the variation of parameters for varying strength $1/f_o$, shown in Fig. 5. Our present solution has $\mu = 0.5 \times \pi$ and actually about the largest value of β^* achievable in this geometry. Turning $1/f_o$ up towards either polarity lowers β^* dramatically. But in that case we can reduce the the distance between the quads to let β^* go back up. In practise, we shorten the distance L_D to 85 m and use DIMAD to fit the section for us. This is straightforward and indeed we are able to fit two solutions, one for $1/f_o$ focusing and the other for $1/f_o$ defocusing, both shown in Fig. 6. Both of these solutions are acceptable; in fact, as we will see in a moment, the focusing on the x and y plane often differs in this fashion.

We now extend the above example to both planes, replacing each quadrupole with a thin-lens doublet with 1 m spacing between quadrupoles. We saw already that this will result in net focusing in both planes. Fitting both planes to $\beta^* = 0.1$ m we get the lattice functions in Fig. 7.

The β maximum in the horizontally focusing quadrupole is almost twice the maximum in the vertically focusing quadrupole. This arises from its longer distance from the IP but is magnified by the defocusing action of the vertically focusing quadrupole. The reason for the observed behavior lies in our using a doublet for an interaction region with equal β in both planes. Replacing the inner doublet by a triplet we can expect much more similarity in both planes.

The quadrupole focal lengths are almost pairwise equal and about $3.5\ldots3.7$ m (outer doublet) and $2.8\ldots3.1$ m (inner doublet), fairly close to what we would expect from the single-plane strengths and evaluation of Eq. (9) for the inner doublet, but less so for the outer doublet.

We will now carry this example further to include the dispersion: Let us assume there is a dipole at 1 m distance from the IP, say, for beam separation. This dipole will generate dispersion, which needs to be cancelled within the straight section of the IR. In order to cancel this dispersion we need a dipole at 180° phase advance. Since the IP dipole will be 90° away from the IP, there is not enough phase advance to go around for this purpose.

To be able to cancel the dispersion we introduce another doublet into the IR lattice, between the outer and the inner doublet. In order to keep apertures small we will place it closer to the outer doublet, with 17 m drift between these (a somewhat arbitrary choice). The dispersion-cancelling dipole $B2$ is placed between the doublets. The two extra quadrupoles bring the number of parameters to be varied up to six, so in principle we can fit the six quantities $(\beta_{x,y}, \alpha_{x,y}, \eta, \eta')$. However, only three focusing magnets are acting mainly horizontally, whereas we have four horizontal quantities to fit, so we may expect difficulties. But we can move $B2$ around within its drift and this will provide another variable (in fact, the location of $B2$ should end up being 180° away from $B1$), so we include the $B2$ location as a fitting parameter. The result of this fit is shown in Fig. 8. The extra cell indeed increases the horizontal phase advance, and inspection of the DIMAD output shows that the two dipoles are indeed 180° apart. We also introduced quadrupoles with finite length (chosen as 1 m). For an electron or positron machine, $B2$ should be a "soft"

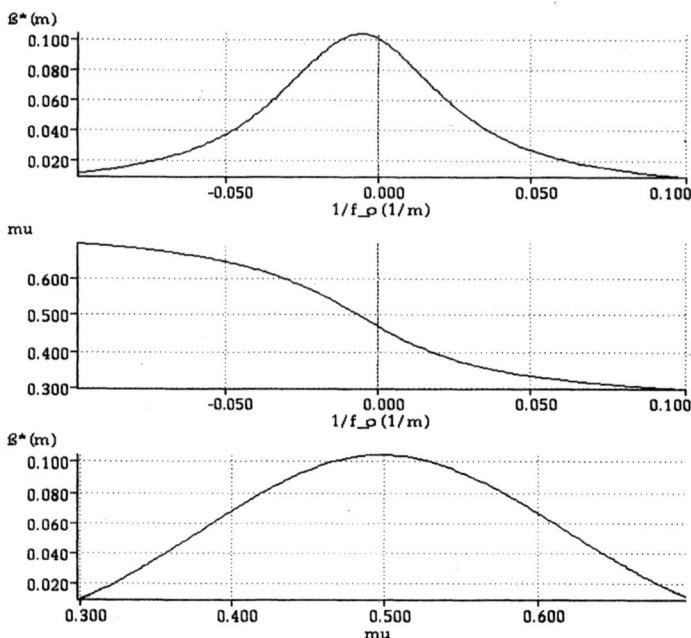

FIGURE 5. Numeric evaluation of the simple IR example

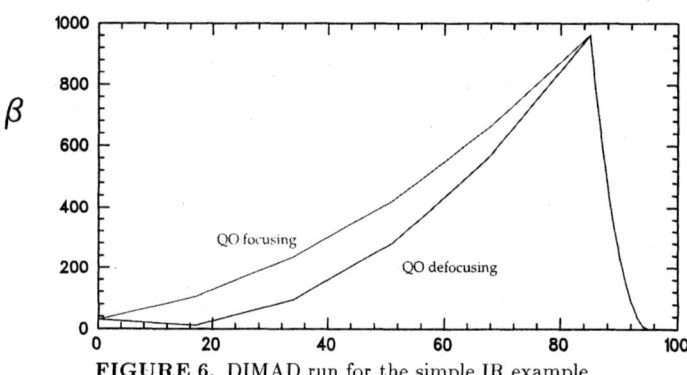

FIGURE 6. DIMAD run for the simple IR example

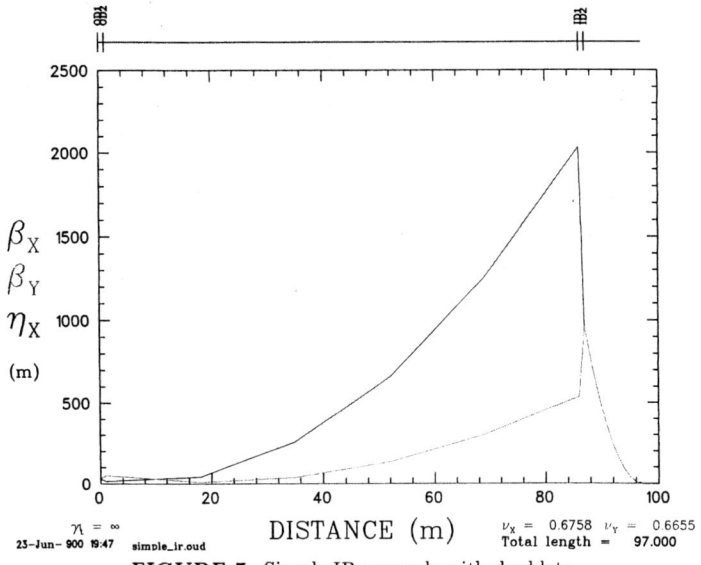

FIGURE 7. Simple IR example with doublets

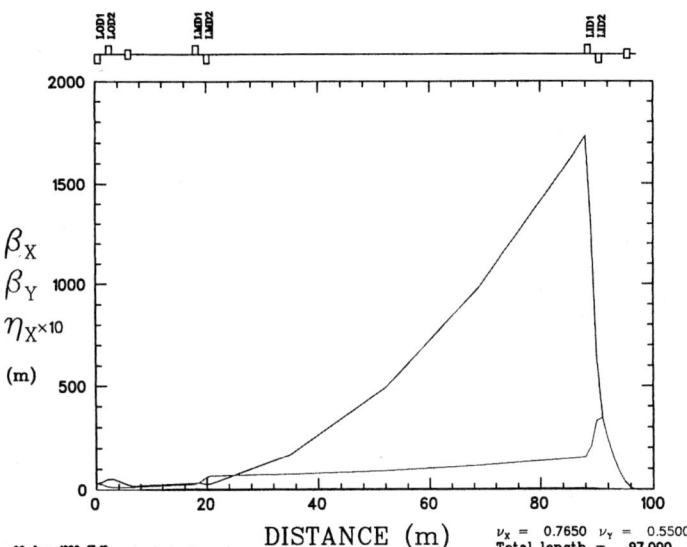

FIGURE 8. Simple IR example with dispersion suppression

(low-field) bend to avoid producing too much synchrotron radiation. The strength of $B2$, incidentally, is chosen for geometry rather than dispersion cancellation; the latter is achieved by the β function the fitter chooses.

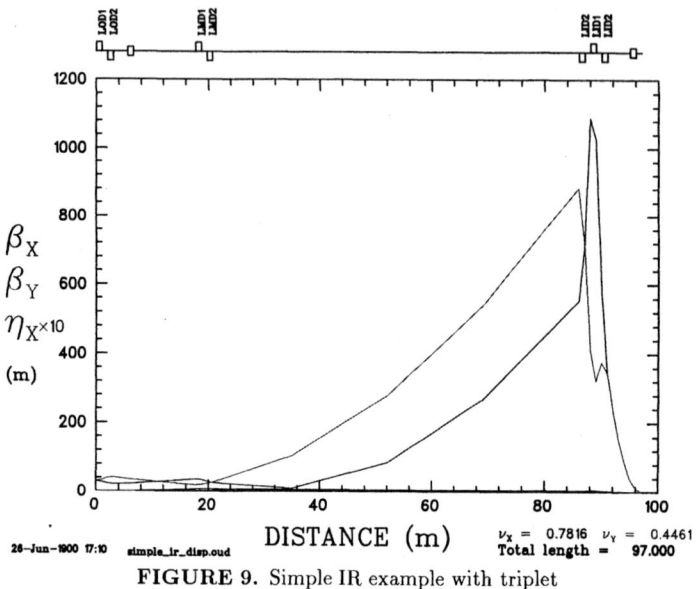

FIGURE 9. Simple IR example with triplet

This begins to look practical except that the β function in the inner doublet is too high. We could move the doublet closer into the IP, but that may interfere with any detector apparatus. It turns out that providing another focusing quadrupole, placed 1 m upstream of the doublet and extending it into a triplet, brings β down significantly, see Fig. 9. Optically, the principal planes of the triplet are much closer together than those of the doublet, which makes it the focusing element of choice for round-beam colliders (*i.e.* most hadron colliders) while in electron colliders with their much lower β_y^* doublets are used, with the D quad invariably being closer to the IP.

V EXAMPLES OF INTERACTION REGIONS

A Electron Machines

PEP-I

Our simplest example is one of the interaction regions of the original PEP ring at SLAC [6]. Its optical layout, shown in Fig. 10, exhibits many of the features of our example IR: dispersion is 0 throughout, and matching and transformation are provided by an outer and an inner doublet. The IP parameters are $\beta^*_{x,y} =$ (2.8, 0.11) m. The design luminosity is up to 10^{32} at 15 GeV. About 7 m of free space on either side of the IP is provided. Being a single-ring collider, PEP did not use

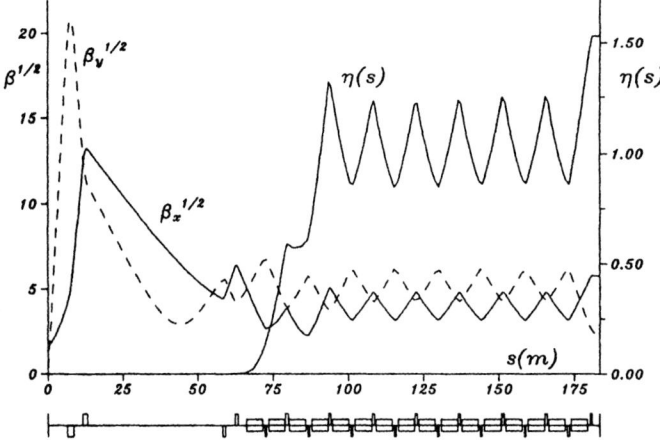

FIGURE 10. PEP-I IR lattice functions

beam-separating magnets, rather, the orbits were separated by using electrostatic deflectors. The dipole magnets closest to the IP are actually low-field dipoles placed at the end of each arc to prevent the strong synchrotron radiation from reaching the experiment while themselves not generating too much synchrotron power.

CESR-III

The CESR-III [7] IR lattice is shown in Fig. 11; a schematic view of the inner IR is in Fig. 12. The insertion quadrupoles are less than 2 m away from the IP, in fact, they are within the CLEO detector. To be able to function in the field of the detector, the insertion magnets have to be either superconducting or permanent magnets (*i.e.* iron-less). At CESR both technologies will be employed. The innermost quadrupole is a permanent quadrupole, the insertion quadrupoles labelled Q01 and Q02 in Fig. 12 are superconducting quadrupoles. In this way, β^*_y can be pushed down to 0.6 and 0.013 m (x and y, resp.). Expected luminosity is about 2×10^{33}.

With beam currents approaching or exceeding 1 A and magnets very close to the detector, synchrotron radiation handling becomes a major issue. A good example of how this is handled is found in the KEK-B and PEP-II *B*-Factories.

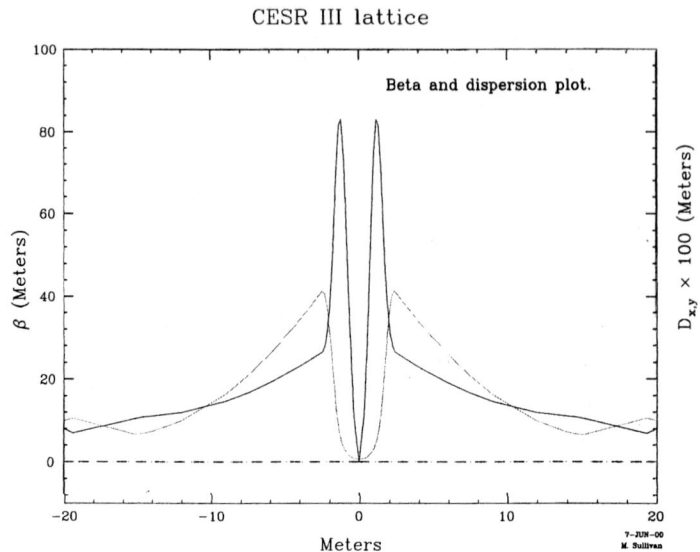

FIGURE 11. CESR III IR lattice functions

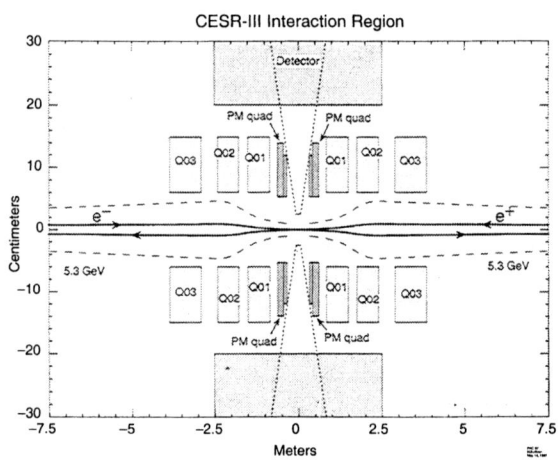

FIGURE 12. CESR III IR layout

PEP-II

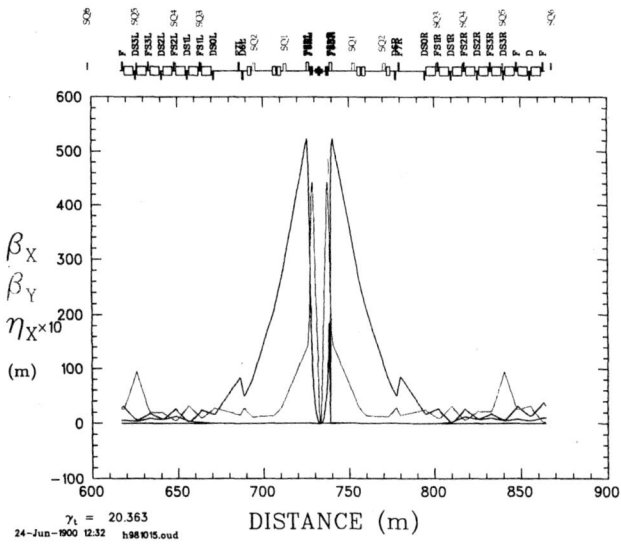

FIGURE 13. PEP-II HER IR lattice functions

The PEP-II B-Factory is an asymmetric 9-GeV electron on 3.1-GeV positron collider. The High Energy Ring (HER) lattice optics, Fig. 13, resembles the doublet version of the simple example: an inner doublet focuses the beam down and an outer doublet matches it to the arc. An extra vertically focusing quad common to both rings is located very close to the IP and provides additional reduction in β_y^*. The skew quadrupoles (SQx, shown in gray) are used to compensate the detector solenoid, cf. Sec. VI B.

The Low Energy Ring IR lattice is shown in Fig. 14. Its layout is quite a bit more complicated because of the need for local chromaticity correction, done by the 180° sextupole pairs SX1 and SX2, and SY1 and SY2, for the two planes.

In order to separate the two beams quickly—needed to allow for many bunches—the beam-separating dipoles (labelled *B1* in Fig. 14) begin at only ±20 cm from the IP, with the first quadrupole (QD1) following at about 1.1 m from the IP. This makes for an extremely crowded IR; the price paid for head-on collisions. The magnets are realized as permanent magnets and are actually buried within the detector solenoid. A layout plot of the IR is shown in Fig. 15.

A major challenge for the B-Factories is the handling of the synchrotron radiation generated by the rather strong dipoles in the detector. The ragged outline of the vacuum chamber in Fig. 15 represents the masking required to shadow the detector from the radiation. These masks have to be able to withstand multi-

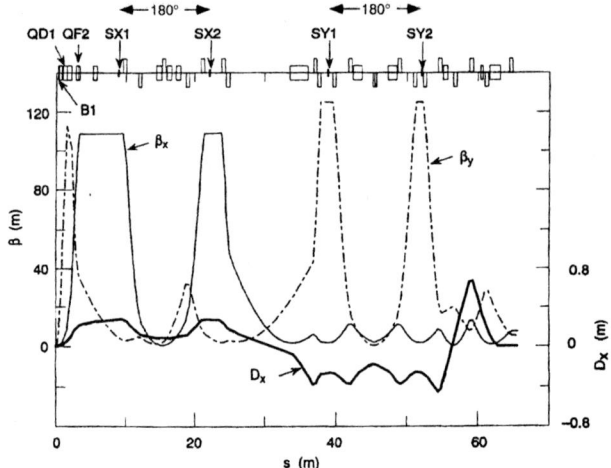

Fig. 4-27. Layout and optics functions of the right half of the IR straight section of the LER. The IP is at the left.

FIGURE 14. PEP-II LER IR lattice functions

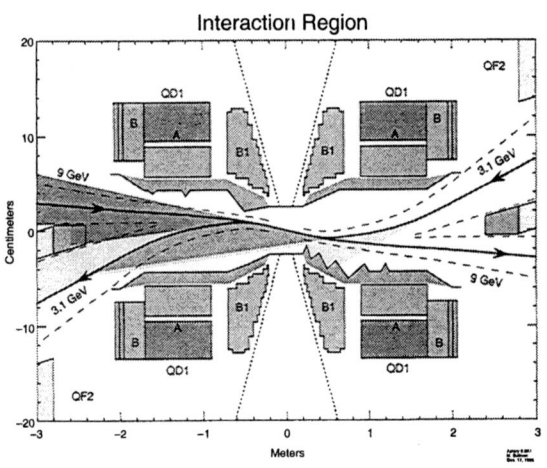

FIGURE 15. PEP-II IR layout

kW of synchrotron radiation power, most of it X-rays. The LER radiation "fans" are indicated as shaded bands, the darkness of which signifies the hardness of the radiation.

KEKB

The amount of synchrotron radiation going through the interaction region is significantly reduced in the design for the B-Factory at KEK [8], by allowing for a 22 mrad beam-crossing angle. This obviates the need for strong beam-separating dipoles in the detector, making room available for the insertion quadrupoles and avoiding the generation of strong synchrotron radiation at this place. This is reflected in the layout of the KEKB interaction region, Fig. 16. As for PEP-II, the

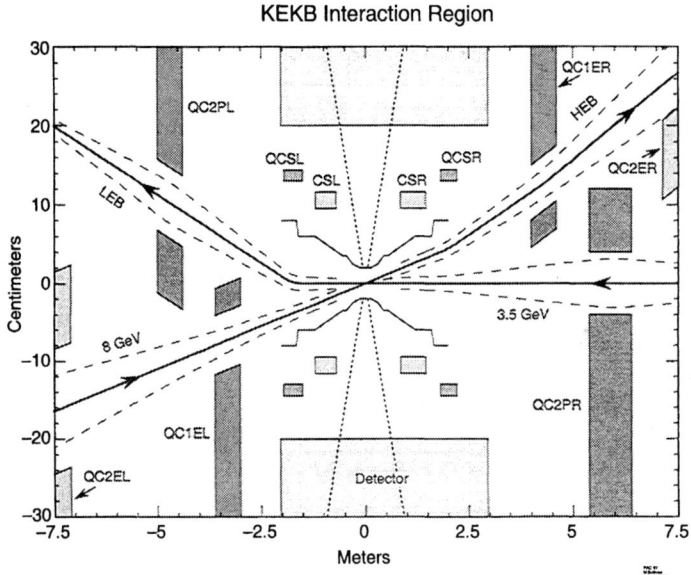

FIGURE 16. KEK B IR layout

insertion quadrupoles are located within the detector field but in this case use superconducting technology. The magnet lattice of the Low Energy Ring is shown in Fig. 17. The innermost quadrupole in the KEKB Low Energy Ring is not as close to the IP as in case of the PEP-II Low Energy Ring, therefore β peaks at about 400 m.

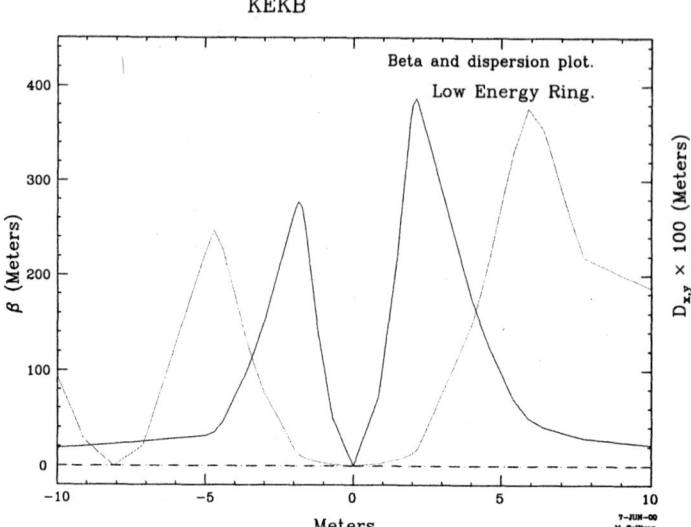

FIGURE 17. KEKB Low Energy Ring IR lattice

B Hadron Machines

SSC

The SCDR design of the low-β (0.5-m) lattice for the interaction region of the now-defunct Superconducting Super Collider, Fig. 18, shows a clear separation of the various functions. The dispersion suppressor at the end of the arc is followed by a 2.5-cell matching section, dipoles, a $-I$ section, more bending and the inner triplet. Because of the round beam at the IP, triplet focusing, as opposed to the doublets in the electron/positron collider examples, is used. There is 20 m of free space on either side of the IP. The $-I$ section changes the sign of the dispersion generated by the inner dipoles and thus matches it to the dispersion generated by the outer bends, for vanishing dispersion in the matching sections. The matching section is varied to vary β^* and thus set up the lattice for injection at a relatively high β^* of 8 m (Fig. 19).

LHC

The rather generous amount of space used for matching in the SSC is not available in the LHC. Its designers saved space by reducing its complexity and not matching dispersion to zero at the entrance to the arcs. Figures 20 and 21 show the high-luminosity insertion lattice of the LHC (V4.1) in its injection ($\beta^* = 6$ m) and its

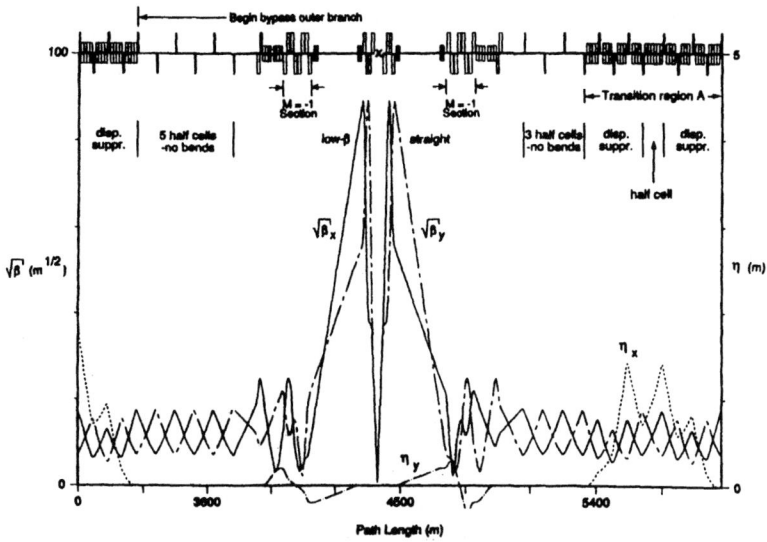

FIGURE 18. SSC low-β IR optics

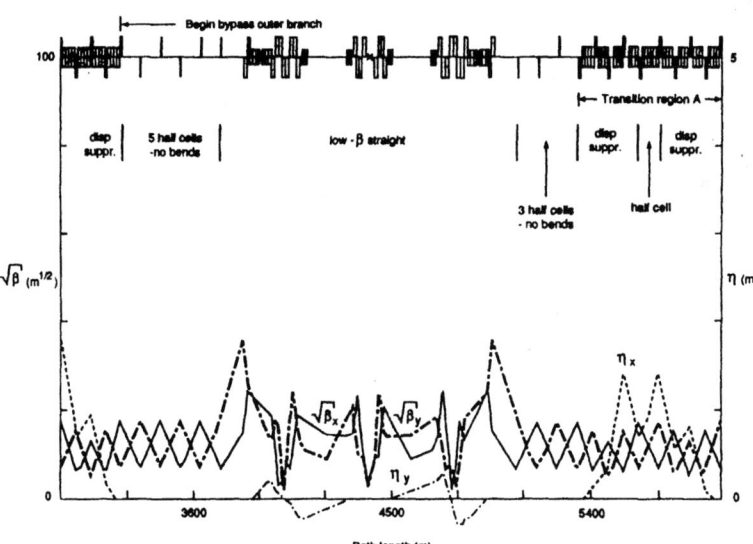

FIGURE 19. SSC low-β IR optics at injection

collision configuration ($\beta^* = 0.5$ m). Detector space is 23 m on either side. The inner triplet of this p-p collider is common to both beams, and the beam-separating dipoles are outside the triplet. Dispersion is matched to the dispersion of the arcs directly.

The β-squeeze from injection to colliding requires rather complicated, nonlinear tracking of the quadrupoles. Fig. 22 shows the program. The inner-triplet magnets (Q1...3) change relatively little while the outer-triplet quadrupoles, in particular Q4 and Q6, vary by a significant amount. A number of dispersion-suppressor quadrupoles is varied as well to ensure proper matching of the dispersion and β functions; however, dispersion at the IP varies, with significant values of η and η' during injection.

RHIC

Unlike the LHC and the SSC, RHIC has no common quadrupoles (Fig. 23); the beam-separating dipoles are closest to the IP. The beta function has a modest minimum of $\beta^* = 3$ m, therefore the extra distance of the inner quadrupole triplet from the IP does not lead to excessive β functions in the insertion quads (Fig. 24). Note that while $\eta = 0$ at the IP, η' is not.

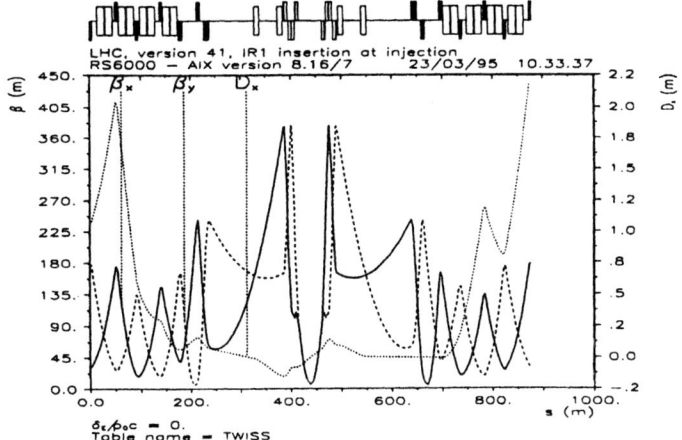

FIGURE 20. LHC IP1 and IP5 lattice, injection optics

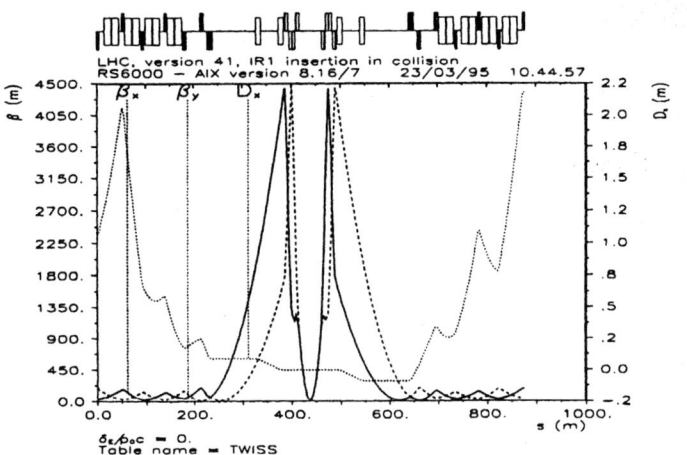

FIGURE 21. LHC IP1 and IP5 lattice, collision optics

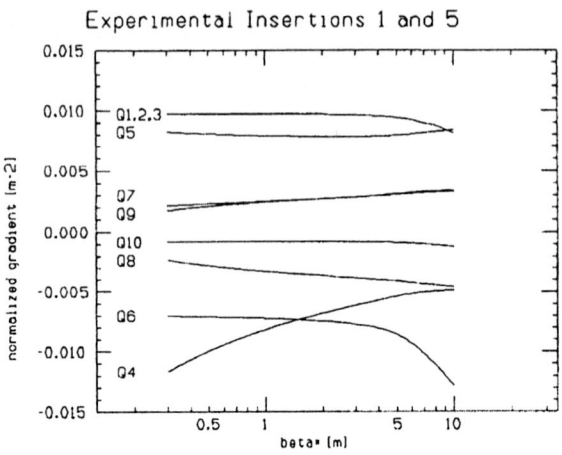

FIGURE 22. Quadrupole tracking for β squeeze, LHC IP1(5) lattice

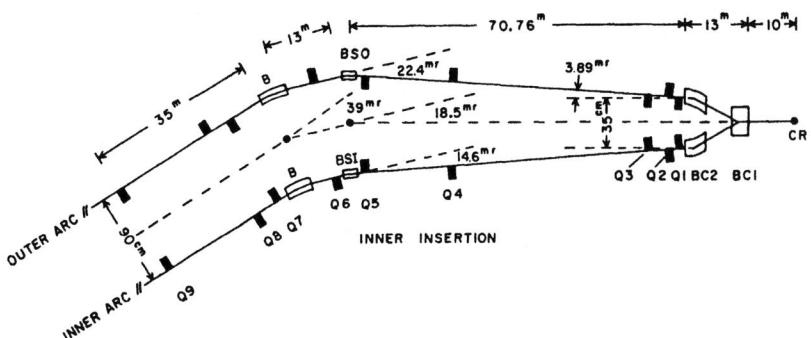

FIGURE 23. IR layout of RHIC

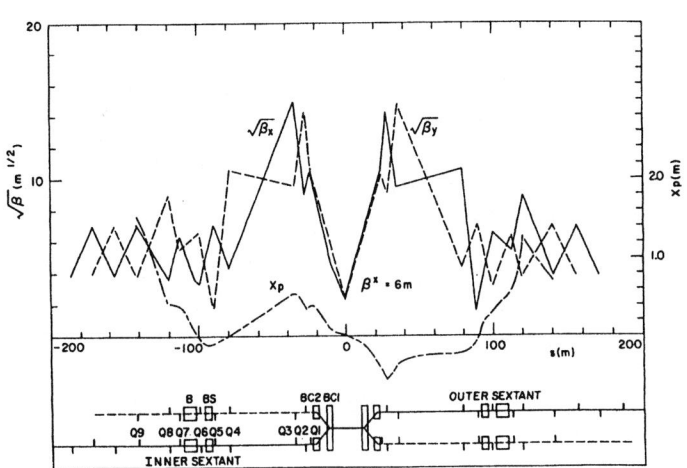

FIGURE 24. IR lattice of RHIC, $\beta^*=6$ m configuration

VI CHROMATICITY AND SOLENOID COMPENSATION

From an engineering point of view, magnet strengths and apertures limit the lowest value for β achievable for a given IR layout. However, besides these engineering limits, chromaticity and optical aberrations limit β^* in a more fundamental way.

A Chromaticity

The chromatic tune shift caused by a single quadrupole is obtained from

$$\delta\nu = \frac{1}{4\pi}\int_0^C \beta(s)\delta k(s)ds, \tag{38}$$

since $\delta k = k \times \delta p/p_0$ and therefore, for a thin quadrupole,

$$\frac{\delta\nu}{\delta p/p_0} = \frac{1}{4\pi}\beta k L \tag{39}$$

with kL its integrated focusing strength. As there are always two quadrupoles, the effective chromatic tune shift generated in the IR is at least twice that.

From Eq. (27) we know that β rises quadratically with distance from the IP and from Eq. (29) that k decreases approximately inversely to the distance, therefore the chromaticity generated in the IR quads rises proportionally to the distance. Detector requirements often dictate a certain amount of drift space to the IP, and one may end up with chromaticity too large to be handled in the lattice. One then applies local chromaticity correction.

The idea is to compensate chromaticity as close as possible to the IR quadrupoles, which requires that dispersion be present close to the IP for sextupoles to be placed there. The dispersion is provided by (sometimes beam-separating) dipoles close to the IP.

The dispersion generated by a parallel-faced dipole followed by a beam-line section with phase advance μ is

$$\eta = \frac{dx}{d\delta} = \sqrt{\frac{\beta_2}{\beta_1}}\rho(1-\cos\Theta)(\cos\mu + \alpha_1\sin\mu) + 2\sqrt{\beta_1\beta_2}\sin\mu\tan\frac{\Theta}{2}, \tag{40}$$

where Θ is the bending angle of the dipole and ρ its bending radius; $\beta_{1,2}, \alpha_{1,2}$ are the Twiss parameters at the beginning and end of the section. To first order in θ this becomes

$$\eta \approx 2\sqrt{\beta_1\beta_2}\Theta\sin\mu \tag{41}$$

and therefore one would like to place the sextupole about 90° away from the dipole. The sextupole will usually be compensated for geometric aberrations by a second sextupole 180° further out. A dipole *has* to be between this pair of sextupoles, else the dispersion will change sign and the chromatic correction will not work. The dispersion achieved is often small and requires fairly high β at the sextupoles to avoid excessive magnet strength.

Two such pairs will be needed if local compensation is required in both planes; these can be interleaved. The dispersion is either matched to 0 by additional dipoles or matched directly into the arcs.

Equation (38) also shows that the field-uniformity requirements for the insertion quadrupoles will be more stringent than for the regular lattice since amplitude-dependent tune shift is generally not tolerable. To make matters worse, the leading field harmonics in quadrupoles usually are 12- and 20-pole terms which generally add up for both quads since they are roughly 180° apart in phase advance.

B Solenoid Decoupling

A significant optical element in the IR will be the momentum-analyzing solenoid of the detector, with strengths often of several Tm. This solenoidal field causes significant coupling between the x and y planes as well as focusing at the edges, which has to be compensated for the machine to achieve its peak performance.

The most straightforward way to compensate the coupling is to include anti-solenoids of just the right strength to cancel the integrated solenoidal field from the outside to the IP and from the IP to the outside again. This compensation has the added benefit of being valid for spin motion as well, in the event that polarized beams are desirable.

A variant of the scheme, applicable if anti-solenoids cannot be placed between the optical elements and the IP, is to place the anti-solenoids further outside and rotate the optical elements closer in by the amount of twist the solenoid introduces. This works perfectly well as long as the solenoid field is fixed unless one is willing to rotate the insertion quadrupoles.

Skew quadrupoles will avoid this restriction. What kind of arrangement will achieve this? We inspect the R matrices for solenoids (ignoring focusing at the ends) and skew quads:

$$R_{sol} = \begin{pmatrix} \cos kL & 0 & \sin kL & 0 \\ 0 & \cos kL & 0 & \sin kL \\ -\sin kL & 0 & \cos kL & 0 \\ 0 & -\sin kL & 0 & \cos kL \end{pmatrix}, \quad (42)$$

$$R_{sq} = \begin{pmatrix} 1 & 0 & 0 & 0 \\ 0 & 1 & kL & 0 \\ 0 & 0 & 1 & 0 \\ kL & 0 & 0 & 1 \end{pmatrix}. \quad (43)$$

We can now introduce phase advance in x and y between the skew quad(s) and the solenoid to determine if we can find a skew-quad arrangement that will compensate the off-diagonal elements of the solenoid. This is done by "projecting" the skew quads to the solenoid location, in which case the matrices all add up and the eight off-diagonal elements shall vanish [14]. This projection is done by

$$R_{proj} = R_{bl} R_{sq} R_{bl}^{-1}. \tag{44}$$

To get a feeling for this we write down the projected matrices for skew quads located $\pi/2$ away in x, y and both x and y:

$$R_{\pi/2,0} = \begin{pmatrix} 1 & 0 & -kL & 0 \\ 0 & 1 & 0 & 0 \\ 0 & 0 & 1 & 0 \\ 0 & kL & 0 & 1 \end{pmatrix}, \quad R_{0,\pi/2} = \begin{pmatrix} 1 & 0 & 0 & 0 \\ 0 & 1 & 0 & kL \\ -kL & 0 & 1 & 0 \\ 0 & 0 & 0 & 1 \end{pmatrix} \quad \text{and}$$

$$R_{\pi/2,\pi/2} = \begin{pmatrix} 1 & 0 & 0 & -kL \\ 0 & 1 & 0 & 0 \\ 0 & -kL & 1 & 0 \\ 0 & 0 & 0 & 1 \end{pmatrix}. \tag{45}$$

Together with the skew quad at 0 (or π) away in both planes this appears to give full control of four off-diagonal elements. Peggs has shown in general that indeed solenoid compensation can be achieved with four skew quadrupole pairs (one up and one downstream each) by deriving the general set of equations

$$\sum_i \Theta_i S_i + \sum_i q_j Q_j = 0. \tag{46}$$

The matrix elements generated at the IP by the skew quads have been written down explicitly by Nosochkov et al. [15]:

$$R_{13} = q\sqrt{\frac{\beta_x^*}{\beta_y^*}} \sin\phi_x \cos\phi_y, \quad R_{14} = q\sqrt{\beta_x^* \beta_y^*} \sin\phi_x \sin\phi_y,$$

$$R_{23} = q\frac{\cos\phi_x \cos\phi_y}{\sqrt{\beta_x^* \beta_y^*}}, \quad R_{24} = q\sqrt{\frac{\beta_x^*}{\beta_y^*}} \cos\phi_x \sin\phi_y, \tag{47}$$

where $q = \sqrt{\beta_x \beta_y} kL$ is the skew-quadrupole strength and $\phi_{x,y}$ the phase advance from the IP in x or y. This shows that our choice of phases above has each skew-quad generating only one specific matrix element, a condition that will hold only approximately in a real machine. Since a "thin" solenoid ($kL \ll 1$) generates only R_{13} and R_{24} elements, one will expect the skew quads at $\pi/2$ phase angle in x or y to be much stronger than the ones at 0 or $\pi/2$ in both planes. For asymmetric interaction regions one expects the skew quadrupoles to be asymmetric.

The HER IR lattice shown in Fig. 13 demonstrates a practical implementation. There are six skew quadrupoles on either side of the IR. Of these, the outer two—located in a dispersive region—are actually a 180° pair used to control vertical dispersion. Therefore five separate skew-quad components are available for solenoid compensation. Five rather than three are used since it was not possible to place the skew-quads at the exact phases given above.

Part 2

VII DISPERSION SUPPRESSORS

A Introduction

The dispersion suppressor is an insertion between the arcs and other parts of a circular accelerator. It is used in most circular accelerators, especially those intended for high energy physics. Its main task is to cancel dispersion generated by arc dipoles. This is necessary in electron machines because non-zero dispersion at the locations of RF cavities leads to excitation of synchro-betatron resonances, which limit machine performance [17,18]. For colliders, zero dispersion greatly simplifies interaction region design. For other machines, straight sections with zero dispersion are useful as well. But "straightforward" cancellation of dispersion produces adverse effects (see Fig. 25). The beta functions lose their periodicity, which might be important. This is the case when nonlinear magnetic elements like sextupoles are located in dispersion suppressors (in addition to their location in arcs). More significant, beta functions have peaks which put strong limits on machine acceptance. Hence, the dispersion suppressor must be prepared in a special way to avoid adverse effects on machine optics.

Some circular accelerators don't have dispersion suppressors. This is the case, for example, for synchrotron radiation sources. The lattice of these accelerators consists of achromats (which provide cancellation of dispersion at their ends) and dispersion-free straight sections.

B Matching of Dispersion in a FODO Lattice

Building an accelerator lattice from FODO cells is popular because it has the advantages of simplicity and economy. The dispersion generated in regular FODO cells can be canceled in a simple way without disturbing other lattice functions [19–21].

Let a cell start and end at the mid-points of its quadrupoles. In the following calculations we assume that dipoles are located in the centers of half-cells and

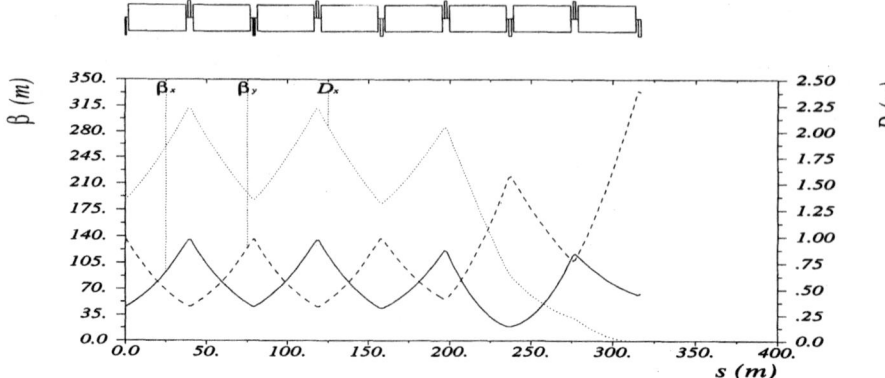

FIGURE 25. Example of a badly prepared dispersion suppressor.

neglect their weak focusing. The notation we use is ϕ_c for the phase advance per cell, $\Delta\phi$ for the phase advance between the symmetry point of a cell and the center of a magnet, l for the length of a magnet and ρ for its bending radius.

Now we apply a general formula for the dispersion and its derivative [22]

$$D(s) = S(s)\int_0^s \frac{C(z)}{\rho}dz - C(s)\int_0^s \frac{S(z)}{\rho}dz, \tag{48}$$

$$D'(s) = S'(s)\int_0^s \frac{C(z)}{\rho}dz - C'(s)\int_0^s \frac{S(z)}{\rho}dz, \tag{49}$$

where the principal trajectories $C(s)$ and $S(s)$ and their derivatives are the components of the transfer matrix

$$M = \begin{pmatrix} \sqrt{\beta_2/\beta_1}\,(\cos\Delta\phi + \alpha_1\sin\Delta\phi) & \sqrt{\beta_2\beta_1}\,\sin\Delta\phi \\ -\frac{(1-\alpha_1\alpha_2)\sin\Delta\phi - (\alpha_2-\alpha_1)\cos\Delta\phi}{\sqrt{\beta_2\beta_1}} & \sqrt{\beta_1/\beta_2}\,(\cos\Delta\phi - \alpha_2\sin\Delta\phi) \end{pmatrix} \tag{50}$$

for the given FODO cell.

To simplify integration in Eqs. (48) and (49), we take the values of the principal trajectories C and S at the mid-points of dipoles, since these functions are linear (in the first order). The transfer matrix between the mid-points of the dipole and the beginning of the FODO cell is

$$\begin{pmatrix} C & S \\ C' & S' \end{pmatrix} = \begin{pmatrix} \sqrt{\beta_m/\beta_c}\,(\cos(\phi_c/2 \pm \Delta\phi) & \sqrt{\beta_m\beta_c}\,\sin(\phi_c/2 \pm \Delta\phi) \\ -\sin(\phi_c/2 \pm \Delta\phi)/\sqrt{\beta_m\beta_c} & \sqrt{\beta_c/\beta_m}\,\cos(\phi_c/2 \pm \Delta\phi) \end{pmatrix}. \tag{51}$$

The transfer matrix for the FODO cell is

$$M_c = \begin{pmatrix} C & S \\ C' & S' \end{pmatrix} = \begin{pmatrix} \cos \phi_c & \beta_c \sin \phi_c \\ -\frac{1}{\beta_c} \sin \phi_c & \cos \phi_c \end{pmatrix}. \tag{52}$$

Substituting Eqs. (51) and (52) into (48) and (49) and using the formulae

$$\cos \alpha + \cos \beta = 2 \cos \frac{\alpha + \beta}{2} \cos \frac{\alpha - \beta}{2}, \quad \cos(\alpha \pm \beta) = \cos \alpha \cos \beta \mp \sin \alpha \sin \beta, \tag{53}$$

we find

$$D(s) = 2\delta_c \sqrt{\beta_m \beta_c} \cos \Delta\phi \sin \phi_c/2, \tag{54}$$

$$D'(s) = 2\delta_c \sqrt{\beta_m/\beta_c} \cos \Delta\phi \cos \phi_c/2, \tag{55}$$

where $\delta_c = l/\rho$. The periodic dispersion function for the FODO cell is found from the equation

$$M_c \begin{pmatrix} D_c \\ D' \\ 1 \end{pmatrix} = \begin{pmatrix} D_c \\ D' \\ 1 \end{pmatrix}. \tag{56}$$

Taking into account $D'_c = 0$ due to the symmetry, we find

$$D_c \cos \phi_c + 2\delta_c \sqrt{\beta_m \beta_c} \cos \Delta\phi \sin \phi_c/2 = D_c, \tag{57}$$

and

$$D_c = \delta_c \sqrt{\beta_m \beta_c} \frac{\cos \Delta\phi}{\sin \phi_c/2}. \tag{58}$$

Now we apply the same procedure to find the dispersion generated in the dispersion suppressor consisting of k FODO cells with dipoles followed by $n - k$ cells without dipoles (n is the total number of cells). The formulae (48) and (49) in this case are modified to give

$$D(s) = S(2nl) \int_0^{2nl} \frac{C(z)}{\rho} dz - C(2nl) \int_0^{2nl} \frac{S(z)}{\rho} dz, \tag{59}$$

$$D'(s) = S'(2nl) \int_0^{2nl} \frac{C(z)}{\rho} dz - C'(2nl) \int_0^{2nl} \frac{S(z)}{\rho} dz. \tag{60}$$

The principal trajectories starting at the beginning of the suppressor (the point with $D = D' = 0$) and ending at the mid-points of a cell with number j are

$$\begin{pmatrix} C & S \\ C' & S' \end{pmatrix} = \begin{pmatrix} \sqrt{\beta_m/\beta_c}\cos[(j-1/2)\phi_c \pm \Delta\phi] & \sqrt{\beta_m\beta_c}\sin[(j-1/2)\phi_c \pm \Delta\phi] \\ -\sin[(j-1/2)\phi_c \pm \Delta\phi]/\sqrt{\beta_m\beta_c} & \sqrt{\beta_c/\beta_m}\cos[(j-1/2)\phi_c \pm \Delta\phi] \end{pmatrix}. \tag{61}$$

For the principal trajectories from the beginning to the end of the suppressor (the point where an arc starts)

$$\begin{pmatrix} C & S \\ C' & S' \end{pmatrix} = \begin{pmatrix} \cos n\phi_c & \beta_c \sin n\phi_c \\ -\frac{1}{\beta_c}\sin n\phi_c & \cos n\phi_c \end{pmatrix}. \tag{62}$$

These expressions are similar to Eqs. (51) and (52). Substituting Eqs. 61) and (62) into (59) and (60), we find

$$D(2nl) = \delta_s\sqrt{\beta_m\beta_c}$$
$$\left[\sin n\phi_c \sum_{j=1}^{k}\{\cos[(j-1/2)\phi_c + \Delta\phi] + \cos[(j-1/2)\phi_c - \Delta\phi]\} - \right.$$
$$\left.\cos n\phi_c \sum_{j=1}^{k}\{\sin[(j-1/2)\phi_c + \Delta\phi] + \sin[(j-1/2)\phi_c - \Delta\phi]\}\right], \tag{63}$$

$$D'(2nl) = \delta_s\sqrt{\beta_m/\beta_c}$$
$$\left[\cos n\phi_c \sum_{i=j}^{k}\{\cos((j-1/2)\phi_c + \Delta\phi) + \cos((j-1/2)\phi_c - \Delta\phi)\} + \right.$$
$$\left.\sin n\phi_c \sum_{j=1}^{k}\{\sin((j-1/2)\phi_c + \Delta\phi) + \sin((j-1/2)\phi_c - \Delta\phi)\}\right]. \tag{64}$$

Using the formulae

$$\sin\alpha + \sin\beta = 2\sin\frac{\alpha+\beta}{2}\cos\frac{\alpha-\beta}{2}, \quad \cos\alpha + \cos\beta = 2\cos\frac{\alpha+\beta}{2}\cos\frac{\alpha-\beta}{2}, \tag{65}$$

we find

$$D(2nl) = 2\delta_s\sqrt{\beta_m\beta_c}\sum_{j=1}^{k}\sin(n-j+1/2)\phi_c, \tag{66}$$

$$D'(2nl) = 2\delta_s\sqrt{\beta_m/\beta_c}\sum_{j=1}^{k}\cos(n-j+1/2)\phi_c. \tag{67}$$

The series in Eqs. (66) and (67) can be calculated by applying the formula

$$\sum_{j=1}^{k} q^{j-1} = \frac{1-q^k}{1-q} \qquad (68)$$

to the series

$$\sum_{j=1}^{k} e^{i(n-k+1/2)\phi_c} = e^{i(n-1/2)\phi_c}\frac{1-e^{-ik\phi_c}}{1-e^{-i\phi_c}} = e^{i(n-k/2)\phi_c}\frac{e^{ik/2\phi_c}-e^{-ik/2\phi_c}}{e^{i/2\phi_c}-e^{-i/2\phi_c}}$$

$$= e^{i(n-k/2)\phi_c}\frac{\sin(k\phi_c/2)}{\sin(\phi_c/2)} \ . \qquad (69)$$

Using Eq. (69), we find formulae for the dispersion and its derivative produced by the dispersion suppressor:

$$D(2nl) = 2\delta_s\sqrt{\beta_m\beta_c}\ \cos\Delta\phi\ \frac{\sin(k\phi_c/2)\sin((n-k/2)\phi_c)}{\sin(\phi_c/2)}, \qquad (70)$$

$$D'(2nl) = 2\delta_s\sqrt{\beta_m/\beta_c}\ \cos\Delta\phi\ \frac{\sin(k\phi_c/2)\cos((n-k/2)\phi_c)}{\sin(\phi_c/2)}\ . \qquad (71)$$

The derivative of the dispersion is equal to zero if

a) $\sin(k\phi_c/2) = 0$, $k\phi_c/2 = \pi m$, $m = 1,2,3,\ldots$,
b) $\cos(n-k/2)\phi_c = 0$, $(n-k/2)\phi_c = \pi(m-1/2)$, $m = 1,2,3,\ldots$. $\qquad (72)$

The first equation in (72) can't help in determining the bending angle δ_s of the suppressor and we don't use it. Using the second equation in (72) and equating Eq. (70) to (58), we find

$$\delta_c = 2\delta_s(-1)^{m+1}\sin\ (k\phi_c/2)\ . \qquad (73)$$

Some of the solutions of Eqs. (72) and (73) are given in Table 1.

TABLE 1. Examples of the missing magnet dispersion suppressor.

n	k	m	ϕ_c	δ_s
2	1	1	60°	δ_c
3	1	1	36°	$\delta_c/(2\sin(\pi/5))$
3	2	1	45°	$\delta_c/\sqrt{2}$
3	2	2	135°	$\sqrt{2}\delta_c$
4	2	1	30°	δ_c
4	2	2	90°	$\delta_c/2$
4	1	2	77°	$-\delta_c/(2\sin(3\pi/14))$

A special solution exists for $k = n$ (no missing magnets). In this case Eq.(72) is modified to give

$$n\phi_c = (2m - 1), \quad m = 1, 2, 3, \ldots, \tag{74}$$

and Eq. (73) is simplified to

$$\delta_s = \delta_c/2. \tag{75}$$

This is the case of the half-field dispersion suppressor. Examples of these solutions are given in Table 2. The particular case of the half-field dispersion suppressor for 60° phase advance in arc is shown in Fig. 26.

TABLE 2. Examples of the half-field dispersion suppressor.

n	m	ϕ_c	δ_s
2	1	90°	$\delta_c/2$
3	1	60°	$\delta_c/2$
4	1	45°	$\delta_c/2$
5	2	108°	$\delta_c/2$

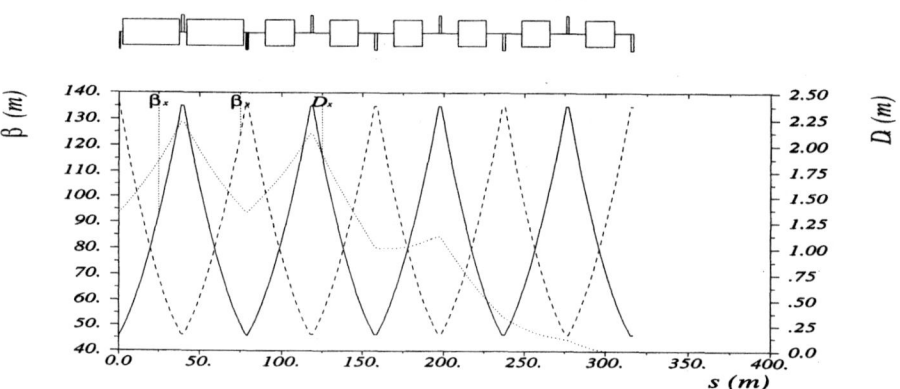

FIGURE 26. The lattice functions in a periodic cell and in a half-field dispersion suppressor with 60° phase advance per cell.

Both solutions are simple and practical. There have simple explanations. For the half-field magnet scheme, the dispersion in the arc is twice as big compared to the full-magnet scheme matched for the dispersion suppression. Hence, it is mismatched and oscillates with an amplitude equal to half of the arc dispersion. In half an oscillation period the dispersion reaches zero, and the dispersion suppressor must start at this point.

For the missing-magnet scheme, we find from Eq. (72b) that the phase advance between the mid-point of k cells generating dispersion and the end of the suppressor is an odd number of $\pi/2$. This means that the k cells produce a kick such that the dispersion reaches its maximum at the end of the suppressor.

The choice of dispersion suppressor for a particular machine depends on general optics design, first of all on the phase advance in an arc cell. The free space in the missing-magnet scheme or in the half-field scheme can be used for other purposes, for example for a wiggler to control machine parameters such as damping time, emittance, bunch length, *etc.*

C More Flexible Dispersion Suppressor: LEP Case

The dispersion suppressor based on the missing-magnet scheme, or the half-field magnet scheme, is most often used. However, it is not flexible. If a machine is designed to operate at different energies, the possibility of working with different phase advances in the arc cell is in strong demand. In this case a more flexible dispersion suppressor can be used.

In the LEP dispersion suppressor [23,24] reduction of the dispersion is achieved in the half cell with a bending magnet angle 1/3 of that in the arc. In half an oscillation period the dispersion reaches zero (see Figs. 27 and 28). The cells in the dispersion suppressor are shorter than the arc cells. This keeps the beta functions in the suppressor smaller than in the arc, thus avoiding an increase of aperture for the optics with 60° phase advance per arc cell. For the optics with 90° phase advance in the arc cells the beta functions are bigger, but this doesn't impose aperture limitation because the beam emittance for high energy electron machines like LEP scales roughly as $\epsilon_x \propto \mu^{-3}$ with phase advance. This dispersion suppressor is so flexible that it is possible to match to the interaction region an arc with 135° phase advance.

The quadrupole strengths in the suppressor are no longer equal to those in the arcs. But eight independent quadrupoles in the dispersion suppressor allow controlling $\beta_x, \beta_y, \alpha_x, \alpha_y$ at its end so the dispersion suppressor can be matched to other insertions; matching D_x, D'_x to zero as well as adjusting the machine tunes Q_x and Q_y over a reasonably wide range.

VIII INJECTION

A Introduction

Injection schemes vary greatly depending on a machine's purpose, its optics, layout, injection energy etc. [21,25]. But all of them are based on the same scheme

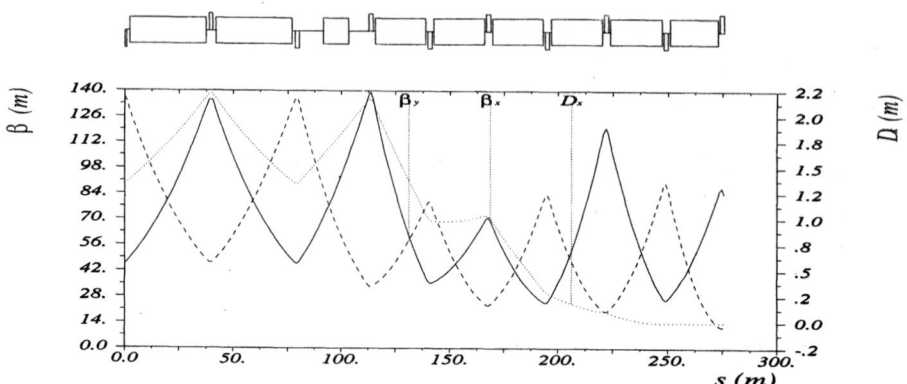

FIGURE 27. The lattice functions in one periodic cell with 60° phase advance and in the dispersion suppressor of LEP.

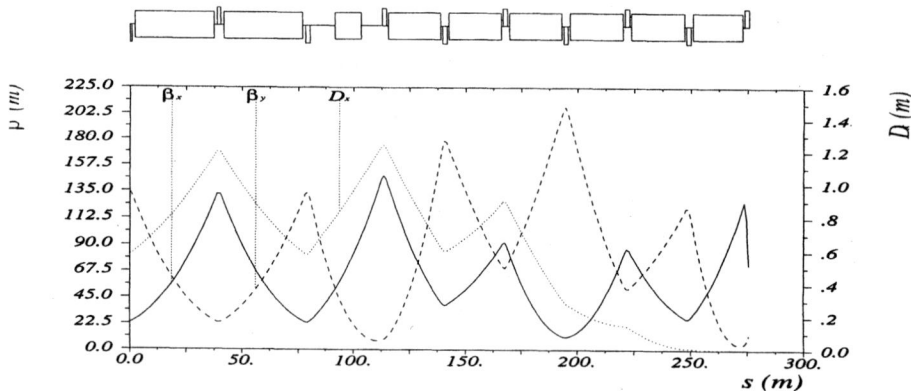

FIGURE 28. The lattice functions in one periodic cell with 90° phase advance and in the dispersion suppressor of LEP.

of optics, which includes a septum magnet and fast kicker(s) or bumper magnet(s) (Fig. 29). A septum is a dipole magnet with a thin plate that isolates the septum's field from the aperture of the circulating beam. When a beam is injected, particles come from a transfer line to a septum and then into the machine. A fast kicker or bumper magnet is used to deflect particles from their injection trajectory onto the central orbit. Below we consider a few typical injection sections.

B Single-Turn Injection

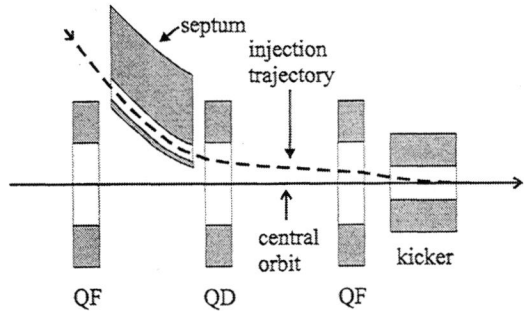

FIGURE 29. Schematic view of a single turn injection.

To clarify the requirements for septum and kicker magnets, we'll calculate their strengths for a given orbit offset at the septum position. The transfer matrix between two points in a machine is given by formula (50). To find the required kicker and septum angles $x'_k = \delta_k$ and x'_s, we apply it to track from septum to kicker with offsets at septum position x_s and at kicker position $x_k=0$:

$$\sqrt{\beta_k/\beta_s}\,(\cos \Delta\phi + \alpha_s \sin \Delta\phi)\,x_s + \sqrt{\beta_k\beta_s}\,\sin \Delta\phi\, x'_s = 0, \tag{76}$$

$$-\frac{1}{\sqrt{\beta_k\beta_s}}[(1 - \alpha_k\alpha_s)\,\sin \Delta\phi - (\alpha_k - \alpha_s)\,\cos \Delta\phi]\,x_s +$$
$$\sqrt{\beta_s/\beta_k}\,(\cos \Delta\phi - \alpha_s \sin \Delta\phi)\,x'_s - \delta_k = 0. \tag{77}$$

Using det $M_{1,2}=1$, we find

$$x_s = -(\delta_k\sqrt{\beta_k})\sqrt{\beta_s}\,\sin \Delta\phi, \tag{78}$$

$$x'_s = (\delta_k\sqrt{\beta_k})/\sqrt{\beta_s}(\cos \Delta\phi + \alpha_s \sin \Delta\phi), \tag{79}$$

hence

$$\delta_k = -\frac{x_s}{\sqrt{\beta_k\beta_s}\,\sin \Delta\phi} \tag{80}$$

$$x'_s = -\frac{x_s}{\beta_s}(\cot \Delta\phi + \alpha_s). \tag{81}$$

The offset x_s between the injected beam and the circulating beam must be large enough to give room for both beams and the septum thickness. To reduce the requirements for the kicker strength, a phase advance between septum and kicker close to $\Delta\phi = 90°$ is desirable, as well as large β_k and β_s. At the same time a large β_k requires a larger aperture for the kicker. Large β_s reduces the septum strength, but this can produce interferences if the next magnet element in the machine upstream to the septum is physically too close to the transfer line. We conclude that the choice of injection system parameters is made for each machine individually, taking into account all constraints and possible solutions.

C Multiturn Injection

Multiturn-injection optics is similar to that for single turn injection, but the circulating beam is moved close to a septum by using a local orbit bump (see Fig. 30). A bump can be made by using fast kickers (electron machines) or bumper magnets (proton and other heavy particle machines) and allows reduction of the oscillation of an injected beam.

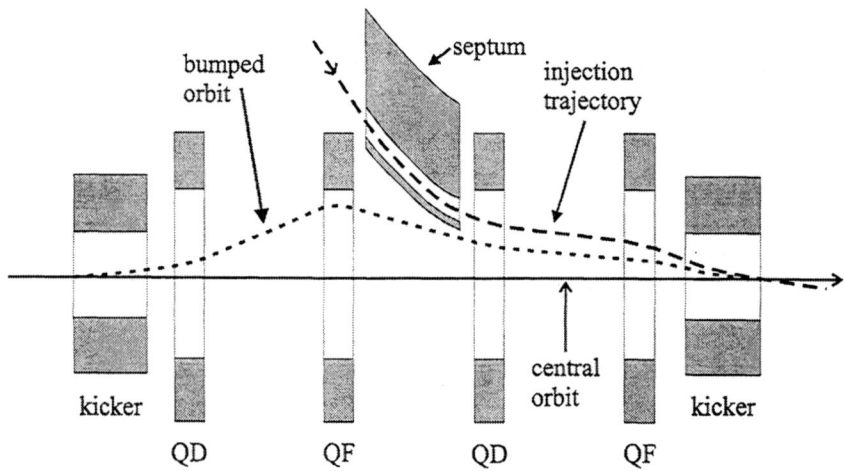

FIGURE 30. Schematic view of a multiturn betatron injection.

To control orbit offset and angle at the septum position, as well as to keep the bump local, in general four kickers (bumpers) are necessary. Under certain conditions a local bump can be implemented using two magnets only.
Exercise 1.
Using formula (50), find conditions for a two-magnet local bump [21,26].
Solution.

For beam trajectories at points 1 and 2 the offsets and their derivatives are $x_1 = 0, x'_1 = \delta_1, x_2 = 0, x'_2 = 0$. We find from formula (50)

$$\sqrt{\beta_2 \beta_1} \sin \Delta\phi = 0, \tag{82}$$

$$\sqrt{\beta_1/\beta_2} \left(\cos \Delta\phi - \alpha_2 \sin \Delta\phi\right) \delta_1 + \delta_2 = 0, \tag{83}$$

hence

$$\sin \Delta\phi = 0, \quad \Delta\phi = \pi(2m - 1), \quad m = 1, 2, 3, ..., \tag{84}$$

$$\delta_2 = (-1)^{m+1} \sqrt{\beta_1/\beta_2} \, \delta_1. \tag{85}$$

From formula (85) we see that two kickers (bumpers) can produce a local orbit bump only if the phase advance between them is a multiple of π. This can be restrictive for injection, and in practice three kickers (or four kickers for a symmetric layout) are used to produce the bump.

Exercise 2.

Find condition for a three-magnet local bump [21,26]. To simplify the calculations, track the beam trajectory forward from point 1 to point 2, and back from point 3 to point 2.

Solution.

Forward tracking from point 1 to point 2 yields

$$x_{2,f} = \sqrt{\beta_1 \beta_2} \sin \Delta\phi_{21} \, \delta_1, \tag{86}$$

$$x'_{2,f} = \sqrt{\beta_1/\beta_2} \left(\cos \Delta\phi_{21} - \alpha_2 \sin \Delta\phi_{21}\right) \delta_1, \tag{87}$$

and back-tracking from point 3 to point 2 yields (here we take into account $\Delta\phi_{23} = -\Delta\phi_{32}, x'_3 = -\delta_3$)

$$x_{2,b} = \sqrt{\beta_3 \beta_2} \sin \Delta\phi_{32} \, \delta_3, \tag{88}$$

$$x'_{2,b} = \sqrt{\beta_3/\beta_2}(\cos \Delta\phi_{32} - (-\alpha_2) \sin \Delta\phi_{32}) \, \delta_3. \tag{89}$$

We find from Eqs. (87) and (89)

$$\sqrt{\beta_1} \delta_1 \sin \Delta\phi_{21} = \sqrt{\beta_3} \delta_3 \sin \Delta\phi_{32}. \tag{90}$$

Kicks at point 2 must be matched to fulfill

$$x'_{2,f} + \delta_2 = -x'_{2,b}. \tag{91}$$

Substituting Eqs. (87) and (89) into (91), we find

$$\sqrt{\beta_2}\delta_2 = -\sqrt{\beta_1}\delta_1 \cos \Delta\phi_{21} - \sqrt{\beta_3}\delta_3 \cos \Delta\phi_{32}. \tag{92}$$

Inspection of Eqs. (90) and (92) shows that $\sqrt{\beta}\delta$ are the sides of a triangle, for which the well known solution is

$$\frac{\sqrt{\beta_1}\Delta_1}{\sin \Delta\phi_{32}} = \frac{\sqrt{\beta_2}\Delta_2}{\sin \Delta\phi_{13}} = \frac{\sqrt{\beta_3}\Delta_3}{\sin \Delta\phi_{21}}. \tag{93}$$

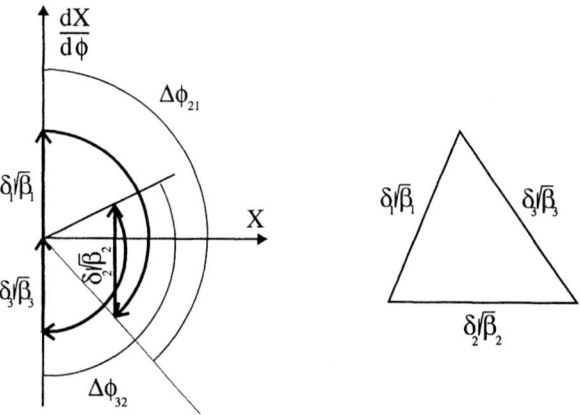

FIGURE 31. Three-magnet local bump in normalized coordinates.

The clear way to see how the three-magnet local bump works is shown in Fig. 31, where the normalized coordinates for transverse motion are

$$X = x(s)/\sqrt{\beta} = A\cos(\phi + B), \tag{94}$$

$$\phi(s) = \int_{s_0}^{s} dz/\beta(z), \quad dX/d\phi = -A\sin(\phi + B). \tag{95}$$

The three-magnet local bump is often used for aperture scans.

Multiturn injection is different for electron and proton machines. For an electron beam radiation damping is employed to accumulate a beam. An injected beam reaches its equilibrium size [27] in a few damping times due to synchrotron radiation. Then new particles can be injected. The procedure is repeated until the accumulated beam reaches its intensity limit. The requirement for high injection

efficiency imposes strong constraints on the dynamic aperture of a machine, the area in phase space where the beam survives for a long time [28,29].
Exercise 3.
Estimate the dynamic aperture required for efficient injection, for an electron machine with parameters given in Table 3. The horizontal beta function at the septum position is 14 m and the septum thickness is 5 mm. Assume for simplicity that the beta function of the transfer line at the septum position is the same as in the machine. Keep in mind that in electron machines natural emittance scales $\propto E^2$.

TABLE 3. Parameters of electron machine.

Full energy, GeV	E	1.2
Injection energy, GeV	E_{inj}	0.8
Natural emittance, nm	ϵ_x	11.1
Emittance of injected beam	ϵ_x^i	18.3
Accumulated current, mA	I	300
Injected cirrent, mA	I_i	0.8
Damping time at 0.8 GeV, msec	τ_x	230

Solution.
For multiturn injection the accumulated beam is moved toward the septum to make injection more efficient. To avoid losses in a circulating beam, the distance between the bumped central trajectory and the septum must be more than a few σ_x. The same rule holds for the distance between the center of injected beam and the septum (see Fig. 32). After injection the new bunches perform oscillations around the central orbit until they damp. For the efficient injection the dynamic aperture must be bigger than

$$x_i \geq n\sigma_c + 2m\sigma_i + d. \tag{96}$$

Typically, $n = 5$ and $m = 4$ are used. For a septum width of 5 mm we find

$$x_i = 5\frac{0.8}{1.2}\sqrt{11.1 \cdot 10^{-9} \cdot 14} + 8\sqrt{18.3 \cdot 10^{-9} \cdot 14} + 0.005 = 10.36 \cdot 10^{-3} \text{ m}. \tag{97}$$

When expressed in the number of standard deviations, N_σ, with σ_x=0.263 mm at injection energy 0.8 GeV, the requirement is N_σ=10.36/0.263=39. This is a tight requirement for the optics design.
Exercise 4.
Do the same for an injection energy equal to the full energy of the machine, 1.2 GeV (in this case the accelerator is called a storage ring). Compare the results.
Solution.
When the beam is accelerated in a linac, its emittance scales inversely with momentum. Repeating the same calculations, we find

$$x_i = 5\sqrt{11.1 \cdot 10^{-9} \cdot 14} + 8\sqrt{12.2 \cdot 10^{-9} \cdot 14} + 0.005 = 10.28 \cdot 10^{-3} \text{ m}. \tag{98}$$

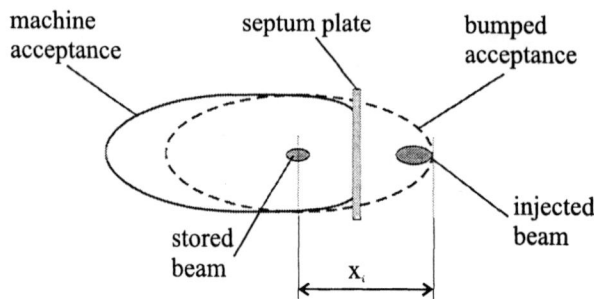

FIGURE 32. Dynamic aperture required for efficient injection.

For a beam size of 0.394 mm this corresponds to $N_\sigma=26$. We conclude that injection at the full energy of a machine provides relaxed requirements to its dynamic aperture, hence better accumulation efficiency. In addition, with weaker requirements for the dynamic aperture we have more flexibility to optimize other machine parameters.

For a proton (or heavy-ion) machine synchrotron radiation is negligible. Nevertheless multiturn injection can be used if the machine acceptance is significantly bigger than the emittance of the injected beam. A schematic view of beam accumulation for this case is shown in Fig. 33 [30].

At the beginning a beam stack is moved close to a septum by creating a closed orbit bump. The value of the closed orbit displacement is linearly reduced to bring the stack back to the center of a vacuum chamber after 10 turns. This can be done by reducing the currents in the bumper magnets. These magnets are slower than the fast kickers and allow changing the orbit typically within 10 to 50 turns. Every injected bunch rotates around the stack in the horizontal phase space (or in the vertical one, if injection is made in the vertical plane). The distance between the injected bunch and the stack increases linearly with the number of bunches.

Exercise 5.
Estimate the non-integer part of the horizontal betatron tune for the multiturn injection scheme shown in Fig. 33. Why is the filling of the available acceptance poor? How can it be improved?

Solution.
A short inspection of the bunch rotation in the horizontal phase space shows that the non-integer part of the horizontal tune is a bit less than 1/3 (exact value is 0.3). To improve filling, one has to use a thin septum of width d_s satisfying the condition

$$d_s \ll \sqrt{\epsilon_x \beta_x}. \tag{99}$$

Also it is worthwhile to partly scrape the bunch by the septum, providing denser packaging of the bunches in the available acceptance, and to use a horizontal beta

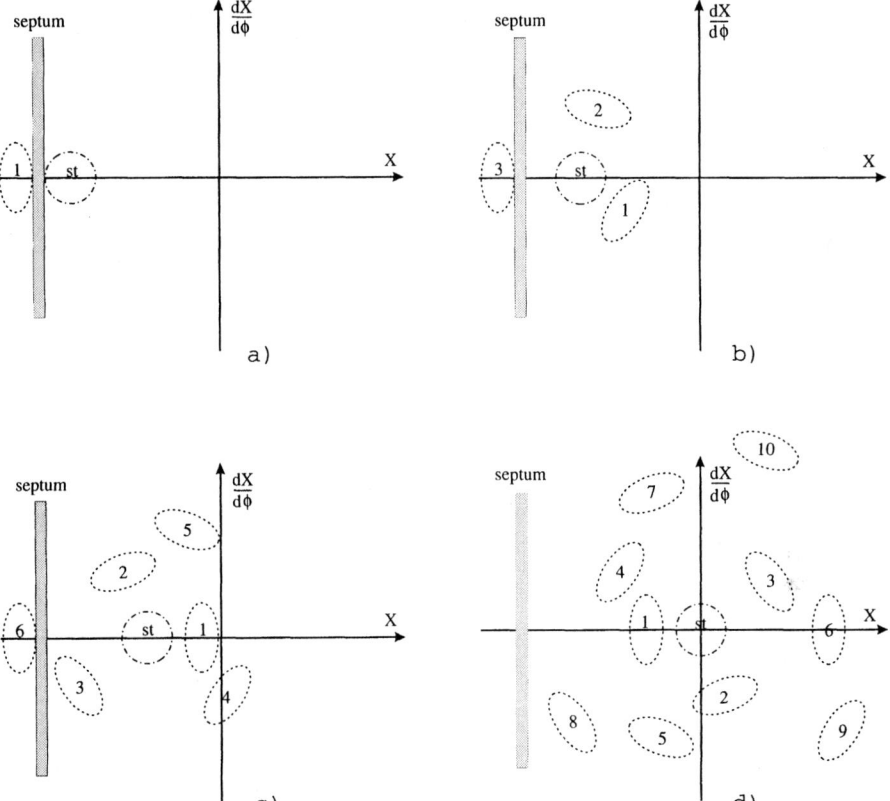

FIGURE 33. Multiturn injection into the horizontal phase space. The injection is shown in the moments of the 1st (a), 3rd (b), 6th (c) and 10th (d) turns. The closed orbit is in the center of the stack and moves together with it to the center of the vacuum chamber when the orbit bump decreases. In the example given it is assumed that the available acceptance is filled in 10 turns.

function in the injection line that is smaller than the one in the machine at the septum position [31]. To avoid losses during injection in betatron phase space, the dispersion at the septum position must be small:

$$D_x \frac{\Delta p}{p} << \sqrt{\epsilon_x \beta_x}. \tag{100}$$

The vertical phase space can also be simultaneously employed for stacking, but the tunes must then be close to the coupling difference resonance [31,32].

D Injection in Synchrotron Phase Space

The injection scheme for this case can include the same elements, but the dispersion function D_x must be non-zero (see Fig. 34). The beam is injected with an energy offset

$$\frac{\Delta p}{p} = \frac{x}{D_x} \qquad (101)$$

and after passing an injection septum and the fast kicker (or bumper) it moves parallel to the circulating beam.

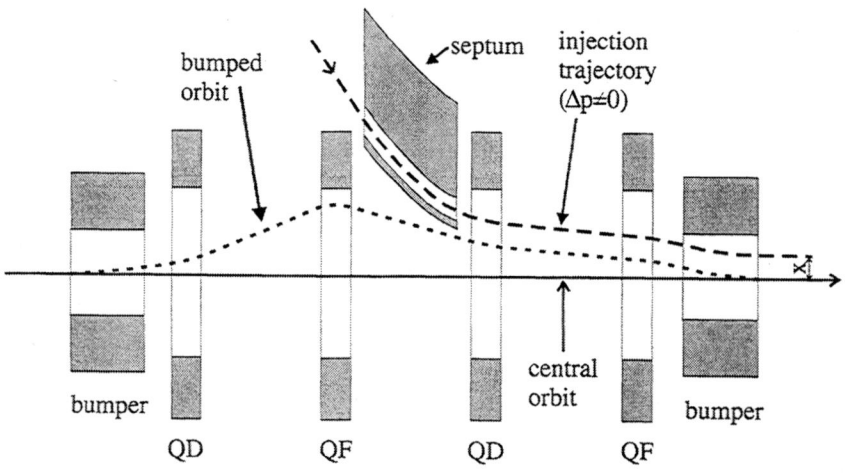

FIGURE 34. Schematic view of a synchrotron injection.

In electron machines the beam oscillates in synchrotron phase space with the synchrotron tune Q_s and damps to the circulating beam. The synchrotron damping time for machines without combined-function magnets is half of the betatron damping time. This provides an opportunity for faster accumulation compared to the betatron injection. Another advantage of synchrotron injection is that the trajectory of the injected beam in the straight sections is flat, providing cleaner radiation conditions for insertions. This is especially important for colliders [33]. The choice of a particular injection scheme depends on machine layout, availability of a section with reasonable dispersion for locating injection elements, acceptance of the transfer line, large momentum acceptance of the machine, *etc.*

For proton and heavy ion machines without radiation damping synchrotron injection still can be used if the momentum spread of an injected beam is significantly smaller than the machine momentum acceptance. An example of multiturn injection into the synchrotron phase space is shown in Fig. 35.

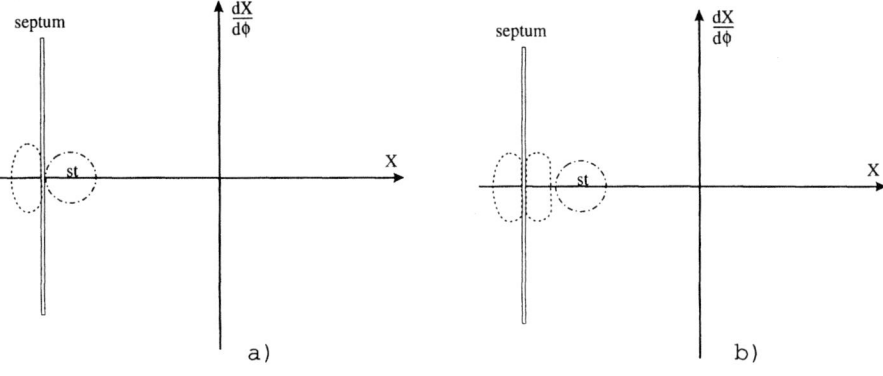

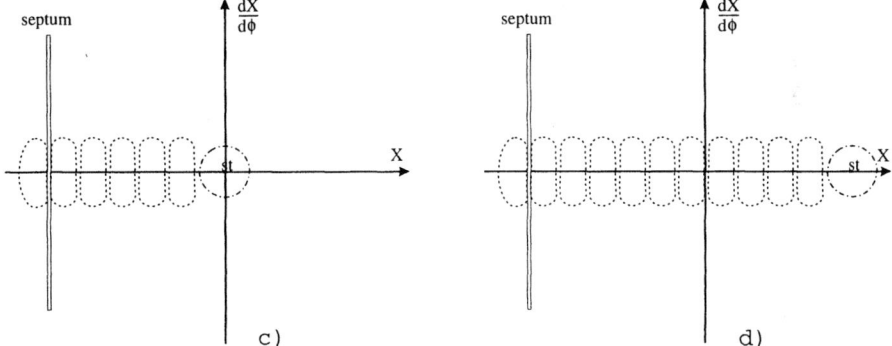

FIGURE 35. Multiturn injection into synchrotron phase space. The injection is shown in the moments of the 1st (a), 2nd (b), 6th (c) and 10th (d) turn. The closed orbit is in the center of the stack at the beginning and moves together with it to the inner side of the vacuum chamber because of the linear decrease in bending field (injection line is located outside the ring). In the given example it is assumed that the available longitudinal acceptance is filled in 10 turns.

At the beginning the central orbit (with the stack on it) is close to the septum (Fig. 35a). Then the orbit is displaced to the inside of the machine by reducing the magnetic field in the bump magnets (Fig. 35b,c,d). The momentum of the injected beam is increased, properly maintaining the orbit of the injected beam in the center of the septum. In contrast to the previous case, the dispersive beam size must be much bigger than the betatron size at the septum position: $D_x \Delta p/p \gg \sqrt{\epsilon_x \beta_x}$.

When the momentum acceptance of the machine is full, cooling can be applied to put all the bunches into one stack again. Note that in Fig. 35 the phase space is filled more densely than in Fig. 33. This is because a part of the injected bunch is scraped by the septum and the septum itself is thinner. The choice between multiturn injection into transverse or longitudinal space depends on the parameters of the injected beam, the acceptance of the machine in the three planes, the Twiss functions at the septum position and possibility for modifying them, machine tunes, etc. [34]. One also has to keep in mind beam cooling times for transverse and longitudinal phase space.

IX INSERTIONS FOR WIGGLERS AND UNDULATORS

A Introduction

Synchrotron radiation sources (SRS) occupy a wide area in the field of accelerators. They are developing rapidly, finding use in a different fields of physics as well as biology, chemistry, material science etc. [35–38]. Depending on the goal, SRS differ in energy, size and devices for synchrotron radiation generation. The photon beams delivered to users vary in terms of their critical wavelength, photon flux, brightness etc. [39,40]. The synchrotron radiation is produced by the bending magnets of the accelerator and by special magnets called insertion devices. These are wigglers and undulators [41–45]. In a simplified way we can consider an insertion device as a sequence of equidistant bending magnets with alternating polarities (Fig. 36).

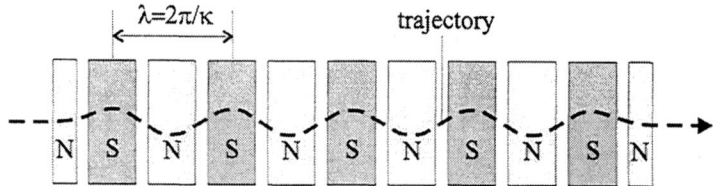

FIGURE 36. Schematic view of a wiggler (undulator).

B Central Orbit in a Wiggler/Undulator Magnet

The expressions for the magnetic field in a wiggler (undulator) are

$$B_y = B_0 \cosh k_x x \cosh k_y y \cos kz,$$
$$B_x = \frac{k_x}{k_y} B_0 \sinh k_x x \sinh k_y y \cos kz,$$
$$B_z = -\frac{k}{k_y} B_0 \cosh k_x x \sinh k_y y \sin kz, \tag{102}$$

where
$$k_x^2 + k_y^2 = k^2 = (2\pi/L)^2 \tag{103}$$

and L is the cell length. Formulae (102) keep only the first harmonics of the field variation along the z-axis and correspond to closely spaced magnets and fulfill Maxwell's equations. The parameter k_x measures the transverse variation of the field due to the limited pole width or curved pole faces.

The equation of motion in the horizontal plane for a particle with momentum p, charge e and mass m is

$$\frac{dp}{dt} = -e\frac{dz}{dt}B_y. \tag{104}$$

Substituting $p = m\gamma dx/dt, s = \beta ct$ and using

$$\frac{p}{e} = B\rho, \tag{105}$$

we find (with cosh $k_x x \simeq 1$)

$$\frac{d^2x}{ds^2} = -\frac{\cos\ kz}{\rho}\frac{dz}{ds}, \tag{106}$$

where $\rho = B\rho/B_0$ is the bending radius in the wiggler magnet. The first integral of Eq. (106) is

$$\theta = \frac{dx}{ds} = -\frac{1}{k\rho}\sin\ kz. \tag{107}$$

The maximum deflection angle of the orbit is

$$\theta_{max} = \frac{dx}{ds} = \frac{1}{k\rho} = K/\gamma, \tag{108}$$

where the K parameter is given by

$$K = \frac{L}{2\pi}\frac{eB}{mc}. \tag{109}$$

Depending on the value of K, the insertion device is called

$$\text{undulator if } K \leq 1, \ (\theta_{max} \leq 1/\gamma), \tag{110}$$

$$\text{wiggler if } K \geq 1, \ (\theta_{max} \geq 1/\gamma). \tag{111}$$

For 1 to 2 GeV machines, K/γ is typically 10^{-3} to 10^{-2} with a maximum of a few 10^{-2} for the strongest wigglers. Examples of insertions devices in operating SRS are given in Ref. [46].

The second integration of Eq, (107) can be done exactly by using the relation

$$\frac{dz^2}{ds} = 1 - \frac{dx^2}{ds} = 1 - \frac{\sin^2 kz}{k^2 \rho^2}, \qquad (112)$$

but with good accuracy

$$x = \frac{\cos kz}{k^2 \rho}, \qquad (113)$$

where we used $s \simeq z$.

C Linear Motion around the Central Orbit

Once the central orbit in an insertion device is known, the equations of motion are derived in the new coordinate system $(\tilde{x}, y, s)$, where s is the distance along the central orbit, $\tilde{x}$ is the displacement in the plane (x, z) perpendicular to the central orbit, and the vertical coordinate is the same (Fig. 37).

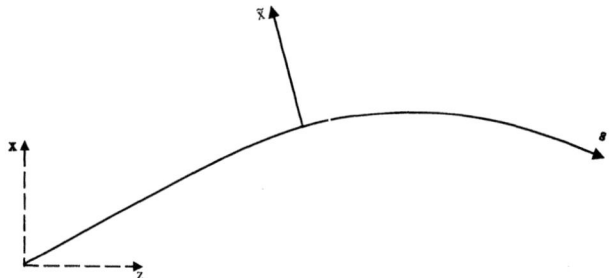

FIGURE 37. The coordinate system on the central orbit.

The old coordinates are related to the new ones by

$$z = z_{c.o.} - x'_{c.o.} \tilde{x}, \qquad x = x_{c.o.} + z'_{c.o.} \tilde{x}, \qquad (114)$$

where the central orbit coordinates $x_{c.o.}$ and $z_{c.o.}$ are defined by formulae (107) and (113). The linear equations of motion can be derived by expanding the fields (102) to first order and averaging over the length of the insertion. They are

$$x'' = -\frac{1}{2\rho^2}\frac{k_x^2}{k^2} x, \qquad y'' = -\frac{1}{2\rho^2}\frac{k_y^2}{k^2} y. \qquad (115)$$

The use of the Hamiltonian formalism allows us to deduce equations of motion keeping non-linear terms which are responsible for the strong influence of insertion devices on beam dynamics [47]. For the typical case of a planar pole face $k_x = 0$

and $k_y = k$, thus a wiggler produces a focusing effect in the vertical plane but doesn't affect the horizontal motion. From Eqs. (115) we find that the focusing of the insertion device is proportional to B^2/E^2 and is much stronger for high-field wigglers in machines with low energy.

D Effect of Insertion Devices on Linear Optics

Perturbation theory is used to estimate the tune shift and beta-function distortion produced by the wiggler (undulator).

Exercise 1.
Suppose a gradient error occurs at position $s = s_1$. Using the transfer matrix of an unperturbed machine (50) and the transfer matrix of a quadrupole kick

$$m = \begin{pmatrix} 1 & 0 \\ -\Delta K ds & 1 \end{pmatrix}, \qquad (116)$$

find the tune shift produced by insertion device.
Solution.
For the one-turn matrix $\beta_2 = \beta_1 = \beta$ and $\Delta\phi = 2\pi Q_0$. The transfer matrix of the perturbed machine is the product Mm of two matrices

$$\begin{pmatrix} \cos 2\pi Q_0 + \alpha \sin 2\pi Q_0 & \beta \sin 2\pi Q_0 \\ -(1-\alpha^2)\sin 2\pi Q_0 & \cos 2\pi Q_0 - \alpha \sin 2\pi Q_0 \end{pmatrix} \begin{pmatrix} 1 & 0 \\ -\Delta K ds & 1 \end{pmatrix} =$$

$$\begin{pmatrix} \cos 2\pi Q_0 + \alpha \sin 2\pi Q_0 + \beta \Delta K ds \sin 2\pi Q_0 & \beta \sin 2\pi Q_0 \\ -(1-\alpha^2)\sin 2\pi Q_0 - \Delta K ds(\cos 2\pi Q_0 - \alpha \sin 2\pi Q_0) & \cos 2\pi Q_0 - \alpha \sin 2\pi Q_0 \end{pmatrix}. \quad (117)$$

Now the tune of the perturbed machine can be found from the formula

$$\cos 2\pi Q_0 = \frac{1}{2}\mathrm{Tr} Mm = \cos 2\pi Q_0 - \frac{1}{2}\beta \Delta K ds_1 \sin 2\pi Q_0. \qquad (118)$$

If the tune shift is small, then

$$\Delta Q = \frac{\beta \Delta K ds}{4\pi}, \qquad (119)$$

where ds is the length of the perturbation. Applying formula (119) to an insertion device with planar pole faces, we find

$$\Delta Q = \frac{\bar{\beta}_y L}{8\pi \rho^2} = \frac{L}{8\pi \rho^2}\left(\beta_y + \frac{L^2}{12\beta_y}\right). \qquad (120)$$

For the case $k_x \neq 0$ the tune shift produced by the insertion device is non-zero in both planes. Formula (120) shows the way to minimize them: one has to choose beta functions at the position of the insertion device equal to

$$\beta_{x,y} = \frac{k_{x,y}^2}{k^2} \frac{L}{2\sqrt{12}}. \tag{121}$$

The distortions of the beta functions can be estimated in a similar way [48]. Let us again consider a gradient perturbation at point s_1. Now we have to check the variation of the beta function throughout the machine, hence we let the observer be at point $s \neq s_1$ (for simplicity $s_1 > s$, but the result does not depend on this assumption). Let us denote by A the transfer matrix of the unperturbed machine from point s to point s_1 and by B the transfer matrix of the perturbed machine from point s_1 to point s. Then the single-turn transfer matrix of the perturbed machine as seen by observer is given by

$$T = \begin{pmatrix} b_{11} & b_{12} \\ b_{21} & b_{22} \end{pmatrix} \begin{pmatrix} 1 & 0 \\ -\Delta K ds & 1 \end{pmatrix} \begin{pmatrix} a_{11} & a_{12} \\ a_{21} & a_{22} \end{pmatrix}. \tag{122}$$

It is sufficient to calculate t_{12} only, which is

$$t_{12} = b_{11}a_{12} + b_{12}a_{22} - \Delta K ds\, a_{12}b_{12}. \tag{123}$$

Furthermore, we are interested in the variation of this term

$$\Delta t_{12} = -\Delta K ds\, a_{12}b_{12}, \tag{124}$$

which we make equal to the variation of the same term given by matrix (50)

$$\Delta(\beta_s \sin 2\pi Q) = -\Delta K ds \beta_s \beta_{s1} \sin(\mu_{s1} - \mu_s) \sin(\mu_s - \mu_{s1})$$
$$= -\Delta K ds \beta_s \beta_{s1} \sin(\mu_{s1} - \mu_s) \sin[2\pi Q - (\mu_{s1} - \mu_s)]. \tag{125}$$

Expanding Eq. (125) to first order and using

$$\sin \alpha \, \sin \beta = -\frac{1}{2}[\cos(\alpha + \beta) - \cos(\alpha - \beta)], \tag{126}$$

we find

$$\Delta \beta_s \sin 2\pi Q_0 + 2\pi \beta_s \Delta Q \cos 2\pi Q_0$$
$$= \frac{1}{2}\Delta K \beta_s \beta_{s1}[\cos 2\pi Q_0 - \cos 2(\mu_s - \mu_{s1}) - \pi Q_0]\, ds. \tag{127}$$

The second term on the left side and the first term on the right side in Eq. (125) cancel due to Eq. (120). We find for the relative beta-beating:

$$\frac{\Delta \beta_s}{\beta_s} = \frac{\Delta K ds\, \beta_{s1} \cos(2|(\mu_s - \mu_{s1}) - \pi|)}{2 \sin 2\pi Q_0}. \tag{128}$$

Here the absolute value of the argument of the cosine is taken when we consider the possibility for s_1 to be upstream of s as well. When applied to insertion devices with $k_x = 0$, this general formula for the beta beating due to quadrupole error yields

$$\frac{\Delta\beta_s}{\beta_s} = \frac{\beta_{s1} L \cos\left(2\left|(\mu_s - \mu_{s1}) - \pi\right|\right)}{4\rho^2 \sin 2\pi Q_0}. \tag{129}$$

The insertion device itself produces dispersion, hence increases the emittance. For SRS the typical requirement for the machine is to keep the emittance as small as possible. The wiggler generates an emittance of approximately the same value as that generated by the rest of the machine bending magnets if its magnetic field is

$$B_w \simeq \pi \left(\frac{15}{2}\right)^{1/3} \left(\frac{\epsilon_0 J_x}{C_q \gamma^2}\right)^{1/3} \frac{B\rho}{(\lambda^2 \bar{\beta}_x)^{1/3}}. \tag{130}$$

As seen from (130), in a strong wiggler the horizontal beta function must be small.

Up to now we have considered the dispersion in the wiggler to be zero. If this is not the case, the wiggler will generate an additional emittance contribution. The emittance generated by dispersion in the wiggler is of the same order as the emittance in the unperturbed machine if the dispersion is

$$D_w \simeq \left(\frac{\epsilon_0 J_x}{C_q \gamma^2}\right)^{1/2} \left(\frac{\beta_x B\rho}{B_w}\right)^{1/2}. \tag{131}$$

With a small horizontal beta function, Eq. (131) imposes a strong restriction to the dispersion in the wiggler to keep emittance of machine small.

We conclude that the beta function in an insertion device must be carefully chosen, especially when the magnetic field is strong, to avoid adverse effects on the machine's linear optics.

E Effect of Synchrotron Radiation on Beam Parameters

The most significant effect of an insertion device is an increase in energy loss.

Exercise 2.

Using the formula for the energy loss in an isomagnetic field,

$$U_0 = C_\gamma \frac{E^4 \text{ (GeV)}}{\rho(\text{m})} \tag{132}$$

and assuming that the dependence of the magnetic field along the z-axis of the insertion device is $B_w \cos kz$, find the energy loss. The length of the insertion device is L, and the bending radius $\rho_w = (B\rho/B_w)$. The value of C_γ is $8.845 \cdot 10^{-5}$ m (GeV)$^{-3}$. Estimate the ratio of the energy loss in the insertion device (B_w=5 T, L=3 m) to that in a machine with 2 GeV energy.

Solution.
According to Eq. (132), the energy loss per unit length is

$$U_0/(2\pi\rho) = C_\gamma \frac{E^4 \cos^2 kz}{2\pi\rho_w}. \tag{133}$$

Integrating Eq. (133) over the length of the insertion device, we find

$$U_w = C_\gamma \frac{L_w E^4}{4\pi\rho_w^2}. \tag{134}$$

The total energy loss in a machine with the insertion device on is given by

$$U_{tot} = U_0 \left(1 + \frac{L_w \rho_0}{4\pi\rho_w^2}\right). \tag{135}$$

Assuming the field in the machine arc dipoles is $B_0=1.5$ T, we find $\rho_0=(B\rho)/B_0=4.4$ m. Following the same way, $\rho_w=1.3$ m. Comparing Eqs. (133) and (135), we find

$$\frac{U_w}{U_0} = \frac{L_w \rho_0}{4\pi\rho_w^2} \simeq 0.6. \tag{136}$$

We can write formula (135) in a general way:

$$U = C_\gamma \frac{E^4}{2\pi} \int \frac{ds}{\rho^2} = C_\gamma \frac{E^4}{2\pi}(I_2^0 + \Delta I_2), \tag{137}$$

where $I_2 = \int ds/\rho^2$ is the so-called second synchrotron integral. The superscript 0 refers to the bending magnets' contribution to the integral, and the second term comes from the contribution of an insertion device which is explicitly seen in formula (135). Other synchrotron integrals are defined in the following way:

$$I_3 = \int \frac{ds}{|\rho^3|}, \quad I_4 = \int \frac{1-2n}{\rho^3} D_x ds, \quad I_5 = \int \frac{H}{|\rho^3|} ds, \tag{138}$$

where

$$H = \gamma D_x^2 + 2\alpha D_x D_x' + \beta D_x'^2, \quad n = \frac{x}{B}\frac{\partial B}{\partial x}. \tag{139}$$

The beam emittance and the energy spread are expressed in the following way via the synchrotron integrals:

$$\epsilon_x = C_q \gamma^2 \frac{I_5}{I_2 - I_4} = C_q \gamma^2 \frac{I_5}{I_2 J_x}, \quad \sigma_E = \left[C_q \gamma^2 \frac{I_3}{2I_2 + I_4}\right]^{1/2}. \tag{140}$$

Here $C_q=3.84 \cdot 10^{-13}$ m. With an insertion device switched on the emittance and the energy spread are modified to

$$\epsilon_x = \epsilon_x^0 \frac{1+\Delta I_5/I_5^0}{1+(\Delta I_2 - \Delta I_4)/(I_2^0 - I_4^0)}, \quad \sigma_E = \sigma_E^0 \frac{1+\Delta I_3/I_3^0}{1+(2\Delta I_2 + \Delta I_4)/(2I_2^0 + I_4^0)}. \quad (141)$$

For an insertion device of N periods with magnetic field $B_y \propto \cos(2\pi z/\lambda)$ and bending radius ρ_w, an approximate formula is valid:

$$\epsilon_x = \epsilon_x^0 \frac{1+\lambda^2/(15\pi^3 \rho_w^5) \cdot \bar{\beta} L/I_5^0}{1+L/(4\pi\rho_0) \cdot (\rho_0/\rho_w)^2}, \quad \sigma_E = \sigma_E^0 \left[\frac{1+2L\rho_0^2/(3\pi^2 \rho_w^3)}{1+L/(4\pi\rho_0) \cdot (\rho_0/\rho_w)^2}\right]^{1/2}. \quad (142)$$

Here the dispersion at the insertion device is assumed to be zero. Inspection of Eq. (142) shows that depending on its magnetic field an insertion device can reduce or increase the emittance and (or) the energy spread. One can calculate the value of the field when this occurs.

Exercise 3.

Using formulae (142), estimate the magnetic field of an insertion device which corresponds to damping of the emittance and reduction of the energy spread with respect to a machine with the insertion device off.

Solution.

The emittance in a machine with the insertion device switched on is equal to that in the machine with the wiggler off when

$$\frac{\lambda^2}{15\pi^3 \rho^5} \frac{\bar{\beta} L}{I_5^0} = \frac{L\rho_0}{4\pi\rho^2}. \quad (143)$$

The fifth synchrotron integral can be found from the first part of formulae (142):

$$I_5^0 = \frac{\epsilon_x^0 I_2^0 J_x}{C_q \gamma^2} = \frac{2\pi \epsilon_x^0}{C_q \rho_0 \gamma^2}. \quad (144)$$

Substituting I_5^0 from Eq. (144) into (143), we get formula (130). For the energy spread we get immediately from the second part of formulae (142)

$$B_w = \frac{3\pi}{8} \frac{B\rho}{\rho_0}. \quad (145)$$

F The Case of a Strong Wiggler

For a strong wiggler the model based on the magnetic field representation (102) is far from reality and can't be used. Computer simulations based on measured magnetic fields are then the only tool available for determining machine optics. After the magnetic measurement the dipole and the sextupole field components are known along the wiggler. The new reference orbit inside the wiggler can then be found numerically. When the reference orbit is known, a field expansion around it is performed to give the dipole and the quadrupole components in the wiggler frame [51]:

$$\tilde{B}_y = B_y + \frac{1}{2}x_{c.o.}^2 \frac{\partial^2 B_y}{\partial x^2}, \tag{146}$$

$$\tilde{G} = \frac{\partial^2 B_y}{\partial x^2}x_{c.o.} - \frac{dB_y}{ds}\theta - \frac{1}{2}\frac{d}{ds}\frac{\partial^2 B_y}{\partial x^2}x_{c.o.}^2\theta, \tag{147}$$

where $\theta = dx_{c.o.}/ds$.
Exercise 4.
Using formula (147), show that, in the absense of a sextupole component, a wiggler produces focusing in the vertical plane only.
Solution.
Integrating (147) by parts, we find

$$\int G ds = -\int \frac{dB_y}{ds}\theta ds = -\int \frac{d(B_y\theta)}{ds}ds = 0 \tag{148}$$

because both B_y and θ are equal to zero at the ends of a wiggler. The focusing in the vertical plane is produced by the dipole component ($1/\rho^2$ term). The effect of a strong wiggler on linear optics is very strong, especially for small machines with low energy. The machine becomes unstable with the wiggler on and must be recovered carefully. The matching technique for insertions is discussed elswhere [46,49,50]. The higher-order field components are found in a similar way. In addition to the "normal" field components others appear, such as "pseudo-sextupole" and "pseudo-octupole" terms, which have strong influence on nonlinear beam dynamics.

ACKNOWLEDGMENTS

One of the authors (U.W.) is indebted to M. Sullivan for providing him with the layout plots of the CESR-III, PEP-II and KEKB interaction regions.

REFERENCES

1. Courant, E.D. and Snyder, H.A., *Annals of Physics* **3**, 1–48 (1958).
2. Wilson, E.J.N., *Circular Accelerators—Transverse*, in Proc. US Particle Accelerator School 1985, AIP Conf. Proc. 153, M. Month & M. Dienes, ed., (1987), p.3.
3. Brown, K.L. and Servranckx, R.V., *Circular Machine Design Techniques and Tools*, ibid. p. 121; SLAC-PUB-3381.

4. Chao, A.W., *Elementary Design and Scaling Considerations of Storage Ring Colliders*, ibid. p. 103.
5. Wiedemann, H., *Particle Accelerator Physics*, Springer, New York, 1993, p. 134.
6. Wiedemann, H., ed. *PEP Design Handbook*, Stanford Linear Accelerator Center and Lawrence Berkeley Laboratory, California, 1977.
7. Peck S.B. and Rubin D.L., *CESR Performance and Upgrade Status*, in Proc. 1999 Part. Accel. Conf., New York, (1999) p. 285.
8. *KEKB B-Factory Design Report*, KEK Report 95-7, Tsukuba, June 1995.
9. *HERA A Proposal for a Large Electron-Proton Colliding Beam Facility at DESY*, DESY-HERA 81/10, Hamburg, 1981.
10. Sanford, J.R. and Matthews, D.M., *SSC Site Specific Conceptual Design of the Superconducting Super Collider*, SSCL-SR-1056 (1990).
11. Zisman, M., ed., *PEP-II An Asymmetric B-Factory*, SLAC-418, LBL-PUB-5379 (1993).
12. JHC Study Group, *Design Study of the Large Hadron Collider (LHC)*, CERN 91-03, Geneva, May 1991.
13. Samios, N.P., *Conceptual Design of the Relativistic Heavy Ion Collider*, BNL-51932, Brookhaven, NY, May 1986.
14. Peggs, S., *The Projection Approach to Solenoid Compensation*, Part. Accel. **12**, 219 (1982).
15. Nosochkov, Y., Cai, Y., Irwin, J., Sullivan, M. and Forest, E. *Detector Solenoid Compensation in the PEP-II B Cactory*, in Proc. 1995 Part. Accel. Conf., Dallas, Tx, 585 (1996).
16. Bryant, P., in *Proc., CAS CERN Accelerator School*, S. Turner, ed., 5th Course, Jyväskylä, Vol. 1, CERN-94-01 (1994).
17. Piwinski A. and Wrulich A., *Excitation of Betatron-Synchrotron Resonances by a Dispersion in the Cavities*, DESY 76/07 (1976).
18. Piwinski A., *Synchro-Betatron Resonances*, in Proc. CAS, Advanced Accelerator Physics, Queen's College, Oxford, 1985; CERN 87-03 (1987).
19. Steffen K., *Periodic Dispersion Suppressors*, Part I, DESY-HERA, 81/19 (1981); Part II, DESY-HERA, 83/02 (1983).
20. Brinkmann R., *Insertions*, in Proc. CAS, Second General Accelerator Physics Course, Aarhus, Denmark, 1986, CERN 87-10 (1987), p. 45.
21. Bryant P. J., *Insertions*, in Proc. CAS, Fifth General Accelerator Physics Course, University of Jyväskyla, Finland, 1992, CERN 94-01 (1994), p. 159.
22. Rossbach J. and Scmüser P., *Basic Course in Accelerator Optics*, ibid, p. 38.
23. Hutton A., Placidi M. and Taylor T., *Low-Beta Insertions and Dispersion Suppressors*, CERN LEP-TH/83-41 (1983), LEP Note 465 (1983).
24. *LEP Design Report*, vol. II, The LEP Main Ring, CERN-LEP/84-01 (1984).
25. Rees G. H., *Injection*, in Proc. CAS, General Accelerator Physics, Gif-sur-Yvette, Paris, 1984, CERN 85-19 (1985), p. 331.
26. Wilson E., *Transverse Beam Dynamics*, in Proc. CAS, Fifth General Accelerator Physics Course, University of Jyväskyla, Finland, 1992, CERN 94-01 (1994), p. 131.
27. Walker R., *Radiation Damping*, ibid., p. 461.
28. Ropert A., *Dynamic Aperture*, in Proc. CAS, Third Advanced Accelerator Physics

Course, Uppsala University, Sweden, 1989, CERN 90-04 (1990), p. 26.
29. Scandale W., *Dynamic Aperture*, in Proc. CAS, Fifth Advanced Accelerator Physics Course, Rhodes, Greece, 1993, CERN 95-06 (1995), p. 109.
30. Maury S. and Möhl D., *Combined Longitudinal and Transverse Multiturn Injection in a Heavy Ion Accumulator Ring*, PS/AR/Note 94-12, May 1994.
31. van der Stok P., *Multiturn Injection into the CERN Proton Synchrotron Booster*, CERN/PS/BR 81-28, December 1981.
32. Schindl K. and van der Stok P., *A Method for Increasing the Multiturn Injection Efficiency in A. G. Synchrotrons by Means of Skew Quadrupoles*, in Proc. 1977 Part. Acc. Conf., IEEE Trans. Nucl. Sci., NS-24, 3, June 1977, p. 1390.
33. *LEP Design Report*, v. 3, LEP2, CERN-AC/96-01 (LEP2), June 1996.
34. Bosser J. et al., *Results on Lead Ion Accumulation in LEAR for the LHC*, in Proc. 6th European Part. Acc. Conf., Stockholm 1998, v. 1, p. 253.
35. Laclare J. L., *Light Source Performance Achievments*, ibid., v. 1, p. 78.
36. Wrulich A., *Future Directions in the Storage Ring Developments for Light Sources*, in Proc. 1999 Part. Acc. Conf., New York, 1999, v. 1, p. 192.
37. Cornacchia M. and Winick H. (eds.), *Workshop on Fourth Generation Light Sources*, 1992, SSRL 92/02.
38. Laclare J. L. (ed.), *Workshop on Fourth Generation Light Sources*, ESRF, Grenoble, 1996.
39. Hubner K., *Synchrotron Radiation*, in Proc. CAS, Synchrotron Radiation and Free Electron Lasers, Chester, United Kingdom, 1989, CERN 90-03 (1990), p. 24.
40. Hofmann A., *Characteristics of Synchrotron Radiation*, in Proc. CAS, Synchrotron Radiation and Free Electron Lasers, Grenoble, France, 1996, CERN 98-04 (1996), p. 1.
41. Elleaume P., *Theory of Undulators and Wigglers*, in Proc. CAS, Synchrotron Radiation and Free Electron Lasers, Chester, United Kingdom, 1989, CERN 90-03 (1990), p. 142.
42. Wille K., *Introduction to Insertion Devices*, in Proc. CAS, Synchrotron Radiation and Free Electron Lasers, Grenoble, France, 1996, CERN 98-04 (1998), p. 61.
43. Walker R., *Wigglers*, in Proc. CAS, Fifth Advanced Accelerator Physics Course, Rhodes, Greece, 1993, CERN 95-06, (1995), p. 807.
44. Walker R., *Insertion Devices: Undulators and Wigglers*, in Proc. CAS, Synchrotron Radiation and Free Electron Lasers, Grenoble, France, 1996, CERN 98-04 (1998), p. 129.
45. Pflüeger J., *Insertion Devices for 4th Generation Light Sources*, in Proc. 1999 Part. Acc. Conf., New York, 1999, v. 1, p. 157.
46. Ropert A., *Lattices and Emittances*, in Proc. CAS, Synchrotron Radiation and Free Electron Lasers, Grenoble, France, 1996, CERN 98-04 (1998), p. 91.
47. Smith L., *Effect of Wigglers and Undulators on Beam Dynamics*, LBL-ESG Tech. Note-24 (1986).
48. Bryant P. J. and Johnsen K., *The Principles of Circular Accelerators and Storage Rings*, Cambridge University Press, 1993.
49. Wrulich A., *Effect of Insertion Devices on Beam Dynamics*, ST/M-87/18 (1987).
50. Katoh M. et al., *The Effect of Insertion Devices on Betatron Functions and their*

Correction in the Low Emittance of the Photon Factory Storage Ring, KEK Report 86-12 (1987).
51. Levichev E. B., Private communication.

Tutorial on Linear Colliders

Frank Zimmermann
CERN, SL Division

Abstract. Proceeding from the collision point towards the source, we discuss purpose and design concepts of the various linear-collider subsystems, as well as important mechanisms of emittance dilution, beam diagnostics, and advanced tuning methods. In particular, we address beamstrahlung, linac emittance degradation due to dispersion and wake fields, scaling of damping-ring parameters with collider energy, fast beam-ion and electron-cloud instabilities, coherent synchrotron radiation, and rf guns. Five case studies are examined in detail.

I PERSPECTIVE

A linear electron-positron collider was first proposed by Tigner in 1965 [1] and later by Amaldi in 1975 [2]. Meanwhile, the Stanford Linear Collider (SLC) [3] has become a successful prototype. During the course of its operation from 1987 to 1998, the SLC performance was continually improved, as numerous advanced beam quality control techniques were invented and implemented [4–7]. We may take for granted that many of the techniques developed for the SLC will also be utilized at the next-generation linear colliders.

All future linear colliders will bring into collision highly energetic electron or positron beams, which are accelerated in two diametrically opposed linear accelerators (linacs). The first and only linear collider so far, the SLC, operated at a center-of-mass energy of about 100 GeV with rms interaction-point (IP) beam spot sizes of several hundred nanometers vertically and more than a micron horizontally. The next generation linear colliders aim for roughly 10 times higher energies and 100 times smaller vertical spot sizes.

Table I compares IP beam parameters for the SLC with those proposed for various future projects. TESLA [8] coordinated by DESY is a superconducting (s.c.) linear collider, NLC [9] designed at SLAC uses a normal-conducting linac at 4 times the SLC rf frequency (11.4 GHz instead of 2.8 GHz), and CLIC [10,11] studied at CERN operates at 30 GHz and its power source is based on two-beam acceleration (see below). The differences between the projects reflect different design choices and emphases.

A linear accelerator (linac) consists of many successive arrays (structures) of coupled rf cavities (cells). In these cells a longitudinal rf electric field accelerates the electron or positron bunches to high energy. A larger accelerating field, or

TABLE 1. Beam and interaction-point parameters for various proposed linear colliders compared with those of the SLC. Note that some numbers may have changed since publication, and that, e.g., TESLA now contemplates an alternative parameter set with higher luminosity. [[†]The SLC spot sizes quoted refer to the 1998 average values.]

Parameter	Symbol	SLC	TESLA	NLC	CLIC
C.m. energy [TeV]	E	0.1	0.5	1	3
Luminosity [10^{34} cm^{-2} s^{-1}]	L	0.0002	0.84	1.3	10
Repetition rate [Hz]	f_{rep}	120	4	120	100
Bunch charge [10^{10}]	N_b	3.7	1.8	1.0	0.4
Bunches/rf pulse	n_b	1	2260	95	154
Bunch separation [ns]	Δ_b	—	354	2.8/1.4	0.67
Av. beam power [MW]	P_b	0.04	13	9	14.8
Bunch length [mm]	σ_z	1	0.5	0.12	0.03
Hor. emittance [μm]	$\gamma\epsilon_x$	50	12	4.5	0.68
Vert. emittance [μm]	$\gamma\epsilon_y$	8	0.03	0.1	0.02
Hor. beta [mm]	β_x^*	2.8	25	12	8
Vert. beta [mm]	β_y^*	1.5	0.5	0.15	0.15
Hor. spot size [nm]	σ_x^*	1700†	783	235	43
Vert. spot size [nm]	σ_y^*	900†	5.5	4	1.0
Upsilon	Υ	2×10^{-3}	0.02	0.3	8.1
Pinch enhancement	H_D	2.0	1.6	1.45	2.24
Beamstrahlung	δ_B [%]	0.06	1.0	10	31
Photons per e$^-$ (e$^+$)	N_γ	1	0.9	1.4	2.3

voltage gradient, is desirable, since it implies a shorter linac length for the same final beam energy. The accelerating gradient G (in units of volts per meter) can be written as

$$G = \sqrt{RP} \qquad (1)$$

where P is the supplied rf power and R the so-called shunt impedance, both per unit length. The shunt impedance in turn may be expressed as $R = (R/Q)Q$, where R/Q is a quantity that depends purely on the geometry of the accelerating cavity. Values of R/Q equal to 200 Ω per cavity length are common. R/Q is reduced for larger iris radii, where the term "iris" refers to the opening hole between successive cavity cells, through which both the beam and the rf wave propagate. The quality factor Q is roughly equal to $V/(S\delta(\omega_{\text{rf}}))$ with V the volume of the cavity, S its surface area, and $\delta(\omega_{\text{rf}})$ the skin depth at the rf frequency [12]. For cavities in a superconducting linac, such as TESLA, the Q value can be extremely high, even larger than 10^{11} [13]. This means it is possible to increase the iris apertures, so as to reduce the beam-induced "wake fields" and, unavoidably, in parallel also the value of (R/Q), but still retain a large shunt impedance R and a modest rf power. On the other hand, the Q value of normal-conducting cavities is much smaller, typically several times 10^3 at 30 GHz [14], and therefore it is important to preserve a large R/Q, even if this implies enhanced wake-field effects.

The higher frequencies chosen for normal-conducting accelerating structures are based on the assumption that the achievable accelerating gradient increases roughly in proportion to the rf frequency, as would be the case if the field gradient is limited by rf breakdown or trapping of dark current generated by field emission [15–17].

In conventional linacs, the rf energy is produced in devices called klystrons, which are powered by other devices called modulators. A klystron uses the bunching of a low-energy electron beam in response to a weak input rf signal to amplify the rf energy. The modulators contain, among other components, an energy storage unit, a fast switch, and a pulse-forming network. Usually the rf pulse generated by the klystrons is a factor of 6–10 longer and weaker than required, and therefore it must be compressed. This is done by properly combining either parts of an rf pulse generated at different times from the same klystron or pulses from adjacent klystrons. The various rf compression schemes are known by their acronyms, such as SLED (SLAC energy doubler), SLED-II, or DLDS (delay-line distribution system). In some of the proposed colliders, several thousands of klystrons and modulators are necessary to reach 1 TeV center-of-mass energy.

The number of klystrons and modulators is reduced drastically in the two-beam approach studied for CLIC. Here a low-energy intense "drive beam" is employed to transport and compress the rf power, which is initially produced in a separate low-frequency drive-beam linac. An attractive feature of CLIC is that there are no active rf components in the main linac, and that the high rf frequency chosen (30 GHz) may allow reaching multiple TeV energies with a linac length comparable to that of the lower-energy lower-frequency designs. The overall layout of a 3-TeV CLIC is shown in Fig. 1. The figure also illustrates the generation of the drive beam and its distribution along the main linac.

The remainder of this report is structured as follows. In the next section, we discuss the constraints arising from beam-beam interaction and the resulting luminosity scaling for linear colliders. We then describe design concepts, beam dynamics, operation, measurement challenges, and tuning methods for the various subsystems, proceeding against the beam direction from interaction point and final focus over collimation and linac towards damping rings and rf gun. Five illustrative case studies are presented in detail.

II BEAMSTRAHLUNG AND LUMINOSITY

Assuming Gaussian bunch distributions and ignoring the variation of beam sizes during collision, *e.g.*, due to the hourglass effect (depth of focus) or to beam-beam forces (pinch enhancement), the luminosity of a linear collider can be written as

$$L = \frac{f_{\text{rep}} n_b N_b^2}{4\pi \sigma_x^* \sigma_y^*} \qquad (2)$$

where f_{rep} denotes the repetition rate at which beam and rf are sent through the linac, n_b the number of bunches per rf pulse, N_b the number of particles per bunch,

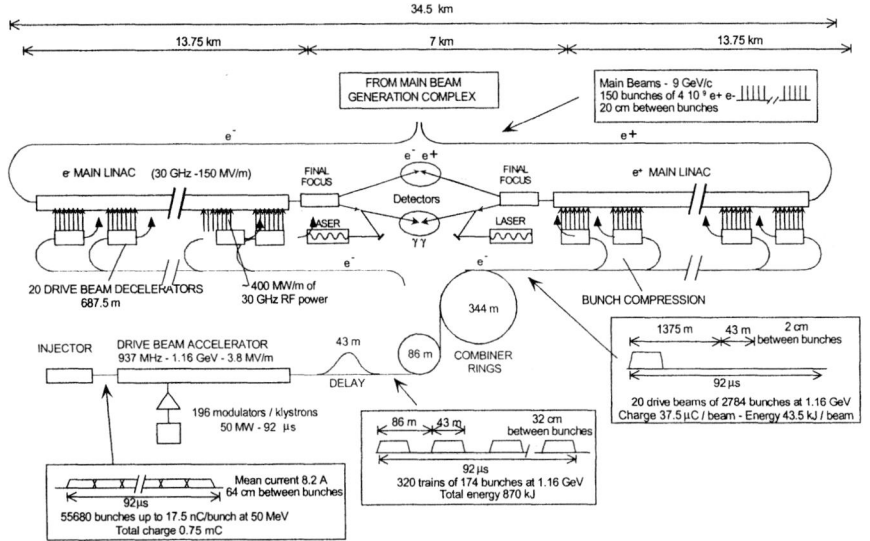

FIGURE 1. Schematic layout of the CLIC 3-TeV linear collider [10,11].

σ_x^* the rms horizontal beam size, and σ_y^* the rms vertical beam size at the interaction point.

During collision, individual electrons or positrons emit synchrotron radiation in the strong field of the opposing beam. This radiation is called beamstrahlung. To preserve an adequate energy spectrum of the luminosity, the number of beamstrahlung photons emitted per electron, N_γ, must be limited to a value of the order of one. Considering this constraint and assuming flat beams with $\sigma_x^* \gg \sigma_y^*$, we can re-express the above luminosity formula as [18]

$$L \approx \left(\frac{5}{r_e}\right) \frac{P_{\text{wall}}}{E_b} N_\gamma \frac{\eta}{\sigma_y^*} \qquad (3)$$

where r_e is the classical electron radius, P_{wall} the wall-plug power, E_b the final beam energy, and η the conversion efficiency of wall-plug power into average beam power ($P_{\text{beam}} = f_{\text{rep}} E_b N_b n_b$). The beam energy is fixed by the physics requirements, and the wall plug power is limited by economic reasons. Hence, there are only two free parameters that can be optimized for maximum luminosity: the conversion efficiency η and the vertical spot size σ_y^*.

At the SLC the parameter η was much smaller than 1%. For all future projects it is raised to roughly 10%, for example, by increasing the number of bunches per rf pulse from 1 to about 100, and by improving the efficiency of all rf components. Clearly, the optimization of η is being pushed to its limits.

The vertical spot size is the second free parameter. In all proposed designs, it is more than 100 times smaller than at the SLC. Such tiny spot sizes are achieved both by much reduced emittances and by interaction-point beta functions that are squeezed down to 150 μm, as shown in Table I.

Many beam-dynamics challenges for a linear collider are related to the small spot size, for example, the design of the final-focus optics, stability tolerances, emittance preservation, and production of a low-emittance beam.

Let us now take a closer look at beamstrahlung. The typical energy of the beamstrahlung photons is characterized by the parameter Υ. This is equal to two thirds of the classical critical energy divided by the beam energy E_b [19],

$$\Upsilon = \frac{2}{3}\frac{\hbar\omega_c}{E_b} \approx \frac{5}{6}\frac{\gamma r_e^2 N}{\alpha\sigma_z(\sigma_x + \sigma_y)}, \tag{4}$$

where α denotes the fine structure constant. For synchrotron radiation emitted from a dipole magnet the critical freqency is $\omega_c = \frac{3}{2}c\gamma^3/\rho$ [20] and, usually, the energy of synchrotron radiation photons is much smaller than the beam energy $\hbar\omega_c \ll E_b$. For the beamstrahlung emitted during the beam-beam collision, this need not be the case. Typical values of Υ at the interaction point are 2×10^{-3} for the SLC, 0.3 for the NLC, and almost 10 for CLIC. If Υ becomes comparable to 1 or larger, a significant portion of beamstrahlung photons convert into real electron-positron pairs in the strong electro-magnetic fields of the two beams. Unfortunately, linear colliders at multi-TeV energies can hardly avoid operating in this regime.

Besides Υ, there is a second parameter of interest, namely the number of beamstrahlung photons emitted per electron. It is [19]

$$N_\gamma \approx \frac{5}{2}\frac{\alpha\sigma_z}{\gamma\lambda_e}\Upsilon\left[\frac{1}{(1+\Upsilon^{2/3})^{1/2}}\right] \approx 2\frac{\alpha r_e N_b}{\sigma_x + \sigma_y}. \tag{5}$$

The last approximation applies if Υ is small. For example, choosing $N = 10^{10}$ and $N_\gamma = 1$, we obtain $(\sigma_x + \sigma_y) \approx 400$ nm, consistent with the NLC parameter set.

Note that by reducing the bunch length σ_z, we can reach a parameter regime where Υ is large and the spot size small, but where we can still ensure that $N_\gamma \leq 1$, thanks to the quantum correction term — the brackets — of Eq. (5). This is sometimes referred to as the quantum suppression of beamstrahlung. It arises, roughly speaking, from the fact that the electrons cannot radiate photons of energy higher than the beam energy. The classical spectrum of synchrotron-radiation photon energies is modified in this extreme quantum regime [19].

Two further quantities characterizing beamstrahlung are the average energy loss per electron [19],

$$\delta_B \approx \frac{1}{2}N_\gamma\Upsilon\left[\frac{(1+\Upsilon^{2/3})^{1/2}}{(1+(1.5\Upsilon)^{2/3})^2}\right], \tag{6}$$

and the fraction of luminosity at the nominal energy [21],

$$\frac{\Delta L}{L} \approx \frac{1}{N_\gamma^2}\left(1 - e^{-N_\gamma}\right)^2, \tag{7}$$

which depends only on N_γ. The value of $\Delta L/L$ drops rapidly for increasing N_γ, e.g., for $N_\gamma = 1$, it is 81%, for $N_\gamma = 2$ only 25%, and for $N_\gamma = 3$ barely 11%.

Introducing the aspect ratio $r \equiv \sigma_y/\sigma_x$, the number of beamstrahlung photons scales as $N_\gamma \propto N_b/(\sigma_x(1+r))$ and the luminosity as $L \propto N_b N_\gamma (1+r)/r$. Hence, in order to maximize the luminosity while at the same time constraining the number of beamstrahlung photons, it is best to operate with flat beams where $r \ll 1$.

Flat-beam parameters have been adopted for all future linear collider designs. Various other methods of overcoming the beamstrahlung problem have been proposed, such as 4-beam collisions [22,23] (2 electron beams colliding with 2 positron beams, so that the net electric and magnetic fields are zero), plasma [24] (where the plasma return current cancels the beam fields), and photon-photon collisions [25,26] (here the beam energy is converted into photon energy by Compton scattering off a high-power laser).

III FINAL FOCUS

The small beta functions at the collision point are achieved by focusing the beam with strong quadrupole magnets located a few meters upstream. The normalized focusing strength of a quadrupole, K (in units of m^{-2}), depends on the particle momentum as

$$K = \frac{B_T}{a(B\rho)} = \frac{eB_T}{ap_0(1+\delta)} \tag{8}$$

where B_T is the pole tip field, a the pole-tip radius, $B\rho \equiv p/e$ the magnetic rigidity of the particle, p_0 the design momentum, and $\delta \equiv (p - p_0)/p_0$ the relative momentum deviation. Typical values for the final quadrupole are $B_T/a \approx 300$–500 T/m and $K \approx 0.1$–0.3 m^{-2}. Since the focusing strength is energy dependent, particles with different momentum deviations will be focused at different distances behind the quadrupole. The change in focal length with particle energy is called the chromaticity of the final focus. It can be computed as an integral over the final quadrupoles $\xi \approx \int ds\, \beta(s) K(s)$. If the chromaticity is not corrected the vertical spot size at the interaction point becomes

$$\sigma_y^* \approx \sigma_{y,0}^* \sqrt{1 + \xi_y^2 \delta_{\rm rms}^2} \tag{9}$$

where $\sigma_{y,0}^*$ $(= \sqrt{\beta_y^* \epsilon_y})$ is the ideal linear design spot size and $\delta_{\rm rms}$ the rms energy spread. Since typically $\xi \approx 30\,000$ and $\delta_{\rm rms} \approx 0.28\%$, the chromaticity, if uncorrected, would increase the IP spot size by several orders of magnitude. Therefore, chromatic correction is indispensable. As in a storage ring, this correction is performed by placing sextupoles at locations with nonzero dispersion. Typically two pairs of sextupoles are used, for horizontal and vertical chromatic correction. The sextupole pairs are placed a multiple of π in betatron-phase advance away from the final quadrupoles. Figure 2 illustrates the basic layout of a final-focus system with chromatic correction.

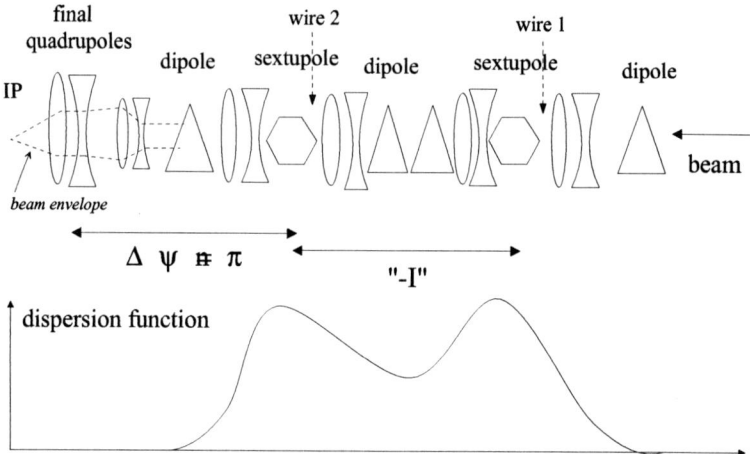

FIGURE 2. Schematic representation of a final-focus system. The beam moves from right to left. It passes the chromatic correction section and then a final demagnifier, before it finally reaches the interaction point, where it collides with an opposing beam. Only one pair of sextupoles is shown. A similar set of sextupoles and bending magnets would be located further on the right. Two pairs of sextupoles (4 sextupoles in total) are necessary in order to compensate the quadrupole chromaticity in both transverse planes and to globally cancel all unwanted low-order nonlinear aberrations induced by a single sextupole (quadratic kicks with amplitude and second-order dispersion) [27]. Wire scanners located near the two sextupoles, as indicated, can measure the energy-position correlation within the bunch; see case study I.

The dispersion at the sextupoles is generated by bending magnets. Thus, we are faced with the undesirable situation that we need bending magnets at the highest energy of the linear collider. Although the bending magnets in the final focus are much weaker than those in a ring collider, the synchrotron radiation in these magnets is important and imposes severe design constraints. In particular, this synchrotron radiation is responsible for a dramatic increase in the length of conventional SLC-type final-focus systems, when extrapolated to TeV energies.

The linear beam transport in a single-pass system is often expressed by a (6×6) matrix relating the initial (subscript i) and final (subscript f) phase-space coordinates [28]:

$$\begin{pmatrix} x \\ x' \\ y \\ y' \\ z \\ \delta \end{pmatrix}_f = \begin{pmatrix} R_{11} & R_{12} & R_{13} & R_{14} & R_{15} & R_{16} \\ R_{21} & R_{22} & R_{23} & R_{24} & R_{25} & R_{26} \\ R_{31} & R_{32} & R_{33} & R_{34} & R_{35} & R_{36} \\ R_{41} & R_{42} & R_{43} & R_{44} & R_{45} & R_{46} \\ R_{51} & R_{52} & R_{53} & R_{54} & R_{55} & R_{56} \\ R_{61} & R_{62} & R_{63} & R_{64} & R_{65} & R_{66} \end{pmatrix} \begin{pmatrix} x \\ x' \\ y \\ y' \\ z \\ \delta \end{pmatrix}_i. \quad (10)$$

In general we can express the position of a particle or of the beam centroid as the sum of betatron oscillation, x_β, dispersion D and higher order dispersion, $D^{(k)}$ ($k = 2, ...$), as

$$x = x_\beta + D\delta + \sum_{k \geq 2} D^{(k)} \delta^k. \quad (11)$$

Often the higher-order dispersion terms ($D^{(k)}$ with $k \geq 2$) are ignored, as we will do in the following analysis.

As already mentioned, in a conventional final focus, two pairs of sextupoles are used for chromatic correction, one pair for the horizontal plane, and the other for the vertical plane. We now consider one such pair in more detail. We assume that at the location of the first sextupole the dispersion is finite, $D_i \neq 0$, but that the slope of dispersion is zero, $D'_i = 0$, and that there are no vertical bending magnets, hence $R_{36} = 0$ and $R_{46} = 0$. The (2×2) submatrices in the vertical and horizontal plane are commonly chosen as $-I$ (minus identity) transformations [27]. The full R matrix between the two sextupoles forming a pair then has the form

$$\begin{pmatrix} -1 & 0 & 0 & 0 & 0 & R_{16} \\ 0 & -1 & 0 & 0 & 0 & R_{26} \\ 0 & 0 & -1 & 0 & 0 & 0 \\ 0 & 0 & 0 & -1 & 0 & 0 \\ 0 & 0 & 0 & 0 & 1 & R_{56} \\ 0 & 0 & 0 & 0 & 0 & 1 \end{pmatrix}. \quad (12)$$

We write the initial position of a particle on a dispersive trajectory as $x_i = x_{\beta,i} + D_i \delta$, and the corresponding final position as $x_f = x_{\beta,f} + D_f \delta$. From Eqs. (10) and (12) it follows that $x_{\beta,f} = -x_{\beta,i}$ and that the dispersion at the second sextupole is

$$D_f = -D_i + R_{16}. \quad (13)$$

This illustrates that without bending magnets (*i.e.*, with $R_{16} = 0$) the dispersion propagates exactly like a betatron oscillation. If there are bending magnets between the initial and final positions, in general the $(1,6)$ matrix element is not zero, *i.e.*, $R_{16} \neq 0$, and, in particular, the strengths of the bending magnets between the sextupoles can be adjusted so that $R_{16} = 2D_i$, whence $D_f = D_i$. In addition, it is possible to design the optics such that $R_{26} = 0$, and hence $D'_i = D'_f = 0$. The last nonzero element, R_{56}, describes the change in path length for different momentum deviations. It is the analogue of the momentum compaction factor in a storage ring. Usually, R_{56} in the final focus is so small, that its effect can be ignored.

What exactly is the idea behind the $-I$ transform? We try to illustrate its merits. For simplicity we consider the horizontal plane only. We assume that the sextupoles are thin, so that their effect may be represented by a single nonlinear deflection, and we denote their integrated strengths by $K_{s,1}$ and $K_{s,2}$. The integrated strength of a sextupole in units of m^{-2} is defined as

$$K_s = l_s \frac{1}{B\rho} \frac{\partial^2}{\partial x^2} B(x)\bigg|_{x=0} = l_s \frac{B_T}{(B\rho)a^2}, \tag{14}$$

where l_s denotes the length of the sextupole, a its inner radius, B_T the pole tip field at the radius a, and $B\rho$ ($= p/e$) the magnetic rigidity. As before, the dispersion at the sextupoles is taken to be $D_i \neq 0$ and the slope to be zero $D'_i = 0$. We denote the particle coordinate just prior to the first sextupole by x_i, the associated trajectory slope by x'_i, and the relative momentum deviation by δ. Behind the first sextupole the slope of the particle trajectory becomes

$$x'_1 = x'_i + \frac{1}{2} K_{s,1} x_i^2 = x'_0 + \frac{1}{2} K_{s,1} x_{\beta,i}^2 + K_{s,1} x_\beta D_i \delta + \frac{1}{2} K_{s,1} D_i^2 \delta^2. \tag{15}$$

In addition to the initial slope, we here recognize three nonlinear dependencies introduced by the thin sextupole. The term proportional to x_β^2 represents a geometric aberration and the component quadratic in δ a second order dispersive term. The mixed product proportional to $x_\beta \delta$ is the chromatic term, which we want to generate in order to compensate the chromaticity of the final-focus quadrupoles.

Now applying the $-I$ transform, Eq. (12), with $R_{26} = 0$ and $R_{16} = 2D_i$, we obtain the particle coordinates and slopes just prior to the second sextupole:

$$x_2 = -x_\beta - D_i \delta + R_{16} \delta = -x_\beta + D_i \delta \tag{16}$$

$$x'_2 = -x'_\beta - \frac{1}{2} K_{s,1} x_i^2. \tag{17}$$

Inserting $x_i = x_{\beta,i} + D_i \delta$ and applying the kick from the second quadrupole, $\Delta x' = K_{s,2} x_2^2$, we obtain

$$x'_2 = -x'_\beta - \frac{1}{2} K_{s,1}(x_\beta^2 + 2x_\beta D_i \delta + D_i^2 \delta^2) + \frac{1}{2} K_{s,2}(x_\beta^2 - 2x_\beta D_i \delta + D_i^2 \delta^2). \tag{18}$$

For equal sextupole strengths, $K_{s,1} = K_{s,2} \equiv K_s$, the geometric aberrations and the second-order dispersion terms cancel exactly, and all that is left is the chromatic component:

$$x'_2 = -x'_\beta - 2K_s x_\beta D_i \delta. \tag{19}$$

The important conclusion is that a $-I$ pair of sextupoles, as considered here, generates only chromaticity and no other low-order aberrations [27]. This conclusion still holds true if the vertical motion is also included in the analysis. Of course, in reality the $-I$ transform is not perfect, but itself varies with the momentum deviation. This gives rise to higher-order chromo-geometric aberrations, which ultimately limit the energy bandwidth of the final-focus system.

In circular accelerators dispersion is usually measured by sampling off-energy orbits with beam-position monitors. In linear colliders, however, varying the energy at some point in the beam line and observing the induced change in orbit measures the R_{16} matrix element between the point of energy change and the BPMs downstream. In general this is not equal to the energy-position correlation within the bunch [29]. The energy-position correlation in the bunch is a result of changes in the individual particle energies,—due to acceleration, synchrotron radiation or wake fields,—and subsequent energy-dependent path lengths all along the beam line, whereas the R_{16} measurement probes the effect of a change in acceleration at one particular location only.

Case Study I: Beam Dispersion

Conceive a scheme by which the horizontal and vertical energy-position correlation in the bunch can be monitored. Hint: one possibility is to use two sets of wire scanners, each with three wires tilted at different angles, e.g., at $90°$, $45°$ and $135°$ with respect to the horizontal plane, and separated by a $-I$ optical transform with bending magnets as in Eq. (12). [29]. A wire scanner is a device that measures the horizontal or vertical or diagonal beam size. It is equipped with thin filaments of, e.g., W or C, which are moved in small steps through the beam. Recording the scattering rate as a function of wire position and correcting for the finite size of the wire, one can determine the rms beam size at the wire location in the direction orthogonal to the wire filament.

Why could it be important to minimize such correlations in the beam?

<u>Solution:</u> *We consider the $-I$ transform of Eq. (12), and place two wire scanners at the initial and final locations. The $90°$ (vertical) filament of the wire measures the horizontal spot size. The latter consists of a betatron part and a dispersive part, added in quadrature:*

$$\sigma_x = \sqrt{\beta_x \epsilon_x + (D_x \delta)^2}. \tag{20}$$

We refer to the first wire by the subscript 1 and to the second wire by 2. The particle position at wire 1 is

$$x_1 = x_\beta + D_x \delta + \Delta D_x \delta. \tag{21}$$

It transforms into

$$x_2 = -x_\beta + D_x \delta - \Delta D_x \delta \tag{22}$$

at the second wire. The term $\Delta D_x \delta$ is the incoming beam dispersion mismatch (or, more precisely, the undesired energy-position correlation) which we want to measure and ultimately correct. This term propagates like a free betatron oscillation.

The rms horizontal beam sizes are computed by averaging over the beam distribution. The beam sizes at the two wire scanners are

$$\sigma_{x1}^2 = \langle x_1 \rangle^2 = \langle x_\beta^2 \rangle + (D_x + \Delta D_x)^2 \langle \delta^2 \rangle, \qquad (23)$$

$$\sigma_{x2}^2 = \langle x_2 \rangle^2 = \langle x_\beta^2 \rangle + (D_x - \Delta D_x)^2 \langle \delta^2 \rangle, \qquad (24)$$

with $\langle x_\beta^2 \rangle = \beta_x \epsilon_x$ and $\langle \delta^2 \rangle = \sigma_\delta^2$. The angular brackets indicate an average over the beam distribution. Without horizontal dispersion mismatch ($\Delta D_x = 0$), the horizontal beam size measured on the two wires is identical. The difference in the squared beam sizes thus provides a measure of the mismatch,

$$\sigma_{x1}^2 - \sigma_{x2}^2 = 4 D_x \Delta D_x \sigma_\delta^2 \qquad (25)$$

or [29]

$$\Delta D_x = \frac{\sigma_{x1}^2 - \sigma_{x2}^2}{4 D_x \sigma_\delta^2}, \qquad (26)$$

where D_x is the (matched) design dispersion at the wire, and the rms momentum spread σ_δ must be obtained from another measurement, typically from an additional wire scan at a location with large dispersion.

Next we look at the situation in the vertical plane. The vertical particle position at the first wire is

$$y_1 = y_\beta + \Delta D_y \delta, \qquad (27)$$

which, at the second wire, transforms into

$$y_2 = -y_\beta - \Delta D_y \delta, \qquad (28)$$

where ΔD_y denotes the vertical dispersion mismatch. Because of the absence of vertical bending, the vertical beam sizes at the two wires are always equal, independent of the amount of vertical dispersion mismatch. The reason is that both terms on the right-hand side of Eq. (27) change sign, and that a constant term, such as D_x horizontally, is missing. However, beam sizes measured on wire filaments tilted at 130° and 35° depend on the product of the vertical dispersion mismatch and the horizontal design dispersion and thereby allow us to determine the vertical dispersion. A wire scanner of this type is sketched in Fig. 3. The tilted wires measure the beam size in the u and v directions defined by

$$u = \frac{x + y}{\sqrt{2}}, \qquad (29)$$

$$v = \frac{x - y}{\sqrt{2}}. \qquad (30)$$

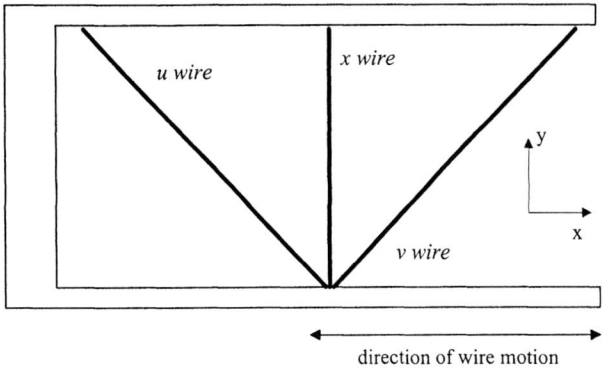

FIGURE 3. Horizontal wire-scanner mount equipped with 3 filaments, which measure the beam size in the x, u and v directions

Computing the rms beam sizes at these tilted wires we find

$$\sigma_u^2 = \langle u^2 \rangle = \frac{1}{2}(\sigma_x^2 + \sigma_y^2) + \langle xy \rangle, \tag{31}$$

$$\sigma_v^2 = \langle v^2 \rangle = \frac{1}{2}(\sigma_x^2 + \sigma_y^2) - \langle xy \rangle, \tag{32}$$

so that, for each wire scanner,

$$\sigma_u^2 - \sigma_v^2 = 2\langle xy \rangle. \tag{33}$$

Inserting the expressions for x and y, Eqs. (21), (22), (27), and (28), the correlation $\langle xy \rangle$ can be written

$$\langle x_1 y_1 \rangle = \langle x_\beta y_\beta \rangle + \Delta D_y (D_x + \Delta D_x) \sigma_\delta^2, \tag{34}$$
$$\langle x_2 y_2 \rangle = \langle x_\beta y_\beta \rangle + \Delta D_y (-D_x + \Delta D_x) \sigma_\delta^2. \tag{35}$$

Subtracting these two expressions, the terms $\langle x_\beta y_\beta \rangle$ (betatron coupling) and the terms proportional to the horizontal mismatch, ΔD_x, cancel, and there remains

$$\langle x_1 y_1 \rangle - \langle x_2 y_2 \rangle = 2 D_x \, \Delta D_y \, \sigma_\delta^2. \tag{36}$$

Re-expressing $\langle xy \rangle$ in terms of $\frac{1}{2}(\sigma_u^2 - \sigma_v^2)$, and using Eq. (33), the vertical dispersion mismatch is found [29]:

$$\Delta D_y = \frac{\sigma_{u1}^2 - \sigma_{v1}^2 - \sigma_{u2}^2 + \sigma_{v2}^2}{4 D_x \sigma_\delta^2}. \tag{37}$$

The dispersion in a transport line is not uniquely defined. The dispersion mismatch we have considered here is related to the energy-position correlation within the initial bunch distribution. This can be made explicit by writing, e.g., the vertical dispersion mismatch as

$$\Delta D_y = \frac{\langle y_1 \delta \rangle}{\langle \delta^2 \rangle}. \tag{38}$$

Correction of the mismatch implies that we remove the correlated component, so as to obtain the new position coordinate

$$y_{\text{cor}} = y_1 - \Delta D_y \delta = y_1 - \frac{\langle y_1 \delta \rangle}{\langle \delta^2 \rangle} \delta. \tag{39}$$

Squaring and averaging over the distribution we find

$$\langle y_{\text{cor}}^2 \rangle = \langle y_1^2 \rangle - \frac{\langle y_1 \delta \rangle^2}{\langle \delta^2 \rangle} \leq \langle y_1^2 \rangle. \tag{40}$$

Hence, the beam size after correction, $\langle y_{\text{cor}}^2 \rangle^{1/2}$, is always smaller than a beam size with some residual correlations, $\langle y_1^2 \rangle^{1/2}$.

In our example, the wire scanners are located a betatron-phase advance of 90° or an integer multiple thereof away from the interaction point (IP). In this case, minimizing the beam size at the wires amounts to minimizing the IP beam divergence.

There are many design constraints for a conventional final focus system:
1) The additional energy spread due to synchrotron radiation emitted in a bending magnet is

$$\delta_{\text{sr}}^2 \approx \frac{55}{24\sqrt{3}} r_e \lambda_e \gamma^5 \frac{\theta^3}{l_b^2} \tag{41}$$

where $\gamma = E/(m_e c^2)$, and r_e is the classical electron radius, θ the bending angle and l_b the length of the dipole magnet. If the radiation occurs after the (first) sextupoles, this energy spread is not chromatically corrected and it increases the IP spot size as

$$\frac{\Delta \sigma_y^*}{\sigma_{y0}^*} = \xi_y \delta_{\text{sr}}. \tag{42}$$

The term in Eq. (42) must be added in quadrature to the design spot size σ_{y0}, yielding a total spot size

$$\sigma_y^* = \sqrt{\sigma_{y0}^{*2} + \Delta \sigma_y^{*2}}. \tag{43}$$

2) Since the chromaticity is proportional to the beta function at the final quadrupoles, it grows roughly inversely with the IP beta function,

$$\xi_y \approx l^*/\beta_y^*, \tag{44}$$

where l^* denotes the effective free length from the last quadrupole to the IP.

3) The condition for the chromatic correction of the final quadrupoles by upstream sextupoles with integrated strength K_s, in units of m^{-2}, is

$$|2D_s K_s \beta_s| = \xi_y \tag{45}$$

where D_s and β_s are the horizontal dispersion and vertical beta function at the two sextupoles.

4) The dispersion scales with bending angle and length as

$$D_s \propto l_b \theta. \tag{46}$$

5) A quadrupole gradient error ΔK in the final focus will shift the waist position longitudinally away from the collision point. This will cause a spot-size increase

$$\frac{\Delta \sigma_y^*}{\sigma_y^*} = \Delta K\, \beta, \tag{47}$$

where β is the beta function at the location of the perturbation and we have assumed a phase advance to the IP equal to an odd multiple of $\pi/2$, as is the case for most final-focus magnets. Again, $\Delta \sigma_y^*$ is added in quadrature; see Eq. (43).

In particular, changes in the horizontal orbit, Δx, at the second sextupole of a pair, e.g. due to vibration or position drifts of quadrupoles located between the two sextupoles, induce a quadrupole component, $\Delta K = K_s \Delta x$, and, thus, increase the IP beam spot size. The tolerance on the orbit stability at the second sextupole, with regard to orbit perturbations generated between the pair of sextupoles, is

$$\Delta x < \frac{1}{5 K_s \beta_s}, \tag{48}$$

corresponding to a 2% increase in the absolute spot size.

Case Study II: Length Scaling of Final-Focus Systems

Assuming that the length of the final focus, $l_{\rm FF}$, increases in proportion to the length of the bending magnets, derive a scaling law for $l_{\rm FF}$ as a function of γ, β^* and Δx.

Solution: Combining Eqs. (41) and (46), we find that

$$\delta_{\rm rms} \propto \frac{D_s^{3/2}}{l_{\rm FF}^{5/2}} \gamma^{5/2}. \tag{49}$$

Together with Eqs. (42) and (44), and limiting the blow-up $\Delta\sigma_y^*/\sigma_y^*$ to, e.g., a value of 0.2, this translates into

$$\frac{D_s^{3/2}\gamma^{5/2}}{l_{FF}^{5/2}}\frac{l^*}{\beta_y^*} < \text{constant}, \tag{50}$$

where the value of the constant on the right-hand side depends on the coefficient relating D_s and $(l_b\theta)$ in Eq. (46). Solving Eqs. (48) and (45) for D_s and using Eq. (44), we find

$$D_s > \frac{5}{2}\Delta x\,\xi_y = \frac{5}{2}\Delta x\,\frac{l^*}{\beta_y^*}. \tag{51}$$

Inserting this into Eq. (50), we obtain the desired scaling law [30]

$$l_{FF} \propto \gamma\Delta x^{3/5}\frac{l^*}{\beta_y^*}. \tag{52}$$

This scaling law predicts many kilometers or even tens of kilometers of final-focus lengths for a few-TeV collider. Therefore, novel final-focus concepts that may allow for a shorter system are now under investigation. One approach is to perform the chromatic correction with sextupoles near the final quadrupole, accepting a nonzero slope of dispersion at the collision point [31]. Another approach is to use a high-frequency rf quadrupole, which can compensate for the correlated energy spread across the bunch [32].

IV COLLIMATION

Collimators are special elements that are positioned closest to the beam. Located somewhere between the linac and the interaction point, their primary function is to remove beam halo at large amplitudes, typically corresponding to 10–20 times the horizontal or 50–80 times the vertical rms beam size. Additional collimation may be required also before or in the linac.

Removal of the halo particles is necessary, since, if lost at aperture restrictions in the final focus or near the collision point, they can produce electromagnetic showers, muons and neutrons, or, if traversing the final quadrupoles with a large transverse offset, they can emit wide-angle synchrotron radiation. In either case, the halo particles may cause unacceptable background in the particle-physics detector. For an ideal linear collider there are only a few unavoidable sources of beam halo, such as scattering off residual gas or thermal photons (blackbody radiation). According to conservative estimates, the known scattering sources result in only about 10^2–10^4 halo particles per bunch train [33,34].

On the other hand, at the SLC, a large number of collimators were added over the first years of operation in an attempt to render the experimental conditions acceptable. Occasionally, more than 10% of the beam had to be scraped at the collimators over periods of hours. There was little quantitave understanding or modeling of the observed halo at the SLC, but the suspected culprits include magnet nonlinearities in the bunch compressor, longitudinal microwave instability in the damping rings, beam dynamics in the linac, and higher-order dispersion.

It is expected that at future colliders an improved design of damping rings and ring-to-linac transfer lines as well as a pre-collimation stage in front of the linac will reduce the halo reaching the end of the linac by several orders of magnitude compared with the SLC value.

The collimators not only remove halo, but also serve a second purpose. In case of a failure (e.g., mis-firing of ring extraction kicker or missing drive beam), they are the elements first hit by a mis-steered beam. Therefore, a common collimator design requirement is that the collimators should survive the single impact of one entire bunch train. After the loss of a mis-steered bunch train and before the next linac pulse, the accelerator can be switched off, and then restarted with a smaller number of low-charge bunches.

For the parameters of all future linear colliders, guaranteeing the collimator survival is a major challenge. Already at the SLC, which should have operated in a safe regime, many collimators have been damaged by the beam [35]. The collimator survival condition can be written as a lower limit on the beam size at the collimator [36],

$$\sigma_x \sigma_y > \frac{\alpha_T Y}{\sigma_{\text{UTS}} C_p} \frac{d\mathcal{E}}{dx} \left(\frac{n_b N_b}{2\pi} \right), \qquad (53)$$

where σ_{UTS} is the ultimate tensile strength, α_T the linear thermal expansion coefficient, C_p the heat capacity, Y the elastic modulus, $d\mathcal{E}/dx$ the specific energy loss per unit length, n_b the number of bunches in the train, and N_b the bunch population.

As an example, taking the material properties of copper, $\alpha = 1.7 \times 10^{-5}$ K^{-1}, $E = 120$ GPa, $C_p = 0.385$ Jg^{-1}K^{-1}, $d\mathcal{E}/dx \approx 1.44$ MeV cm^2/g, and $\sigma_{UTS} = 300$ MPa, we find

$$(\sigma_x \sigma_y)^{1/2} \geq 200 \ \mu\text{m} \quad \text{or} \quad \beta_{x,y} \geq 1000 \text{ km},$$

where, in the last step, we assumed normalized emittances of $\gamma \epsilon_x = 0.68$ μm and $\gamma \epsilon_y = 0.02$ μm at 1.5-TeV beam energy. Such enormous beta-functions imply a long system and tight tolerances.

Fortunately, the above estimates are too pessimistic. For many conceivable failure modes, the emittance will blow up considerably before the beam hits the collimator. An example is shown in Fig. 4. In the simulation of a mis-steered beam, rapid filamentation due to large energy spread leads to an emittance increase by two or three order of magnitude. This suggests that the nominal beta-functions at the collimators could be reduced accordingly. The required values then would appear more reasonable.

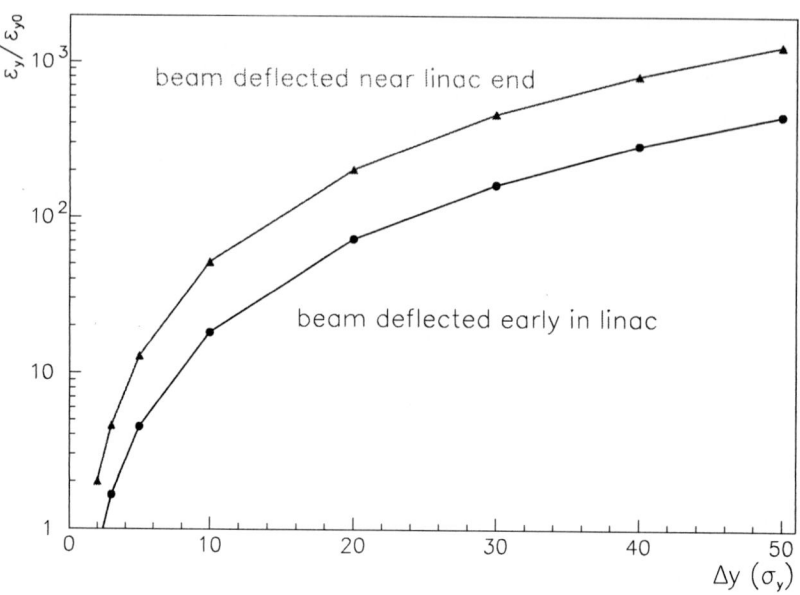

FIGURE 4. Emittance growth of a beam deflected in the early (bottom curve) or late part (top curve) of the linac as a function of induced oscillation amplitude in units of the rms beam size [37]. Because of rapid filamentation an unstable beam is accompanied by significant emittance blow-up.

The development of a workable collimator system is still a matter of active research. The final design will depend on the answers to the following questions:

- How large is the beam halo?

- How many muons produced per bunch train are acceptable for the detector?

- Which are the failure modes that can mis-steer the beam, and what is the resulting emittance increase for each of these?

The NLC design is considering the deployment of replaceable or renewable collimators [38].

V LINAC

The linac is the heart of a linear collider. It should not only accelerate the beam to high energy, but in addition preserve the transverse emittance and also supply the beam stably to the downstream final-focus and collimation systems. Many perturbation sources can cause pulse-to-pulse variation in the linac trajectory, for example, mechanical vibration of magnet supports, ground motion, small changes in the initial beam distribution from the damping ring, or drifts or fluctuations in the rf systems. Most dangerous are changes that occur from one rf pulse to the next, because it is extremely difficult to counteract these by feedback.

A Dispersion

We will now look more closely at the beam dynamics in a linac, in order to understand how emittance growth can occur. Both the beam centroid and the individual beam particles perform betatron oscillations along the linac. If we ignore the energy change due to acceleration, the betatron oscillation for arbitary momentum error δ is a solution of Hill's equation

$$x'' + \frac{K(s)}{1+\delta} x = 0, \tag{54}$$

where the quadrupole focusing strength $K(s)$ in units of m^{-2} is a function of the longitudinal position s, and the prime denotes the derivative with respect to s. In the presence of dipole magnets or quadrupole misalignments, the right hand-side of this equation would contain an inhomogeneous term $1/\rho$, but we will ignore this for the moment. We next use a smooth approximation and replace the s dependent force by a constant average. Instead of Eq. (54), we thus write

$$x'' + \frac{K}{1+\delta} x = 0 \tag{55}$$

where we may identify $K = \langle K(s) \rangle_s$ with $1/\beta^2$, the inverse square of the smoothed beta function. The solution can be written as an expansion in δ:

$$x = x_\beta + D\delta + D^{(2)}\delta^2 + ... \tag{56}$$

where D is the dispersion and $D^{(2)}$ the second-order dispersion. Inserting this solution into Eq. (55) and equating terms with equal powers of δ, we obtain an equation for the on-energy betatron motion,

$$x_\beta'' + Kx_\beta = 0, \tag{57}$$

and, from the terms of first order in δ, an equation for the dispersion,

$$D'' + KD = Kx_\beta. \tag{58}$$

According to Eqs. (57) and (58) a betatron oscillation propogating through the linac resonantly drives the dispersion: the natural oscillation frequencies of x_β and D are identical and x_β enters as an excitation on the right-hand side of Eq. (58).

Dispersion in a linac arises from deflections by misaligned quadrupoles or by wake fields excited by a beam passing off-center through an accelerating structure. We can estimate the magnitude of these effects by considering a two-particle model, where the bunch is represented by a leading and a trailing (macro-)particle, denoted by 1 and 2, each with half the bunch charge and separated by, e.g., twice the rms bunch length. For simplicity, we will ignore acceleration and use a smooth approximation for the betatron motion. We first look at the leading particle.

Case Study III: Linac Dispersion

(1) Dispersion for free betatron oscillation. Consider a beam deflected by an angle θ at $s = 0$. The betatron motion of the leading particle is described by

$$x_1''(s) + \frac{k_\beta^2}{1+\delta_1} x_1(s) = \frac{\theta}{1+\delta_1} \delta(s) \tag{59}$$

where $\delta(s)$ is the Dirac delta function, indicating a single deflection of strength θ at location $s = 0$, and $k_\beta = \sqrt{K}$ is the wave number of the betatron oscillation. Equation (59) describes the motion of an individual off-momentum particle in the head of the bunch, as well as the centroid motion of the bunch head if the latter experiences a centroid momentum offset δ_1. The equality of the particle and centroid motion (for small bunch charges) is an advantage, since the single-particle dispersion can be measured by observing the response of the centroid motion to an energy error.

Solve Eq. (59), linearize the solution in δ_1 and determine the dispersion for $k_\beta s \delta_1 \ll 1$.

Solution: This is the equation of a harmonic oscillator. The solution is [39]

$$x_1(s) = \frac{\theta}{k_\beta \sqrt{1+\delta_1}} \sin \frac{k_\beta s}{\sqrt{1+\delta_1}} \approx \frac{\theta}{k_\beta \sqrt{1+\delta_1}} \left[\sin k_\beta s - \frac{1}{2} k_\beta s \delta_1 \cos k_\beta s \right] \tag{60}$$

or

$$x_1(s) \approx \frac{\theta}{k_\beta} \sin k_\beta s - \frac{1}{2} \left[\theta s \cos k_\beta s + \frac{\theta}{k_\beta} \sin k_\beta s \right] \delta_1 + \mathcal{O}(\delta_1^2). \tag{61}$$

From the term linear in δ_1 we infer the dispersion at the bunch head,

$$D_1(s) = -\frac{1}{2} \left[\theta s \cos k_\beta s + \frac{\theta}{k_\beta} \sin k_\beta s \right]. \tag{62}$$

The solution is illustrated in Fig. 5. The linear increase with s reflects that the dispersion is resonantly driven [39].

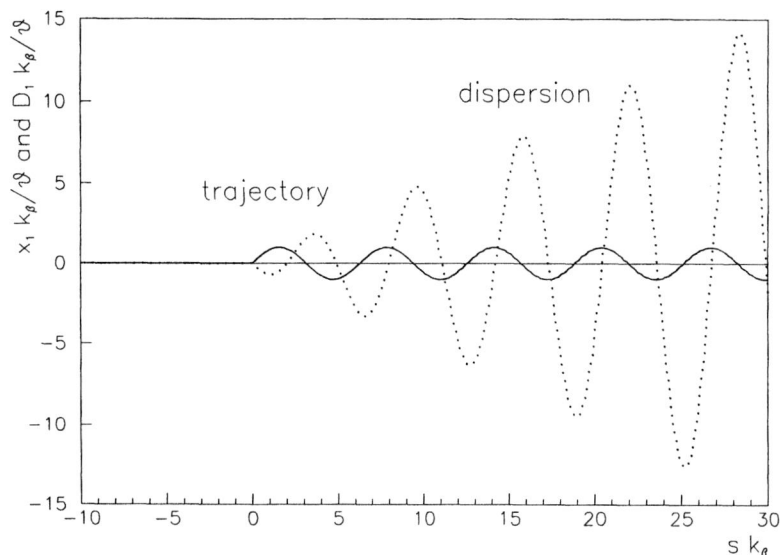

FIGURE 5. Trajectory oscillation, $x_1 k_\beta/\theta$ for $\delta_1 = 0$, and resonantly growing dispersion at the bunch head, $D_1 k_\beta/\theta$, induced by a deflection at $s = 0$, according to Eqs. (61) and (62).

(2) Dispersion behind π bump. Orbit correction can be thought of as a superposition of π bumps. Calculate the dispersion generated by a bump, represented by two kicks θ, at $s_1 = 0$ and $s_2 = \pi/k_\beta$.

<u>Solution:</u> The dispersion generated by a single kick at $s = 0$, $D_1(s)$, was computed in Eq. (62). The dispersion generated by the second kick is obtained by simply shifting the argument by s_2, i.e., it is given by $D_1(s - \pi/k_\beta)$. The dispersion arising from the π bump is then the sum of the terms generated by the two kicks [39]:

$$D_\pi = D_1(s) + D_1(s - \pi/k_\beta) = -\frac{\theta\pi}{2k_\beta} \cos k_\beta s. \qquad (63)$$

The solution is illustrated in Fig. 6. While the orbit after the π bump is zero, the dispersion propagates at a constant amplitude. A perfectly centered orbit in the downstream linac section does not imply that the dispersion is zero as well.

The linearly growing dispersion component in a linac arises from the energy dependence of the oscillation frequency (chromaticity). Note that the definition of the linac dispersion D_1 differs from that of the periodic dispersion in a storage ring. Nevertheless, also the latter can be efficiently controlled by 'resonant' orbit bumps [40].

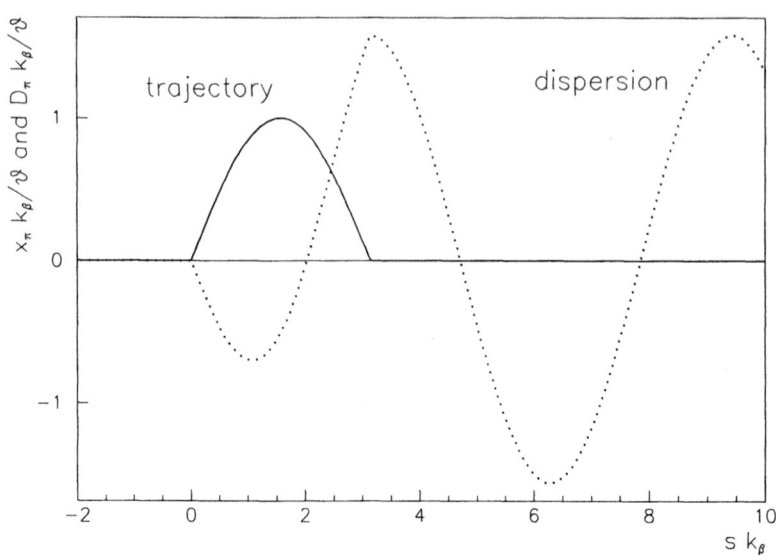

FIGURE 6. Trajectory perturbation, $x_\pi k_\beta/\theta$, and subsequent constant dispersion, $D_\pi k_\beta/\theta$, induced by a π bump, according to Eq. (63).

B Pulse-to-Pulse Stability

Consider the situation where the vertical positions of the linac quadrupoles vary from one pulse to the next. Each of the misaligned quadrupoles will induce a betatron oscillation travelling down the linac. These betatron oscillations in turn give rise to dispersion. Because of the resonant excitation the dispersion grows much faster than the betatron motion. For this reason, the blow up of the final beam size at the end of the linac due to energy spread and accumulated dispersion may cause a larger luminosity loss than the beam-beam separation induced by the centroid betatron motion.

If the ith quadrupole at location i is misaligned by an amount Δy, the beam is deflected by $\theta_i = K_q^i \Delta y$, where K_q denotes the integrated strength of the quadrupole (in units of m^{-1}). From the above solution, the amplitude of the resulting dispersion at the end of the linac is $D_i = \frac{1}{2}\theta_i(L - s_i)$, with L denoting the full linac length and s_i the position of the quadrupole. If a total of N_q quadrupoles randomly shift in position by an rms value $\Delta y_{\rm rms}$, the expectation value for the final dispersion is the incoherent sum of the individual contributions, and a rough estimate for the rms dispersion at the end of the linac is [43]

$$D_{\rm rms} = \frac{1}{4\sqrt{2}}\theta_{\rm rms} L\sqrt{N_q} = \frac{1}{4\sqrt{2}} K_q L \sqrt{N_q} \Delta y_{\rm rms}. \tag{64}$$

The factor $1/\sqrt{2}$ is the rms value of the cosine function in Eq. (62), where we have assumed a uniform distribution of phase advances between the quadrupoles and the IP. In order to re-express K_q in terms of the average beta function, we note that the betatron phase advance over a linac FODO cell of length $L_{\rm cell}$ is $\phi_{\rm cell} \approx L_{\rm cell}/\beta \approx \pi/2$. This relation is strictly true for a 90° lattice. In addition, the cell length and the quadrupole strength are roughly related as $L_{\rm cell} \approx 2\sqrt{2}/K_q$, and we thus approximate $K_q \approx 2/\beta$. Note that again we have considered a constant beam energy and a smooth beta function.

Typically we require that the absolute beam size increase due to the additional dispersion generated between two linac pulses,

$$\Delta \sigma_y = \left(\sqrt{\sigma_y^2 + (D_{\rm rms} \delta_{\rm rms})^2} - \sigma_y \right), \tag{65}$$

should be less than 2%, or $D_{\rm rms}\delta_{\rm rms} < \sigma_y/5$, where $\sigma_y = \sqrt{\epsilon_y \beta_y}$ is the rms beam size. We then obtain the following tolerance on the pulse-to-pulse quadrupole stability:

$$\Delta y_{\rm rms} < \frac{2\sqrt{2}\sigma_y \beta}{5\sqrt{N_q}\delta_{\rm rms} L}. \tag{66}$$

Inserting parameters that approximately correspond to the CLIC linac, $\beta_y \approx 7$ m, $\gamma\epsilon_y \approx 5\times 10^{-9}$ m, $L = 15$ km, $\delta_{\rm rms} \approx 1\%$, $\gamma \approx 2\times 10^4$, and $N_q = 1500$, we find a tight tolerance: $\delta y_{\rm rms} \lesssim 1$ nm, i.e., between two pulses the rms quadrupole motion in the CLIC linac should be less than 1 nanometer! The example's parameters were chosen such that this estimate agrees with the result of an elaborate simulation [41], which includes the actual energy profile along the bunch, wake fields, and acceleration. Refined analytical formulae and many other details on emittance preservation for linear colliders can be found in Ref. [42].

C Wake Fields

Until now we have looked only at dispersion. Dispersion increases the emittance because of the nonzero energy spread across the bunch. A second important effect degrading linac emittance is transverse wake fields. The effect of a single-bunch wake field can be illustrated using the two-particle model [12]. For simplicity, we consider only the case without momentum deviation ($\delta_1 = \delta_2 = 0$). We have seen already that the bunch head performs the usual betatron motion

$$x_1'' + k_\beta^2 x_1 = 0 \tag{67}$$

and that the solution of Eq. (67) is $x_1 = \hat{x} \cos k_\beta s$. In the two-particle model, the equation of motion for the trailing macro-particle is

$$x_2'' + k_\beta^2 x_2 = -\frac{N_b r_e W_1}{2\gamma} x_1. \tag{68}$$

The term on the right-hand side represents the wake field excited by the leading particle with a charge of $N_b/2$. The strength of the wake, W_1, in units of inverse cubic length, depends on the distance of the bunch head from the tail, hence on the rms bunch length. Typically, for short bunches, $W_1 < 0$ and the wake is "defocusing," *i.e.*, the tune shift for a rigid coherent betatron oscillation is negative.

Since the oscillation frequencies of x_1 and x_2 are the same, the trailing particle is resonantly driven. The solution of Eq. (68) becomes

$$x_2 = \hat{x} \cos k_\beta s - \Upsilon_{\text{BBU}} \frac{s}{L} \hat{x} \sin k_\beta s \tag{69}$$

where

$$\Upsilon_{\text{BBU}} = -\frac{N_b r_e W_1 L}{4\gamma \sqrt{K}} \tag{70}$$

is called the beam break up parameter, and L denotes the total length of the linac. We have chosen the initial conditions $x_1(0) = x_2(0)$ and $x_1'(0) = x_2'(0) = 0$.

D Acceleration

So far we have completely ignored the main task of a linac, which is to accelerate, *i.e.*, to increase the beam energy. If we include acceleration, Eq. (67) for the head particle is replaced by

$$\frac{d}{ds}\left[\gamma(s)\frac{dx_1}{ds}\right] + k_\beta^2 x_1 = 0 \tag{71}$$

where $\gamma(s) = \gamma_i(1+\alpha s)$ is the relativistic factor with $\gamma_i = \gamma(0)$, and α is a constant proportional to the acceleration gradient. Here we again assume $\delta_1 = \delta_2 = 0$. The solution to Eq. (71) depends on the way the betatron wavenumber varies along the linac. Most future designs assume a scaling close to

$$k_\beta \approx \text{constant} \tag{72}$$

which implies

$$\beta \approx \gamma^{1/2}. \tag{73}$$

This scaling can be realized by increasing the lengths of all elements as $\sqrt{\gamma}$ and keeping the quadrupole pole-tip fields constant. Introducing the new independent variable $z = \sqrt{1+\alpha s}$, Eq. (71) is rewritten as

$$z^2 \frac{d^2}{dz^2} x_1 + z \frac{d}{dz} x_1 + 4 \frac{k_\beta^2}{\alpha^2} z^2 x_1 = 0. \tag{74}$$

This is a Bessel equation. For the initial condition $x_1(0) = \hat{x}$, $x_1'(0) = 0$ the solution for the head of the bunch is

$$\begin{aligned} x_1 = -\frac{\pi k_\beta}{\alpha} \hat{x} \Bigg[& Y_1\left(\frac{2k_\beta}{\alpha}\right) J_0\left(\frac{2k_\beta}{\alpha}\sqrt{1+\alpha s}\right) \\ & - Y_0\left(\frac{2k_\beta}{\alpha}\right) J_1\left(\frac{2k_\beta}{\alpha}\sqrt{1+\alpha s}\right) \Bigg]. \end{aligned}$$

A useful relation exists between the modified Bessel functions $Y_{0,1}$ and $J_{0,1}$,

$$Y_0(w) J_1(w) - Y_1(w) J_0(w) = \frac{2}{\pi w}, \tag{75}$$

and, for large arguments, $w \gg 1$, these Bessel functions may be further approximated as

$$J_0(w) \approx -Y_1(w) \approx \sqrt{\frac{2}{\pi w}} \cos\left(w - \frac{\pi}{4}\right), \tag{76}$$

$$J_1(w) \approx Y_0(w) \approx \sqrt{\frac{2}{\pi w}} \sin\left(w - \frac{\pi}{4}\right). \tag{77}$$

Using the above relations, we can rewrite our solution as

$$x_1(s) \approx \hat{x} \left[\frac{1}{(1+\alpha s)^{1/4}} \cos\left(\frac{2k_\beta}{\alpha}(1 - \sqrt{1+\alpha s})\right) \right]. \tag{78}$$

Note that the oscillation amplitude decreases as $|x_1| \propto \sqrt{\beta/\gamma} \sim \gamma^{-1/4}$, due to the combined effect of adiabatic damping and variation in focusing strength along the linac.

The equation of motion for the tail particles again includes the wake field:

$$x_2'' + \frac{d\gamma/ds}{\gamma} x_2' + k_\beta^2 x_2 = -\frac{Nr_0 W_1}{2\gamma} x_1 \tag{79}$$

with $z = \sqrt{1+\alpha s}$. The solution is

$$x_2(z) = x_1(z) - \frac{2Nr_0 W_1}{\gamma \alpha^2} \int_1^z dz' G(z, z') y_1(z') \tag{80}$$

with the Green function

$$G(z, z') = \frac{\pi}{2} z' \left\{ J_0\left[\frac{2k_\beta}{\alpha} z'\right] Y_0\left[\frac{2k_\beta}{\alpha} z\right] - Y_0\left[\frac{2k_\beta}{\alpha} z'\right] J_0\left[\frac{2k_\beta}{\alpha} z\right] \right\}. \tag{81}$$

In the limit $\alpha \ll k_\beta$ this simplifies to

$$G(z, z') \approx \sqrt{\frac{z'}{z} \frac{\alpha}{2k_\beta}} \sin\left(\frac{2k_\beta}{\alpha}(z - z')\right) \qquad (82)$$

where again $z = \sqrt{1 + \alpha s}$. The final solution is

$$x_2(s) \approx \hat{x} \frac{1}{(1+\alpha s)^{1/4}} \left[\cos\left(\frac{2k_\beta}{\alpha}(\sqrt{1+\alpha s} - 1)\right) \right.$$
$$\left. - \frac{Nr_0 W_1(\sigma_z)}{2\gamma_i \alpha k_\beta}(\sqrt{1+\alpha s} - 1) \sin\left(\frac{2k_\beta}{\alpha}(\sqrt{1+\alpha s} - 1)\right) \right].$$

Hence, for $s = L$ and $\alpha L \gg 1$, the beam break up parameter with acceleration is

$$\Upsilon_{\rm BBU} = -\frac{NW_1 r_0 L}{2k_\beta} \frac{1}{\sqrt{\gamma_f \gamma_i}}. \qquad (83)$$

It can be obtained from the no-acceleration result by simply replacing $1/\gamma$ with $2/\sqrt{\gamma_f \gamma_i}$.

Assuming a constant beta-function along the linac instead of $\beta \propto \gamma^{1/2}$, a similar expression is found; see Ref. [12]. In this case, the factor $1/\gamma$ in the no-acceleration version of $\Upsilon_{\rm BBU}$ must be replaced by $1/\gamma_f \ln(\gamma_f/\gamma_i)$.

Thus, it is often convenient to perform calculations first without acceleration and then to include it at the end, using the appropriate substitution rule.

For the Stanford Linear Collider (SLC), $\Upsilon_{\rm BBU} \approx 15$, which means that if the beam was injected with a certain betatron oscillation, the oscillation of the tail increased by a factor of 15 along the linac. For the proposed 3-TeV CLIC linac, we compute $\Upsilon_{\rm BBU} \approx 5$.

(3) Dispersion with wake field. The equation of motion for the tail of the bunch includes the effect of the wake generated by the bunch head as an additional excitation term on the right-hand side,

$$x_2''(s) + \frac{k_\beta^2}{1+\delta_2} x_2(s) = \frac{\theta}{1+\delta_2}\delta(s) - \frac{Nr_0 W_1}{2\gamma(1+\delta_2)} x_1(s). \qquad (84)$$

Solve this equation for $\delta_1 = 0$, determine the trajectory for $\delta_2 = 0$, and, by linearizing in δ_2, the additional dispersion arising from the wake field, possibly using the beam break up parameter $\Upsilon_{\rm BBU}$. Acceleration can be approximately included 'a posteriori' by inserting the correctly modified value of $\Upsilon_{\rm BBU}$. The motion of the head particle was calculated above. For $\delta_1 = 0$ it is

$$x_1(s) = \frac{\theta}{k_\beta} \sin k_\beta s. \qquad (85)$$

Solution: The solution is the sum of an oscillation at the natural frequency of the tail particle, $k_\beta/\sqrt{1+\delta_2}$, and a response to the head driving force at frequency k_β. We thus make the ansatz

$$x_2(s) = A\sin k_\beta s + B\sin\frac{k_\beta}{1+\delta_2} + C\cos\frac{k_\beta}{1+\delta_2}s. \tag{86}$$

Inserting this ansatz into Eq. (84) and also considering the two initial conditions $x_2(0) = 0$ and $x'_2(0) = x'_1(0) = \theta$, we can solve for the three constants of integration, A, B and C. We then obtain the solution of Eq. (84) as [39]

$$x_2(s) = \frac{\theta}{k_\beta}\left(\frac{1}{\sqrt{1+\delta_2}}\sin\frac{k_\beta s}{\sqrt{1+\delta_2}} - \frac{\Upsilon_{\mathrm{BBU}}2}{Lk_\beta\delta_2}\left(\sin k_\beta s - \sqrt{1+\delta_2}\sin\frac{k_\beta s}{\sqrt{1+\delta_2}}\right)\right).$$

For $k_\beta s\delta_2 \ll 1$ we can again expand in δ_2. We find

$$\begin{aligned}x_2(s) = &\frac{\theta}{k_\beta}\sin k_\beta s - \frac{\Upsilon_{\mathrm{BBU}}\theta}{k_\beta L}\left(s\cos k_\beta s - \frac{1}{k_\beta}\sin k_\beta s\right) \\ &- \frac{\Upsilon_{\mathrm{BBU}}\theta}{4k_\beta L}\left(k_\beta s^2\sin k_\beta s - s\cos k_\beta s + \frac{1}{k_\beta}\sin k_\beta s\right)\delta_2 \\ &- \frac{1}{2}\left(\theta s\cos k_\beta s + \frac{\theta}{k_\beta}\sin k_\beta s\right)\delta_2.\end{aligned} \tag{87}$$

We identify the term linear in δ_2 with the tail dispersion

$$\begin{aligned}D_2(s) = &-\frac{\Upsilon_{\mathrm{BBU}}\theta}{4k_\beta L}\left(k_\beta s^2\sin k_\beta s - s\cos k_\beta s + \frac{1}{k_\beta}\sin k_\beta s\right) \\ &- \frac{1}{2}\left(\theta s\cos k_\beta s + \frac{\theta}{k_\beta}\sin k_\beta s\right).\end{aligned} \tag{88}$$

The solution is sketched in Fig. 7. Note that the wake-induced dispersion is doubly resonantly driven, and grows quadratically with distance!

At the end of the linac the amplitude ratio of the wake-induced dispersion and the regular dispersion (without the wake-field effect) generated by a deflection early in the linac is $\Upsilon_{\mathrm{BBU}}/2$. Hence, at the SLC, with $\Upsilon_{\mathrm{BBU}} \approx 15$, the wake-induced dispersion was about 7 times larger.

(4) BNS damping is one way of avoiding the beam break up caused by wake fields. It is named after its inventors Balakin, Novokhatsky and Smirnov [44], who showed that the effect of the wake field on the tail of the bunch can be partially compensated by increasing the focusing strength for the tail particles, from k_β to $k_\beta + \Delta k_\beta$ [44]. For a free betatron oscillation propagating through the linac, the additional focusing counteracts the wake-field "defocusing" for a coherent betatron

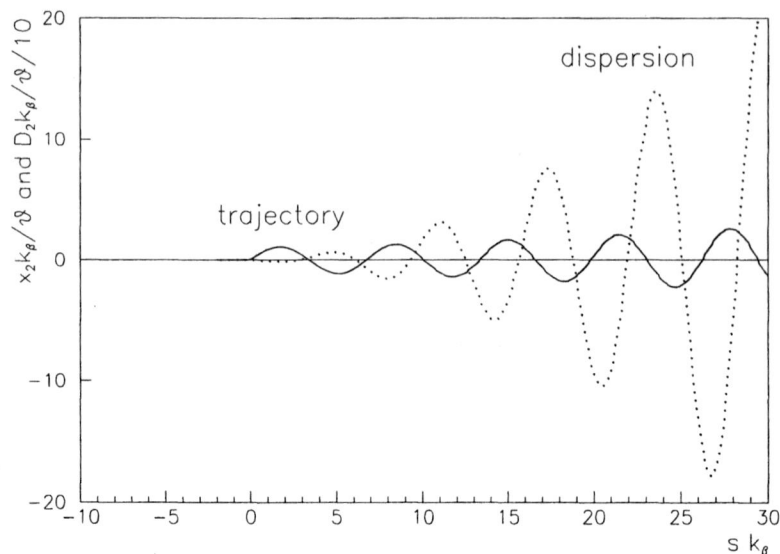

FIGURE 7. Growing oscillation of the trailing macroparticle, $x_2 k_\beta/\theta$, due to beam break up, and the doubly resonantly driven tail dispersion, $D_2 k_\beta/\theta/10$, induced by a deflection at $s = 0$ according to Eq. (87), for $\Upsilon_{\text{BBU}}/(k_\beta L) = 0.08$ (SLC value).

oscillation. To this end either rf quadrupoles with rapidly varying field can be used [45] or the position of the bunch can be adjusted with respect to the crest of the rf wave so that the tail acquires less energy than the head.

With BNS damping the equation of motion for the trailing macroparticle is

$$x_2''(s) + \frac{(k_\beta^2 + \Delta k_\beta^2)}{1 + \delta_1} x_2(s) = \frac{\theta}{1 + \delta_2} \delta(s) - \frac{N r_0 W_1}{2\gamma(1 + \delta_2)} x_1(s) \tag{89}$$

where again $x_1(s) = (\theta/k_\beta) \sin k_\beta s$ assuming $\delta_1 = 0$. BNS damping is achieved if

$$\left(1 + \frac{\Delta k_\beta}{k_\beta}\right)^2 = 1 + \frac{2\Upsilon_{\text{BBU}}}{k_\beta L_0}. \tag{90}$$

Show that under this condition and with $\delta_2 = 0$ the orbit of a trailing particle is identical to that of the bunch head: $x_2(s) = x_1(s)$.

Evaluate the dispersion of the trailing particle x_1 with BNS damping, assuming $\Delta k_\beta \ll k_\beta$, $\delta \ll 1$, $k_\beta s \delta \ll 1$, $\Upsilon_{\text{BBU}}/(k_\beta L) \ll 1$, and $\delta \ll 2\Upsilon_{\text{BBU}}/(k_\beta L_0)$.

<u>Solution:</u> The solution to Eq. (89) is [39]

$$x_2(s) = \frac{\theta}{(k_\beta + \Delta k_\beta)\sqrt{1+\delta_2}} \sin\frac{(k_\beta + \Delta k_\beta)s}{\sqrt{1+\delta_2}} - \frac{2\Upsilon_{\rm BBU}\theta}{k_\beta L\left(1+\delta_2 - \left(1+\frac{\Delta k_\beta}{k_\beta}\right)^2\right)}$$

$$\times \left[\frac{1}{k_\beta}\sin k_\beta s - \frac{\sqrt{1+\delta_2}}{k_\beta + \Delta k_\beta}\sin\frac{(k_\beta + \Delta k_\beta)s}{\sqrt{1+\delta_2}}\right]. \tag{91}$$

Eliminating $(k_\beta + \Delta k_\beta)$ using the BNS condition, Eq. (90), this solution simplifies to [39]

$$x_2(s) = x_1(s) + \frac{\theta\delta_2}{k_\beta\left(\delta_2 - \frac{2\Upsilon_{\rm BBU}}{k_\beta L}\right)}\left[\sqrt{\frac{1+\frac{2\Upsilon_{\rm BBU}}{k_\beta L}}{1+\delta_2}}\sin\left(k_\beta s\sqrt{\frac{1+\frac{2\Upsilon_{\rm BBU}}{k_\beta L}}{1+\delta_2}}\right) - \sin k_\beta s\right]$$

$$= x_1(s) + D_2(s)\delta_2 + \mathcal{O}(\delta_2^2) \tag{92}$$

with

$$D_2(s) = -\frac{\theta L}{2\Upsilon_{\rm BBU}}\left[\sin\left(k_\beta s + \frac{\Upsilon_{\rm BBU}}{L}s\right) - \sin k_\beta s\right]. \tag{93}$$

Note that with BNS damping the tail particle experiences no regular dispersion, and the wake-induced dispersion is suppressed by a factor $1/\Upsilon_{\rm BBU}$. The solution is illustrated in Fig. 8.

(5) Dispersion from a misaligned structure. The kick θ discussed above affected both head and tail particles. If a structure, e.g., an rf accelerating cavity, is misaligned, only the tail receives a kick θ_s while the head orbit is unperturbed. We can write the centroid motion of head and tail as $x_1(s) = 0$, and $x_2(s) = (\theta_s/k_\beta)\sin k_\beta s$, assuming that the centroid momentum errors are zero. Orbit correction will apply a kick $\theta = -\theta_s/2$ to both the head and tail particles in order to correct the centroid motion.

Compute the dispersion for a particle in the bunch tail after orbit correction, assuming that the BNS condition is fulfilled. Compare the result with the dispersion generated by a π bump through a misaligned quadrupole given by Eq. (63).

<u>Solution:</u> *After correcting the centroid orbit behind the misaligned structure, the particle motion for the head and tail of the bunch is roughly given by (see Eq. (60)):*

$$x_1(s) \approx -\frac{\theta_s}{2k_\beta\sqrt{1+\delta_1}}\sin\frac{k_\beta s}{\sqrt{1+\delta_1}}, \tag{94}$$

$$x_2(s) \approx \frac{\theta_s}{k_\beta\sqrt{1+\delta_2}}\sin\frac{k_\beta s}{\sqrt{1+\delta_2}} - \frac{\theta_s}{2k_\beta\sqrt{1+\delta_2}}\sin k_\beta s. \tag{95}$$

Dispersion generated by the second term in the equation for x_2 will not be significant if the BNS condition is fulfilled; see Eq. (93). Thus the dispersion in the tail comes mainly from the first term on the right hand side, which represents the deflection by the wake field. It is approximately given by Eq. (62) with $\theta = \theta_s/2$ [39]:

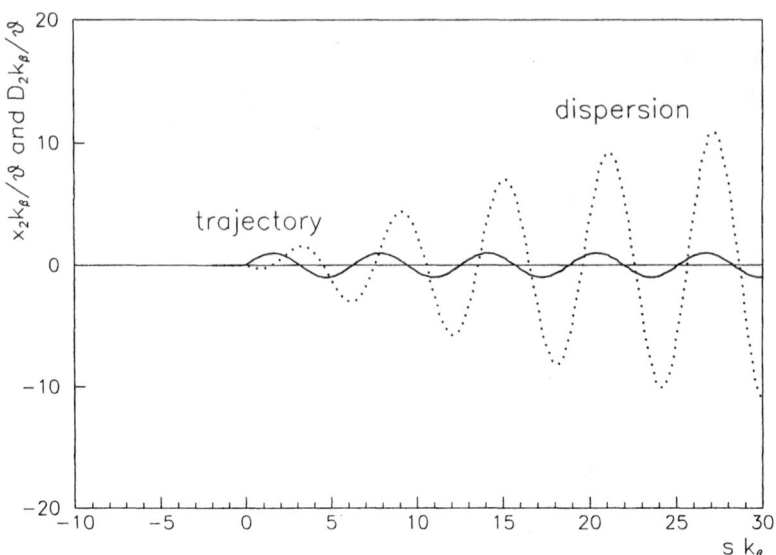

FIGURE 8. Oscillation of the trailing macroparticle, $x_2 k_\beta/\theta$, with wake fields and BNS damping, and the tail dispersion, $D_2 k_\beta/\theta$, induced by a deflection at $s = 0$, according to Eqs. (92) and (93), for $\Upsilon_{\text{BBU}}/(k_\beta L) = 0.08$ (SLC value).

$$D_2(s) \approx -\frac{\theta_s}{4}\left[s \cos k_\beta s + \frac{1}{k_\beta} \sin k_\beta s \right]. \quad (96)$$

The dispersion for the head particles is also obtained from Eq. (62), but with $\theta = -\theta_s/2$. It thus is of the same magnitude, but of opposite sign: $D_1(s) = -D_2(s)$. Therefore, even after perfectly steering the orbit through the center of all beam-position monitors (BPMs) and quadrupoles, there can still be significant dispersion across the bunch. This solution is illustrated in Fig. 9.

E Quadrupole and Structure Misalignment

Note that, for a quadrupole misalignment, orbit correction via π bumps leads to a constant residual dispersion, whereas in case of a misaligned structure the orbit is corrected essentially by a single deflection and as a result the dispersion grows resonantly.

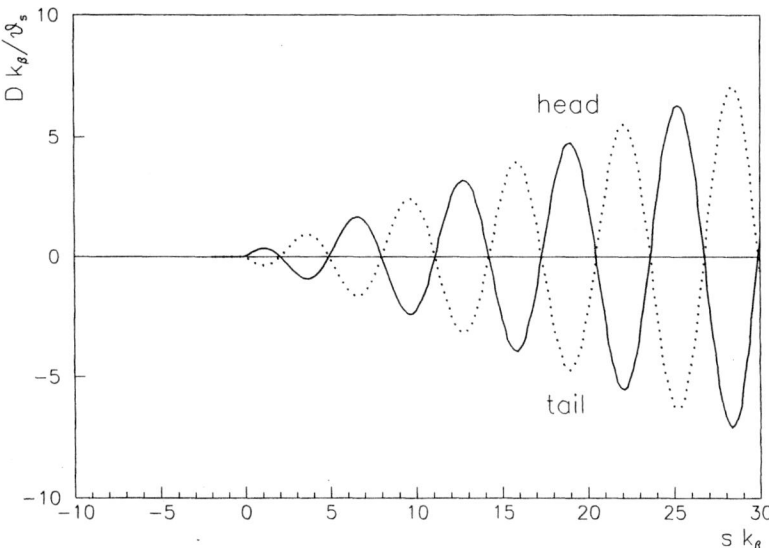

FIGURE 9. Dispersion for particles in bunch head and tail, Dk_β/θ_s, induced by a misaligned structure at $s = 0$ deflecting the tail by θ_s and subsequent orbit correction, according to Eq. (96). The dispersion functions at the head and the tail are of opposite sign and both grow linearly along the linac, after Ref. [43].

We can evaluate the magnitude of the deflection expected from a single structure misalignment, using again the two-particle model. Traversing a structure misaligned by $\Delta x_{\rm acc}$, the tail will experience a kick

$$\theta_s = \frac{N_b r_e (W_1 l_{\rm acc})}{2\gamma} \Delta x_{\rm acc} \tag{97}$$

where $l_{\rm acc}$ denotes the length of the misaligned accelerator structure, and W_1 the transverse wake field per unit length in the structure, at a distance of 1–2 σ_z. With $\gamma = 2 \times 10^4$, $N = 4 \times 10^9$, $W_1 \approx 4 \times 10^6$ m^{-3}, $l_{\rm acc} = 0.5$ m, we find $\theta_s \approx 6 \times 10^{-4}\,\Delta x_{\rm acc}$ [m].

On the other hand, approximating the integrated quadrupole strength as $K_q \approx 2/\beta$, for a quadrupole misaligned by $\Delta x_{\rm quad}$ we estimate the deflection from a displaced quadrupole as

$$\theta = K_q \Delta x_{\rm quad} \approx \frac{2}{\beta} \Delta x_{\rm quad}, \tag{98}$$

or $\theta \approx 2 \times 10^{-1} \Delta x_{\text{quad}}$ [m]. For equal displacement ($\Delta x_{\text{acc}} = \Delta x_{\text{quad}}$), the deflections from a quadrupole are 300 times stronger.

F Dispersion-Free Steering

To detect and correct the residual dispersion and the wake field effects, special steering algorithms have been developed. The main idea is that not only the nominal orbit is corrected, but simultaneously also the orbits measured for different bunch charges, bunch lengths or linac-quadrupole strengths. Since the orbit difference is proportional to the dispersion or wake-field effects, the latter are then minimized as well.

One of the most important and fruitful algorithms is *dispersion-free steering* [46]. Look again at the equation of betatron motion in a linac:

$$x'' + \frac{K}{1+\delta}x = \sum_i \frac{K_i \Delta x_i}{1+\delta}\delta(s - s_i), \tag{99}$$

where on the left we used a smooth approximation for the focusing, and on the right we introduced the deflections from off-center quadrupoles. The displacement of the quadrupole with respect to the nominal beam position is denoted Δx_i.

If we change the strength of all quadrupoles by ΔK, or ΔK_i, the on-momentum trajectory ($\delta = 0$) becomes

$$x'' + K\left(1 + \frac{\Delta K}{K}\right)x = \sum_i \left(1 + \frac{\Delta K_i}{K_i}\right)\Delta x_i\, \delta(s - s_i). \tag{100}$$

We observe that Eq. (99) for a particle trajectory with $\delta \neq 0$ is formally identical to Eq. (100) for a trajectory with $\delta = 0$ and modified quadrupole strength

$$\frac{\Delta K}{K} = \frac{\delta}{1+\delta}. \tag{101}$$

The idea of dispersion free steering is then, rather than to change the beam energy, to vary the relative strength of all the quadrupoles, and to minimize the resulting change in beam orbit. In addition, as for conventional orbit correction, also the absolute orbit readings are reduced.

The principle and benefit of dispersion-free steering is illustrated schematically in Fig. 10 [43]. Standard orbit correction will steer the beam through the center of all BPMs, which are assumed to be fixed to the quadrupole centers. A misaligned quadrupole will provoke a bump. Although the orbit perturbation is localized, this bump will generate dispersion. The dispersive trajectory becomes visible as a change in beam orbit, when the strengths of all dipole and quadrupole correctors are scaled. Subsequent minimization of the difference trajectory results in a more efficient compensation of the misalignment, which no longer generates dispersion.

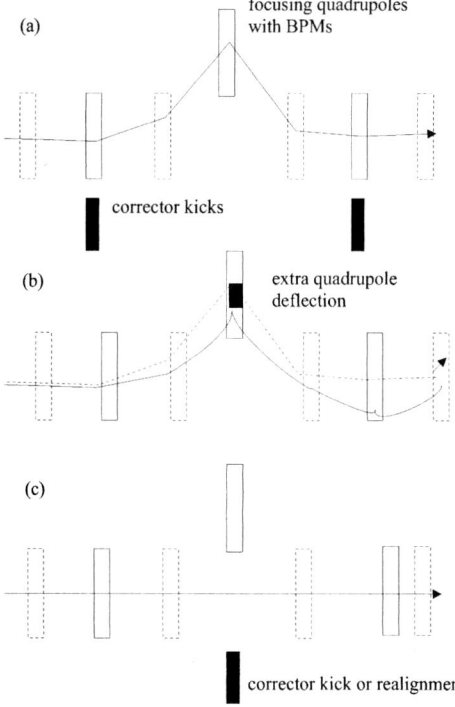

FIGURE 10. Schematic of dispersion-free steering, after Ref. [43]: (a) standard orbit correction that minimizes BPM readings at a misaligned quadrupole, thereby generating dispersion; (b) an extra deflection induced by the misaligned quadrupole which becomes apparent when quadrupoles and dipole correctors are scaled; (c) the result of the dispersion-free correction, which eliminates or compensates the kick from the misaligned quadrupoles.

In the following we briefly outline the mathematical algorithm of dispersion-free steering [43]. The orbit generated at the jth BPM by all upstream deflections θ_i (due to misalignments and steering coils) is

$$x^j = \sum_{i=1}^{j-1} R_{12}^{i \to j} \theta_i \qquad (102)$$

where

$$R_{12}^{i \to j} = \sqrt{\frac{E^i}{E^j}} \sqrt{\beta_x^i \beta_x^j} \sin\left[\psi_x^i - \psi_x^j\right] \qquad (103)$$

and we used the beam energies, beta-functions, and betatron phases at corrector i and monitor j.

Scaling all quadrupoles and dipoles in the lattice by $\kappa = 1 + \Delta K/K$ results in the orbit change

$$\Delta x^j(\kappa) = x^j(0) - x^j(\kappa) = \sum_{i=1}^{j-1}(R_{12}^{i \to j} - R_{12,\kappa}^{i \to j})\theta_i \qquad (104)$$

where $R_{12,\kappa}^{i \to j}$ denotes the R matrix for the scaled lattice.

We introduce vectors for the BPM measurements and the associated weights:

$$\vec{B} = \begin{pmatrix} x^1 \\ \Delta x^1(\kappa_1) \\ \Delta x^1(\kappa_2) \\ \Delta x^1(\kappa_3) \\ x^2 \\ \Delta x^2(\kappa_1) \\ \Delta x^2(\kappa_2) \\ \Delta x^2(\kappa_3) \\ \ldots \\ \Delta x^n(\kappa_3) \end{pmatrix} \quad \text{and} \quad \vec{W} = \begin{pmatrix} W^1 \\ \Delta W_\Delta^1(\kappa_1) \\ \Delta W_\Delta^1(\kappa_2) \\ \Delta W_\Delta^1(\kappa_3) \\ W^2 \\ \Delta W_\Delta^2(\kappa_1) \\ \Delta W_\Delta^2(\kappa_2) \\ \Delta W_\Delta^2(\kappa_3) \\ \ldots \\ \Delta W_\Delta^n(\kappa_3) \end{pmatrix}. \qquad (105)$$

The weights for the nominal orbit are a combination of the statistical variation, $\sigma^2(x^j)$, and the BPM misalignment or electronic offset, σ_{BPM}^2:

$$W^j = \frac{1}{\sigma^2(x^j) + \sigma_{\text{BPM}}^2}. \qquad (106)$$

The weight for the difference orbit includes the statistical contributions, $\sigma^2(x^j)$ and $\sigma^2(xj, \kappa)$, for the two orbits whose difference is computed, as well as a systematic error representing orbit or BPM drifts between the measurements, σ_{sys}^2:

$$W_\Delta^j(\kappa_i) = \frac{1}{\sigma^2(x^j) + \sigma^2(x^j, \kappa_i) + \sigma_{\text{sys}}^2}. \qquad (107)$$

After averaging over many orbits the statistical contributions become negligible and the χ^2 of the measurement reads

$$\chi^2 \approx \sum_j \left[\frac{x^{j\,2}}{\sigma_{\text{BPM}}^2} + \sum_{\kappa_i} \frac{\Delta x^{j\,2}(\kappa_i)}{\sigma_{\text{sys}}^2} \right]. \qquad (108)$$

Finally we define a correlation matrix

$$\mathbf{A} = \begin{pmatrix} R_{12}^{1 \to 1} & 0 & \ldots & 0 \\ R_{12,\kappa_1}^{1 \to 1} & 0 & \ldots & 0 \\ R_{12,\kappa_2}^{1 \to 1} & 0 & \ldots & 0 \\ R_{12,\kappa_3}^{1 \to 1} & 0 & \ldots & 0 \\ R_{12}^{1 \to 2} & R_{12}^{2 \to 2} & \ldots & 0 \\ R_{12,\kappa_1}^{1 \to 2} & R_{12,\kappa_1}^{2 \to 2} & \ldots & 0 \\ \ldots & & & \end{pmatrix} \qquad (109)$$

where the R_{12}s are the transport-matrix elements between the correctors and BPMs. The dispersion-free trajectory is obtained by solving for the vector $\vec{X}$ of corrector settings which simultaneously minimizes the trajectory offsets and the dispersion:

$$\min_{X} ||\vec{W}(\vec{B} + \mathbf{A}\vec{X})||_2. \tag{110}$$

Several variations of this method have been tried successfully at the SLC. For example, instead of changing the strength of the quadrupoles, one could compare the orbits of electron and positron beams (the opposite charge is equivalent to a -200% change in focusing strength), and the least-squares minimization was replaced by a singular-value decomposition [47,48]. In addition, it is possible to extend the dispersion-free algorithm so that it also minimizes the wake-field effect for coherent betatron oscillations [49].

G Computer Simulations

Complementing the analytical treatment described above, dedicated computer programs like LIAR [50], MUSTAFA [51], or PLACET [52] can provide improved estimates of emittance growth and beam stability in linear-collider linacs. These computer simulations include acceleration, magnet and structure misalignments, BPM errors, realistic steering procedures, correction algorithms, orbit-feedback systems, and both single and multi-bunch wake fields. It is reassuring that the measured emittance growth along the SLAC linac was reproduced in a simulation [53].

VI DAMPING RINGS

The purpose of damping rings is to reduce the phase-space volume of a beam produced by a positron source or an electron gun so that the design beam emittances are obtained. The transverse emittances typically must be decreased by several orders of magnitude. The damping ring accomplishes this via radiation damping. The latter is characterized primarily by two parameters: the damping time and the equilibrium emittance.

A Synchrotron Radiation

The equation describing horizontal emittance evolution in a damping ring is

$$\tau_x \frac{d\epsilon_x}{dt} = -2(\epsilon_x - \epsilon_{x,\infty}) \tag{111}$$

with the solution

$$\epsilon_x(t) = \epsilon_{x,0} e^{-2t/\tau_x} + \epsilon_{x,\infty}\left(1 - e^{-2t/\tau_x}\right) \tag{112}$$

where τ_x is the horizontal radiation damping time, $\epsilon_{x,0}$ the initial emittance of the injected beam, and $\epsilon_{x,\infty}$ the equilibrium emittance. If the ring is large enough, several bunch trains can be stored simultaneously, and an individual bunch train may stay in the ring for a correspondingly longer time. What matters for the performance is the effective damping time $\tau_{\text{eff}} = \tau_x/n_{\text{train}}$ with n_{train} denoting the number of trains stored. If the effective damping time is $1/(3f_{\text{rep}})$ or smaller—where f_{rep} is the repetition rate of the collider—the initial emittance can be reduced by a factor $e^{-6} \approx 0.002$ (depending on the values of $\epsilon_{x,0}$ and $\epsilon_{x,\infty}$), or more.

The expression for the horizontal damping time is

$$\tau_x = \frac{1}{C_d J_x E^3 \langle 1/\rho^2 \rangle} \tag{113}$$

where E is the beam energy, ρ the bending radius, and $J_x \approx 1$ the damping partition number; and $C_d = c r_0/(3m_e^3 c^6) \approx 2 \times 10^3$ m^2 GeV^{-3} s^{-1} where $m_e \approx 9.11 \times 10^{-31}$ kg denotes the electron mass. For example, with $E = 2$ GeV and $\rho \approx 10$ m, we find $\tau_x \approx 6$ ms. We will see that for obtaining smaller equilibrium emittances it is desirable to increase the bending radius and hence the circumference. In order to keep the effective damping time $\tau_{\text{eff}} \propto \rho/E^3$ constant, the beam energy then must also be increased as $E \propto \rho^{1/3}$. This constraint could be relaxed if most of the energy loss occurs in special damping wigglers and not in the arcs [54]. We will not make that assumption.

After a sufficiently long storage time the beam loses the memory of the injection conditions, and its emittance acquires an equilibrium value [20,55]

$$\epsilon_{x,\infty} = \frac{C_q}{J_x} \gamma^2 \theta^3 F \tag{114}$$

where θ denotes the bending angle per dipole, assuming all dipoles are identical, and $C_q = 3.83 \times 10^{-13}$ m. The function F is determined by the lattice,

$$F \equiv \frac{\rho^2}{l^3} \langle \mathcal{H} \rangle_\rho, \tag{115}$$

with l the length of a dipole, and

$$\langle \mathcal{H} \rangle_\rho = \frac{1}{l} \int_0^l (\gamma_x D_x^2 + 2\alpha_x D_x D'_x + \beta_x D'^2_x) ds. \tag{116}$$

Here D_x is the dispersion, β_x the beta function, and $\alpha_x = -\frac{1}{2}\beta'_x$, and $\gamma_x = (1 + \alpha_x^2)/\beta_x$. In the following exercise we drop the subscript x. The normalized emittance, $\gamma \epsilon_\infty$, increases as the third power of energy and bending angle. If the length per cell is held constant, and we assume $E \propto \rho^{1/3}$ so as to maintain a constant effective damping time, the normalized emittance decreases inversely with the square of the radius, *i.e.*, $\gamma \epsilon \propto \rho \theta^3 \propto 1/\rho^2$, where we considered $F \approx$ constant and $\theta \propto 1/\rho$.

Case Study IV: Minimum Emittance Lattice

Assume that $D = D' = 0$ at one end of a dipole, so as to produce a dispersion-free straight section, and that the beta-function has a minimum β_0 at some position s_0. Determine the values of β_0 and s_0 that minimize the function F. Recall that inside the dipole magnet the equations describing beta function and dispersion are $2\beta\beta'' - \beta'^2 - 4 = 0$, and $D'' = 1/\rho$.

Solution: the general form of Twiss parameters and dispersion functions in a region without quadrupoles and with bending radius ρ (ignoring the weak focusing from the dipole) is

$$\beta = \beta_0 + \frac{(s - s_0)^2}{\beta_0}, \tag{117}$$

$$\alpha = -\frac{1}{2}\beta', \tag{118}$$

$$\gamma = \frac{1}{\beta_0}, \tag{119}$$

$$D' = \frac{s}{\rho}, \tag{120}$$

$$D = \frac{1}{2}\frac{s^2}{\rho}. \tag{121}$$

$$\tag{122}$$

Inserting this into Eq. (116), we find [55]

$$\langle \mathcal{H} \rangle_\rho = \frac{1}{l}\int_0^l \left(\frac{1}{\beta_0}\frac{s^4}{4\rho^2} - \frac{(2-s_0)}{\beta_0}\frac{s^3}{\rho^2} + \frac{(s-s_0)^2}{\beta_0}\frac{s^2}{\rho^2} + \beta_0\frac{s^2}{\rho^2}\right) ds$$

$$= \frac{1}{3}\frac{l^3}{\rho^2}\left[\frac{\beta_0}{l} + \frac{l}{\beta_0}\left(\frac{s_0^2}{l^2} - \frac{3}{4}\frac{s_0}{l} + \frac{3}{20}\right)\right]. \tag{123}$$

This expression becomes minimum for $s_0/l = 3/8$ and $\beta_0/l = \sqrt{3/320}$, where $F = 0.065$ [55].

As an example, consider a damping ring accommodating 100 bending magnets, each with angle $\theta = 2\pi/100 \approx 0.06$ rad, length $l = 0.6$ m, $\rho = 10$ m, and $E = 2$ GeV ($\gamma \approx 4000$). For $F = 0.065$ the equilibrium emittance would be

$$\gamma \epsilon_x \approx 3.4 \times 10^{-7} \text{ m}, \tag{124}$$

close to the requirements of a future multi-TeV collider (see Table I).

However, in practice it is not easy to design a lattice with the optimum parameters calculated above. For a more realistic lattice, composed of a symmetric

TABLE 2. Example of damping ring parameters for a 3-TeV linear collider, scaled from the 1-TeV lattice of Ref. [56]

Variable	Symbol	Value
Energy	E	3.0 GeV
Circumference	C	700 m
Hor./vert. emit.	$\gamma\epsilon_{x,y}$	$5/0.1 \times 10^{-7}$ m
Hor. beam size	σ_x	30 μm
Vert. beam size	σ_y	4 μm
Hor./vert. half ap.	$h_{x,y}$	2 cm
Av. beta function	$\beta_{x,y}$	~ 10 m
Bunch length	σ_z	1.8 mm
Train length	l_{train}	154 ns
No. of bunches	n_b	154
Bunch population	N_b	4×10^9
Bunch spacing	L_{sep}	0.2 m

achromatic bend connecting two dispersion-free straight sections, the minimum value of F turns out to be 0.1054 [55]. Then, in order to still achieve the desired horizontal emittance, in our example the number of bending magnets must be increased by a factor $(0.1054/0.065)^{1/3}$, or 20%, namely from 100 to 120.

Table 2 lists a crudely scaled set of tentative parameters for the damping ring of a 3-TeV CLIC collider. Later on we will use these parameters for estimating various instability growth rates. Note that the rms bunch length in the ring is a few millimeters, whereas in the linac and at the collision point bunch lengths of the order of 30–150 μm are required. A similar situation is encountered for NLC and TESLA. Therefore, all linear-collider designs foresee bunch compression after extraction from the damping ring and before injection into the main linac.

B Intrabeam Scattering

Up to now, we have calculated only the ideal horizontal emittance due to synchrotron radiation alone. For small beam sizes, multiple scattering of particles inside the bunch off each other will increase all three emittances. This effect is known as "*intrabeam scattering*".

The equilibrium emittance with intrabeam scattering can be obtained from a balance of two excitations and one damping term as [57]

$$\epsilon_x = \frac{1}{4}\tau_x \left[Q_x^{SR} + Q_x^{IBS} \right] \qquad (125)$$

where

$$Q_x^{SR} = \frac{55}{24\sqrt{3}} \frac{r_e^2 c}{\alpha} \gamma^5 \left\langle \mathcal{H}_x/\rho^3 \right\rangle \qquad (126)$$

refers to the quantum excitation and

$$Q_x^{IBS} = \left\langle \frac{Nr_e^2 c \beta_x}{8\pi \gamma^3 \sigma_x^2 \sigma_y \sigma_z} f(\chi_m) \mathcal{H}_x \right\rangle \qquad (127)$$

to the intrabeam scattering. The latter occurs all around the ring, whereas the former comes only from the dipoles (since $1/\rho \approx 0$ at other places, if the beam has no large offsets in the quadrupoles). In Eq. (127) we used the following symbols [57]: $\chi_m = r_e \beta_x^2/(b_{max} \gamma^2 \sigma_x^2)$, $b_{max} = (N/((2\pi)^{3/2} \sigma_x \sigma_y \sigma_z))^{-1/3}$, $\mathcal{H}_x = (1/\beta_x) \left[D^2 + (\beta_x D' - \frac{1}{2}\beta_x' D)^2 \right]$, and $f(\chi_m) \approx 50 - 200$ for $\chi_m \approx 10^{-5} - 10^{-9}$. The angular brackets indicate an average over the ring: $\langle (...) \rangle \equiv 1/C \oint (...) ds$.

Combining the above expressions we obtain

$$\epsilon_x = \frac{\tau_x}{4} \left[\frac{55}{24\sqrt{3}} \frac{r_e^2 c}{\alpha} \gamma^5 \left\langle \frac{\mathcal{H}_x}{\rho^3} \right\rangle + \frac{Nr_e^2 c f(\chi_m)}{8\pi \gamma^3 \epsilon_x^{3/2} \sqrt{\kappa} \sigma_z} \left\langle \frac{\mathcal{H}_x}{\beta_y^{1/2}} \right\rangle \right] \qquad (128)$$

with $\kappa = \epsilon_y/\epsilon_x$, and σ_z the increased bunch length [57]

$$\sigma_z \approx \sigma_{z0} \left(1 + \frac{Nr_e^2 c \tau_\delta f(\chi_m)}{\sigma_{\delta,0}^2 32\pi \gamma^3 \sigma_{x0}^2 \sigma_{y0} \sigma_{z0}} \right)^{1/2}. \qquad (129)$$

The subscript 0 denotes beam sizes and rms energy spread without intrabeam scattering, and τ_δ is the longitudinal damping time.

The equilibrium emittance with intrabeam scattering can be estimated by solving Eqs. (128) and (129) numerically, or it can be calculated more accurately by using computer codes such as ZAP [58,59]. Typically, intrabeam scattering increases the equilibrium emittance of a linear-collider damping ring by 20–50%.

C Emittance Measurements

How can one verify that the expected emittance has been achieved? The Accelerator Test Facility (ATF) [60] at KEK in Japan is a prototype damping ring, which was built to stably produce a low emittance beam as required by future 1-TeV linear colliders, and to develop adequate tuning techniques. Over the last few years a large part of the ATF machine studies were devoted to measuring the vertical emittance. Unlike the horizontal emittance, the vertical emittance is not determined by the design accelerator optics, but arises from residual vertical dispersion and linear coupling.

In light sources often the image from a synchrotron light monitor is used to infer the beam size σ_y (or the beam divergence $\sigma_{y'}$) and from this the emittance via $\epsilon_y = \sigma_y^2/\beta_y$ (or $\epsilon_y = \sigma_y'^2 \beta_y$), taking into account possible corrections for depth of field and diffraction [61]. This approach is not directly feasible for the small emittances in a linear collider damping ring. Using standard deviations for a Gaussian distribution, the diffraction-limited photon beam emittance is $\epsilon_\gamma \sim \lambda_\gamma/(4\pi\gamma)$, from Heisenberg's uncertainty principle, where λ_γ denotes the photon wavelength. If no other information is available, it is not possible by conventional imaging techniques to resolve electron beam emittances smaller than the diffraction-limited photon-beam emittance. However, often, e.g., when the photon divergence is much larger than the beam divergence, the beam emittance can still be determined from a photon image, using the known optical functions at the light emission point.

For soft x-rays with $\lambda \approx 5$ nm as monitored at the LBNL ALS, the photon-beam diffraction limit is reached at $\epsilon_\gamma \approx 400$ pm, still much larger than the ATF design vertical emittance of 10 pm. Two possibilities for measuring the ATF beam emittance would be x-ray imaging with compound refractive lenses (CRL) or using a pin-hole camera at sub-Angstrom wavelengths [62]. At the ESRF, CRLs provide a beam-size resolution of 4 μm for a photon energy of 23 keV [63], and an x-ray pin-hole camera may diagnose emittances down to 5 pm, at typical photon energies of 40 keV [64].

One can also take advantage of the closeness to the diffraction limit and infer the beam size from the visibility of an interference pattern [65,66]. At the KEK ATF a stellar interferometer is employed for this purpose. A simplified schematic is shown in Fig. 11. Quasi-monochromatic synchrotron light at $\lambda = 500$ nm is sent through a double slit, and the interference pattern observed on a screen behind. An ideal point source creates a perfect interference pattern. If the source has a finite vertical extent, the contrast of the interference is reduced. This not only can be used to measure the emittance, but also has proven to be a valuable online tuning tool for optimizing the emittance, e.g., by varying bumps or skew quadrupoles in the ring. The most precise emittance values are obtained by measuring the interference contrast as a function of the distance between the two slits. The resolution at present is limited by vibrations of the interferometer support platform and the required exposure time (several ms).

In addition to the interferometer, the ATF team has developed three other procedures for measuring the vertical beam emittance in the damping ring.

The rms energy spread can be inferred from the beam size at a high-dispersion point, either in the ring or after extraction. Because of intrabeam scattering the energy spread increases with bunch current. The intrabeam-scattering blow up depends on the transverse and longitudinal emittances. If horizontal and longitudinal emittances are known, the vertical emittance may be obtained by fitting the measured increase in energy spread to theoretical expressions [57,67,68].

A similar technique can be applied to the beam lifetime, which for small emittances is limited by the Touschek effect [69]. This refers to binary collisions of particles within a bunch, by which so much energy is transferred from transverse into longitudinal phase space that the scattered particles leave the stable rf bucket. Since it is caused by a particle-particle collision, the loss rate due to the Touschek effect is quadratic in the bunch population, and inversely proportional to the bunch

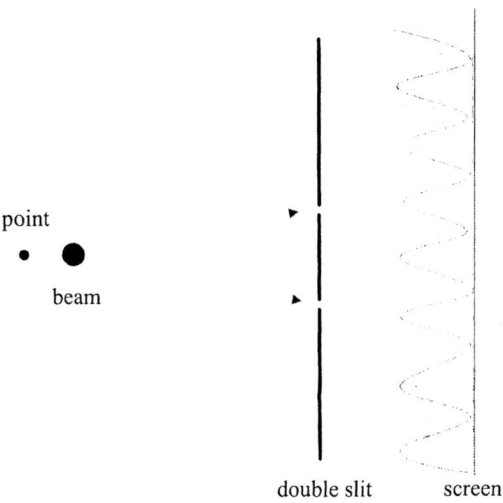

FIGURE 11. Schematic of ATF beam size measurement using stellar interferometer [65,66].

volume. Again, if horizontal emittance and bunch length are known, the measured beam lifetime as a function of current can be used to estimate the vertical emittance by fitting to the analytical expressions [70]. Since the theory of the Touschek lifetime is simpler than the theory for intrabeam scattering, and since, in addition, an increase of the energy spread could also be caused by longitudinal instabilities, the lifetime method appears to be the more reliable of the two.

The clearest verification of the ring emittance is to extract the beam from the ring and to measure the beam sizes at various wire scanners. Note that conventional wire scanners cannot be employed inside the ring, as the beam would break the wires within a few turns. Therefore, for the ATF ring an unbreakable laser wire is under development [71].

It is straightforward to compute the beam emittance from wire-scanner measurements. We define a reference point s_0 and denote the R matrix between s_0 and wire number i by $R^{(i)}$; see Fig. 12.

Considering purely horizontal or vertical motion, the R matrix is

$$R^{(i)} = \begin{pmatrix} R^{(i)}_{11} & R^{(i)}_{12} \\ R^{(i)}_{21} & R^{(i)}_{22} \end{pmatrix}. \tag{130}$$

The measured squared beam sizes, the Twiss parameters at point s_0 and the emittance are related as follows:

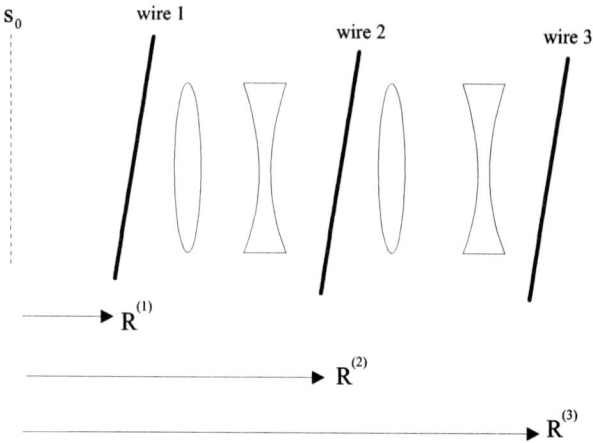

FIGURE 12. Schematic of emittance measurement using multiple wires.

$$\begin{pmatrix} \sigma_x^{(1)\,2} \\ \sigma_x^{(2)\,2} \\ \dots \\ \sigma_x^{(n)\,2} \end{pmatrix} = \begin{pmatrix} R_{11}^{(1)\,2} & -2R_{11}^{(1)}R_{12}^{(i)} & R_{22}^{(1)\,2} \\ R_{11}^{(2)\,2} & -2R_{11}^{(2)}R_{12}^{(i)} & R_{22}^{(2)\,2} \\ \dots & \dots & \dots \\ R_{11}^{(n)\,2} & -2R_{11}^{(2)}R_{12}^{(n)} & R_{22}^{(n)\,2} \end{pmatrix} \begin{pmatrix} \beta_0 \epsilon \\ -\alpha_0 \epsilon \\ \gamma_0 \epsilon \end{pmatrix}. \quad (131)$$

The procedure is now to determine a least-squares solution for the vector $(\beta_0 \epsilon, -\alpha_0 \epsilon, \gamma_0 \epsilon)$, and then to solve for the initial Twiss parameters and the emittance. Note that $\gamma_0 = (1 + \alpha_0^2)/\beta_0$, and hence all variables can be calculated. Evidently, for a meaningful measurement, at least three wire scanners at different betatron phases are required. This is called a multiwire emittance measurement.

Alternatively, if only one wire is available, one can measure the beam size at this wire for different strengths of an upstream quadrupole magnet. This is known as a quadrupole scan. Mathematically it can also be described by Eq. (131). The only difference is that now the R matrices do not refer to different wires but to different settings of the quadrupole.

Timing and pulse-amplitude jitter of the extraction kicker may affect the beam stability and hence impair the quality of the wire scans, since the latter are not single-shot measurements but require of the order of 50 beam pulses during which the wire is moved across the beam. It took some effort to stabilize the ATF extraction sufficiently, but in spring 2000 an extremely small vertical emittance close to the design value was demonstrated with good scan quality. Figure 13 shows recent wire scans in the ATF extraction line and illustrates the high degree of stability that has been achieved.

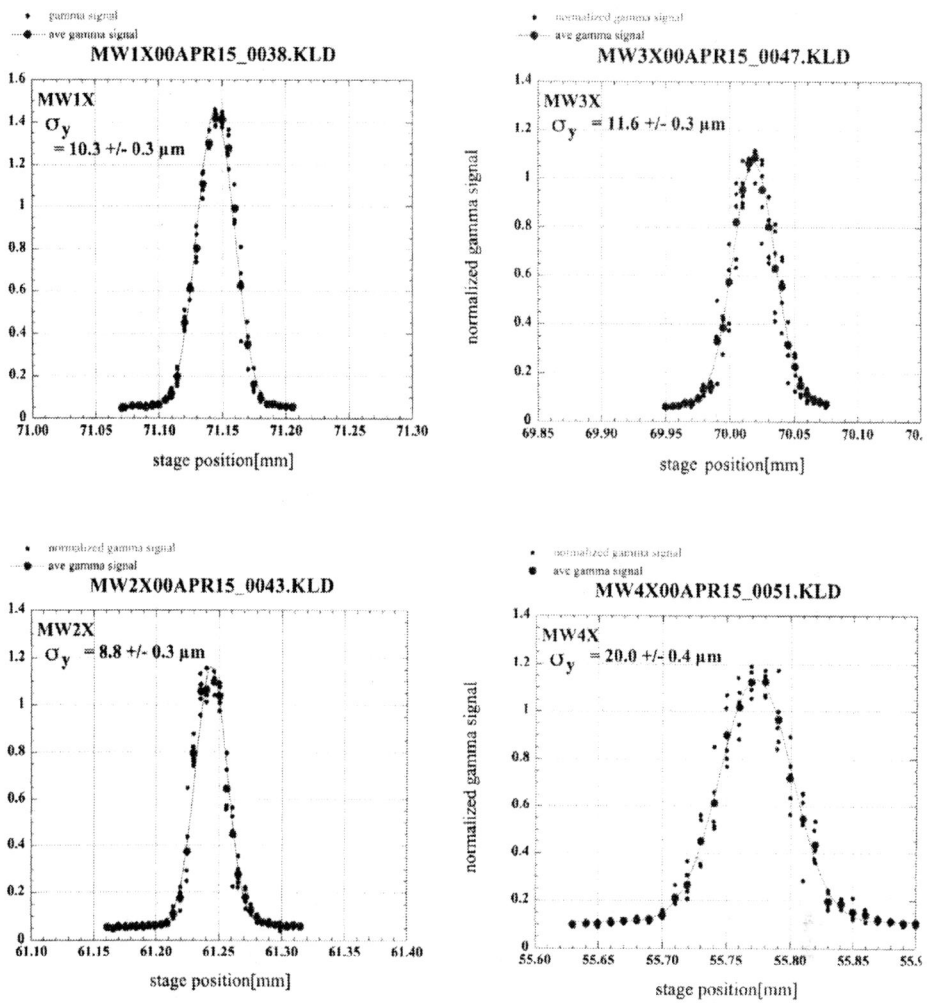

FIGURE 13. Vertical wire scans in the ATF extraction line, using W filaments with 10 μm diameter. (Courtesy J. Urakawa.)

TABLE 3. Momentum compaction factor for existing and proposed rings.

Ring	α_C
SLC	1.8×10^{-2}
ATF	2.1×10^{-3}
NLC	4.7×10^{-4}
TESLA	3.4×10^{-4}
1-TeV CLIC	2.4×10^{-4}
LEP	1.9×10^{-4}
3-TeV CLIC	4.0×10^{-5}

D Momentum Compaction Factor

Many other effects must be carefully considered in the design of damping rings. Some are related to the small value of the momentum compaction factor. This factor, denoted α_C, quantifies the change in electron path length, C_e, as a function of relative energy deviation,

$$\alpha_C = \frac{\Delta C_e/C_e}{\Delta p/p}, \qquad (132)$$

or, vice versa, if the rf frequency and thus the electron path length C_e stay constant, α_C gives the change in electron momentum as a function of the circumference C,

$$\Delta p/p = \frac{1}{\alpha_C}\frac{\Delta C}{C}. \qquad (133)$$

The momentum compaction factor can be calculated from the dispersion function D_x and the local bending radius ρ as

$$\alpha_C = \frac{1}{C}\oint \frac{D}{\rho}ds. \qquad (134)$$

It is often convenient to approximate the momentum compaction factor as [20] $\alpha_C \approx 1/\nu_x^2$, suggesting that $\alpha_C \propto 1/\rho^2$ for constant cell length.

In future damping rings, a small equilibrium emittance is achieved by reducing the dispersion in the bending magnets. According to Eq. (134), we expect that the momentum compaction factor will also decrease. Table 3 shows that this is indeed the case. The values of α_C in future designs are two orders of magnitude smaller than at the SLC.

If the ambient temperature changes by an amount ΔT, the floor and magnet supports will expand or contract by $\Delta l/l \propto \alpha_T \Delta T$, where α_T denotes the thermal expansion coefficient of the tunnel-floor or magnet-support material. The average path length of the electron beam is determined by the rf frequency, and unchanged by the temperature variation. However, if the quadrupole magnets are moved inwards or outwards by thermal expansion, the beam will experience additional bending fields from the resulting off-center orbit in the quadrupoles, and its energy will change.

Combining the two equations above, the beam energy change is given by

$$\frac{\Delta p}{p} = \frac{\alpha_T}{\alpha_C}\Delta T. \quad (135)$$

For example, taking a typical value of $\alpha_T \approx 10^{-5}$, the CLIC damping ring temperature should be stabilized to 0.5 mK, in order to maintain a constant beam energy to within 10% of the rms energy spread. This tight tolerance could be relaxed by means of an automatic path-length feedback employing chicanes. Such a scheme might be tested at the ATF damping ring [72]. It is noteworthy that, despite an extremely small momentum compaction factor, temperature variation does not appreciably affect the LEP circumference. The largest relative energy excursions at LEP, of the order of 10^{-4}, are caused by tidal effects [73].

Aside from increased sensitivitiy to temperature, low momentum compaction also increases the likelihood of longitudinal single-bunch instabilities. The current threshold for longitudinal microwave instability, with growth times much shorter than a synchrotron period, can be estimated from the Boussard criterion [74,75],

$$\frac{Ne^2 c|Z/n|}{(2\pi)^{3/2}\alpha_C E \sigma_\delta^2} \approx 1. \quad (136)$$

The term Z/n refers to the longitudinal impedance, with Z denoting the impedance at $n\omega_0$ (ω_0 is the angular revolution frequency), and n the number of revolution harmonics.

Assuming a constant number and size of strong inductive impedance sources, such as rf cavities, the longitudinal impedance decreases inversely with the bending radius [12], $Z/n \propto 1/\rho$. On the other hand, α_C decreases more strongly, as $\alpha_C \propto 1/\rho^2$ (since $D/\rho \propto 1/\rho^2$). Recalling the assumed energy scaling $E \propto 1/\rho^{1/3}$, we can stay below the threshold, if the current per bunch decreases as $1/\rho^{2/3}$.

Again, we consider as examples some parameters for CLIC. Using Eq. (136) with $\alpha_C = 2.4 \times 10^{-3}$, $N = 4 \times 10^9$, $\sigma_z = 1.8$ mm, $E = 2.15$ GeV, and $\sigma_\delta = 8.2 \times 10^{-4}$, an upper bound for the acceptable impedance is $|Z/n| < 0.05$ Ω. This appears feasible, since for the smaller damping ring of the NLC [9] a more detailed evaluation predicts $|Z/n| \approx 0.03$ Ω.

E Novel Instabilities

The performance of the recently commissioned B factories, at SLAC and KEK, is affected by new types of instabilities, in which electron beams interact with ions created from the residual gas, and positron beams with photo- and secondary electrons.

1 Fast Beam-Ion Instability

The *fast beam ion instability* [76,77] is a single-pass coupled-bunch instability occurring in electron beams; see Fig. 14. It is driven by ions created from the residual gas during the single passage of a bunch train. It was first observed at the LBL ALS [78], and shortly thereafter confirmed at the Pohang Light Source [79]. Recently, it has been observed also at the ESRF when operated with low-emittance optics and degraded vacuum pressure [80].

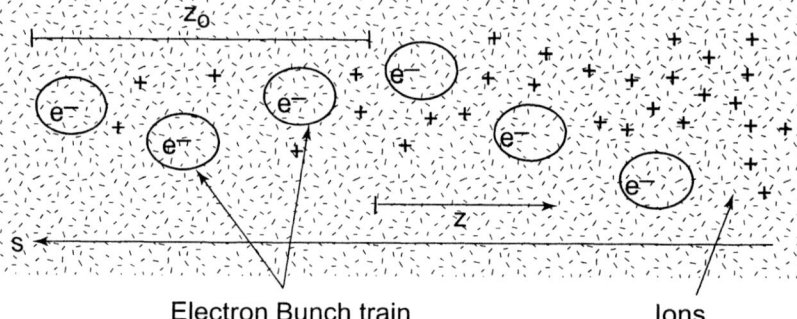

FIGURE 14. Schematic of fast beam-ion instability [76,77], which can arise due to ion trapping in an e^- bunch train.

The linear theory [76] predicts a quasi-exponential rise time, $y \propto \exp\sqrt{t/\tau_c}$, with a characteristic growth rate

$$\frac{1}{\tau_c} = \frac{4 d_{\text{gas}} \sigma_{\text{ion}} \beta_y N_b^{3/2} n_b^2 r_e r_p^{1/2} L_{\text{sep}}^{1/2} c}{\sqrt{33} \gamma \sigma_y^{3/2} (\sigma_x + \sigma_y)^{3/2} A^{1/2}} \quad (137)$$

at the end of the bunch train. A refined theory [81], taking into account the decoherence due to ion frequency spread within the bunch and around the ring, predicts exponential growth, $y \propto \exp(t/\tau_e)$, with a growth rate of

$$\frac{1}{\tau_e} \approx \frac{1}{\tau_c} \frac{5c}{\sqrt{2} l_{\text{train}} \tilde{\omega}_i} \quad (138)$$

for the last bunch in the train, where

$$\tilde{\omega}_i \equiv c \left(\frac{4 N_b r_p}{3 A L_{\text{sep}} \sigma_y (\sigma_x + \sigma_y)} \right)^{\frac{1}{2}} \quad (139)$$

denotes the coherent angular ion oscillation frequency, l_{train} the length of the bunch train (in meters), L_{sep} the bunch spacing in meters, d_{gas} the gas density (in molecules per cubic meter), n_b the number of bunches, N_b the bunch population, σ_{ion} the ionization cross section, and $\sigma_{x,y}$ the average horizontal and vertical rms beam sizes.

Assuming an ionization cross section of $\sigma_{\text{ion}} \approx 2$ Mbarn, e.g., for carbon monoxide, a gas density of $d_{\text{gas}} \approx 3 \times 10^{13}$ m^{-3}, and an atomic mass $A = 28$ (carbon monoxide or nitrogen), a pressure $p = 1$ nTorr and the 3-TeV ring parameters discussed above, we find $\tilde{\omega}_i \approx 7 \times 10^8$ s^{-1}, $\tau_c \approx 1$ μs, and $\tau_e \approx 20$ μs. The latter growth rate corresponds to about 10 turns, so that a bunch-by-bunch feedback system may be effective.

2 Electron-Cloud Instability

Positron or proton beams can suffer from a different kind of instability. Because of their opposite charge they may interact with the electron cloud created by photoemission or secondary emission. Figure 15 illustrates electron cloud build-up during the passage of a bunch train in the LHC beam pipe.

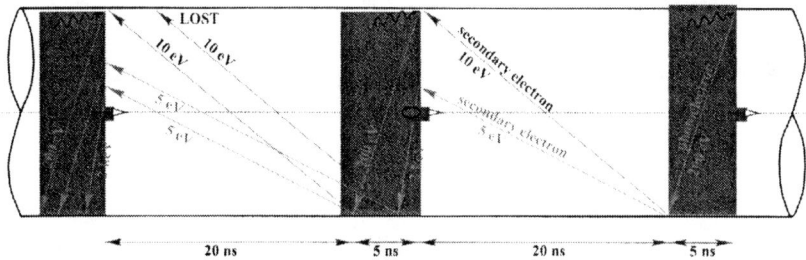

FIGURE 15. Schematic of electron-cloud build up in the LHC beam pipe. (Courtesy Francesco Ruggiero.)

So far two different manifestations of electron-cloud instabilities have been observed with multi-bunch positron beams. The *Ohmi effect* [82] refers to a coupled-bunch instability caused by an electron cloud. This instability was first seen at the KEK photon factory [82,83] and later verified at BEPC [84].

The second electron instability, potentially more harmful, is a *head-tail instability driven by an electron cloud*, where the cloud produced by previous bunches acts like a short-range wake field [85]. The resulting single-bunch instability was observed with positron beams at KEKB LER [86] and SLAC PEP-II, and possibly with the LHC proton beam at the CERN SPS [87].

We can derive a rough analytical estimate for the strength of this instability [85]. According to simulations, after a few bunches the electron cloud reaches a saturation density which is roughly equal to the neutralization density, defined as

$$\rho_e \approx \frac{N_b}{\pi h_x h_y L_{\text{sep}}}. \tag{140}$$

At this density the time average of the electric field on the chamber wall vanishes. Here h_x and h_y are the horizontal and vertical chamber half-apertures and L_{sep} is the bunch spacing (in meters). Note that, unlike the beam, the electrons are distributed almost uniformly across the entire vacuum chamber, and only a small portion are in the vicinity of the beam at the moment of a bunch passage. This fraction of electrons is responsible for the single-bunch instability. Considering a flat beam, and using a two-particle model (because of the peculiar nature of the electron-cloud "wake," the head particle is not chosen to be pointlike but to have a finite longitudinal extent of the order of the bunch length), an approximately constant vertical wake function can be derived [85]:

$$W_1 \approx \frac{8C}{h_x h_y L_{\text{sep}}}. \qquad (141)$$

This transverse wake function has the remarkable property that it depends only on the vacuum chamber dimensions and on the bunch spacing.

The wake function of Eq. (141) is a good approximation as long as the bunches are longer than $\sigma_x \sigma_y / (N_b r_e)$, with r_e the classical electron radius, and N_b the bunch population. For short bunches the formula must be modified [85].

By inserting the wake function W_1 into the standard expressions for instability growth rates [12], we can estimate the growth rates for fast beam break up (ignoring the synchrotron motion) as

$$\frac{1}{\tau} \approx \frac{2\pi \rho r_e c \langle \beta_y \rangle}{\gamma}, \qquad (142)$$

and the growth rate for the $l=1$ head-tail mode of the regular head-tail instability as

$$\frac{1}{\tau^{(1)}} \approx \frac{64}{3} \frac{\rho \langle \beta_y \rangle r_e \sigma_z Q'_y}{T_0 \alpha_C \gamma}. \qquad (143)$$

In addition, we can compute the threshold density for the strong head-tail instability. It is

$$\rho_{\text{thr}} = \frac{2\gamma Q_s}{r_e C \beta_y} \qquad (144)$$

where β_y is the average beta function, weighted with the local electron-cloud density.

Using the parameters from our example for the positron damping ring and applying these approximate relations, the saturated electron density is about $\rho_{e,\text{neutr}} \approx 1.6 \times 10^{13}$ m^{-3}, the wake function $W_1 \approx 7 \times 10^7$ m^{-2}, the growth rate for beam break-up $1/\tau_{\text{BBU}} \approx 1.4 \times 10^5$ s^{-1}, the growth rate for the $l=1$ head-tail instability $1/\tau^{(1)} \approx 1.6 \times 10^4 Q'_y$ s^{-1}, and the threshold density for the strong head-tail instability (also called TMCI instability) $\rho_{\text{thr}} \approx 3 \times 10^{12}$ m^{-3}, assuming a synchrotron tune $Q_s = 0.005$. Note that the estimated neutralization electron density, $\rho_{e,\text{neutr}}$ is about 5 times higher than the TMCI threshold, a clear warning sign.

The electron cloud also induces coherent and incoherent tune shifts of about $\Delta Q = 2r_e \beta \rho/\gamma$, which, for our parameters, evaluates to $\Delta Q \approx 0.1$. This can be useful for diagnostics purposes, since measuring the tune shift allows us to monitor the electron-cloud build up and the average electron density [86].

F Coherent Synchrotron Radiation

Another potentially harmful effect is coherent synchrotron radiation (CSR). At wavelengths longer than the bunch length, the bunch may radiate like a single macroparticle of charge $N_b e$. Since the synchrotron radiation is proportional to the charge squared, at these wavelengths the radiated power is enhanced by a factor N_b compared with normal synchrotron radiation. This enhanced radiation is coherent synchrotron radiation. The shorter the bunch, the larger the frequency range in which it is observed and the larger the overall effect.

The rms energy spread induced by CSR in a bend of length L_d and radius R is [88]

$$\Delta \delta_{\text{rms}}^{\text{CSR}} \approx 0.2 \frac{N r_e L_d}{\gamma \rho^{2/3} \sigma_z^{4/3}}. \tag{145}$$

The CSR is unimportant at ultra-high energies. *E.g.*, for a bending section in the CLIC final focus ($N_b = 4 \times 10^9$, $E = 1.5$ TeV, $L_d \approx 176$ m, $\theta_d \approx 244$ μrad, $\sigma_z \approx 30$ μm) we have $\Delta \delta_{\text{rms}}^{\text{CSR}} \approx 2 \times 10^{-8}$, much smaller than the energy spread due to incoherent synchrotron radiation, $\Delta \delta_{\text{rms}}^{\text{SR}} \approx 10^{-5}$.

However, CSR can be important for a damping ring. CSR effects are larger for smaller bending radius and lower beam energy. In order to examine the worst case, we take numbers typical of the damping-ring lattice for a 1-TeV collider [56] rather than a 3-TeV one. Specifically, we assume $E = 2.15$ GeV, $\sigma_z = 1.8$ mm, $B = 1.4$ T, and $L_{d,\text{tot}} \approx 30$ m. We find an additional induced energy spread of $\Delta \delta_{\text{rms}}^{\text{CSR}} \approx 2.6 \times 10^{-5}$ per turn. Further assuming $C = 283$ m, $\alpha = 2.4 \times 10^{-4}$, $\tau_\delta \approx 10.4$ ms, $\sigma_{\delta 0} \approx 8.2 \times 10^{-4}$, and considering regular synchrotron radiation as the only source of damping, we estimate the equilibrium energy spread due to CSR alone:

$$\sigma_\delta^{\text{CSR}} \approx \sqrt{\frac{1}{4} \tau_E \frac{(\Delta \delta_{\text{rms}}^{\text{CSR}})^2 c}{C}} \approx 1.3 \times 10^{-3}, \tag{146}$$

where τ_δ is the longitudinal damping time. In this example, the energy spread induced by CSR is larger than the natural energy spread!

The low-frequency CSR will be shielded by the vacuum chamber, if the latter has a full aperture smaller than the *critical aperture* [89]

$$h_{\text{crit}} = \left(\pi \sigma_z \sqrt{R}\right)^{2/3}. \tag{147}$$

An empirical formula for the shielding efficiency obtained by fitting to a large number of computer calculations [90] is [91]

$$\delta_{\text{rms}}^{\text{CSR}}(h) \approx \left(1 - e^{-\frac{2h}{h_{\text{crit}}}+0.8}\right) \delta_{\text{rms}}^{\text{CSR}}(\infty), \tag{148}$$

where $\delta_{\text{rms}}^{\text{CSR}}(\infty)$ denotes the energy spread induced without the shielding, Eq. (145).

As an example, with a bending radius of $R = 5$ m and a full chamber aperture $h = 60$ mm ($a = h/2 = 30$ mm), the CSR is shielded for bunch lengths $\sigma_z > 2$ mm. If the full aperture can be reduced to $h = 20$ mm, the CSR is shielded already for bunch lengths $\sigma_z > 400$ μm. However, small apertures like this may be impractical with respect to vacuum, beam lifetime, or impedance. Perhaps one may want to control the bunch length in the ring via the inductive impedance (without entering the turbulent bunch-lengthening regime) [92].

VII RF GUN AND POSITRON SOURCE

Challenges for the source design are posed by all the beam parameters required: small emittances, high charge, repetition rate, and electron polarization.

Electron beams can be generated in a variety of ways. Accordingly a number of different devices exist which can serve as electron sources for linear colliders: thermionic guns, dc guns with laser photocathodes (used at the SLC), or rf guns. In the future, also polarized rf guns may become available.

For example, in a laser-driven rf gun, or rf photoinjector, a high-power pulsed laser illuminates a photocathode placed on the end wall of an rf cavity. The emitted electrons are accelerated immediately in the rf field. The time structure of the electron beam is controlled by the laser pulse, and the rapid acceleration minimizes the effect of space-charge repulsion.

Several effects contribute to the normalized emittance attainable by such an rf gun [93]:

(1) The *thermal* emittance is determined by the initial transverse momenta of the electrons at the moment of their emission,

$$\gamma \epsilon_{x,y}^{th} \text{ [mm mrad]} \approx \frac{1}{4} \sqrt{\frac{k_B T_\perp}{m_e c^2}}\, \sigma_{x,y} \text{ [mm]},$$

where $k_B T_e \approx 0.1$ eV represents the thermal emission temperature.

(2) An *rf emittance* arises from the time-dependent transverse focusing at the exit of the cavity,

$$\gamma \epsilon_{x,y}^{\text{rf}} \text{ [mm mrad]} \approx \frac{e E_{\text{rf}}}{\sqrt{8} m_E}\, \sigma_{x,y}^2 \sigma_z^2 \omega_{\text{rf}}^2,$$

where E_{rf} is the peak accelerating field.

(3) The residual space-charge emittance is due to the repelling force between the equally charged beam particles [94],

$$\gamma \epsilon_{x,y}^{sc} \text{ [mm mrad]} \approx \frac{2N_b r_e}{7\sigma_{x,y} W} \exp\left(-3\sqrt{W\sigma_y}\right) \sqrt{\frac{\sigma_y}{\sigma_z}},$$

where $W = eE_{rf} \sin \phi_0/(2m_e c^2)$ and ϕ_0 is the rf phase at the beam center. Since the transverse space-charge force depends on the local charge density of the bunch, it disorients in phase space the transverse slices located at different longitudinal positions along the bunch. For round beams this dilution can be almost fully inverted by properly placed solenoids [95].

Linear colliders require flat electron beams at the collision point, in order to maximize the luminosity for a certain amount of beamstrahlung. However, electron guns usually produce round beams.

Case Study V: Flat RF Gun

Conceive a scheme by which one can transform a round beam ($\epsilon_x = \epsilon_y$) into a flat beam ($\epsilon_x \gg \epsilon_y$). Hint: one possibility starts with the beam from an rf gun immersed in a solenoid field, which is followed by a set of linear transformations.

Solution: A scheme for flat-round conversion was proposed by Brinkmann et al. in 1999 [96,97]. We describe the idea following Edwards [98]. The basic scheme consists of two parts:

(1) the beam from a cathode immersed in a solenoidal field develops an angular momentum at exit from the solenoid;

(2) subsequently this beam is passed through a quadrupole (or skew quadrupole) channel with 90° phase advance difference between the two planes, and length scale defined by the solenoid field.

Consider electrons moving parallel to a solenoid field whose axis is oriented in the z direction. Maxwell's equations imply the presence of a radial magnetic field at the exit of the solenoid. This radial field gives rise to a transverse deflection, which depends on the distance from the solenoid axis. For example, the vertical deflection at the solenoid exit is

$$\Delta y' = \frac{1}{B\rho} \int B_x dz = \frac{1}{B\rho} \frac{x_0}{2} B_z, \qquad (149)$$

where B_z is the longitudinal field inside the solenoid and x_0 the horizontal offset. A similar expression holds for $\Delta x'$. Abbreviating, we write $\Delta y' = kx_0$, $\Delta x' = -ky_0$ with $k = B_z/(2B\rho)$. After leaving the solenoid, the beam takes on a clock-wise rotation

$$\begin{pmatrix} x \\ x' \\ y \\ y' \end{pmatrix}_0 = \begin{pmatrix} x_0 \\ -ky_0 \\ y_0 \\ kx_0 \end{pmatrix}.$$

We have neglected any initial uncorrelated momenta, assuming that these are much smaller than kx_0 or ky_0. Actually, these terms are important, as they do determine the final flat-beam emittance. We will see this below.

Suppose now that the quadrupole channel behind the solenoid produces an I matrix in x and an additional $90°$ phase advance in y:

$$\begin{pmatrix} x \\ x' \\ y \\ y' \end{pmatrix}_1 = \begin{pmatrix} 1 & 0 & 0 & 0 \\ 0 & 1 & 0 & 0 \\ 0 & 0 & 0 & \beta \\ 0 & 0 & -1/\beta & 0 \end{pmatrix} \begin{pmatrix} x_0 \\ -ky_0 \\ y_0 \\ kx_0 \end{pmatrix} = \begin{pmatrix} x_0 \\ -ky_0 \\ k\beta x_0 \\ -\frac{1}{\beta}y_0 \end{pmatrix}.$$

If we choose $\beta = 1/k$, the final phase-space vector becomes

$$\begin{pmatrix} x \\ x' \\ y \\ y' \end{pmatrix}_1 = \begin{pmatrix} x_0 \\ -ky_0 \\ x_0 \\ -ky_0 \end{pmatrix}.$$

This is a flat beam inclined at $45°$. If one uses a skew quadrupole channel instead of quadrupole channel, the beam can be made flat in the vertical plane, as shown next.

The 4×4 transport matrix from the end of the solenoid through the skew quadrupole channel can be written as

$$M = R^{-1}TR \quad \text{with} \quad R = \frac{1}{\sqrt{2}}\begin{pmatrix} I_2 & I_2 \\ -I_2 & I_2 \end{pmatrix},$$

where I_2 is 2×2 identity, and the matrix T represents a normal quadrupole channel:

$$T = \begin{pmatrix} A & 0 \\ 0 & B \end{pmatrix}.$$

Combining the above, we write M as

$$M = \frac{1}{2}\begin{pmatrix} A+B & A-B \\ A-B & A+B \end{pmatrix}.$$

The initial state after the solenoid exit is

$$X \equiv \begin{pmatrix} x_0 \\ -ky_0 \end{pmatrix} \quad \text{and} \quad Y \equiv \begin{pmatrix} y_0 \\ kx_0 \end{pmatrix},$$

which we write more elegantly as

$$Y = SX \quad \text{using} \quad S \equiv \begin{pmatrix} 0 & -\frac{1}{k} \\ k & 0 \end{pmatrix}.$$

The final state is then

$$\begin{pmatrix} X \\ Y \end{pmatrix}_1 = \frac{1}{2} \begin{pmatrix} \{A+B+(A-B)S\}X \\ \{A-B+(A+B)S\}X \end{pmatrix},$$

and the condition for a flat beam is $Y_1 = 0$, or $I = -(A-B)^{-1}(A+B)S$.

Using the Courant-Snyder parametrization [100] $A = \exp(J\mu)$, $B = \exp(J(\mu + \Delta))$, where J denotes the matrix

$$J = \begin{pmatrix} \alpha & \beta \\ -\gamma & -\alpha \end{pmatrix},$$

the flat-beam condition becomes

$$I = -\frac{\cos(\Delta/2)}{\sin(\Delta/2)} \begin{pmatrix} k\beta & \alpha/k \\ -k\alpha & \gamma/k \end{pmatrix}.$$

This is fulfilled for $\Delta = -\pi/2$, $\alpha = 0$ and $\beta = 1/k$.

Finally, adding a random component to the slope to the initial vector $(x, x', y, y')_0 = (x_0, -ky_0 + x'_0, y_0, kx_0 + y'_0)$ we can apply the same transformation M and, assuming that the initial beam is round with $\sigma_{x0} = \sigma_{y0}$ and $\sigma'_{x0} = \sigma'_{y0}$, we find [96]

$$\epsilon_{y,1} = \frac{1}{2} \frac{{\sigma'_{x,y}}^2}{k}$$

and

$$\epsilon_{x,1}/\epsilon_{y,1} = 1 + 4k^2 \frac{\sigma^2_{x,y}}{{\sigma'_{x,y}}^2}.$$

The larger the k, i.e., the stronger the solenoid field, the flatter the beam becomes.

First experimental tests of a flat beam electron source at Fermilab have demonstrated the viability of this scheme [99].

The conventional approach to producing positron beams is to hit a high-Z target with a several-GeV e⁻ beam. An electro-magnetic shower of bremsstrahlung and pair creation develops, in the course of which a large number of positrons are produced. In order to get enough positrons, a thick high-Z material is chosen as a target. This method was used at the SLC and it is the preferred option for NLC.

The TESLA project considers an alternative approach to generating the positron beam. Here, a high-energy electron beam passing through a wiggler emits hard synchrotron-radiation photons which impact on a thin target downstream and produce positrons via pair creation [8,101,102]. A schematic of the TESLA positron source is shown in Fig. 16. In this design, a thin low-Z target is sufficient to produce the desired number of positrons. Its main advantages are, first, the lower heat capacity C_p, and hence the smaller target temperature rise, and, second, the reduced scattering. The latter implies smaller positron divergence and better capture

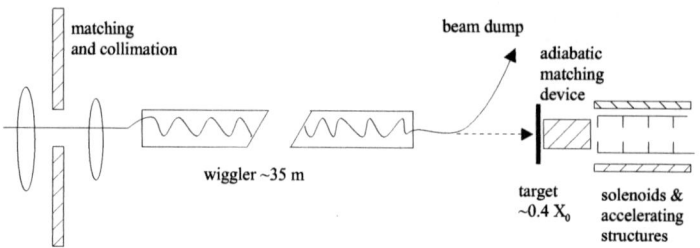

FIGURE 16. Schematic of TESLA positron source, consisting of a wiggler, a thin low-Z target, and a capture section with adiabatic matching [8].

efficiency. The TESLA source is even capable of producing polarized positrons if the wiggler is replaced by a helical undulator.

The positron capture section behind the photon target consists of acceleration units embedded in a strong solenoid field. The acceptance of the solenoid channel is well suited for a large spot size and small angles, whereas the positrons emerge from the target with a small spot size and large angles. A tapered solenoid with adiabatically increasing field strength [103] provides the optical matching between the two regions.

ACKNOWLEDGEMENTS

I am grateful to R. Assmann, M. Minty, F. Ruggiero, F. Schmidt, D. Schulte, and J. Urakawa for various helpful discussions and generous contributions. I thank F. Ruggiero, M.-P. Zorzano, H. Burkhardt, and E. Keil for careful reading of the manuscript and valuable comments. I also owe many thanks to the students who joined this tutorial. Without their active participation it would have been much less exciting. Their questions greatly inspired the written document. Finally, I thank the organizers of the Joint Accelerator School, in particular E. Wilson, S.Y. Lee, J. Miles, J. Murphy, and E. Perevedentsev, for their constant encouragement and for providing the possibility for this tutorial, and my co-teacher P. Logachev for his cheerful collaboration.

REFERENCES

1. M. Tigner, "A Possible Apparatus for Electron-Clashing Experiments," Nuovo Cimento, **37** (3), 1228 (1965).
2. U. Amaldi, "A Possible Scheme to Obtain e^+e^- Collisions at Energies of Hundreds of GeV," Physics Letters **61B** (3), 313 (1976).
3. R. Erickson (ed.) et al., "SLAC Linear Collider Design Handbook," Stanford Linear Accelerator Center (1984).

4. J. Seeman, K.L. Bane, T. Himel, W.L. Spence, C. Adolphsen, "Observation and Control of Emittance Growth in the SLC Linac," in 14th Int. Conf. on High Energy Accelerators, Tsukuba, Japan; Part. Accel. 30, 97 (1990).
5. R. W. Assmann, "Beam dynamics in SLC," in 17th IEEE Particle Accelerator Conference (PAC 97), Vancouver, Canada; SLAC-PUB-7576 (1997).
6. R. W. Assmann et al., "Accelerator Physics Highlights in the 1997/98 SLC Run," in 1st Asian Particle Accelerator Conference (APAC 98), Tsukuba, Japan; SLAC-PUB-7782 (1998).
7. P. Raimondi et al., "Recent Luminosity Improvements at the SLC," in 17th Int. Conf. on High-Energy Accelerators (HEACC 98), Dubna, Russia; SLAC-PUB-7955 (1998).
8. R. Brinkmann et al. (eds.), "Conceptual Design of a 500 GeV e^+e^- Linear Collider with Integrated X-ray Laser Facility," DESY 1997-048, ECFA 1997-182 (1997).
9. C. Adolphsen et al., "NLC Zeroth Order Design Report for the Next Linear Collider," SLAC Report 474 (1996).
10. J.P. Delahaye and I. Wilson, "CLIC a Multi-TeV e+e- Linear Collider," CERN/PS 99-062 (LP).
11. The CLIC Study Group, G. Guignard (ed.), "General Description of a 3 TeV Linear Collider Based on the CLIC Technology," CERN Report 2000-008 (2000).
12. A. Chao, Physics of Collective Beam Instabilities in High Energy Accelerators, Wiley (1993).
13. H. Padamsee, J. Knobloch, T. Hays, RF Superconductivity for Accelerators, Wiley (1998).
14. E. Jensen, private communication (2000).
15. M. Chodorow et al., Rev. Sci. Instrm. **26**, 134 (1955).
16. A.E. Vlieks, et al., "Breakdown phenomena in High-Power Klystrons," SLAC-PUB-4546 (1988).
17. D. Whittum, H. Henke, P. Chou, "High-Gradient Cavity Beat Wave Accelerator at W Band," in PAC 97; SLAC-PUB-7805 (1998).
18. A similar formula was presented by R. Brinkmann around 1992.
19. K. Yokoya, P. Chen, "Beam-beam phenomena in linear colliders," Lecture at 1990 US-CERN School on Particle Accelerators, Hilton Head, SC (1990).
20. M. Sands, "The Physics of Electron Storage Rings," SLAC Report 121 (1970).
21. P. Chen, "Differential Luminosity under Multi-Photon Beamstrahlung," Physical Review **D46**, 1186 (1992).
22. V.E. Balakin, N.A. Solyak, in Proc. 13th Int. Conf. on High Energy Accelerators, Novosibirsk (1986).
23. J.B. Rosenzweig, B. Autin, P. Chen, "Instability of Compensated Beam-Beam Collisions," in 1989 Lake Arrowhead Workshop on Advanced Accelerator Concepts (1989).
24. D. H. Whittum, A. M. Sessler, J. J. Stewart, S.S. Yu, "Plasma Suppression of Beamstrahlung," Part. Accel. **34**, 89 (1990).
25. I. F. Ginzburg, G. L. Kotkin, V. G. Serbo, V.I. Telnov, "Colliding Gamma e and Gamma Gamma Beams Based on the Single Pass Accelerators (of VLEPP Type)," Nucl. Instrum. Meth. **205**, 47 (1983).

26. V. Telnov, "Principles of Photon Colliders," Nucl. Instrum. Meth. **A355**, 3 (1995).
27. K. L. Brown, "Basic Optics of the SLC Final Focus System," SLAC-PUB-4811; presented at Workshop on Physics of Linear Colliders, Capri, Italy, June, 1988 and at Int. Workshop on the Next Generation of Linear Colliders, Stanford, CA, Nov 28 - Dec 9, 1988.
28. K.L. Brown, "A First and Second Order Matrix Theory for the Design of Beam Transport Systems and Charged Particle Spectrometers," Adv. Part. Phys. **1**, 71 (1968).
29. P. Emma, D. McCormick, M.C. Ross, "Beam Dispersion Measurements with Wire Scanners in the SLC Final Focus System," in PAC93, Washington; SLAC-PUB-6208 (1993).
30. F. Zimmermann, "New Final-Focus Concepts at 5 TeV and Beyond," in 8th Workshop on Advanced Acceleration Concepts, Baltimore, MD; SLAC-PUB-7883 (1998).
31. P. Raimondi, A. Seryi, "A Novel Final-Focus Design for High-Energy Linear Colliders," in EPAC 2000, Vienna (2000).
32. S. Fartoukh, J.B. Jeanneret, "Using Microwave Quadrupoles to Shorten the CLIC Beam Delivery Section," in EPAC 2000, Vienna (2000).
33. H. Burkhardt, "Background in Future e^+e^- Linear Colliders," CERN-SL-057-AP; CLIC Note 416 (1999).
34. R. Brinkmann, Talk at International Linear Collider Workshop LC99, Frascati (1999).
35. F. J. Decker et al., "Design and Wakefield Performance of the New SLC Collimators," in LINAC96, Geneva; SLAC-PUB-7261 (1996).
36. R. Brinkmann et al., TESLA 95-25 (1995).
37. R. Assmann et al., "Design Status of the CLIC 3-TeV Beam Delivery System and Damping Rings," in EPAC 2000, Vienna (2000).
38. J. Frisch, E. Doyle, K. Skarpaas VIII, "Advanced Collimator Systems for the NLC," in Proc. Linac 2000, Monterey (2000).
39. R. Assmann, A. Chao, "Dispersion in the Presence of Strong Transverse Wakefields," in PAC97, Vancouver; SLAC-PUB-7581 (1997).
40. F. Ruggiero, A. Zholents, "Resonant Correction of Residual Dispersion in LEP," CERN SL MD Note 26 (1992).
41. D. Schulte, "Emittance Preservation in the Main Linac of CLIC," in EPAC98, Stockholm; CERN-PS-98-018-LP (1998).
42. T. Raubenheimer, "The Generation and Acceleration of Low Emittance Flat Beams for Future Linear Colliders," Ph.D. Thesis, Stanford; SLAC Report 387 (1991).
43. R. Assmann, T. Chen, F.J. Decker, M. Minty, P. Raimondi, T.O. Raubenheimer, R. Siemann, "Simultaneous Trajectory and Dispersion Correction in the SLC Linac," Unpublished.
44. V. Balakin, A. Novokhatsky, V. Smirnov, in 12th Int. Conf. on High Energy Accelerators, Fermilab (1983).
45. W. Schnell, "Microwave Quadrupoles for Linear Colliders," CLIC Note 34 (1987).
46. T. Raubenheimer, R. D. Ruth, "A Dispersion Free Trajectory Correction Technique for Linear Colliders," Nucl. Instrum. Meth. **A302**, 191 (1991).
47. V. Ziemann, "Corrector Ironing," SLAC-CN-393 (1992)

48. P. Raimondi implemented the SVD algorithm for dispersion-free steering at the SLC.
49. T. Raubenheimer, R. D. Ruth, "A New Method of Correcting the Trajectory in Linacs," in IEEE PAC91, San Francisco (1991).
50. R. Assmann et al., "The Computer Program LIAR for the Simulation and Modeling of High Performance Linacs," in 17th IEEE PAC 97, Vancouver, Canada; SLAC-PUB-7577 (1997).
51. G. Guignard, J. Hagel, "Mustafa Environment Description and Users' Guide with Applications to CLIC," CERN-SL-98-002-AP; CLIC-Note-349 (1998).
52. D. Schulte, "PLACET: A Program to Simulate Drive Beams," CERN-PS-2000-028-AE (2000).
53. R.W. Assmann, "Beam dynamics in SLC," IEEE PAC 97, Vancouver, Canada; SLAC-PUB-7576 (1997).
54. J. Jowett, Private communication (2000).
55. L. Teng, "Minimizing the Emittance in Designing the Lattice of an Electron Storage Ring," FNAL TM-1269 (1985).
56. J.P. Potier and L. Rivkin, "A Low Emittance Lattice for the CLIC Damping Ring," in IEEE PAC 1997, Vancouver (1997).
57. J. Le Duff, "Single and Multiple Touschek Effects," in CERN Accelerator School, West Berlin and Rhodes, CERN-95-06 (1995).
58. M. S. Zisman, S. Chattopadhyay, J.J. Bisognano, "Zap User's Manual," LBL-21270 (1986).
59. T. O. Raubenheimer, "The Core Emittance with Intrabeam Scattering in e+/e− Rings," Part. Accel. **45**, 111 (1994).
60. F. Hinode et al., "ATF Design and Study Report," KEK Internal 95-4 (1995).
61. A. Hofmann, F. Meot, "Optical Resolution of Beam Cross-Section Measurements by Means of Synchrotron Radiation," Nucl. Instrum. Meth. **203**, 483 (1982).
62. O. Chubar, "Novel Applications of Optical Diagnostics," in EPAC 2000, Vienna (2000).
63. T. Weitkamp et al., "Electron Beam Profile Measurements with Refractive X-Ray Lenses," in EPAC 2000 Vienna (2000).
64. P. Elleaume, C. Fortgang, C. Penel, E. Tarazona, "Measuring Beam Sizes and Ultra-Small Electron Emittances Using an X-Ray Pinhole Camera," J. Synchrotron Rad. **2**, 209 (1995).
65. T. Mitsuhashi, "Beam Profile and Size Measurement by SR Interferometers," in Joint US-CERN-Russia-Japan School on Particle Accelerators: Beam Measurement, Montreux (1998).
66. T. Mitsuhashi, T. Naito, "Measurement of Beam Size at ATF Damping Ring with the SR Interferometer," in EPAC Stockholm; ATF Report 98-17 (1998).
67. A. Piwinski, "Intrabeam Scattering," in 9th Int. Conf. on High Energy Accelerators, Stanford 1974, Springfield (1975); see also CERN Accelerator School, Oxford 1985 (1985).
68. J. D. Bjorken, S. K. Mtingwa, "Intrabeam Scattering," Part. Accel. **13**, 115 (1983).
69. C. Bernadini et al., Phys. Rev. Letters **10**, 407 (1963).
70. R.P. Walker, "Calculation of the Touschek Lifetime in Electron Storage Rings," in

PAC 1987 (1987).
71. H. Sakai, Private communication (2000).
72. J. Urakawa, M. Ross, T. Raubenheimer, Private communications (1998).
73. L. Arnaudon et al., "Effects of Terrestrial Tides on the LEP Beam Energy," Nucl. Instrum. Meth. **A357**, 249 (1995).
74. D. Boussard, CERN Lab II/RF/Int 75-2 (1975).
75. E. Keil, W. Schnell, CERN Report TH-RF/69-48 (1969); V.K. Neil, A.M. Sessler, Rev. Sci. Instr. **36**, 429 (1965).
76. T. O. Raubenheimer, F. Zimmermann, "A Fast Beam-Ion Instability in Linear Accelerators and Storage Rings," Phys. Rev. **E52**, 5487 (1995).
77. G.V. Stupakov, T.O. Raubenheimer, F. Zimmermann, "Fast Beam-Ion Instability. 2. Effect of Ion Decoherence," Phys. Rev. **E52**, 5499 (1995).
78. J. Byrd, A. Chao, S. Heifets, M. Minty, T. O. Raubenheimer, J. Seeman, G. Stupakov, J. Thomson, F. Zimmermann, "First Observations of a Fast Beam-Ion Instability," Physical Review Letters **79**, 1, 79 (1997).
79. J. Y. Huang, M. Kwon, T.-Y. Lee, I.S. Ko, Y. H. Chin, H. Fukuma "Direct Observation of the Fast Beam-Ion Instability," Phys. Rev. Letters **81**, 4388 (1998).
80. R. Nagaoka, J.L. Revol, J. Jacob, "Observation, Analysis and Cure of Transverse Multibunch Instabilities at the ESRF," in EPAC 2000 Vienna (2000).
81. G.V. Stupakov, "A Fast Beam Ion Instability," in Proc. Int. Workshop on Collective Effects and Impedance for B Factories (CEIBA95); KEK Proc. 96-6, 243 (1996).
82. K. Ohmi, "Beam Photoelectron Interactions in Positron Storage Rings," Phys. Rev. Lett. **75**, 1526 (1995).
83. M. Izawa, Y. Sato, T. Toyomasu, "The Vertical Instability in a Positron Bunched Beam," Phys. Rev. Lett. **74** (25), 5044 (1995).
84. Z.Y. Guo et al., "Study of the Beam-Photoelectron Instability in BEPC," in 1st APAC Tsukuba (1998).
85. K. Ohmi, F. Zimmermann, "Head-Tail Instability Caused by Electron Cloud in Positron Storage Rings," Physical Review Letters **85** (18), 3821 ; CERN-SL-2000-015 AP (2000).
86. H. Fukuma et al., "Observation of Vertical Beam Blow-Up in KEK Low Energy Ring," presented at EPAC 2000, Vienna (2000).
87. Proceedings of Chamonix X, in particular contributions by W. Hoefle, M. Jimenez, and G. Arduini; CERN-SL-2000-007 (DI) (2000).
88. Ya.S. Derbenev, J. Rossbach, E.L. Saldin, V.D. Shiltsev, "Microbunch Radiative Head-Tail Interaction," DESY TESLA-FEL 95-09 (1995).
89. R.L Warnock, "Shielded Coherent Synchrotron Radiation and Its Possible Effect in the Next Linear Collider," in Proc. IEEE PAC 1991, San Francisco; SLAC-PUB-5523 (1991).
90. R.L. Warnock and P. Emma, Unpublished (1996).
91. P. Emma, Private communication (1996).
92. F. Ruggiero, Private communication (2000).
93. K.-J. Kim, "RF and Space-Charge Effects in Laser-Driven RF Electron Guns," Nucl. Instr. Methods **A275**, 201 (1989).
94. J.B. Rosenzweig, E. Colby, G. Jackson, T. Nicol, "Design of a High Duty Cy-

cle, Asymmetric Emittance RF Photoinjector for Linear Collider Applications,", in Proc. IEEE PAC 1993, Washington, DC (1993).
95. M.E. Jones, B.E. Carlsten, "Space-Charge Induced Emittance Growth in the Transport of High-Brightness Electron Beams," in IEEE PAC, Washington (1987).
96. R. Brinkmann, Ya. Derbenev, K. Flöttmann, "A Flat Beam Electron Source for Linear Colliders," TESLA 99-09 (1999).
97. R. Brinkmann, Y. Derbenev, K. Flöttmann, "A Low Emittance, Flat-Beam Electron Source for Linear Colliders," in Proc. EPAC 2000, Vienna (2000).
98. D. Edwards, "Notes on the Production of Flat Beams," Unpublished, dated January 17, 2000.
99. D. Edwards, H. Edwards, N. Holtkamp, S. Nagaitsev, J. Santucci, R. Brinkmann, K. Desler, K. Floettmann, I. Bohnet, M. Ferrario, "The Flat Beam Experiment at the FNAL Photoinjector," Presented at LINAC 2000, Monterey (2000).
100. E.D. Courant and H.S. Snyder, "Theory of the Alternating Gradient Synchrotron," Ann. Phys. **281**, 360 (1958).
101. V.E. Balakin, A.A. Mikhailichenko, "The Conversion System for Obtaining High Polarized Electrons and Positrons," Preprint INP 79-85 (1979).
102. K. Flöttmann, "Investigations Toward the Development of Polarized and Unpolarized High Intensity Positron Sources for Linear Colliders," DESY-93-161 (1993).
103. R. Helm, "Adiabatic Approximation for Dynamics of a Particle in the Field of a Tapered Solenoid," SLAC-4 (1962).

Beam Quality Control for Linear Colliders

P.V. Logatchov

Budker Institute of Nuclear Physics
Novosibirsk 630090, Russia

In this part of the tutorial we present, on one side, a detailed description of some new beam diagnostic devices, and, on the other side, a set of problems regarding the basic operational principles of these devices. The recent developments in accelerator technology tend toward high intensity beams of very small size. This requires nondestructive beam diagnostic tools with high space and time resolution. It is the main reason for returning to the old idea of using an electron beam as a nondestructive beam diagnostic tool [1, 2].

Exercise 1: Interaction of low energy electron beam with an intense ultra-relativistic bunch.

The aim of this exercise is to show the basic operational principles of the electron beam probe. A very thin electron beam with low current moves across the trajectory of an intense ultra-relativistic bunch with offset parameter ρ (see Fig. 1). An ultra-relativistic bunch moves in the Z direction with the velocity of light c ($\beta=1$). A probe electron beam moves in the X direction, perpendicular to the direction of motion of an ultra-relativistic bunch, with velocity βc. The particles of a probe electron beam are deflected in the electromagnetic fields of an ultra-relativistic bunch. Each particle of a

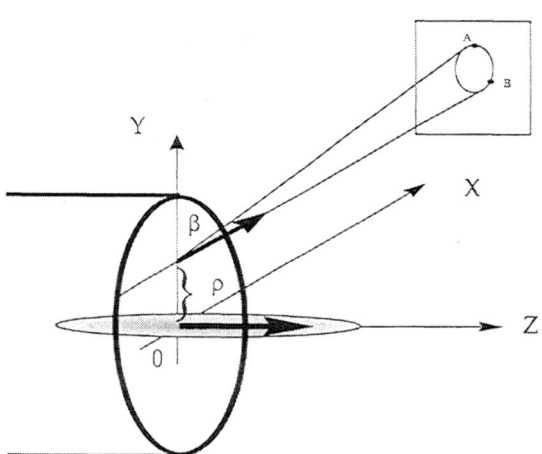

FIGURE 1. Deflection of a probe particle by the fields of an ultra-relativistic bunch.

probe beam has θ_y and θ_z deflection angles after passing the interaction region. The transverse sizes of both beams are much less than the offset parameter ρ. The longitudinal distribution of charge density in an ultra-relativistic bunch is given as a function $n(z)$ at time moment $t = 0$. At the same time moment ($t = 0$) each particle of a probe beam has its own value of coordinate x. Find the functions $\theta_y(x)$ and $\theta_z(x)$ if the vacuum chamber walls are absent.

Solution:
The total number of particles in an ultra-relativistic bunch can be found as

$$Ne = \int_{-\infty}^{\infty} n(z)\,dz.$$

A small kick for the probe beam electron can be expressed as

$$d\theta_z = \frac{dP_z}{P_x} = \frac{1}{\gamma mc\beta} \cdot \frac{e^2 2n(z)\beta}{\sqrt{\rho^2 + (x+\beta ct)^2}} \cdot \frac{x+\beta ct}{\sqrt{\rho^2 + (x+\beta ct)^2}} \cdot dt$$

$$= \frac{1}{\gamma mc} \cdot \frac{e^2 2n(z)}{\sqrt{\rho^2 + (x+\beta z)^2}} \cdot \frac{x+\beta z}{\sqrt{\rho^2 + (x+\beta z)^2}} \cdot \frac{dz}{c} = \frac{2r_e}{\gamma} \cdot \frac{n(z)(x+\beta z)\,dz}{\rho^2 + (x+\beta z)^2}.$$

We can use also the simple correlation between t and z for an ultra-relativistic bunch ($z = ct$). For the Y direction it is easy to find

$$d\theta_y = \frac{dP_y}{P_x} = \frac{1}{\gamma mc\beta} \cdot \frac{e^2 2n(z)}{\sqrt{\rho^2 + (x+\beta ct)^2}} \cdot \frac{\rho}{\sqrt{\rho^2 + (x+\beta ct)^2}} \cdot dt$$

$$= \frac{1}{\gamma mc\beta} \cdot \frac{e^2 2n(z)}{\sqrt{\rho^2 + (x+\beta z)^2}} \cdot \frac{\rho}{\sqrt{\rho^2 + (x+\beta z)^2}} \cdot \frac{dz}{c} = \frac{2\rho r_e}{\gamma\beta} \cdot \frac{n(z)\,dz}{\rho^2 + (x+\beta z)^2}.$$

Finally the dependencies can be written as

$$\theta_y(x) = \frac{2\rho r_e}{\beta\gamma} \int_{-\infty}^{+\infty} \frac{n(z)\,dz}{\rho^2 + (x+\beta z)^2},$$

$$\theta_z(x) = \frac{2r_e}{\gamma} \int_{-\infty}^{+\infty} \frac{(x+\beta z)n(z)\,dz}{\rho^2 + (x+\beta z)^2}. \qquad (1)$$

The use of infinity in the integration limits is justified for a vacuum chamber diameter significantly larger than the ρ value. Otherwise we should use the correct value for a particular vacuum chamber design. It is necessary to emphasize that the simple calculations presented here do not take into account the influence of the vacuum chamber walls. This approach is perfect only for ultra-relativistic bunch motion on the axis of the vacuum chamber pipe. Another restriction comes from the assumption that γ and β values are constant for the probe beam. The perturbation of the probe beam longitudinal motion by the electric field of an ultra-relativistic bunch should be small; therefore the value of θ_y^{MAX} should be significantly smaller than unity. But the test beam traces the closed curve on the screen (see Fig.1). By assuming a constant current I of a probe beam, we can derive a simple correlation between the x-coordinate and the charge distribution $q(l)$ along the indicated curve on the screen from point A to point B (see Fig. 1):

$$x = \frac{\beta c}{I} \int_A^B q(l)\, dl. \qquad (2)$$

Integrating the charge along the curve from point A to point B (Fig. 1) we can find the x-coordinate, Eq. (2), and its correspondence to the angles $\theta_y(x)$ and $\theta_z(x)$ at point B. Since the dependencies $\theta_y(x)$ and $\theta_z(x)$ are determined, it is possible, by using any of these functions, to restore the dependence $n(z)$.

Exercise 2: Using the function $\theta_y(x)$ try to find the function $n(z)$.

Solution:

Let us define $\theta_y(k)$ as

$$\theta_y(k) = \int_{-\infty}^{+\infty} \theta_y(x) e^{-ikx}\, dx.$$

Using Eq. (1) it is easy to find

$$\theta_y(k) = \frac{2\rho r_e}{\gamma\beta} \cdot \int_{-\infty}^{\infty} n(z)\, dz \cdot \int_{-\infty}^{\infty} \frac{e^{-ikx}}{\rho^2 + (x + \beta z)^2}\, dx.$$

Taking into account the result of integration,

$$\int_{-\infty}^{\infty} \frac{e^{-ikx}}{\rho^2 + (x+\beta z)^2}\, dx = \pi \frac{e^{-|k|\rho}}{\rho} e^{ik\beta z},$$

we can express

$$\theta_y(k) = \frac{2\pi r_e}{\gamma\beta} e^{-|k|\rho} n(-k\beta),$$

$$\theta_y(k) = \frac{2\rho r_e}{\gamma\beta} \cdot \int_{-\infty}^{\infty} n(z)\, dz \cdot e^{ik\beta z} \cdot \pi \cdot \frac{e^{-|k|\rho}}{\rho} = \frac{2\rho r_e}{\gamma\beta} \cdot \pi \cdot \frac{e^{-|k|\rho}}{\rho} \cdot \int_{-\infty}^{\infty} n(z) e^{ik\beta z}\, dz.$$

Finally the dependence of $n(z)$ can be written as follows:

$$n(-k\beta) = \frac{\gamma\beta}{2\pi r_e} e^{|k|\rho} \theta_y(k),$$

$$n(z) = \int_{-\infty}^{\infty} n(-k\beta) e^{ik\beta z} \frac{d(k\beta)}{2\pi} = \frac{\gamma\beta^2}{4\pi^2 r_e} \int_{-\infty}^{\infty} \theta_y(k) e^{(ik\beta z + |k|\rho)}\, dk.$$

Thus, if an experiment can give us the function $\theta_y(x)$, we can restore the dependence $n(z)$.

Exercise 3: In the case of $n(z) = N_e \delta(z)$ find the equation for the closed curve on the screen (see Fig. 1).

Solution:

Using the expressions for deflecting angles, Eq. (1), we can find

$$\theta_y(x) = \frac{2 N_e r_e \rho}{\gamma \beta (\rho^2 + x^2)},$$

$$\theta_z(x) = \frac{2 N_e r_e x}{\gamma (\rho^2 + x^2)}.$$

Simple calculations give the result

$$\left(\theta_y \beta - \frac{r_e N_e}{\gamma \rho}\right)^2 + \theta_z^2 = \left(\frac{r_e N_e}{\gamma \rho}\right)^2.$$

Thus, it is possible to monitor the value N_e / ρ by measuring the vertical loop size on the screen.

All these simple ideas and solutions were taken as the basis for the electron beam probe prototype briefly presented here.

We tested the electron beam probe at the VEPP-3 storage ring at bunch energies of 350, 1200 and 2000 MeV. We placed the device in the straight section of the ring between two RF cavities. The probe system was evacuated to a typical storage ring vacuum level of 10^{-9} Torr. A schematic diagram of the layout is shown in Fig. 2. The probe electron gun had a flat diode geometry with a 0.2-mm-diameter anode diaphragm. We used a 4-mm dispenser cathode with an emission ability of 3 A/cm². The maximum pulse current of the probe electron beam was 1 mA at an energy of 60 keV. An axial magnetic focusing lens formed a transverse probe beam of minimal size at the interaction region and on the screen. Transverse correction coils were installed to adjust the position of the probe beam on the screen. We directed the probe beam to the thin strip placed just before the 20-mm-diameter Micro Channel Plate (MCP). This allowed us to avoid MCP saturation by the 5-µs, 1-mA probe beam and to measure the beam pulse current. We also measured the probe beam energy. These two parameters made it possible for us to restore the x value, Eq. (2), or time, from the charge distribution at the MCP entrance. The relativistic bunch duration was in the range of one nanosecond, so all voltages on MCP, screen, and gun could be considered as constant during that time period. The shortest pulse on the MCP served as a gate pulse. It helped make a single-bunch picture on the phosphor screen (the revolution frequency of the bunch in the ring is 4.03 MHz). To digitize the screen image we used a conventional black and white CCD video camera and a special ADC-based grabber of standard video signal with external start. Synchronous start of the camera is absolutely necessary to have stable brightness of the screen image for brightness-to-charge conversion.

Since a built-in brightness-to-charge calibration system was not ready for the first set of experiments, we used, for that purpose, longitudinal charge distribution data obtained by a dissector [12]. This stroboscopic device works properly for operation with a stable bunch at a time resolution level of 100 ps.

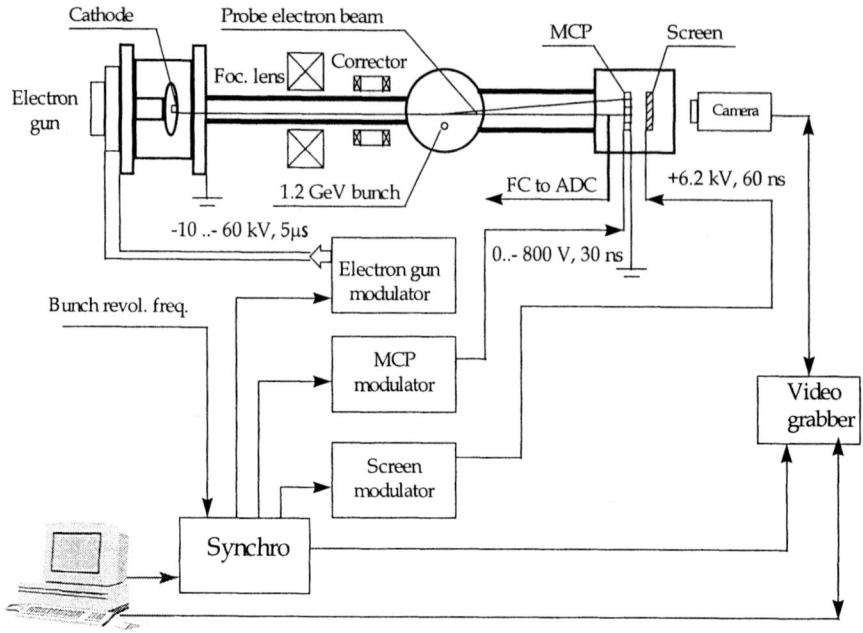

FIGURE 2. Electron beam probe installation.

The maximum repetition rate for our system was limited by the screen luminescence time (5 ms) and video data acquisition time (500 ms). Therefore all the measurements presented were made at a repetition rate of 0.5 Hz.

The probe beam focusing system, all modulators, and the video and synchronous start system were controlled by computer from the main control room.

EXPERIMENTAL RESULTS

First we adjusted the synchronous start system and modulators to reach reliable operation with good time stability (maximum long-time jitter was less than 1 ns). The pulse-to-pulse voltage stability at the moment of relativistic bunch passing was better than 2% for each modulator. Then we checked the surface uniformity of the MCP-screen-camera conversion system; the non-uniformity was less than 3%. All the measurements presented were made with a 60-keV, 1-mA probe electron beam. The probe beam size was 0.5 mm at the interaction point and 1 mm on the screen. To restore the bunch shape according to Eq. (4) below, we need the ρ value. It was measured directly by moving the probe beam up to its crossing point with the relativistic bunch trajectory. We could recognize the crossing picture very clearly.

After that, we performed the calibration measurements with a stable bunch in the ring in order to have a real longitudinal charge distribution in the bunch from the dissector (the longitudinal bunch shape is very close to Gaussian). Using these data we could calculate the brightness-to-charge conversion coefficient. Since the range of brightness changing was not very big in comparison with the dynamic range of the video signal (less than 10%), and the MCP-screen system can be considered as linear within our range of parameters, we could use that conversion coefficient for most of our measurements. Figure 3 shows an example of a calibration measurement with a stable bunch state.

After calibration we used our device to monitor longitudinal bunch instability. In this case the signal from the dissector was very wide and unstable and it had two peaks with flashing amplitudes. Figure 4 shows typical single-bunch pictures for that instability.

This example shows the potential of the method, but for detailed instability analysis we need at least a few pictures: for example after the 10th, 20th, 30th turn and so on. This is possible, and a multi-shot device is now being installed on the VEPP-4 collider at BINP.

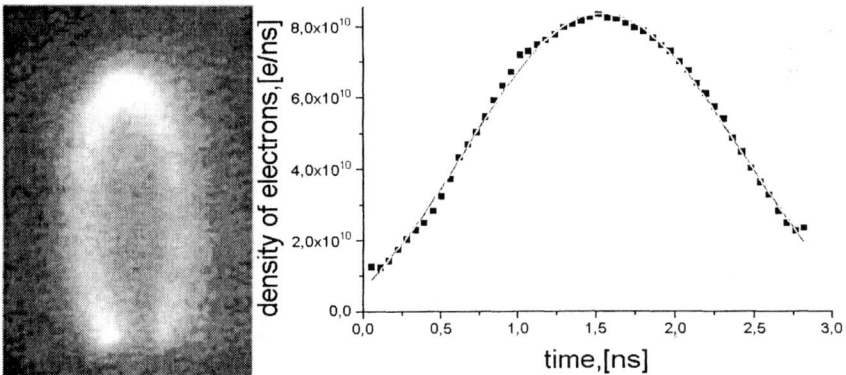

FIGURE 3. Calibration measurement with the dissector at a stable bunch state. The solid line is the best Gaussian fitting to the dissector data. Dots correspond to our measurements.

Exercise 4: Evaluate the time resolution of this method with the assumption that an ultra-relativistic bunch has a Gaussian longitudinal charge distribution with parameter σ and total number of particles Ne.

Solution:

The basic idea of this diagnostic tool looks simple, but we should be very careful in evaluating the time resolution of this method. At first, we have a finite angle resolution $\Delta\theta_Y = d_{scr}/L$ due to the probe beam size on the screen or the spatial resolution of the electron detecting system. Here d_{scr} is the maximum of two values:

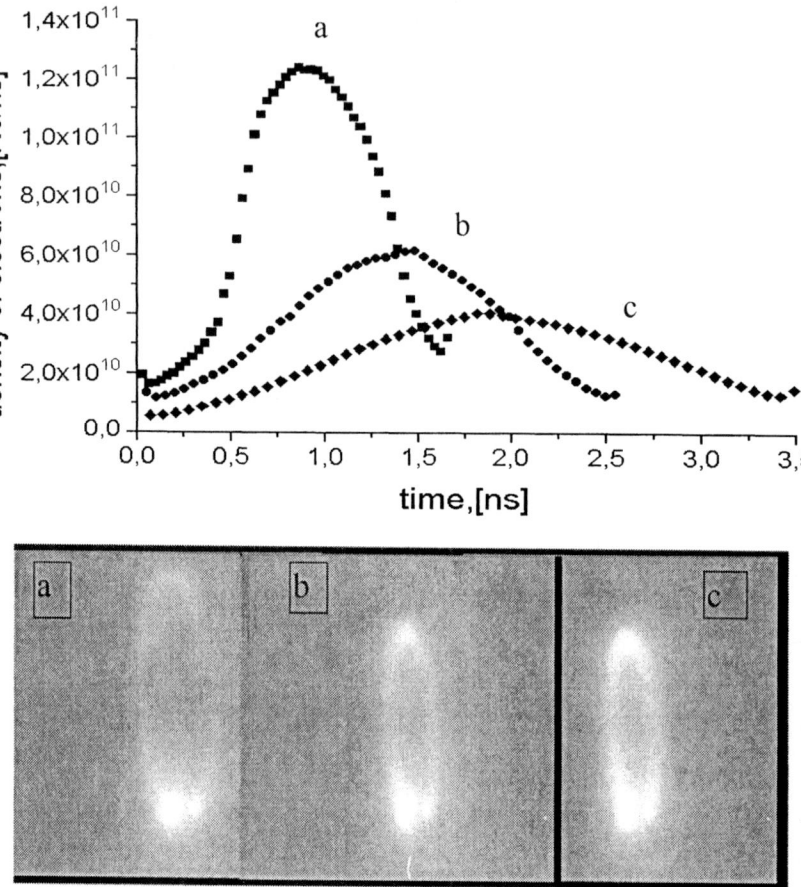

FIGURE 4. Typical single-bunch pictures for longitudinal instability: (a) – minimum bunch length, (b) – intermediate state, (c) – maximum.

the probe beam spot size on the screen and the spatial resolution of the screen, and L is the distance between the interaction point and the screen. Anglular resolution can be recalculated to time resolution as follows:

$$\Delta t = \frac{\Delta \theta_Y}{[d\theta_Y(x)/dx]_{MAX}} \cdot \frac{1}{\beta c}.$$

On the other hand, modulation of the longitudinal probe beam velocity due to the longitudinal component of the bunch electric field increases the time error. The total time resolution is given by both effects:

$$\tau = \frac{\Delta \theta_Y}{[d\theta_Y(x)/dx]_{MAX}} \cdot \frac{1}{\beta c} + \frac{\rho}{\beta c} \cdot \frac{\Delta \beta}{\beta} = \frac{\Delta \theta_Y}{[d\theta_Y(x)/dx]_{MAX}} \cdot \frac{1}{\beta c} + \frac{\rho}{\beta c} \cdot \theta_y^{max}. \quad (3)$$

The maximum vertical deflecting angle can be calculated for a Gaussian bunch as follows:

$$\theta_y^{max} = \frac{r_e N_e \sqrt{2\pi}}{\beta^2 \sigma} e^{\frac{\rho^2}{2\beta^2 \sigma^2}} \left(1 - \mathrm{erf}\left(\frac{\rho}{\beta \sigma \sqrt{2}}\right)\right). \qquad (4)$$

Assuming that $n(z) = \frac{Ne}{\sigma \sqrt{2\pi}} e^{-\frac{z^2}{2\sigma^2}}$ and using Eqs. (3) and (4), we can express:

$$\tau = \frac{\sqrt{2\pi} d_{scr} \sigma_z \rho \beta}{4 r_e N_e L c} / \int_{-\infty}^{\infty} \frac{z dz}{(1+z^2)^2} e^{-\frac{\rho^2}{2\beta^2 \sigma^2}\left(z - \frac{x}{\rho}\right)^2} + \frac{\rho \cdot r_e \cdot N_e \sqrt{2\pi}}{\beta^3 c \cdot \sigma} e^{\frac{\rho^2}{2\beta^2 \sigma^2}} \left(1 - \mathrm{erf}\left(\frac{\rho}{\beta \sigma \sqrt{2}}\right)\right).$$

We can evaluate the time resolution in two limiting cases:

$$\text{for } \frac{\sigma \cdot \beta}{\rho} \gg 1, \ \tau \approx \frac{r_e N e \rho}{c \sigma \beta^3} + \frac{d_{scr} \sigma^2 \beta^2}{L c r_e N e}, \qquad (5)$$

$$\text{and for } \frac{\sigma \cdot \beta}{\rho} \ll 1, \ \tau \approx \frac{d_{scr} \rho^2}{L c r_e N e} + \frac{2 r_e N e}{c \beta^2} e^{-\sqrt{\frac{\rho}{\beta \cdot \sigma \sqrt{2}}}}. \qquad (6)$$

Our experiment fits the first case, Eq. (5); the expression (6) is suitable for linear accelerators. Taking into account the final size of the screen, we can evaluate, from Eq. (5), the time resolution value for our experiment (50 ps for total bunch duration 1 ns).

An electron beam probe can be used also as a precise beam position monitor. Further information can be found in Refs. [1] and [3].

Analysis of optical and electrical signals in picosecond and subpicosecond ranges is important in accelerator [4], laser [5], and free-electron laser [6] technologies, and in plasma physics [7]. Measuring such short pulses can be performed in either the time or the frequency domain. Although operation in the frequency domain results in very good performance [8], operation in the time domain is sometimes preferable because the duration of an event is directly measurable with no need for conversion. Streak cameras (SCs) are commonly acknowledged to be the most rapid instruments operating in the time domain. These devices are capable of measuring the duration of optical signals down to the picosecond and subpicosecond ranges. The transformation of a short electron bunch into an optical signal is usually done by two methods. The first uses transition radiation [9] and the second, synchrotron radiation light [10].

We will discuss the performance of a radio-frequency-based streak camera. This method is based on magnetic deflection of the electron beam incoming from the gun with a photocathode in a radio-frequency cavity. The device provides time resolution down to the sub-ps domain. Further advantages of this method are simplified triggering, weaker dependence on space-charge effects within the electron beam, high modularity, and ease of implementation.

However, SCs have some disadvantages: the demand for an ever lower resolution limit contrasts with the short—but not negligible—rise time of high-voltage on deflecting plates, which is no faster than some nanoseconds at best. In addition, linearity should be preserved, further limiting the slope of the sweep transient.

Usually, the deflecting force has a constant direction and varies linearly in time over a sufficiently long interval T. Hence, the pulse being measured and the deflecting force must be synchronized with an accuracy better than T, which means picosecond accuracy for a device designed for subpicosecond resolution. In the case of stable and periodic pulses or pulses phase-locked to a reference signal, stroboscopic methods can easily be accomplished. Indeed, when single or sporadic pulses are being measured, triggering may become a serious problem.

Once the sweep has been accomplished as fast as possible, the resolution limit of a SC is established by the finite spatial resolution (about 50 µm) of the elements of the microchannel plate image intensifier (MCP) and by space-charge effects within the electron beam in the streak tube. All these circumstances limit the time resolution of a SC.

The purpose of the following exercises is to illustrate the performance of an RF-based streak camera (RFC), which will be shown to be working with sub-ps resolution, and to simplify some problems that are typical of normal SCs, i.e. space-charge and triggering.

Exercise 5:

The circular deflection of an electron beam can be performed by a transverse magnetic field of TM_{110} mode with circular polarization in a cylindrical cavity. Circular polarization is provided by exciting two orthogonal modes shifted by $\pi/2$ rad in phase. Find the dependence of the radius of the circle on the screen upon the magnetic field strength on the axis of the cavity and the velocity of the streak camera electron beam.

Solution:

Let us consider the motion of an electron, undergoing a rotating magnetic field with a step-like distribution along the z-axis:

$$H_x = H_0 \cos\omega t, \quad H_y = H_0 \sin\omega t, \quad \text{for} \quad 0 \leq z \leq d,$$
$$H_x = H_y = 0, \quad \text{for} \quad z < 0, \quad \text{and} \quad z > d,$$

where d is the length of the cavity gap, $\omega = 2\pi\nu$, ν being the frequency of the electromagnetic wave. Under the assumption that the deflection angle α is small, the longitudinal momentum p_z and velocity β_z can be regarded as constants. In this case the equations of motion can be written as

$$\dot{p}_x = eH_0\beta_z \sin\omega t,$$
$$\dot{p}_y = eH_0\beta_z \cos\omega t.$$

After integration over time from the entrance time of an electron in the cavity τ_0 to the exit time $\tau = \tau_0 + d/\beta_z c$, we obtain the deflection angles:

$$\alpha_x = \frac{eH_0\lambda}{\pi\gamma mc^2}\sin\mu \sin(\omega t + \mu),$$

$$\alpha_y = \frac{eH_0\lambda}{\pi\gamma mc^2}\sin\mu \cos(\omega t + \mu),$$

where $\mu = \pi d/\lambda\beta_z$ is the transit angle and λ is the wavelength of the electromagnetic wave. The transverse position of an electron after a drift distance L can be parameterized by radius R and azimuth angle θ:

$$R = \frac{eH_0\lambda L}{\pi\gamma mc^2}\cdot\sin\mu$$

$$\theta = \omega\tau_0 + \mu - \frac{\pi}{2}.$$

These are the polar coordinates of a point over which an electron would impinge on a screen placed perpendicularly to the z-axis at a distance L from the cavity. Maximum deflection is attained when the transit angle is $\pi/2$ rad; in this case

$$R = \frac{eH_0\lambda L}{\pi\gamma mc^2}$$

$$\theta = \omega\tau_0.$$

The radius of deflection does not depend on time, and the azimuth angle is linearly proportional to time. If an electron bunch of finite duration $\Delta\tau$ coming from the photocathode passes through the cavity, it comes out as an arc of a circumference drawn by electrons over the screen, whose angular width is $\Delta\theta$; by measuring this angle, we can determine $\Delta\tau$:

$$\Delta\tau = \frac{\Delta\theta}{\omega}$$

This measurement is absolute if the frequency of the electromagnetic field is known exactly. The only limitation to this system is determined by the full-bunch duration, which should be necessarily less than one RF period. The next exercise is suggested in order to overcome this limitation.

Exercise 6:

Using the results of the previous exercise, suggest a scheme that can enable spiral scanning on the screen of the streak camera (hint: you can use two cavities).

Solution:
To obtain a good spiral on the screen with a step equal to few diameters of the beam, we should have a fast drop of the RF power in the cavity: 20 times faster than the original RF decay in a high-quality-factor cavity. It is not a good idea also to reduce the Q value of the cavity, because then we would need an unreasonably powerful and very expensive RF source. One possible solution involves two cavities

slightly different frequencies ω_1, ω_2 : each cavity makes a circular scanning. The cavities are placed one just after the other. In this case the scanning angle amplitude will oscillate with frequency $(\omega_2 - \omega_1)$. We can choose the difference in the frequencies in order to reach the spiral parameters we need. We can use MCP as a gate device to choose the time period in which good spiral scanning is possible.

We give below some experimental results [11] that illustrate how the RF streak camera operates. The setup of the experiment is sketched in Fig. 5. It basically consisted of three parts: electron gun, chopper-system, and RFC.

In order to test the time resolution of the RFC, we had to provide a sub-picosecond electron bunch. Since the existing GaAs photo-gun operated at 50 kV was unable to deliver electron bunches shorter than 60 ps, a system to form a sub-ps electron bunch was implemented. This was a chopper consisting of deflecting cavity $C1$, drift tube, and slit S (see Fig. 5a). An iris with a 400-µm aperture was installed to reduce the transverse beam size at the entrance. The slit width of the chopper could be varied through micrometric control.

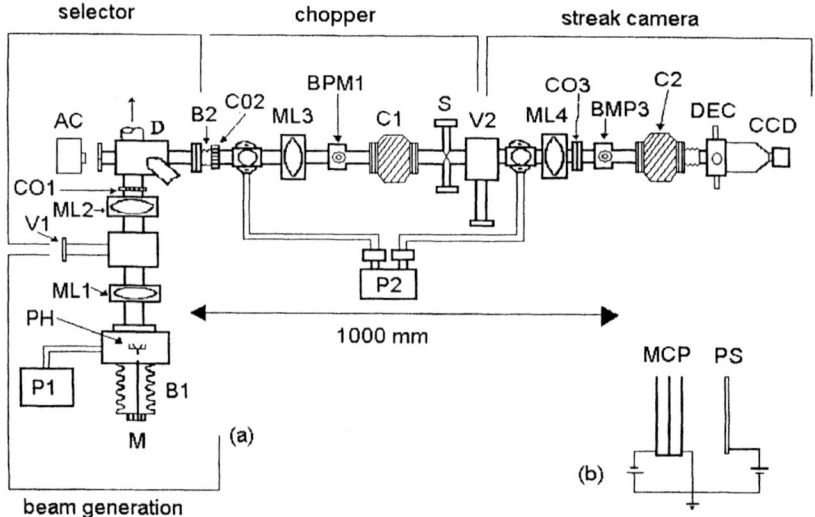

FIGURE 5. The experimental setup.

The deflecting cavity was axially symmetric and made of OFHC copper. The electromagnetic field of the TM_{110} mode inside the cavity was excited through a magnetic loop. The bunch coming out of the cavity was swept onto the horizontal plane and hit onto a vertical slit. In our case a minimum beam size of about 200 µm provided a minimum bunch length of 0.7 ps at the exit of the chopper system. This way the electron bunch passing through the chopper cavity was sufficiently short to calibrate the RFC.

The measuring part of the experimental apparatus (strictly speaking, the RFC) consisted of deflecting cavity $C2$, drift tube, detector DEC and magnetic lens $ML4$.

Cavity $C2$ had the same design as the chopper cavity ($C1$) but was fed by two orthogonal TM_{110} modes, resulting in a rotating magnetic field orthogonal to the cavity axis. Each mode was excited by a separate magnetic loop. The two loops were geometrically at right angle and fed by two RF amplifiers, reciprocally shifted by $\pi/2$ rad in phase. The resonant frequency of each mode could be separately adjusted by means of two piston tuners.

Deflected electrons passed through the drift tube and were collected by a position-sensitive detector. This was a two-stage MCP coupled to a circular phosphor screen, 28 mm in diameter (see Fig. 5b). High voltage of 4 kV was applied between the backside of the MCP and the screen. The image on the screen was read out by a charge-coupled device (CCD) camera and acquired by computer. The overall resolution of the detector was about 60 µm (rms) limited mainly by the MCP.

A pulsed RF-power amplifier with three separate channels was used to feed both the chopper and the measuring cavities. Output power up to 800 W could be independently controlled on each channel. The phase difference between the two channels could be varied by a coaxial phase shifter. The signal of the mode-locking RF generator of the laser was multiplied by 64 in frequency and used as input for the amplifier. This scheme provided synchronization between electron bunch and chopper phase.

We firstly adjusted the optics to focus the beam as much as possible onto the screen, then measured an rms size of about 100 µm for the beam spot—limited by the beam emittance.

Then the RF power was fed into the cavity, and the phase shift between the two modes was optimized to achieve an arc of a circumference for the beam trace, as shown in Fig. 6. The arc was not a complete circumference because the laser pulse was shorter than one RF period. The radius of deflection drawn by electrons on the screen is an important parameter since it is bounded by the resolution of the instrument. Measuring the radius allowed us to calibrate the RFC: here a diameter of 314 pixels was measured, corresponding to 18.5 mm.

The response of the RFC to a short electron bunch was tested using the chopper system and the adjustable slit. As RF power was supplied to the chopper cavity, short electron bunches entered the RFC and were analyzed, as shown in Fig. 7, where a beam-trace image and its cross-section are shown. The resolution of the instrument was sufficiently high to follow the profile of the electron bunch.

The ultimate capability of the RFC in terms of resolution can be determined by probing through even shorter electron bunches. As the RF power to the chopper cavity was raised, the electron bunch coming out became shorter, as shown in Fig. 8. Error bars are rms values of the distribution obtained after a large number of measurements. The dashed line represents the resolution limit achievable by the instrument in its present configuration, i.e. 700 fs (rms).

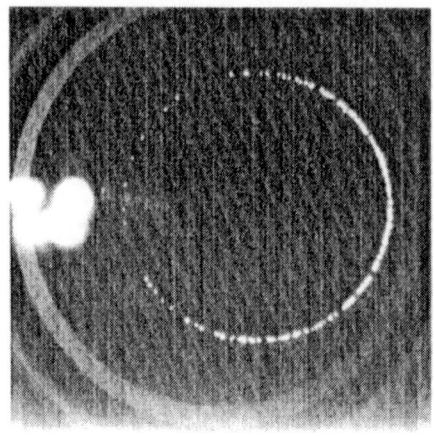

FIGURE 6. Image of a circularly scanned beam on the phosphor screen. The chopper cavity is switched off.

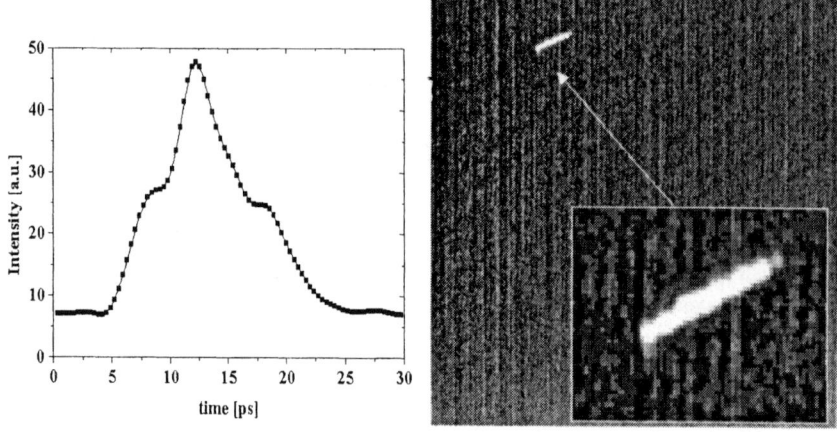

FIGURE 7. Image of the electron beam on the phosphor screen (magnified in the inset) and its cross-section when the chopper cavity is switched on.

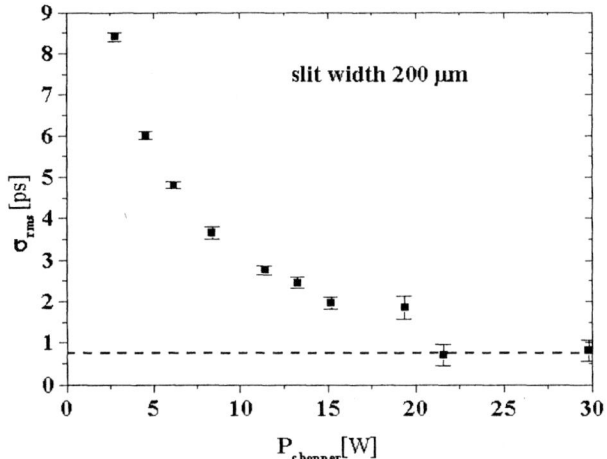

FIGURE 8. Dependence of short bunch length on power coming into the chopper cavity.

REFERENCES

1. John A. Pasour, Mai T. Ngo, "Nonperturbing Electron Beam Probe to Diagnose Charged-Particle Beams," *Rev. Sci. Instrum.* **63** (5), 3027 (1992).
2. P.V. Logatchov, P.A. Bak, A.A. Starostenko, N.S. Dikansky, V.S. Tupikov, K.V. Gubin, S.M. Mishnev, M.B. Korabelnikov, M.G. Fedotov, "Nondestructive Single-Pass Monitor of Longitudinal Charge Distribution in an Ultra-Relativistic Electron Bunch," in *Proc. 1999 PAC*, New York 1999, p. 2167.
3. A.A. Starostenko, P.A. Bak, Ye.A. Gusev, N.S. Dikansky, P.V. Logatchov, A.R. Frolov, "Non-Destructive Single-Pass Bunch Length Monitor: Experiments at VEPP-5 Preinjector Electron Linac," in *Proc. EPAC 2000*, Vienna 2000, p. 1720.
4. M. Uesaka, K. Tauchi, T. Kozawa, T. Kobayashi, T. Ueda, K. Miya, *Phys. Rev.* E **50**, 3068 (1994).
5. T. Starczewski, J. Larsson, C.-G. Wahlström, M.H.R. Hutchinson, J.E. Muffett, R.A. Smith, J.W.G. Tisch, *Journal of Physics* B **27**, 3291 (1994).
6. H. A. Schwettman, T. I. Smith, R. L. Swent, *Nucl. Instrum. Methods in Physics Research* A **358**, 19 (1994).
7. J. Workman, A. Maksimchuk, X. Liu, U. Ellenberger, J.S. Coe, C.-Y. Chien, D. Umstadter, *Phys. Rev. Letters* **75**, 2324 (1995).
8. H. Lihn, P. Kung, C. Settakorn, H. Wiedemann, D. Bocek, *Phys. Rev.* E **53**, 6413 (1996).
9. J.D. Jackson, *Classical Electrodynamics*, Wiley, New York, 1975.
10. T. Mitsuhashi, "Beam Profile and Size Measurement by SR Interferometers," in *Proc. Joint US-CERN-Japan-Russia School on Particle Accelerators*, World Scientific, 1999, p. 399.
11. A.V. Aleksandrov, N.S. Dikansky, V. Guidi, G.V. Lamanna, P.V. Logatchov, S.V. Shiyankov, L. Tecchio, "Setting Up and Time-Resolution Measurement of a Radio-Frequency-Based Streak Camera," *Rev. Sci. Instrum.* **70** (6), 2622 (1999).
12. E.I. Zinin, "Stroboscopic Method of Electron-Optical Pico-Second-Resolution Chronography and Its Application in Synchrotron Radiation Experiments," *Nucl. Instrum. Methods* **208**, 439 (1983).

Future Colliders

E. Keil

CERN, Geneva, Switzerland

Abstract. Future high-energy colliders are discussed, in particular linear e^+e^- colliders, $\mu^+\mu^-$ colliders and hadron colliders. The energies of the e^+e^- colliders are larger than that of the existing e^+e^- collider LEP. The energies of the very large hadron colliders are larger than that of the LHC being constructed at CERN. Two $\mu^+\mu^-$ colliders at 100 and 3000 GeV are included. For all these colliders, parameter lists, topics of R&D, and progress with component tests are discussed.

I INTRODUCTION

In this talk, I shall restrict myself to future high-energy colliders, in particular linear e^+e^- colliders, muon colliders, and hadron colliders. The e^+e^- colliders are discussed in Section II. Their energies are larger than that of the existing e^+e^- collider LEP. The two $\mu^+\mu^-$ colliders at 100 and 3000 GeV are discussed in Section III. The hadron colliders are discussed in Section IV. Their energies are larger than that of the LHC being constructed at CERN. My conclusions are in Section V.

II LINEAR ELECTRON-POSITRON COLLIDERS

Linear e^+e^- colliders are being studied at several laboratories that have joined together in the Inter-Laboratory Collaboration for R&D on TeV-Scale Linear Colliders. The report of the International Linear Collider Technical Review Committee (ILC-TRC) [1] was the first attempt to gather in one document the current status of every major e^+e^- linear collider project in the world. Table 1 is a brief summary of typical parameters, taken from the WWW page of the ILC-TRC, last updated in October 1999 [2]. I include only colliders that are currently being pursued actively, i.e. TESLA, JLC/NLC and CLIC, and leave out the S and C band colliders and VLEPP. The listed parameters are for 1TeV nominal energy in the center of mass. The main linac technologies are quite different: TESLA is super-conducting, the others are at room temperature, and the RF frequencies of the main linac range from 1.3 to 30 GHz between TESLA and CLIC. The RF power sources are klystrons for all machines except CLIC, which uses a second beam. The two figures for the accelerating gradient are the unloaded and loaded values, respectively. The beamstrahlung parameters [3] $\Upsilon_{\text{effective}}$, δ_B, n_γ/e, the number of e^+e^- pairs, hadrons and jets in a crossing N_{pairs}, N_{hadrons}, N_{jets}, relevant to particle physics, have converged towards similar values over the last few years. Up to date information is on WWW pages for CLIC [4], JLC [5], NLC [6], and TESLA [7].

TABLE 1. Linear collider parameters for 1 TeV in the center of mass, including the dates of the latest update. The accelerator physics parameters are in the upper part, those related to the beam-beam collisions and particle physics in the lower part. The horizontal and vertical beam radii at the IP are taken before the pinch.

	TESLA 8/98	JLC/NLC 12/98	CLIC 9/99
Energy in center of mass (TeV)	0.8	1	1
RF frequency of main linac (GHz)	1.3	11.4	30
Actual luminosity $(\text{nbs})^{-1}$	12.8	12.9	27.2
Linac repetition rate (Hz)	5	120	150
Bunch separation (ns)	189	2.8	0.67
Beam power/beam (MW)	12.2	8.7	7.4
Unloaded/loaded gradient (MV/m)	34/34	72/55	172/150
Total two-linac length (km)	30	18	10
Hor./vert. beam radii (nm) at IP	391/2	234/3.9	115/1.75
$\Upsilon_{\text{effective}}$		0.3	1.02
δ_B (%)	4.7	10.3	11.2
n_γ/e		1.4	1.1
N_{pairs}	28	18.4	11.1
N_{hadrons}/crossing	0.19	0.33	0.28
$N_{\text{jets}} \cdot 10^{-2}$	0.66	2.3	4.4

Figure 1 is a schematic drawing of a linear e^+e^- collider, using CLIC as an example. The main e^+e^- beams are generated by a sequence of linear accelerators, a target for e^+ production, one or more damping rings for achieving low emittances in all three degrees of freedom, and one or more bunch compressors for shortening the bunches. The main e^+e^- beams are accelerated in two long linear accelerators, and delivered to the interaction point by two final focus systems.

A Component Tests for Linear e^+e^- Colliders

Components of future linear e^+e^- colliders are being tested in several laboratories [2]. The Final Focus Test Beam (FFTB) at the end of the SLAC linac has achieved and measured spot sizes close to the design goal, 1 μm horizontally and 60 nm vertically through 1998. In the TESLA Test Facility (TTF) at DESY, three modules of eight superconducting RF cavities were tested. Between 1997 and 1999, the average gradients increased from about 16, through 20, to about 22MV/m, well above the TTF design gradient 15MV/m. The Accelerator Test Facility (ATF) at KEK consists of an S-band injector linac, a damping ring and a bunch compressor. Its goal is to demonstrate the small emittances and bunch lengths needed in the X-band linear collider designs. It has been operated at 1.28 (1.54) GeV, and achieved tunes Q_x, Q_y, damping times τ_i, and a

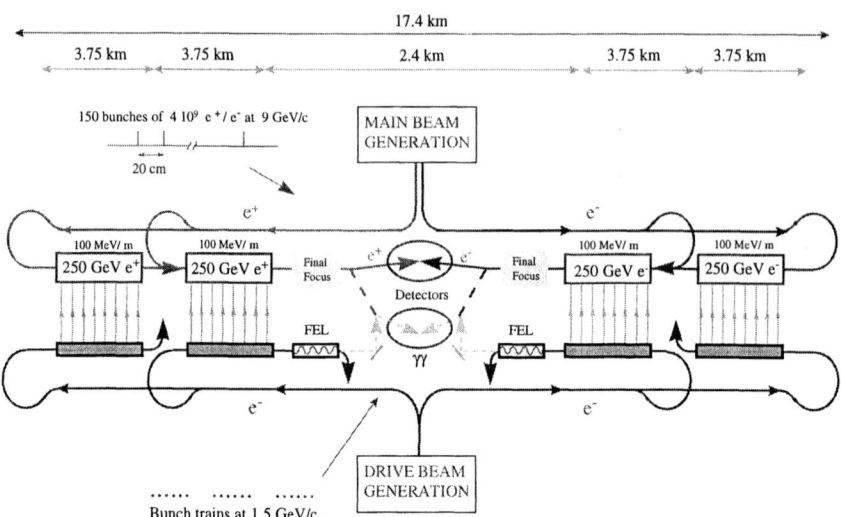

FIGURE 1. Schematic linear e^+e^- collider CLIC at 1 TeV in the center of mass as envisaged in 1998. The main beams are accelerated in accelerating modules, operating at a gradient of 100MeV/m and increasing the energy by 250 GeV each, and focused at the center of the detector by a final focus system. The RF power for the accelerating modules is generated in transfer structures from a drive beam. The latter is generated in a linear accelerator, operating at 1.5 GeV with trains of bunches at a pulse length of 92 μs.

horizontal emittance ϵ_x close to the design values, shown in parentheses. However, the bunch current is only about 60% of the design value, and the vertical emittance ϵ_y is (1.5 ± 0.25) times design, possibly caused by intra-beam scattering. The Next Linear Collider Test Accelerator (NLCTA) at SLAC, a prototype high-gradient X-band linac, has accelerated a 120 ns pulse of 0.34 A through 6 structures to 305MeV, corresponding to a loaded gradient of 34MV/m in September 1997. The beam loading was compensated to 0.5% absolute energy variation instead of 15% without compensation. The second CLIC Test Facility (CTF2) is a prototype two-beam 30 GHz linac. It has generated 27 (71) MW of RF power, and accelerated beam at 59 (95) MV/m. Power and gradient are limited by the drive beam charge through the decelerator 0.37 (0.64) μC. Recently, damage of the accelerating structures was found at CERN and SLAC, presumably due to local heating caused by RF breakdown.

Since the beginning of the test facility construction, the concepts of several linear e^+e^- collider modules have been modified. At SLAC, it is foreseen to modify NLCTA, incorporating these changes. At CERN, a new test facility (CTF3) has been approved. Its construction will start after the final shutdown of LEP in autumn 2000. Results from these facilities are expected in a few years.

B Linear e^+e^- Collider Schedules

Zero-th order design reports have been published for the NLC [8], TESLA [9] and JLC [10]. An updated NLC design and a cost estimate were presented in 1999. The estimated cost was considered rather high. Since then, the NLC design has been modified in order to reduce its cost. A conceptual description of a 3TeV e^+e^- linear collider based on CLIC technology was published recently [11]. A conceptual design report for TESLA is scheduled for 2001.

When I gave a similar lecture in 1998 [12], I thought that a linear e^+e^- collider might be under construction before the LHC is completed in 2005. This now appears to be unlikely.

C Prospects for Higher-Energy e^+e^- Colliders

Once a first linear e^+e^- collider with a center of mass energy between 0.5 and 1TeV is completed, its operation at the initial energy and upgrades to a higher energy might run in parallel. Energy upgrades to 1.5TeV are included in the design of the NLC, and to 1.6TeV for TESLA. The nominal CLIC energy is 3TeV. A collaboration between SLAC and the CLIC group at CERN has published a pilot design for a 5 TeV collider [13]. CERN and SLAC experts agree that higher frequencies are more favorable for energies above 2 TeV, because of the higher accelerating gradient, made possible by the higher threshold gradient for dark-current capture. The CLIC group is making a significant contribution to the world-wide linear-collider effort, and is working successfully at the frontier of high frequencies and high energies. The drive-beam concept has recently evolved considerably. The two-beam high-frequency approach of the CLIC will undoubtedly be an essential aspect of any future high-frequency, high-energy machine.

III MUON COLLIDERS

The possibility of muon colliders was introduced by Budker [14], Parkhomchuk and Skrinsky [15], and Neuffer [16], and they have been developed intensively over the past three years [17–20]. A feasibility study for a 4 TeV muon collider was presented at Snowmass [21,22]. Since then, the Muon Collider and Neutrino Factory Collaboration has been set up [23], and an updated status report published [24]. More recently, attention has turned to neutrino factories and away from $\mu^+\mu^-$ colliders. It was realized [25] that a muon storage ring with long straight sections, filled with a muon beam from a muon source rather similar to that of a $\mu^+\mu^-$ collider, and operating at about 20 to 50 GeV, is a source of an intense, well collimated, flavor-pure neutrino beam. The cooling requirements of such a neutrino source are much less stringent than those of a $\mu^+\mu^-$ collider, since the emittance of the μ beam must be reduced by a factor 10 only in both transverse planes, and no longitudinal cooling is needed. Design studies of such neutrino factories have been published [26,27].

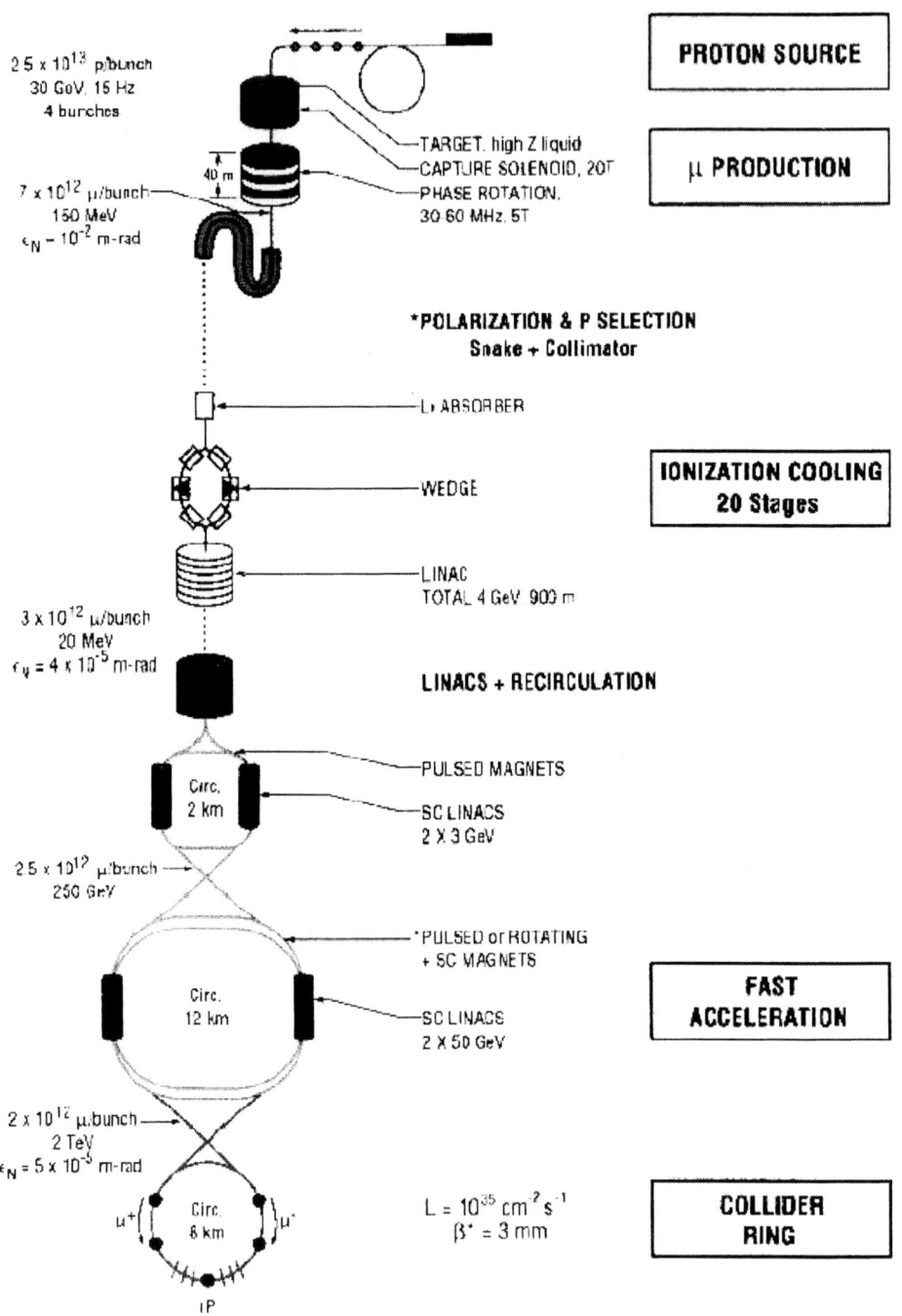

FIGURE 2. Schematic 4TeV $\mu^+\mu^-$ collider. The main components are the proton source, $\mu^+\mu^-$ production, ionization cooling, fast acceleration, and the collider ring. The diagram is not to scale.

A Muon Collider Components

Figure 2 shows a schematic layout of a $\mu^+\mu^-$ collider complex. A powerful proton synchrotron accelerates a few short proton bunches that hit a small target for copious π production. The experiment E951 at BNL will test targetry concepts [28]. Particle production will be measured down to about 2 GeV proton energy in the HARP experiment at CERN [29]. A system of solenoids and RF cavities then captures and accelerates the π's and μ's. An ionization cooling channel reduces the six-dimensional phase space volume of the $\mu^\pm$ beams by a factor between 10^5 and 10^6. Such a channel consists of many cooling cells. Each cell contains an absorber, typically liquid hydrogen, and an RF cavity. The energy loss in the absorber reduces both transverse and longitudinal muon energies. The RF cavity compensates the longitudinal energy loss. Multiple scattering and straggling in the absorber heat the muon beam, and result in a lower limit of the emittances that can be achieved in the cooling channel. The muon scattering experiment MUSCAT at TRIUMF [31] has taken its first data. The simulation tools DPGeant, [1] ICOOL [32] and PATH [33] allow study of the particle dynamics in the cooling channel. An ambitious μ cooling experiment MUCOOL [34] at Fermilab was proposed. A cascade of recirculating linear accelerators similar to CEBAF rapidly accelerates the $\mu^+\mu^-$ to the collision energy. They are finally stored in a collider ring. The $\mu^+\mu^-$ collisions occur in a fancy low-β insertion.

TABLE 2. Energy E, repetition rate f and beam power P of projected and existing high-intensity proton sources

Machine	E/GeV	f/Hz	P/MW
New Fermilab booster	16	15	4
KAON	30	10	3
JHF	50	0.3	0.5
ESS	1.33	50	5
SNS	1	60	2
ISIS	0.8	50	0.16
Fermilab booster	8	15	0.067
CPS	28	0.4	0.056

Table 2 compares the parameters of a typical proton source for a $\mu^+\mu^-$ collider, the new Fermilab booster, with the proton source projects KAON [35], JHF [36], ESS [37], SNS [38], and existing proton synchrotrons. Some of them provide the proton energy E, some the repetition frequency f, and some the beam power P, but only the new Fermilab booster achieves the combination of E, f and P needed for a $\mu^+\mu^-$ collider.

---

[1]) DPGeant is based on Geant3 [30]. Tracking uses 64-bit precision. Classical electro-magnetic fields are implemented.

TABLE 3. Parameters of a "Higgs factory" at 100 GeV and a high-energy $\mu^+\mu^-$ collider at 3TeV in the CoM

Centre of mass energy (GeV)	100		3000
No. of $\mu^\pm$ bunches/sign	1		2
$\mu^\pm$ bunch population (10^{12})	4		2
Collider circumference (m)	300		6000
Free space ℓ^* at IP (m)	5		6.5
RMS momentum spread $\Delta p/p$	0.12%	0.003%	0.16%
β^* at IP (cm)	4	13	0.3
Bunch length σ_z (cm)	4	13	0.3
RMS beam radius at IP σ_r (μm)	82	270	3.2
Beam-beam tune shift ξ	0.05	0.015	0.043
Luminosity (nbs)$^{-1}$	0.12	0.01	50

B Muon Collider Ring

Because of their larger mass, muons produce much less synchrotron radiation than electrons of the same energy, and can be recirculated and stored in circular machines at high bending field. Hence, $\mu^+\mu^-$ colliders are much more compact than circular e^+e^- colliders at the same energy. The 3TeV $\mu^+\mu^-$ collider is about the size of the SPS at CERN with $C = 6911$ m, while the e^+e^- collider LEP at CERN with 200 GeV in the center of mass has $C = 26659$ m.

The most important parameters of a $\mu^+\mu^-$ collider ring appear in the following equation for the luminosity $\bar{L}$, averaged over the relativistic μ lifetime:

$$\bar{L} = \left(\frac{\tau_0}{e\mu_0}\right) \dot{N}_\mu \left(\frac{\xi \gamma B}{\beta_\perp}\right) \left(\frac{2\pi\rho}{C}\right). \tag{1}$$

The first parenthesis contains the natural constants μ lifetime at rest $\tau_0 = 2.19703 \pm 0.00004$ μs, μ charge e, and permittivity of free space μ_0. Then comes the $\mu^+\mu^-$ storage rate $\dot{N}_\mu$, determined by the $\mu^+\mu^-$ source. The second parenthesis contains the initial beam-beam tune shift parameter ξ, the relativistic factor γ, the dipole field in the arcs B, and the value of the β-function at the interaction point $\beta_\perp$. The last parenthesis is a filling factor < 1, the ratio of the total length of dipoles in the arcs $2\pi\rho$ and the ring circumference C. Only two assumptions enter into Eq. (1): The interval between fills is long compared to the relativistic μ lifetime, and the beams are round at the interaction point. For good average luminosity $\bar{L}$, the quantities in the numerator should be large, and those in the denominator small. The derivation of Eq. (1) is left to the reader.

Table 3 shows the parameters of $\mu^+\mu^-$ colliders at two energies, a "Higgs factory" at 100 GeV, and a high energy one with 3TeV in the center of mass. A $\mu^+\mu^-$ collider is particularly interesting for Higgs production since the cross section is $(m_\mu/m_e)^2 \approx 4\cdot 10^4$ times larger than in an e^+e^- collider. The width of the Higgs is believed to be very narrow, and the momentum spread in the CoM should not be larger than the width.

It might be possible to achieve this with a larger momentum spread in the beams by monochromatization [39], which has often been included in τ-charm factory designs [40]. The arcs of a $\mu^+\mu^-$ collider have a high dipole field in order to maximize the number of turns in the $\mu^+\mu^-$ lifetime. They are nearly isochronous in order to achieve a short bunch length σ_s, which in turn allows low values of the β-function at the interaction point IP, but still keeping $\beta_\perp \geq \sigma_s$, avoiding a luminosity drop due to the "hourglass" effect. The low-β insertion uses techniques from final-focus systems of linear e^+e^- colliders to achieve a low value of $\beta_\perp$ at the IP where the detector is installed. The projected beam-beam tune shift ξ is higher than in hadron colliders, and comparable to that of e^+e^- colliders.

C Critical $\mu^+\mu^-$ Collider Issues

Many of the components of a $\mu^+\mu^-$ collider system are at the limit of what is technologically possible. These components should be tested one by one. Some of these tests could be done within the ongoing studies of neutrino factories.

At one time, $\mu^+\mu^-$ colliders were believed to be a more compact and hence more economic method for achieving multi-TeV lepton collisions than linear e^+e^- colliders. Since then, it has turned out that there is an upper limit on the muon energy at a few TeV. Neutrinos from muons with energy E, decaying in the arcs, form a fan in the median plane of the $\mu^+\mu^-$ collider with an opening angle of order $1/\gamma = E_\mu/E$ perpendicular to the median plane. Here E_μ is the rest energy of the muon. Similarly, muons in the straight sections decay into a cone of neutrinos with the same opening angle. Unless these neutrino beams are sufficiently diluted, they are a radiation hazard [41] if they leave the Earth in an uncontrolled area. Dilution is achieved in two ways, by the opening angle leading to an upper limit on E, and by the distance that the neutrinos travel through the Earth leading to a lower limit of the depth of the $\mu^+\mu^-$ collider.

IV FUTURE LARGER HADRON COLLIDERS (VLHC)

Future larger hadron colliders beyond the LHC and the discontinued SSC were discussed in 39 Eloisatron studies at Erice, Italy [42], since 1986, and in "VLHC" workshops at Bloomington, Indiana, in 1994 [43] and Snowmass, Colorado, in 1996 [44]. Exploratory studies continue in several laboratories [45–47]. A National VLHC Organization was set up in the US with a steering committee, working groups on magnet technology, accelerator technology and accelerator physics, working group meetings, annual plenary meetings and reports, etc. [48]. Practically achievable $\bar{p}$ production rates are much smaller than the consumption rates at the luminosities listed in Table 4. Hence, future high-energy, high-luminosity hadron colliders will be proton-proton colliders, and neither $\bar{p}p$ nor heavy ion colliders.

A VLHC Studies

Table 4 compares the parameters of the LHC with those of three larger machines, a 50TeV collider (Low B) with 1.8T dipoles [44], a 50TeV collider (High B) with 12.6T dipoles [44], and a 100TeV collider E12T with 12T dipoles [42,47]. The combination of a high and a low dipole field B, and of two energies E, and the comparison with the LHC, clearly show how the choice of B and E changes the parameters. The stored energy is given in tons of TNT. The conversion factor is 1 t TNT $\equiv$ 4.7 GJ.

TABLE 4. Comparison of LHC and VLHC Parameters

	LHC	Low B	High B	E12T
Beam energy E/TeV	7	50	50	100
Dipole field B/T	8.4	2.0	11.6	12
Circumference C/km	27	520	95	229
Luminosity $L/(\text{nbs})^{-1}$	10	10	10	10
Events/collision n_c	19	21.5	21.5	21
Damping time τ_z/h	26	81	2.6	1.5
Particles/beam $N/10^{14}$	3	27.6	5.0	5.5
Radiation power P/MW	0.0037	0.048	0.189	1.08
Stored energy G/t TNT	0.07	4.7	0.85	0.63
Debris power D/kW	0.8	4.8	4.8	9.6

Contrary to a $\mu^+\mu^-$ collider system, the scale of a VLHC system is dominated by the collider ring proper, not by its injectors. The number of events in a collision, $n_c = L\sigma_{\text{inel}}s$, expresses the ease or difficulty of analyzing the events in the detector, with inelastic cross section $\sigma_{\text{inel}} \approx 120$ mb. Keeping n_c at values comparable to those for LHC essentially imposes an upper limit on the bunch spacing s in seconds. The power in the debris $D = L\sigma_{\text{inel}}E$ of the collisions, which must be absorbed by shielding to prevent the cascades from heating super-conducting coils, is given by Nature and the performance parameters, leaving no choice for the designer.

B Critical VLHC Issues

The product of damping time τ_z and damping partition number J_z for synchrotron radiation in the z-plane is

$$\tau_x J_x = \left(\frac{3E_0^3}{e^2 c^3 r_c}\right)\frac{1}{EB^2}\left(\frac{C}{2\pi\rho}\right). \tag{2}$$

Here z may be horizontal, vertical or along the beam; E_0 is the rest mass of the particle and r_c its classical radius. Numerically the factor in the left parenthesis is 16644 h TeV T^2. The right parenthesis is the reciprocal of the filling factor appearing in Eq. (1). By using Eq. (2), the synchrotron radiation power P, which may cause a considerable heat load when it gets absorbed in a vacuum chamber at cryogenic temperatures, and the stored energy in one beam G, which must be absorbed by a beam dump without destroying it, can be brought into the following forms:

$$P = \left(\frac{4\pi r_c^{3/2}}{\sqrt{3cE_0}}\right) \frac{E^{3/2} L \beta_{\rm IP}}{\xi \sqrt{J_z \tau_z}} \sqrt{\frac{C}{2\pi\rho}} \qquad (3)$$

$$G = \left(\frac{2\pi r_c^{3/2}}{\sqrt{3cE_0}}\right) \frac{E^{3/2} L \beta_{\rm IP} \sqrt{J_z \tau_z}}{\xi} \sqrt{\frac{C}{2\pi\rho}}. \qquad (4)$$

The central fraction in Eqs. (3) and (4) contains the design parameters E, L, $\beta_{\rm IP}$, and ξ, and opposite powers of $\sqrt{J_z \tau_z}$. By choosing B, and hence τ_z, one can trade a reduction of P against an increase of G, and vice-versa.

Various collective effects may limit beam current and luminosity [47]. One dominant collective effect is the transverse resistive-wall instability. Its growth rate is a function of the conductivity, and hence depends on the material and temperature of the vacuum chamber. By choosing the temperature, one can trade the growth rate of the instability, and the feedback system needed to damp it, against the heat load caused by the synchrotron radiation absorbed by the vacuum chamber. A second important collective effect is the coherent synchrotron tune shift, driven by the longitudinal broad-band impedance. In order to ensure longitudinal Landau damping, the bunch length must be increased. All collective effects get worse with small vacuum chamber radii. In the high B colliders, the growth times caused by intra-beam scattering are only a little larger than the damping times.

C VLHC Magnet Development

Simply scaling the VLHC cost from LEP, LHC and/or SSC would result in exorbitant figures. Therefore, R&D programs, aiming at significant reductions of unit prices, have been launched. In the US, they are coordinated by the National VLHC Organization. The proceedings of the Magnet Technology Workshops at Port Jefferson in 1998 and Fermilab in 2000 [48] contain a wealth of information. Two lines of magnet R&D specifically address this issue for the Low B and High B colliders respectively, in Table 4.

Figures 3 and 4 show the concepts of the double-aperture low-field "transmission line" dipole [49] and of the high-field dipole with "common coils" [50]. Since the protons in the two rings must circulate in opposite direction, the dipole fields in the arcs must have opposite signs. Figures 3 and 4 show how this is achieved in both cases. Figure 5 shows the test facility for low-field dipoles at Fermilab.

The design of the high-field dipole has flat coils and large radii of curvature at the ends, and is much simpler and better adapted to brittle materials such as Nb_3Sn and high-temperature super-conductors than the traditional design of $\cos\theta$ coils with three-dimensional ends and tight radii of curvature. The simple shape of the coils also simplifies the design of the steel yoke and the clamping. A model of a common coil dipole

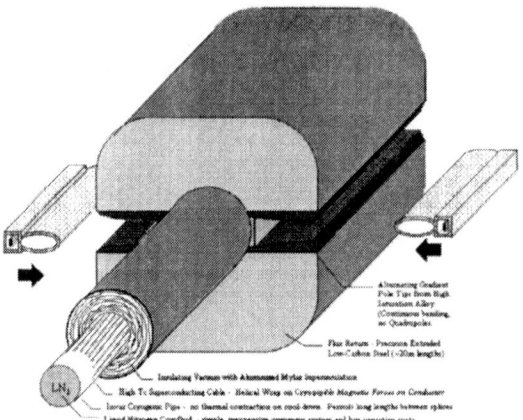
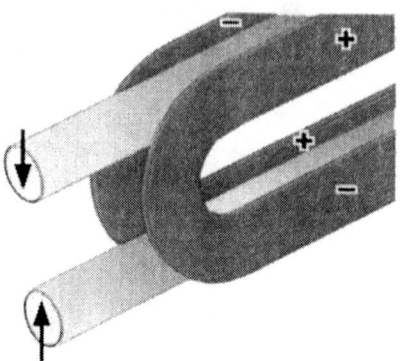

FIGURE 3. Schematic low-field transmission line magnet. The two beam pipes are arranged side by side. Two C shaped yokes above and below the beam pipes carry the magnetic flux. The pole tips create an alternating gradient, and define the field quality. The aluminum vacuum chambers are extruded. The magnet is excited by a NbTi cable, helically wrapped around an Invar cryogenic pipe. Return lines for both helium and current are below the magnet.

FIGURE 4. Concept of the high B dipole with common coils. The two beam apertures are arranged one above the other. Two racetrack coils with currents flowing in opposite directions excite vertical dipole fields with opposite directions in the beam apertures as required.

was constructed, and tested at LBNL in September 1998 with Nb_3Sn cable left over from the SSC [51]. Figure 6 shows the second model RT-1 that was built with state-of-the art superconductor and achieved a field of approximately 12.2T in March 2000.

V CONCLUSIONS

Possible future colliders for e^+e^-, $\mu^+\mu^-$ and pp are discussed. Their energies are well beyond those of e^+e^- and pp colliders such as LEP, Tevatron and LHC which exist or are under construction. The engineering and performance parameters of the future colliders are tabulated. They are all large and expensive machines, if their cost is extrapolated from the existing machines. All future colliders need R&D, both in accelerator physics and in engineering. The chief goal of the engineering R&D is to reduce the unit prices by half an order of magnitude at least. Methods of keeping the cost under control are keeping the magnets simple and making the aperture small. Foster [52] has argued that the cost of a tunnel, fully equipped with magnets, etc., must be of order 2000 USD per running metre in a machine with 500km circumference, in order to keep the total cost at about 10^9 USD. The aim of accelerator physics should be a demonstration that future colliders can be made to work with small apertures. Further studies of future high-energy colliders, adequately funded, should provide many opportunities for students at schools like JAS'2000 to develop bright ideas. I have indicated promising directions for R&D.

FIGURE 5. Fermilab transmission line magnet test facility. The conventional magnet is configured as a DC transformer; the secondary is a superconducting loop. This facility has been excited to more than 100kA. It has a 4 metre long replacable section for testing various conductor and splice configurations.

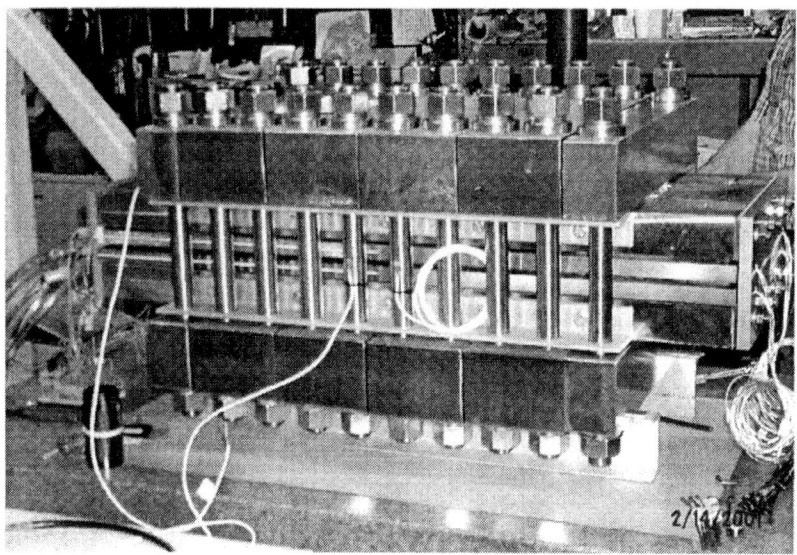

FIGURE 6. Common Coil Nb_3Sn Magnet Model RT-1 at LBNL that was built with state-of-the art superconductor and achieved a field of approximately 12.2T in March 2000.

REFERENCES

1. G.A. Loew and T. Weiland (eds.), *International Linear Collider Technical Review Committee ILC-TRC Report 1995*, SLAC-R-95-471 (1995)
2. www.slac.stanford.edu/xorg/ilc-trc/ilc-trchome.html, October 1999
3. P. Chen and K. Yokoya, Lecture Notes in Physics **400** (Springer, New York 1992) 415
4. http://www.cern.ch/CERN/Divisions/PS/CLIC/
5. http://lcdev.kek.jp/
6. http://www-project.slac.stanford.edu/lc/nlc.html
7. http://tesla.desy.de/
8. NLC Design Group, LBNL 5424, SLAC R 474, UCRL-ID 124161, UC-414 (1996)
9. R. Brinkmann et al. (eds.), DESY 1997-048 (1997)
10. N. Toge (ed.), KEK Report 97-1, April 1997
11. G. Guignard (ed.), CERN 2000-008 (2000)
12. E. Keil, CERN-SL-98-067 AP (1998)
13. J.P. Delahaye et al., Proc. PAC'97 (Vancouver 1998) 482
14. G.I. Budker, *Accelerators and Colliding Beams, Proc. 7th Internat. Accelerator Conference* (Erevan, 1969) Vol. I, 33, Talk at the International High Energy Physics Conference (Kiev, 1970)
15. V.V. Parkhomchuk and A. N. Skrinsky, Proc. 12th Int. Conf. on High Energy Accelerators, F. T. Cole and R. Donaldson (eds.) (1983) 485; A. N. Skrinsky and V.V. Parkhomchuk, Sov. J. Part. Nucl. **12** (1981) 223; D. Cline, AIP Conf. Proc. **352** (1996) 3
16. D. Neuffer, IEEE Trans. **NS-28** (1981) 2034
17. D. Cline, Nucl Inst. and Meth. **A350** (1994) 24−56; *Proc. of the Muon Collider Workshop*, February 22, 1993, Los Alamos National Laboratory Report LA- UR-93-866 (1993); *Physics Potential & Development of $\mu^+\mu^-$ Colliders 2nd Workshop*, Sausalito, California, D. Cline (Ed.), AIP Conf. Proc. **352** (1996)
18. Transparencies at the 2 + 2 TeV $\mu^+\mu^-$ Collider Collaboration Meeting, Feb 6-8, 1995, BNL, compiled by Juan C. Gallardo; www.cap.bnl.gov/mumu/ Transparencies at the 2 + 2 TeV $\mu^+\mu^-$ Collider Collaboration Meeting, July 11-13, 1995, Fermilab, compiled by R. Noble; *Proc. of the 9th Advanced ICFA Beam Dynamics Workshop*, ed. J.C. Gallardo, AIP Conf. Proc. **372** (1996)
19. D.V. Neuffer and R.B. Palmer, Proc. EPAC'94 (London, 1994) 52; M. Tigner, AIP Conf. Proc. **279** (1993) 1
20. R.B. Palmer et al., AIP Conf. Proc. **352** (1996) 108; R. B. Palmer *et al.*, Nucl. Phys. B (Proc. Suppl.) **51A** (1996)
21. $\mu^+\mu^-$ *Collider: A Feasibility Study*, BNL-52503, Fermilab-Conf-96/092, LBNL-38946 (1996)
22. R.B. Palmer, A. Tollestrup and A. Sessler, Proc. New Directions for High-Energy Physics (Snowmass, 1996) 203
23. http://www.cap.bnl.gov/mumu/
24. C.M. Ankenbrandt et al., Phys. Rev. ST Accel. Beams **2** (1999) 081001
25. S. Geer, Phys.Rev. **D57** (1998) 6989 and **59** (1999) 039903
26. R.B. Palmer, C. Johnson and E. Keil, CERN SL/99-070 (AP) (1990)

27. N. Holtkamp and D. Finley (eds.), FERMILAB-PUB-00-108-E (2000)
28. K. McDonald, Targetry Experiment at BNL, Proc. NuFact'00, Monterey
29. M.G. Catanesi et al., Proposal to study hadron production for the neutrino factory and for the atmospheric neutrino flux, CERN-SPSC/99-35 (1999)
30. Geant Manual V3.2.2.1, CERN Program Library W5013 (1994), cf. http://wwwinfo.cern.ch/asdoc/geant_html3/geantall.html
31. R. Edgecock, *Muon Scattering Experiment*, Proc. NuFact'00, Monterey
32. R.C. Fernow, Proc. PAC'99 (New York, 1999) 3020
33. R. Tracz, CERN Document PS/HP-note-99 (1999) (unpublished)
34. C.M. Ankenbrandt et al., Ionization Cooling Research and Development for a High Luminosity Muon Collider (Proposal P904, Fermilab 1998)
35. *KAON Factory Study – Accelerator Design Report*, TRIUMF, Vancouver B.C., Canada (1990)
36. Y. Mori et al., Proc. EPAC'96 (Sitges, 1996) 569
37. H. Lengeler et al. (ed.), *The European Spallation Source, Vol.III, The ESS Technical Study*, ESS-96-53-M (1996)
38. J.R. Alonso, Proc. PAC'99 (New York, 1999) 574
39. A. Renieri, Frascati Preprint INF-75/6 (R) (1975)
40. J.M. Jowett, Lecture Notes in Physics **425** (1994) 79
41. B.J. King, Proc. PAC'99 (New York, 1999) 318
42. A.G. Ruggiero (ed.), *Hadron Colliders at the Highest Energy and Luminosity* (World Scientific, Singapore 1998)
43. Proc. Workshop on Future Hadron Facilities in the U.S., Bloomington, Indiana, 6–10 July 1994, Fermilab-TM-1907
44. G. Dugan, P. Limon, M. Syphers, Proc. New Directions for High-Energy Physics (Snowmass, 1996) 251
45. http://www-ap.fnal.gov/VLHC
46. S. Peggs, M. Harrison, F. Pilat, M. Syphers, Proc. PAC'97 (Vancouver 1998) 95
47. E. Keil, Proc. PAC'97 (Vancouver 1998) 104
48. http://www.vlhc.org
49. G.W. Foster et al., Proc. PAC'97 (Vancouver 1998) 3392
50. R. Gupta, Proc. PAC'97 (Vancouver 1998) 3344
51. S.A. Gourlay et al., Proc. PAC'99 (New York 1999) 171
52. G.W. Foster, VLHCPub-93 (1999)

High Power DC Electron Accelerators of the ELV Type

R.A. Salimov

*Budker Institute of Nuclear Physics,
Novosibirsk 630090 Russia*

Abstract. The parameters of powerful electron accelerators capable of continuous operation are given. The main systems of the accelerator and a wide set of supplementary devices extending its application range are described. Some directions of further development are noted.

MAIN PARAMETERS OF ELV-ACCELERATORS

Beginning in 1971, the Budker Institute of Nuclear Physics (BINP) of the Siberian Branch of the Russian Academy of Science (SB RAS) started the development and manufacturing of electron accelerators of the ELV-type for use in industrial and research radiation-technological installations. ELV-type accelerators are designed with unified systems and units enabling them to be adapted to the specific requirements of the customer by varying the main parameters such as energy range, beam power, length of extraction window, etc. The design and schematic solutions provide long-term and round-the-clock operation of accelerators under the conditions of industrial production processes. The specific features of ELV-accelerators are simplicity of design, convenience and ease of control, and reliability in operation.

BINP proposed a series of electron accelerators of the ELV-type covering the energy range from 0.2 to 2.5 MeV with a beam of accelerated electrons of up to 200 mA and maximum power of up to 160 kW. By now, more than 70 accelerators had been delivered inside our country and abroad and the total operation time exceeds 500 accelerator-years.

The basic parameters of the ELV-type accelerators are given below.

	Energy range, MeV	Beam power, kW	Max. beam current, mA
ELV-mini	0.2 – 0.4	20	50
ELV-0.5	0.4 – 0.7	25	40
ELV-1	0.4 – 0.8	25	40
ELV-2	0.8 – 1.5	20	25
ELV-3	0.5 – 0.7	50	100

ELV-4	1.0 – 1.5	50	100
ELV-6	0.8 – 1.2	100	100
ELV-8	1.0 – 2.5	90	50
ELV-6M	0.75 – 0.95	160	200
Torch	0.5 – 0.8	500	800
ELV-12	0.6 – 1.0	400	400

The last three lines refer to accelerators of a new generation. At a relatively low energy they have quite large power. Their designs are oriented mainly to the solution of ecology problems.

DESIGN

Figure 1 shows the overall dimensions of ELV-type accelerators. A general view of an ELV-type accelerator with foil extraction is given in Fig. 2. Inside the tank filled with SF_6 gas are located: primary winding, high voltage rectifier with a built-in accelerating tube, high voltage electrode, and injector control unit. The location of the accelerating tube inside the column of the high voltage rectifier makes ELV-accelerators the most compact among devices of this class. The vacuum system components and extraction device are fixed to the bottom of the tank. Electrons emitted by

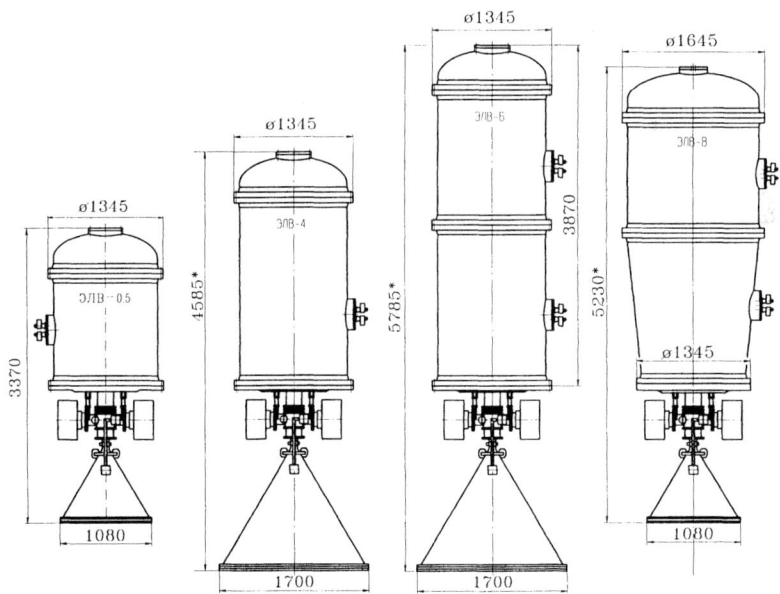

FIGURE 1. Overall dimensions of ELV-type accelerators.

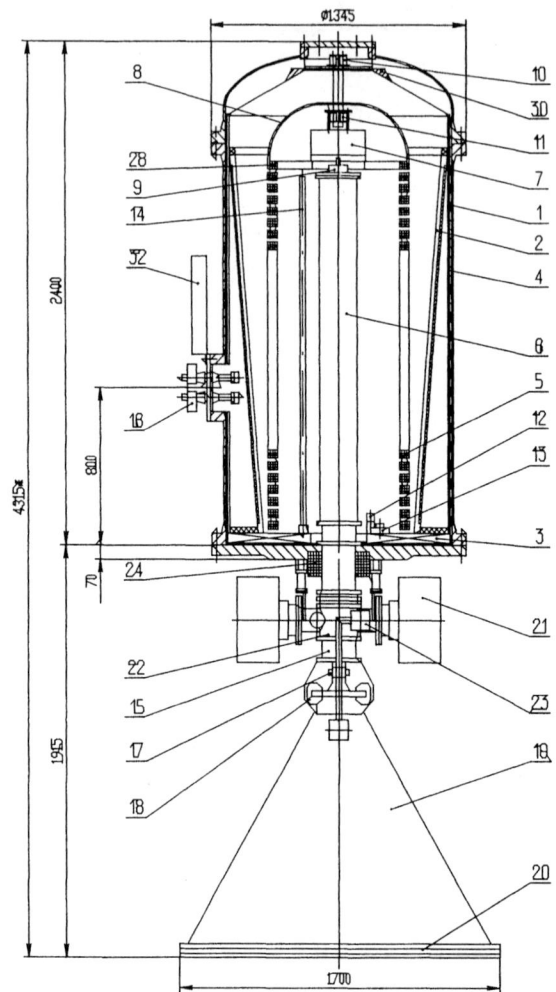

FIGURE 2. General view of ELV-4 accelerator: 1 – vessel, 2 – primary winding; 3,4 – magnetic guides; 5 – rectifier sections; 6 – accelerating tube; 7 – injector control unit; 8 – high voltage electrode; 9 – injector; 10,11 – optical channels for injector control; 12 – section divider; 13 – capacitor unit; 14 – energy divider; 15 – vacuum gate; 16 – primary winding terminals; 17,18 – scanning coils; 19 – extraction device; 20 – extraction window frame; 21 – vacuum pumps; 22 – cross head; 23 – vacuum gate; 28 – base of high voltage electrode; 29 – magnetic lens; 30 – high voltage shield; 32 – clamp set.

the cathode, placed on the upper end of the accelerating tube, have total energy eU_0 at the output of the accelerating tube. Passing through the vacuum system they reach the extraction device, where they are homogeneously distributed along the foil by the scanning electromagnets and then extracted into air. The irradiated material is transported under the frame of the extraction window. A simplified electric circuit of the accelerator is given in Fig. 3.

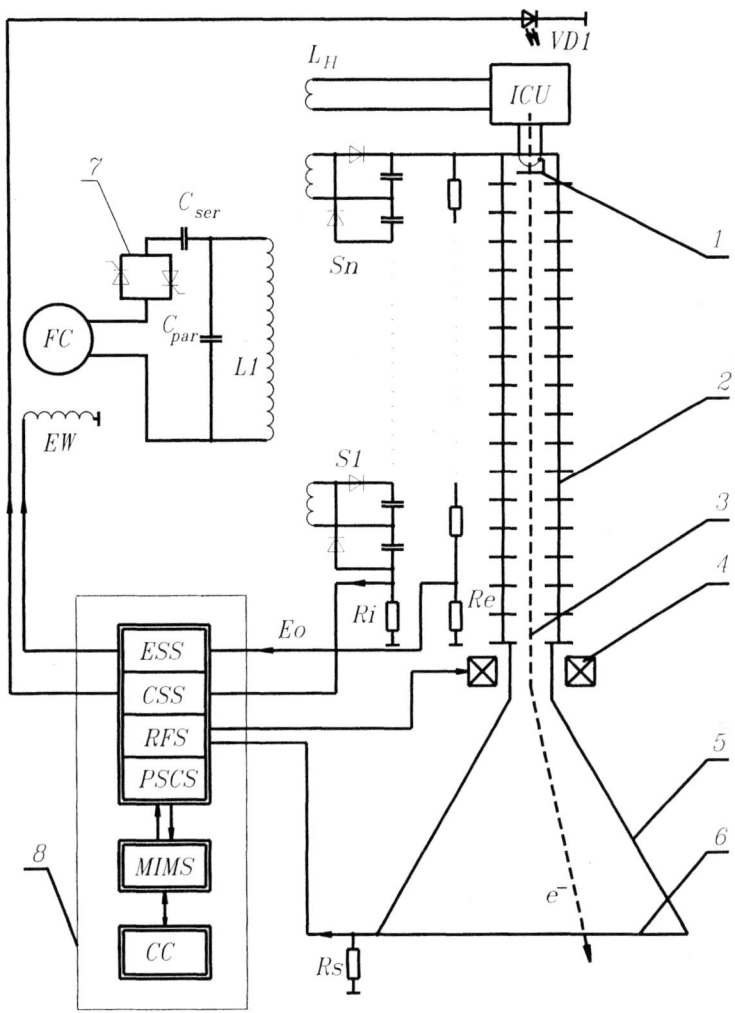

FIGURE 3. Simplified electric circuit of ELV accelerator: 1 – cathode of electron gun; 2 – accelerating tube; 3 – electron beam; 4 – coils of the raster formation system; 5 – extraction device; 6 – titanium foil; 7 – thyristor switch; 8 – control system (ESS – energy stabilizing system, CSS – current stabilizing system, RFS – raster formation system, MIMS – module information–measuring system, CC – Control computer, PSCS – power supply control system, FC – frequency converter, ICU – injector control unit).

HIGH VOLTAGE RECTIFIER

The source of high voltage is a cascade generator with parallel inductive coupling. The rectifier section column is installed inside the primary winding. The electric circuit of the section is given in Fig. 4. The coil of the secondary winding has 3000 turns and the maximum voltage induced on its ends is 20 kV. This voltage is rectified by the voltage doubling scheme. Thus, the output voltage of the rectifying section is 40 kV. The rectifying sections are connected either in series (Fig. 4a) or in series-parallel (Fig. 4b). The rectifiers with series connection are of higher voltage and those with the series-parallel connection are of higher current. The rectifier section column is terminated with a high voltage electrode, inside which are the injector control unit and a special coil for its power supply.

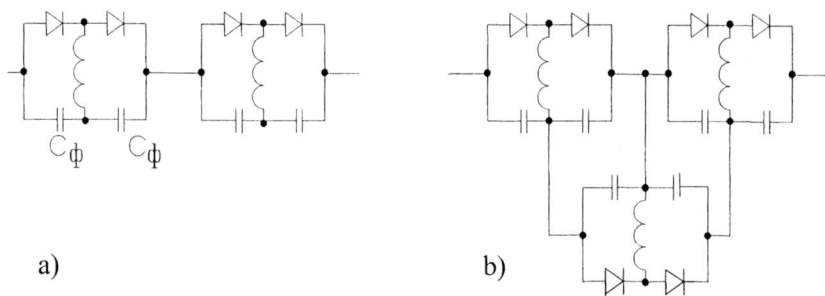

FIGURE 4. ELV rectifying section circuit, and ways of connecting sections.

Note that, unlike "conventional" transformers, our design has no central magnetic core. This substantially simplifies the high voltage source design but has no practical influence on the operational characteristics of the rectifier because of the successful design of the primary winding, the high quality of the energy stabilizing system, and the low turn voltage (6 V/turn). The transformer specific power in ELV-type accelerators is about 40 kW/m.

The use of low inductive capacitors K-15-10, the original scheme of intrasection connections, and the presence of damping resistances provide reliable protection of components of the high voltage rectifier against overvoltages during breakdowns of both vacuum and gas insulation. Generally speaking, breakdown in ELV-type accelerators is an exceptionally rare event; however, during the design stage (and this principle is always followed) we assumed that even a great number of breakdowns (hundreds or thousands) should not lead to damage of the high voltage rectifier.

Practice has proved the high reliability of the high voltage rectifier. The only reason for its malfunction is malfunction of capacitors K-15-10 caused by micro-charges. Therefore, the preliminary selection of capacitors can provide a substantial increase in rectifier reliability.

ACCELERATING TUBE

A general view of the accelerating tube is shown in Fig. 5. The channel aperture is 100 mm. This provides good vacuum conditions in the cathode region and, consequently, long lifetime. The outer diameter of the insulator is 205 mm, its inner diameter is 180 mm. A distance between the electrodes is 21 mm. The 20-mm ceramic rings UF-46 are connected to the electrodes either with the high-molecular glue PVA or by thermodiffusion welding. In order to avoid the influence of alternating magnetic fields, the tube is shielded by short-circuited copper rings. To shield the transverse component of the magnetic field, a few layers of transformer iron are placed inside the rings.

The potential distribution among the electrodes is produced by the high-Ohmic divider. The typical current value of the tube divider is up to 50 mA. The voltage distribution is homogeneous except for the upper part where the resistor value is determined by the conditions of the maximum electric strength of a tube. The divider resistors are fixed directly on the electrodes.

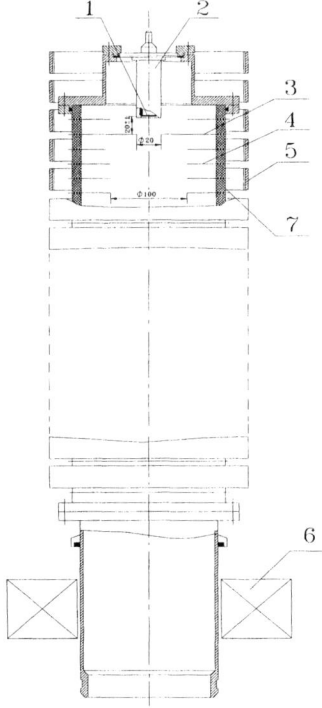

FIGURE 5. General view of accelerating tube: 1 – cathode; 2 – cathode heater; 3,4 – electrodes; 5 – shielding rings; 6 – magnetic lens; 7 – ceramic.

The maximum operation gradient in the tube is 10 kV/cm, but for long-term and round-the-clock operation its value does not exceed 8 kV/cm; therefore no vacuum breakdowns occur in accelerating tubes.

The cathode, in the form of a LaB_6 tablet either 6 or 10 mm in diameter, has indirect heating. For heating the injector, a power of about 50 W is required. The beam current value is controlled by the cathode temperature, i.e. the gun is operated in a regime of full emission current take-off. To this end, and injector control unit is envisaged which is located under high potential inside the high voltage electrode, and a beam current stabilizing unit is placed in the control cabinet. The stabilizing system provides stability at a level no worse than ±0.3 mA, which does not exceed 1% of the maximum beam current value.

For lossless passing of the beam through the vacuum system and extraction device, a magnetic lens is installed on the lower end of the accelerating tube. The lens current is regulated automatically without an operator in case of an energy change.

VACUUM SYSTEM

An operational vacuum in the accelerating tube is provided by two magneto-discharge pumps, each of 400 l/s capacity. The preliminary start is provided by a fore-vacuum aggregate AVZ-20 with a nitrogen trap. As a rule, this aggregate is used only at the first start after opening the vacuum system (assembly, cathode or filter replacement). Furthermore, in normal operation, intervals in operation of up to two days do not require initial vacuum pumping for starting the magneto-discharge pumps. In the design of the vacuum system, simple and efficient rubber seals are used and thereby the operation vacuum value is limited to a level of $10^{-4}-10^{-5}$ Pa. This vacuum level is quite acceptable since the cathode lifetime is not reduced; particle scattering on the residual gas is of the order of 10^{-6} of the beam current, i.e. the particle flux to the tube electrodes is insignificant and therefore the electric strength is not reduced during operation with a beam.

The vacuum value is measured by the current of the magneto-discharge pumps. The interlock system is a two-step one: at vacuum of 5×10^{-4} Pa, a warning signal for the operator is sent to the terminal, and at a vacuum of 10^{-3} Pa the accelerator is de-energized. The vacuum gate (pos. 15 in Fig. 2) enables cathode replacement with no loss of vacuum in the extraction device or, correspondingly, foil replacement without loss of vacuum in the accelerating tube.

EXTRACTION DEVICE WITH A FOIL WINDOW

A schematic diagram of the device designed for beam extraction into air through the foil is given in Fig. 6. An electron beam is scanned over the foil in two mutually perpendicular directions with the use of two electromagnets. The scanning frequencies have the ratio 251/15. Due to this, there is no overlapping of beam trajectories and the foil is filled completely. The low frequency beam is scanned along the foil and the high frequency beam is scanned across the foil. The scanning frequency along the foil

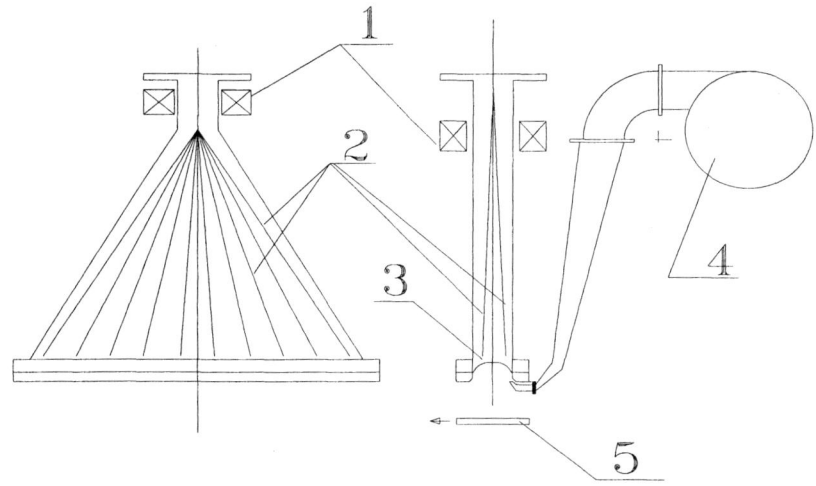

FIGURE 6. Schematic diagram of extraction device with a foil window: 1 – scanning electromagnets, 2 – beam trajectories; 3 – foil of extraction window; 4 – foil cooling fan; 5 – movable target.

is about 50 Hz if there are no special technological requirements. The maximum deflection angle of a beam is 30°.

The foil is cooled with an air jet. To this end, a high pressure fan is used with an initial jet rate of 180–200 km/h. At this rate, the average current density on the foil does not exceed 100 mA/cm^2, i. e. the maximum extracted current is 70 mA/m. This is approximately half the maximum admissible value for the current density on the foil for this jet rate. This double reserve of current density throughout the foil makes its lifetime practically limitless.

Figure 7 shows the distribution of linear current density at a distance of 50 mm from the frame of the extraction window. The linear current density is the part of a beam measured by a long probe installed across the extraction window. The value of the absorbed dose in the irradiated material is proportional to this parameter. Usually, we guarantee an inhomogeneity of the linear current density of no worse than ±10% at a distance of no more than 50 mm with 90% use of the beam current. The current loss on the distribution tails is caused by electron scattering on the extraction window foil and in the air.

TWO-WINDOW EXTRACTION DEVICE

As mentioned above, the maximum current extracted through the foil window is 70 mA/m. The use of radiation technologies in large-scale industrial production (flue gas treatment, metallurgy; waste water treatment, etc.) requires an increase in accelerator power up to a few hundred kilowatts. The electron beam optimum energy for most of these application is in the range of 0.7–1.5 MeV. Therefore, in order to achieve the

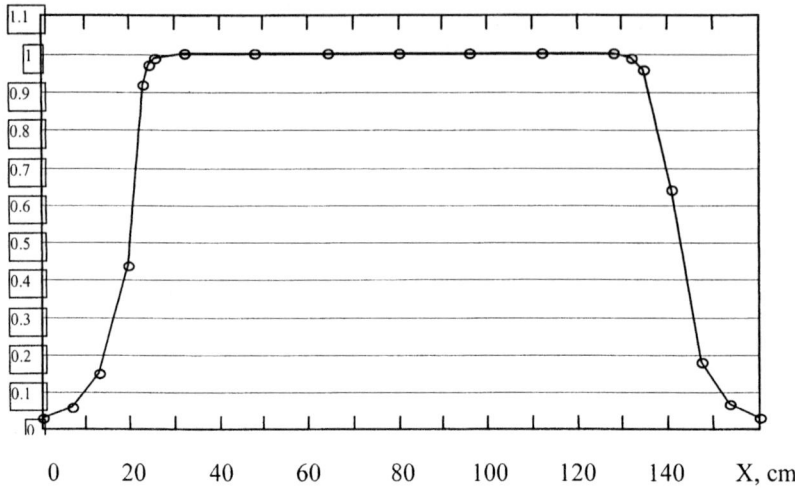

FIGURE 7 Typical distribution of linear current density under the frame of the extraction window at an electron energy of 1.5 MeV.

required power one has to extract into air an electron beam with a current of a few hundred mA at quite a low current density, i.e. the extraction window area should be enlarged. The use of support grids in the given range of energies is not reasonable since their transparency is 80–90%. If a single window is used, its width is determined by the mechanical strength of the foil and it does not exceed 7–10 cm. Therefore, it was decided to develop a new extraction device with two extraction foils enabling the doubling of the extraction window area with no substantial change in the device overall dimensions.

The new extraction device (Fig. 8a,b) was developed on the basis of the existing design with an extraction window length of 1600 mm. Two parallel foils are used in the device. An approximate trajectory of beam movement is shown in Fig. 8c. Beam scanning both along and across the window is produced with the help of standard deflecting magnets with a frequency scanning ratio of 15/251. The beam is transported from one window to the other with a switching magnet. The transport moments are synchronized with the scanning frequency along the foil in such a way that the change of beam current polarity in the switching magnet is made when the maximum deflection is achieved (on the window ends). To protect the foil fixing and its seals against the direct action of a beam during beam transport, a water-cooled cylinder is installed (Fig. 8d) which also increases the rigidity of the structure. Two additional magneto-discharge pumps, installed for better vacuum, pump directly from the extraction device.

The diaphragm on the extraction device input is an element of vacuum resistance. As a result, a simple system of differential pumping is produced, and the accelerating tube vacuum during operation with a beam is two or three times as good as that in the extraction device. At a beam current of 200 mA and an energy of 0.8 MeV the accelerating tube vacuum is $1-2\times10^{-4}$ Pa. Thus, high beam current does not influence

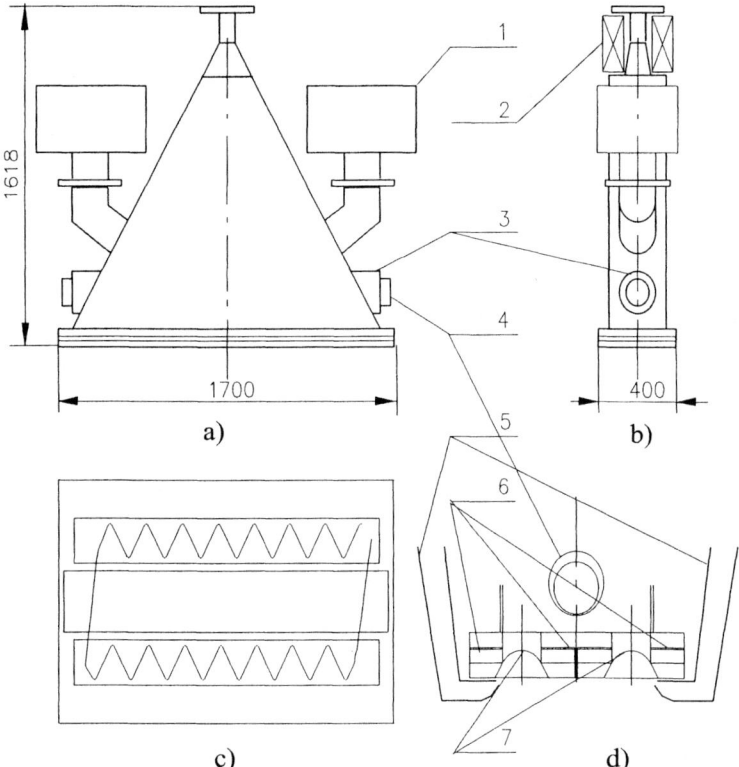

FIGURE 8. Two-window extraction device: 1 – ion pumps, 2 – coils and cores of the beam scanning system, 3 – protection cylinder flange, 4 – protection cylinder, 5 – air jet cooling, 6 – frame for fixing the foil, 7 – extraction foils.

the vacuum conditions of the accelerating tube. In the development and adjustment of the extraction device two major problems had to be solved:
- to provide a sufficiently short time of beam transport from one window to the other;
- to correct an inhomogeneity of the transverse deflection of a beam along the window caused by the difference in beam path lengths from the deflection point to the foil surface and also by the influence of edge focusing of the longitudinal deflection magnet i.e. to compensate for raster distortion of the "sabre" type.

The minimum time for beam transport is determined by the wall thickness of the extraction device, since the switching magnet is installed on the outer side of the extraction device, and also by the coil inductance of the switching magnet. With a 3-mm wall thickness of the stainless mouth, the minimum time for changing the field polarity of the switching magnet is about 0.25 ms, which provides a beam loss at a level of 3%.

The correction of raster distortion is achieved by providing the required shape of current in the coils of the switching magnet. The power source of the switching magnet is computer controlled with a special circuit on the base of the programmable DAC made in the CAMAC standard. Control of the beam transport system is fully automated and produced by a standard computer in the framework of the general accelerator control program.

An electron beam with a 200-mA current at energies of 750–800 keV is extracted with use of this extraction device. This device passed the bench test. Long operation of this device is planned in a pilot industrial plant.

CONTROL SYSTEM

The control system for the industrial accelerator determines its operation characteristics such as its reliability, continuous operation, repair fitness, and level of personnel qualification.

The operator of a technological installation communicates with the accelerator through the computer. The accelerator control system comprises a set of software and hardware covering all the accelerator units requiring operating control and diagnostics. The multifunctional control system enables one:

- to have automated control of the accelerator. Algorithms introduced into the accelerator control program solve the problems of preparing the accelerator for operation (speed-up of frequency converter, switch-on of the foil blow motor, switch scanning, and, if necessary, technological equipment); watch the status of interlocks; and, after switching on of the accelerator, provide the energy and current of the electron beam for a given regime.
- to stabilize safely the main parameters of an electron beam (electron energy, beam current, size and position of the raster on the foil of the extraction window) which provides high quality radiation treatment.
- to provide continuous diagnostics of the high voltage rectifier and self-testing of the other accelerator systems during operation of the accelerator.
- to synchronize accelerator operation and technological equipment; in this case, operation of the accelerator integrated into the technological line in a completely automated regime is possible.
- provides the personnel with a wide choice of preliminary commands for regimes of testing and adjusting the accelerator.

Figure 3 shows a functional diagram of the ELV-accelerator connections to the control system.

The high voltage rectifier column, which consists of rectifying sections $S_1 \ldots S_n$ connected in series through the primary winding L_1, takes power from the frequency converter FC. The high voltage is measured with the help of the energy resistive divider, and signal E_0 is applied to the input of the energy stabilizing system ESS. The control action proportional to the difference between the requested and real energy is applied to the excitation winding EW of the frequency converter FC. The long-time instability of the energy is in the range of 1–2%.

A positive output of the high voltage rectifier is connected with the "ground" through the current-measuring resistor R_1. The measuring signal of rectifier current I_0 is applied to the input of the current-stabilizing system CSS. The error signal from the output of the system through the optical connection line controls the output voltage of the injector control unit ICU applying the current into the heater of the cathode indirect filament. The beam current control by the filament current regulation increases the cathode lifetime since, in this mode, the heater uses the minimum current. The beam current instability lies within the limits of 1% of its maximum value.

The raster formation system RFS coil 4 (see Fig. 3) scans the beam along the foil of extraction window 6. The raster formation system provides the power supply of these coils by the saw-tooth current, power supply of the lens, automatic correction of the position (centering) of the beam raster on the extraction foil, de-energizing the accelerator if the scanning current or lens current values decrease below an admissible value. The raster centering on the foil is done by the beam position stabilizer. The beam position signal is formed by an analog treatment of the beam sediment signal (less than 1% of the full current) to the walls of the extraction device measured with resistor R_S. If the raster is not centered on the foil, the beam position stabilizer applies the correcting current to the corresponding correction coil.

These three systems (ESS, CSS, RFS) added to the power supply control system of PSCS make up the lower level of the control system – the level of the end control units. All the control commands for the setting of modes, etc., are formed on the next level of the control system: the module information-measuring system MIMS.

The CAMAC section comprising the set of measuring and control modules required for a specialized ELV-accelerator control station, based on a microprocessor, is used as the module information-measuring system. The station comprises the 64-channel input and output registers, 12-bit ADC with 64-channel analog multiplexer on the input, and 16 12-bit DACs.

The third level of the control system is the control computer loaded with packages of specialized software. This software provides a user-friendly system of dynamic menus, and text and graphic visualization of the accelerator operation run.

POWER SUPPLY

The power supply of the primary winding of the accelerator is provided at a frequency of 400 HZ from the frequency converter. Rotary frequency converters have been used which were attractive because of low price, simplicity, and reliability. Their only deficiency is low efficiency (65–80% depending on power). Attempts are now being made to replace them by static frequency converters (both the thyristor and transistor types). This is expected to increase the total efficiency up to 85% for machines of up to 100 kW and up to 92% for more powerful accelerators.

The thyristor switch is designed for fast ($<10^{-3}$ s) de-energizing of the accelerator in case of an emergency such as vacuum or gas insulation breakdown, decrease in currents of scanning and lens, or loss of foil cooling.

All the operative switching in the power supply is done by the control program automatically without action by the operator.

DEVICES FOR EXTRACTION INTO AIR OF THE CONCENTRATED ELECTRON BEAM

When extracting an electron beam into air through the foil the maximum current density does not exceed 100 mA/cm^2. However, a number of beam technologies require higher current densities. To this end, devices for extracting a concentrated electron beam into air have been developed. The electron beam current density on the output of these devices may reach 10 A/cm^2, and power density 10 MW/cm^2. The beam is extracted through a system of holes in diaphragms. The holes are burnt by the beam and their diameters are within the limits of 1–2 mm. The operation vacuum in the accelerating tube is provided by the continuous operation of differential pumping. Two versions of accelerators with this kind of extraction device have been developed.

In the first one the beam is focused by two magnetic lenses. This version is used in the ELV-6 accelerator with maximum power of 100 kW at an energy of 1.5 MeV. (See Fig. 9). Its operation is briefly described below. An electron beam, after passing through the accelerating tube, is focused by the magnetic lens. A diaphragm is placed in the lens crossover. Further on, the expanded beam reaches a second lens with a smaller focal length. Two diaphragms are located in the lens crossover. The gas going through the holes in the diaphragms is evacuated with vacuum pumps. The maximum value of the extracted beam current is limited by the ripple of accelerating voltage which cause an increase in the holes in the diaphragms, admissible at a level of 2–3%. The ELV-2, ELV-3, ELV-4 and ELV-6 accelerators can be equipped with this device for the extraction of focused electron beams.

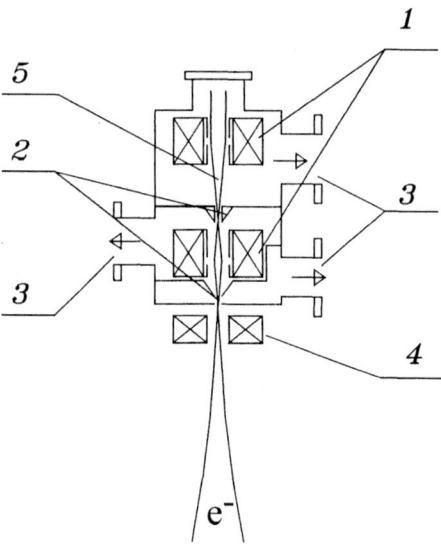

FIGURE 9. Schematic diagram of extraction device: 1 – magnetic lens; 2 – diaphragms; 3 – vacuum lines; 4 – scanning magnets; 5 – beam envelope.

For experiments having no requirements regarding the maximum densities of power, the device is equipped with magnets for scanning in two mutually perpendicular directions. In this case, the beam is deflected in air. This system of scanning also enables one to provide different configurations of the dose field (Fig. 10) according to the technology requirements.

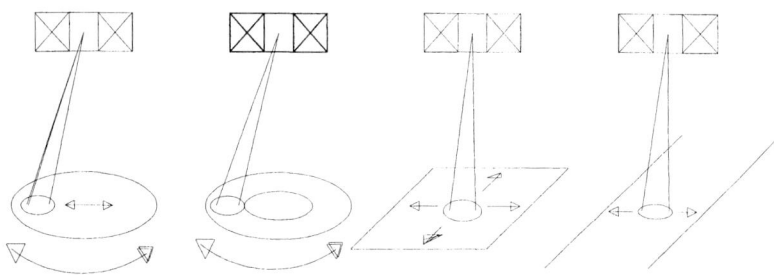

FIGURE 10. Possible configurations of irradiation fields.

Another way to decrease the beam diameter in the output diaphragms is to compress the beam by adiabatically increasing the longitudinal magnetic field. This method is used in the 500-kW Torch accelerator. The main advantage of adiabatic compression is the low sensitivity of the beam size to changes of electron energy, which is especially important during the development of accelerators with a power of a few hundred kilowatts where the problem of ripple and instabilities of the accelerating voltage is vital.

The accelerating tube and extraction device are placed on the same axis, and the magnetic field smoothly increases from 100 G on the cathode to 10000 G in the region of the output diaphragms. In this case, the beam size decreases inversely proportionally to the square root of the magnetic field value The longitudinal magnetic field is produced by a system of solenoids and coils. To increase the field value only in the region of the output diaphragms, a steel concentrator is used. A schematic diagram of this extraction device is given in Fig. 11. The power consumed by the magnetic system is 50 kW. The vacuum system consists of the diaphragms and tubes of the high vacuum resistance and is built into the magnetic system. Figure 12 shows the magnetic and vacuum systems of the accelerator.

At first sight, it seems that the efficiency of this extraction device is lower than that of the device where the beam is extracted into air through the foil. However, the electron energy loss in 50-mm Ti foil is 35 kW and, in addition, a small fraction of the beam current (1–2%) reaches the walls of the extraction device. This leads to the finding that, if a 1-A beam were extracted into air through the foil, the beam powerloss would be 50 kW, i.e. even at a current of 1 A the efficiency of the device with adiabatic beam compression is not lower than that of the foil extraction device. As the beam current increases, the losses during extraction through the foil increase proportionally, but in the device with adiabatic compression they remain constant.

The maximum extracted current value was 0.8 A.

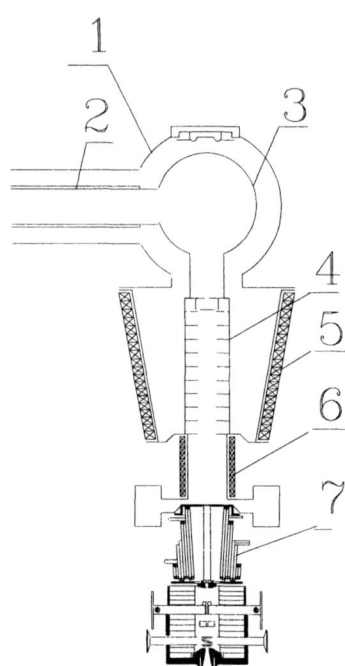

FIGURE 11. Device for the extraction into air of an adiabatically compressed intense electron beam: 1 – tank, 2 – high voltage gas feeder connecting the device with the source of an accelerating voltage, 3 – high voltage electrode inside of which the components of the injector power supply system are placed, 4 – accelerating tube, 5,6 – solenoids producing an increasing longitudinal magnetic field, 7 – coils of the magnetic system with a built-in system for differential pumping.

SYSTEMS FOR CIRCULAR AND TWO-SIDED IRRADIATION

In order to extend the technological capabilities of accelerators, systems for two-sided and circular irradiation are being developed and manufactured which provide efficient use of a beam extracted into air through a foil for irradiating cables and tubes of up to 60-mm diameter as well as for two-sided irradiation of bands of up to 300-mm width. The action of bending magnets changes the motion of the electrons extracted through the foil and makes it possible to irradiate the whole object, as shown in Fig. 13.

The device for two-sided irradiation is operated similarly. The systems are efficiently operated at an electron energy above 1.2 MeV. They are also operable at low energies, but the beam utilization factor is reduced because of electron scattering in the foil and air. The systems are supplied as supplementary equipment for typical machines and are easy installed and removed at the replacement of technology.

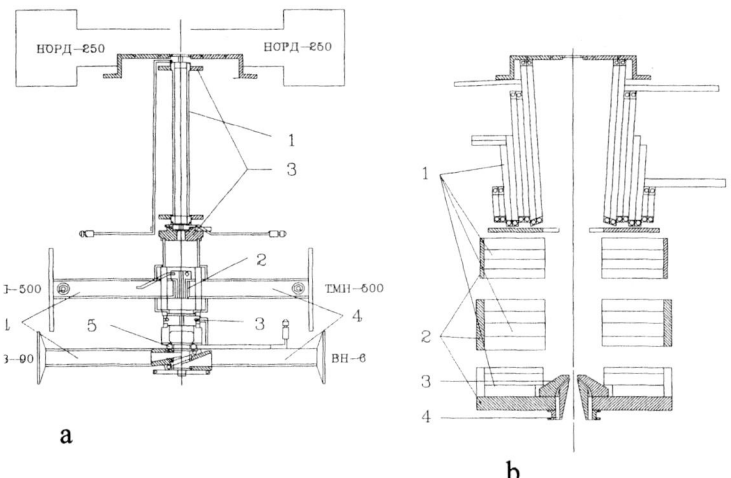

FIGURE 12. a) Vacuum system of an extraction device with adiabatic compression of the beam: 1,2 – tubes of high vacuum resistance, 3,5 – diaphragms, 4 – vacuum lines. b) Magnetic system of extraction device: 1 – coils, 2 – magnetic guides, 3 – concentrator, 4 – cooling shield.

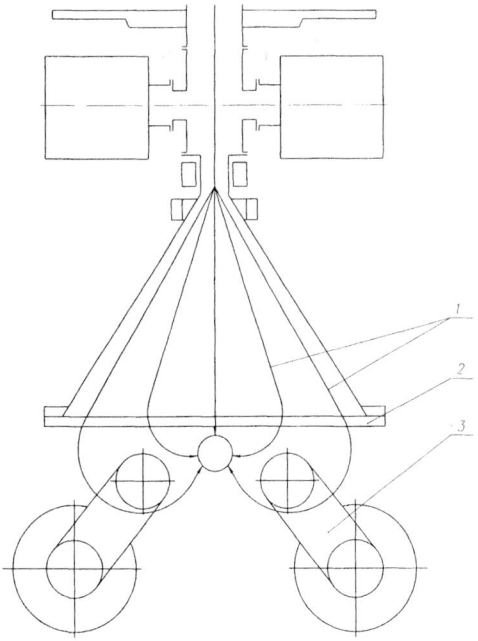

FIGURE 13. System of circular irradiation: 1 – electron trajectory, 2 – extraction window, 3 – bending magnets.

ACCELERATORS WITH EXTRACTED BEAM POWER OF A FEW HUNDRED KILOWATTS

Available accelerators mainly have power of up to 100 kW and cannot satisfy the needs of power-intensive radiation technologies (mainly ecological) which require accelerators with total electron beam power of megawatts and tens of megawatts. The modules of such complexes must have unit powers of hundreds kilowatts as a minimum.

At the Budker INP, a new generation of high voltage accelerators is being developed which produce an extracted beam of the required power. Representatives of this new family of accelerators are the ELV-6M with an energy of 0.75–1.0 MeV and a power of 160 kW; the "TORCH" accelerator with an energy of 0.5–0.8 MeV at a power of 500 kW, and the ELV-12 accelerator with a power of 400 kW at an energy of 0.6–1.0 MeV.

The ELV-6M Accelerator

A schematic diagram of the ELV-6M accelerator is shown in Fig. 14a. It resembles the ELV-6 accelerator where two rectifying columns operate in parallel to the common load. The columns are located vertically one over another. The accelerating tube is placed in the lower column. Each of the columns consists of 38 rectifying sections connected in series-parallel as is shown in Fig. 14b. Note that the similar connection diagram allows operation with no filter capacitors in the rectifying section, as shown in Fig. 14. In this accelerator a new scheme is used to supply power to the primary windings: the windings are connected to different phases of a three-phase frequency converter and the output voltage of each column has a phase shift. This provides additional smoothing of ripple of the output voltage.

The two-window device for extracting beam currents of up to 200 mA into air was specially developed for this accelerator and tested on it. This accelerator has been manufactured and tested successfully. Upon completion of the development of the pilot installation for electron-beam purification of gases, it will be delivered to the Slavyansk HES.

"Torch" Accelerator

In the "Torch" accelerator (Fig. 15), the high voltage rectifier is placed in a separate tank and is connected with an accelerating tube through the gas feeder. The rectifier consists of two parallel columns with an output of high voltage in between. The sections of each column are connected in series-parallel and have no filter capacitors. The primary winding with the central magnetic guide is located inside the column of the high voltage rectifier. The accelerator operation frequency is 1000 Hz and it is supplied from the converter PPFV-500.

The accelerator is equipped with a device (described above) for extracting the adiabatically compressed electron beam into air. The maximum parameters obtained on this accelerator are the following: beam current = 0.8 A at an energy of 0.5 MeV, energy = 0.8 MeV at a current of 0.5 A, beam power = 500 kW (0.7 MeV × 0.7 A).

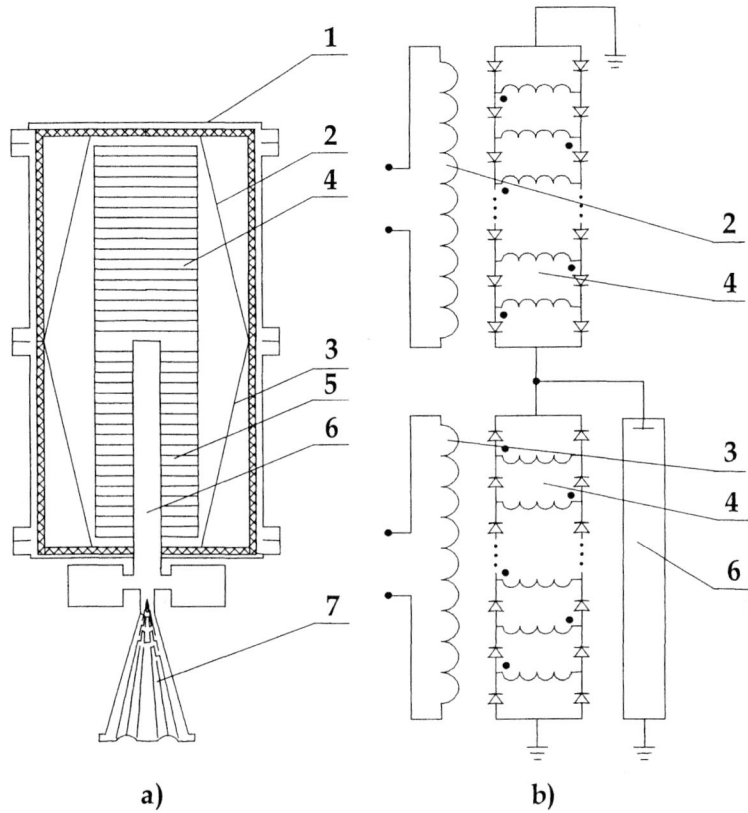

FIGURE 14. a) Schematic diagram of the ELV-6M; b) electric circuit of the ELV-6M: 1 – tank; 2,3 – primary windings; 4,5 – columns of rectifying sections; 6 – accelerating tube; 7 – extraction device.

ELV-12 Accelerator

The ELV-12 accelerator (Fig. 16) unites design solutions tested in the process of developing previous accelerators. The high voltage power supply source, which consists of two parallel columns, is placed in a separate tank and is connected by the gas feeders with two accelerating modules. The rectifying sections in the columns are connected in series-parallel and they have no filter capacitors since the output capacity is quite large because of the presence of gas feeders and operation frequency increased up to 1000 Hz. New coils of the secondary windings were developed for this accelerator. They have a larger diameter, and the winding voltage is increased to 20 V/turn. The power supply from the thyristor frequency converter is planned.

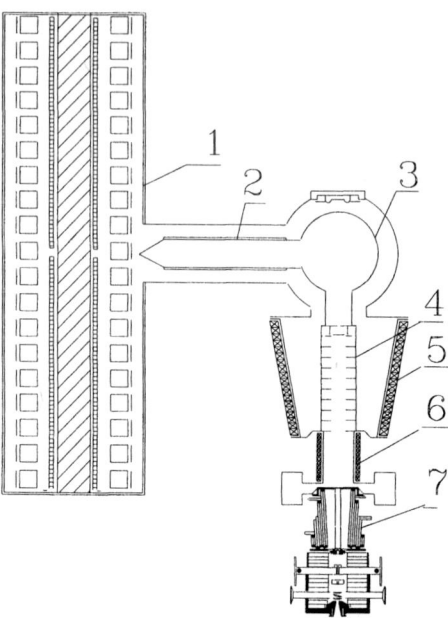

FIGURE 15. "Torch" accelerator: 1 – the source of accelerating voltage, 2 – gas feeder, 3 – injector control system, 4 – accelerating tube, 5,6,7 – the system of raster formation and vacuum system.

The accelerating module is an accelerating tube placed in a separate tank and connected to the two-window extraction device. In addition, one tube may be located inside the accelerator tank. The beam current in each module is regulated independently; this enables regulation in a wide range of power distributions of an absorbed dose. The maximum current value of an individual module is 200 mA. In this case, the maximum total accelerator current is 400 mA. In principle, the rectifier under construction (if necessary) would provide power up 1 MW with an increased number of irradiators.

The ELV-12 accelerator is designed mainly for operation in electron beam devices for the purification of gas and sewage water. At present, the accelerator is under construction.

CONCLUSION

Our accumulated experience in the design, development, and manufacture of the ELV series of accelerators enables us to propose to the customer machines which, by their parameters, do not rank below but in the majority of cases even surpass the best world samples of such machines.

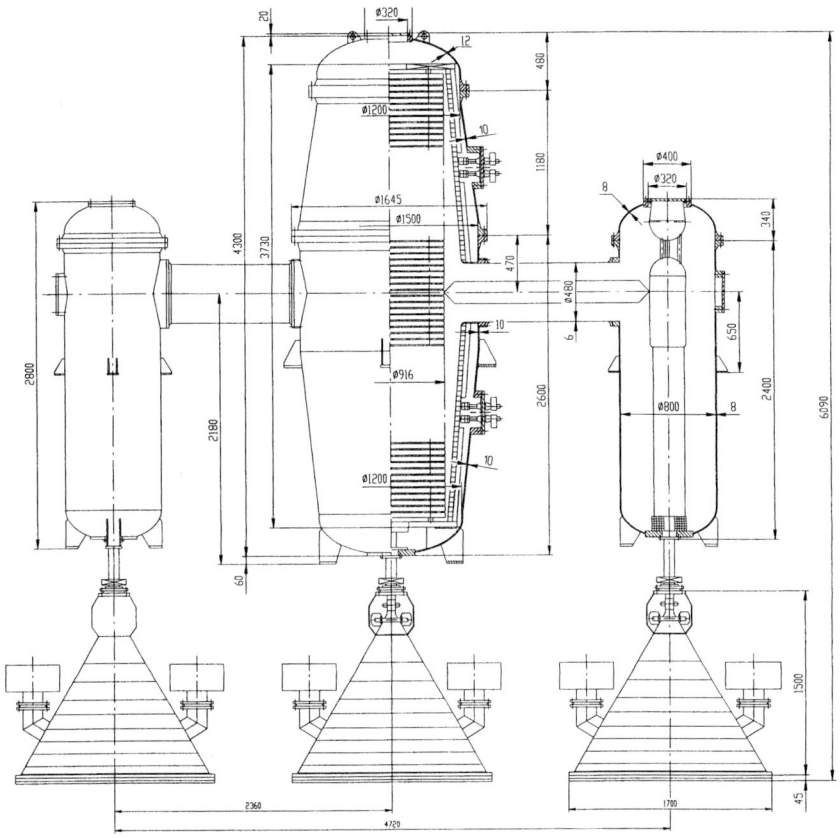

FIGURE 16. General view of ELV-12 accelerator (the version with three accelerating tubes)

SUMMARY: MAIN APPLICATIONS OF ELV-ACCELERATORS

	Type of technology	Country	Number
	Modification of polymer products		
1	Modification of the polyethylene insulation in the production of thermoresistant wires and cables 0.5–120 mm^2 with a capacity of up to 200 m/min	Russia	5
		Byelorussia	5
		Ukraine	5
		Czechia	1
		China	10
		Korea	3
2	Production of thermo-shrinkable pipes, films and bands with a capacity of up to 1000 kg/h	Russia	3
		Moldavia	2
		China	3
		Korea	1
3	Production of gel	Russia	2
4	Production of artificial leather and rubber technical products with a capacity of up to 1000 m/h	Russia	4
5	Curing lacquer-paint coatings on various bases for the building industry with a capacity of up to 500 m^2/h	Russia	2
		Uzbekistan	1
	Ecology		
6	Sewage water treatment	Russia	4
		Korea	1
7	Purification of flue gases by removing sulphur oxides and nitrogen oxides with a capacity of 20000 m^3/h	Russia	1
		Poland	2
		Japan	2
	Other applications		
8	Surface build-up and hardening of metal; production of catalysts for ammonia synthesis; development of high-temperature chemical technologies	Russia	5
9	Disinsectization of grain with a capacity of up to 200 t/h	Russia	1
		Ukraine	2
10	Research	Russia	5
		Bulgaria	1
		Germany	1
		Korea	2

In total, over 70 accelerators operate in technological lines and research centers

Muon Collider(s):
Basics, Status, Problems, Prospects

A. Skrinsky

Budker Institute of Nuclear Physics
Novosibirsk 630090, Russia

Abstract. The aim of this paper is to present the main specific features of muon colliders, and to show the (very personal) ways to solve the most difficult problems of their design and construction.

WHY AND WHEN DO WE NEED MUON COLLIDER(S)?

In this article I shall consider the general scheme of a muon collider complex presented in Fig. 1, initiated long ago [1,2], which is now the subject of many efforts, including those of the international Muon Collider Collaboration [3]. But why (and at which stage of elementary particle physics development) do we need to consider the muon collider as an important and necessary step?

First of all – it is much easier to construct a hadron (proton) collider of the same energy and the same luminosity. But hadrons bring into the collision their complex structure (quarks and gluons), whereas in lepton colliders (involving electrons/positrons or muons) we deal directly with fundamental (unstructured up to now) incident particles. As a consequence of hadron complexity, also, only a small fraction of the total hadron-hadron interaction energy transfers into the fundamental interaction – less than one tenth, on average. Hence, for a hadron collider we need ten times higher energy just to compete in energy. Additionally, the fundamental interactions are not monochromatic at all (100% energy spread in collision!) and each of them is accompanied by many interactions of remnants of hadrons, producing the given fundamental interaction.

Of course, a hadron collider – even for 10 times higher energy – is cheaper (especially before the muon collider concept matures). And the non-monochromaticity of the fundamental interaction has some advantage, when we enter a new, unexplored energy domain and need to discover something of unknown mass.

If we compare different lepton-(anti)lepton colliders, we need to consider the electron-positron *linear* collider as a reference to the muon collider – the cyclic collider road to electron-positron collisions came to an end with LEP-200. Brilliant success in raising the LEP energy is an explicit demonstration of such an end (2% of full electron energy per turn to compensate synchrotron radiation energy losses for a 30 km circumference collider!). Transfer to linear colliders (as was understood long ago at Novosibirsk [1,2] where the first self-consistent physics project was presented

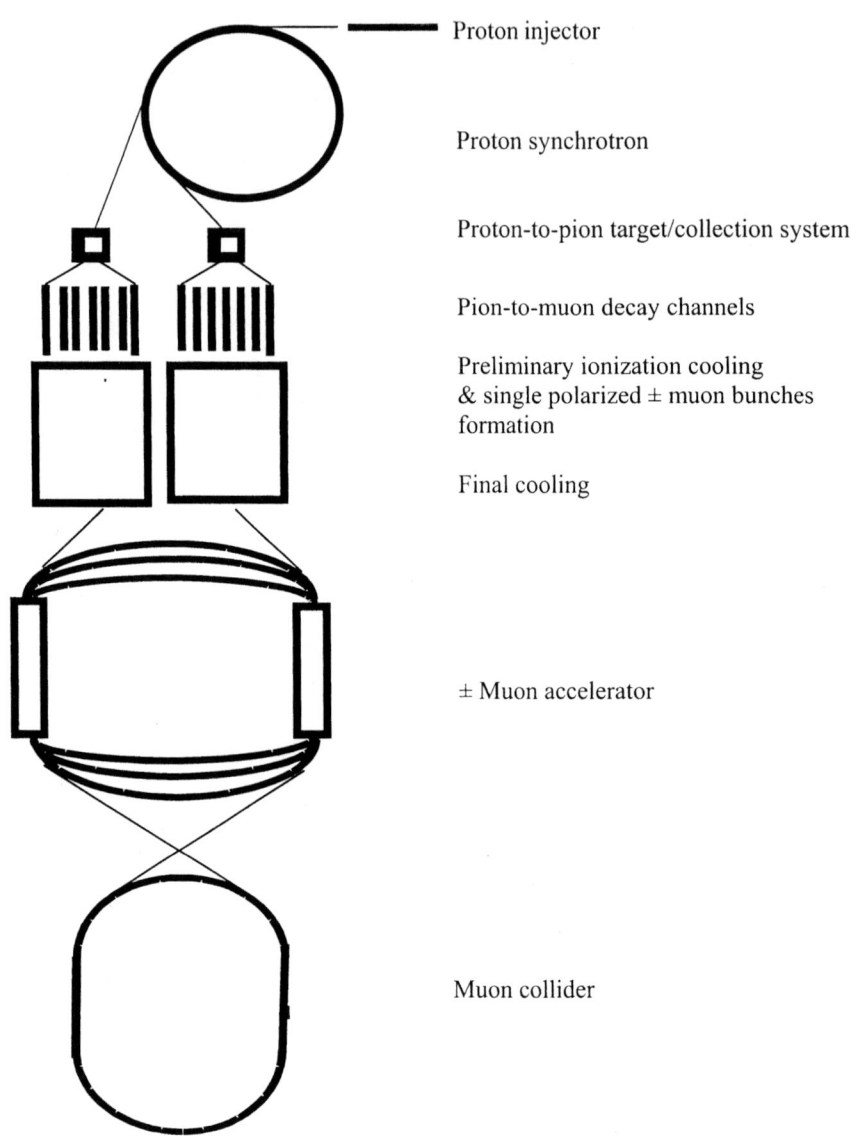

FIGURE 1. Schematic layout of muon collider complex.

in 1978 [4]) removes the problem of synchrotron radiation energy losses practically completely at all the stages of the collider cycle – except in the process of the collisions themselves! To reach high enough luminosity, we need high enough density of the colliding bunches, and if the collision spot is axially symmetric, the coherent electromagnetic field of opposing bunch is so high that it induces absolutely unacceptable radiation losses – in a single pass. Even for modest hundreds of GeV per beam we need to change the shape of the collision spot so that the horizontal dimension is very large, while the vertical one is proportionally smaller – so as not to lose bunch density (luminosity). The need to operate with a nanometer vertical size of bunches practically eliminates the possibility of reaching more than, say, 1 TeV per electron/positron beam for a linear collider. But still the synchrotron radiation losses remain reasonably big, and instead of simple monochromatic electron-positron collisions we get collisions with a large energy spread, accompanied by large numbers of photon-photon and photon-electron parasitic collisions.

(I do not consider here the very interesting photon-photon and electron-photon options of linear collider operation, when electrons just before the collision region are converted into full energy photons by passing through a dense "laser target".)

As a result of the above considerations, we need to conclude that the muon collider(s) is a natural and almost inevitable step in elementary particle physics development while studying (total) energies higher (and much higher) than 5 TeV, starting from 1 or 2 TeV.

LOW LIFE-TIME PARTICLE COLLIDER

The main obvious specificity of muon accelerators/colliders applies to the need to deal with particles that live in their rest frame only 2.2 microseconds. Of course, muon life-time grows proportionally to $E_\mu/E_{\mu 0}$, but it remains short.

Hence, the cooling of muon beams and their acceleration to the energy required should be fast enough. To evaluate the necessary acceleration gradient in the relativistic case is quite easy: if the acceleration is linear in time (the acceleration gradient $\dfrac{dE_\mu}{ds}$ is constant), the number of particles N_μ accelerated up to energy $E_{\mu\text{fin}}$, starting from $N_{\mu 0}$ muons of energy $E_{\mu\text{ini}}$, will be

$$N_\mu = N_{\mu 0} \left(\frac{E_{\mu\text{ini}}}{E_{\mu\text{fin}}} \right)^{\frac{E_{\mu 0}}{c\tau_{\mu 0}} \Big/ \frac{dE_\mu}{ds}}.$$

The requirement for muon acceleration gradient is not very difficult: $\dfrac{dE_\mu}{ds} = 0.16$ MeV/m, hence the average accelerating gradient of 3 MeV/m is enough to lose 30% of muons only, upon 1000 times acceleration.

The limited muon lifetime enters into the requirement to use in the collider itself the highest possible magnetic field B_{coll}. The number of turns useful for muon-muon luminosity in the collider is

$$N_{lumi} = \frac{ec\tau_{\mu 0}}{2\pi E_{\mu 0}} \cdot B_{coll} ,$$

and does not depend on the muon energy. For an average collider field of 10 Tesla

$$N_{lumi} = 3000.$$

IONIZATION COOLING

The next important aspect in reaching high luminosity muon collisions is to arrange effective cooling of muon beams – without beam compression satisfactory luminosity can not be reached. And the only way, in my opinion, is to use ionization cooling – the ionization energy loss plus its compensation by an external electrical longitudinal RF-field [1,2]. Fortunately, ionization cooling is appropriate for muons beams – and for muon beams only [5, 6]! And it could be made fast enough to prevent muon intensity losses.

Consequently, the general scheme should look like the one presented in Fig. 1.

For the case of cooling due to "full energy losses" of any origin, the increment of six-dimensional density (or the sum of decrements for all three emittances) is

$$\delta_{\Sigma 0} := \frac{2 \cdot P_{fr}}{p_\mu \cdot v_\mu} \cdot \left(1 - \frac{P_{\mu long}}{2 \cdot v_\mu} \cdot \frac{d}{dp_{\mu long}} v_\mu \right) + \frac{1}{v_\mu} \cdot \frac{d}{dp_{\mu long}} P_{fr}$$

(the "equilibrium particle" energy losses are considered to be compensated by an external source).

The power of the ionization energy loss by a charged particle is

$$P_{fr}(\beta_\mu) := 4\pi \cdot r_e^2 \cdot m_e \cdot c^2 \cdot N_e \cdot \frac{c}{\beta_\mu} \cdot \left[\ln \left[\frac{2 \cdot m_e \cdot c^2 \cdot \beta_\mu^2}{I \cdot \left[1 + \frac{2}{\sqrt{1-\beta_\mu^2}} \cdot \frac{m_e}{M_\mu} + \left(\frac{m_e}{M_\mu}\right)^2 \right] \cdot (1-\beta_\mu^2)} \right] - \beta_\mu^2 \right]$$

Consequently, the decrements sum (now in cm^{-1} of cooling matter, see Fig. 2) is

$$\delta_{\Sigma 0}(\beta_\mu) := \frac{2 \cdot P_{fr}(\beta_\mu) \cdot \sqrt{1-\beta_\mu^2}}{(M_\mu \cdot c^3) \cdot \beta_\mu^3} \cdot \left(\frac{1+\beta_\mu^2}{2}\right) + \frac{(1-\beta_\mu^2)^{\frac{3}{2}}}{(M_\mu \cdot c^3 \cdot \beta_\mu^2)} \cdot \left(\frac{d}{d\beta_\mu} P_{fr}(\beta_\mu)\right)$$

where P_{fr} is the power of ionization losses:

$$P_{fr} = 4\pi \cdot m_e c^3 r_e^2 N_e \beta_\mu^{-1} \cdot \left(\ln \frac{2 m_e c^2 \beta_\mu^2 \gamma_\mu^2}{I\left(1 + 2\gamma_\mu \frac{m_e}{M_\mu}\right)} - \beta_\mu^2 \right) .$$

In the above formulae all constants are the usual ones, and β_μ and γ_μ are the ratios, during the cooling process, of muon velocity to the velocity of light, and of muon full energy to muon rest frame energy.

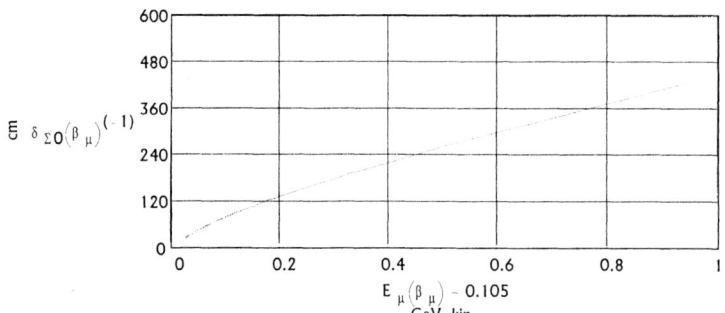

FIGURE 2. The "cooling length" for the product of all three emittances in lithium.

Hence, to cool the three-emittance a million times, say, at 200 MeV (kinetic), we need to travel through the lithium for about 15 meters (with continuous energy recovery).

But real life is more complicated: the transverse decrements are positive and of the order of the six-dimensional decrement, but the longitudinal decrement is very small at "high" energy (Fig. 3), and at lower energies becomes negative (Fig. 4) – heating instead of cooling!

Hence, it is obligatory to <u>redistribute</u> effectively the decrements sum in favor of longitudinal degree of freedom (κ_{long} = fraction of the full three-emittance decrement transferred to the longitudinal degree of freedom).

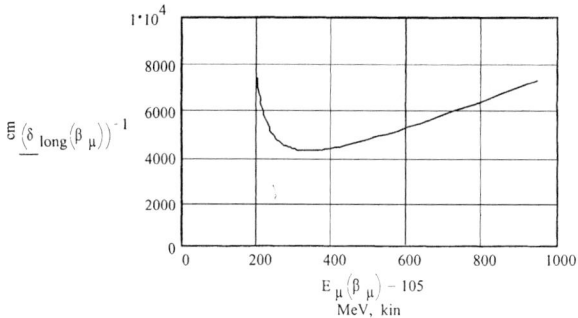

FIGURE 3. "Pure" longitudinal cooling length at high energies.

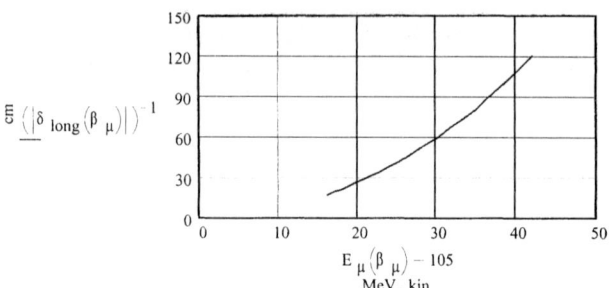

FIGURE 4. "Pure" longitudinal anti-damping length at low energies.

In addition to ionization *cooling*, the pass through some substance causes multiple scattering of muons and fluctuation of their ionization losses. The interplay of cooling and scattering gives the muon beam the equilibrium angles and energy spread.

Transverse cooling asymptotically shrinks the muon angular spread down to the equilibrium one, acquired due to multiple scattering at one transverse emittance cooling length (Fig. 5):

$$\vartheta_{x,y}^2 = 4\pi \cdot r_\mu^2 N_e (Z+1) L_c \frac{1}{\gamma_\mu^2 \beta_\mu^4} \cdot \left(\frac{1-\kappa_{long}}{2} \cdot \delta_\Sigma \right)^{-1},$$

and the corresponding equilibrium relative energy spread would be (Fig. 6)

$$\Delta_E^2 = \frac{2\pi \cdot r_\mu^2 N_e \left(2 - \beta_\mu^2 \right)}{\kappa_{long} \delta_\Sigma}.$$

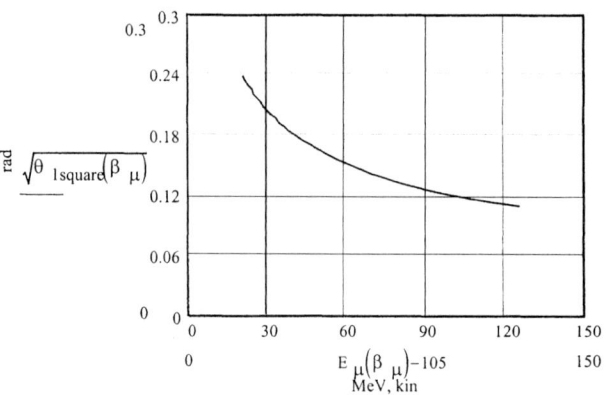

FIGURE 5. Equilibrium muon angles (κ_{long} is assumed optimal, around 0.25).

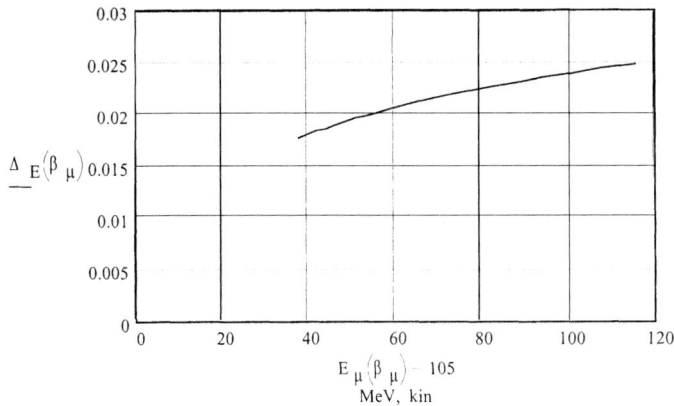

FIGURE 6. Equilibrium relative energy spread (κ_{long} is assumed optimal, around 0.25).

As you see, the equilibrium angles and energy spread at interesting energies are not small at all, and to reach small enough emittances at the final stage of cooling – and, hence, to reach acceptable collider luminosity – we need to use at this stage very strong focusing in all three directions – as strong as practically possible!

To arrange the strongest possible transverse focusing, my preference is to use current-carrying lithium rods, which focus muon beams via a magnetic field gradient, limited by the magnetic field on the surface (10 Tesla or somewhat higher – pulsed). The whole device is just a very long lithium lens, developed at Novosibirsk [7] for positron and antiproton collection decades ago, and still in use now (INP, CERN, FNAL). The radius of the rod and the surface field at the final stage should provide acceptance, say, 2-3 times more than the final muon emittance (the final lithium rod diameter is about 6 mm in this case), and for the necessary high repetition rate we need to use liquid lithium to extract the Ohmic heat [8].

While using lithium rods at the final stage, the normalized equilibrium transverse emittance ($\varepsilon_{neqtran} = \dfrac{a_{\mu eq}^2}{\beta_{local}} \cdot \beta_{\mu cool} \gamma_{\mu cool}$) should shrink to 0.04–0.05 cm·rad (Fig. 7).

To arrange simultaneous cooling in the transverse and longitudinal directions is a more troublesome task. The necessary transfer of the summary cooling decrement to the longitudinal degree of freedom is best arranged, to my current understanding, by converting the cylinder into a helix (the local helical curvature radius of about 10 cm is produced by a corresponding external magnetic field H_{helix}), and by a positive electron density gradient (Fig. 8). The latter aim can be reached by combining lithium and beryllium in the cooling cylinder – more beryllium at the outer side of the helix – with a teeth-like boundary between the lithium and beryllium. The helical parts of the final cooling device, of alternating direction of curvature to "extract" a fraction of the decrement from both transverse directions, are separated by RF accelerating sections – to compensate the average ionization energy losses; in these RF sections, a proper and

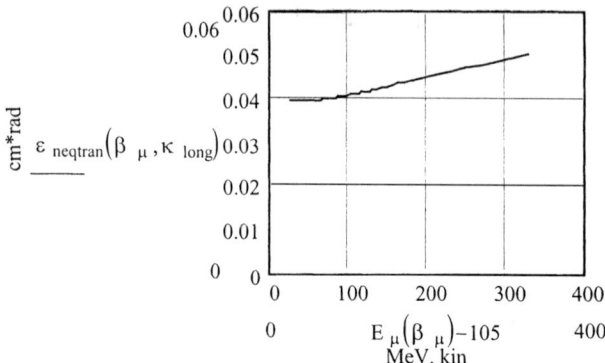

FIGURE 7. Normalized transverse emittance upon ultimate cooling.

strong enough focusing should be arranged to match emittances in the neighboring cooling sections in spite of very high momentum spread [11]. To make the longitudinal emittance, and consequently the three-emittance, minimal at the final stage of ionization cooling, we need an RF system that operates at the shortest possible wave-length (10 cm?) and a high accelerating gradient (30 MeV/m on average?). In this case, it is possible to reach the emittances presented in Figs. 9 and 10.

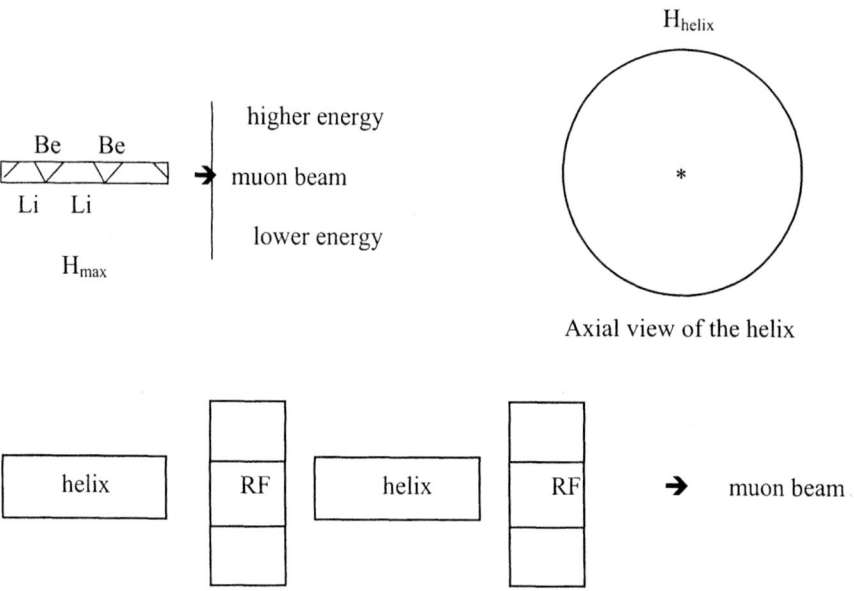

FIGURE 8. The schematics for simultaneous transverse and longitudinal cooling.

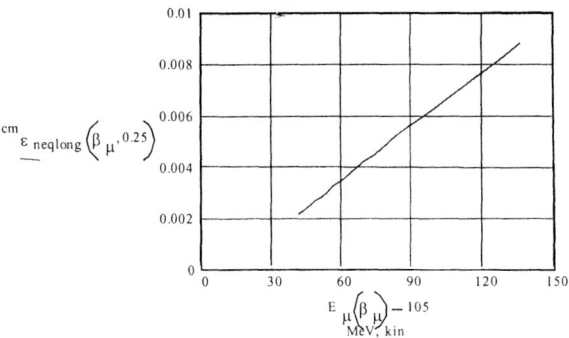

FIGURE 9. Equilibrium normalized longitudinal emittance.

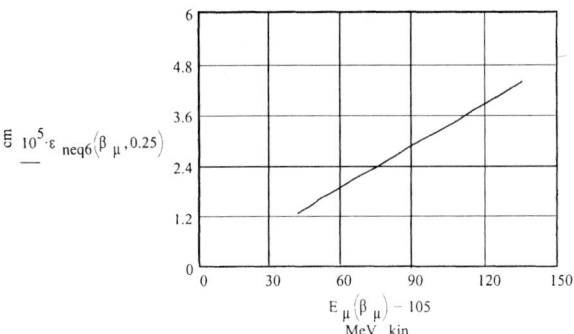

FIGURE 10. Equilibrium normalized three-emittance.

ULTIMATE LUMINOSITY

If at the cooling stage the normalized transverse emittance $\varepsilon_{neqtran}$ and the longitudinal emittance $\varepsilon_{neqlong}$ were reached and were kept constant at all the stages including the collision stage, the bunch length at collision is limited by $\varepsilon_{neqlong}$ and by the maximal acceptable energy spread ΔE_{max}, transverse beta-functions are made equal to the bunch length, and the collider magnetic field H_{coll} provides N_{life} effective turns for muons before they decay, then the ultimate luminosity would be [10]

$$L_{\mu\mu\, max} = \frac{N_\mu^2}{4\pi} \cdot \frac{\gamma_{coll}^2}{\varepsilon_{neqtran} \cdot \varepsilon_{neqlong}} \cdot \frac{\Delta E_{max}}{E} \cdot N_{life} \cdot f_0 \ .$$

The luminosity of an "optimal" collider (E_μ= 2 TeV + 2 TeV, N_μ=1·10^{12}, H_{coll}=10 T, f_0=15 s^{-1}, $\Delta E_{max}/E = 0.3\%$, the fraction of the sum of cooling decrements transferred to the longitudinal direction $\kappa_{long}= 0.25$) with the equilibrium emittances reachable as the ultimate limit in the cooling process (see above), as a function of cooling kinetic energy, is shown in Fig. 11.

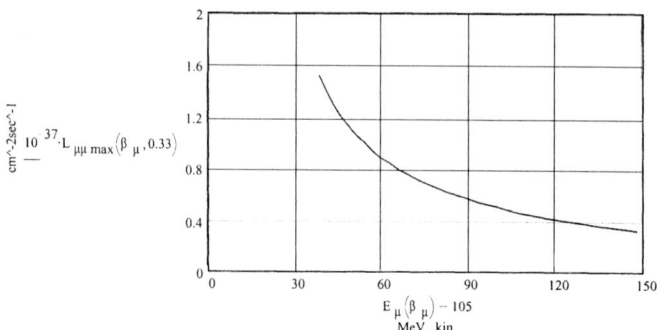

FIGURE 11. "Maximum" collider luminosity (without "emittance gymnastics" and with no limitation for final focusing and energy spread).

But if we calculated the beta-value at collision, assumed here as always, to be equal to the muon bunch length, we would get 5 microns! Such a bunch length and beta-function look impractical. If we limit these values as $\sigma_{longcoll}$, the following formula for "<u>practi-cal</u>" maximum luminosity should be valid [10]:

$$L_{\mu\mu} = \frac{N_\mu^2}{4\pi} \cdot \frac{\gamma_{coll}^{\frac{3}{2}}}{\varepsilon_{neq6}^{\frac{1}{2}} \sigma_{longcoll}^{\frac{1}{2}}} \cdot \left(\frac{\Delta E_{max}}{E}\right)^{\frac{1}{2}} N_{life} f_0 .$$

For a reasonable limit of $\sigma_{longcoll} = \beta_{coll} = 300$ microns, the luminosity dependence on cooling energy is shown in Fig. 12.

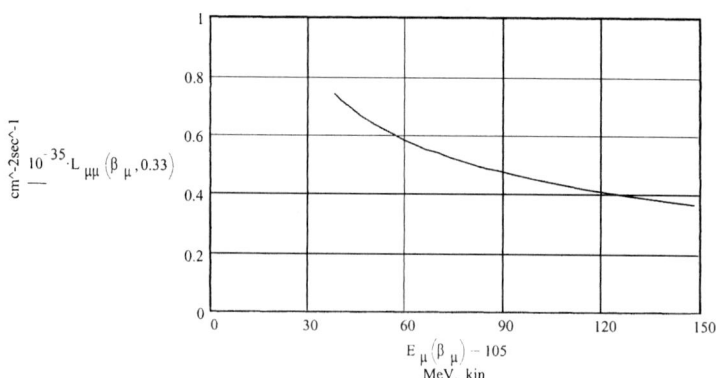

FIGURE 12. "Practically achievable" collider luminosity.

Hence, "under practical circumstances," the equilibrium normalized three-emittance ε_{neq6} enters the maximum luminosity directly. And the aim of final ionization cooling is to reach a minimum of

$$\varepsilon_{neq6} = (\varepsilon_{neqtran})^2 \varepsilon_{neqlong}.$$

In a special section we discuss the way to rearrange the structure of this emittance properly – to reach the highest luminosity, and the necessary monochromaticity and/or degree of polarization.

EMITTANCE GYMNASTICS

As we discussed earlier, to "extract" full usefulness of the low three-emittance reached at the final cooling stage, we need to rearrange this emittance according to the acceptable muon bunch length in the collider, taking into account the desired muon-muon monochromaticity (or the acceptable energy spread in the collider) and the necessity to keep the degree of polarization at an acceptable level. For this emittance gymnastic purpose, we need [10] to use a combination of dispersive and septum elements, RF accelerating/decelerating structures, and delay lines (but not ionization components, which damage six-density!). Such a transformation is worth arranging at any convenient energy. The schematic layout of such a device for maximizing collider luminosity is shown in Figure 13. The "monochromatic" collider option (as in the case of the "low energy Higgs Factory") could require rearrangement in the opposite direction.

In-coming muon bunch: "wide, but short"

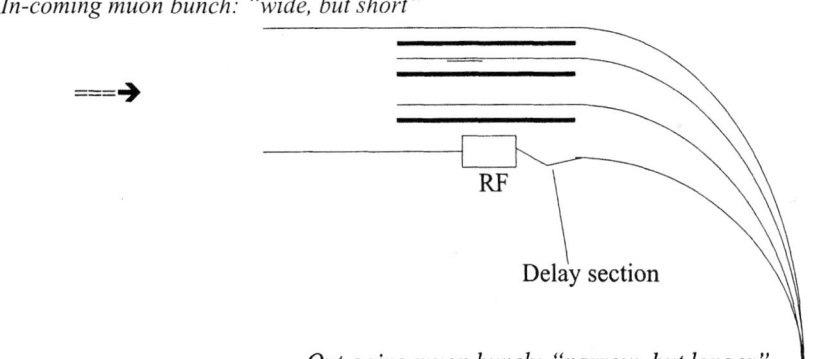

Out-going muon bunch: "narrow, but longer"

FIGURE 13. Transformation of a bunch with big transverse emittance into a bunch with a smaller one but with proportionally higher momentum spread and longitudinal emittance (an option).

POLARIZED MUONS

A high degree of polarization is very important for extracting full physics information from muon collider experiments. Hence, first of all, it is worthwhile to find a way to produce highly polarized intense muon beams [6,9]. We assume that positive and negative pions are collected, being generated by different proton bunches.

A possible option for a proton-to-pion multi-channel conversion system, followed by multi-channel pion-to-muon decay channels, is sketched in Figure 14. It may be reasonable to arrange a sectioned target (using additional channels). This could be especially useful at high proton energy – around 100 GeV.

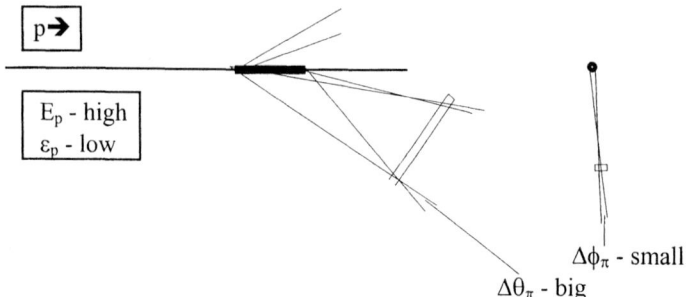

FIGURE 14. Schematics for a multi-channel proton-to-pion conversion system.

In each pion-collecting straight channel, with the use of one-dimensional "thin surface-current-carrying lens" doublets for the initial matching focusing, it is necessary to direct a wide spectrum of pions into many independent channels. In each channel, in the θ-direction the beam transverse emittance is big, but in the ϕ-direction it is quite small. These beams can easily be transported straight out of the target area, and the following channel gymnastics will happen in reasonably free space.

The next step is to arrange the energy dispersion in this smaller emittance direction in each channel, and then to direct each of the ±5% momentum spread pion beams into additional separate strong-focusing decay channels.

Pion beams with such a narrow momentum spread (with a very small emittance in one direction), upon passing about two decay lengths (proportional to the pion energy in each channel, around $15\beta_\pi\gamma_\pi$ meters), generate muon beams of about ±30% momentum spread (Fig. 15), with a strong correlation of the muon's spin direction to its momentum.

Consequently, for every particular muon beam, we cut away the middle 30% of the muon spectrum and direct the upper and lower parts with opposite helicities into two separate sub-channels. At the next phase, we shift the energy in each muon channel by RF acceleration/deceleration to the energy optimal for ionization cooling. Then, upon preliminary cooling, we combine all the "upper sub-channel" muons in one longitudinally polarized bunch, and all the "lower sub-channel" muons in another bunch of opposite helicity, each with a 70% degree of polarization. If it were found useful, this procedure could be arranged in a few stages. Then all four bunches (μ^+ and μ^-) wouldl be cooled down to the lowest three-emittance.

Afterwards, we can reverse the helicity of the "lower" bunches at a later stage upon acceleration to 45 GeV (by an additional non-accelerating full turn) and then combine the two bunches into one (one μ^+ bunch + one μ^- bunch) with the three-emittance twice as high as that reachable at the final cooling.

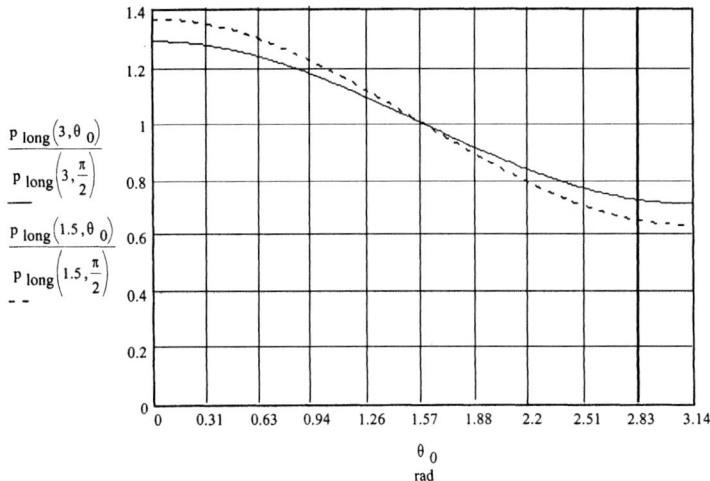

FIGURE 15. The dependence of the relative muon momentum in the lab frame on the polar angle of decay θ in the CoM frame, for pion kinetic energies of 300 MeV (solid line) and 60 MeV (dotted line).

The helicity reversing of muons is due to their anomalous magnetic moment. Positive spin-to-velocity relative rotation is very slow at low energy (e.g. at the cooling stage) and thus does not damage the initial degree of muon beam polarization, but it becomes faster proportionally to the muon energy, and at 45 GeV each full turn of the muon trajectory results in reversing of muon helicity. Keep in mind that all the muon spin motion proceeds in the median plane of the collider.

Helicities of colliding bunches are modulated at relative frequency

$$v_{spin} = \frac{\mu_{anom}}{\mu_0} \gamma_\mu = \frac{E_{GeV}}{90} .$$

Because of this modulation, at integer spin resonances the helicity remains the same throughout a collision process. At half-integer resonances the helicity reverses at consecutive turns. At intermediate energies the spin-at-collision modulation proceeds with a non-integer fraction of v_{spin}.

Relative helicities of muon bunches at the interaction region (from ±± to ±∓) can be controlled by choosing a proper injection path (e.g. by an additional non-accelerating turn of one beam at, say, 45 GeV).

At high energy, when $v_{spin} \gg 1$, incomplete coherence of spin rotation becomes important, and this effect can lead to lowering of the degree of polarization due to beam energy spread. The loss becomes significant if the spin frequency difference in the beam reverses the relative spin orientation at a half synchrotron oscillation period. The effective loss factor of the degree of polarization ζ_{eff} (compared to the initial degree) can be expressed as

$$\zeta_{eff} = 1 - \frac{1}{2} \cdot (2\pi \cdot \frac{v_{spin}}{v_{synch}} \cdot \Delta_{Ecoll})^2 ,$$

where ΔE_{coll} is the muon beam energy spread in the collider and v_{synch} is the relative synchrotron frequency (this evaluation is meaningful if ζ_{eff} is not very far from 1; otherwise, the degree of polarization goes to zero).

MUON COLLIDER OPTIONS

Let us present now [10] the potential luminosity $L_{\mu\mu}$ and polarization factor ζ_{eff} for a muon collider at different energies (the actual degree of polarization at the collision stage would be close to the product of ζ_{eff} and the degree of polarization at muon production).

For all the options, we assume the "ultimately" low beta-value at the interaction $\beta_0 = 3$ mm, equal to the muon bunch length $\sigma_{longcoll}$. The average magnetic field in the collider H_{coll} is assumed always to be equal to 10 Tesla, except for the "Higgs Factory" for which it is assumed to be lower (7 Tesla) since at this small collider the bigger fraction should be used for interaction regions, etc. The relative betatron frequency in the collider v_R, RF wavelength λ_{RF}, and total RF voltage V_{RF} (in MV) are chosen to correspond to the relative energy spread in the collider $\Delta E_{max}/E$.

TABLE: Collider options for $5 \cdot 10^{12}$ muons per bunch (luminosity per detector).

$2E_{coll}$	N_μ 10^{12}	R_{coll} m	$\Delta E_{max}/E$	v_R	V_{0coll} MV	ζ_{eff}	$L_{\mu\mu}$ cm^{-2}s^{-1}	
100 GeV	5	25	$3 \cdot 10^{-5}$	20	0.1	1	$7 \cdot 10^{32}$	Higgs Factory
1 TeV	5	180	$3 \cdot 10^{-4}$	30	100	0.8	$2 \cdot 10^{35}$	
4 TeV	5	720	$2.5 \cdot 10^{-4}$	30	1000	0.8	$7 \cdot 10^{35}$	
10 TeV	5	1800	$1 \cdot 10^{-4}$	30	1000	0.8	$1.5 \cdot 10^{36}$	
10 TeV	**5**	**1800**	**$1 \cdot 10^{-3}$**	**300**	**1000**	**0**	**$6 \cdot 10^{36}$**	Ultimate collider

The average muon current at $H_{coll} = 10$ T and $f_0 = 15$ Hz for $N_\mu = 5 \cdot 10^{12}$ per bunch is the same for all the energies, around 40 mA. For energies of 10 TeV (total) and higher, the muon current becomes almost constant at this level in spite of decay, and the muon generation/acceleration repetition rate should go down.

OTHER COLLISIONS

Let us underline that the muon collider provides almost automatically the same full luminosity for μ^--proton collisions. Additional care and arrangements are needed to get longitudinally polarized protons at collision. To reach this, we need to install proper proton spin rotators on both sides of the interaction region(s) in order to transform vertically polarized protons in the arc to longitudinally polarized at collision, and to restore transverse polarization before entering the next arc. The influence of these spin rotators on muon polarization is negligibly small.

To arrange (if we are interested) same-sign collisions, we need a double-ring collider; the other components of the complex remain almost unaltered.

ACKNOWLEDGEMENTS

It is my great pleasure to thank V. Parkhomchuk, G. Silvestrov, and T. Vsevolozhskaya, for long-lasting and fruitful collaboration, and R. Palmer, D. Cline, and J. Gallardo for interesting and useful discussions.

REFERENCES

1. Skrinsky A.N., "Colliding Beams Program in Novosibirsk," presented at International Seminar on Prospects in High Energy Physics; Morges, CERN, 1971.
2. Skrinsky A.N., "Accelerator and Instrumentation Prospects of Elementary Particle Physics," in *Proc. 20th Intern. Conf. on High Energy Physics*, Madison, 1980, v. 2, pp. 1056-1093; *Uspekhi Fiz. Nauk* Moscow **138** (1), 3-43 (1982).
3. Muon Collider Collaboration, "Muon Collider Design," in *Proc. Symp. on Physics Potential and Development of mu^+mu^- Colliders*, San Francisco, 1995, *Nuclear Physics B, Proceedings Supplement*, **51A**, 61-84 (1996).
3a. Palmer R.B. and Gallardo J.C., "Muon-Muon and Other High Energy Colliders," *BNL Report 64148*, 1997.
4. Balakin V.E., Budker G.I. and Skrinsky A.N., "On Feasibility of Super-High Energy Electron-Positron Collider," in *Proc. 6th All Union Accelerator Conf.*, 1978, v. 1, pp. 27-34.
5. Parkhomchuk V.V. and Skrinsky A.N., "Methods of Cooling of Charged Particles Beams," in *Physics of Elementary Particles and Atomic Nuclei*, Moscow, 1981, v. 12, pp. 557-613; *Soviet Journal on Particles and Nuclei*, **12** (3), 223-232 (1981).
6. Skrinsky A.N., "Ionization Cooling and Muon Collider," in *Proc. 9th ICFA Beam Dynamics Workshop: Beam Dynamics and Technology Issues for Muon-Muon Colliders*, Montauk, NY, 1995, AIP Conference Proceedings 372, pp. 133-139; *Nuclear Instruments and Methods*, **A391**, 188-195 (1997).
7. Silvestrov G.I., "Problems of Intense Secondary Particle Beams Production," in *Proc. 13th Intern. Conf. on High Energy Accelerators*, Novosibirsk, 1986, v. 2, pp. 258-263.
8. Silvestrov G.I., "Lithium Lenses for Muon Colliders," in *Proc. 9th ICFA Beam Dynamics Workshop: Beam Dynamics and Technology Issues for $\mu^+\mu^-$ Colliders*, Montauk, NY, 1995, AIP Conference Proceedings 372, pp. 168-177.
9. Skrinsky A.N., "Polarized Muon Beams for Muon collider," in *Proc. Symp. on Physics Potential and Development of mu^+mu^- Colliders*, San Francisco, 1995, *Nuclear Physics B, Proceedings Supplement*, **51A**, 201-203 (1996).
10. Skrinsky A.N., "Towards Ultimate Luminosity Polarized Muon Collider (problems and prospects)," in *Proc. 4th Intern. Conf. on Physics Potential and Development of mu^+mu^- Colliders*, San Francisco, 1997, AIP Conference Proceedings 441, pp. 249-264.
11. Skrinsky A.N., "Remarks on High Energy Muon Collider," *in Proc. Workshop on Physics Potential and Development of mu^+mu^- Colliders*, Montauk, 1999.

The Antihydrogen and Positronium Problem in Particle Physics

I.N. Meshkov

Joint Institute for Nuclear Research, Dubna, 141980, Russia

Abstract. The importance of experimental studies of antihydrogen and positronium at extremely high precision relates to the fundamental problems of the properties of matter – symmetry (CPT theorem) and the nature of basic interactions (quantum electrodynamics). Generation of antihydrogen and positronium atoms "in-flight," especially in the form of rather intense and well directed streams, opens new opportunities for experiments, as discussed in this report.

I INTRODUCTION

The physics of antihydrogen $\overline{H}^0$ is of great interest from the viewpoint of understanding the fundamental properties of matter, and, in particular, its symmetry properties. A specific problem which was suggested by the authors of the first studies on antihydrogen production [1], is that of verifying the CPT theorem. This has become quite feasible since the first synthesis of antihydrogen atoms in 1995 at the LEAR antiproton ring at CERN [2] and in 1996 at Fermilab. Although these experiments were essentially just a demonstration, it can now be claimed that antihydrogen exists as a physics object.

Currently, there are two quite different approaches in antihydrogen physics. The first is the production of antihydrogen atoms one by one in antiproton and antipositron traps at ultraslow energies, followed by confinement of the atoms in magnetic traps with the minimum magnetic field and cooling down to a temperature of order 1° K by means of laser radiation [3].

The second approach is the use of antiproton and positron storage rings [1], [4]. A recently suggested variant of this idea [5] would allow the production of intense pencil beams of 30 to 3×10^4 antihydrogen atoms per second with velocities in the range 0.03 to 0.3 times the speed of light (antiproton energies of 0.5 to 50 MeV). This setup would simultaneously produce pencil beams of positronium (Ps) with an intensity of $2 \cdot 10^4$ to 30 sec^{-1}, which are of independent interest as an object of study.

The purpose of the present report is to describe the methods of $\overline{H}^0$ and Ps generation and some possibilities offered by experiments with them.

II PROBLEMS IN THE PHYSICS OF ANTIHYDROGEN AND POSITRONIUM

The principle of CPT invariance (CPT theorem) can be verified by comparing particle and antiparticle parameters: their masses, absolute values of electric charges and magnetic moments, and gyromagnetic ratios. Of course, it is interesting to perform measurements at a level of accuracy exceeding that attained at present. The latter can be judged from the data, given in Table 1.

TABLE 1. Parameters of the fundamental particles.

Parameter	Value	Accuracy
Electron and positron		
Electron mass, MeV	0.510 099 906(15)	$3 \cdot 10^{-7}$
Mass difference $|m^+ - m^-|/m^-$	$< 4 \cdot 10^{-8}$	$< 4 \cdot 10^{-8}$
Charge inequality $|e^+ - e^-|/e^-$	$< 4 \cdot 10^{-8}$	$< 4 \cdot 10^{-8}$
Difference of the charge to mass ratios	$< 3 \cdot 10^{-8}$	$< 1 \cdot 10^{-8}$
Electron magnetic moment (in Borh magnetons)	1.011 159 652193(10)	$1 \cdot 10^{-11}$
Gyromagnetic ratios $|g^+ - g^-|/g^-$	$(-0.5 \pm 2.1) \cdot 10^{-12}$	$2.1 \cdot 10^{-12}$
Proton and antiproton		
Proton mass, MeV	938.2723(28)	$3 \cdot 10^{-7}$
Mass difference, $\Delta M/M$	$< \pm 4 \cdot 10^{-8}$	$< 4 \cdot 10^{-8}$
Inequality of the proton and electron charges, $|e_p - e^-|/e^-$	$< 1 \cdot 10^{-21}$	$< 1 \cdot 10^{-21}$
Inequality of the proton and antiproton charges, $|e_p - e_a|/e_p$	$< 2 \cdot 10^{-5}$	$< 2 \cdot 10^{-5}$
Inequality of the proton and antiproton charge-to-mass ratios	$< 1.5 \cdot 10^{-9}$	$< 1.1 \cdot 10^{-9}$
Proton magnetic moment (in nuclear magnetons)	2.792 847 39(6)	$2 \cdot 10^{-8}$
Antiproton magnetic moment (in nuclear magnetons)	$-2.8005(90)$	$3 \cdot 10^{-3}$

Although at present there is no experimental evidence calling the validity of CPT invariance into question, there is also no reason not to verify CPT invariance. In this respect, CPT invariance is just as much an axiom of modern physics as are other axioms. The high accuracy attained by indirect comparison of the masses of neutral kaons

$$\left| \frac{m(K^0) - m(\overline{K^0})}{m(K^0)} \right| \leq 5 \cdot 10^{-19}$$

cannot serve as absolute proof of the theorem, because, in general, any particle "has the right" to be asymmetric relative to its antiparticle. Therefore, the experimental verification of the symmetry of each of the known particles is of independent importance. It is most important for the "most fundamental" particles, i.e., the proton and electron.

The use of antihydrogen atoms ($\overline{H^0}$) as a test object allows direct and highly accurate comparison of the electric charges of the antiproton and positron (See 4.1). The precise measurement of the hypefine structure and the Lamb shifts of the optical spectrum of antihydrogen is important because, if these values were found to differ for atoms and antiatoms, this would in itself be an indication of symmetry breaking in the fundamental interactions. In particular, the Lamb shift, which is calculated in quantum electrodynamics including radiative corrections, is a parameter of the atomic spectrum measured with a high degree of accuracy. Accordingly, by comparing its value for hydrogen and antihydrogen it becomes possible to judge the symmetry of the interactions. Moreover, a precise measurement of the hyperfine splitting would allow the determination of the magnetic moment – both its absolute value and, with much higher accuracy, the relative difference between the proton and antiproton magnetic moments.

Positronium (Ps), being a very simple quantum system, has been described quite well theoretically and perhaps plays the same role in quantum electrodynamics as the hydrogen atom in nonrelativistic quantum mechanics. Experiments involving positronium, which so far have been carried out under rather complicated conditions in which it is not easy to isolate the effect of the target, have, as a rule, produced rather inaccurate results, and even results which in some cases are ambiguous (see details in Ref. 6). A precise measurement of the positronium parameters is therefore of great interest.

Among experiments of this type are the following:

- the measurement of *the orthopositronium* (o-Ps) *lifetime and the parapositronium* (p-Ps) *lifetime;*
- *the positronium spectrum;*
- the search for orthopositronium annihilation with violation of momentum conservation and charge invariance: o-Ps$\rightarrow 2n\gamma$, where n is an integer;
- the search for exotic and rare decay channels of parapositronium:

$$p\text{-}Ps \rightarrow n\gamma, \quad n > 2;$$

- search for a light, neutral, short-lived boson, via which o-Ps annihilation can occur

$$\begin{array}{c} o\text{-}Ps \rightarrow b + \gamma; \\ \downarrow \\ 2\gamma \end{array}$$

Perhaps, the most intriguing problem in positronium physics is the search for the *"Mirror Universe"* (see Section VII D).

III THE FIRST EXPERIMENT PRODUCING ANTIHYDROGEN

In the experiment [2] performed at the LEAR antiproton storage ring at CERN, a circulating antiproton beam of energy 1.2 GeV (momentum 1.94 GeV/c) interacted with the xenon atoms of an internal cluster (jet) target. The target thickness reached 3×10^{13} atoms/cm^2, and the intensity of the antiproton beam was 1.7×10^{10} particles for a lifetime of 3 min. The integrated luminosity for September and October of 1995 was 5×10^4 ($\pm 50\%$) cm^{-2}.

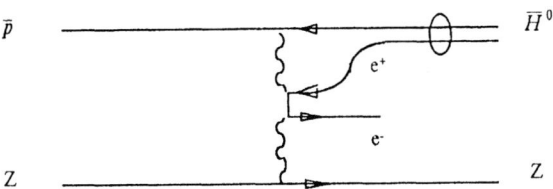

FIGURE 1. Graph of two-proton e^+e^- production pair in the interaction of an antiproton with a nucleus and production of an antihydrogen atom.

Atoms of antihydrogen appeared inside the target when antiprotons interacted with xenon nuclei: an antiproton interacting with a nucleus generates an e^+e^- pair (Fig. 1) and captures a positron if the momentum of the latter is close in magnitude and direction to that of the antiproton (an energy difference of less than 13.6 eV). The cross section for this process is

$$\sigma \sim 2Z^2 \text{pb} \sim 6 \cdot 10^{-33} \text{ cm}^2,$$

where $Z = 54$ is the atomic number of xenon. Therefore, for the given luminosity, it could be expected that 30 atoms of $\overline{H^0}$ would be obtained. The fast, neutral $\overline{H^0}$ atoms (antiproton energy equal to 1.217 GeV and positron energy equal to 0.663 MeV), which were not deflected in the magnetic field of the storage ring, traveled into the detection channel (Fig. 2). Passing through two of the three silicon counters Si (of thickness 700 μm each), the $\overline{H^0}$ atoms lost the positron, which was stopped in one of the counters, producing a γ pair. The third counter recorded dE/dx from the remaining antiproton. The annihilation γ pair was recorded by a cylindrical NaI calorimeter (with energy resolution 14 %), which covered 91 % of the entire solid angle, thereby ensuring a total efficiency of 82 %.

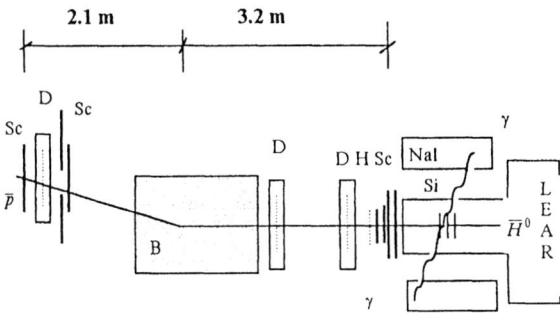

FIGURE 2. Scheme for antihydrogen generation: Si – three silicon counters; Sc – silicon counter-trigger and scintillators; D – proportional chambers; NaI – 6-section calorimeter; H – scintillation hodoscope; B – dipole magnet.

The antiprotons appearing after the stripping of the $\overline{H}^0$ atom passed through three start scintillators Sc (each 4 mm thick) and a hodoscope H of 16 fibers (2×2×32 mm), and then a group of four stop scintillators Sc (Fig. 2). The three drift chambers D with corresponding readout delays recorded the passage of antiprotons deflected in the dipole magnetic field B. The needed calibrations were performed using cosmic rays.

The experiment recorded 11 $\overline{H}^0$ atoms for a background contribution of no more than 2 ± 1 with 95 % probability.

The main result of the experiment, namely, the proof of the "existence theorem," i.e., of the possibility of synthesizing antihydrogen atoms under terrestrial conditions, is certainly of fundamental importance.

Later, in 1996, a similar experiment was performed at Fermilab and 30 $\overline{H}^0$ atoms were recorded.

Thus, *"The Theorem of Existence"* was proved for $\overline{H}^0$!

IV ANTIHYDROGEN PRODUCTION AND STORAGE IN MAGNETIC TRAPS; AD AND ATHENA/ATRAP PROJECTS

The first experiment in which physics parameters of $\overline{H}^0$ will be measured is in preparation now at CERN. After the LEAR accelerator was shut down in December 1996, specialists and "users" of antiproton beams at CERN began to look for a chance to reactivate a $\overline{p}$ source and proposed the Antiproton Decelerator (AD) project [7], which was approved in 1997 and soon realized: the new $\overline{p}$ storage ring was assembled on the base of elements of the "old" AC (Antiproton Collector) storage ring (Fig. 3).

AD accumulates and decelerates antiprotons generated on an existing target system: the proton beam from the Proton Synchrotron at 26 GeV/c proton momentum produces $\overline{p}$ bunch on the target, and $5 \cdot 10^7$ antiprotons with momentum 3.57 GeV/c are collected by a pulsed lens (a so-called magnetic horn) and injected into AD. After deceleration (in several steps) to 100 MeV/c momentum (about 5.9 MeV kinetic energy) and cooling, antiprotons are extracted in one or several bursts with a length ranging from 200 to 500 ns. Such a pulsed beam will be used in several experiments, among which the ATHENA project (AnTiHydrogEN Apparatus) [8] and ATRAP (Antihydrogen Trap Collaboration) [9] are dedicated to antihydrogen generation and study.

ATHENA (Fig. 4) consists of several traps, where antiprotons and e^+ are to be stored, cooled down and recombined to form $\overline{H}^0$ atoms. Antiprotons and positrons meet and recombine in the Central (Recombination) Trap (Fig. 4), but only after a long passage and preparation procedure.

Antiprotons extracted from AD, at kinetic energy 5.88 MeV, enter the Antiproton Trap, and are decelerated in the Degrader by ionization energy losses (Fig. 5, stage 1). The Trap has a longitudinal magnetic field up to 6 T. After a bunch passes the Degrader, a negative potential is applied to the Degrader foil and antiprotons are trapped in the potential well, oscillating in a space between Ring Electrodes with negative potential and the Degrader (Fig. 5, stages 2, 3). A preliminary electron cloud

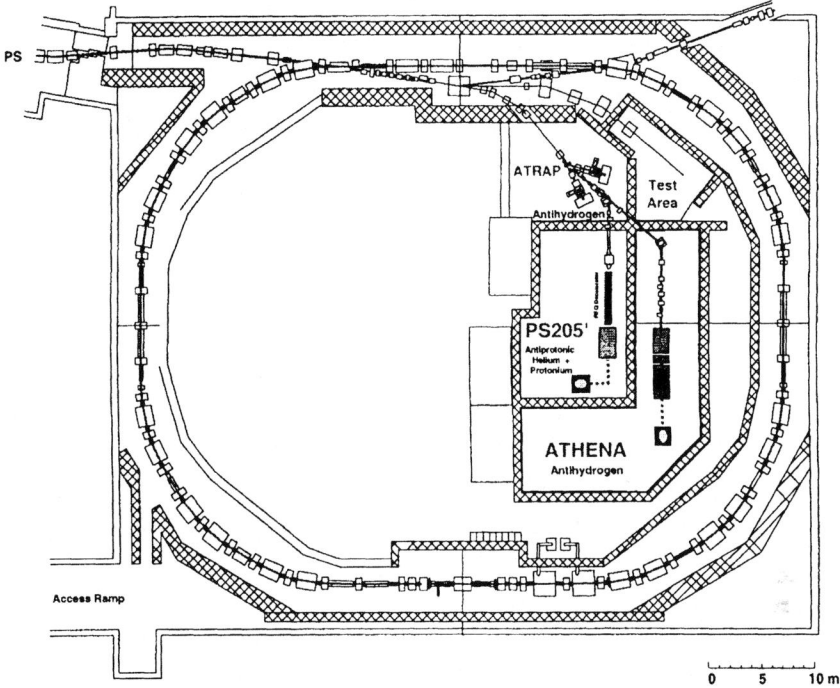

FIGURE 3. Layout of the AD storage ring.

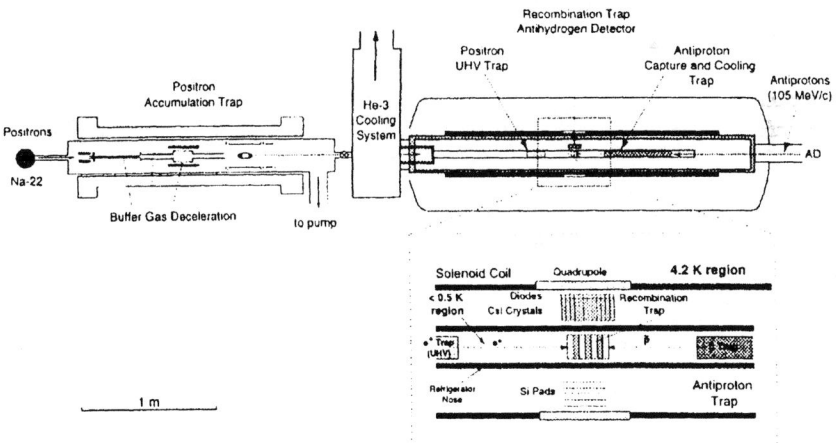

FIGURE 4. Athena schematic overview.

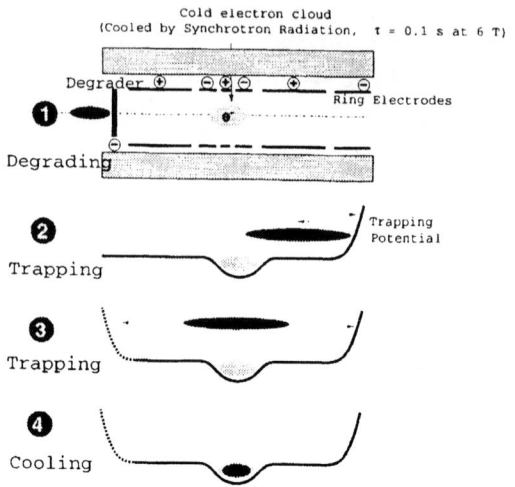

FIGURE 5. Antiproton capture and cooling.

is injected from an electron gun in the central part of the Trap, caught in another potential well formed by additional rings, and cooled via emission of synchrotron radiation due to rotation in a longitudinal magnetic field applied parallel to the trap axis. This radiation (which actually is classic dipole radiation, or "cyclotron radiation" at low electron energy) cools the electrons for a characteristic time of about 0.1 s at B = 6 T. Penetrating the cloud of cold electrons, the trapped antiprotons reduce their kinetic energy until it equalizes with that of the electrons. Then the $\bar{p}$ bunch is transferred (by pulsing of ring potentials) into the Recombination Trap.

Positrons are generated by a Na^{22} radioactive source and collected in the Positron Accumulation Trap (Fig. 4), losing their energy in collisions with Buffer Gus molecules. Such a cooling method provides low positron temperature and accumulation. After storing and cooling, the positron bunch is transferred through the channel by differential vacuum pumping to the Positron Ultra High Vacuum Trap, where several positron bunches are to be stored to get necessary number of positrons. Both bunches, e^+ and $\bar{p}$, merge in the Recombination Trap, where they recombine and form *antihydrogen atoms*. Some ways to intensify this process are under study, for instance, stimulation of recombination by laser radiation or a pulsed electric field.

The Recombination Trap has a system of current coils, which form special nonhomogeneous magnetic field with a minimum at the central point of the Trap – so-called "Ioffe-trap." It consists of a pair of ring coils, which generate magnetic fields of opposite directions, and the windings of a transverse quadrupole field. Antiproton atoms experience the focusing action of the field gradient and can not leave the Trap. Thus, some certain number of $\overline{H^0}$-atoms can be stored and kept in the Trap.

The Trap walls are cryogenically cooled below 4.2 K, so ultrahigh vacuum, of the order of 10^{-12} Torr or better, is expected in the central zone. Being in thermoequilibrium with the walls (via thermoradiation) $\overline{H^0}$ atoms are at the same low temperature, which brings significant gain in the precision of spectroscopic measurements.

The final goal of the experiment is "Doppler-free spectroscopy" of the two-photon 1S-2S transition in antihydrogen with an expected precision better than

$$\frac{\Delta\omega}{\omega} \leq 1\cdot 10^{-12}.$$

(see details in Ref. 6).

V GENERATION OF ANTIHYDROGEN IN-FLIGHT

The idea of producing beams (fluxes) of antihydrogen atoms is closely related to the technique of electron cooling [1]. For antihydrogen generation, the antiproton source, like the antiproton installations at CERN or Fermilab, must be provided with two storage rings. The first is used to store low-energy antiprotons and has a traditional strong-focusing system. The second is a positron storage ring of the racetrack type with four straight sections. The storage rings are combined such that in one of the straight sections the beams pass through each other (Fig. 6). This is where the $\overline{p}$ and e^+ particles recombine, forming $\overline{H^0}$ atoms. Each storage ring has its own electron cooling system, allowing dense, cold beams of recombining particles to be obtained. *Positronium production* also occurs in the section for electron cooling of positrons. Its long-lived component, orthopositronium, like the $\overline{H^0}$ atoms, can be extracted from the focusing system into the detection channel.

The positron storage ring scheme proposed recently [5] has a special focusing system with longitudinal quasi-uniform and helical quadrupole magnetic fields. A bending magnetic field consistent with the positron energy is also applied to the toroidal segments of this storage device. Such a focusing system of the "stellarator" type ensures stability of the beam of circulating electrons.

The fundamental feature of this scheme, distinguishing it from those proposed earlier for low antiproton energies [1], [4], is the magnetization of the positrons, whose source is also immersed in the longitudinal field. This complicates the injection of positrons into the ring, making it necessary to use a special injection system. However, the magnetization leads to several important advantages (see details in Ref. 10).

The design with a magnetized beam is only weakly sensitive to retuning of the particle energy, which suggests that it might also be applicable for stabilizing the positron beam down to very low positron energies of the order of hundreds of eV (accordingly, antiproton energies of hundreds of keV) at rather high beam intensity. Such a storage ring is equivalent to a ring with very strong focusing, because the role of the betatron function in it is played by the Larmor helix, the period of which is of the order of several centimeters for the highest energy planned.

The electron cooling of the two beams of recombining particles, antiprotons and positrons, ensures low temperatures of each (small spread and velocity), thereby

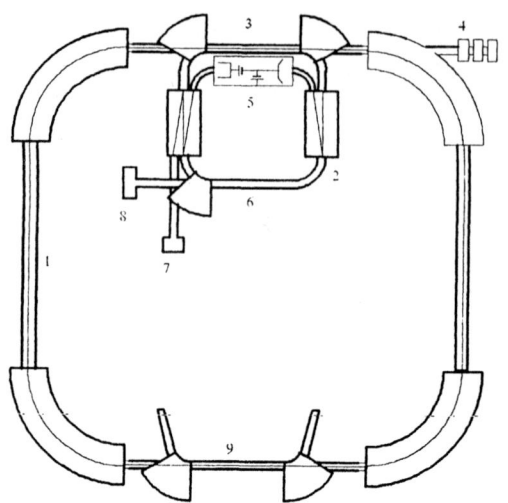

FIGURE 6. Scheme for generating antihydrogen and positronium: 1 – antiproton storage ring; 2 – positron storage ring; 3 – $\overline{p}e^+$ – recombination section; 4 – $\overline{H}^0$ detection channel; 5 – source and collector of the electron beam; 6 – section for electron cooling of positrons; 7 – positron injection channel; 8 – positron detection channel; 9 – system for electron cooling of antiprotons.

leading to a high recombination rate. Therefore, the low temperature of cooled beams makes it possible to obtain rather intense *fluxes of* $\overline{H}^0$ *and Ps atoms with small angular and velocity spreads.*

Positron production and storage is a special problem, and the proposed schematics allows to resolve it effectively (see Ref. 5).

Table 2 lists the design values of the parameters of storage rings and estimates of the intensities of antihydrogen and orthopositronium fluxes made in the radiative recombination approximation.

Yet another extremely important advantage of systems with electron cooling, which follows from the quality of the average velocities of the cooling and cooled particles, is the possibility of *absolute calibration* of the $\overline{H}^0$ atom velocity (using the voltage at the cathode of the electron gun) and the possibility of smooth, controlled *tuning* of this velocity within a broad range (using the same voltage or the potential of a "suspended" cooling segment).

To obtain very slow $\overline{H}^0$ atoms the proposed scheme is not suitable because, when the positron energy is of the order of a hundred eV, the lifetime of circulating positrons is too short (or, in other words, the vacuum conditions are nonrealistic) - about 10 s, when vacuum pressure is of the order of 10 pTorr. As a result, the lowest energy of antihydrogen atoms available in such a scheme is of the order of a few hundred keV. One can avoid this problem and produce very slow antihydrogen atoms by using electrostatic deceleration of the antiprotons and positrons in the recombination section [5]. Several limitations do not allow one to get an intense $\overline{H}^0$ flux with this scheme.

However, one can hope to generate several atoms per second at an energy of the order of 10 eV (see details in Ref. 6).

The project dedicated to creation of a Low Energy Positron Storage Ring (LEPTA) with parameters close to those listed in Table 2 is under development now at JINR (Fig. 7).

TABLE 2. Antiproton and positron storage ring parameters

Antiproton storage ring		
Circumference, m	80	
Antiproton energy, MeV	50	0.5
Density of cooling electron beam, A/cm^2	1.0	0.2
Electron longitudinal temperature, μeV	120	70
Number of stored antiprotons	$1 \cdot 10^{11}$	$1 \cdot 10^9$
Antiproton current, mA	20	2.0
Positron storage ring		
Circumference, m	20	
Positron energy, MeV	27.2	0.272
Longitudinal magnetic field, T	0.1	0.05
Number of stored positrons	$1 \cdot 10^9$	$1 \cdot 10^8$
Positron beam current, μA	800	80
Density of cooling electron beam, A/cm^2	1.0	0.002
Antiproton flux		
Intensity, sec^{-1}	$3 \cdot 10^4$	30
Angular spread, μrad	1.1	8.5
Velocity spread, 10^{-6}	1.1	8.5
Doppler spread, $\Delta v/c$, 10^{-7}	3.5	2.7
Orthopositronium flux		
Intensity, sec^{-1}	$1.7 \cdot 10^4$	35
Angular spread, mrad	1.5	16
Velocity spread, 10^{-5}	5.1	40
Doppler spread, 10^{-5}	1.5	1.2

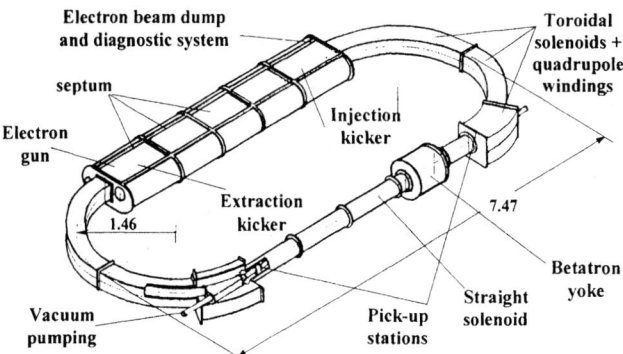

FIGURE 7. Schematic of the LEPTA ring.

VI EXPERIMENTS WITH $\overline{H^0}$ IN-FLIGHT

Generation of an intense $\overline{H^0}$ flux opens up the feasibility of experimental studies in this field and, simultaneously, requires a specific approach to the experimental setups. A few examples are considered below (see details in Ref. 6).

A Direct Comparison of Particle Electrical Charges

This experiment can be done by measuring the displacement of the beam of atomic particles – antihydrogen, hydrogen, and positronium – emitted from the storage ring and deflected in the transverse magnetic field. It is possible to obtain an upper limit on the difference between the electric charges δe [6] at a resolution of $(\delta e/e)_{\overline{H^0}} \sim 2 \cdot 10^{-9}$.

For positronium the value of $B_\perp$ is limited owing to interference between its ortho- and para-states accompanied by rapid annihilation of the p-Ps component. Nevertheless, even in fields of the order of 2 T we can count on $(\delta e/e)_{Ps} \sim 4 \cdot 10^{-10}$. This value decreases the present limit $(4 \cdot 10^{-8})$ by two orders of magnitude. Nevertheless, by performing an experiment of this design with antihydrogen and positronium and using the high accuracy with which the charges e_p and e^- coincide, it is possible to "close" (via positronium!) the chain of charges of all four particles with an accuracy of at least 10^{-8}, i.e., to improve the inequality of e_p and e_a by three orders of magnitude. Further improvement is determined by the limit on the inequality of e^+ and e^-.

We should stress the fact that the proposed experiment gives the *difference between the electric charges* of the particles, and therefore it differs from those where e/m is measured. The results of these two experiments allow the upper limit on the inequality of m_p and m_a to be improved.

B Microwave Spectroscopy of the 2S-2P States of Antihydrogen

The methods of radiospectroscopy and atomic interferometry developed in measurements of the hyperfine structure and the Lamb shift of the hydrogen spectrum can be used in these experiments. These methods are based on the interference of two nearby states of an atom in an external electromagnetic field.

The first method, developed in precision measurements of the hyperfine structure of the hydrogen atom, can also be used in experiments involving an antihydrogen flux. The idea of the method is that an external perturbation can be used to excite transitions between hyperfine levels of the metastable $2^2 S_{1/2}$ state (whose lifetime is 1/7 sec), and then the resonance frequency of the transition can be measured. The atoms that have undergone this transition are detected by exciting the next transition from the 2S state to one of the short-lived $2^2 P_{1/2}$ states (whose lifetime is about 1.5 nsec), with the decay of this state detected by using the $2P \rightarrow 1S$ transition (Fig. 8). The frequencies of 2S-2P transitions lie in the centimeter wavelength region, and those of $2P \rightarrow 1S$ transitions lie in the vacuum ultraviolet (10.2 eV).

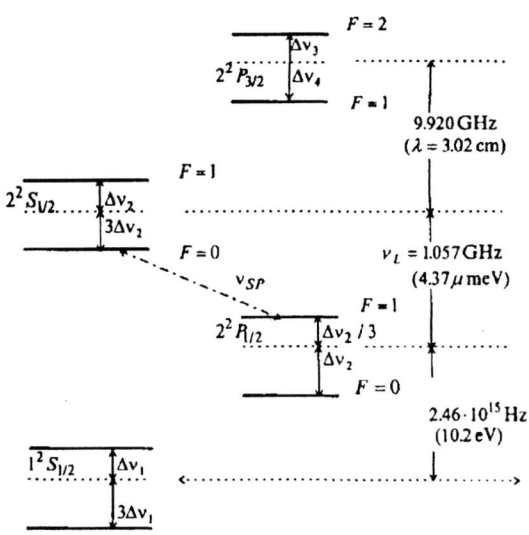

FIGURE 8. Structure of the low-lying levels of the hydrogen atom. $\Delta v_1 = 355.1014$ MHz, $\Delta v_2 = \Delta v_1/8$, $\Delta v_3 = \Delta v_2/10$, $\Delta v_4 = (13/30)\Delta v_2$, $v_L = 1057.8514(19)$ MHz is the Lamb shift.

The main difference between the experiment discussed here and the pioneering one (see Ref. 6) is the relatively high velocity of the atoms, v_0. This affects the resolution of the method, and also makes it necessary to take into account the Doppler frequency shift:

$$\omega_{Lab} = \frac{\omega'}{\gamma(1-\beta\cos\theta)}, \quad \gamma = (1-\beta^2)^{-1/2}, \quad \beta = v_0/c.$$

Here ω' is the frequency in the atomic rest frame and θ is the angle between the directions of the atomic velocity and the cavity axis.

The accuracy (resolution) can be improved by many orders of magnitude by using the technique of the atomic interferometer.

The operating principle of such an interferometer is the following. Two short electromagnetic field pulses, separated in space and time by a relatively long interval L, are applied, where the atom moves in free space. In the radiospectroscopic version of this interferometer these pulses are generated by the electromagnetic field of two cavities P and A (Fig. 9) excited at the same frequency ω. The first cavity P induces, as before, transitions between initial "lower" $2^2 S_{1/2}$ ($F=0$) states and final "upper" $2^2 S_{1/2}$ ($F=1$) states. The phase of the wave function at the output of C_1 depends on the cavity length ℓ and the frequency. After passing through the free segment L, the atom enters the second cavity A, where it again undergoes transitions between the same states. The atoms leaving the interferometer in the "upper" state undergo

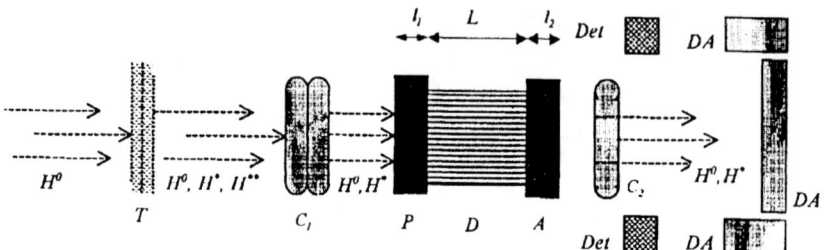

FIGURE 9. Scheme for atomic interferometer: $\overline{H}^0$, $\overline{H}^*$, $\overline{H}^*$ are antihydrogen atoms in the ground and excited states; T is the target; C_1 and C_2 are cavities for purification and analysis. The interferometer: ℓ is the length of cavities P and A; L is the length of free space (D); Det are UV-radiation detectors; DA are annihilation detectors.

transitions to the 2P state in the cavity C_2 and transit to the 1S state with some probability $P(\omega)$. The photons of the UV radiation are recorded by the detectors D.

The best accuracy achieved with this method was of order 50 Hz, which allows measurement of the hyperfine structure of the $2S_{1/2}$ state with an accuracy of order $\delta\omega/\omega \sim 3 \cdot 10^{-7}$. Zeeman splitting and double fitting were used for this.

Determination of the values of the hyperfine splitting of the $2S_{1/2}$ level of hydrogen and antihydrogen allows comparison of the proton and antiproton magnetic moments at a level 10^{-6} to 3×10^{-7}.

A variant of the atomic interferometer with static electric fields (the v_{SP} transition, Fig. 8) was successfully used in experiments involving hydrogen atoms of low energy of order 20 keV. The first measurements of the Lamb shift with a flux of fast ($v_0/c \approx 0.35$) hydrogen atoms were performed at a proton storage ring with electron cooling NAP-M (Budker INP, Novosibirsk) [6].

C The Atomic Interferometer and the Stern-Gerlach Method; Spectroscopy of the 1S State

Use of the atomic interferometer in the classical Stern-Gerlach method made it possible to perform high-precision measurements of the hyperfine structure of the ground $1^2S_{1/2}$ state of the hydrogen atom. This scheme can be applied nearly without change in experiments involving antihydrogen fluxes. For this the polarizer and analyzer of the atomic interferometer must be realized as ordinary EPR spectrometers: each must consist of a dipole magnet with a uniform field $B_\perp$ directed across the beam, and a cavity in which an electromagnetic field is produced. Its magnetic component $B_g(t) = B_g \sin \omega t$ excites EPR transitions. In addition, a sorting system composed of two gradient magnets, one before and one after the analyzer, is provided. The magnets separate atoms with different polarization (the Stern-Gerlach method), and guide atoms in a given state to the detector. The transition from one state to another is accompanied by change of the polarization, as a result of which the atom is sent to the detector by the second sorting magnet. In the current version, the gradient sorting magnets are made in the form of sextupoles as, for example, in the hydrogen maser - the time standard.

It can be hoped that this method will allow the Doppler-broadening limit to be exceeded by an order of magnitude, giving $\delta\omega/\omega_0 \leq 3\cdot 10^{-8}$. Knowledge of the frequency of transitions between hyperfine levels of the ground state makes it possible to determine the value of the antiproton magnetic moment, which, for this accuracy, gives $\delta\mu_a/\mu_a \sim 2\times 10^{-5}$. Higher accuracy, 1×10^{-7} at least, can be obtained for the difference between μ_p and μ_a by comparing the values of the hyperfine splitting for H^0 and $\overline{H^0}$.

D The Laser Spectroscopy of Fast Antihydrogen Atoms

The most interesting measurement can be done with *the 1S-2S two-proton transition* because the small width of the metastable $2S_{1/2}$ level (a lifetime of 1/7 sec), makes it possible, in principle, to measure the transition frequency with an accuracy

$$\frac{\delta\omega}{\omega} \leq \frac{\Gamma}{\omega} \sim 10^{-17}.$$

The Doppler-free scheme can be used in this experiment if it is somehow possible to solve the laser problem. The scheme assumes the excitation of a two-photon transition in an atom bombarded by two colliding beams of the same laser, one beam directed along the atomic flux, and the other opposite to it. Therefore, the Doppler shifts have different signs for each beam. This is actually the main idea of this scheme: the total energy (frequency) of two colliding photons in the frame of the atom is independent, in a first approximation, of the angular spread of the beam and of the velocity spread and direction:

$$\omega_1' + \omega_2' = 2\gamma\omega_{laser} + \Delta\omega', \quad \frac{\Delta\omega'}{\omega'} \sim \beta^2 \frac{\Delta v}{v} + \beta\cdot\Delta\theta(\theta_0 + \delta\theta),$$

where Δv and $\Delta\theta$ are the velocity and angular spreads of the $\overline{H^0}$ flux, $\delta\theta$ is the angular spread of the laser beam, and θ_0 is the crossing angle of both laser beams with the $\overline{H^0}$ flux. Thus, the angular spread of the laser beam $\delta\theta$ does not significantly influence the resolution level when $\theta_0 \gg \delta\theta$. Reduction of the $\overline{H^0}$ angular and velocity spreads (which is possible by using special electron guns in the electron cooling system), allows one to reach the resolution

$$\Delta\omega'/\omega' \sim 10^{-11}.$$

One should underline the basic advantage of experiments with $\overline{H^0}$ in-flight compared to the experiments with $\overline{H^0}$ in traps: using a directed flux of atoms and coincidence schematics of detection, one can clearly distinguish the transitions in $\overline{H^0}$ from those in hydrogen atoms of residual gas.

VII EXPERIMENTS USING POSITRONIUM FLUXES

Use of a positron storage ring like LEPTA (Fig. 7) as a generator of a directed flux of orthopositronium (o-Ps) opens new possibilities for experimental studies of this physics object. The first is *the comparison of e^+e^- charges*, considered in Sec. 6.1.

A Positronium Spectroscopy

When o-Ps fluxes of high intensity and small velocity spread are used, a level of precision of at least 10^{-6} can be expected. Both radiospectroscopic and laser-spectroscopic realization of these experiments are possible. The latter may also allow measurements on two-photon transitions with cancellation of the Doppler broadening. Among such spectroscopic experiments one can point out the measurements of the fine structure of the ground state, o-Ps – p-Ps transitions in a magnetic field, the transition energy of different states, and the fine structure of excited states. All these experimental data are of great importance for QED, where positronium is used as a test particle for theory.

B e^+e^- recombination

This is another experiment of similar significance if high precision in measurement of the parameters of the electron-positron recombination process with formation of Ps atoms can be reached. Pure and well controlled conditions are provided, when this process occurs in the electron cooling section of the positron storage ring.

C Search for Exotic Decay Channels of positronium

This is a way of experimental testing of the fundamental physics law of momentum conservation. The forbidden decay modes

$$\text{o-Ps} \to 4\gamma, \quad \text{p-Ps} \to 3\gamma$$

can be detected with a probability of the order of 10^{-8} using pure conditions of o-Ps in-flight and o-Ps $\to$ p-Ps transition in a magnetic field under well controlled conditions. The same level of experimental resolution can be obtained for allowed (but never observed!) decay modes

$$\text{p-Ps} \to 4\gamma, \quad n > 2.$$

D The Lifetime of Positronium; the Search for "Mirror Universe"

The problem of o-Ps *life time* has some history, since the discrepancy between its theoretical and experimental values (Fig. 10) gave rise to the hypothesis of short-lived neutral bosons (see details in Ref. 6 and 11). Another fascinating idea related to this problem is the hypothesis of a so-called Mirror Universe, proposed by I.Yu. Kobzarev, L.B. Okun' and I.Ya. Pomeranchuk (see Ref. 6 and 12). They showed that the "usual" world, which consists of "left" (L) particles, can interact with the "mirror" world of

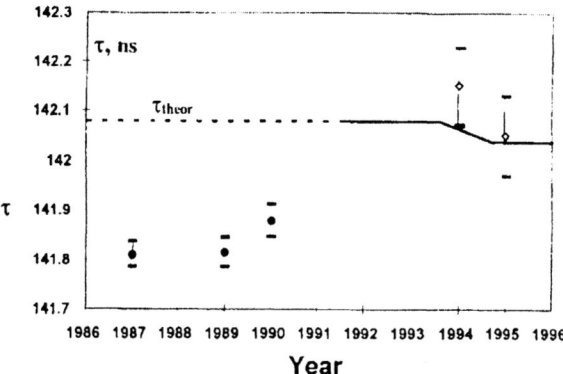

FIGURE 10. Theoretical and experimental values of orthopositronium life time. 1987-1990: studies by Michigan group, 1994-1995: studies by Tokio group (see Ref. 11).

right (R) particles only by exchange with photons or gravitons. In more recent times S. Glashow proposed the use of positronium as a test system, considering the process of exchange of L- and R-positronium atoms by photons (Fig. 11a). Such an interaction gives a "coupling" of o-Ps and o-Ps' (R) with splitting of the coupled state into two single ones (Fig. 11 b).

What can be observed? By definition, in our L-world one can detect o-Ps (L-system) but one cannot detect o-$\overline{\text{Ps}}$ (R-system). Then, because of coupling, some the left particles can "escape" into the R-system (Mirror Universe). This will shorten the o-Ps lifetime τ measured in our L-world.

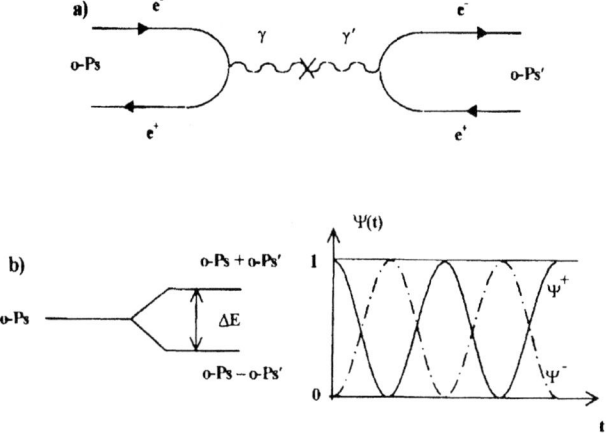

FIGURE 11. (a) diagram of the interaction between o-Ps (L) and o-Ps' (R) particles. (b) coupling of two states of positronium.

The analysis of statistics errors (defined by o-Ps flux intensity!) and systematic errors related to o-Ps velocity spread and the dependence of the decay rate Γ on the velocity (via Lorentz-factor), the accuracy of measurement of the decaying positronium position, etc., shows that a level of resolution

$$(\Delta\tau/\tau)_{exp} \sim 10^{-6}$$

is achievable. This allows to detection of a difference (if it exists!) between τ values predicted by QED and by the theory of the Mirror Universe.

CONCLUSIONS

Experiments on directed monochromatic atoms of antihydrogen and positronium reveal new possibilities for experimental testing of the CPT theorem and quantum electrodynamics. The accuracies of measuring the parameters of the fundamental particles and the simplest atoms that can be expected in the future on the basis of the described methods significantly exceed the present level of precision of particle data.

This work is supported by the Russian Foundation for Basic Research, grant No. 99-02-17716.

REFERENCES

1. Budker G.I. and Skrinsky A.N., *Usp. Fiz. Nauk* **124**, 561 (1978) [*Sov. Fiz. Usp.* **21**, 277 (1978)].
2. Baur G., Boero G., W. Oelert W. et al., *Phys. Lett. B* **368**, 251 (1996).
3. Charlton M., Eades J., Horvath D. et al., *Phys. Rep.* **241**, 67 (1994).
4. Herr H., Moehl D., and Winnacker A., in *Physics at LEAR with Low-Energy Cooled Antiprotons*, Plenum Press, Mew York, 1984, p. 659.
5. Meshkov I. N., and Skrinsky A.N., *NIM A* **379** 41-49 (1996).
6. Meshkov I.N., *Fiz. Elem. Chastits At. Yadra* **28**, 198 (1997) [*Sov J. Part. Nucl.* **28** 198 (1997).
7. http://www.cern.ch/PSdoc/acc/ad
8. http://www.cern.ch/Athena
9. http://hussle.harvard.edu/~atrap
10. Meshkov I.N., *Fiz. Elem. Chastits At. Yadra* **25** 1487 (1994) [*Sov. J. Part. Nucl.* **25** 631 (1994)].
11. Khriplovich I.B., Meshkov I.N., Milstein A.I., *JINR Rapid Communications*, **83** 68-71 (1997).
12. Meshkov I.N., "Generation of Directed Flux of Positronium and Experimental Studies with Positronium In-Flight", in *Hadronic Atoms and Positronium in the Standard Model-1998*, edited by E.A. Kuraev et al., JINR, Dubna, 1998, pp. 176-182.

AUTHOR INDEX

B

Beloshitsky, P., 435

G

Gareyte, J., 24, 260

K

Kasuga, T., 44
Keil, E., 566

L

Logatchov, P., 552

M

Machida, S., 405
Meshkov, I. N., 616
Migliorati, M., 231
Minty, M., 118

P

Palumbo, L., 231
Parkhomchuk, V. V., 53
Perevedentsev, E. A., 6
Pestrikov, D. V., 317, 339, 356

S

Salimov, R. A., 580
Seeman, J. T., 163
Shatunov, Y. M., 279
Skrinsky, A., 1, 601
Struckmeier, J., 66
Stupakov, G. V., 205

T

Tobiyama, M., 374

V

Vinokurov, N. A., 390

W

Wienands, U., 435
Willeke, F., 85

Y

Yokoya, K., 185

Z

Zimmermann, F., 494